T0375834

Rheology
Concept, Methods, and Applications
4th Edition

Prof. Dr. Alexander Ya. Malkin
Academy od Sciences, Institute of Petrochemical Synthesis
Moscow, Russia

Prof. Dr. Avraam I. Isayev
The University of Akron, Department of Polymer Engineering
Akron, Ohio, USA

ChemTec Publishing

Toronto 2022

Published by ChemTec Publishing
38 Earswick Drive, Toronto, Ontario M1E 1C6, Canada

© ChemTec Publishing, 2005, 2012, 2017, 2022
ISBN 978-1-927885-93-2 (hard cover); ISBN 978-1-927885-94-9 (epub)

Cover design: Anita Wypych

Library and Archives Canada Cataloguing in Publication

Title: Rheology : concepts, methods, and applications /
Prof. Dr. Alexander Ya. Malkin, Prof. Dr. Avraam I. Isayev.
Names: Malkin, A. IA. (Aleksandr Iakovlevich), author.
| Isayev, Avraam I., 1942- author.
Description: 4th edition. | Includes bibliographical references and index.
Identifiers: Canadiana (print) 20210194707
| Canadiana (ebook) 2021019474X | ISBN 9781927885932
  (hardcover) | ISBN 9781927885949 (PDF)
Subjects: LCSH: Rheology-Textbooks. | LCGFT: Textbooks.
Classification: LCC QC189.5 .M33 2022 | DDC 531/.1134-dc23

Printed in Australia, United Kingdom and United States of America

TABLE OF CONTENTS

PREFACE

A number of books devoted to different aspects of both theoretical and applied rheology were published in the last 20 years. The keyword in the last sentence is "different aspects." Rheology has a unique structure with its own language, fundamental principles, original concepts, rigorous experimental methods, and a set of well-documented observations with inherent interrelations between various branches of natural science and numerous practical applications.

By examining the enormous volume of rheological literature and meeting various people interested in rheology (university teachers, students, applied scientist, and engineers), the authors felt the need for a systematic presentation of the subject matter in one book – a book that includes all components of rheology and presents them as an independent branch of natural science.

However, it became apparent from the early planning stages that some information would need to be omitted to provide a clear presentation of the concepts, methods, and applications which constitute the essence of efforts that created this science. The wish to present all aspects of rheology will inevitably lead to a book of enormous size. Every attempt to write a scientific book is personal and objective; objective because science benefits from objective assessments and personal because our experiences make us feel that certain aspects are more important than others. In our case, we are university teachers and researchers primarily in the field of applied rheology. An attentive reader will most likely find some reflections of our personal preferences.

Considering the book's goals and the tasks, the authors tried to limit the choice of references to the first publications on a particular subject, including reviews and papers providing the most expressive examples and illustrations of the topics under discussion. Accordingly, a significant number of original publications are not mentioned. It is regrettable because any serious publication is worth mentioning.

The authors hope that the readers of the book will benefit from our presentation of rheology as an interrelated system of concepts, principal phenomena, experimental methods, and directions of their application. Our rheology is also a science interwoven with other branches of theoretical and applied sciences. We take many opportunities to emphasize these links because they enrich science, make it easier to understand and apply, and this also helps to fulfill our goals concisely expressed in the book title. To amplify its usefulness as a teaching tool, all chapters of the book contain questions to be used by readers to assess their knowledge of a particular subject. Answers to these questions are included in the last part of the book.

Finally, the authors are glad to fulfill their pleasing duty to thank Dr. Andrei Andrianov (Moscow State University) for his technical assistance in preparing the computer versions of many figures and realizing the liaison between the authors. The authors are also grateful to Dr. Sayata Ghose for painstaking proofreading and making corrections for this book.

Special gratitude goes to Professor J.L. White, who read the book's manuscript and made many valuable comments, which helped enrich the presented text.

We express our deep gratitude to publishers of various journals (Advances in Polymer Science, Colloid Journal, European Polymer Journal, International Journal of Polymeric Materials, Journal of Applied Polymer Science, Journal of Macromolecular Science, Journal of Non-Newtonian Fluid Mechanics, Journal of Polymer Science, Journal of Rheology, Macromolecular Chemie, Polymer, Polymer Engineering and Science, Polymer Science USSR, Reviews of Scientific Instruments, Rheologica Acta) and books (Rheology of Elastomers by P. Mason and N. Wookey (eds), A Practical Approach to Rheology and Rheometry by G. Schramm for permission to use figures from their publications.

Alexander Ya. Malkin,
Moscow, Russia
Avraam I. Isayev
Akron, Ohio, USA
July, 2005

PREFACE TO THE 2ND EDITION

In preparing the Second Edition of this book, the book's general structure is maintained, and some necessary corrections and additions are made. The most important recent results published in periodicals till the middle of 2011 are added. In particular, Section 2.8.1 of Chapter 2, Section 3.5.2 of Chapter 3, and Subsection 5.8.1.2 of Chapter 5 are modified. A new Subsection 5.6.2.6 on Capillary breakup in elongational rheometry is added. Furthermore, Section 3.2.3 on the Viscosity of anisotropic liquids and Section 3.6.3 on Instabilities of the flow of elastic fluids of Chapter 3 are completely rewritten. Many other modifications in the text are made, and some new figures are added. Also, all the detected misprints and errors found by ourselves or pointed out by colleagues are corrected.

After the publication of the First Edition of the book, a lot of comments and advice from our friends and colleagues were received. We are very grateful to all of them for constructive criticism and valuable comments.

We are also grateful to our Editor, Dr. G. Wypych, for his hard work in improving the manuscript and making it ready for publication.

> Alexander Ya. Malkin,
> Moscow, Russia
> Avraam I. Isayev
> Akron, Ohio, USA

August 2011

PREFACE TO THE 3RD EDITION

In the 3rd Edition of the book, we updated the material paying special attention to the issues which have become the hot spots of rheology during the last decade. These are such topics as the rheology of polymeric materials containing fillers, the concept of heterogeneity of the flow, including the effect of shear-banding, and new ideas in understanding the visco-plastic media. We have added references to the mostly up-to-date publications in these fields.

We are grateful to readers who were kind to bring to our attention some misprints and not quite clear explanations used in the former editions. Necessary corrections have been implemented.

We hope that this book is continued to be used by students and young researchers who only start their careers in the intriguing world of the rheology of real materials surrounding us in our life.

Alexander Ya. Malkin,
Moscow, Russia
Avraam I. Isayev
Akron, Ohio, USA

December 2016

PREFACE TO THE 4TH EDITION

Rheology is a living and developing science. Every month. various journals, publish dozens of papers devoted to general problems and foundation of rheology as well as application of the rheological methods for solving different applied tasks in the oil industry and food production, road building, and polymer technology, etc. It is not possible and not necessary to collect all this information, but among these publications, one can find new approaches and conceptions which have value for different aspects of rheology. In preparing the new edition of the book, we tried to catch this new knowledge or put emphasis on some old works that obtained modified reflections. First of all, the corrections touched on understanding the nature of instabilities in the flow of viscoelastic liquids, the concept of plasticity as the part of elastoplastic behavior of highly concentrated suspensions, some details in the capillary viscometry.

We are grateful to our colleagues who spent their time and were kind to point out some unclear places in the text and made advice concerning a more rigorous description of some theoretical aspects of rheology.

This book is addressed to students and young researchers interested in expanding their theoretical background and applied science-based knowledge and plan to continue their carrier in physics and technology of complex liquids related to rheological measurements.

Alexander Ya. Malkin
    Moscow, Russia

Avraam I. Isayev
    Akron, OH, USA

December 2021

INTRODUCTION.
RHEOLOGY: SUBJECT AND GOALS

Rheology, as an independent branch of natural sciences, emerged more than 70 years ago. It originated from observations of the "strange" or abnormal behavior of many well-known materials and difficulties in answering some "simple" questions. For example:

- paints are evidently liquids because they can be poured into containers, but why do they remain on vertical walls without sagging down, unlike many other liquids?
- clays look solid, but they can be molded into a shape; they may occupy vessels the way any liquid does; why do clays behave like many liquids?
- yogurt does not flow out of a container (it has high viscosity), but after intensive mixing, its viscosity decreases and then increases again when left to rest, so which value of viscosity should be considered?
- concrete mix appears to be solid and rigid, but when subjected to an external force, it changes its shape similar to liquids; what are the reasons for such behavior?
- parts made out of polymeric materials (plastics) look solid and hard, similar to parts made out of metal, but they are noticeably different: when force is applied to a metallic part, it slightly changes its shape and maintains its new shape for a long time; this is not the case with plastics which also change their shape after force is applied, but they continue to change shape; if this material is solid, why does it "creep"?
- pharmaceutical pastes (for example, toothpaste or body lotion) must be "liquids" when applied, and they should immediately become "solids" to remain on the skin; are they liquids or solids?
- sealants widely used in construction must be fluid-like to seal all spaces and to fill cavities, but then sealant must rapidly "solidify" to prevent sagging; is sealant liquid or solid?
- metals are definitely solids, but how is it possible to change their initial form by punching and stamping as if metal was liquid?

These are just a few examples. It is common for them to represent many real materials' properties and exhibit a mixture of *liquid-like* and *solid-like* properties. This shows that the commonly used words "liquid" and "solid" are insufficient to describe their properties, and *new concepts* are needed to understand the properties of many real materials. A *new terminology* emerges from discoveries and description of new features of materials.

New methods are needed to characterize and measure their properties. *New fields of appli-cation* can be expected from the application of new concepts and the results of studies. All these are the essence of *rheology*.

Superposition of liquid-like and solid-like features in the behavior of technological materials is directly regarded as the consequence of *time effects*, i.e., the results of observations depend on a *time scale*. Possibly, this is the most common feature of the materials which were listed above. Time by itself has no meaning, but time is a reflection of changes in material structure taking place during the period of observation (or experiment).

The primary method of rheology consists of constructing *models*, which are useful in qualitative or (better) quantitative descriptions of experimental results of different materials' mechanical behavior. Any natural science pretends to deal with reality and does so by means of *phenomenological models*. Any model is created not to reflect all but the most important characteristic features of an object. The concepts of *liquids* and *solids* are also models, and their formal (mathematical) representation originated from the classical works by Isaac Newton and Robert Hooke.

Newton (1687) reflected upon resistance of liquids to a cylinder rotating in a vessel. His ideas were converted to a more accurate form by Stokes, who formulated a general law of liquid-like behavior, known as the Newton-Stokes law. According to this concept, the deformation rate is expected to be proportional to stress, and the constant coefficient of proportionality is called viscosity, which is a material parameter of liquid. This law assumes that, in flow of liquids, a force (or resistance to flow) is proportional to a velocity (of movement).

Hooke (1676) formulated a similar proposal concerning the properties of solids. The law, named after him, was translated to modern form by Bernoulli and then by Euler. *Hooke's law* states that in the deformation of solids, stress is proportional to deformation. The coefficient of proportionality is called *Young's modulus*.

Both models represent properties of many real materials and work well in describing their behavior with a considerably high degree of accuracy. However, there are many other materials that *are not described* by the Newton-Stokes and the Hooke laws. Rheology relies on the concept that *non-Newtonian* and *non-Hookean* materials exist in reality. These materials are interesting from both theoretical and applied aspects, and that is why such materials must be the objects of investigation.

It is important to emphasize that every model describes the properties of real materials with a different degree of approximation. The Newton-Stokes and Hooke laws are not exceptions, and more strict and complex laws and equations give a much better approximation of reality than the classical Newton-Stokes and Hooke laws known from school years.

Both basic phenomenological (i.e., taken as probable *assumptions*, but only assumptions) relationships (the Newton-Stokes and Hooke laws) do not include the inherent structure of matter. Because matter consists of molecules and intermolecular empty spaces, every material is heterogeneous. At the same time, an observer sees a body as a homogeneous continuous mass without holes and empty spaces. The obvious way out of these contradictory evidence lies in the idea of the *space scale of observation*. This scale can be small enough to distinguish individual molecules or their parts. Then, molecules can be combined in regular arrangements, such as crystals, and then crystals can be organized in super-crystalline (or super-molecular) arrangements. All this leads to the concept of material *structure*, i.e., more or less well-organized and regularly spaced shapes. The structure might be well determined. This is the case, for example, of reinforced plastics and monocrystals. In other cases, "structure" can mean complex intermolecular interactions, which cannot be observed by direct methods. Rheology is especially interested in structured materials because their properties change due to the influence of applied forces on the structure of matter.

The definition of rheology as a branch of natural science and the subject of rheological studies can be formulated based on the following argument. Traditionally, rheology is defined as "the study of deformations and flow of matter" (College Dictionary). However, this definition is ambiguous. The definition is close to the mechanics of continuum and does not distinguish special features of rheology.

The following points should be emphasized:

- Rheological studies are not about "deformation and flow" but about *properties* of matter determining its *behavior*, i.e., its reaction to deformation and flow.
- Rheology deals with materials having properties not described by the models of Newton-Stokes and Hooke. It is a negative statement (the rule of contraries). The positive statement is that rheology studies materials are having properties described by *any* relationship between force and deformation. In this sense, the Newton-Stokes and Hooke laws are limiting cases formally lying on the border of rheology. The subject of rheology is not about all matters, but only those for which *non-linear* dependencies between forces and deformations or rates of deformations are the main characteristics.
- Rheology is interested in materials, deformation of which results in *superposition* of *viscous and elastic* effects.
- Rheology studies materials with *structure changes* under the influence of applied forces.

One of the keywords in rheology is the behavior of various real continuous media. What is the meaning of "*behavior*"? For a body of finite size, it is a relationship between external action (forces applied to a body) and internal reaction (changes of a body shape). For continuous media, this approach is extrapolated to a point, and the relationship between forces and deformations at this point (i.e., changes of distance between two arbitrary points in a body) is examined. Thus, by discussing what happens in a body *at a point of reference*, the problem of a geometrical form of a body as such is avoided, but the subject of investigations are only its substantial, inherent properties.

The *first* main *goal* of rheology consists of establishing the relationship between applied forces and geometrical effects induced by these forces at a point. The mathematical form of this relationship is called the *rheological equation of state*, or the *constitutive equation*. The Newton-Stokes and Hooke laws are the simplest examples of such equations. Rheological equations of state can be (and they are!) very different for numerous real materials. The rheological equations of state found for different materials are used to solve macroscopic problems related to the continuum mechanics of these materials. Any equation is just a model of physical reality.

The independent *second goal* of rheology consists of establishing relationships between the rheological properties of the material and its molecular structure (composi-

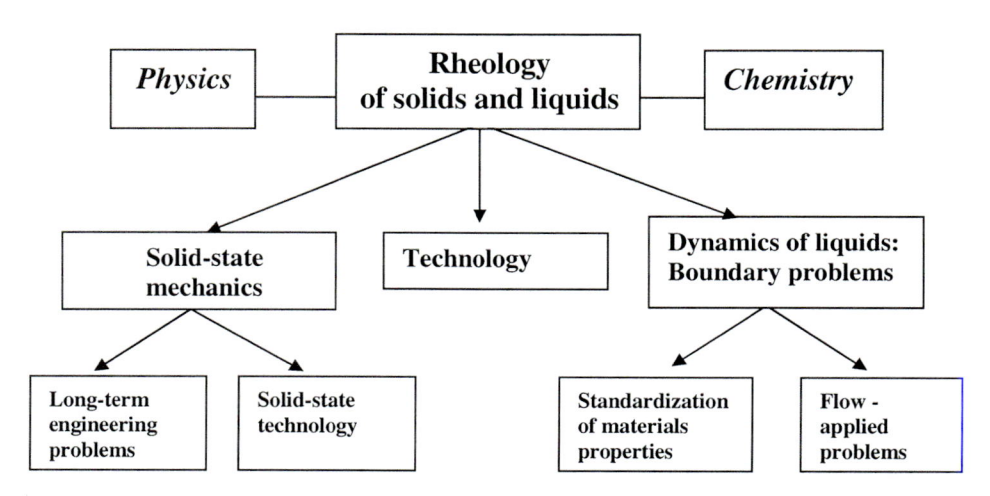

Fig. 1. Rheology as an interdisciplinary science – its place among other sciences.

tion). This is related to estimating the quality of materials, understanding laws of molecular movements and intermolecular interactions. The term *microrheology*, related to classical works by Einstein (1906, 1911), devoted to viscous properties of suspensions, is sometimes used in this line of thought. It means that the key interest is devoted not only to movements of physical points but also to what happens inside a point during the deformation of the medium. Therefore, it is a search for rheological equations of the state of different materials based on the basic physical concepts. Then, the constants entering these models are related to various molecular parameters of the material.

In its origin, the term "rheology" applies to flowing media since the main root of the word means "to flow" (*rheo* in Greek). But it is very difficult to classify the material as solid or fluid, and, therefore, this term is now used for any material. As a result, many analytical methods used for solids and liquids are very similar. The majority of publications devoted to rheology and, consequently, the largest part of this book deal with flowing materials, whereas rheology of solids is frequently treated as a part of the mechanics of solids and discussed as a separate branch of science.

The place of rheology among other natural sciences and applied problems is shown in Figure 1. Rheology is a multi-disciplinary science having many relationships with fundamental physics and chemistry, as well as many applications in technology and engineering of materials and many fields of biological sciences.

Indeed, the connection between *rheology* and *physics* consists of explanation and predictions of rheological properties based on knowledge of the molecular structure and fundamental laws of physics (molecular physics, statistical physics, thermodynamics, and so on). The connection between *rheology* and *chemistry* consists of existing experimental evidence of a direct correlation between chemical parameters (molecular mass and molecular mass distribution, chemical structure, intermolecular interactions, etc.) and rheological properties; therefore, it is possible to synthesize materials with desirable properties.

The second layer of interrelations consists of the connection between *rheology* and *mechanics of continuum*. The results of rheological studies give the background for the formulation of boundary problems in solid-state mechanics as well as in dynamics of liquids. This includes governing equations and their solutions to find numerical values of macro-parameters, such as pressure, forces, displacements, etc.

In the framework of solid-state mechanics, such effects as long-term behavior, engineering properties of materials as well as their technological properties are the direct objects of rheological analysis. Solutions of boundary problems in dynamics of liquids are used for the analysis of flow in technology (calculation of pressure and output, resistance to movement of solid bodies in liquids, and so on), as well as for standardization of the methods of quality control in the technology of liquid products.

It is worthwhile to list materials for which the rheological analysis is the most important. In fact, such a list should include all materials because the Newton-Stokes and Hooke laws are the limiting cases only. However, the following list gives an impression of fields that cannot be developed without the participation of rheological studies:

- metals, alloys, and composites at large deformations and different technological operations
- concrete, ceramics, glass, and rigid plastics (including reinforced plastics) at long periods of loading
- polymer melts and solutions, including filled composites, rubbers
- foodstuffs
- lubricants, greases, sealants
- pharmaceuticals and cosmetics
- colloid systems, including emulsions and detergents of any types
- paints and printing inks
- mud, coal, mineral dispersions, and pulps
- soils, glaciers, and other geological formations
- biological materials such as bones, muscles, and body liquids (blood, saliva, synovial liquid, and others).

It is worth mentioning that polymers and plastics continue to be the main object of rheological studies, and at least half of the publications on rheology are devoted to them.

To recapitulate, it seems useful to point out that the main ideas of the *Introduction* permit us to compose a dictionary of rheology.

Rheology is a science concerned with the mechanical properties of various solid-like, liquid-like, and intermediate technological and natural products. It accomplishes its goals by means of *models* representing principal peculiarities of *behavior* of these materials. The behavior of the material is a *relationship* between forces and changes of shape. A model gives a mathematical formulation of such a relationship. *Rheological properties* are expressed by the model structure (i.e., its mathematical image), and values of constants included in the model are characteristics of the material.

Rheological models are related to *a point*, which is a physical object including a sufficient number of molecules to neglect the molecular structure of matter and treat it as a continuum. The rheological analysis is based on the use of *continuum* theories, meaning that the following is assumed:

- there is no discontinuity in the transition from one geometrical point to another, and the mathematical analysis of infinitesimal quantities can be used; discontinuities appear only at boundaries

- properties of the material may change in space (due to the gradient of concentration in multicomponent mixtures, temperature distribution, or other reasons), but such changes occur gradually; these changes are reflected in space dependencies of material properties entering equations of continuum theories which must be formulated separately for any part of material surrounded by the boundary surfaces at which discontinuity takes place
- continuity theories may include an idea of anisotropy of properties of material along with different directions.

The rheological behavior of material depends on *time* and *space scales* of observation (experiment). The former is important as a measure of the ratio of the (time) rate of inherent processes in a material to the time of experiment and/or observation; the latter determines the necessity to treat material as homo- or heterogeneous.

The results of the macroscopic description of the behavior of real engineering and biological media, based on their rheological properties, are used in numerous applications related to the technology of synthesis, processing, and shaping various materials listed above.

LITERATURE

Many books devoted to rheology were published during the 70-year history of this science. It is not possible to list all of them. The most important books which gave the input to the development of rheology and reviewed results of principal rheological schools are cited in the text of this book. The same concerns many historically important and original publications.

The modern history of rheology began with the publication of several books in the 1940s, which had a great impact on the education of future generations of rheologists. Among them, it is necessary to recall:

T. Alfrey, **Mechanical Behavior of High Polymers**, *Interscience*, N.Y. 1948.
M. Reiner, **Twelve Lectures on Theoretical Rheology**, *North Holland Publ. Co.,* Amsterdam. 1949.
G.W. Scott Blair, **A Survey of General and Applied Rheology**, *Pitman*, London, 1949.

Those who are interested in special aspects of rheology may like to pay attention to the following books published more recently.

1   Deepak Doraiswamy, The origin of rheology: For a short historical excursion see
     http://sydney.edu.au/engineering/aeromech/rheology/Origin_of_Rheology.pdf
     The development of rheology including sketches devoted to people who played the most important role in rheology can be found in a monograph:
     R.I. Tanner, K. Walters, **Rheology: A Historical Perspective**, *Elsevier*, Amsterdam, 1998.

2   The state of rheology and the main results obtained until 1969 were summarized in a five-volume book
     Rheology. Theory and Applications, Ed. F.R. Eirich, v. 1-5, *Academic Press*, London, 1956-1959.
     One can find a comprehensive analysis of rheology in the following books:
     J. D. Ferry, **Viscoelastic Properties of Polymers**, *Wiley*, NY, 1980.
     C.W. Macosko, **Rheology: Principles. Measurements and Applications**, *VCH*, New York, 1993.
     F.A. Morrison, **Understanding Rheology**, *Oxford University Press*, New York, 2001.

3   The following books are devoted to the general theoretical background of rheology:
     G. Astarita, G. Marucci, **Principles of Non-Newtonian Fluid Mechanics**, *McGraw-Hill*, New York, 1974.
     R.B. Bird, R.C. Armstrong, O. Hassager, **Dynamics of Polymeric Liquids**, v. 1-2, *Wiley*, New York, 1987.
     M. Doi, S.F. Edwards, **The Theory of Polymer Dynamics**, *Oxford Science Publisher*, Oxford, 1986.
     J. Furukawa, **Physical Chemistry of Polymer Rheology**, *Springer*, Berlin, 2003.
     H. Giesekus, **Phänomenologische Rheologie: Eine einführung**, *Springer*, Berlin, 1995.
     R.R. Huilgol, N. Phan-Thien, **Fluid Mechanics of Viscoelasticity**, *Elsevier*, Amsterdam, 1997.
     G.D.C. Kuiken, **Thermodynamics of Irreversible Processes: Application to Diffusion and Rheology**, *Wiley*, New York, 1994.
     R.G. Larson, **Constitutive Equations for Polymer Melts and Solutions**, *Butterworths*, Boston, 1988.
     R.G. Larson, **The Structure and Rheology of Complex Fluids**, *Oxford University Press*, New York, 1999.
     A.I. Leonov, A.N. Prokunin, **Nonlinear Phenomena in Flows of Viscoelastic Polymer Fluids**, *Chapman and Hall*, London, 1994.
     A.S. Lodge, **Elastic Liquids**, *Academic Press*, London, 1960.

H.C. Öttinger, **Stochastic Processes in Polymer Fluids**, *Springer*, Berlin, 1996.

W. R. Schowalter, **Mechanics of Non-Newtonian Fluids**, *Pergamon*, 1978.

N.W. Tschoegl, **The Phenomenological Theory of Linear Viscoelasticity**, *Springer*, Berlin, 1989.

A.S. Wineman, K.R. Rajagopal, **Mechanical Response of Polymers: Introduction**, *Cambridge University Press*, Cambridge, 2000.

4　The following monographs were devoted to different materials extensively studied by rheological methods:

D. Acierno, A.A. Collyer (Eds.), **Rheology and Processing of Liquid Crystal Polymers**, *Chapman and Hall*, London, 1996.

J.M.V. Blanshard, P. Lillford, **Food Structure and Behavior**, *Academic Press*, London, 1987.

O.E. Briskoe, **Asphalt Rheology: Relationship to Mixture**, *ASTM*, Philadelphia, 1987.

P. Coussot, **Mudflow Rheology and Dynamics**, *A.A. Balkema*, Rotterdam, 1997.

D.A. Drew, D.D. Joseph, S.L. Pasman (Eds.), **Particulate Flows: Processing and Rheology**, *Springer*, New York, 1998.

H.A. Faradi, J.M. Faubion (Eds.), **Dough Rheology and Baked Products Texture**, *Van Nostrand Reinhold*, New York, 1990.

Y.C. Fung, **Biomechanics: Mechanical Properties of Living Tissues**, *Springer*, New York, 1993.

R.K. Gupta, **Polymer and Composite Rheology**, *Marcel Dekker*, New York, 2000.

B.G. Higgins, **Coating Fundamentals: Suspension Rheology for Coating**, *TAPPI Press*, Atlanta, 1988.

M.J. Keedwell, **Rheology and Soil Mechanics**, *Elsevier Science*, London, 1984.

D. Laba (Ed.), **Rheological Properties of Cosmetics and Toiletries**, *Marcel Dekker*, New York, 1993.

G.D.O Lowe (Ed.), **Clinical Blood Rheology**, *CRC Press*, Boca Raton, 1988.

J.M. Mazumdar, **Biofluids Mechanics**, *World Scientific*, Singapore, 1992.

W.B. Russel, D.A. Saville, W.R. Schowalter, **Colloidal Dispersion**, *Cambridge University Press*, Cambridge, 1989.

A.V. Shenoy, **Rheology of Filled Polymer Systems**, *Kluwer*, Dordrecht, 1999.

G.V. Vinogradov, A.Ya. Malkin, **Rheology of Polymers**, *Springer*, Berlin, 1980.

5　Applied problems of rheology are presented (in addition to the above-listed monographs) in the books:

V. Capasso, **Mathematical Modeling of Polymer Processing**, *Springer*, Berlin, 2003.

J.M. Dealy, K.F. Wissbrun, **Melt Rheology and Its Role in Plastics Processing**, Van Nostrand, New York, 1990.

C.D. Han, **Rheology in Polymer Processing**, *Academic Press*, NY, 1976.

C.D. Han, **Multiphase Flow in Polymer Processing**, *Academic Press*, NY, 1981.

A.I. Isayev (Ed.), **Injection and Compression Molding Fundamentals**, *Marcel Dekker*, New York, 1987.

A.I. Isayev (Ed.), **Modeling of Polymer Processing. Recent Developments**, *Hanser*, Munich, 1991.

J.M. Piau, J.F. Agassant (Eds), **Rheology for Polymer Melt Processing**, *Elsevier*, Amsterdam, 1996.

R.I. Tanner, **Engineering Rheology**, 2nd Edition, *Oxford University Press*, Oxford, 2000.

M.R. Kamal, A.I. Isayev and S.-J. Liu (Eds.), **Injection Molding Technology and Fundamentals**, *Hanser*, Munich, 2009.

J.F. Steffe, **Rheological Methods in Food Process Engineering**, 2nd Edition, *Freeman Press*, East Lansing, 1996.

J.L. White, **Principles of Polymer Engineering Rheology**, *Wiley*, New York, 1990.

6　Experimental methods of rheology are discussed in the following books:

H.A. Barnes, J. F. Hutton, K. Walter, **An Introduction to Rheology**, *Elsevier*, Amsterdam, 1989.

D.V. Boger, K. Walters, **Rheological Phenomena in Focus**, *Elsevier*, Amsterdam, 1993.

A. Collyer, D.W. Clegg (Eds), **Rheological Measurement**, *Chapman and Hall*, London, 1998.

A. Collyer (Ed.), **Techniques in Rheological Measurement**, *Chapman and Hall*, London, 1993.

J.M. Dealy, **Rheometers for Molten Plastics. A Practical Guide to Testing and Property Measurement**, *Van Nostrand Reinhold*, New York, 1982.

J.M. Dealy, P.C. Saucier, **Rheology in Plastics Quality Control**, *Hanser*, Munich, 2000.

G.F. Fuller, **Optical Rheometry of Complex Liquids**, *Oxford University Press*, New York, 1995.

H. Janeschitz-Kriegl, **Polymer Melt Rheology and Flow Birefringence**, *Springer*, Berlin, 1983.

W.-M. Kulicke, B.C. Clasen, **Viscometry in Polymers and Polyelectrolytes**, *Springer*, Berlin, 2004.

V.I. Levitas, **Large Deformation of Materials with Complex Rheological Properties at Normal and High Pressures**, *Nova Science*, New York, 1996.

A.Ya. Malkin, A.A. Askadsky, V.V. Kovriga, A.E. Chalykh, **Experimental Methods of Polymer Physics**, *Prentice-Hall*, Englewood Cliffs, 1983.

K. Walters, **Rheometry**, *Chapman and Hall*, London, 1975.

R.W. Whorlow, **Rheological Techniques**, *Ellis Harwood*, New York, 1992.

7　Computational methods are very important in applications of rheology because of the rather complicated equations used in the calculation. However, although many papers in journals have been published, only a very limited number of books are devoted to this problem, for example:

T. J. Chung, **Computational Fluid Dynamics**, *Cambridge University Press*, New York, 2002.

M. J. Crochet, A. R. Davies, K. Walters, **Numerical Simulation of Non-Newtonian Flow**, *Elsevier*, Amsterdam, 1984.

R.G. Owens, T.N. Phillips, **Computational Rheology**, *Imperial College Press*, London, 2002.

J.R.A. Pearson, S.M. Richardson (Eds), **Computational Analysis in Polymer Processing**, *Applied Science Publishers*, London, 1983.

C. Pozdrikidis, **Fluid Dynamics. Theory, Computation and Numerical Simulation**, *Kluwer Academic Publisher*, Dordrecht, 2001.

C.L. Tucker (Ed.), **Computer Modeling for Polymer Processing**, *Hanser*, Munich, 1989.

Papers devoted to rheology appear in a large number of journals, but there are six special journals devoted to rheology, as follows:

Journal of Rheology

Journal of Non-Newtonian Fluid Mechanics

Nihon Reorogi Gakkaishi (Journal of Society of Rheology Japan)

Rheologica Acta

Applied Rheology

Korea-Australia Rheology Journal

New results in different branches of rheology can be found in these journals.

Continuum Mechanics as a Foundation of Rheology

Rheology is a science dealing with the deformation and flow of matter. Relationships between *stresses* and *deformations* are the fundamental concepts of continuum mechanics, which are discussed in this chapter. The modern history of rheology was marked by the publication of several books[1] in the 1940s, which impacted the education of future generations of rheologists.

1.1 STRESSES

Internal *stresses* are directly related to *forces* applied to a body regardless of their origin. Only in special cases do internal stresses exist in the absence of external forces. These are, for example, thermal stresses caused by temperature inhomogeneity throughout a body or frozen stresses stored as a result of the thermal and/or mechanical history of a body treatment caused by its heterogeneity.

1.1.1 GENERAL THEORY

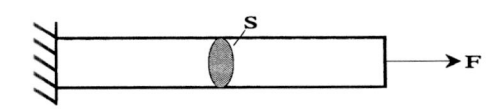

Figure 1.1.1. A bar loaded with a normal force.

Any external force applied to a body leads either to a movement of the body as a whole or to a change of its initial shape. Both may occur simultaneously. The movement of a body in space and/or its rotation around its center of gravity, with no change to its shape, is a subject of study by mechanics, and as such is not discussed in this book. The principal focus of our discussion here are changes that occur inside a body on the application of an external force. The applied forces create *dynamic reactions* at any point of a body, which are characterized by a physical factor called stress.

Stress can be explained using a simple example. Let us consider a body (a bar). The area of its normal cross-section is S (Fig. 1.1.1). The force, F, is normal to the surface, S. The specific force at any point of the cross-section equals F/S. The ratio is *normal stress* or *tensile stress*, σ_E:

$$\sigma_E = F/S \qquad\qquad [1.1.1]$$

i.e., *stress is the force per unit of the surface area*. The force at any surface may not be constant, i.e., be a function of coordinates. For example, a train moving on rails presses rails at local zones (where wheels touch the rail). The force is then distributed within the rail according to a complex pattern of stress distribution.

In our case, we do not consider force distribution because we have selected a small surface area, ΔS. A relative (specific) force, ΔF, acting on the area of ΔS is used to calculate the ratio $\Delta F/\Delta S$. By decreasing the surface area, we eventually come to its limiting form, as follows:

$$\sigma = \lim \frac{\Delta F}{\Delta S}, \text{ at } \Delta S \to 0, \text{ i.e., } \sigma = \frac{dF}{dS} \qquad [1.1.2]$$

This is a more general and exact definition of stress than given by Eq. 1.1.1 because it is related to a reference point, such as the surface area, ΔS. However, the definition is still not complete. A force at the area ΔS can have any direction, therefore a force is, in fact, a *vector* **F**. This vector can be decomposed into three components along three coordinate axes, in particular, it can be decomposed to one perpendicular and two tangential components. The perpendicular component is *normal stress*, and the tangential components are *shear stresses*.

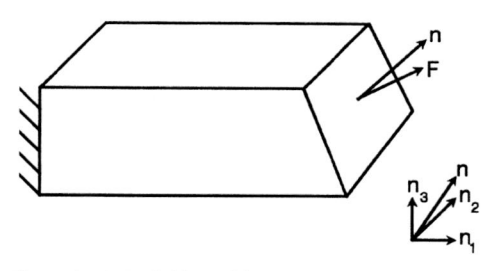

The selection of a small area ΔS is arbitrary, therefore it is better expressed by vector **n**, which determines the *orientation* of ΔS to the normal orientation. The stress is described by a combination of two vectors, **F** and **n**, defined at any reference point. This is shown in Fig. 1.1.2. The stress is a derivative dF/dn, and it is independent of the coordinate axis. Any vector is a physical object existing regardless of the choice of a coordinate system.

Figure 1.1.2. Definition of the stress tensor – two vectors: force, **F**, and orientation of a surface, **n**.

In practical applications, it is more convenient to operate with its projections on the coordinate axes rather than with a vector itself. Any vector can be decomposed into its three projections on the orthogonal coordinate axes; let it be Cartesian coordinates, x_1, x_2, and x_3.[2]

It follows from the above explanations that a complete characterization of stress as a physical object requires the identification of two vectors: a force and a normal orientation to the surface to which this vector is applied. The physical objects determined in such a manner are called *tensors*, and that is why stress is a value of tensor nature.

Let both vectors, **F** and **n**, defined at any reference point, be represented by their three projections along the orthogonal coordinate axes:

$$\mathbf{n} = n(n_1, n_2, n_3)$$

$$\mathbf{F} = F(F_1, F_2, F_3)$$

Nine values can be obtained from three projections of force on the surfaces determined by the three coordinate vectors. All values of force, F_i (i = 1, 2, 3), must be divided by the surface area to give the components of a stress tensor, σ_{ij}. The first index gives the orientation of a force, and the second index designates the orientation of a surface.

The result is written in the table form (*matrix*), as follows:

$$\sigma = \begin{bmatrix} \sigma_{11} & \sigma_{12} & \sigma_{13} \\ \sigma_{21} & \sigma_{22} & \sigma_{23} \\ \sigma_{31} & \sigma_{32} & \sigma_{33} \end{bmatrix} \qquad\qquad [1.1.3]$$

where the components of the stress tensor, σ_{ij}, mean the following: the first column represents components of a force (a force vector) that acts on the plane normal to the x_1 axis, the second column gives the same for x_2 axis, and the third for x_3 axis. The directions to the normal are indicated by the second indices.

The matrix contains all components (projections) of a force vector applied to different planes at an arbitrary point inside a body. In order to emphasize that this set of parameters presents a single physical object, i.e., stress tensor,[3] it is usual to put the table between the brackets.

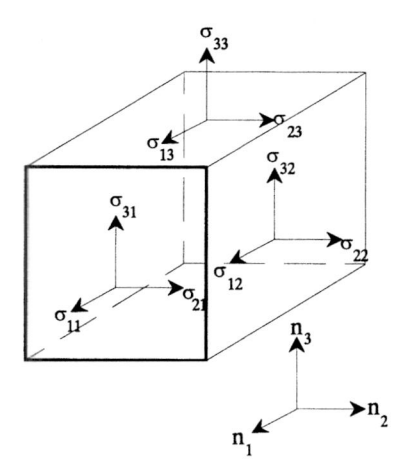

Figure 1.1.3. Three-dimensional stress state, a definition of the stress tensor components.

Fig. 1.1.3 shows all components of the stress tensor acting on a selected point. The components having the same numbers in their index are normal stresses, which are equivalent to the initial definition of the normal stress in Eq. 1.1.1, and all values with different numbers in the index are shear stresses.

All components of the stress tensor are determined at a point, and they can be constant or variable in space (inside a medium). It all depends on the distribution of external forces applied to a body. For example, the force field is homogeneous in Fig. 1.1.1, and thus a stress tensor is constant (inside a body). But the stress field (or stress distribution) is very complex in many other cases, for example, in a liquid flowing inside a channel or in the case of a roof covered with snow.

The values of the stress tensor components depend on the orientation of coordinate axes. They change with the rotation of the coordinate axes in space, though the stress state at this point is the same. It is important to remember that, regardless of the choice of the coordinate axes, this is the same physical object, invariant to the choice of the coordinate axis.

Some fundamental facts concerning the stress tensor (and any other tensor) are discussed below.

Comments – operations with tensors

There are several general rules concerning operations with objects of a tensor nature.[4] Two of them will be used in this chapter.

The first is the rule of summation of tensors. Tensor $\mathbf{A}$ is the sum of tensors $\mathbf{B}$ and $\mathbf{C}$ if components of $\mathbf{A}$, a_{ij}, are the sum of components b_{ij} and c_{ij} with the same indices, i.e., the equality $\mathbf{A} = \mathbf{B} + \mathbf{C}$ means that $a_{ij} = b_{ij} + c_{ij}$.

The second is the rule of multiplication of a tensor by a constant. Tensor $\mathbf{A}$ equals the product of a scalar constant k and a tensor $\mathbf{B}$ if components of $\mathbf{A}$, a_{ij}, are equal to kb_{ij}, i.e., the equality $\mathbf{A} = k\mathbf{B}$ means that $a_{ij} = kb_{ij}$. The unit tensor (also called the Kronecker delta, δ_{ij}) will be used below. This object is defined as a tensor, for which all diagonal (normal) components are equal to 1 and all shear components (when $i \neq j$) are equal to zero.

1.1.2 LAW OF EQUALITY OF CONJUGATED STRESSES

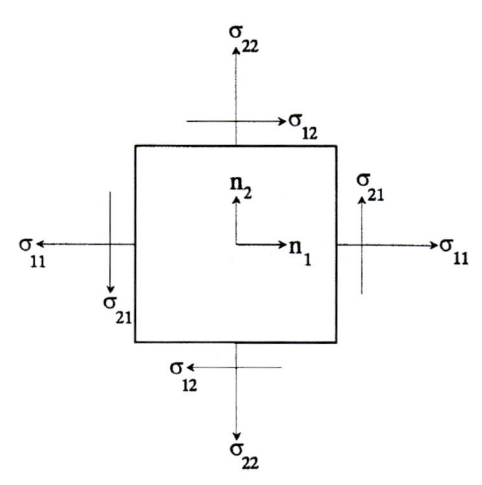

Figure 1.1.4. Two-dimensional (plane) stress state.

Let us consider a plane section of a unit cube in Fig. 1.1.3. The section is shown in Fig. 1.1.4. The rotational equilibrium condition about the central point of the square immediately gives equality

$$\sigma_{12} = \sigma_{21}$$

The same is true for any other pair of shear stresses. The general rule can be formulated as

$$\sigma_{ij} = \sigma_{ji} \qquad [1.1.4]$$

These equalities are known as the *Cauchy rule.*[5]

The result means that only three independent shear components of the stress tensor exist, and the stress state at a point is completely defined by six independent values: three normal, σ_{11}, σ_{22}, and σ_{33}, and three shear stresses, σ_{12}, σ_{13}, and σ_{23}.

However, it is necessary to mention that some special materials may exist, for which the Cauchy rule is invalid. It can happen if there is an inherent *moment of forces* acting inside any element of a medium.

1.1.3 PRINCIPAL STRESSES

The concept of *principal stresses* is a consequence of the dependence of stresses on the orientation of a surface. If stress components change on the rotation of the coordinate axes, there must be such orientation of axes, at which the numerical values of these components are extreme (maximum or minimum).

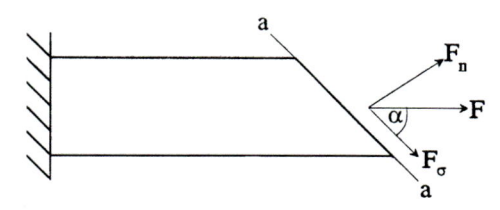

Figure 1.1.5. Stresses on the inclined section of a bar – decomposition of a normal force at the arbitrary oriented surface.

This idea is illustrated by a simple two-dimensional example generated from Fig. 1.1.1. Let the bar be cut at some angle α, as shown in Fig. 1.1.5, and let the force **F** act at this angle α to the plane *aa*. Then, it is easy to calculate two components of the vector **F** – normal and tangential forces, F_n and F_σ, respectively:

$$F_n = F\sin\alpha\,; \quad F_\sigma = F\cos\alpha$$

Then, the stress tensor components can be found, taking into account that the surface area of the inclined cross-section is $S/\sin\alpha$. Stress components, related to the force per unit of surface area, are found as follows:

normal stress, σ_E

$$\sigma_E = \frac{F_n}{S}\sin\alpha = \frac{F}{S}\sin^2\alpha = \sigma_0\sin^2\alpha \qquad [1.1.5]$$

shear stress, σ

$$\sigma = \frac{F_\sigma}{S}\sin\alpha = \frac{F}{S}\sin\alpha\cos\alpha = \frac{\sigma_0}{2}\sin 2\alpha \qquad [1.1.6]$$

where $\sigma_0 = F/S$.

The following special orientations can be found in a body:

at $\alpha = 90°$ the normal stress $\sigma_E = \sigma_0$ is at maximum, and the shear stress $\sigma = 0$;
at $\alpha = 45°$ the normal stress $\sigma_E = \sigma_0/2$ and the shear stress $\sigma = \sigma_0$ is at maximum;
at $\alpha = 0°$ both σ_E and σ are equal zero, i.e., this plane is free from stresses.

The discussion shows that, at any arbitrary orientation (or direction) of a body, both normal and shear stresses may exist. Moreover, there is always a direction at which the normal or the shear stresses are at maximum. The last observation is very important because various media resist the application of *extension* (normal) force or *shear* (tangential) force in different manners. For example, it is difficult to *compress* liquid (compression is achieved by the application of negative normal stresses) but it is very easy to shear liquid (to move one layer sliding over another). Another case: when a thin film is stretched, it breaks as a result of action of normal stresses, but shear stresses are practically negligible in this case.

The above examples are only illustrations of a general idea that all components of a stress tensor depend on the orientation of a surface because the size of a vector projection depends on the orientation of axes in space.

The theory of operations with tensor objects gives the general rules and equations for calculating components of the stress tensor in a three-dimensional stress state. The simpler equations for a two-dimensional stress state ("plane stress state") are given by Eqs. 1.1.5 and 1.1.6. Theoretical analysis shows that for any arbitrary stress tensor it is possible to find three orthogonal (i.e., perpendicular to each other) directions, at which normal stresses are extreme and shear stresses are absent (see Fig. 1.1.5). The normal stresses σ_{ii} are maximum along with the directions at which shear stresses are absent, $\sigma_{ij} = 0$ ($i \neq j$). These normal stresses are called the principal stresses.

The existence of the principal stresses constitutes a general law for any stress tensor. In fact, it is a particular case of a more general statement: the existence of three principal values is the general law for any tensor.

The concept of the principal stresses permits finding a minimal number of parameters that characterize the stress state at any point. It is more difficult to compare stress states at different points of a body or in different bodies operating with six independent components of the stress tensor acting along different directions. It is much easier to do so dealing with only three normal principal stresses. All numerical values of the components of a stress tensor depend on the choice of the coordinate axes, while the principal stresses do not.

The "threshold" effects on the material behavior can be treated in an unambiguous manner using the principal stresses as a criterion of an event, but not separate components of a stress tensor. It means that physical phenomena caused by the application of mechanical forces can be considered in terms of principal stresses. The examples include phase transition induced by applied forces; heat dissipation in the flow; storage of elastic energy; non-sag properties of some semi-liquid materials; rupture of solid bodies; the slow movement of snow with a sudden transition to avalanche; sand or mud on slopes, etc. The observed physical effects are usually caused by the principal stress which attains maximum value.

1.1.4 INVARIANTS OF A STRESS TENSOR

Knowledge of the principal stresses allows us to distinguish between different stress states of matter (e.g., three different values of the principal stresses or all principal stresses having the same value, etc.).

The principal stresses are characteristic of the stress state of a body (at a given point). They are not influenced by orientation. In other words, they are *invariant* to the choice of orientation.

How to calculate principal stresses if all components of the stress tensor are known for some arbitrary coordinate system is thus an essential practical question. The theory of tensors gives an answer to this question in the form of a cubic algebraic equation:

$$\sigma^3 - I_1\sigma^2 + I_2\sigma - I_3 = 0 \qquad [1.1.7]$$

and principal stresses denoted as σ_1, σ_2, and σ_3 appear to be the three roots of this equation. These roots are evidently expressed through coefficients of Eq. 1.1.7 – I_1, I_2, and $_3$. These coefficients are constructed by means of all components of a stress tensor for arbitrary orthogonal orientations in space as:

$$I_1 = \sigma_{11} + \sigma_{22} + \sigma_{33} \qquad [1.1.7a]$$

$$I_2 = \sigma_{11}\sigma_{22} + \sigma_{11}\sigma_{33} + \sigma_{22}\sigma_{33} - (\sigma_{12}^2 + \sigma_{13}^2 + \sigma_{23}^2) \qquad [1.1.8a]$$

$$I_3 = \sigma_{11}\sigma_{22}\sigma_{33} + 2\sigma_{12}\sigma_{13}\sigma_{23} - (\sigma_{11}\sigma_{23}^2 + \sigma_{22}\sigma_{13}^2 + \sigma_{33}\sigma_{12}^2) \qquad [1.1.9a]$$

The principal stresses σ_1, σ_2, and σ_3 do not depend on the orientation of axes of a unit cube (at a point) in space but they are expressed by values of I_1, I_2, and I_3. This leads to the conclusion that I_1, I_2, and I_3 are also invariant with respect to the choice of directions of orientation and that is why they are usually called *invariants of a stress tensor* at a point. According to its structure (the power of the components), I_1 is the first (linear), I_2 is the second (quadratic), and I_3 is the third (cubic) invariant. The invariants can also be expressed *via* the principal stresses only. These formulas are easily written based on Eqs. 1.1.7a -1.1.9a.

$$I_1 = \sigma_1 + \sigma_2 + \sigma_3 \qquad [1.1.7b]$$

$$I_2 = \sigma_1\sigma_2 + \sigma_1\sigma_3 + \sigma_2\sigma_3 \qquad [1.1.8b]$$

$$I_3 = \sigma_1\sigma_2\sigma_3 \qquad [1.1.9b]$$

Any combination of invariants I_1, I_2, and I_3 is also invariant with respect to the orientation of axes in space. Various mathematical structures of invariants can be derived but it is a fundamental result that three and only three independent values of such kind exist.

Invariants are characteristics of the physical state of matter under the action of forces, either internal or external. This means that neither any stress by itself nor its arbitrary combination but only invariants determine a possibility of occurrence of various physical effects and threshold phenomena, some of which were mentioned above.

The fundamental principle says that the *physical effects must be independent of the choice of a coordinate system* which is quite arbitrary, and that is why invariants (as the combination of principal stresses), which are values independent of a coordinate system, govern physical phenomena, which occur because of the application of mechanical forces.

In many practical applications, a two-dimensional (also known as plane stress) state exists with stress in the third direction being absent. Thin-walled articles having stress-free outer surfaces (e.g., balloons, membranes, and covers) are typical examples. The analysis of the two-dimensional stress state is an adequate solution in these cases. "Thin" means that the dimension in the direction normal to the surface is much smaller than in the other two directions.

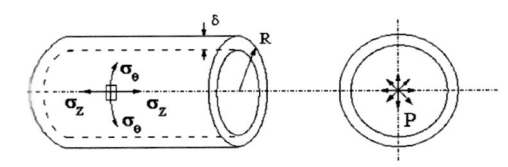

Figure 1.1.6. A thin-wall cylinder loaded by the inner pressure – stresses in the wall.

Examples. Stresses in a thin-wall cylinder

Internal pressure, p, in a thin-wall cylinder, closed by lids from both sides is typical of vessels working under pressure (chemical reactors, boilers, tubes in tires, and so on). Only normal stresses, σ_θ and σ_z, act inside a wall, where σ_θ is the stress acting in the circumferential direction and σ_z is the longitudinal stress (see Fig. 1.1.6). The values of these stresses are calculated from

$$\sigma_\theta = \frac{pR}{\delta}, \text{ and } \sigma_z = \frac{pR}{2\delta}$$

where R is the radius of a cylinder, and δ is the thickness of its wall.

This is a typical two-dimensional (plane) stress state, where both components of normal stresses are principal stresses and shear stresses are absent.

A long thin cylinder is twisted by applying a torque, T.

Torque is produced by relative turning, rotating, or twisting of a cylinder (inner or outer with no effect on the result). Shear stress, σ, can be calculated from

$$\sigma = \frac{2T}{\pi(2R + \delta)^2 \delta}$$

Generalization of plane stress state causes all components containing the index "3" to vanish, which gives the full stress tensor as in Eq. 1.1.10 instead of Eq. 1.1.3:

$$\sigma = \begin{bmatrix} \sigma_{11} & \sigma_{12} & 0 \\ \sigma_{21} & \sigma_{22} & 0 \\ 0 & 0 & 0 \end{bmatrix}$$ [1.1.10]

In this case, one principal stress, σ_3, is zero and two other, σ_1 and σ_2, are the roots of a quadratic (but not cubic) algebraic equation as follows:

$$\sigma_{1,2} = \frac{\sigma_{11} + \sigma_{22}}{2} \pm \sqrt{\left(\frac{\sigma_{11} + \sigma_{22}}{2}\right)^2 + \sigma_{12}^2}$$ [1.1.11]

There are two simple cases of the plane stress state: simple (or unidimensional) tension and simple shear. In the first case: $\sigma_1 = \sigma_{11} = \sigma_E$ and $\sigma_2 = 0$. In the second case: $\sigma_{11} = \sigma_{22} = 0$, $\sigma \equiv \sigma_{12}$ and therefore $\sigma_1 = \sigma$ and $\sigma_2 = -\sigma$. The last example demonstrates that, even when only shear stresses are applied, there are two orthogonal planes in a matter where only normal stresses act.

1.1.5 HYDROSTATIC PRESSURE – SPHERICAL TENSOR AND DEVIATOR

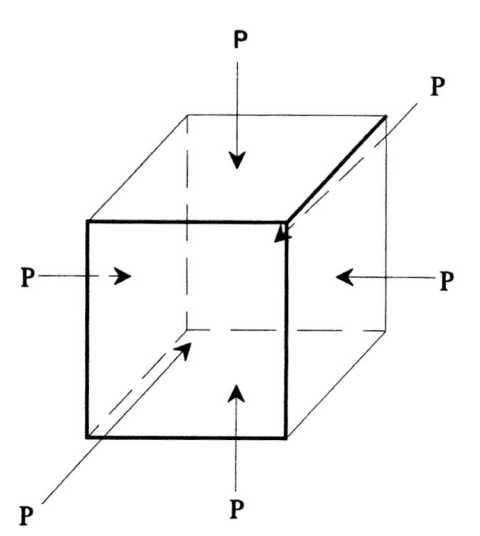

It seems pertinent that only normal stresses can change the volume of a body, while shear stresses may distort its form (shape). For this reason, it appears to be reasonable to divide a stress tensor into two components.

Fig. 1.1.7 shows the compression of a body under hydrostatic pressure. The main feature of *hydrostatic pressure* is the absence of shear stresses; hence, all stress components with exception of the normal stresses are equal to zero, and the stress tensor can be written as follows:

$$\sigma = \begin{bmatrix} -p & 0 & 0 \\ 0 & -p & 0 \\ 0 & 0 & -p \end{bmatrix}$$ [1.1.12]

Figure 1.1.7. Hydrostatic pressure, p – all-directional (tri-axial) compression of a unit cube.

All principal stresses are the same and equal to -p:

$$\sigma_1 = \sigma_2 = \sigma_3 = -p$$ [1.1.13]

and the "minus" sign shows that the force is directed inside an element of a matter.

Eq. 1.1.12 can be written in a short form using the above-discussed rules of operation with tensors:

$$\sigma_{ij} = -p\delta_{ij}$$ [1.1.14]

It means that $\sigma_{ii} = -p$ (the same two indexes of σ) and $\sigma_{ij} = 0$ (if $i \neq j$).

This stress tensor shows that shear stresses are absent in any direction in space. The tensor explains hydrostatic pressure shown in Fig. 1.1.7. The hydrostatic pressure is expressed as

$$p = -\frac{\sigma_{11} + \sigma_{22} + \sigma_{33}}{3} = -\frac{I_1}{3} \qquad [1.1.15]$$

The last definition of hydrostatic pressure is true for any stress state, even when σ_{11}, σ_{22}, and σ_{33} are not equal to each other. Eq. 1.1.15 is considered as a general definition of *pressure*, and the stress tensor, Eq. 1.1.12, is called the *spherical stress tensor*.

However, one intriguing question arises: whether the value $-I_1/3$, calculated according to Eq. 1.1.15, and called pressure, has the same physical meaning as the pressure used in thermodynamic relationships. Certainly, it is true for hydrostatic pressure when all normal stress components are the same, but this equivalence is assumed to be valid for an arbitrary stress state, though, possibly, it needs separate experimental evidence.

For the plane shear stress state, as was shown above, $\sigma_1 = -\sigma_2 = \sigma$ and $\sigma_3 = 0$. The same conclusion is correct for all other shear components of the stress tensor. It means that in *simple shear*, $I_1 = 0$, i.e., hydrostatic pressure is absent ($p = 0$). This shows that *shear stresses do not influence the volume of a body but may change its shape*.

It is now possible to write down a general expression for any stress tensor by separating the hydrostatic component. In this approach, all shear stresses remain untouched and each diagonal member of the tensor becomes equal to $(\sigma_{ij} - p)$.

This part of the stress tensor (complete tensor minus hydrostatic component) is called a *deviator* or deviatoric part of the stress tensor. It is thought that this part of the tensor is responsible for changes in the shape of a body but not its volume.

Uniaxial extension

The idea of splitting a stress tensor into spherical and deviatoric parts is well illustrated by the example of uniaxial stretching. It results in a body extension and it most likely leads to a volume change of a body. The question arises if the uniaxial extension is equivalent to negative hydrostatic pressure? Stress tensor for a uniaxial extension is written as

$$\sigma = \begin{bmatrix} \sigma_E & 0 & 0 \\ 0 & 0 & 0 \\ 0 & 0 & 0 \end{bmatrix} \qquad [1.1.16]$$

Similar to Fig. 1.1.1, all other forces except normal force, F, are absent. Therefore, there is no reason for other stress components, except for σ_{11}, and that is why all components in the matrix 1.1.16 equal zero (in particular $\sigma_{22} = \sigma_{33} = 0$), except for $\sigma_{11} = \sigma_E$.

It is now possible to split this tensor into hydrostatic and deviatoric parts, separating hydrostatic pressure (the remaining part of the stress tensor is the deviator). The deviator is simply a difference between full stress tensor and hydrostatic pressure. Then, the stress tensor for a uniaxial extension can be written as:

$$\sigma = -p\delta_{ij} + \begin{bmatrix} \frac{2}{3}\sigma_E & 0 & 0 \\ 0 & -\frac{1}{3}\sigma_E & 0 \\ 0 & 0 & -\frac{1}{3}\sigma_E \end{bmatrix} = -p\delta_{ij} + \frac{\sigma_E}{3}\begin{bmatrix} 2 & 0 & 0 \\ 0 & -1 & 0 \\ 0 & 0 & -1 \end{bmatrix} \qquad [1.1.17]$$

Any component of the full stress tensor, σ, equals the sum of the components of both addenda with the same indices; for example:

$$\sigma_{11} = -p + \frac{2}{3}\sigma_E = \frac{1}{3}\sigma_E + \frac{2}{3}\sigma_E = \sigma_E$$

and

$$\sigma_{22} = \sigma_{33} = -p - \frac{1}{3}\sigma_E = \frac{1}{3}\sigma_E - \frac{1}{3}\sigma_E = 0$$

The last rearrangements prove that Eqs. 1.1.16 and 1.1.17 are equivalent.

Comparison of Eqs. 1.1.12 and 1.1.17 shows that the *uniaxial extension is not equivalent to hydrostatic pressure* (the sign is not essential in this discussion) as the former leads to the appearance of a deviatoric component of the stress tensor too. In particular, it means that in the case of uniaxial extension it is possible to find such directions in a body where the shear stress exists – contrary to hydrostatic pressure where the shear stresses are principally absent (see discussion of Fig. 1.1.7).

The interpretation of the uniaxial extension as the sum of hydrostatic pressure and deviator explains that one-dimensional tension creates not only negative pressure ("negative" means that stresses are oriented outward of unit areas inside a body) but also different normal stresses acting in all directions. This is the physical reason why all dimensions of a body change in the uniaxial tension (increase along the direction of the extension but decrease in perpendicular (lateral) directions).

Some comments and examples

The uniaxial extension is important for many technical applications. In particular, this mode of loading is frequently used in material testing. Stress calculations in uniaxial extension need to be done very accurately. At first glance, the problem is well represented by Eq. 1.1.1. However, two limitations are essential. First, this equation is valid only far from the ends, because stress distribution near the ends of a sample is determined by details of force application, which is usually not uniform. Thus, the stresses can be calculated from Eq. 1.1.1 only for long samples. Second, the cross-sectional area of the sample changes with extension. Therefore Eq. 1.1.1 only describes the initial state of a sample. In technical applications, stress is often calculated without considering such effects. It is correct then to consider it as some "conventional" or "engineering" stress.

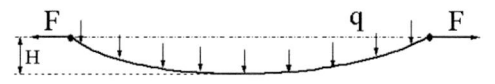

Figure 1.1.8. A sagging fiber loaded by a distributed load.

A loading under own weight of a sample suspended at one end is a special case of uniaxial extension. The normal (extensional) stress is caused by the gravitational force. The maximum stress, σ_{max}, acts at the cross-section at which it is suspended. This stress equals $\sigma_{max} = \rho g L$ (where ρ is density, g is the gravitational acceleration and L is the length of the sample). The σ_{max} increases as the sample length increases.

There exists a length of the sample at which σ_{max} exceeds the material strength. This limiting length, correspondng to σ_{max} can be used as a measure of the material's strength. Such measure is used in engineering practice for characterizing the strength of fibers expressed by a "breaking length".

Analysis of uniaxial loading can be useful in solving many practical problems. For example, let us consider a horizontal fiber (string, rope, etc.) loaded by a distributed load, q (Fig.1.1.8). This distributed load can be its own weight, snow cover, strong wind, etc. Load provokes the sagging of a flexible material. The application of the extension force, F, is one possibility of counteracting too extensive, unacceptable sagging. The length between the anchoring points at both sides is L. Then the height of the maximum sagging, H, is calculated from:

$$H = \frac{qL^2}{8F}$$

There is a direct correspondence between extending force and sagging height. The increase of extension force results in the decrease in sagging. However, the force cannot be too large because the increase in stress may eventually exceed the strength limit of a fiber.

1.1.6 EQUILIBRIUM (BALANCE) EQUATIONS

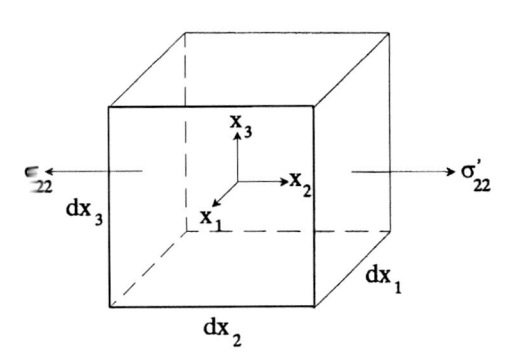

Figure 1.1.9. Components of the stress tensor in Cartesian coordinates – illustration of the stress difference at parallel cube faces along the infinitesimal distance.

The distribution of stresses throughout a body is described by equilibrium (or balance) equations as formulated by Navier,[6] Poisson,[7] and Cauchy[8] in their classical studies. In essence, it is a form of Newton's second law written for a continuum because the sum of all forces at a point equals to the product of mass (of this point) multiplied by acceleration.

A "point" in the theoretical analysis is an elementary (infinitesimal) space with sides oriented along the orthogonal coordinate axes (in Fig. 1.1.9 this space is a cube in the Cartesian coordinates). The idea of the analysis consists of a projection of all the external forces on the faces of the cube along three coordinate axes and their sum equals zero.

Forces are continuously changing at infinitesimal distances along the axis. If there is no special case having discontinuities in force, it is reasonable to think that, for example, a force on the left-hand face of the cube (Fig. 1.1.9) equals to σ_{22}, and on the parallel right-hand face it equals

$$\sigma'_{22} = \sigma_{22} + \frac{\partial \sigma_{22}}{\partial x_2} dx_2$$

The last relationship supposes that stress σ_{22} changes by infinitesimally small value to σ'_{22} at a small distance dx_2.

Other forces also exist and need to be taken into account in the formulation of the balance equations. These are forces, presented by the vector $\mathbf{X}(X_1, X_2, X_3)$ per unit volume, and inertia forces equal (per unit volume) to $\rho\mathbf{a}$, where ρ is the density of matter and $\mathbf{a}(a_1, a_2, a_3)$ is a vector of acceleration.

Then, writing the sum of projections of all forces (stresses are multiplied by the unit areas of the cube face) and dividing them all by $dx_1 dx_2 dx_3$ (which is an infinitesimally small value of the higher order), one comes to *equilibrium* (or *balance*) *equations* or *equations of momentum conservation*. For all three coordinate axes, this gives the following system of equations:

$$\frac{\partial \sigma_{11}}{\partial x_1} + \frac{\partial \sigma_{12}}{\partial x_2} + \frac{\partial \sigma_{13}}{\partial x_3} + X_1 = \rho a_1$$

$$\frac{\partial \sigma_{21}}{\partial x_1} + \frac{\partial \sigma_{22}}{\partial x_2} + \frac{\partial \sigma_{23}}{\partial x_3} + X_2 = \rho a_2 \qquad [1.1.18]$$

$$\frac{\partial \sigma_{31}}{\partial x_1} + \frac{\partial \sigma_{32}}{\partial x_2} + \frac{\partial \sigma_{33}}{\partial x_3} + X_3 = \rho a_3$$

The system of Eqs. 1.1.18 includes pressure gradients of normal components of the stress tensor. Sometimes, the pressure gradient is written separately and in this case, σ_{11}, σ_{22}, and σ_{33} must be regarded as deviatoric components of the stress tensor.

For many rheological applications, it is reasonable to treat problems restricted to static equilibrium, and in such cases $a = 0$. The existence of a body force is important, for example, if movement occurs because of the action of gravity (e.g., sagging paints or sealants from vertical or inclined surfaces, the flow of glaciers, etc.). However, in many cases the influence of these forces is negligible and it is possible to assume that $X = 0$.

Then it is possible to omit the last two members of the balance equations and to simplify the system of Eqs. 1.1.18. This simplified (and usually used) system of balance equations is written as follows:

$$\frac{\partial \sigma_{11}}{\partial x_1} + \frac{\partial \sigma_{12}}{\partial x_2} + \frac{\partial \sigma_{13}}{\partial x_3} = 0$$

$$\frac{\partial \sigma_{21}}{\partial x_1} + \frac{\partial \sigma_{22}}{\partial x_2} + \frac{\partial \sigma_{23}}{\partial x_3} = 0 \qquad [1.1.19]$$

$$\frac{\partial \sigma_{31}}{\partial x_1} + \frac{\partial \sigma_{32}}{\partial x_2} + \frac{\partial \sigma_{33}}{\partial x_3} = 0$$

Equilibrium can be considered with respect to different coordinate systems but not restricted to a Cartesian system. The choice of coordinates is only a question of convenience in solving a specific boundary problem. The choice of the coordinate system depends, generally, on the shape and the type of symmetry of a geometrical space which is the most convenient for an application. For example, if the round shells or tubes with one axis of symmetry are discussed, the most convenient coordinate system is cylindrical polar coordinates with r, z, and θ axes.

Components of the stress tensor in the cylindrical (polar) coordinates are shown in Fig. 1.1.10. The static balance equations in the absence of inertia forces, $a = 0$, and volume – body – forces, $X = 0$, for the point (or infinitesimal volume element), shown in Fig.

1.1.10, represent equilibrium conditions with respect to r, z, and θ directions. These equations can be written as:

$$\frac{\partial \sigma_{rr}}{\partial r} + \frac{\partial \sigma_{rz}}{\partial z} + \frac{1}{r}\frac{\partial \sigma_{r\theta}}{\partial \theta} + \frac{\sigma_{rr} - \sigma_{\theta\theta}}{r} = 0$$

$$\frac{\partial \sigma_{\theta r}}{\partial r} + \frac{1}{r}\frac{\partial \sigma_{\theta\theta}}{\partial \theta} + \frac{\partial \sigma_{\theta z}}{\partial z} + \frac{2\sigma_{\theta r}}{r} = 0 \qquad\qquad [1.1.20]$$

$$\frac{\partial \sigma_{zr}}{\partial r} + \frac{1}{r}\frac{\partial \sigma_{z\theta}}{\partial \theta} + \frac{\partial \sigma_{zz}}{\partial z} + \frac{\sigma_{zr}}{r} = 0$$

The meaning of the stress tensor components is explained in Fig. 1.1.10.

Some practical cases are symmetrical to the z axis so that all terms containing $\partial/\partial z$ become zero as does the shear stress $\sigma_{r\theta}$. In some cases, the cylindrical bodies can be very long and variations of stresses along the axis of symmetry are absent (or can be taken as negligibly small). This allows us to continue simplification of the balance equations written in the cylindrical coordinates. In this case, the balance equations reduce to

$$\frac{d\sigma_{rr}}{dr} + \frac{\sigma_{rr} - \sigma_{\theta\theta}}{r} = 0 \qquad [1.1.21]$$

All shear components of the stress tensor are absent.

All systems of balance equations, Eqs. 1.1.18 - 1.1.20, contain 6 unknown space functions (formally they contain 9 stress tensor components, but the use of the Cauchy rule decreases this number to 6), i.e., the stress components depend on space coordinates in an inhomogeneous stress field. In order to close the system of equations – to make it complete –

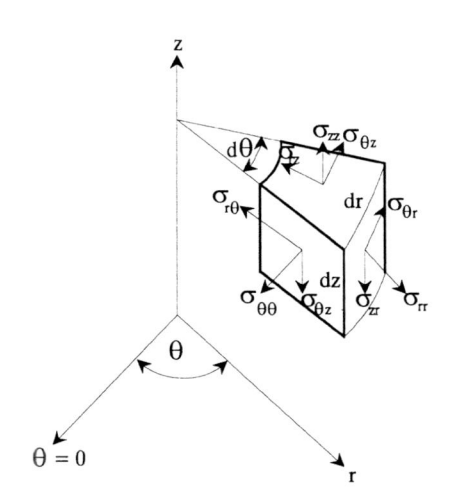

Figure 1.1.10. Components of the stress tensor in cylindrical (polar) coordinate system.

it is necessary to add the constitutive equations, connecting stress components with deformations. This is the central problem of rheology because these equations express the rheological properties of matter.

1.2 DEFORMATIONS

1.2.1 DEFORMATIONS AND DISPLACEMENTS

1.2.1.1 Deformations

The result of the action of external forces can either be the movement of a body in space or a change to its shape. Continuum mechanics is interested in changes occurring inside a body. The change of body shape is essentially the change of *distances* between different

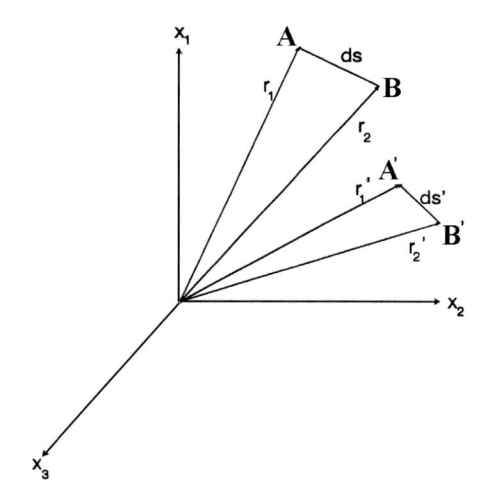

Figure 1.2.1. Displacements of two points in a body –
the origin of deformations.

sites inside the material, and this phenomenon is called *deformation*. Deformation is just a geometrical concept and all interpretations of this concept have clear geometrical images.

The change of distances between points inside a body can be monitored by following changes of very small (infinitesimally small) distances between two neighboring points.

Let the initial distance between two points A and B in the material be ds (Fig. 1.2.1). For some reason, they both move and their new positions become A' and B'. Their movement in space is not of interest by itself for continuum mechanics. Only the new distance between them, which becomes, ds', is of interest. Moreover, the absolute value of the difference (ds' – ds) is also not important for continuum mechanics, because the initial length ds might be arbitrary. Only the relative change of the distance between two points is relevant, and it is determined as

$$\varepsilon = \frac{ds' - ds}{ds} \qquad\qquad [1.2.1]$$

The distance between points A and B is infinitesimal. Assuming that a body after deformation remains continuous (between sites A and B), the distance between points A' and B' is still infinitesimal.

The definition, Eq. 1.2.1, is not tied to any coordinate system, and it means that ε is a scalar object. However, this value can be expressed through components of the tensor of deformation (or strain).[9]

Square of ds is calculated as:

$$(ds)^2 = (dx_1)^2 + (dx_2)^2 + (dx_3)^2 \qquad\qquad [1.2.2]$$

Here, three Cartesian coordinates are defined as x_1, x_2, and x_3.
The square of the length $(ds')^2$ is calculated from

$$(ds')^2 = (dx'_1)^2 + (dx'_2)^2 + (dx'_3)^2 \qquad\qquad [1.2.3]$$

The coordinates of the new position of the A'B' length are calculated from

$$dx'_1 = \left(1 + \frac{\partial u_1}{\partial x_1}\right)dx_1 + \frac{\partial u_1}{\partial x_2}dx_2 + \frac{\partial u_1}{\partial x_3}dx_3$$

$$dx'_2 = \frac{\partial u_2}{\partial x_1}dx_1 + \left(1 + \frac{\partial u_2}{\partial x_2}\right)dx_2 + \frac{\partial u_2}{\partial x_3}dx_3 \qquad\qquad [1.2.4]$$

$$dx'_3 = \frac{\partial u_3}{\partial x_1}dx_1 + \frac{\partial u_3}{\partial x_2}dx_2 + \left(1 + \frac{\partial u_3}{\partial x_3}\right)dx_3$$

With neglecting the terms of higher orders than dx, it is easy to calculate the difference $(ds')^2 - (ds)^2$, which equals to

$$(ds')^2 - (ds)^2 = 2[\varepsilon_{11}(dx_1)^2 + \varepsilon_{22}(dx_2)^2 + \varepsilon_{33}(dx_3)^2] +$$

$$+ 4[\varepsilon_{12}dx_1 dx_2 + \varepsilon_{13}dx_1 dx_3 + \varepsilon_{23}dx_2 dx_3] \qquad [1.2.5]$$

The change of length is expressed by six values of ε_{ij}, which can be expressed in symmetric form as

$$\varepsilon_{ij} = \frac{1}{2}\left(\frac{\partial u_i}{\partial x_j} + \frac{\partial u_j}{\partial x_i}\right) + \frac{1}{2}\left(\frac{\partial u_1}{\partial x_i}\frac{\partial u_1}{\partial x_j} + \frac{\partial u_2}{\partial x_i}\frac{\partial u_2}{\partial x_j} + \frac{\partial u_3}{\partial x_i}\frac{\partial u_3}{\partial x_j}\right) \qquad [1.2.6]$$

The indices i and j are used instead of 1, 2, and 3 for brevity to not repeat Eq. 1.2.6 for six components of ε_{ij}.

The values ε_{ij} are not equal to the change of length of the distance AB but are only the *measures* of this change.

It is possible to prove that the values are components of a tensor, and this tensor is called the *tensor of large deformations*.[10] The complete expression for ε_{ij} consists of linear (first term in Eq. 1.2.6) and quadratic (second term in Eq. 1.2.6) terms.

If derivatives in Eq. 1.2.6 are small (<< 1) and their pairs of products, which enter into the second right-hand side term in Eq. 1.2.6, are negligibly smaller than derivatives, relationships can be further simplified. The derivatives can be omitted, and only the first term of the equation remains. This only holds true for small deformations, and that is why the tensor consisting of only of the first derivatives is called a *tensor of small* or *infinitesimal deformation* (or *strain*). This tensor, d_{ij}, can be written as follows:

$$d = \begin{bmatrix} \dfrac{\partial u_1}{\partial x_1} & \dfrac{1}{2}\left(\dfrac{\partial u_1}{\partial x_2} + \dfrac{\partial u_2}{\partial x_1}\right) & \dfrac{1}{2}\left(\dfrac{\partial u_1}{\partial x_3} + \dfrac{\partial u_3}{\partial x_1}\right) \\[3mm] \dfrac{1}{2}\left(\dfrac{\partial u_2}{\partial x_1} + \dfrac{\partial u_1}{\partial x_2}\right) & \dfrac{\partial u_2}{\partial x_2} & \dfrac{1}{2}\left(\dfrac{\partial u_2}{\partial x_3} + \dfrac{\partial u_3}{\partial x_2}\right) \\[3mm] \dfrac{1}{2}\left(\dfrac{\partial u_3}{\partial x_1} + \dfrac{\partial u_1}{\partial x_3}\right) & \dfrac{1}{2}\left(\dfrac{\partial u_3}{\partial x_2} + \dfrac{\partial u_2}{\partial x_3}\right) & \dfrac{\partial u_3}{\partial x_3} \end{bmatrix} \qquad [1.2.7]$$

The first row of the tensor d_{ij} represents the projections of deformations along the x_1 axis, and so on.

It is worth repeating that the components of ε_{ij} tensor, as defined by Eq. 1.2.1, were calculated based on pure geometrical arguments. The final result of these calculations is given by Eq. 1.2.6, or in the case of small deformations, by Eq. 1.2.7.

In the separate sections, the small deformation tensor, d_{ij}, and the complete (large deformation) tensor, ε_{ij}, will be further discussed.

1.2.1.2 DISPLACEMENTS

As shown in section 1.1.1, any tensor can be defined by two vectors. It is similar to deformations. The position of any site (point) in a body is characterized by its radius vector, $\mathbf{r}$. Two sites are involved in the definition of deformation, A and B. Therefore it is necessary to introduce two vectors: $\mathbf{r}_1$ for point A and $\mathbf{r}_2$ for the point B.

The value u_i, entering Eq. 1.2.6 via expressions for ε_{ij}, characterizes projections of the displacement vector, $\mathbf{u}$, which represents the movement of the site A into its new position A'.

The quantitative determination of deformation can be accomplished by following *displacement*, $\mathbf{u} = (\mathbf{dr}_1 - \mathbf{dr}_2)$. The result of subtraction of two vectors is also a vector, and it can be expressed by its three projections: $\mathbf{u}(u_1, u_2, u_3)$. Relative displacement is expressed as $(\mathbf{dr}_1 - \mathbf{dr}_2)/\mathbf{dr}_1$. This object – contrary to the vector $\mathbf{u}$ – is characterized not only by its length but also by its orientation in space. Since two vectors, $\mathbf{u}$ and $\mathbf{x}(x_1, x_2, x_3)$, describe the relative displacement, the latter is of tensorial nature. Indeed, deformation and relative displacement are tensors, and components of both tensors can be calculated through the derivative $d\mathbf{u}/d\mathbf{x}$. It is also pertinent that there are nine such values (three projections of vector $\mathbf{u}$ and three of vector $\mathbf{x}$), as it could be expected for a tensor.

The values of all derivatives are dimensionless and they are expressed in absolute numbers or percents.

The tensor of relative displacement, $\mathbf{g}$, is, by definition,

$$\mathbf{g} = \text{grad } \mathbf{u} \qquad\qquad [1.2.8]$$

and it can be written via the components, g_{ij}, of this tensor

$$\mathbf{g} = \begin{bmatrix} \dfrac{\partial u_1}{\partial x_1} & \dfrac{\partial u_1}{\partial x_2} & \dfrac{\partial u_1}{\partial x_3} \\[2ex] \dfrac{\partial u_2}{\partial x_1} & \dfrac{\partial u_2}{\partial x_2} & \dfrac{\partial u_2}{\partial x_3} \\[2ex] \dfrac{\partial u_3}{\partial x_1} & \dfrac{\partial u_3}{\partial x_2} & \dfrac{\partial u_3}{\partial x_3} \end{bmatrix} \qquad\qquad [1.2.9]$$

The first row includes derivatives of the u_1-component of displacement along the three coordinate axes, the second row is the same for the u_2-component, and the third, for the u_3-component of the vector $\mathbf{u}$.

The displacement tensor, defined by Eq. 1.2.9, is *not* deformation.

It is quite evident that the tensors d_{ij} and g_{ij} are not equivalent. The difference between them becomes clear if one decomposes the components of the tensor g_{ij} into two parts in the following manner:

$$g_{ij} = \frac{\partial u_i}{\partial x_j} = \frac{1}{2}\left(\frac{\partial u_i}{\partial x_j} + \frac{\partial u_j}{\partial x_i}\right) + \frac{1}{2}\left(\frac{\partial u_i}{\partial x_j} - \frac{\partial u_j}{\partial x_i}\right) \qquad\qquad [1.2.10]$$

The first, the so-called symmetrical, part of the tensor g_{ij}, coincides with the deformation tensor d_{ij}, but it is evident that the deformations are something different than the displacements.

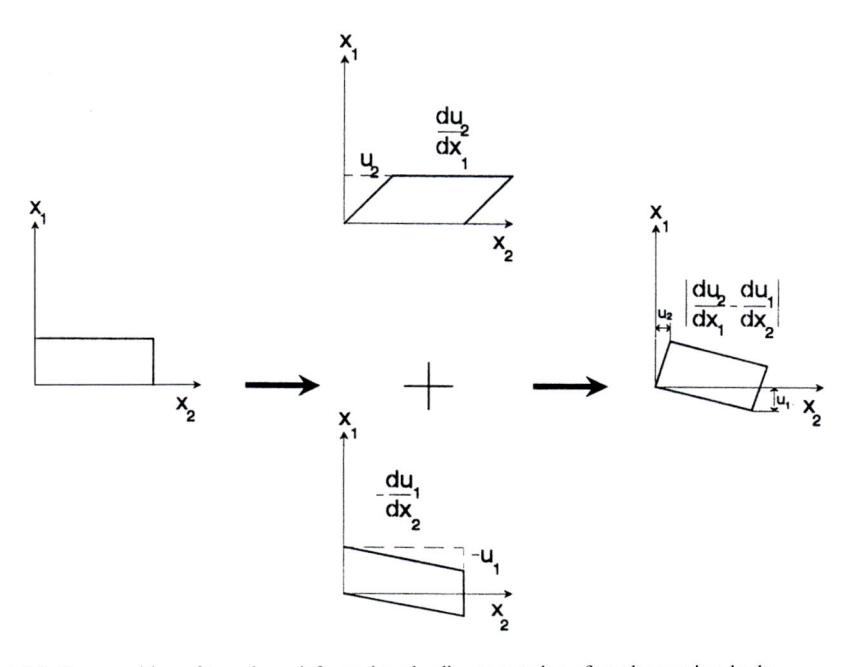

Figure 1.2.2. Superposition of two shear deformations leading to rotation of an element in a body.

The sense of this difference or the meaning of the second, the so-called antisymmetrical, part of the displacement tensor is explained in Fig. 1.2.2. Let us follow the deformation of an infinitesimal two-dimensional (plane) body element drawn here as a rectangle in the left diagram. Two displacements, u_2 and u_1, having gradients du_2/dx_1 and $-du_1/dx_2$, may occur as shown in the central part of Fig. 1.2.2. Now, let us superimpose these two displacements, as shown in the right diagram. It is evident from Fig. 1.2.2 that the summation of du_2/dx_1 and $-du_1/dx_2$ does not lead to deformation but to the rotation of the body element. It means that the second term in Eq. 1.2.10 represents rotation, but not deformation. It can be written in the following manner:

$$g_{ij} = d_{ij} + \theta_{ij} \qquad [1.2.11]$$

where θ_{ij} is given by the following equation:

$$\theta_{ij} = \frac{1}{2}\left(\frac{\partial u_i}{\partial x_j} - \frac{\partial u_j}{\partial x_i}\right) \qquad [1.2.12]$$

These values are the components of the *tensor of rotations* (*vorticity*) of infinitesimal volumes inside a body. Thus displacement at any point of a body is a sum of deformation and rotation.

1.2.2 INFINITESIMAL DEFORMATIONS: PRINCIPAL VALUES AND INVARIANTS

The pure geometrical analysis demonstrates that the diagonal components of tensor d_{ij} expressed by Eq. 1.2.7 are equivalent to relative elongations (extension ratios) and non-diagonal components are shear or changes of angles between two orthogonal lines at a point.

    The tensor of small (infinitesimal) deformations has all the general features of any other tensor, for example, the stress tensor discussed above. In particular, it is possible to calculate the principal values and invariants of this tensor using the same equations as for the stress tensor, only with changes in symbols. However, the invariants of the deformation tensor have a definite geometrical interpretation.

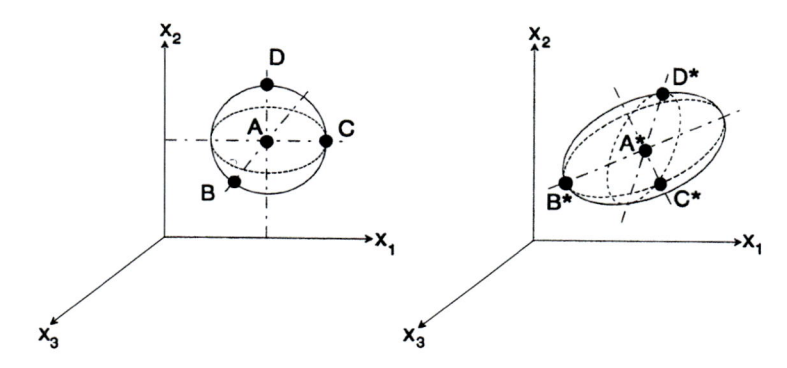

Figure 1.2.3. Transformation of a sphere into an ellipsoid as a consequence of three principal deformations along their axes.

    For the infinitesimal deformation tensor, the principal deformations, d_1, d_2, and d_3, are extensions in three orthogonal directions. It can be illustrated by deformations in the vicinity of some arbitrary point. Let us represent an infinitely small volume in a body as a sphere (Fig. 1.2.3) with a center positioned at a point A and radius of the sphere dr (infinitesimal small length). The coordinates of the central point A are x_1, x_2, and x_3. As a result of movements and displacements, the following changes have taken place in a body: point A has moved to a new position A*, the directions of the radii AB, AC, and AD have changed to the directions A*B*, A*C*, and A*D*, respectively. As a result, the sphere itself has transformed into an ellipsoid with semi-axes of length $(1 + d_1)$dr, $(1 + d_2)$dr, and $(1 + d_3)$dr, respectively.

    The deformations characterize the change in the shape of a volume element of the body on the transition from a sphere to an ellipsoid. Besides they determine the relative change in volume, ε_V, which can be written as follows:

$$\varepsilon_V = \frac{V_{ell} - V_{sph}}{V_{sph}}$$

A simple calculation shows that

$$\varepsilon_V = (1 + d_1)(1 + d_2)(1 + d_3) - 1 \qquad [1.2.13]$$

It is very easy to show that ε_V is expressed by invariants of the deformation tensor. The change in volume must not be associated with a choice of the coordinate system, and the invariants do not depend on the coordinate axes. Therefore, only invariants can determine the change of volume. The relationship between invariants is simple if deformations are small and it is possible to neglect quadratic terms in Eq. 1.2.13. Eq. 1.2.13 gives the following result:

$$\varepsilon_V = d_1 + d_2 + d_3 \qquad [1.2.14]$$

i.e., *volumetric changes are equal to the first (linear) invariant* of the tensor of infinitesimal deformations and that is its physical meaning.

The volumetric changes in deformation can also be represented by extension ratios, λ_i. For this purpose, let us (conditionally) cut out a small rectangular parallel slab, at some site in a body, oriented along the principal axes. Let the length of its edges be a, b and c before deformation, and let them become a*, b*, and c* as a result of deformation. Then, the extension ratios are:

$$\lambda_1 = a^*/a; \ \lambda_2 = b^*/b; \ \lambda_3 = c^*/c$$

The volume change is calculated as

$$\frac{V^* - V}{V} = \frac{\Delta V}{V} = \frac{a^* b^* c^*}{abc} - 1 = \lambda_1 \lambda_2 \lambda_3 - 1 \qquad [1.2.15]$$

The last equation shows a very simple rule of the constancy of volume in deformations of any type:

$$\lambda_1 \lambda_2 \lambda_3 = 1 \qquad [1.2.16]$$

Like any other tensor, the deformation tensor, d_{ij}, can be decomposed into spherical and deviatoric parts. The first invariant is the volume change. It is possible to write:

$$d_{ij} = \frac{\varepsilon_V}{3} \delta_{ij} + d_{ij}^{(dev)} \qquad [1.2.17]$$

The second term, on the right-hand side of the equation, is a deviatoric part, $d_{ij}^{(dev)}$ of the d_{ij} tensor which describes shape transformations occurring without changes in volume. Splitting the deformation tensor, d_{ij}, into spherical and deviatoric parts corresponds to separating the complete deformation into changes of volume and shape.

1.2.3 LARGE (FINITE) DEFORMATIONS

The difference between small (*infinitesimal*) and large (*finite*) deformations depends on the values of derivatives in Eq. 1.2.6. If all derivatives are much smaller than 1, the quadratic terms, i.e., products of derivatives (in parentheses), can be neglected and the tensor d_{ij} is used instead of ε_{ij}.

In the discussion of the concept of large deformations, it is always assumed that a *reference state* of deformation can be established. In this sense, the flow of liquid may not be considered as a deformation because all states are equivalent. The liquid does not have an initial (or reference) state. That is why only materials having *memory* of their initial state are important in determining deformations. Having such an approach, it is very easy to illustrate the essential difference between small and large deformations, using the simplest model of uniaxial extension from Fig. 1.2.4. Let a fiber (or a bar) of the length l_0 be stretched by Δl. The simple question is: what is the deformation in this case?

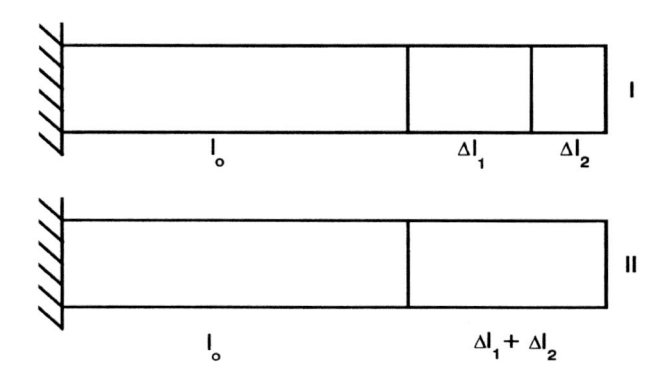

Figure 1.2.4. Two ways of realization of large deformation – two-step extension (I) or one-step extension (II).

In the first case, let $l_0 = 1$ and $\Delta l = 0.1$. The so-called *engineering measure* of deformation is

$$\varepsilon^* = \frac{\Delta l}{l_0} \qquad [1.2.18]$$

and in the example under discussion $\varepsilon^* = 0.1$ or 10%.

The reasoning becomes more complex if Δl is comparable with l, for example, let Δl be equal 1. An engineering measure of deformation is the characteristic of the change of specimen length, and $\varepsilon^* = 1$ (or 100%). But this approach to the definition of deformation contains an inherent contradiction. Let us compare two situations, drawn in Fig. 1.2.4. In the first case (case I), the increase in the length occurs in two consequent steps: initially by Δl_1 and then, separately, Δl_2. Then, the deformation in the first step is $\varepsilon_1^* = \Delta l_1/l_0$, and, in the second step, it is $\varepsilon_2^* = \Delta l_2/l_1$, because the initial length of the sample in the second step is l_1. The total deformation, ε_{total}, is the sum of both deformations, as follows

$$\varepsilon_{total}^{I} = \varepsilon_1^* + \varepsilon_2^* = \frac{\Delta l_1}{l_0} + \frac{\Delta l_2}{l_1} = \frac{l_0(\Delta l_1 + \Delta l_2) + \Delta l_1^2}{l_0 l_1}$$

where l_0 is the initial length of the sample and $l_1 = l_0 + \Delta l_1$.

In the second case (case II), the increase of the length is achieved just in one step. This increase equals $(\Delta l_1 + \Delta l_2)$ and the total deformation is calculated from

$$\varepsilon_{total}^{II} = \frac{\Delta l_1 + \Delta l_2}{l_0} = \frac{l_0(\Delta l_1 + \Delta l_2) + \Delta l_1^2 + \Delta l_1 \Delta l_2}{l_0 l_1}$$

If elongations are small ($\Delta l_1 \ll 1$, and $\Delta l_2 \ll 1$), the difference between ε^I_{total} and ε^{II}_{total} is negligible. However, if it is not so, then the above-written formulas clearly demonstrate that $\varepsilon^I_{total} \neq \varepsilon^{II}_{total}$, i.e., the final results of the extension are different. This contradicts the physical meaning of the experiment interpretation: in reality, the final result is the same in both cases and the sample does not "know" which way it was brought to the final state, whereas calculations show a difference. This contradiction appears only as a result of large deformations because if deformations are small, the quadratic terms in formulas for ε_{total} are negligible in comparison to the linear terms.

It becomes apparent that there is a need to introduce such a measure of deformation that does not depend on the sequence of operations. Such measure is called a logarithmic or the Hencky strain measure,[11] ε^H, which is defined by:

$$\varepsilon^H = \ln\left(\frac{l_0 + \Delta l}{l_0}\right) \qquad [1.2.19]$$

It is easy to prove that large deformations analyzed by this measure obey the law of additivity. Therefore, in the example discussed above, the resulting deformation, determined by the Hencky strain measure, does not depend on the history of deformation, as required:

$$\varepsilon^I_{total} = \ln\left(\frac{l_0 + \Delta l_1}{l_0}\right) + \ln\left[\frac{l_0 + (\Delta l_1 + \Delta l_2)}{l_0 + \Delta l_1}\right] = \ln\left[\frac{l_0 + (\Delta l_1 + \Delta l_2)}{l_0}\right]$$

and

$$\varepsilon^{II}_{total} = \ln\left[\frac{l_0 + (\Delta l_1 + \Delta l_2)}{l_0}\right]$$

Some other measures of large deformations are also used in the rheological literature. For example, let a fiber (or a bar) of the initial length l_0 be stretched and increase its length by Δl. The extension ratio, λ, equals $(\Delta l + l_0)/l_0$. Then

$$\frac{du_1}{dx_1} = \lambda - 1$$

and according to the definition given by Eq. 1.2.6

$$\varepsilon_{11} = \frac{du_1}{dx_1} + \frac{1}{2}\left(\frac{du_1}{dx_1}\right)^2 = (\lambda - 1) + \frac{1}{2}(\lambda - 1)^2 = \frac{1}{2}(\lambda^2 - 1) \qquad [1.2.20]$$

This measure of large deformations was introduced by George Green.[12] The large deformation tensor, used in continuum mechanics and based on the definition expressed by Eq. 1.2.6, is called a *Cauchy-Green tensor*, C_{ij}, and it is defined as

$$C_{ij} = \delta_{ij} + 2\varepsilon_{ij} \qquad [1.2.21]$$

where δ_{ij} is the Kronecker delta.

Similar to the tensor defined by Eq. 1.2.6, a Cauchy-Green tensor of finite deformations characterizes the change in distance between two arbitrary sites at a "point":

Another tensor of large deformations is also frequently used. This is the inverse (or reciprocal) tensor to the Cauchy-Green tensor, C_{ij}, named the *Finger tensor*, C_{ij}^{-1}.[13] According to the definition, the relationship between both tensors is

$$C_{ij} C_{ij}^{-1} = \delta_{ij} \qquad [1.2.22]$$

The principal components of the large deformation tensor, ε_i, are expressed by equation equivalent to Eq. 1.2.20:

$$\varepsilon_i = \frac{1}{2}(\lambda_i^2 - 1) \qquad [1.2.23]$$

where λ_i are the principal elongation ratios.

The principal values of the tensors C_{ij} and C_{ij}^{-1} are also expressed *via* the principal elongation ratios:

$$C_i = \lambda_i^2 \; ; \text{ and } C_i^{-1} = \lambda_i^{-2} \qquad [1.2.24]$$

The first invariants of both tensors, $C_{I,inv}$, are as follows:

$$C_{I, inv} = \lambda_1^2 \lambda_2^2 \lambda_3^2 \; ; \text{ and } C_{I, inv}^{-1} = \lambda_1^{-2} \lambda_2^{-2} \lambda_3^{-2} \qquad [1.2.25]$$

Introducing different measures of deformations does not exclude the main question regarding the initial state – point of reference of the deformed state. The importance of this question has already been demonstrated by the example of large deformations in uniaxial extensions. For static states, this problem can be solved by introducing the Hencky measure of deformations. The same problem appears and becomes more pertinent for a continuously moving medium where the position of deformed elements of a body is changing in time and it is necessary to describe the process or the rate of deformation. This problem will be discussed in more detail in Section 1.3.2.

1.2.4 SPECIAL CASES OF DEFORMATIONS – UNIAXIAL ELONGATION AND SIMPLE SHEAR

1.2.4.1 Uniaxial elongation and Poisson's ratio

Experiments show that a sample being stretched in the axial direction changes dimensions in the lateral direction. The relation between relative changes of dimensions in the lateral and the axial directions cannot be established on the basis of pure geometrical arguments because this relation reflects an inherent, independent property of the material. The ratio of relative lateral contraction to the relative longitudinal extension is the quantitative characteristic of material property. This property of material is called *Poisson's ratio*.[7]

Let the radius of the cross-section of a bar in the initial state be r_0, and the length, l_0. If its elongation is Δl, and, as a result of stretching, the radius is decreased by Δr, then, by definition, Poisson's ratio, μ, is:

$$\mu = \frac{\Delta r / r_0}{\Delta l / l_0} \qquad [1.2.26]$$

It is now easy to calculate the volume change, resulting from uniaxial stretching. The relative change of volume, $\Delta V / V_0$, is

$$\frac{\Delta V}{V_0} = \frac{(r + \Delta r_0)^2(l_0 + \Delta l) - r_0^2 l_0}{r_0^2 l_0} \qquad [1.2.27]$$

where $V_0 = \pi r_0^2 l_0$ is the initial volume of a sample (in a non-deformed state).

For small deformations $\Delta l \ll l_0$ and consequently $\Delta r \ll r_0$. In this case, Eq. 1.2.27 gives

$$\frac{\Delta V}{V_0} = 1 - 2\mu \qquad [1.2.28]$$

Poisson's ratio is a measure of volume changes during small deformations. From Eq. 1.2.28, one can see that deformations occur without volume changes when $\mu = 0.5$. For solid materials, $\mu < 0.5$, (for many solid materials $\mu \approx 0.3 - 0.35$). This means that their elongation is accompanied by an increase in specific volume. Only some rubbers and polymer melts $\mu \cong 0.5$ deform without volume changes.

The concept of Poisson's ratio allows one to use the general method of decomposing the deformation tensor, d_{ij}, into spherical and deviatoric terms for a uniaxial extension. If $\lambda \ll 1$, and deformation in uniaxial extension equals ε^*, the tensor of infinitesimal deformations, Eq. 1.2.7, for such cases can be written as:

$$d = \begin{bmatrix} \varepsilon^* & 0 & 0 \\ 0 & -\mu\varepsilon^* & 0 \\ 0 & 0 & -\mu\varepsilon^* \end{bmatrix} = \frac{1-2\mu}{3}\varepsilon^*\delta_{ij} + \frac{1+\mu}{3}\varepsilon^*\begin{bmatrix} 2 & 0 & 0 \\ 0 & -1 & 0 \\ 0 & 0 & -1 \end{bmatrix} \qquad [1.2.29]$$

The structure of this sum is very similar to the structure of the stress tensor decomposed into two parts (compare with the analogous procedure in section 1.1).

A more precise analysis of Eq. 1.2.27 shows, however, that for large deformations Eq. 1.2.28 is not valid, and the rule of $\mu = 0.5$, as the condition for maintaining the constant volume at stretching, has no general meaning. Indeed, preserving a formal definition of Eq. 1.2.26 for Poisson's ratio, according to Eq. 1.2.27, the condition for $V = 0$ is:

$$1 - 2\mu(1+\varepsilon) + \mu^2\varepsilon(1+\varepsilon) = 0 \qquad [1.2.30]$$

If $\varepsilon \ll 1$, Eq 1.2.30 is converted to an ordinary condition $\mu = 0.5$, but in the more general case it is not true.

Example – Poisson's ratio in finite deformations

Let the bar be stretched by 9 times (e.g., rubber ribbon or melted fiber). It means that $l/l_0 = 8$ and the volume can remain unchanged if the final radius becomes equal to 1/3 of its initial value. Then $\Delta r/r_0 = 2/3$. In this case, according to Eq 1.2.30, and following the formal definition, Eq. 1.2.25, $\mu = 1/12$.

This example shows that adaptation of infinitesimal deformation mechanics ($\mu = 0.5$ as a necessary condition for the constant volume at extension) to the domain of large deformations must not be done in a straightforward manner.

1.2.4.2 Simple shear and pure shear

The movement of all fluids and liquid-like materials is based on the model of sliding of neighboring layers relative to each other. This is a case of *simple shear*. Simple shear is also realized in several modes of deformations of solids, such as, for example, twisting long tubes or wires.

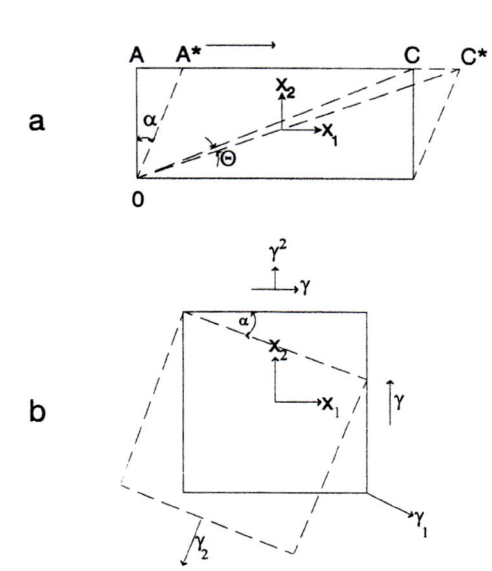

a

b

Figure 1.2.5. Small (a) and large (b) deformations in simple shear.

The schemes of two-dimensional (plane) simple shear for an element of a body in small deformation, and for a general case of arbitrary deformation, are shown in Figs. 1.2.5 a and b, respectively. Along the direction of shear marked by an arrow, a displacement, u_1, takes place. Its gradient, du_1/dx_2, is determined by the slope which is denoted as:

$$\gamma = \tan\alpha = \frac{du_1}{dx_2} \qquad [1.2.31]$$

Since the length of linear elements, which were directed before deformation in the x_2 direction, is changed in shear, one more displacement component, u_2, appears. It is related to the change in the length of the segment OA, which, after displacement, becomes equal to OA*:

$$\frac{OA^* - OA}{OA} = (1 + \gamma^2)^{1/2} - 1 \qquad [1.2.32]$$

The value of $\gamma = du_1/dx_2$ in a simple shear determines all components of the tensor at large deformations. According to the definition of the ε_{ij} tensor, its components are:

$$\varepsilon_{12} = \varepsilon_{21} = 0.5\gamma; \ \varepsilon_{22} = 0.5\gamma^2 \qquad [1.2.33]$$

This tensor is graphically illustrated in Fig. 1.2.5 b in which the components of the tensor, ε_{ij}, are marked by arrows (factor 0.5 is omitted in drawing this figure). The appearance of a diagonal component in the deformation tensor in a simple shear is a direct consequence of large deformations. It is a second-order effect because ε_{22} is proportional to γ^2 and its value becomes negligible if $\gamma \ll 1$. This phenomenon is known as the *Poynting effect*,[14] which is observed in wire twisting (their length slightly changes). Twisting is an example of shear deformation, and the observed change of the length is regarded as relative to the ε_{22} component of the deformation tensor.

Shear produces a shift between the direction of shear, x_1, and the orientation of the principal axis. The shift is denoted by an angle, α, as shown in Fig. 1.2.5 b. The angle can be calculated from:

$$\alpha = \frac{1}{2}\arctan(\gamma/2) \qquad [1.2.34]$$

The main components of the deformation tensor may be written as follows:

$$\varepsilon_{11} = \cot\alpha; \ \varepsilon_{22} = \tan\alpha; \ \varepsilon_{33} = 1 \qquad [1.2.35]$$

The results obtained from Eqs. 1.2.35 indicate that in simple shear no volume change occurs because the product $\varepsilon_1\varepsilon_2\varepsilon_3 = 1$

Expressions for the components of the Cauchy-Green and the Finger tensors in shear are important for future discussion concerning rheological models of elastic bodies of different types. Direct calculations give the following expressions for C_{ij} and C_{ij}^{-1}:

$$C = \begin{bmatrix} 1 & \gamma & 0 \\ \gamma & 1+\gamma^2 & 0 \\ 0 & 0 & 1 \end{bmatrix} \text{ and } C^{-1} = \begin{bmatrix} 1+\gamma^2 & -\gamma & 0 \\ -\gamma & 1 & 0 \\ 0 & 0 & 1 \end{bmatrix} \quad [1.2.36]$$

In a simple shear, not only the lengths of linear elements change (e.g., along the principal directions), but the rotation of the elements of a body also takes place. This effect is well seen in Fig. 1.2.5, where the angle of rotation, θ, of the diagonal element from OC to OC* position, is shown.

Shear deformation in this figure is due to displacement, AA*, and α is its gradient. Any gradient of displacement consists of deformation and rotation, which in general form is expressed by Eq. 1.2.11. For a small displacement, the angle of rotation, $\theta = \alpha/2$, is used, unlike for large deformations, where the general Eq. 1.2.10 is applicable.

It is possible to find such shear conditions where no rotation occurs. This case, called *pure shear*, is based on the definition of θ_{ij} from Eq. 1.2.12. $\theta_{ij} = 0$ if all differences of the displacement gradients equal zero. For the simple shear, this condition has the following form: $du_1/dx_2 = du_2/dx_1$.

A geometrical image of pure shear is drawn in Fig. 1.2.6. In pure shear, the diagonal AB of the small square (at some point) moves, due to deformation, into new position A*B*, parallel to its initial position, and the diagonal OM does not change its position at all, being only extended to OM*. Therefore, no element of the body undergoes rotation.

Fig. 1.2.6 can be obtained in a different way. It is quite evident that transition from the square OAMB to the rhomb OA*M*B* can be achieved by pressing the square along the direction AB, with simultaneous stretching along the direction OM. It means that pure shear can be realized through the superposition of two uniaxial extension deformations (with different signs).

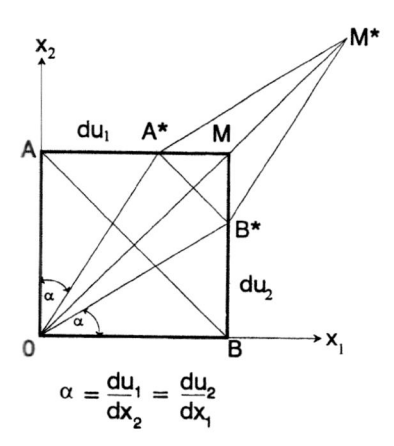

$$\alpha = \frac{du_1}{dx_2} = \frac{du_2}{dx_1}$$

Figure 1.2.6. Pure shear of an element of a body.

The difference between simple shear and pure shear is the same as the difference between deformation and displacement. This difference may appear important in formulations of constitutive equations describing rheological properties and behavior of real materials.

1.3 KINEMATICS OF DEFORMATIONS

1.3.1 RATES OF DEFORMATION AND VORTICITY

The motion of a body is characterized by *velocity*, which is a vector. If velocity at any given point of a body is the same, it means that the body moves as a whole and no deformation takes place. The deformation appears only as a consequence of a velocity gradient at a "point", which means that two neighboring locations (the distance between them being infinitesimal) move with different velocities. If velocity is **v** (a vector value), the components of its gradient, a = d**v**/d**r**, are calculated as

$$a_{ij} = \frac{dv_i}{dr_j} \qquad [1.3.1]$$

The space coordinates are described by radius-vector, **r**. Thus **a** is a tensor with components a_{ij} determined by two vectors (**v** and **r**). The velocity is the rate of displacement, i.e., **v** = d**u**/dt. The relationship between the gradient of velocity, a_{ij} = d**v**/d**r**, and gradient of displacement, g_{ij} = d**u**/d**r**, can be established from the following rearrangements:

$$a_{ij} = \frac{dv}{dr} = \frac{d}{dr}\left(\frac{du}{dt}\right) = \frac{d^2u}{drdt} = \frac{d}{dt}\left(\frac{du}{dr}\right) = \frac{dg_{ij}}{dt} \qquad [1.3.2]$$

In section 1.2.1, it was established that the whole gradient of displacement is not controlling deformation, only its symmetric part. The same is true for the deformation rate. The reasoning is the same as above. Differentiation with respect to a scalar − time, d/dt, adds nothing new to the result. By decomposing tensor a_{ij} into symmetrical and antisymmetrical components,

$$a_{ij} = \frac{1}{2}\left(\frac{\partial v_i}{\partial x_j} + \frac{\partial v_j}{\partial x_i}\right) + \frac{1}{2}\left(\frac{\partial v_i}{\partial v_j} - \frac{\partial v_j}{\partial v_i}\right) \qquad [1.3.3]$$

one obtains

$$a_{ij} = D_{ij} + \omega_{ij} \qquad [1.3.4]$$

where D_{ij} is the *rate of deformation* tensor, and ω_{ij} is the so-called *vorticity tensor*.

As in the previous case, the rate of deformation tensor characterizes local changes of shape. The deformation is related to the first term of Eq. 1.3.4, while the vorticity tensor describes the rate of rotation of local elements of a body without their deformation.

The difference between the tensors a_{ij} and D_{ij} (which is quite similar to the difference between the tensors g_{ij} and d_{ij}) can be illustrated by a simple example. Let us analyze the rotation of a solid (non-deformable) body around some axes. The velocity, **v**, at a point located at the distance, **r**, from the axis of rotation equals ωr, where ω is the constant angular velocity. Thus, v = ωr, and the gradient of velocity, grad v = dv/dr, is evidently equal ω. It means that, during rotation of a solid body, the gradient of velocity does exist and there is no deformation because (as initially assumed) the body is non-deformable.

This example is also valid for any rotational movement, for example, for a circular movement of liquid placed between the stationary inner and the rotating outer cylinders (for any liquid point the difference between the gradient of velocity and rate of deformation exists). In the latter case:

$$\frac{dv}{dr} = \frac{d(\omega r)}{dr} = \omega + \frac{rd\omega}{dr} \qquad [1.3.5]$$

Rate of deformation equals the second member of the sum, $r(d\omega/dr)$, whereas the first member, ω, represents superimposed rotation which does not influence the deformation of matter placed between the cylinders. Indeed, it is possible to add a constant angular velocity of rotation, Ω, to both cylinders (to force both to rotate with the same constant angular velocity added to the rotation of the outer cylinder with the angular velocity, ω). It will increase the velocity gradient by this value but will not change the deformation rate.

1.3.2 DEFORMATION RATES WHEN DEFORMATIONS ARE LARGE

Some difficulties in the calculation are encountered in the case of large deformations. At the end of section 1.2.3 it was pointed out that the description of large (finite) deformations requires special attention and monitoring of a continuously moving medium because positions of deformed elements are changing with time.

Substantial derivative

Similar solutions are required in classical hydrodynamics, when, for example, temperature effect due to heat exchange is included or material is transformed by chemical reaction. Such processes happen in media in motion. The problem is solved by using the so-called *material* or *substantial derivative*, D/Dt, which can be written for an arbitrary variable, Y, which depends on time and a site position moving in space, as:

$$\frac{DY(x_i, t)}{Dt} = \frac{\partial Y}{\partial t} + \sum_{k=1}^{3} v_k \frac{\partial Y}{\partial x_k} \qquad [1.3.6]$$

The first term represents local changes of the Y value, whereas the second term describes the movement of this local site in three-dimensional space.

In the theory of large deformations, it is important to know the rate of deformation in a fixed and moving coordinate system. Changes occur in a traveling element of material, which deforms along with its replacement. This is called the *principle of material indifference*, which states that all physical phenomena must not depend on a coordinate system used for their mathematical formulation.[15]

As a result of large deformations, material elements can travel far away from their initial position, and that is why it is important to apply proper rules of transition from the reference state. Similar to the discussion of large uniaxial extension (see section 1.2.3), it is also important to choose different reference states in such a manner that they will not lead to an ambiguous estimation of deformation. An observer who measures the properties of a material is always positioned in a fixed (unmovable) coordinate system. Hence, the general approach consists of formulating ideas concerning the possible rheological behavior of a material for a moving (and deforming) element of a medium, recalculating them into a fixed coordinate system, and then comparing the results with an experiment.

This is true for the rate of deformation. There are many mathematical avenues to transform the rate of deformation tensor into a fixed coordinate system and, depending on selection, various forms of time derivatives were proposed, one of them is Eq. 1.3.6. In some theoretical studies, kinematic tensors of higher order were introduced, which are time derivatives of the Cauchy-Green or the Finger tensors. They are used when the rheological behavior of a material depends on higher derivatives of deformation. In Chapter 2, devoted to properties of viscoelastic materials, it will be demonstrated that their behavior can be modeled by equations containing a sum of n-th order time derivatives of deformation (the so-called rheological equations of a differential type).

The physical meaning of substantial time derivative, D/Dt, requires that the derivative is calculated for a moving medium in which a material point follows time changes and leaves its initial position. The most popular are the *Rivlin-Ericksen*, $A_n(t)$,[16] and the *White-Metzner*, $B_n(t)$,[17] *tensors* of the n-th order. They are determined as

$$A_n(t) = \frac{D^n C_{ij}(t)}{Dt^n} \qquad\qquad [1.3.7]$$

and

$$B_n(t) = -\frac{D^n C_{ij}^{-1}(t)}{Dt^n} \qquad\qquad [1.3.8]$$

where C_{ij} and C_{ij}^{-1} are the Cauchy-Green and the Finger tensors, respectively.

The use of various measures of large deformations and different types of their time derivatives permits us to make qualitative predictions concerning all possible effects in the mechanical behavior of different materials. It is the global task of an experiment to evaluate possible models and to find the simplest of them which can adequately describe numerous physical phenomena observed in real materials in an unambiguous manner.

The practical application of the above-discussed approaches for formulating constitutive equations for different materials and using them in solving dynamic (boundary) problems are considered in more detail in Chapters 2 and 6 of this book.

1.4 HETEROGENEITY ON FLOW

The general concept formulated in the title of the Section "Continuum Mechanics as a Foundation of Rheology" is certainly valid. However, there are some cases in which it is necessary to treat this concept with caution or special comments. This applies to the *heterogeneity* of the materials having a large size of inherent structure elements.[18] In this case, the term "continuum" cannot be applied.

Three different situations can exist. First, heterogeneity was present in the initial structure of a material. This includes multi-component materials, such as filled plastics and rubbers, reinforced structures, and colloidal systems.

Second, heterogeneity can appear as a result of the inner reaction in the initially homogeneous medium. This is the case, for instance, of polymerization leading to the formation of a new phase.

Third, heterogeneity appears as the result of deformation-inducing structure rearrangements of a material. The third case is of special interest to rheology.

1.4.1 PARTICLE DISTRIBUTION IN DISPERSE SYSTEMS

Let the initial spatial distribution of dispersed particles be uniform (Fig. 1.4.1, a). Will it remain the same during the flow? It is not obligatory. Two effects are known. First, particles move from a solid wall and a layer of the low viscosity fluid free of solid particles appears (Fig. 1.4.1, b). Second, particles form self-arranged necklace structures (Fig. 1.4.2).[19]

The multi-component system in the process of shearing cannot be treated as homogeneous. If a low-viscous layer exists at the wall (Fig.1.4.1), dominating shear takes place in this layer and this radically changes the boundary conditions in dynamic problems. After the formation of the necklace structure (Fig. 1.4.2), the system becomes not only heterogeneous but also anisotropic (its properties along the particle chain and in the transversal direction become different).

An interesting case of the formation of the layered structure is shown in Fig. 1.4.3.[20] Such a structure appears in multi-component systems in the range of the so-called lowest Newtonian viscosity. This part of the flow curve was usually treated as the flow of the uniform material with a completely destroyed structure.

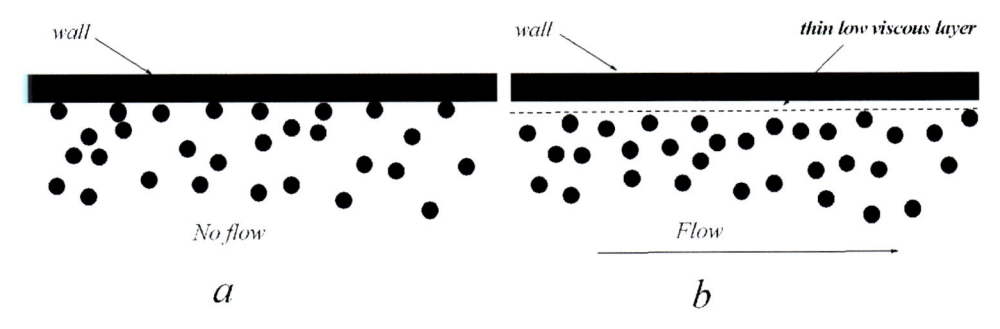

Figure 1.4.1 Formation of a thin fluid layer near a wall free of solid particles.

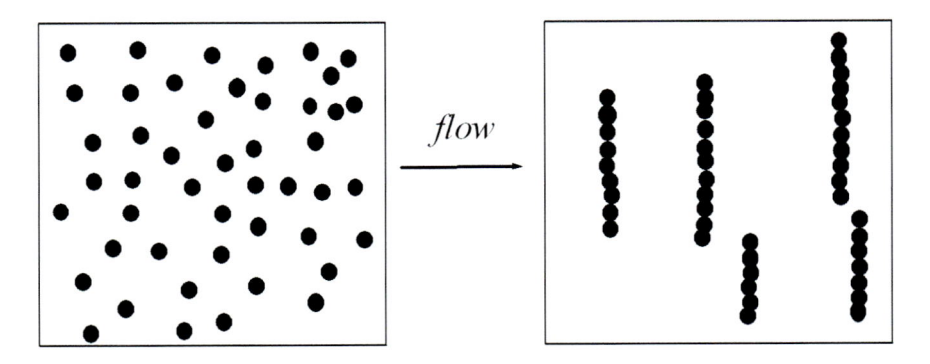

Figure 1.4.2. Self-organization in shearing a dispersion of solid particles – transition from a random distribution in the initial state (left) to the necklace structure in shearing (right).

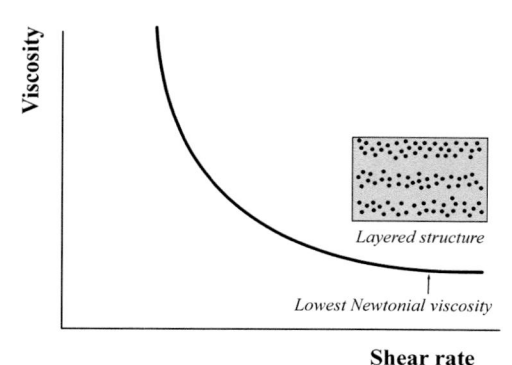

Figure 1.4.3. Layered structure instead of "completely destroyed" state at the lowest Newtonian viscosity domain of a flow curve.

Actually, this is flow through fluid interlayers between regularly organized layers of solid particles.

1.4.2 PHASE SEPARATION

Shearing can promote phase separation in a flow. This phenomenon is discussed in detail in Section 3.5. Here, it is worth mentioning that this effect leads to changes in the rheological properties of polymeric and colloidal systems, especially at high deformation rates. Then the system becomes heterogeneous and it is necessary to take into account the spatial-temporal distribution of the rheological properties of such materials in the flow. As the limiting situation, the breaks in the continuum can appear with different laws of flow on both sides of these lines.

1.4.3 FLOW OF THE LARGE-SCALE STRUCTURE ELEMENTS

In the general concept of continuum mechanics, it is assumed that the movement takes place on the molecular size scale and the super-molecular structure can be neglected. This approach is wrong, if a multi-component system initially consists of large-scale volume elements, as, e.g., in emulsions or non-colloidal suspensions. In some cases, the procedure of averaging allows us to neglect this factor but not always. For example, it is impossible to do this if the size of the channel is comparable to the size of structural elements. This is a case of the flow of blood in vessels and many others. So, it appears necessary to understand *"How does a concentrated emulsion flow?"*[21] The answer to this question is: such objects flow (or more correct, move) on the super-molecular size level.

Direct observations of a highly concentrated emulsion flow showed that there were two mechanisms of droplet displacement (assuming that just droplet movement is the flow).[22] At low stresses, droplets do not deform and large droplets roll on small droplets, like balls in ball bearings. At high stresses, droplets deform and all droplets move (Fig. 1.4.4).

Moreover, it was shown that the movement of droplets in concentrated emulsions can be cooperative, i.e., elementary droplets can aggregate in cylindrical flocks which roll together.[23]

Heterogeneous flow is rather typical for such complex fluids as concentrated disperse systems and though this effect was called *"unexpected"*.[24] Such a type of movement can be accompanied by a breakup in the continuity and appearance of the sliding planes. This phenomenon characteristic for the concentrated suspensions will be discussed below (see Subsection 3.3.4.3 and illustrated in Fig. 3.3.9).

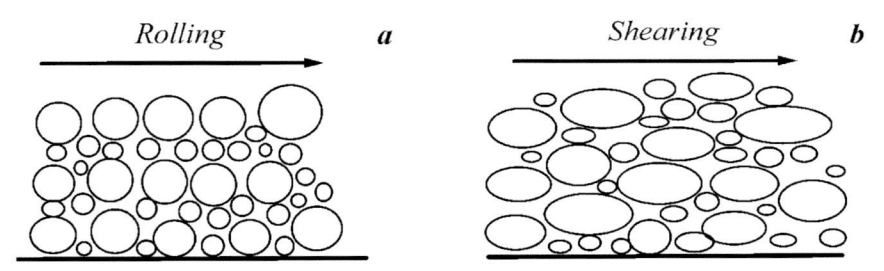

Figure 1.4.4. Mechanisms of movement of highly concentrated polydisperse emulsions at low (a) and high (b) shear stresses.

This type of displacement can hardly be called "flow". It is closer to the movement of granular media.

1.5 SUMMARY – CONTINUUM MECHANICS IN RHEOLOGY

1.5.1 GENERAL PRINCIPLES

Classical continuum mechanics is one of the milestones of rheology. Rheology, dealing with *properties of matter*, regards these properties as relationships between *stresses* and *deformations*, which are the fundamental concepts of continuum mechanics.

The idea of a continuum, as well as mathematical operations used in mechanics, assumes that there are a continuous transition and movement from point to point. A "*point*" is understood as a mathematical object of infinitesimally small size. However, it is necessary to accept the following contradiction: a "physical" point is different than a mathematical point.

Almost everybody is convinced that matter consists of molecules and intermolecular empty spaces, which means that in reality, any material body is heterogeneous. At the same time, an observer is sure that he "sees" a body of matter as a homogeneous continuous mass without holes and empty spaces. The obvious way out of these contradictory evidence lies in the idea of the *geometrical scale of observation*.

This scale must be large enough to distinguish individual molecules or their segments. The characteristic order of the size of a molecule (its cross-section or length of several bonds) is 1 nm. Then, only when dealing with the sizes of the order of at least 10 nm, one may neglect molecular structure and treat a body as *homogeneous*. It means that a characteristic volume is of an order larger than 10^3 nm^3. This is a real size of a physical "point", which is quite different from a philosophical or geometrical point. The latter is an infinitely small object of zero size. The physical "point" contains 10^4 molecules or segments of the macromolecule, and throughout its volume, all molecular size fluctuations are averaged. The number of molecules in such a point is large enough for smoothing and averaging procedures.

In many cases, especially when discussing properties of a single-component material, it is possible to neglect the inherent structure of the medium and the difference between ideas of the "physical" and "mathematical" concepts of a point as immaterial.

Having in mind the real scale of a physical point, it is supposed that it is permissible to apply methods of mathematical analysis of infinitesimal quantities (which formally relate to a geometrical point) to a physical medium. The formal extrapolation of physics-based analysis to infinitely small sizes tacitly avoids the incorrectness of this operation, and the only justification for this is the fact that in almost all practical applications, nobody is interested in what really happens in a very small volume.

However, there are at least three important principal exceptions.

1. A central physical problem exists in the explanation of macro-observations of the molecular structure of matter. One would like to understand what happens to a molecule or how intermolecular interactions occur; then going through micro-volumes containing numerous molecules and averaging molecular phenomena, one would come to the macro properties of a body.

2. In some applications, we use "zero" size. If geometrical shapes under consideration have sharp angles and the size at the corner of any angle (formally) equals zero, extrapolation of calculations results in such "zero" volume and sometimes leads to infinite values, and this is out of the realm of physical meaning. The analysis of problems of this kind requires special methods.

3. There are many materials that cannot be considered homogeneous in principle. It is, therefore, necessary to consider their *structure*, i.e., such materials are *heterogeneous* by definition. For example, a medium can be a statistical or a regular mixture of some components with step-like transitions between them. Typical examples of such heterogeneity are suspensions and filled polymers, sometimes with well-arranged (in reinforced plastics) structures. In some applications, the structure of heterogeneous materials may be out of interest and it is possible to continue to treat the medium as homogeneous, averaging inner differences in relation to a much higher geometrical scale. For example, for many astronomic observations, the Sun and the Earth are regarded as quite homogeneous and moreover can be treated as "points". In other cases, the role of heterogeneity can be important and it may become a determining factor (for example, for reinforced plastics), but, in any case, the scale of such heterogeneity has to be much larger than the characteristic molecular sizes.

1.5.2 OBJECTS OF THE CONTINUUM AS TENSORS

Continuum mechanics of any material operates with some fundamental concepts characterizing dynamic (*stresses*), geometrical (*deformations*), and kinematic (*deformation rates*) situation at a point (*site*). In this approach, a "point" is always understood in a mathematical sense as infinitesimal small objects. All these concepts are physical objects existing regardless of the choice of the coordinate system.

Stress is a measure of forces acting on a point and it is defined as a relative force or a force related to the unit area. Stress values depend on the direction of the applied force and the orientation of a surface on which acting forces are considered. Stress is an object of the *tensor* nature.

Stresses determine the deformation of matter, and, in limiting cases, when they overcome some threshold, they lead to *transitions* and eventually to rupture of the material.

Stresses can be *normal* (perpendicular) and *shear* (tangential) to the surface where they act.

The stress tensor is written via its *components* – projections of the force on the coordinate axes.

It is always possible to calculate components of the stress tensor for any direction and to find such *principal* directions and principal normal stresses, which are extreme with shear stresses absent in those directions.

There are three particular combinations of any arbitrary stress tensor which do not depend on the choice of axes or their orientation in space. These combinations are called *invariants*. The independence of these combinations of the stress components on the choice of the coordinate system is evidence of the existence of stress as a physical quantity regardless of the coordinate system.

The stress tensor can be divided into two parts, one of which (known as *spherical*), being hydrostatic pressure, is responsible for the volume changes, and the other part (called *deviatoric*) is responsible for shape (or form) changes of a body (at a point). The spherical part of the stress tensor (and the first invariant of this tensor) determines the hydrostatic *pressure* (all-directional, tri-axial, compression acting on a body).

Calculation of stresses throughout a body is realized by solving differential equations with appropriate boundary conditions. These equations represent the law of equilibrium (or balance) of all forces applied at a point.

Due to different reasons and, in particular, due to the action of external forces, the points in a body can move in space and this is known as *displacement*. If displacements are inhomogeneous throughout a body (i.e., different at different points) *relative* displacements appear and they lead to *deformations*, which are determined as the changes of infinitesimal distances between points inside a body.

Displacement is a *vector*, but relative displacements, as well as deformations, similar to stress, are the quantities of tensor nature because their presence is controlled by two vectors. The relative displacement is described by radius vectors of two points for which the displacement is considered, and the deformation is characterized by means of a vector of displacement and a radius vector at a point, where the displacement occurs.

Deformation is only a part of the relative displacement, the latter also includes rotation of elements of a body as a whole.

Deformations can be small (or *infinitesimally small*) or large (or *finite*). The boundary between them is determined by the value of relative displacement (or gradient of displacement), which is a dimensionless value. If this value is small ($\ll 1$), it is reasonable to neglect the square of this value in comparison with the value itself. One can thus neglect all quadratic terms included in the definition of deformation. In this case, deformations can be treated as infinitesimally small.

The deformation tensor can be divided into two parts: the spherical part, which represents volume changes, and the deviatoric part, which is a characteristic of the shape transformations.

If *large deformations* are considered, some new effects appear. First of all, deformations occur at a site that moves, as a result of displacement, and vacates its initial position. Description of all occurrences (including deformation itself) must be done in relation to a moving point. An observer, carrying out experiments, follows its behavior and treats the results of measurements in a fixed coordinate system. Hence, it is necessary to know the rules of transformations and the tensor values used for projecting deformations from moving to a fixed coordinate system.

Large deformations are characterized by special measures of deformation, such as the *Hencky measure* (a logarithmic measure subjective to additivity rule), and the *Cauchy-Green* and the *Finger tensors* of large deformations.

The tensors of deformations, similar to any other tensors, have principal axes along which the principal values of this tensor are calculated. Besides, three invariants of the deformation tensors are calculated by the standard rules of operations with tensors. The geometrical sense of the first invariant of the deformation tensor is volume change caused by deformation.

The kinematic picture of the relative movement of points of a continuum is characterized by the time derivative of displacement of a point (*its velocity*), the time derivative of relative displacement (*gradient of velocity*), and time derivative of deformation (*rate of deformation*). Time derivatives of tensors are also tensors. For calculation of the rate of deformation, special rules exist which take into consideration large deformations and movements of a deforming site in space. The gradient of velocity is the sum of the rate of deformation and *vorticity* tensors of elements of a body, which – due to displacements – can rotate simultaneously with deformation.

Two special cases of deformations are of interest: uniaxial longitudinal extension and simple shear. In the process of extension, a body undergoes lateral compression. The ratio of relative changes of lateral and longitudinal sizes is called *Poisson's ratio*, which is an inherent property of a material. For the range of small deformations, the volume of a body remains unchanged if Poisson's ratio equals 0.5. In simple shear, volume changes are not taking place at all. However, at large shear deformations, diagonal components of the deformation tensor appear, and they lead to some second-order effects.

Simple shear is accompanied by rotation of elementary volumes in space. In order to exclude rotation, it is necessary to apply *pure shear* in which rotation does not exist. This type of deformation is equivalent to a two-dimensional superposition of extension and compression in mutually perpendicular directions.

REFERENCES

1　　T. Alfrey, **Mechanical Behavior of High Polymers**, *Interscience*, N.Y. 1948.
　　　M. Reiner, **Twelve Lectures on Theoretical Rheology**, *North Holland Publ. Co.*, Amsterdam. 1949.
2　　Sometimes the coordinate system used in this book will be marked as x, y, and z, which are equivalent to x_1, x_2, and x_3, respectively.
3　　Stress tensor is a particular case of a more general concept of tensors. This particular case is called tensor of the second rank.
4　　Those who are interested to understand more about tensor analysis may use textbooks, such as for example, R. Aris, **Vectors, Tensors and the Basic Equations of Fluid Mechanics**, *Prentice Hall*, Englewood Cliffs, 1962; or E.C. Young, **Vector and Tensor Analysis**, *Marcel Dekker*, New York, 1993.
5　　A.L. Cauchy (1789-1857) – French mathematician and physicist. One of the founders of the modern mathematical analysis and mechanics. His main publication on the subject under discussion: A. L. Cauchy, De la pression ou tension dans un corps solide, *Ex. de Math.*, 1827.
6　　C.L.M.H. Navier (1785-1836) – French engineer and physicist, an author of fundamental works on the theory of strength, fluid dynamics, and the theory of elasticity. The balance equations were published in: C.L.M.H. Navier, Mémoire sur les lois de l'équilibre et du mouvement des corps solides élastique, *Mém. Acad. Sci. Inst. France*, 1827.
7　　S.D. Poisson (1781-1840) – French mathematician and physicist, an author of fundamental works in the field of mathematical analysis and the theory of elasticity. His publication on the subject under discussion: S.D. Poisson, Mémoire sur l'équilibre et le mouvement des corps élastiques, *Mém. Acad. Sci. Inst. France*, 1829.
8　　A.L. Cauchy, Sur les équations differentielles d'équilibrium ou de mouvement pour le points matériels, *Ex. de Math.*, 1829.
9　　Some authors try to distinguish between "strain" and "deformation". The College Dictionary has the following definition: "strain is a deformation of a body or a structure as a result of an applied force". This shows that both terms have the same meaning.
10　　A.J.B. de St Venant was the first who realized the necessity to consider the concept of large (finite) deformations. A. J. B. de St. Venant, Mémoire sur équilibre des corps solides, dous les limits de leur élasticité, et sur les conditions de lèur resistance quond es désplacements êpouvés par leurs points ne sont par trés petit, *C. R. Acad. Sci.*, Paris, **24**, 1847. Later the fundamental ideas of the theory of large

deformations were formulated by F. D. Murnaghan, *Ann. L. Mech.*, **59**, 235 (1937). Also the following paper contained many ideas, which became commonly accepted and used afterward: M. Biot, Theory of elasticity with large displacements and rotations, Proc. 5 th Int. Congr. Appl. Mech., 1939.

11 H. Hencky, *Ann. Physik*, **2**, 617 (1931).

12 A.E. Green published (together with R.S. Rivlin) a number of important works in the field of continuum mechanics of solids capable of large deformations. See: A. E. Green, R. S. Rivlin, *Arch. Ration. Mech. Anal.*, **1**, 1 (1957) and **4**, 387 (1960). His general approach to the theory of large deformations was summarized in the book: A.E. Green, J.E. Adkins, **Large Elastic Deformations and Non-linear Continuum Mechanics**, *Clarendon Press*, Oxford, 1960. George Green, *Cambridge Phil. Soc. Trans.*, **7** (1839) (see **Mathematical Papers of the Late George Green**, edited by N.M. Ferrers, London, 1871)

13 J. Finger, *Sitzungsberichte Acad. Wiss. Wien*, (IIa), **103**, 163 (1894).

14 J.H. Poynting, *Proc. Roy. Soc. London*, A **82**, 546 (1909); A **86**, 534 (1912). This is, probably, the first experimental evidence of the second-order effect in finite deformations.

15 The fundamental principles which must be fulfilled in formulating the rheological constitutive equations in the case of finite deformations were advanced in a very clear and closed form in an important paper of J.G. Oldroyd, *Proc. Roy. Soc.*, A**200**, 523 (1950). This publication stimulated a lot of investigations on the mechanics of continuum of non-linear materials. However, it is worth mentioning the earlier precursors in this field: pioneering works of S. K. Zaremba, *Bull. Acad. Sci. Cracovie*, **85**, 380 and 594 (1903) and G. Jaumann, **Grundlagen der Bewegungslehre**, *Springer*, Leipzig, 1905.

16 R.S. Rivlin, J.L. Erickson, *J. Rat. Mech. Anal.*, **4**, 323 (1955); see also the review: R.S. Rivlin, Proc. IX Intern. Cong. Rheol., B. Mena, A. Garcia-Rejon, C. Rangel-Nafaile, Part 1, 1 (1984).

17 J.L. White, A.B. Metzner, *J. Appl. Polym. Sci.*, **7**, 1867 (1963).

18 A.Ya. Malkin, V.G. Kulichikhin, Spatial-temporal phenomena in the flows of multi-component materials, *Appl. Rheol.*, **25**, 3, 35358 (2015).

19 There are many publications devoted to this type of structure formation, See the latest ones: A. Mirsepassi, Bh. Rajaram, A. Mohraz, D. Dunn-Rankin, *J. Non-Newton. Fluid Mech.*, **179-180**, 1-8 (2012); R. Pasquino, D. Panariello, N. Grizzuti, *J. Coll. Interface Sci.*, **394**, 49-54 (2013).

20 N.B. Ur'ev, *Coll. Surf. A: Physicochem. Engng. Aspects*, **87**, 1-14 (1994).

21 L. Bécu, P. Grondin, A. Colin, S. Manneville, *Coll. Suf. A: Physchem. Engng. Aspects*, **263**, 146-152 (2005).

22 I. Masalova, M. Taylor, E. Kharatiyan, A.Ya. Malkin, *J. Rheol.*, **49**, 839-849 (2005).

23 A. Monsenti, A.A. Peña, P. Pasquali, *Phys. Rev. Lett.*, **92**, 058303 (2004).

24 J. Miller, A complex fluid exhibits unexpected heterogeneous flow, *Phys. Today*, **63**, 18-19 (2010).

QUESTIONS FOR CHAPTER 1

QUESTION 1-1
What is the equilibrium state of liquid and solid in the absence of stresses?

QUESTION 1-2
What are the possible limits of Poisson's ratio, μ? Can its value exceed 0.5? Can it be negative?

QUESTION 1-3
What are the pressure and the shear stresses in the stress state created by the following normal stresses: $\sigma_{11} = \sigma_0$; $\sigma_{22} = -\sigma_0$ and $\sigma_{33} = 0$? What are shear stresses in this case?

QUESTION 1-4
Calculate stresses acting on a thread being suspended by its end and stretched by its own weight.

QUESTION 1-5
Analyze a situation where a horizontal long flexible engineering element (fiber, bar, etc.) is loaded along its length by a distributed force, q (i.e., force, normal to the bar, per the unit of a length).

QUESTION 1-6
In section 1.3.1, the difference between the gradient of velocity and the rate of deformation is explained. What is the situation with these values for a uniaxial extension?

QUESTION 1-7
Calculate the stresses in a hemispherical cup loaded by its own weight. Such a case is part of many engineering designs, for example, in a spherical roof covering a large area of a stadium or a warehouse.

QUESTION 1-8
Let liquid be placed between two coaxial cylinders with radii R_o (outer) and R_i (inner). The gap between cylinders $\Delta = R_o - R_i$ is small in comparison with the cylinder radii. Let the outer cylinder rotate with an angular velocity, Ω. Then, the assembly of both cylinders begins to rotate with the same angular velocity, ω. What are the shear rates and gradients of velocity in these two cases?

QUESTION 1-9
A cylindrical thread of length l_0 is fixed at one end and stretched at the other end. What must be the time dependence of velocity, $v(t)$, of stretching that is sufficient to maintain a constant deformation rate, $\dot{\varepsilon}_0 = const$?

QUESTION 1-10
Put-forth your arguments proving the possibility to neglect shear stresses in a thin-wall cylinder as in the Example in section 1.1.4.

Answers can be found in a special section entitled Solutions.

VISCOELASTICITY

2.1 BASIC EXPERIMENTS

Two simple and easy to grasp concepts (or *models*) describing mechanical properties of materials originated in the XVII century. One is the Newton law of liquids:[1]

$$\dot{\gamma} = \frac{\sigma}{\eta} \qquad\qquad [2.1.1]$$

and another is Hooke's law of solids:[2]

$$\varepsilon = \frac{\sigma_E}{E} \qquad\qquad [2.1.2]$$

where ε – deformation, $\dot{\gamma}$ – the rate of shear deformation, σ – shear stress, σ_E – tensile stress, E – elastic (or Young's) modulus, and η – viscosity.

Both equations are the simplest *rheological equations of state* or *constitutive equations* of material. These two concepts and their further developments are discussed in detail in the subsequent chapters of this book (liquids in Chapter 3 and solids in Chapter 4). Here, we would like to present the main features of these fundamental laws, which are:
- linearity of $\varepsilon(\sigma_E)$ dependence for solids and $\dot{\gamma}(\sigma)$ dependence for liquids
- absence of time effects in $\varepsilon(\sigma_E)$ dependence for solids; deformation strictly corresponds to stress and changes immediately following stress evolution
- absence of any fixed deformation of liquids because at any stress deformations develop unlimitedly and they increase linearly with time at constant stress
- existence of a single constant characterizing properties of the material – viscosity, η, for liquids and elastic modulus, E, for solids.

Many independent experimental pieces of evidence are well-known that cannot be explained within the framework of the classic theories of fluid dynamics and elasticity, based on Eqs. 2.1.1 and 2.1.2, respectively, and these are discussed in the subsequent sections.

2.1.1 CREEP (RETARDED DEFORMATION)

Let a bar of a length, l, be stretched by the force, F, as shown in Fig. 1.1.1. What happens afterward?

Time dependencies of deformation, $\varepsilon(t)$, for the two above-mentioned classical models are shown in Fig. 2.1.1a. Deformation of liquid (Eq. 2.1.1) increases linearly with time (line N) and deformation of solid (Eq. 2.1.2) instantly increases until it reaches a definite level and then stays constant (line H) as long as an observer wishes to follow it. If the

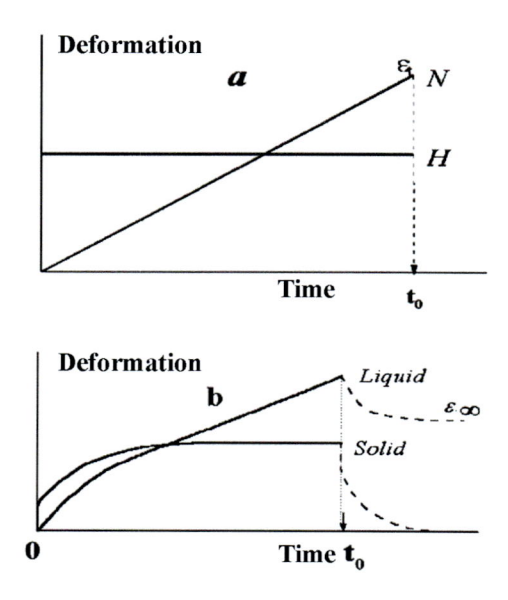

Figure 2.1.1. Development of deformation and retardation in materials of different types.

force is taken away at the time t_0, the deformation of liquid remains at the level reached before t_0, at the point ε_t, and the deformation of solid instantly disappears.

These relationships appear to be very simple, and, in fact, many real materials behave as predicted by these models. However, there are also many other materials which have different behavior. A typical illustration of such behavior is given in Fig. 2.1.1b for both liquid and solid. The main peculiarity of these curves is the delay of deformation with instantaneous application and release of forces. It is related to the initial time where a constant force acts (until the point t_0) and to *retardation*, which is the reverse of deformations (after the point t_0 at which force is removed). The difference between the pairs of curves N–Liquid and H–Solid is remarkable.

The difference between the two curves in Fig. 2.1.1a leaves no doubt about the difference in the behavior of liquid and solid. This difference is not so evident in Fig. 2.1.1b because the initial parts of the curves are similar. In practice, this initial behavior continues for a long time. The difference also appears in the final parts of these curves. The comparison of the two curves in Fig 2.1.1b helps to point out differences between liquids and solids. In *liquids*: the final part of the $\varepsilon(t)$ dependence is a line inclined to the coordinate axes and the liquid viscosity is responsible for its slope; after the force is removed, some residual deformation, ε_∞, can always be found. In *solids*: the final part of the $\varepsilon(t)$ dependence is a horizontal straight line, which is then stopped. After force is removed, deformation disappears completely and no residual deformation remains.

This difference between liquids and solids, which is very clear in the model picture, may be less definite in real practice because the processes of deformation and retardation might be very slow and an observer is never certain whether he waited long enough to reach an unambiguous conclusion.

The phenomenon of the slow development of deformations is called *creep*,[3] and the effect of retardation is sometimes called *elastic recoil*.

2.1.2 RELAXATION

Let the bar drawn in Fig. 1.1.1 be stretched rapidly by a force, F, to some length and then be fixed at this new length. What is the force necessary to maintain deformation?

The answer for the two simplest models of liquid and solid is pertinent: if there is no continuation of deformation rate, the force cannot exist in liquid (Eq. 2.1.1), i.e., the force instantly drops to zero. If the deformation of a solid is fixed (ε = const), in accordance with Eq. 2.1.2 the force remains constant during any long time observation. Again, these conclusions can be very easily grasped and they are true for many real materials.

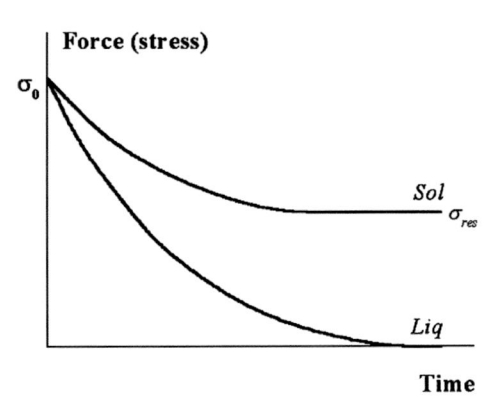

Figure 2.1.2. Stress relaxation in solids (marked as Sol) and liquids (marked as Liq).

However, many other materials have different behavior. It is possible to note a slow decay of forces (or stresses) resulting from *stress relaxation*.[4,5] Two possible modes of relaxation are shown in Fig. 2.1.2. Stresses in liquids, after unloading, relax (sooner or later) from its initial value, σ_0, until zero is reached because liquids cannot store stresses if they do not flow. Stresses in solids decrease after unloading from their initial value of σ_0, but always some residual (equilibrium) stresses, σ_{res}, remain regardless of how long the relaxation continues.

This difference in the behavior of materials is characteristic of definitions of liquids and solids. The model presents some uncertainties in practice because very slow relaxation may not be noticed by some impatient observers. The rate of the relaxation process is quantitatively characterized by the relaxation time.

Relaxation time - general concept in physics
The concept of relaxation has a general meaning for many physical phenomena. It is a reflection of an idea of the restoration of equilibrium state from a non-equilibrium condition, regardless of the reasons which caused the departure from equilibrium. For example, this can be concentration fluctuation caused by purely statistical reasons as was considered by Maxwell. Let the equilibrium value of some physical parameter be X_0, the current value of this parameter be X, and let it be supposed that the rate of approach of equilibrium is proportional to the distance from the equilibrium. This assumption immediately leads to the following first-order kinetic equation:

$$\frac{dX}{dt} = -k(X - X_\infty) \qquad [2.1.3]$$

where k is a kinetic rate constant with the dimension of reciprocal time.

The parameter X in the initial state equals to X_0. Then, the solution of this equation is

$$\frac{X(t) - X_\infty}{X_0 - X_\infty} = e^{-kt} \qquad [2.1.4]$$

Now, if $X_\infty = 0$, then the simplest form of this equation is

$$X(t) = X_0 e^{-kt} \qquad [2.1.4a]$$

The last two equations describe the relaxation process and the value of

$$\theta = k^{-1} \qquad [2.1.5]$$

is called *relaxation time*. Its value characterizes the rate of approach of the equilibrium (but not the complete time necessary to reach this equilibrium because it is infinitely large according to Eq. 2.1.4).

The exponential equation is the simplest way of quantitative description of the relaxation process, directly related to the basic concept of relaxation. In reality, the relaxation process is more complex. For example, if one places a small hot ball into a cold liquid bath, the temperature of a ball will approach the temperature of a liquid bath but the temperature evolution will be described by quite different (non-exponential) law. The general way of treating real relaxation processes using the superposition of several individual exponential modes is discussed in section 2.2.

A large number of empirical equations were proposed to describe $\sigma(t)$ functions during relaxation. The Kohlrausch function[6] is frequently used. It can be written as

$$\sigma(t) = \sigma_0 \exp[-(t/\theta_K)^n] \qquad\qquad [2.1.6]$$

where σ_0 is the initial stress, θ_K is a characteristic time of the relaxation process which must not be confused with a relaxation time as defined by Eqs. 2.1.4 and 2.1.5, and n is an empirical constant. The equation of this structure is frequently used in different branches of physics for fitting experimental data.

2.1.3 FADING MEMORY

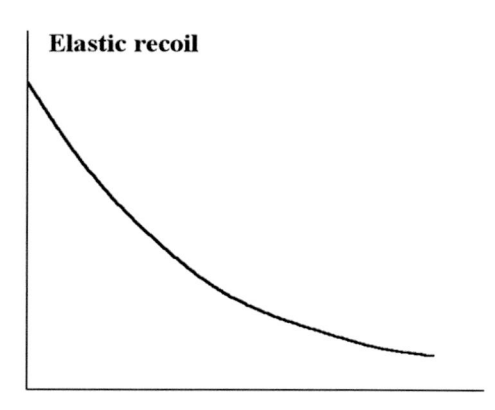

Elastic recoil

Time of rest

Figure 2.1.3. Fading memory – decrease of elastic recoil as a function of time delay.

Fading memory is a more general concept than creep or relaxation because it unites them both. This concept consists of the following: Let a bar be stretched, as in the case of creep and elastic recoil, but the sample s not released instantly but held for some time before the elastic recoil. In reality, the term "instantly" always is realized with some delay, depending on the skills of the experimentalist and the capabilities of the experimental technique.

What is the influence of this delay on the elastic recoil? In the case of liquid, according to Eq. 2.1.1, the material instantly "forgets" its previous deformation, any current state is equilibrium, elastic recoil is absent, and, as a result, liquids are considered to have no memory. In the case of elastic material, according to its model, Eq. 2.1.2, the time of rest has no influence on its behavior after releasing stresses. Such materials do not forget the pre-history of their deformations at all. So, their memory does not fade.

However, the behavior of many real materials is intermediate. It is possible to observe a phenomenon of retarding elastic recoil by increasing the delay time (Fig. 2.1.3), i.e., it is reasonable to treat this effect in terms of the "fading memory" of the pre-history of their deformations. The reasons are quite evident: in the period preceding the elastic recoil, a relaxation partly takes place and stresses responsible for the post-effect decrease. For an observer, it appears as if the material was forgetting the pre-history of its deformations, or in other words, this is the phenomenon of fading memory.

The concept of viscoelasticity comes from the above-described experimental evidence.[7] They demonstrate that in the deformation of many real materials it is necessary to consider a combination of viscous and elastic behavior. This is very clear in a creep experiment: deformation consists of flow and an elastic (retarded) part. The same is observed from a relaxation curve because this curve reflects features of viscous liquids and elastic solids. The concept of fading memory also assumes that the superposition of elastic behavior and viscous flow explains experimental facts.

2.2 RELAXATION AND CREEP – SPECTRAL REPRESENTATION. DYNAMIC FUNCTIONS

2.2.1 RETARDATION AND RELAXATION SPECTRA – DEFINITIONS

Experimental *linear* relaxation and creep functions are demonstrated here by their spectral representations. The description of the viscoelastic behavior of the material is the same for any geometrical mode of deformation, e.g., extension, shear, or volume changes. Therefore, no special definition concerning the geometry of deformations and type of stress are considered below.[8] The creep and the relaxation phenomena are compared side-by-side in the table below.

Creep	Relaxation
The typical creep functions are represented in Fig.2.1.1. The *linear* viscoelastic behavior in creep means that deformation, γ, at any time is proportional to the applied stress, σ_0. Based on the typical form of experimental data, the following equation for the creep curve can be written as: $$\gamma(t) = \sigma_0 \left[J_0 + \psi(t) + \frac{t}{\eta} \right] \qquad [2.2.1]$$ where σ_0 is the constant stress-producing creep.	The typical relaxation curves are represented in Fig. 2.1.2. The *linear* viscoelastic behavior in relaxation means that the stress, σ, at any time, is proportional to deformation created before relaxation. Then, based on the typical form of experimental data, the following equation for the relaxation curve can be written: $$\sigma(t) = \gamma_0 [G_\infty + \varphi(t)] \qquad [2.2.2]$$ where γ_0 is constant deformation, at which relaxation takes place.
The linearity of viscoelastic behavior is expressed by positioning the stress out of square brackets. So, the function called *viscoelastic compliance* $$J(t) = \frac{\gamma(t)}{\sigma_0} = J_0 + \psi(t) + \frac{t}{\eta} \qquad [2.2.3]$$ is independent of stress and can be considered as a rheological property of matter.	The linearity of viscoelastic behavior is expressed by positioning the deformation out of square brackets. So, the function called *viscoelastic* or *relaxation modulus* $$G(t) = \frac{\sigma(t)}{\gamma_0} = G_\infty + \varphi(t) \qquad [2.2.4]$$ is independent of deformation and can be considered as a rheological property of matter.
J_0 is *instantaneous compliance*, η is viscosity, and $\psi(t)$ is a *creep function*, which is a representation of the viscoelastic behavior of a sample in an experiment. These two constants, J_0 and η, and the creep function, $\psi(t)$, are characteristic parameters of the material. The creep function, $\psi(t)$, is also called the *creep compliance*, J_c. It is a ratio of (creep) deformation to stress.	G_∞ is an *equilibrium modulus*, and $\varphi(t)$ is a *relaxation function*, which is a representation of the viscoelastic behavior of a sample in an experiment. The constant G_∞ and the *relaxation function*, $\varphi(t)$, are individual parameters of the material. The relaxation function, $\varphi(t)$ is also called the *relaxation modulus*, G_r. It is a ratio of (relaxing) stress to deformation.
The difference between solids and liquids is as follows. For a liquid $\eta < \infty$, i.e., the material can flow. For solid: $\eta = \infty$, i.e., irreversible deformations are absent. This difference is clearly seen in Fig. 2.1.1. The creep function has a limit at $t \to \infty$; $\psi(\infty) < \infty$. The equilibrium elastic deformations are expressed via *the equilibrium shear compliance*, J_e $$J_e = J_0 + \psi(\infty) \qquad [2.2.5]$$	The difference between solids and liquids is as follows. For liquid $G_\infty = 0$ and $\varphi(\infty) = 0$ because, by definition, it is assumed that $\varphi(\infty) = 0$. For solids $G_\infty > 0$. This difference is clearly seen in Fig. 2.1.1. The initial value of relaxation modulus (at $t = 0$) is called an *instantaneous modulus*, G_0, and its value is found as $$G_0 = G_\infty + \varphi(0) \qquad [2.2.6]$$

Creep	Relaxation
The creep function, by its physical meaning, is an increasing function of time having a definite limit. The functions of such type can *always* be represented by the following integral	The relaxation function, by its physical meaning, is a decreasing function of time having zero limit at $t \to \infty$. The functions of such type can be *always* presented by the following integral
$$\psi(t) = \int_0^\infty J(\lambda)[1 - e^{-t/\lambda}]d\lambda \qquad [2.2.7]$$	$$\varphi(t) = \int_0^\infty G(\theta)e^{-t/\theta}d\theta \qquad [2.2.8]$$
where λ is called a retardation time, and $J(\lambda)$ is a function of distribution of retardation times, or a *retardation time spectrum*.	where θ is called a relaxation time, and $G(\theta)$ is a function of distribution of relaxation times, or *relaxation time spectrum* as measured in shear.
Retardation for many materials takes place in a wide time scale, and therefore it is important to know the retardation spectrum in a wide time scale. The *logarithmic retardation spectrum*, $k(\ln\lambda)$, is commonly used and Eq. 2.2.7 is written as	Relaxation for many materials takes place in a wide time scale, and therefore it is important to know the relaxation spectrum in a wide time scale. The *logarithmic relaxation spectrum*, $h(\ln\theta)$, is commonly used and Eq. 2.2.8 is written as
$$\psi(t) = \int_0^\infty k(\ln\lambda)[1 - e^{-t/\lambda}]d\ln\lambda \qquad [2.2.7a]$$ and evidently $$J(\lambda) = \frac{k(\ln\lambda)}{\lambda}$$	$$\varphi(t) = \int_0^\infty h(\ln\theta)e^{-t/\theta}d\ln\theta \qquad [2.2.8a]$$ and evidently $$G(\theta) = \frac{h(\ln\theta)}{\theta}$$
The integral expression can be approximated by the sum of the limited number of members:	The integral expression can be approximated by the sum of the limited number of members:
$$\psi(t) = \sum_{i=1}^N J(\lambda_i)[1 - e^{-t/\lambda_i}] \qquad [2.2.9]$$	$$\varphi(t) = \sum_{i=1}^M G(\theta_i)e^{-t/\theta_i} \qquad [2.2.10]$$
In this formula, λ_i is a set of *retardation times* and J_i are "weights" or *partial compliances*.	In this formula, θ_i is a set of *relaxation times* and G_i are "weights" or *partial moduli*.
If	If
$$J(\lambda) = \sum_{i=1}^N J_i\delta(\lambda - \lambda_i)$$ Then	$$G(\theta) = \sum_{i=1}^M G_i\delta(\theta - \theta_i)$$ Then
$$\int_0^\infty J(\lambda)[1 - e^{-t/\lambda}]d\lambda = \sum_{i=1}^N J_i[1 - e^{-t/\lambda_i}]$$ and the validity of Eq. 2.2.9 is evident.	$$\int_0^\infty G(\theta)e^{-t/\theta}d\theta = \sum_{i=1}^M G_ie^{-t/\theta_i}$$ and the validity of Eq. 2.2.10 is evident.

Creep	Relaxation
The discrete approximation helps us to understand the meaning of spectral representation of the creep function. This function is a sum (or linear superposition) of several retardation modes, each of which is characterized by its own retardation time. The distribution of the retardation times can be continuous, as written in Eq. 2.2.7 or discrete, as in Eq. 2.2.9. In both cases, the limits of distribution are not known beforehand and need to be determined from an experiment or a theoretical model.	The discrete approximation helps us to understand the meaning of the spectral representation of the relaxation function. This function is a sum (or linear superposition) of several relaxation modes, each of which is characterized by its own relaxation time. The distribution of the relaxation times can be continuous, as written in Eq. 2.2.8, or discrete, as in Eq. 2.2.10. In both cases, the limits of distribution are not known beforehand and need to be determined from an experiment or a theoretical model.
The simplest case of creep function is described by the single retardation time:	The simplest case of the relaxation function is described by the single relaxation time:
$$\psi(t) = J(1 - e^{-t/\lambda}) \qquad [2.2.11]$$	$$\varphi(t) = Ge^{-t/\theta} \qquad [2.2.12]$$
This equation for the creep function is known as the Kelvin-Voigt model.[10] The methods for finding a retardation spectrum will be discussed below, in section 2.5.3.	This is the Maxwell model of relaxation. The methods for finding a relaxation spectrum will be discussed below, in section 2.5.3.

The transition from continuous spectrum to the set of discrete values, i.e., the transition for Eq. 2.2.7 to Eq. 2.2.9 or from Eq. 2.2.8 to Eq. 2.2.10, is based on the Dirac[9] concept or the delta function.

Comments – delta-function

The delta-function, $\delta(x)$, belongs to the class of special singular functions. This is a "line" with zero width and infinite length, defined by the following equality:

$$\int_{-\infty}^{\infty} \delta(x)dx = 1$$

The main feature of this function used in many applications is its definition *via* the following functional:

$$\int_{-\infty}^{\infty} \delta(x - x_0)f(x)dx = f(x)$$

here f(x) is any "regular" function. It means that $\delta(x - x_0)$ is defined through the above written functional which puts into conformity $\delta(x - x_0)$ to some number $f(x_0)$.

It is easily seen that presenting $J(\lambda)$ or $G(\theta)$ as a sum of delta functions defined at a set of points λ_i or θ_i, respectively, one converts the integrals as in Eqs. 2.2.7 and 2.2.8, to sums as in Eqs. 2.2.9 and 2.2.10, respectively.

It is not necessary for creep and relaxation functions, or, consequently, relaxation and retardation spectra to be the same or to be mirror reflections of each other. In fact, they are not, though there is a mathematical relation between them, which will be discussed in section 2.4.

2.2.2 DYNAMIC FUNCTIONS

In order to find rheological characteristics of the material it is preferable to carry out an experiment using the simplest and the easiest to interpret conditions. In section 2.2.1, the regimes of constant stress or deformation were discussed.

It is possible to propose another simple experimental scheme such as *periodic oscillation*.[12] The form of an oscillation can be arbitrary, but it is preferable to use harmonic oscillation as the basic stress (or deformation) mode. The mathematical analysis of harmonic functions is very well developed. Also, any periodic function can be represented by a sum of harmonic functions (by the Fourier integrals).

Harmonic oscillation may continue for the duration of the experiment design. The changes taking place in the material are measured as a function of time. It is also important for modern experimental techniques to use harmonic oscillation in a very wide frequency domain, exceeding many (at least 6-8) decimal orders.

The discussion in this section will be limited to small deformations within the range of linear mechanical behavior of the material. Such a regime of deformation is known as the dynamic mechanical studies of materials.

Similar to static experiments (e.g., measurement of creep and relaxation), dynamic experiments can be conducted in stress- or deformation-controlled modes of deformation. The interpretation of both experimental modes is similar and it will be compared in the table below.

In operation with harmonic functions, it is convenient to use the mathematics of complex numbers, because it simplifies calculations.

Comments – the Euler equality
The central operation in this calculation uses the Euler theorem:

$$e^{i\omega t} = \cos(\omega t) + i\sin(\omega t)$$

where $\omega = 2\pi f$ is frequency of oscillation in rad/s, f is the frequency expressed in Hz (s⁻¹), and t is time. Writing periodic function as an exponent indicates that the real or imaginary part of the sum is involved in an experiment.

Stress-controlled experiment	**Deformation-controlled experiment**
Let stress change according to the harmonic law as $$\sigma(t) = \sigma_0 e^{i\omega t} \qquad [2.2.13]$$ where σ_0 is the amplitude of harmonic oscillation of stresses and ω is frequency. It leads to deformation, which (in a general case) changes as $$\varepsilon(t) = J_0\sigma(t) + \gamma_0 e^{i(\omega t - \delta)} - i\frac{1}{\omega\eta}\sigma(t) \quad [2.2.15]$$	Let deformation change according to the harmonic law as $$\gamma(t) = \gamma_0 e^{i\omega t} \qquad [2.2.14]$$ where γ_0 is the amplitude of harmonic oscillation of deformations and ω is frequency. It leads to stresses which (in a general case) change as $$\sigma(t) = G_0\varepsilon(t) + \sigma_0 e^{i(\omega t + \delta)} \qquad [2.2.15]$$

Stress-controlled experiment	Deformation-controlled experiment
This equation is equivalent to Eq. 2.2.1 with the same meaning of members. The new factor here is the angle, δ, which reflects the possible delay of deformation changes that follows the oscillation of stresses. The value σ_0 is the amplitude of periodically changing deformations.	This equation is equivalent to Eq. 2.2.2 with the same meaning of members. The new factor here is the angle, δ, which reflects the possible delay of stress changes which follow the oscillation of deformations. The value σ_0 is the amplitude of periodically changing stresses.
Dividing $\gamma(t)$ by $\sigma(t)$ the equation is obtained which is equivalent to Eq. 2.2.3	Dividing $\sigma(t)$ by $\gamma(t)$ the equation is obtained which is equivalent to Eq. 2.2.4
$$J^* = \frac{\gamma(t)}{\sigma(t)} = \left(J_0 + \frac{\gamma_0}{\sigma_0}\cos\delta\right) -$$ $$-i\left(\frac{1}{\omega\eta} + \frac{\gamma_0}{\sigma_0}\sin\delta\right) \qquad [2.2.17]$$	$$G^* = \frac{\sigma(t)}{\gamma(t)} = \left(G_0 + \frac{\sigma_0}{\gamma_0}\cos\delta\right) +$$ $$+i\left(\frac{\sigma_0}{\gamma_0}\sin\delta\right) \qquad [2.2.18]$$
The value J^* is called *complex shear compliance* and J^* can be written as a sum:	The value G^* is called *complex shear modulus* and G^* can be written as a sum:
$$J^* = (J_0 + J') - i\left(\frac{1}{\omega\eta} + J''\right) \quad [2.2.19]$$	$$G^* = (G_0 + G') + iG'' \qquad [2.2.20]$$
The components of the complex compliance, J' and J'', are called real and imaginary parts of the complex compliance, respectively, and are expressed as:	The components of the complex modulus, G' and G'', are called real and imaginary parts of the complex modulus, respectively, and are expressed as:
$$J' = \frac{\gamma_0}{\sigma_0}\cos\delta \text{ and } J'' = \frac{\gamma_0}{\sigma_0}\sin\delta \quad [2.2.21]$$	$$G' = \frac{\sigma_0}{\gamma_0}\cos\delta \text{ and } G'' = \frac{\sigma_0}{\gamma_0}\sin\delta \quad [2.2.22]$$
The J' and J'' components of complex compliance represent deformations changing in-phase and out-phase along with stress. The angle δ is calculated as	The G' and G'' components of complex modulus represent stresses changing in-phase and out-phase along with deformation. The angle δ is calculated as
$$\tan\delta = \frac{J''}{J'} \qquad [2.2.23]$$	$$\tan\delta = \frac{G''}{G'} \qquad [2.2.24]$$

In some applications, changes in the deformation rate are followed instead of deformation at a stress-controlled regime of deformations.

Formal rearrangements give the following result:

$$\eta^* = \frac{\sigma(t)}{\gamma(t)} = (\eta + \eta') - \eta'' \qquad [2.2.25]$$

where η^* is called *complex dynamic viscosity* and its components η' and η'' are real and imaginary parts of complex dynamic viscosity, respectively. These factors are expressed as

$$\eta' = \frac{\sigma_0}{\gamma_0\omega}\sin\delta \text{ and } \eta'' = \frac{\sigma_0}{\gamma_0\omega}\cos\delta \qquad [2.2.26]$$

The following relationships follow:

$$\eta' = \frac{G''}{\omega} \text{ and } \eta'' = \frac{G'}{\omega} \qquad [2.2.27]$$

The last central equation of the theory of periodic oscillation in studies of material properties is the consequence of expressions for J* and G*. If constants are neglected, the following simple relationship is valid:

$$J*G* = 1 \qquad\qquad [2.2.28]$$

Two parameters (in addition to constants) are measured in a dynamic regime of deformation at any frequency. They are components of the dynamic modulus, or components of dynamic compliance, or any of the components and the angle δ.

Dynamic measurements are carried out in the *linear* range of mechanical properties of material *when* and *if* the ratio of amplitudes of stress and deformations does not depend on amplitudes and therefore these two functions (J', J" or G', G") do not depend on σ_0 and consequently on γ_0. This happens when deformations are small. However, the last term (linear) must be understood not in the geometrical meaning only. The amplitude of deformations may be small ($\gamma_0 \ll 1$) but sufficient to influence the structure of the material. It may result in the observed changes of J* or G*. The latter is true, for example, in disperse systems. The deformation may exceed the characteristic size of dispersed particles or structural elements in material (which might be very small), and in this case deformations are not small in relation to the physical (structural) size, though the geometrical condition $\gamma_0 \ll 1$ is strictly fulfilled.

Frequency dependence of the dynamic modulus, G'(ω), may span over a very wide frequency range, covering many decimal orders of magnitude. Values of G' also vary in a wide range. For many polymeric substances, there is a frequency range where G' = const. This is known as *plateau modulus* (in extension, E_N^0, and in shear deformations, G_N^0), which plays an important role in the mechanical characterization of materials.

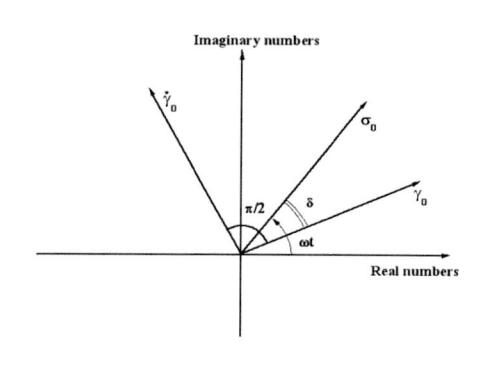

Figure 2.2.1. Graphic interpretation of oscillations in dynamic measurements of properties of viscoelastic materials.

The above-formulated theory of dynamic measurements can be interpreted in a graphical form as in Fig. 2.2.1. Let stresses and deformations be presented by the vectors with the lengths γ_0 and σ_0, respectively. The vector $\dot{\gamma}$ representing the rate of deformation is also shown in Figure 2.2.1. The vectors rotate counter-clockwise with an angular velocity of ω. The angles between γ_0 and σ_0 and the axis of real numbers at any moment, t, equal to ωt and $\omega t - \delta$, respectively.

Linearity of viscoelastic properties of materials means that stretching any vector results in stretching all other vectors by the same ratio. The angle between both vectors, σ_0 and γ_0, remains unchanged.

The projection of the stress vector on the γ_0-vector is $\sigma_0 \cos\delta$ and the projection in the perpendicular direction is $\sigma_0 \sin\delta$. G' and G" are the ratios of these projections on γ_0, i.e., these ratios are the in- and out-of-phase components of G*. The same is true for $\eta*$ with changes of terms.

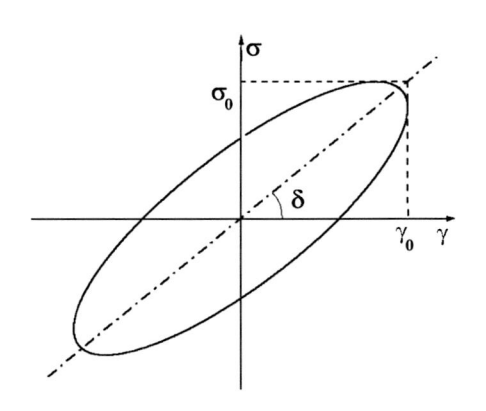

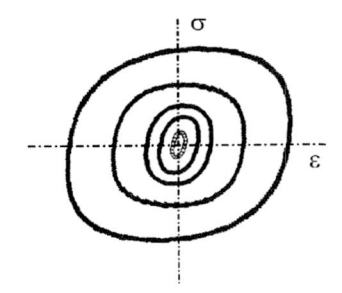

Figure 2.2.3. Several examples of Lissajous figures at different amplitudes of deformations obtained in study of concentrated emulsions. (deformation is designated, ε, as in the original publication). [Adapted, by permission, from T.G. Mason, P.K. Rai, *J. Rheol.*, **47**, 513 (2003)].

Figure 2.2.2. Graphical representation of the stress-deformation relationship in dynamic measurements of viscoelastic properties.

Another useful graphical interpretation of dynamic experiment in the measurement of viscoelastic properties is based on a combination of equations for stresses and deformations by excluding time as a parameter. Then, the following relationship between stress and deformation is obtained:

$$\left(\frac{\sigma}{\sigma_0}\right)^2 + \left(\frac{\gamma}{\gamma_0}\right)^2 = (\sin\delta)^2 + 2\left(\frac{\sigma}{\sigma_0}\right)\left(\frac{\gamma}{\gamma_0}\right)\cos\delta \qquad [2.2.29]$$

This equation is that of an ellipse with the principal axis inclined to the abscissa by angle, δ, as represented in Fig. 2.2.2.

The exclusion of argument, t, in both dependencies, $\sigma(t)$ and $\gamma(t)$, permits building the closed curves called the *Lissajous figures* (Fig.2.2.3). These curves are not necessarily ellipses, and their shapes differ depending on deformation. The non-elliptic shape of the Lissajous figures is a reflection of the non-linear behavior of the material in oscillation (see section 2.8).

The area, A, of an ellipse equals:

$$A = \pi\sigma_0\gamma_0\sin\delta \qquad [2.2.30]$$

Two limiting cases are of special interest.

1.　$\delta = \pi/2$. In this case, the ellipse transforms into a circle with coordinates (σ/σ_0) and (γ/γ_0) and its area is $A = \pi\sigma_0\gamma_0$. This case corresponds to G' = 0 and it means that the material is a purely viscous liquid without elasticity.

2.　$\delta = 0$. In this case, the ellipse degenerates into a straight line. This means that G" = 0 and the material becomes a purely elastic body. The area of the ellipse equals zero, A = 0.

The representation of experimental results of stress-deformation relationships by means of an elliptic figure permits use of a simple method for calculation of angle, δ. The area of a rectangle circumscribed around an ellipse equals

$$S = 4\sigma_0\gamma_0$$

Then, $\sin\delta$ is expressed by the ratio of the ellipse surface area to the surface area of a circumscribed rectangle as

$$\sin\delta = \left(\frac{4}{\pi}\right)\left(\frac{A}{S}\right) \qquad [2.2.31]$$

This approach is used in some standards and experimental devices because the value of angle δ has important applications by itself (for example, in estimating the damping characteristics of rubbers, rubber compounds, and porous materials). The angle δ can be found from a hysteresis loop (i.e., a surface area of an ellipse, as in Fig. 2.2.2) with no need to calibrate stress and deformation scales and to find σ_0 and ε_0.

The graphic representation of oscillating deformation gives a physical interpretation of parameters introduced for quantitative description of results of dynamic experiment.

The work of entire oscillation cycle (gained and lost) is calculated as

$$W = \int_0^T \sigma(t)d\gamma \qquad [2.2.32]$$

where $T = 2\pi/\omega$ is the duration of a single cycle of oscillations.

Direct calculation shows that this work is as follows

$$W = \pi\sigma_0\gamma_0\sin\delta \qquad [2.2.33]$$

Eqs. 2.2.30 and 2.2.33 are identical, which means that the surface area of an ellipse is directly interpreted as the work dissipated during the cycle of oscillation. The angle δ is a relative measure of work losses. For this reason, it is called the *loss angle*. For an elastic body $\delta = 0$ (an ellipse degenerates into a straight line) and losses are absent. For a viscous liquid $\delta = \pi/2$, and consequently the losses are at maximum. By decreasing δ, and consequently decreasing viscous losses, material transits from pure viscous to pure elastic.

The introduction of values J" and G" into Eq. 2.2.33 leads to the following relationship:

$$W = \pi\sigma_0^2 J" = \pi\gamma_0^2 G" \qquad [2.2.34]$$

Values of J" and G" are also measures of losses (heat dissipation in periodic oscillation) and therefore they are called *loss compliance* and *loss modulus*, respectively.

It is possible to show that real components of dynamic compliance and modulus, J' and G', are measures of elasticity because the energy stored (and then returned) during the cycle of oscillation is proportional to these values. Therefore J' and G' are called *storage compliance* and *storage modulus*, respectively.

In real practice, viscoelastic materials can be used in many forms and applications. Engineering products (e.g., springs) must be highly elastic (losses must be low) and they should be made out of materials having low loss angle values. This is also true of church bells. A bell continues to sound as long as the mechanical losses of material, from which it is made, are low. Shock absorbers, sound insulators, and materials for many other similar applications must possess a high dissipative function (energy), meaning that the loss angle of such materials must be as close to $\pi/2$ as possible.

The above-formulated parameters are used to describe viscoelastic effects and characterize properties of real materials. However, they are not constants but functions. Just as creep and relaxation functions depend on time, the dynamic properties depend on the frequency of oscillation. In representing viscoelastic properties of the material, it is necessary to consider the functions $J'(\omega)$ and $J''(\omega)$, or $G'(\omega)$ and $G''(\omega)$ as permitted by equation 2.2.28. Other parameters can also be included (loss angle or dynamic viscosity). The principal conclusion is as follows: the viscoelastic properties of the material, measured in a dynamic experiment, are represented by two frequency-dependent functions.

2.3 MODEL INTERPRETATIONS

2.3.1 BASIC MECHANICAL MODELS

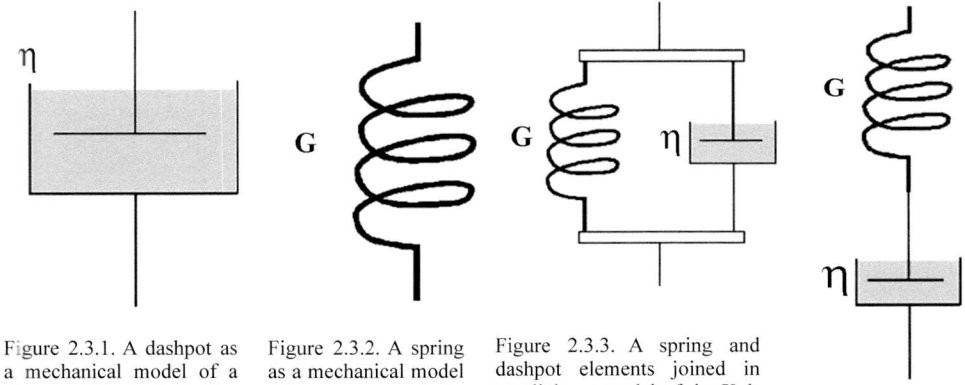

Figure 2.3.1. A dashpot as a mechanical model of a viscous (Newtonian) liquid.

Figure 2.3.2. A spring as a mechanical model of an elastic (Hookean) solid.

Figure 2.3.3. A spring and dashpot elements joined in parallel – a model of the Kelvin-Voigt viscoelastic liquid.

Figure 2.3.4. A spring and a dashpot element joined in series – a model of the Maxwellian viscoelastic liquid.

It is instructive and useful for better understanding the ideas of theory of viscoelasticity to illustrate some typical behaviors of viscoelastic materials by simple mechanical models.

The simplest model of viscous behavior of the material is a dashpot, a piston moving inside (Fig. 2.3.1) a cylinder filled with liquid. It is supposed that the speed of movement, $V = dX/dt$, of the piston is proportional to the applied force, F:

$$F = \eta V \tag{2.3.1}$$

where X is displacement, t is time, and η is the coefficient of proportionality.

This equation is formally analogous to the basic equation for Newtonian liquid and one assumes that the speed is an analogue of the rate of deformation, the force F is an analogue of stress, σ, and the coefficient η is an analogue of viscosity.

The simplest mechanical model of an elastic body is a spring (Fig. 2.3.2). The displacement X is proportional to the applied force, F:

$$X = \frac{F}{G} \tag{2.3.2}$$

In order to compare this equation with the standard formulation of Hooke's law, Σ can be treated as a relative deformation, F as an analogue of stress, and G as an analogue of the elastic modulus.

Both equations are quite trivial and do not add anything new to the initial concepts of viscous (Newtonian) liquid and elastic (Hookean) material. However, it is instructive to analyze the following mechanical models: let spring and dashpot elements be connected in parallel (Fig. 2.3.3) or in series (Fig. 2.3.4). These two mechanical models are compared below.

Let the constant force, $F = F_0 = $ const, be suddenly applied to the end of the jointed two-component model shown in Fig. 2.3.3 and continue to be applied for an unlimited time. It is easy to anticipate that both components will resist the movement, with a spring stretching and the piston slowly pulling out of a cylinder. This process continues until the spring is stretched to the length corresponding to the applied force. The movement of the piston stops at this state because the force is balanced by a spring and the force is absent at the piston. The mathematical representation of the above-mentioned statement is as follows.	Let the constant displacement, $X = X_0 = $ const, be suddenly created at the end of the jointed two-component model shown on Fig. 2.3.4 and then fixed for an unlimited time. It is easy to anticipate that this displacement immediately stretches the spring. Then, the extended spring will pull a piston out of a cylinder and this process will continue for some time because the movement in a viscous liquid is not very fast. This process continues until the spring comes to the equilibrium state (its initial length is restored). The mathematical representation of the above-mentioned statement is as follows.
Total force, F, of the two-component model F is the sum of forces acting on both elements of a model: $$F_{sp} + F_{pist} = F$$ where F_{sp} is the force acting on the spring branch and F_{pist} is the force acting on the piston.	Total displacement, X, of the two-component model is the sum of displacements of both components $$X_{sp} + X_{pist} = X$$ where X_{sp} is the displacement of the spring and X_{pist} is the displacement of the piston. The same equality is true for the derivatives of the components and their sum $$\dot{X}_{sp} + \dot{X}_{pist} = \dot{X}$$
Now, using Eqs. 2.3.1 and 2.3.2, the last sum can be rearranged as $$\eta\dot{X} + GX = F_0 \qquad [2.3.3]$$	Now, using Eqs. 2.3.1 and 2.3.2, the last sum can be rearranged as $$\frac{\dot{F}}{G} + \frac{F}{\eta} = 0 \qquad [2.3.4]$$
The solution of this equation gives the time dependence of the displacement X Eq. [2.3.5] below: $$X(t) = X_\infty\left(1 - e^{-\frac{t}{\eta/G}}\right) = EF_0\left(1 - e^{-\frac{t}{\eta/G}}\right)$$	The solution of this equation gives the time dependence of the force F: $$F(t) = F_0 e^{-\frac{t}{\eta/G}} = X_0 E e^{-\frac{t}{\eta/G}} \qquad [2.3.5]$$

The last expression is the analogue of Eq. 2.2.11. The ratio (k/G) has the meaning of a *retardation time*. The combination of components as shown in Fig. 2.3.3 behaves like Kelvin-Voigt viscoelastic material and this is why it is called the Kelvin-Voigt model. The Kelvin-Voigt material is solid because the application of a constant force leads to a limited displacement as for any solid body (though with delayed elasticity). The relaxation time of the Kelvin-Voigt model is absent (or to be more exact is equal to infinity). The Kelvin-Voigt model stretched by a constant force supports this force unlimitedly long due to stretching a spring that does not relax.	The last expression is the analogue of Eq. 2.2.12, where the ratio (k/G) has the meaning of a *relaxation time*. The combination of components as shown in Fig. 2.3.4 behaves like the Maxwell viscoelastic material and this is why it is called the Maxwell model. It is easy to see that Maxwellian material is liquid because the application of a constant force leads to unlimited movement of the piston, i.e., this is a model of flow. The retardation time of the Maxwell model equals zero. The Maxwell model immediately follows the applied force due to the reaction of a spring acting without any delay.
Let us now join N the Kelvin-Voigt models in series to each other as shown in Fig. 2.3.5. Formal calculations show that this model's behavior can be described by Eq. 2.2.9 with N retardation times.	Let us now join M the Maxwell models in parallel to each other as shown in Fig. 2.3.6. Formal calculations show that this model's behavior can be described by Eq. 2.2.10 with M relaxation times.

The use of the mechanical analogue models illustrates the behavior of the material in different modes of deformations. As an example, dynamic functions of both models are calculated by changing the model parameters, such as the rheological characteristics, viscosity, η, and modulus, E, of components of the Kelvin-Voigt and Maxwell models.

The parameters of the model and the components of the dynamic functions of the Kelvin-Voigt model are expressed as: retardation time $\lambda = \eta/G$ relaxation time $\theta = \infty$	The parameters of the model and the components of the dynamic functions of the Maxwell model are expressed as: relaxation time $\theta = \eta/G$; retardation time $\lambda = 0$
Then $$G'(\omega) = G \ \& \ G''(\omega) = \omega\eta$$ $$J'(\omega) = \frac{1}{G}\frac{1}{1+(\lambda\omega)^2} \ \& \ J''(\omega) = \frac{1}{G}\frac{\omega\lambda}{1+(\lambda\omega)^2}$$ $$\tan\delta = \omega\lambda \qquad\qquad [2.3.7]$$	Then $$G'(\omega) = \frac{G(\omega\theta)^2}{1+(\omega\theta)^2} \ \&$$ $$G''(\omega) = \frac{G(\omega\theta)}{1+(\omega\theta)^2}$$ $$J'(\omega) = G^{-1} \ \& \ J''(\omega) = (\omega\theta)^{-1}$$ $$\tan\delta = (\omega\theta)^{-1} \qquad\qquad [2.3.8]$$

The comparison of the above-presented results in both columns shows that the Kelvin-Voigt and the Maxwell models predict different behavior of the material, in respect of retardation and relaxation times, frequency dependencies of loss tangent, etc. The Maxwell material is liquid and the Kelvin-Voigt material is solid. Using mechanical analogue models it is easy to transit from describing solid-like to liquids-like behavior and *vice versa*.

Let us assume that in a multi-component Kelvin-Voigt model one partial modulus equals zero. Then, this component degenerates into a single viscous element, and finally, it means that such multi-component system models do not have solid but liquid-like behavior, because unlimited deformations become possible.	Let us assume that in a multi-component Maxwell model one partial viscosity is infinite. Then, this component degenerates into a single elastic element, and finally, it means that such a multi-component system model is not a liquid but it has solid-like behavior, because the spring stores non-relaxing stress.

For combinations of either the Kelvin-Voigt (Fig. 2.3.5) or the Maxwell (Fig. 2.3.6) models, the expressions for the components of the dynamic functions include sums of the same structure as the equations for the single elements.

Frequency dependencies of the components of the dynamic functions for multi-element models are

$$J'(\omega) = \sum_{i=1}^{N} \frac{1}{G_i} \frac{1}{1+(\omega\lambda_i)^2}$$ $$J''(\omega) = \sum_{i=1}^{N} \frac{1}{G_i} \frac{(\omega\lambda_i)}{1+(\omega\lambda_i)^2} \qquad [2.3.9]$$ where N is the number of elements in the model in Fig. 2.3.5 (retardation times).	$$G'(\omega) = \sum_{i=1}^{M} G_i \frac{(\omega\theta_i)^2}{1+(\omega\theta_i)^2}$$ $$G''(\omega) = \sum_{i=1}^{M} G_i \frac{(\omega\theta_i)}{1+(\omega\theta_i)^2} \qquad [2.3.10]$$ where M is the number of elements in the model in Fig. 2.3.6 (relaxation times).

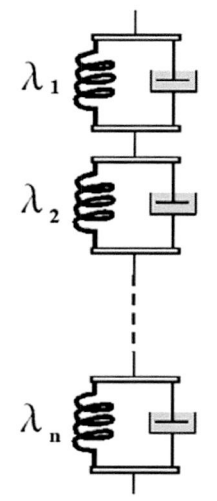

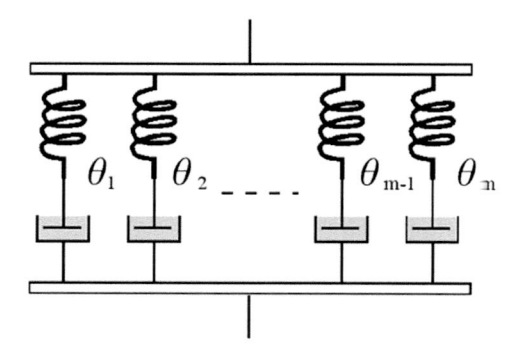

Figure 2.3.6. A combination of the Maxwell models joined in parallel – a model of viscoelastic liquid with a set of relaxation times θ_i.

The use of the model representation opens a good potential for analysis of material behavior in various deformation-stress modes.

Figure 2.3.5. A combination of the Kelvin-Voigt models joined in series – a model of viscoelastic solid with a set of retardation times, λ_i.

2.3.2 COMPLICATED MECHANICAL MODELS - DIFFERENTIAL RHEOLOGICAL EQUATIONS

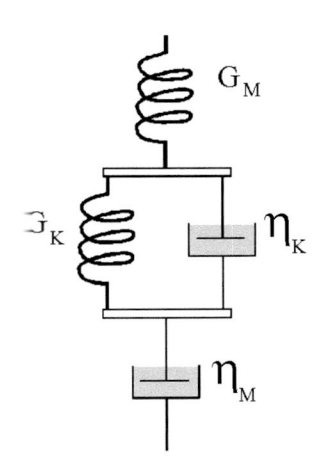

Figure 2.3.7. The Burgers model – predicting superimposed relaxation and retardation.

The Maxwell and the Kelvin-Voigt models can be joined in parallel, in series, or combinations thereof. In many applications, these models are also joined in their combinations. The combination of the Maxwell model and the Kelvin-Voigt model joined in series (known as the Burgers model[13]) is shown in Fig. 2.3.7. This combination represents a popular quantitative model of the behavior of polymeric materials.

The central peculiarity of this model is the possible combination of relaxation and retardation phenomena in one material. Let us suppose that at low temperatures the "Maxwell viscosity", η_M, is very high – a material is "frozen" – and it is possible to neglect deformations of this element. Then, the model under discussion presents a solid-like behavior due to the spring G_M. With temperatures increasing, viscosity, η_M, decreases and relaxation become possible. Then it is typical relaxation-retardation behavior of a viscoelastic material. At even higher temperatures, the viscosity of both components, η_M and η_K, becomes very low and the model represents the behavior of primarily viscous liquid, such as polymer melt (with a slight retardation).

The mathematical description of the behavior of a model represented in Fig. 2.3.7 is based on the summation of deformations of three components of the model

$$\gamma_{M, sp} + \gamma_K + \gamma_{M, pist} = \gamma$$

where γ is the total deformation, which is the sum of deformation of a Maxwellian spring, $\gamma_{M,sp}$, Maxwellian viscous element, $\gamma_{M,pist}$, and deformation of a Kelvin-Voigt element, γ_K. The direct substitutions lead to two equations:

$$\begin{cases} \dfrac{\dot{\sigma}}{G_M} + \dfrac{\sigma}{\eta_M} + \dot{\gamma}_K = \dot{\gamma} \\ \sigma = G_K \gamma_K + \eta_K \dot{\gamma}_K \end{cases}$$

Then, after excluding $\dot{\gamma}_K$ from both equations, the final rheological equation for $\gamma(\sigma)$ dependence is obtained. The structure of this dependence is of special interest. This is an equation including higher derivatives of variables:

$$k_2 \frac{d^2\sigma}{dt^2} + k_1 \frac{d\sigma}{dt} + k_0 \sigma = l_1 \frac{d\gamma}{dt} + l_0 \gamma \qquad [2.3.11]$$

where the coefficients of members of sums, k_i and l_i, are constructed from *four* rheological constants of the Burgers model.

The relaxation behavior of the Burgers model can be determined. $\sigma(t)$ is found as the solution of the *second-order differential equation*, as follows:

$$\sigma(t) = C_1 e^{-t/\theta_1} + C_2 e^{-t/\theta_2}$$

where C_1 and C_2 are constants expressed *via* k_i (or four rheological parameters of the Burgers model) and θ_1 and θ_2 are *two* independent relaxation times also expressed *via* four parameters of the Burgers model. This means that the Burgers model is equivalent to two Maxwell models joined in series. However, it is evident that the Burgers model predicts a single retardation time only.

The construction of rheological models by joining the simplest the Kelvin-Voigt and the Maxwell models in various (sometimes rather whimsical) combinations leads, in a general case, to the following *operator equation*:

$$\sum_{n=0}^{N} k_n \frac{d^n}{dt^n} \sigma = \sum_{m=0}^{M} l_m \frac{d^m}{dt^m} \gamma \qquad\qquad [2.3.12]$$

where k_n and l_m are individual rheological parameters of the material. Relaxation and retardation times are expressed *via* these parameters.

If the material is liquid $N > M$. If it is solid $N = M$. The differential rheological model predicts the existence of a set of discrete relaxation times. Consequently, it is possible to prove that, for a model of liquid, the number of relaxation times, m, equals the number of retardation times, n, plus 1.

Differential equations are considered to be old-fashioned and they are used infrequently for presenting rheological properties of the material. More popular are the integral equations derived from the superposition principle (see section 2.4).

2.3.3 NON-MECHANICAL MODELS

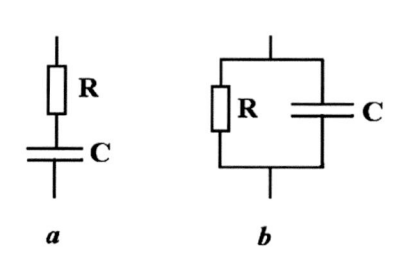

Any combination of physical elements leading to the same mathematical predictions, primarily exponential decay and growth of some variables, can be treated as an analogue model of the viscoelastic rheological behavior of material.

Among them, the most interesting and useful are electrical analogue systems,[14] because electrical analogue models can be easily built in a laboratory and their behavior can be followed in detail using simple experimental techniques.

a   *b*

Figure 2.3.8. Combination of resistor and capacitor joined in series as an analogue of the Kelvin-Voigt solid (a) and in parallel as an analogue of the Maxwell liquid (b).

In order to illustrate the electrical analogue, let us consider a circuit consisting of two elements: a resistor with resistance, R, and a capacitor with capacitance, C. Electrical charge, Q, is an analogue of mechanical deformation, current, $J = dQ/dt$, is an analogue of deformation rate and voltage, U, is an analogue of mechanical stress. Both elements can be joined in series and parallel as shown in Fig. 2.3.8.

The starting relationships for the simplest elements are:

for a resistor: $J = \dot{Q} = \dfrac{1}{R}U$

for a capacitor: $Q = CU$

Then the analysis of Fig. 2.3.8 gives the following results.

For a capacitor and a resistor joined in series, the total voltage, U_0, is a sum of voltages at both components: $$R\dot{Q} + \frac{1}{C}Q = U_0$$ and it is easy to show that charge increases according to the following equation: $$Q = Q_\infty(1 - e^{-t/RC})$$ where $Q_\infty = CU_0$ is the equilibrium charge of the capacitor.	For a capacitor and a resistor joined in parallel, the total current, J, is a sum of currents in both components: $$\frac{U}{R} + CU = J$$ and it is easy to show that at $Q = $ const $(J = 0)$, the voltage relaxes according to the following equation: $$U = U_0 e^{-t/RC}$$ where $U_0 = Q_0/C$, and Q_0 is the initial charge of the capacitor.
This is a direct analogue of the Kelvin-Voigt model and the product (RC) has the meaning of the retardation time.	This is the direct analogue of the Maxwell model and the product (RC) has meaning of the relaxation time.
The set of models shown in Fig. 2.3.8a joined in parallel presents an analogue of the multi-retardation time of the Kelvin-Voigt model.	The set of models shown in Fig. 2.3.8b joined in series presents an analogue of the multi-relaxation time the Maxwell model.

Electrical analogue modeling permits the construction of even more complicated versions of relaxation and retardation behavior and examination of their behavior by varying the model parameters.

2.4 SUPERPOSITION – THE BOLTZMANN-VOLTERRA PRINCIPLE

2.4.1 INTEGRAL FORMULATION OF THE SUPERPOSITION PRINCIPLE

The phenomenon of fading memory can be formulated in the following way: the longer the time interval between events and their observed consequences, the weaker the influence of these events on the observed material behavior.

Mathematical formalization of this assumption means that functions connecting deformations and stresses must be decreasing and written *via* an argument (t - t'), where t is the actual (current) time, and t' is the time at which some event took place. This formalism leads to the general formulation of a relationship between stresses and deformations.

Two principal ideas are the basis of this approach:
- the response to any event is linear
- all consequent events lead to independent responses.

This is the *principle of linear superposition* of stresses and/or deformations.[15] It means that material reacts to the next action as if no former action took place. In other words, the structure and properties of the material are not changed, regardless of its deformation, and the last statement is a real physical meaning of the principle of linear superposition.

Now, let us write the above-stated concept in the form of mathematical formalism. Let the initial stress, acting from the time t = 0, be equal σ_0. Then, deformations immediately begin to change according to Eq 2.2.1. At some point in time, t', let stress change by $\Delta\sigma$. The principle of linear superposition assumes that in this case, deformation changes accordingly:

$$\gamma(t) = \sigma_0\left[J_0 + \psi(t) + \frac{t}{\eta}\right] + \Delta\sigma\left[J_0 + \psi(t - t') + \frac{t - t'}{\eta}\right] \qquad [2.4.1]$$

Stress can change at any given time. For any such moment of time and any corresponding change of stress, one can add an independent term in the last equation for $\gamma(t)$.

So, $\gamma(t)$ is written as a sum:

$$\gamma(t) = \sum_i \Delta\sigma_i\left[J_0 + \psi(t - t'_i) + \frac{t - t'_i}{\eta}\right] \qquad [2.4.2]$$

where $\Delta\sigma_i$ is the new change of stress added at the time t_i.

This equation shows the following:
- all stress inputs are independent, they do not interact with each other and the deformation is proportional to stress (the principle of linear superposition)
- the influence of stress changes on deformation decreases with time. If stress was applied earlier, as determined by the argument (t - t') (fading memory) more changes were observed compared with the recently applied stresses.

Stress may change continuously, this leads to the final integral (instead of sum) formulation of the principle of linear superposition:

$$\gamma(t) = \int_{-\infty}^{t}\left[J_0 + \psi(t - t') + \frac{t - t'}{\eta}\right]d\sigma(t') \qquad [2.4.3]$$

or

$$\gamma(t) = \int_{-\infty}^{t}\frac{d\sigma}{dt'}\left[J_0 + \psi(t - t') + \frac{t - t'}{\eta}\right]dt' \qquad [2.4.4]$$

The lower limit of integration ($-\infty$) reflects an idea that the *whole* history of deformation influences deformation at any chosen moment, t.

The analogous line of arguments can be used to describe changes in stresses, and in this case, Eq 2.2.2 is a starting point. The final result is quite similar to Eq 2.4.4 and can be written as

$$\sigma(t) = \int_{-\infty}^{t}\frac{d\gamma}{dt'}[G_\infty + \varphi(t - t')]dt' \qquad [2.4.5]$$

A pair of symmetrical Eqs 2.4.4 and 2.4.5 is called the *Boltzmann-Volterra equations*.[16] They form a complete mathematical formulation of the principle of linear superposition.

The integrals representing the Boltzmann-Volterra superposition principle contain the difference (t - t') as an argument. The integrals of such structure are called the *hereditary integrals*, because they reflect the fading influence of pre-history of deformations on the current state of material.

The hereditary part of Eq. 2.4.5 is

$$\sigma(t) = \int_{-\infty}^{t} \dot{\gamma}(t')\varphi(t-t')dt'$$ [2.4.5a]

This expression is conveniently used in the discussion of the viscoelastic behavior of liquids because it is characterized by the rate of deformation. For solids, deformation by itself is the determining factor. Therefore, the Boltzmann-Volterra superposition principle is more useful in the alternative form

$$\sigma(t) = \int_{-\infty}^{t} \gamma(t')m(t-t')dt'$$ [2.4.6]

where m(t - t') is called a *memory function*. Its physical meaning is equivalent to the meaning of the relaxation function in Eq. 2.4.5 because both characterize the fading influence of pre-history of deformations on the current stress state of the material.

The relationship between functions φ (t - t') and m(t - t') is established from comparison of Eqs 2.4.5 and 2.4.6:

$$m(t) = \frac{d\varphi(t)}{dt}$$ [2.4.7]

The behavior of viscoelastic material according to the principle of superposition can be illustrated by the following example for elastic recoil (retardation) after forced deformation of a body.

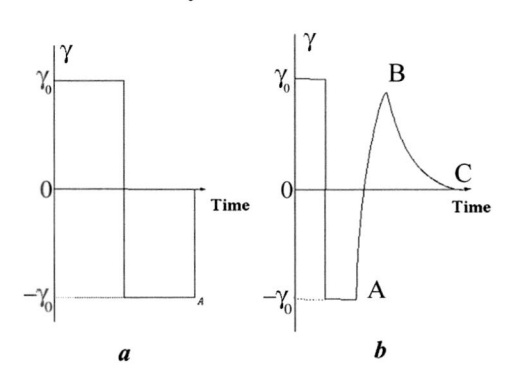

Let the history of deformations be as shown in Fig. 2.4.1a: the external force created deformation, γ_0, and then (very rapidly) the same deformation but with the opposite sign, $-\gamma_0$, is realized. When the force was acting during two short periods of time, one could neglect partial relaxation at deformations γ_0 and $-\gamma_0$. Now, we follow what happens if at the moment A the external force is removed. An ideal elastic body immediately returns to its initial state, as shown by the vertical line from point A

Figure 2.4.1. Deformation history created by external force (a) and the post-reaction (b) of viscoelastic material after removal of the force at the point A.

in Fig. 2.4.1a. The behavior of a viscoelastic body is quite different, as illustrated by the line ABC in Fig. 2.4.1b. Seemingly, the shape of the curve looks rather strange. Indeed, why does the deformation cross the zero line and reach point B? The first part of this line, AB, is the retardation from the second deformation, $-\gamma_0$, but the sample "remembers" that the first deformation was γ_0 and strives to restore itself to a state determined by the first deformation, γ_0. Only after that slow (delayed) action does a return to the zero state occurs.

Another very interesting (and important for technological applications) example of the influence of deformational prehistory on the behavior of the material is related to polymer processing (thermoplastic and rubber compounds). During extrusion of the continuous profile, a molten material moves between the screw and the barrel of an extruder, then it passes through transient channels. Finally, it is shaped in an outlet section of a die. It is desirable that the shape of the final profile is equivalent to the shape of the outlet section of the die. But the material continues to react to all deformations, which took place before the outlet section of the die. As a result, distortion of its shape occurs; therefore, the final section of the part can be different than expected. The related effects are discussed in Chapter 3.

Memory effects become complicated and sometimes rather unexpected when temperature changes during the process of deformation. In these cases, if temperature decreases, the memory of former deformations can be frozen and the material looks stable (retardation times become too long for an observer). However, upon heating of an article, the frozen deformations release, and one observes effects that are unusual. Many examples of memory of the previous deformation states of such kind are characteristic of applications of viscoelastic materials.

The following example can be considered to illustrate the critical case related to frozen stresses. A polymer block stores large internal (frozen) stresses due to its previous temperature-deformation pre-history. This block appears stable. However, when additional stresses are imposed during its machining (drilling, cutting, or other operations) the block may disintegrate into small pieces, may even become powder. This is typical of some materials (polymers and inorganic glasses, large crystallized blocks) especially if they are cooled rapidly.

Viscoelastic materials have a fading memory of the history of previous deformations. The integrals in Eqs. 2.4.4 and 2.4.5 are called hereditary integrals because they summarize events that took place before the current moment of time and are responsible for the stress (or deformation) state of the material at the current moment.

The relaxation function, $\varphi(t)$, is a decreasing function. Therefore, its values are higher when the argument is smaller. It means that the changes of deformation, which happened earlier, influence stress to a lesser degree than later changes. In the first case, the value of the argument $(t - t')$ in Eq 2.4.5, for a fixed moment of time, t, is smaller than for events that happened later because values of t' are smaller. In other words, a material continuously "forgets" what happened before, and in this sense the integrals in Eqs. 2.4.4 and 2.4.5 form a model of material with "fading memory".

It is interesting to outline the limiting cases. They are:
- liquid which "forgets" everything immediately (energy of deformation completely dissipates); in this case, the integral Eq. 2.4.5 transforms to the Newton law
- solid which "remembers" everything (energy of deformation is completely stored), and in this case, the integral Eq. 2.4.3 transforms to Hooke's law.

2.4.2 SUPERPOSITION PRINCIPLE EXPRESSED VIA SPECTRA

Relaxation or retardation spectra can be inserted into the general integral of Eqs. 2.4.4 and 2.4.5 by substitution of Eqs. 2.2.7 and 2.2.8 into Eqs. 2.4.4 and 2.4.5, respectively.

Below, this procedure is illustrated by calculations of stresses, $\sigma(t)$, developing in liquid ($E_\infty = 0$); the same will be done for deformations, $\gamma(t)$.

Substitution and the subsequent rearrangements lead to the following result:

$$\sigma(t) = \int_{-\infty}^{t} \frac{d\gamma}{dt'}[\varphi(t-t')]dt' = \int_{-\infty}^{t} \frac{d\gamma(t-t')}{dt'}\left[\int_{0}^{\infty} G(\theta)e^{-\frac{t-t'}{\theta}}dt'd\theta\right] =$$

$$= \int_{0}^{\infty} G(\theta)\left[\int_{-\infty}^{t} \frac{d\gamma(t-t')}{dt'}e^{-\frac{t-t'}{\theta}}dt'\right]d\theta \qquad [2.4.8]$$

The expression in the square brackets is a function of t only. It permits us to analyze some typical cases of stress development in various deformation modes.

2.4.3 SIMPLE TRANSIENT MODES OF DEFORMATION

The superposition principle can be illustrated by several simple examples related to transient modes of deformation. The examples given below are based on Eqs. 2.4.5 or 2.4.6 and they relate to shear deformation.

2.4.3.1 Relaxation after sudden deformation

First of all, it is necessary to define the word "sudden" for the purposes of mathematical language. This deformation mode can be formulated as

$$\sigma(t) = \gamma_0 \varphi(t) \qquad [2.4.9]$$

Evidently, this is nothing else than the definition of the relaxation function expressed by Eq. 2.2.2.

From Eq. 2.4.6, the following equation for $\sigma(t)$ can be derived

$$\sigma(t) = \gamma_0 \int_{0}^{\infty} G(\theta)d\theta \qquad [2.4.10]$$

This equation is also already known: it is Eq. 2.2.8. The initial value of the $\sigma(t)$ function divided by γ_0 equals $\varphi(0)$:

$$\frac{\sigma(t=0)}{\gamma_0} = \varphi(0) = \int_{0}^{\infty} G(\theta)d\theta \qquad [2.4.11]$$

and this is a part of the instantaneous modulus (see Eq. 2.2.4).

2.4.3.2 Developing stresses at a constant shear rate

In this deformation mode $\dot{\gamma} = \dot{\gamma}_0 = \text{const}$, deformation is suddenly created at the time $t = 0$. Then it is easy to show that Eq. 2.4.5 leads to the following expression:

$$\sigma(t) = \dot{\gamma}_0 \int_{0}^{\infty} \varphi(x)dx \qquad [2.4.12]$$

Calculations based on Eq. 2.4.8 result in the following expression for stress evolution in material with the known relaxation spectrum

$$\sigma(t) = \dot{\gamma}_0 \int_{0}^{\infty} \theta G(\theta)(1 - e^{-t/\theta})d\theta \qquad [2.4.13]$$

The equilibrium value of σ (at $t \to \infty$) divided by $\dot{\gamma}_0$ equals

$$\frac{\sigma(t \to \infty)}{\dot{\gamma}_0} = \int_0^\infty \varphi(x)dx = \int_0^\infty \theta G(\theta)d\theta \qquad [2.4.14]$$

Viscosity at the steady-state regime of flow is a physical meaning of this value.

2.4.3.3 Relaxation after steady shear flow

Let the deformation rate be $\dot{\gamma} = \dot{\gamma}_0 = \text{const}$ at $t < 0$, and, at the time $t' = 0$, the deformation rate suddenly drops to zero. Here, the deformation rate is described by two members: $\dot{\gamma} = \dot{\gamma}_0 = \text{const}$ at $t' < 0$ and by means of the negative delta-function at the point $t' = 0$. After some formal rearrangements, Eq. 2.4.5 leads to the final expression for stress relaxation:

$$\sigma(t) = \dot{\gamma}_0 \left\{ \int_0^\infty \varphi(t)dt - \int_0^t \varphi(t)dt \right\} = \dot{\gamma}_0 \int_t^\infty \varphi(t)dt \qquad [2.4.15]$$

The first integral term in this equation represents the equilibrium value of stress at steady flow (see Eq. 2.4.14).

The difference between Eqs. 2.4.9 and 2.4.15 (both written for "relaxation") demonstrates that the relaxation process can be very different depending on the history of deformation prior to relaxation.

Calculations using Eq. 2.4.8 give the following result:

$$\sigma(t) = \dot{\gamma}_0 \int_0^\infty \theta G(\theta) e^{-t/\theta} d\theta \qquad [2.4.16]$$

The difference between Eqs. 2.4.10 and 2.4.16 is evident.

It is also worth mentioning that Eqs. 2.4.12 and 2.4.15 as well as Eqs. 2.4.13 and 2.4.16 are mirror relations of each other.

Eqs. 2.4.4 and 2.4.5 can be used for quantitative analysis of any other arbitrary regimes of stress or deformation evolution.

2.4.3.4 Relationship between relaxation and creep functions

Eqs. 2.4.4 and 2.4.5 contain deformation and stress, and each of them can be treated as an equation for either stress or deformation. Eq. 2.4.4 determines the development of deformation for the known evolution of stresses. It can be considered as an integral equation for $\sigma(t)$ if the function $\gamma(t)$ is known. The same is true for Eq. 2.4.5. Therefore, it is possible to exclude these functions by substituting, for example, the function $\gamma(t)$ from Eq. 2.4.4 to the right side of Eq. 2.4.5. After some formal mathematical rearrangements, the relationship between rheological parameters is obtained which does not contain $\gamma(t)$ or $\sigma(t)$. The resulting equation includes only constants and creep and relaxation functions in the following form:

$$G_\infty J_0 + J_0 \varphi(t) + G_\infty \left[\frac{t}{\eta} + \psi(t) \right] + \int_{-\infty}^t \varphi(t') \left[\frac{1}{\eta} + \frac{d\psi(t-t')}{d(t-t')} \right] dt' = 1 \qquad [2.4.17]$$

where G_∞ is the equilibrium modulus, J_0 is the instantaneous compliance, $\psi(t)$ is the creep function, $\varphi(t)$ is the relaxation function and η is viscosity.

Eq. 2.4.17 shows that the relaxation and creep functions are not independent but related to each other by the integral equation. If one of these functions is known (measured, calculated, or assumed), the other can be found from Eq. 2.4.17. This equation formally, and rigorously, confirms that the behavior of the material, in different modes of deformation, is governed by the same inherent properties.

Eqs. 2.4.4 and 2.4.5 give the mathematical ground for calculation of stress-deformation relationship at any arbitrary path of material loading. The only, but very essential, limitation in the application of these equations is the requirement of linearity of rheological behavior of medium, i.e., independence of all material constants and functions entered into these equations (instantaneous compliance, equilibrium modulus, viscosity, relaxation, and creep functions) on stresses and deformations.

Eqs. 2.4.4 and 2.4.5 are general rheological equations of state, or constitutive equations, for viscoelastic materials. The differences in properties of various materials are hidden in the values of constants and types of viscoelastic (creep or relaxation) functions.

It is essential to remember that the above properties are related to a "point" as adapted for Newtonian liquid and Hookean solid and in general for any rheological equation of state. In order to find stress deformation distribution throughout a body, it is necessary to combine these equations with equilibrium conditions (equations of conservation, introduced in Chapter 1) and appropriate boundary conditions.

2.4.3.5 Relaxation function and large deformations

The concept of large deformation (as discussed in section 1.2) requires the treatment of any deformation as three-dimensional. In simple shear, it results in the appearance of normal stresses, N_1 (see section 3.4.2 for more details concerning the normal stress effect in shear flow).

Shear stress evolution in simple shear, $\sigma(t)$, proceeding with constant shear rate, $\dot{\gamma} = \text{const}$, created at the time $t = 0$ is described by Eq. 2.4.13, which is a direct consequence of Eq. 2.4.5 in which the large deformation measure is used.

The analysis of the three-dimensional problem gives the following result:[17] the direct calculations lead to the following expression for the time evolution of the first difference of normal stresses:

$$N_1^+(t) = 2\dot{\gamma}^2 \int_0^t x\varphi(x)dx \qquad [2.4.18]$$

where $\varphi(t)$ is a relaxation function.

Another form of the equation for the transient increasing function $N_1^+(t)$ is obtained from a relaxation spectrum (instead of a relaxation function) as the characteristic of rheological properties of the material. In this case, the following result is obtained:

$$N_1^+(t) = 2\dot{\gamma}^2 \int_0^\infty \theta^2 G(\theta)\left[1 - e^{-t/\theta}\left(1 + \frac{t}{\theta}\right)\right]d\theta \qquad [2.4.19]$$

The difference in the development of shear and normal stresses (Eqs. 2.4.13 and 2.4.18 and 2.4.19, respectively) is pertinent. It is also possible to prove that regardless of the relaxation spectrum, the development of normal stresses proceeds slower than shear stresses.

The equilibrium value of the coefficient of the first difference of normal stresses[15] corresponding to the regime of steady-state flow is expressed *via* relaxation spectrum as

$$\Psi_1 = \frac{N_1^+(t \to \infty)}{\dot{\gamma}^2} = 2\int_0^\infty \theta^2 G(\theta)d\theta \qquad [2.4.20]$$

Relaxation of normal stresses is described by an equation which is the mirror image of Eq. 2.4.19, i.e.:

$$N_1^-(t) = 2\dot{\gamma}^2\int_0^\infty \theta^2 G(\theta)e^{-t/\theta}\left(1 + \frac{t}{\theta}\right)d\theta \qquad [2.4.21]$$

It is possible to prove that the relaxation of normal stresses always proceeds slower than shear stresses.

The most important physical result of the above-mentioned derivations is that the transient behavior of normal stresses is determined by the same viscoelastic functions (relaxation function or relaxation spectrum) as the evolution of one-dimensional shear deformations (stresses).

With regard to dynamic functions, oscillating normal stresses also appear in periodi-cally changing shear deformations.[19] But deformations are changing according to the har-monic law:

$$\gamma(t) = \gamma_0 e^{i\omega t}$$

In addition to shear stresses oscillating with the same frequency and described by dynamic functions (as discussed in Section 2.2.2), normal stresses appear and are oscillating with *double* frequency. The time dependence of the first difference of normal stresses in har-monic oscillations of deformations is:

$$N_1(\omega, t) = N_{1,c}(\omega) + N_{1,osc}(\omega)e^{2i(\omega t - \delta)} \qquad [2.4.22]$$

The last equation shows that normal stresses appearing in shear oscillations consist of three components: a constant steady-state component, $N_{1,c}$, depending on a frequency only, in-phase and out-of-phase components of N_1 having the amplitude, $N_{1,osc}$, changing with double frequency and characterized by the retardation angle, δ.

The same formalism as for dynamic moduli in shear deformation can be introduced for oscillating normal stresses. In-phase and out-of-phase components of dynamic normal stresses can be characterized by frequency-dependent "moduli of normal stresses". In the general case, in a *quasi-linear* mode[20] of shear deformations, normal stresses are described by means of the frequency-dependent coefficients

$$\Psi(\omega) = \Psi_0 + \Psi'(\omega) + i\Psi''(\omega) \qquad [2.4.23]$$

The members of the sum in Eq. 2.4.21 reflect the components of the sum in Eq. 2.4.22.

It was theoretically proven and experimentally confirmed[21] that these coefficients are not independent parameters of material but are directly related to a standard set of vis-coelastic functions of materials. It means that the effect of oscillating normal stresses, as well as transient changes in normal stresses, are not separate properties of the material but only the consequence of its general viscoelastic behavior, whereas normal stresses appear as a second-order effect due to large deformations.

More complicated modes of deformation can also be studied on the basis of the general relationships discussed above. As an example of such complex deformation modes, where the viscoelastic behavior of the material is a dominating factor, sometimes the superposition of low-amplitude oscillations on steady shear flow is considered. It was demonstrated that the superposition of flow changes viscoelastic behavior.[22] This and other complicated cases must be treated in the framework of more general non-linear theories of viscoelastic behavior.

2.5 RELATIONSHIPS AMONG VISCOELASTIC FUNCTIONS

2.5.1 DYNAMIC FUNCTIONS – RELAXATION, CREEP, AND SPECTRA

The main goal of the theory of viscoelasticity is to establish a relationship among all functions mentioned above and used it in the interpretation of experimental data obtained in different deformation-stress modes.[23]

In the most general form, it is done by the principle of superposition. However, it is useful to obtain some simpler relationships that are more convenient in practice.

The connection between dynamic functions, creep, and relaxation functions can be found on the basis of Eqs 2.4.4 and 2.4.5 by substituting the harmonic functions into these equations. Subsequent calculations lead to the following equations:

For the components of dynamic compliance	For the components of dynamic modulus
$$J'(\omega) = \int_0^\infty \frac{\partial \psi}{\partial t} \cos \omega t \, dt$$	$$G'(\omega) = \omega \int_0^\infty \varphi(t) \sin \omega t \, dt$$
$$J''(\omega) = \int_0^\infty \frac{\partial \psi}{\partial t} \sin \omega t \, dt \qquad [2.5.1]$$	$$G''(\omega) = \omega \int_0^\infty \varphi(t) \cos \omega t \, dt \qquad [2.5.2]$$

These types of equations are well known in mathematics: they are called Fourier transforms.[24] The structure of the inverse transformation is the same as the direct one. It is possible to rearrange Eqs. 2.5.1 and 2.5.2, treating them as equations of functions $\psi(t)$ and $\varphi(t)$. The solutions are

$$\frac{\partial \psi}{\partial t} = \frac{2}{\pi} \int_0^\infty J'(\omega) \cos \omega t \, d\omega =$$	$$\varphi(t) = \frac{2}{\pi} \int_0^\infty \frac{G'(\omega)}{\omega} \sin \omega t \, d\omega =$$
$$= \frac{2}{\pi} \int_0^\infty J''(\omega) \sin \omega t \, d\omega \qquad [2.5.3]$$	$$= \frac{2}{\pi} \int_0^\infty \frac{G''(\omega)}{\omega} \cos \omega t \, d\omega \qquad [2.5.4]$$

The following conclusions can be drawn from Eqs. 2.5.1-2.5.4.
- there are direct relationships between time-dependent (relaxation and creep) functions and frequency-dependent dynamic functions, thus, any of them can be calculated if the others are known

- the pair of dynamic functions (either J' and J" or G' and G") are, in fact, not independent but can be expressed by each other. The resulting relationships, known as the *Kronig-Kramer equations*,[25] are:

$$J'(\omega) = \frac{2}{\pi} \int_0^\infty \frac{\alpha J''(\alpha)}{\alpha^2 - \omega^2} d\alpha$$
$$J''(\omega) = \frac{2}{\pi} \int_0^\infty \frac{\omega J'(\alpha)}{\omega^2 - \alpha^2} d\alpha \qquad [2.5.5]$$

$$G'(\omega) = \frac{2}{\pi} \int_0^\infty \frac{\omega^2 G''(\alpha)}{\alpha(\omega^2 - \alpha^2)} d\alpha$$
$$G''(\omega) = \frac{2}{\pi} \int_0^\infty \frac{\omega G'(\alpha)}{\alpha^2 - \omega^2} d\alpha \qquad [2.5.6]$$

The last pairs of relationships are the functional representations of dynamic functions *via* a retardation (or relaxation) spectrum. They are also based on the superposition principle, Eqs. 2.4.4. and 2.4.5, and the path of rearrangements is the same as for Eq. 2.4.8. The final results omit the constants:

$$J'(\omega) = \int_0^\infty J(\lambda) \frac{1}{1 + (\omega\lambda)^2} d\lambda$$
$$J''(\omega) = \int_0^\infty J(\lambda) \frac{\omega\lambda}{1 + (\omega\lambda)^2} d\lambda \qquad [2.5.7]$$

$$G'(\omega) = \int_0^\infty G(\theta) \frac{(\omega\theta)^2}{1 + (\omega\theta)^2} d\theta$$
$$G''(\omega) = \int_0^\infty G(\theta) \frac{\omega\theta}{1 + (\omega\theta)^2} d\theta \qquad [2.5.8]$$

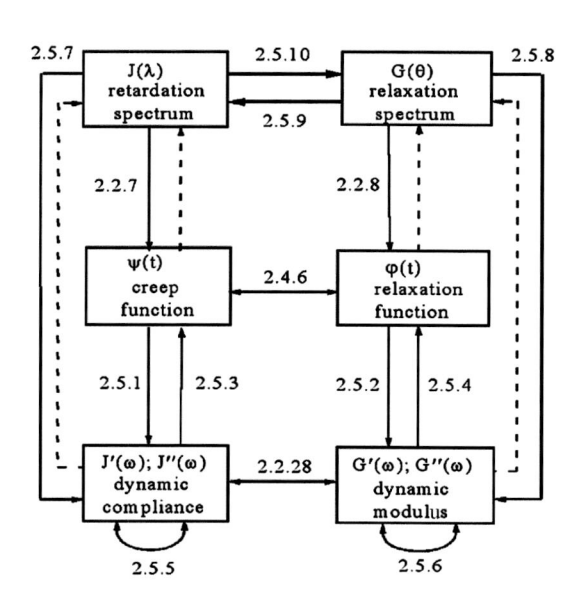

It is interesting to show the evident analogy between these equations and Eqs. 2.3.9 and 2.3.10 obtained for mechanical analogues, i.e., representing a discrete retardation (relaxation) spectrum.

The main sense of all equations given in this chapter is, first of all, to demonstrate the existence of relationships for all viscoelastic characteristics under discussion (they can be used for mutual calculations), and secondly, to emphasize the fact that all these relationships are represented by the integral equations with infinite limits.

The first point can be represented as in Fig. 2.5.1. The second point will be discussed in more detail below. The scheme in Fig.

Figure 2.5.1. Interrelations between different functions used to characterize viscoelastic behavior of material in shear.

2.5.1 represents the general structure of the theory of viscoelasticity. The solid lines are relationships and the numbers of suitable equations, which were discussed above are clearly marked.

Double lines in the upper part of the scheme indicate the relationship between relaxation and retardation spectra. This relationship can be obtained by substituting the equations for the components of J' and J", on one hand, and G' and G", on the other hand, and expressed *via* spectra based on Eq. 2.2.28.

The final pair of relationships, known as the *Cross equations*, are as follows:

$$J(\lambda) = \frac{G(\theta)}{\left[G_\infty - \int_0^\infty \frac{G(x)}{(\lambda/x)-1}dx\right]^2 + [\pi\theta G(\theta)]^2} \qquad [2.5.9]$$

$$G(\theta) = \frac{J(\lambda)}{\left[J_0 - \int_0^\infty \frac{J(x)}{1-(x/\theta)}dx - \frac{\theta}{\eta}\right]^2 + [\pi\lambda J(\lambda)]^2} \qquad [2.5.10]$$

The functions describing one spectrum can be found from the other spectrum at $\theta = \lambda$.

The dashed lines, i.e., the methods of calculation of a retardation (relaxation) spectrum, are worth special discussion, and section 2.5.3 is specifically devoted to this subject.

2.5.2 CONSTANTS AND VISCOELASTIC FUNCTIONS

An experiment gives characteristics of material behavior. The characteristics do not change with time nor depend on frequency. These parameters, as well as viscoelastic functions, can be determined for different geometrical modes of deformations. Below, they will be related to shear.

The following constants[26] were introduced in the theory of viscoelasticity:

Newtonian viscosity $\qquad\qquad \eta = \sigma/\dot\gamma$

steady-state compliance $\qquad\qquad J_s^0 = \sigma/\gamma_\infty$

instantaneous elastic modulus $\qquad\qquad G_0$

where σ is steady shear stress, $\dot\gamma$ is the shear rate in a steady flow, and γ_∞ is equilibrium (stored in the state of steady flow) elastic (or recoverable) shear deformation, or *elastic recoil*.[27]

The theoretical analysis made within the framework of the linear theory of viscoelasticity leads to the following relationships between a relaxation spectrum and the above-listed constants.

The *zero moment* of a relaxation spectrum is an instantaneous modulus

$$G_0 = \int_0^\infty G(\theta)d\theta \qquad [2.5.11]$$

(compare with Eq. 2.4.11).

The *first moment* of a relaxation spectrum is viscosity:

$$\eta_0 = \int_0^\infty \theta G(\theta)d\theta \tag{2.5.12}$$

(compare with Eq. 2.4.14).

Steady-state shear compliance is calculated *via* the moments of a relaxation spectrum in the following way:

$$J_s^0 = \frac{\int_0^\infty \theta^2 G(\theta)d\theta}{\left[\int_0^\infty \theta G(\theta)d\theta\right]^2} \tag{2.5.13}$$

The last equation is directly related to the expression of the coefficient of normal stresses (see section 3.4.2) calculated as the *second moment* of the relaxation spectrum

$$\Psi_1 = 2\int_0^\infty \theta^2 G(\theta)d\theta \tag{2.5.14}$$

(compare with Eq. 2.4.20].

Then, comparing the last two equations, one comes to the *Lodge equation:*[28]

$$J_s^0 = \frac{\Psi_1}{2\eta_0^2} \tag{2.5.15}$$

where η_0 and Ψ_1 are initial values of viscosity and the coefficient of the first difference of normal stresses, respectively, i.e., these values are determined in the domain of linear viscoelastic behavior of the material.

Eq. 2.5.15 can also be written in another form:

$$\sigma_{11} - \sigma_{22} = 2\sigma\gamma_\infty \tag{2.5.16}$$

where γ_∞ is ultimate recoil (complete elastic deformation stored during steady shear flow).

All three principal characteristics of steady rheological behavior of liquid are expressed *via* different moments of the relaxation spectrum, *zero for instantaneous modulus, the first for Newtonian viscosity, and the second for elastic properties* (equilibrium compliance and normal stresses).

The theory also gives some useful "limiting" expressions. If $G'(\omega)$ and $G''(\omega)$ have been measured in a wide frequency range, it could be proven that, at sufficiently low frequencies, G' is expected to be proportional to ω^2, and G'' to ω. Then the following limits are valid:

$$\eta_0 = \lim\frac{G''(\omega)}{\omega} \text{ at } \omega \to 0 \tag{2.5.17}$$

and

$$J_s^0 = \lim \frac{G'(\omega)}{[G''(\omega)]^2} \text{ at } \omega \to 0 \qquad [2.5.18]$$

Sometimes another equation is used instead of Eq. 2.5.18:

$$J_s^0 = \lim \frac{G'(\omega)}{[G'(\omega)]^2 + [G''(\omega)]^2} \text{ at } \omega \to 0 \qquad [2.5.18a]$$

Because at low frequencies $G'' >> G'$, Eq. 2.5.18a degenerates to Eq. 2.5.18.

The last useful relationship is the consequence of the equation for Ψ_1. It is easily seen that

$$\Psi_1 = \lim \frac{2G'(\omega)}{\omega^2} \text{ at } \omega \to 0 \qquad [2.5.19]$$

The above formulated integral expressions, Eqs. 2.5.1-2.5.8, can also be written in adequate forms if integrals are replaced by sums. This is reasonable if a continuous spectrum, entering the integral equations, is replaced by a discrete spectrum (discrete distribution of relaxation modes), found by treating experimental data (see section 2.5.3). In this case, a function $G(\theta)$ is given by the set of pairs $\theta_i - g_i$.

One more interesting relationship is from a relaxation curve measured after steady shear flow, $\sigma(t)$, as expressed by Eq. 2.4.2 (with the consequent change in symbols). The area under a relaxation curve, S, is calculated as

$$S = \int_0^\infty \sigma(t)dt = \dot{\gamma} \int_0^\infty \theta^2 G(\theta)d\theta \qquad [2.5.20]$$

where $\dot{\gamma}$ is the shear rate at a steady flow.

The last integral is exactly the expression for the coefficient of the first normal stress difference, Ψ_1, (Eq. 2.4.20), and therefore the following equality is valid:

$$\Psi_1 = \frac{2S}{\dot{\gamma}} \qquad [2.5.21]$$

The physical meaning of this equation is evident: the area under the relaxation curve, as well as the coefficient of normal stresses, are the measures of stored elastic energy.

2.5.3 CALCULATION OF A RELAXATION SPECTRUM

2.5.3.1 Introduction – general concept

A relaxation spectrum,[29] by definition, is a function defined by Eq. 2.2.8, which is well known in mathematics as the *Laplace transform*.[30] The theory says that any decreasing function (such as a relaxation function) can be represented by the Laplace integral. The theory also permits finding G(t) if a function $\varphi(t)$ is known in an *analytical form in the whole range* of its argument, i.e., from zero to infinity. However, if one refers not to mathematics but to a *practical* determination of a relaxation spectrum based on *experimental grounds*, at least two principal difficulties appear and the roots of these difficulties are as follows:

- first, no experiment can be performed "from zero to infinity", neither in time scale nor in frequency scale; and therefore what the contribution of absent ranges of the experimental scale (outside of the "*experimental window*") is to the results of calculations is unknown and cannot be estimated
- second, no experiment gives the absolutely correct measured value, but only within some experimental confidence, and the influence of the experimental error on the results of calculations, especially considering that the integral transforms are non-linear, is uncertain and can be large.

Therefore, it is necessary to refuse attempts to find an unambiguous and rigorous answer to the question as to how to find the function G(t) if a function $\varphi(t)$ has been measured. Then, it is necessary to state the meaning of the determined relaxation spectrum. The answer depends on a goal: why are we are interested in finding a relaxation spectrum?

First, a relaxation spectrum appears as a mathematical image and only this is the complete and full definition of a spectrum. In this case, a spectrum is treated as some *fitting function* reflecting viscoelastic properties of the material. This function, found in one experiment, can be used in solving different problems related to any arbitrary deformation modes.

According to this concept, a relaxation spectrum can be found from experimental data by some procedures, based on the integral equations discussed in the previous sections of this chapter. This is essentially the *inverse problem*.

Problems of such kind are usually treated as *ill-posed* or incorrect because the determination of a function by solving an integral equation leads (in principle) to ambiguous results strongly and uncertainly depending on slight perturbation of the experimentally measured function. It is quite evident that any limited set of experimental data, known within some limits of error, can be fitted by many independent ways.

There are several approximate methods useful in applications. These methods will be considered below. This attitude to a relaxation spectrum as some fitting function is the only one close to today's understanding of the problem. Indeed, the final result of numerous attempts to find the "true" relaxation spectrum was summarized in the following way: the problem of line spectrum determination is essentially a curve fitting procedure, and "no line spectrum – produced by whatever method – is ever the *true* spectrum".[31] In this sense, the choice of an algorithm for determining a relaxation spectrum is a "personal preference rather than an objective definition".[32] Moreover, it was proven that different approximations, based on various fitting procedures and resulting in different forms of a relaxation spectrum, lead to very close predictions concerning the viscoelastic behavior of the material.[33]

Second, another attitude to a relaxation spectrum determination is based on the modeling molecular movements in material (consisting, for example, of individual macromolecules, their aggregates, or supermolecular structures) and treating these movements in terms of a set of relaxation times (i.e., a relaxation spectrum).

Obviously, a relaxation spectrum reflects molecular movements. However, this statement is nothing more than the general idea in the first approach, but it leads to unique and unambiguous predictions in the second approach, and, within the framework of an accepted model, any relaxation mode has definite physical meaning. This concept will be discussed in more detail in section 2.7.

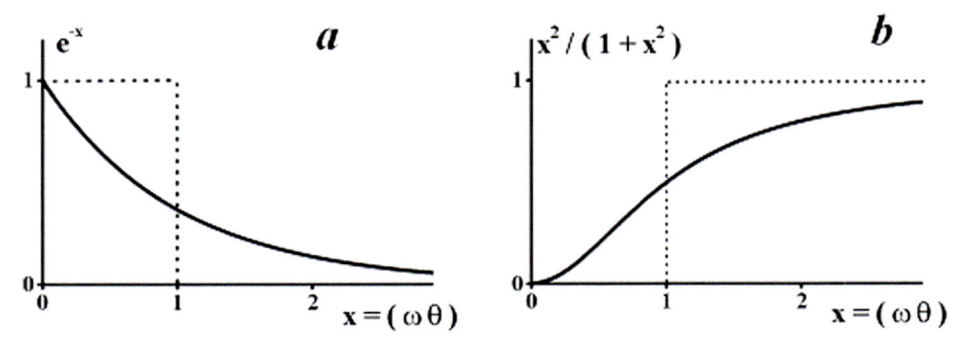

Figure 2.5.2. Approximations of some kernels of integral transforms by step functions: for a relaxation function (a) and for dynamic modulus (b).

Then, if a form of relaxation mode is known beforehand (from a molecular model), it is easy to calculate any viscoelastic function which is then compared with independent experimental data.

In this approach, a relaxation spectrum is not a pure mathematical image but any mode of a spectrum has a quite clear physical meaning, and the problem of relaxation spectrum determination transforms from an inverse mathematical problem to a *semi-inverse* problem.

2.5.3.2 Kernel approximation – finding a continuous spectrum

It is noticeable from any of the integral transforms including a relaxation spectrum that a kernel of all equations is a quickly decreasing function of its argument. The exponential function, e^{-x}, entering Eq. 2.2.18, decreases in comparison with its initial value (at $x = 0$) by 10 times already at $x = 2.3$. A function $x^2/(1 + x^2)$ entering Eq. 2.5.8, increases from its initial value (at $x = 0$) by 10 times at $x = 3$. Analogous conclusions can be made for other kernels of integral transforms. The character of some kernel functions is shown in Fig. 2.5.2.

It is possible to make the following approximations of the kernel functions:

$$e^{-x} = \begin{cases} 1, x \le 1 \\ 0, x > 1 \end{cases}; \quad \frac{x^2}{1 + x^2} = \begin{cases} 0, x \le 1 \\ 1, x > 0 \end{cases} \qquad [2.5.22]$$

These approximations are also shown in Fig. 2.5.1 by dotted lines.

It can be anticipated that small values of kernels strongly diminish the input of the part of an integral with small values of a kernel and it is reasonable to neglect this part in calculating a viscoelastic function. Using these approximations, the integral transforms are easily solved in an analytical form and the final results are as follows.

The first level approximation of Eq. 2.2.8 is:

$$G(\theta) \cong -\frac{d\varphi(t)}{dt} \text{ at } \theta = t \qquad [2.5.23]$$

An analogous approach to Eq. 2.5.8 gives the following final result:

$$G(\theta) \cong \omega^2 \frac{dG(\omega)}{d\omega} \quad \text{at } \theta = \omega^{-1} \qquad\qquad [2.5.24]$$

The higher level approximations can also be obtained by the same method. Similar ideas are used for calculations of a retardation spectrum.

The approach based on kernel approximations was very popular in the pre-computer era because they are quite simple to handle and do not require time-consuming calculations.[34] However, today the latter argument is not important and the methods based on kernel approximations are practically not used, being replaced by computer methods utilizing a spectrum represented by a set of discrete lines.

2.5.3.3 Computer-aided methods for a discrete spectrum

2.5.3.3.1 The direct method[35]

According to this method, it is assumed that a unique continuous spectrum describes the viscoelastic behavior of the material. However, this spectrum is discretized and it is presented as a set of independent relaxation modes, though both (continuous and discrete spectra) are considered equivalent.

This is the direct method for searching a relaxation spectrum, which is supposed to exist in the form of a set of discrete lines, $G_i - \theta_i$. The values of $G_i - \theta_i$ are used for calculating the experimentally observed functions (let it be $G'_{exp}(\omega)$ and $G''_{exp}(\omega)$). These values are varied and the calculated values of $G'_{cal}(\omega)$ and $G''_{cal}(\omega)$ are compared with experimental values. The deviation of the results of approximation (in varying $G_i - \theta_i$) from experimental data is estimated by the *functional of errors*, E. It can be calculated in different manners but the simplest and the most evident form of this functional is

$$E = \frac{1}{2M} \sqrt{\sum_{n=1}^{M} \left[\left(\frac{G'_{exp,n} - G'_{calc,n}}{G'_{exp,n}} \right)^2 + \left(\frac{G''_{exp,n} - G''_{cal,n}}{G''_{exp,n}} \right)^2 \right]} \qquad [2.5.25]$$

The subscript symbol, n, shows that this result relates to the n-th experimental point and the total number of these points is M.

The general computer-aided procedure consists of minimization of functional of errors, which, according to Eq. 2.5.25, is nothing else than a standard average deviation. This procedure results in a set of parameters $G_i - \theta_i$ for the best fit of experimental data.

Two important points are taken into account in an algorithm of calculations:
- initial independence of 2M parameters; the search for a minimum is a non-linear problem by its nature
- obtaining the best fit of experimental data with a minimum number of modes (the latter allows authors to call their approach a *parsimonious model*).

Authors say that the discrete relaxation modes are not meaningful by themselves and can be replaced by other sets of pairs of $G_i - \theta_i$. So, the method does not give a unique or unambiguous solution to the problem. However, a continuous spectrum obtained from this set of parameters "certainly is a meaningful representation of the macromolecular dynamics".

2.5.3.3.2 Method of linearization[36]

This method is based on the representation of experimental dependencies (either $G'(\omega)$ and $G''(\omega)$, or a relaxation curve determined inside an experimental window) by the expansion series using increasing powers of ω. The unknown values are the coefficients of the series. Contrary to the procedure described above, the search for these parameters by minimizing the functional of errors (such as Eq. 2.5.25 or any other) is a linear problem by itself, and therefore this problem has a unique solution.

The result of the computer-aided calculations is also a set of discrete modes. The set of relaxation modes obtained by this method is different than that obtained by the direct method discussed above. None of these modes has any definite physical meaning, but they adequately represent the viscoelastic properties of materials.

The advantage of this method is in unambiguity of results since there is only a single minimum of linear functional of errors.

2.5.3.3.3 Semi-inverse method[37]

The freedom of choice of relaxation modes permits presetting the distribution of relaxation times along the frequency (or time) scale. This distribution can be arbitrary. However, for simplicity, it is preferable to establish certain rules for the choice of relaxation times. As an example, the relaxation time distribution can be expressed as

$$\log \theta_n = \log \theta_{max} + Cn \qquad [2.5.26]$$

where n is an ordinary number of a relaxation mode and C is a step. The equidistant distribution in log-scale is proposed. The other possible distribution of relaxation times is proposed in the form of the power law:

$$\theta_n = 3\theta_{max}/n^{-\alpha} \qquad [2.5.27]$$

where θ_{max} is the initial value of the distribution, n is an ordinary number of relaxation modes, and α is an arbitrary factor.

Experimental data are always known for a limited "window" of the frequency range, from ω_{min} to ω_{max}. However, it does not mean that the distribution must also be limited to the same range. For computation purposes it is preferable to take the maximum value of the relaxation time beyond the upper limit of frequencies, i.e., it is reasonable to take $\theta_{max} > \omega_{min}^{-1}$ and θ_{min} can be less than ω_{max}^{-1}.

This does not mean that we may determine the relaxation spectrum beyond the boundaries of the experimental window: the choice of θ_{max} is important for the method of fitting the experimental data.

The preliminary choice of the relaxation time distribution makes the calculations much easier because the determination of the spectrum degenerates into a linear problem that is necessary to find the weights of relaxation modes (partial moduli) only, satisfying the condition of minimizing the standard deviation or other measures of fitting errors.

2.5.3.3.4 Regularization method[38]

This is a particular case of a more general approach to solving the Fredholm integral equations of the first kind. All inverse problems of searching for a relaxation (retardation) spectrum belong to these equations. One of the most popular general methods is based on the Tikhonov regularization.[39] The method is based on minimization of the following functional $V(\lambda)$ (omitting the constants), which is somewhat more of a general expression than the standard average deviation as in Eq. 2.5.25:

$$V(\lambda) = \sum_{i=1}^{N} \frac{1}{\sigma_i^2} \left\{ g_i^{\sigma} - [\int K(t-\tau)G(\ln\theta)d\ln\tau] \right\}^2 + \lambda(L|E|)^2 \qquad [2.5.28]$$

Here g_i^{σ} is a measured value of the experimental function (let it be a relaxation function or elastic modulus) determined with some error, or noise, σ; $K(t-\tau)$ is a kernel as in formulas for an experimental function; $G(\theta)$ is a relaxation spectrum (the argument is used in a logarithmic scale because an experimental window covers several decimal orders of time or frequency changes); N is the number of experimental points.

The second, new term in Eq. 2.5.28, $\lambda L|E|$, reflects the idea of smoothing the calculated function. Here, λ is a *regularization parameter* and $L|E|$ is an arbitrary operator, usually, it is the second derivative of the function $G(\ln\theta)$. Introducing this term leads to the minimization of undesirable oscillations of unknown function.

The regularization method is a rigorous way of searching for a relaxation spectrum. Appropriate software gives a powerful method of the computer-aided solution of the problem under discussion.

However, the above-cited remark that "no line spectrum – produced by whatever method – is ever the true spectrum" is still valid. And again, it is necessary to stress that different approximations based on various fitting procedures can lead to very close predictions concerning the viscoelastic behavior of the material as described in this chapter.

2.6 VISCOELASTICITY AND MOLECULAR MODELS

2.6.1 MOLECULAR MOVEMENTS OF AN INDIVIDUAL CHAIN

Viscoelastic (or relaxation) properties are very important characteristics of polymeric materials in the form of dilute solutions, melts, or solids. The origin of viscoelasticity is attributed to the molecular movement of polymeric chains. Some macro-models were constructed to explain the nature of the relaxation phenomenon in polymer substances and to predict the distribution of relaxation times in real materials.

The relaxation phenomenon is caused by the molecular movement of a polymer chain. These movements change the conformations of flexible macromolecular chains. The transition between different conformations proceeds in time and these transitions are of a relaxation nature.

The model of macromolecular movement is visualized by simple viscoelastic elements described in section 2.3. Here, basic ideas and principal conclusions, use-

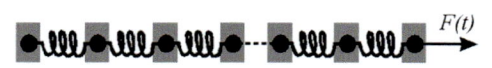

Figure 2.6.1. A spring-and-bead model (free-draining coil).

ful in rheological applications, are briefly discussed. Mathematical calculations are not included because they belong to the field of polymer physics rather than rheology.[40]

2.6.1.1 A spring-and-bead model ("free-draining chain")

Fig. 2.6.1 shows a model of a macromolecular chain called a *spring-and-bead model*.[41] This model predicts viscoelastic or relaxation behavior due to the combination of viscous resistance and elastic recoil. The elements of the model are not directly related to atoms in a macromolecular chain: the model represents an image of some peculiarities of chain deformation. This model is also called a model of a free-draining chain because an effect of intramolecular interactions (between different elements of a model) is neglected.

A chain contains $(N + 1)$ identical beads and N identical springs. Resistance to the displacement of the n-th bead is expressed as the product of viscosity of a surrounding medium and its velocity, $\eta(du_n/dt)$. In this case, "viscosity" is some measure of the intermolecular interaction between a macromolecule and a surrounding liquid. Force acting on every i-th string is the product of its modulus, G, by relative displacement of its ends: $G(u_n - u_{n-1})$. Analogous equations are written for every chain element. Then, as a result of accurate calculations, the expression for displacement of the chain end under constant force, F_0, is given by:

$$U(t) = \frac{F_0 t}{\eta(N + 1)} + F_0 \sum_{n=1}^{N} G_n(1 - e^{-t/\lambda_n}) \qquad [2.6.1]$$

The first term expresses a continuous movement of a chain under constant force ("flow"), and the second term represents the retarded deformation as in any viscoelastic model element discussed in section 2.3. The partial moduli, G_n, are calculated via constants of the model.

$$G_n = \frac{1}{2G(N + 1)} \left[\cot \frac{n\pi}{2(N + 1)} \right]^2 \qquad [2.6.2]$$

The constants λ_n in Eq. 2.6.1 are retardation times expressed by

$$\lambda_n = \frac{\eta}{4G} \left[\sin \frac{n\pi}{2(N + 1)} \right]^2 \qquad [2.6.3]$$

The existence of a set of identical elements joined in a chain leads to the appearance of a retardation (and relaxation) spectrum by itself. The number of lines in a spectrum (the number of retardation times) is determined by the number of elements in the chain.

A long chain, consisting of many elements $(N \gg 1)$ can be analyzed. The range of retardation times is determined by the boundary values of minimum, λ_{min}, and maximum, λ_{max}, retardation times, respectively. Calculations show that

$$\lambda_{min} \approx \frac{1}{4} \left(\frac{\eta}{G} \right) \qquad [2.6.4]$$

and

$$\lambda_{max} \approx \frac{(N + 1)^2}{\pi^2} \left(\frac{\eta}{G} \right) \qquad [2.6.5]$$

The minimum retardation time is practically independent of the full length of the chain because quick relaxation movements occur inside the chain segments. The maximum retardation time is proportional to N^2, i.e., it increases with its molecular mass.

There is only a single independent retardation time and all others can be expressed by it. The n-th retardation time (in the range of long retardation times, n < 5) is expressed as

$$\lambda_n = \lambda_{max} n^{-2} \qquad\qquad [2.6.6]$$

A retardation spectrum represented by the pairs G_n-λ_n is discrete.[42] If N is suffi ciently large, summation can be replaced by integration in order to obtain a continuous spectrum. Then, the following formula for a relaxation spectrum can be obtained:

$$G(\theta) = K\theta^{-3/2} \qquad\qquad [2.6.7]$$

This formula is valid within the limits from λ_{min} to λ_{max}. Sometimes a logarithmic form of a relaxation spectrum, $h(\ln\theta)$, is used (see Eq. 2.2.8a). Then

$$h(\ln\theta) = K\theta^{-1/2} \qquad\qquad [2.6.8]$$

The constant K is found from any integral characteristic of the spectrum (see section 2.5.2).

Molecular theory helps to express the model constants by parameters that have physical meaning. The elasticity of macromolecules originates from their Brownian movement and therefore the elastic modulus, G_e, is written as

$$G_e = akTN \qquad\qquad [2.6.9]$$

where k is the Boltzmann constant, T is the absolute temperature, N is a factor depending on concentration (number of chains in a unit volume) and a is some front-factor close to one. It is easy to show that

$$G_e = \frac{3G}{N}$$

and finally

$$G = \frac{1}{3}akTN^2 \qquad\qquad [2.6.10]$$

A value of "viscosity" entering the molecular model is expressed as

$$\eta = \frac{\eta_0 - \eta_S}{N} \qquad\qquad [2.6.11]$$

where η_0 is the viscosity of the solution, η_S is the viscosity of the solvent, i.e., the value of η is decrement of viscosity (an increase of viscosity) caused by the presence of a macromolecule in a solution related to a unit volume.

After some evident rearrangements, the following formula for maximum retardation time is obtained:

$$\lambda_{max} \approx \frac{3(\eta_0 - \eta_S)}{\pi^2 NkT} \qquad\qquad [2.6.12]$$

and all other times can be calculated from a spectrum using Eq. 2.6.6. Eq. 2.6.12 relates relaxation times of macromolecular movements to macroscopic parameters of a polymer system.

The following features of a spring-and-bead model need to be mentioned:

- this model relates to a very dilute solution without taking into account intermolecular interaction but based on the analysis of behavior of an individual polymer chain
- this is a linear model, i.e., it predicts linear viscoelastic behavior of the solution
- relaxation properties of dilute polymer solutions can be predicted from macroscopic properties of liquid
- the whole relaxation spectrum is defined through a single relaxation time, i.e., all relaxation modes are not independent but are the consequence of the movement of identical "sub-molecules".

2.6.1.2 Model of a non-draining coil[43]

This is the same model as the spring-and-bead model discussed above, but it is supplemented by intramolecular interactions inside the chain. It means that in contrast to the above-discussed models, this model takes into account perturbations of the flow field in liquid due to the presence of foreign particles in it. This concept leads to another method of calculation of a force acting on every bead in a chain. Without discussion of details of calculations, the following main theoretical results for a model of a non-draining coil are of interest:

- maximum relaxation time in this model, θ_{max}^{Z}, is expressed in the same manner as in the spring-and-bead-model, though with a different front-factor
- all other relaxation times are not independent but can be found within the framework of a model
- the distribution of relaxation times is different than in the spring-and-bead-model (more narrow), though the differences are not very large
- new relaxation time distribution results in slightly different predictions concerning experimentally observed functions, primarily $G'(\omega)$ and $G''(\omega)$.

The difference in predictions of the $G'(\omega)$ and $G''(\omega)$ functions in both theories permits us to compare them with experimental data in a wide frequency range. The following theoretical predictions are obtained from both models. In the low-frequency range, both models give practically identical predictions for $G'(\omega)$ and $G''(\omega)$ dependencies. However, in the high-frequency range, a spring-and-bead model predicts that $G'(\omega)$ must coincide with $[G'' - \omega\eta_S]$ where η_S is the viscosity of solvent and the slope of both functions (in log-log scale) equals 1/2. The model of non-draining chains predicts that $[G'' - \omega\eta_S]$ exceeds G' by $\sqrt{3}$ times and the slope of both functions is 2/3.

The intermediate case of partial draining was examined by many authors.[44] This model, regardless of the details of the calculation, leads to predictions lying in between the results of calculations of the two above-mentioned theories.

It is difficult to make accurate experimental measurements that permit us to evaluate different theoretical predictions. This is because measurements have to be made for very dilute solutions (formally for infinitely dilute solutions), the use of "monodisperse" polymers, and the use of a correct and unique measure of molecular interaction in such solutions. However, several very accurate experimental studies were carried out and they

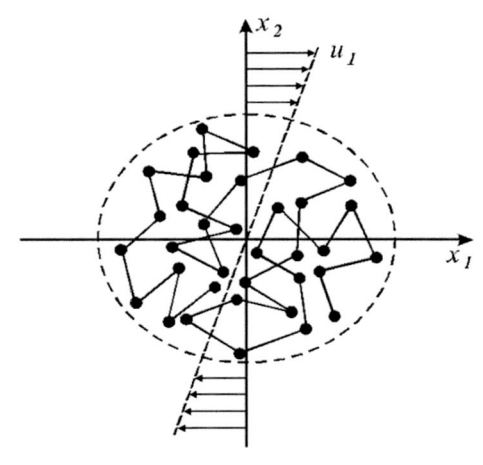

Figure 2.6.2. A model of rotating macromolecular coil.

confirmed the principal conclusions of the model predictions.[45] If a "good" solvent is taken, the viscoelastic behavior of the solution is described by a model of a free-draining chain, and if a "theta-solvent" is taken, the results of the measurements are much closer to predictions of the model of a non-drained chain.[46]

2.6.1.3 Model of a rotating coil[47]

A polymer molecule in a solution forms a statistical coil. If this coil is placed in a shear field, forces appear which tilt a position of any element of a coil from its equilibrium state (Fig. 2.6.2). As a result, a restoring force appears and this force tends to push an element back to its equilibrium state. This effect is modeled by a spring resistance and resistance of a bead to a medium. The behavior of a macromolecular coil is modeled by a set of Maxwellian elements, attached to the center of mass of a coil.

Rotation of a macromolecular coil with respect to the center of mass in a shear field results in some additional loss of energy (energy dissipation) and this is equivalent to an increase in apparent viscosity.

Calculations based on this model lead to the following conclusions:

- the maximum retardation time calculated from this model equals $2\lambda_{max}$, which is found using Eq. 2.6.12
- the dependence of apparent viscosity on the shear rate is the same as the dependence of dynamic viscosity on frequency, $\eta'(\omega)$, assuming that shear rate equals frequency,[48] i.e., the theory predicts linear viscoelastic behavior with simultaneous shear-rate dependence of viscosity in shear flow
- as in any other molecular theory, all other relaxation times can be expressed by the maximum relaxation time, i.e., all relaxation modes are interrelated.

2.6.2 RELAXATION PROPERTIES OF CONCENTRATED POLYMER SOLUTIONS AND MELTS

The intermolecular interaction, which is certainly present in concentrated polymer solutions and melts, is simulated by the friction of beads moving through a viscous medium in all models concerning individual chains. This may be insufficient in modeling relaxation properties of concentrated solutions and melts where every chain interacts with other long chains. There are several more or less realistic models which represent these cases.

2.6.2.1 Concept of entanglements

The concept of *entanglements* assumes that every long-chain interacts with other chains, and it is necessary to account for restrictions to motion of an individual chain caused by molecular movements of other chains. This idea was first introduced in the model of entangled ropes:[49] pulling one rope from a bundle inevitably leads to the movement of all other ropes and the force of pulling depends on the length of ropes and their number.

Though the idea of this approach was rather clear, quantitative predictions of theory were not so adequate.

Later, the idea of entanglements was developed on the basis of a model of a single free-draining chain, and subsequent principal steps were related to the law of frictions of the beads of a chain. There were numerous versions of the friction law, and some of the most popular are discussed below.

One can assume that the space distribution of entanglements is random. However, deformation changes the situation and leads to the formation of concentrated zones (as shown by the model – Fig. 2.6.3).

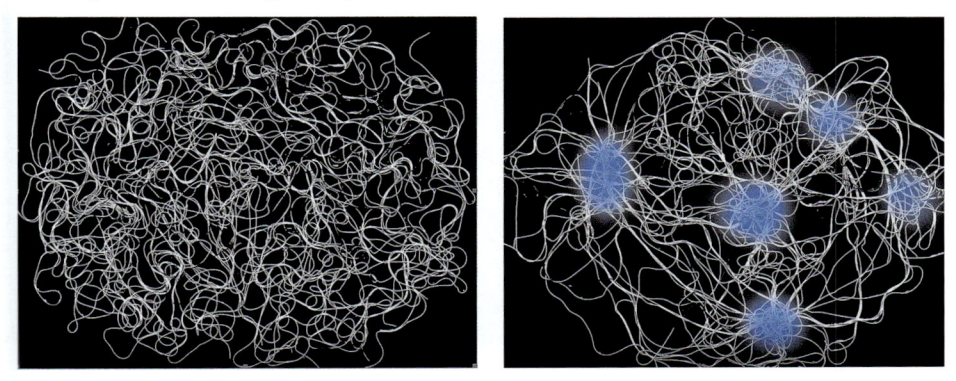

Figure 2.6.3. Model illustrating random entanglements (left) and the formation of knots created by concentrated entanglements (right). [Adapted, by permission, from A.Ya. Malkin, A.V. Semakov, V.G. Kulichihkin, *Rheol. Acta*, **50**, 485 (2011); *Appl. Rheol.*, **22**, 32575 (2012).]

Then there are two possibilities. At low velocity (low deformation rate), chains can slip out of the entanglements (knots) and it is a case of flow. If the velocity exceeds some threshold, knots become tight and irreversible movement becomes impossible. This change in the behavior of the intermolecular entanglement network corresponds to the liquid-to-rubbery state transition and appearance of instabilities (see Section 3.6.3).

The disentanglement and interchain slip are due to Brownian movement while the formation of stable knots can be attributed to the elasticity of macromolecules and storage of elastic energy. Then the following dimensionless criterion can be proposed to characterize the limit of the flow:[50]

$$M^* = \frac{\sigma_s^2}{2G_{term}\rho RT} M_c$$

where σ_s is the shear stress at the liquid-to-rubbery state transition, G_{term} is the elastic modulus at the terminal zone, M_c is the molecular weight of the chain segment between neighboring entanglements, ρ is density.

The ratio $\rho RT/M_c$ is the value of the elastic modulus on the viscoelastic plateau. So, the criterion of the loss of fluidity is characterized by the threshold of the criterion

$$M^* = \frac{\sigma_s^2}{2G_eG_{term}}$$

Different theories based on the model of macromolecular entanglements are discussed below.

2.6.2.2 Two-part distribution of friction coefficient

The coefficient of friction, f, of beads moving in a viscous medium depends on the length of a chain. The law of friction was assumed to be as follows:[51]

$$f = f_0 = const \qquad [2.6.13a]$$

for short portions of a chain, where f_0 is the same constant as in the model of a free-draining chain, i.e., short segments of chain between entanglements are treated as free-draining chains, and

$$f = f_0(M/M_e)^{2.4} \qquad [2.6.13b]$$

for long chains, where M is the molecular mass of an entire chain and M_e is the average molecular mass of a chain segment between equivalent entanglements.

The idea of increasing resistance to movement due to the presence of macromolecular entanglements is expressed in this model very clearly and leads to the appearance of a spectrum of long relaxation times.

The power factor 2.4 is chosen arbitrarily because of the need to use a power law for the dependence of viscosity on molecular mass (it is well known that the universal rule $\eta \propto M^{3.4}$ is applicable – see Chapter 3).

Further development of this model can be found in many publications, and in particular in a model of two chains sliding at the points of junctions.[52] The restriction to movement is due to interaction in some entanglement points, where the friction is determined by a new parameter of the model – coefficient of sliding, δ. For a chemically cured network (with permanents crosslinks) $\delta = 0$; for free-draining chains moving independently from each other $\delta = 1$.

The concept of a two-part relaxation spectrum is very useful and was explored in many other molecular models.

2.6.2.3 Non-equivalent friction along a chain

It is possible that the coefficient of friction of beads is changing along a macromolecular chain. The coefficient of friction distribution can also be represented by the following rule:[53]

$$f = f_0\frac{L}{N}(1 + |q|^b) \qquad [2.6.14]$$

where L is the chain length, N is the number of entanglement points, q and b are constants determining the character of friction coefficient distribution along the chain. The power factor $b = 2.4$ is assumed to fit the standard viscosity vs. molecular mass dependence ($\eta \propto M^{3.4}$).

Other versions of the same approach were also discussed based on a different distribution of the friction law along a chain.[54]

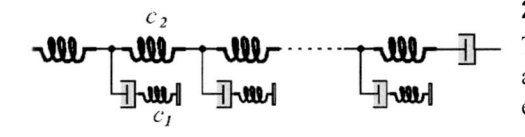

Figure 2.6.4. A model of viscoelastic interaction in an entanglement.

2.6.2.4 Viscoelastic entanglements

The basic idea in this approach is the assumption that a surrounding medium exerts (*via* entanglements) not only a viscous drag but an elastic resistance as well.[55] Each entanglement is treated by means of a spring-bead interaction as shown in Fig. 2.6.4. The original model of such kind was proposed in some versions differing in magnitudes of elastic spring rigidity C_1 and C_2.[56]

This model predicts some interesting relationships between viscoelastic parameters of materials. The following expression for steady-state (equilibrium) compliance, J_s^0, gives a good correlation with experimental data:

$$J_s^0 = \frac{\alpha_1 E}{1 + \alpha_2 E} \qquad [2.6.15]$$

where α_1 and α_2 are constants, and E is the number of entanglements per macromolecule, i.e., its value is proportional to the full length of a chain.

The model of viscoelastic junctions was developed in several publications.[57] It gives realistic predictions of the viscoelastic behavior of polymer melts.

2.6.2.5 Rubber-like network

It is reasonable to suppose that macromolecules form temporary junctions (entanglements) with different characteristic lifetimes and/or propensities to slide at these junctions. A portion of the chain between two neighboring knots can be long enough to treat the molecular movement of each portion as a free-draining chain, as discussed above. The junctions of chains can also be permanent. This is the case of cured rubber because no sliding occurs in crosslinks. In the other limiting case, chains are moving without "noticing" each other: this is a case of very dilute solutions. However, in numerous intermediate cases, it is necessary to take into account interactions between macromolecules, which form temporary junctions. It is a case of concentrated solutions and melts.

The central equation of the model of the rubber-like network is an expression for the concentration of junctions between two different chains a and b:[58]

$$N(t - t') = \sum_{a, b} L_{a, b} e^{-(t - t')/\theta_{a, b}} \qquad [2.6.16]$$

where $L_{a,b}$ is the rate of formations of junctions between chains a and b; $\theta_{a,b}$ is the rate constant characterizing the breakdown of junctions; this constant can be treated as the characteristic relaxation time of the corresponding junctions. The argument of the function N(t - t') is typical for all hereditary processes discussed in the theory of viscoelasticity.

The stresses acting on the network chains at each instance of time are proportional to the value of N changing with time. This is the reason why the memory function, entering the constitutive equation, can have very different forms depending on the kinetic parameters of the model.

It is important that the values $L_{a,b}$ and $\theta_{a,b}$ are not specified in the original theory because it gives freedom of selection of these factors for fitting experimental data.

The initial version of the rubber-like network can be modified in different directions. It is possible to assume that the distance between neighboring junctions is not constant but described by Gaussian or some other distribution. It is also possible to suppose that the kinetics of formation and breakdown of knots depend on stress; this is a natural way for introducing non-linear effects in viscoelasticity.

2.6.2.6 "Tube" (reptation) model

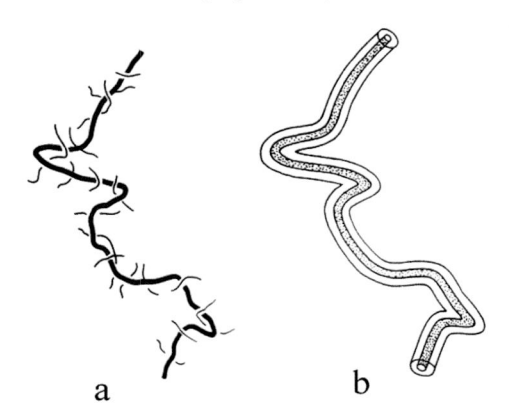

A modern approach to the modeling of relaxation behavior of concentrated polymer solutions (and melts) also utilizes the concept of restriction of molecular movement of an individual chain due to intermolecular entanglements. However, the model presents these restrictions in a different way without the localization of these entanglements. Fig. 2.6.5a shows a singled out macromolecule (solid line) and numerous other macromolecules which prevent its movement in the direction perpendicular to the chain backbone, creating strong resistance to movements at distances longer than the characteristic cross-section of

Figure 2.6.5. A "tube" (reptation) model.

a macromolecule. This can be presented by a slightly modified scheme as shown in Fig. 2.6.5b, and this is a *"tube" model.*[59] The effect of topological constraints is similar to that of a macromolecule placed into a tube of the same configuration as the chain. According to this model, it is assumed that the long-range motions of a chain are allowed essentially along its own length only. This type of motion resembles a displacement of a snake and was called reptation, and the model is also called a *reptation model.*[60]

As is seen from Fig. 2.6.5b, the reptation model does not include special points of interactions (e.g., "beads" as in some of the above-mentioned models) but operates with characteristic dimensions: length of a chain, L, and the diameter of a tube, d.

The theory determines some characteristic relaxation times that determine the time scales of different kinds of molecular motion. The first of them is the Rouse relaxation time, $\theta_{R,0}$, which is related to the movement of the segment of a chain of the minimal length which moves independently of the chain as a whole. In terms of the entanglement model, a segment is a distance between neighboring knots of a part of a chain with molecular weight M_e. This parameter describes local molecular motions at short times, $t << \theta_{R,0}$ corresponding to high-frequency motions. This molecular parameter does not depend on the full length of a macromolecule. The second time scale is related to the diffusion movement of a chain as a whole and characterizing the wriggle motions along the contour of the chain. This relaxation time reflects the Rouse movements of an individual chain, and actually, there is a set of relaxation times between $\theta_{R,0}$ and the longest Rouse relaxation time, $\theta_{R,max}$. Finally, the third relaxation time scale is related to the interaction of a macromolecule with its surroundings, or "walls' of a tube. This relaxation time θ_{tube} describes the time required for a chain to leave its initial tube (its surrounding) and appear inside a new surrounding (tube). This time is often referred to as the *tube renewal time*. Actually, this relaxation time is related to the flow of concentrated solutions and melts because it reflects the long-distance displacement of a chain. This time scale strongly depends on the molecular weight of a chain. Usually, the molecular weight of linear chains is much larger than Me and therefore $\theta_{tube} \gg \theta_R$.

The characteristic relaxation times can be found from the frequency dependence of the storage modulus measured in a wide frequency range.[61] The usual method for defining the longest Rouse relaxation time, $\theta_{R,max}$ is based on the Rouse model predicting that the storage modulus G'(ω), is proportional to $\omega^{1/2}$ over a range of angular frequency, ω. This takes place at rather high frequencies beyond the elasticity plateau. In the terminal zone, in the limit of low frequencies, the storage modulus is a quadratic function of frequency (G'=Aω^2) and the loss modulus is a linear function of frequency (G''=Bω) the maximal relaxation time θ_d describing the viscoelastic behavior of a liquid is found as (A/B). This value can be identified with θ_{tube}.

According to the initial version of the reptation model, the characteristic (terminal) relaxation time, θ_d, and viscosity are proportional to the cube of the molecular mass. This high value of the exponent reflects strong intermolecular interactions (entanglements), though even this exponent's value is lower than the experimental value of the exponent close to 3.5.

It was also shown[62] that the theory permits calculation of characteristic relaxation time, θ_d, using measured rheological parameters:

$$\theta_d = \frac{10}{\pi^2}\eta_0 J_s^0 \qquad\qquad [2.6.17]$$

where η_0 is Newtonian viscosity and J_s^0 is the steady-state (equilibrium) compliance.

The reptation model is the most widely used (in different modifications) for interpretation of experimental results in studies of viscoelastic properties of concentrated polymer solutions and melts. However, the predictions of the theory, though they correctly reflect some principal features of relaxation properties of long-chain polymers, are more qualitative than quantitative. Possibly, it is due to some oversimplifications of the real molecular movements. In particular, the theory predicts a very narrow relaxation spectrum, while real relaxation spectra are much wider.

The attempts to improve the reptation model were based on the concept of "*double reptation*"[63] or including the entanglements (in addition to a contour tube) as superimposing restrictions to molecular movements.[64] It is known that the presence of long branching in a macromolecule leads to significant problems in the framework of a tube model, because reptations, as they are described in the basic model, become impossible. This special case of a tube model required further development and was discussed in the so-called "pom-pom" macromolecules.[65] It was shown that branching leads to some special features of viscoelastic properties, and it is especially important for extension.[66]

2.6.2.7 Some conclusions

Summarizing the above discussion of molecular models, it is possible to make the following conclusions regarding numerous mechanical macromolecular models proposed for the description of viscoelastic properties of concentrated polymer solutions and melts.

- all models are based on sets of mechanical elements, which can be joined in numerous arbitrary combinations; this allows one to vary theoretical predictions concerning experimentally observed functions; indeed, all models give a possibility to calculate viscoelastic properties, in particular, $G'(\omega)$ and $G''(\omega)$ dependencies, which can be compared with experimental data
- relaxation spectra of concentrated solutions or melts are assumed to consist of the sum of two parts – rapid relaxation time, obeying the same distribution as for a free-draining chain, and a slow relaxation time distribution; the latter reflects the existence of intermolecular interactions ("entanglements"); any model includes some arbitrary ("free") scaling parameters.

A complete molecular model should describe the following principal and general experimental facts, such as:

- viscosity depends on the molecular mass (MM) as $\eta \sim MM^\alpha$, where α is of the order of 3.5
- existence of the plateau value, G_N^0, on the frequency dependence of dynamic storage modulus, the length of this plateau depends on MM
- independence of steady-state compliance on MM
- the empirical rule: $J_s^0 G_N^0 \approx 2$
- it is very desirable that the theory correctly predicts the frequency dependencies of G' and G" in a wide frequency range.

2.6.3 VISCOELASTICITY OF POLYDISPERSE POLYMERS

Any molecular theory initially operates with molecules of equal length. But real polymers are polydisperse, i.e., all polymers are mixtures of molecules of different lengths and the latter is characterized by molecular-mass distribution (MMD).

About MMD

MMD is characterized by the function w(M), where dw is the mass share of the fraction with molecular mass, MM, from M to (M + dM). MMD, by its physical sense, is discrete because the values of MM change discretely. However, it is convenient to neglect this and to treat MMD as continuous due to small steps in the argument of a distribution. In many cases, it is useful to operate with some average values of MMD:

number-average MM:
$$\overline{M}_n = \int_0^\infty \frac{w(M)}{M} dM$$

weight-average MM:
$$\overline{M}_w = \int_0^\infty w(M) dM$$

z-average MM:
$$\overline{M}_z = \int_0^\infty M w(M) dM$$

z+1-average MM:
$$\overline{M}_{z+1} = \int_0^\infty M^2 w(M) dM$$

and so on. The lower boundary in these integrals is rather formal because MM cannot equal zero.

According to the tube model, the θ_d for monodisperse polymers is determined by Brownian motion (or diffusion) of a macromolecule as a whole. However, if a contour tube is formed by shorter molecules (as in a polydisperse polymer) the tube renewal happens in a shorter time scale and it provides an additional relaxation mechanism with relaxation times being different in comparison with θ_d.[676] It means that a relaxation spectrum of a polydisperse polymer is not a simple mixture of relaxation times of the fractions forming MMD of a polydisperse polymer.

The central problem, when passing from a monodisperse polymer to polydisperse samples, is in the construction of a *"mixing rule"*, i.e., formulation of a law of summarizing inputs of different fractions that will match observed viscoelastic properties. It relates to the integral constants, as well as to the relaxation spectrum itself.

The principal and applied interests were concentrated on the most easily measured parameters, such as Newtonian viscosity, η_0, steady-state compliance, J_s^0, plateau modulus, G_N^0, coefficient of normal stresses, Ψ, the crossover frequency, ω_c (the frequency at which G' = G"), the curvature of a flow curve (see Chapter 3), and so on.

Numerous publications have proven that in viscosity calculations the following mixing rule is valid in many cases:[68]

$$\eta_0 = K\overline{M}_w^\alpha \qquad [2.6.18]$$

where the scaling parameter, α, is commonly accepted as $\alpha = 3.4 - 3.5$, though it is possible to meet other values, from 3.2 to 3.9. The presence of low-MM fractions and branching affects the value of α. The prefactor K depends on the molecular structure and temperature, and it is not the subject of this discussion.

This equation is assumed to be valid for high-molecular-mass polymers, where the whole MMD lies above some critical value of MM, M_c.

A very impressive illustration of the $\eta_0(\overline{M}_w)$ dependency is shown in Fig. 2.6.6 and some other examples are presented in

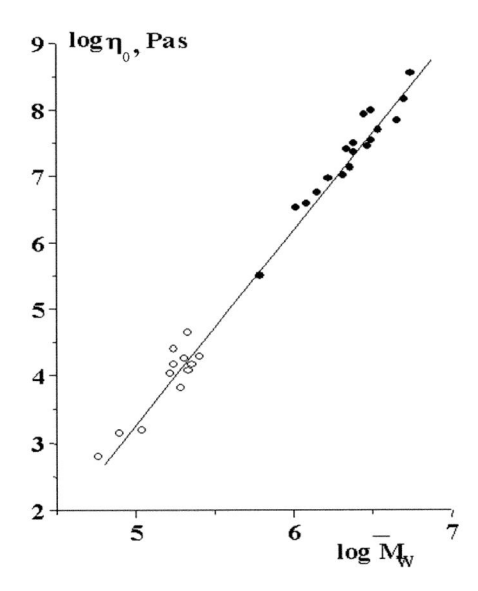

Figure 2.6.6. Viscosity-MM dependence for polyethylenes. The unfilled circles refer to the data taken from literature. The filled circles are the original data of the author of the publication. [Adapted, by permission, from M.T. Shaw, *Polym. Eng. Sci.*, **17**, 266 (1977)].

Chapter 3 (Fig. 3.3.2) and in Chapter 6 (Figs. 6.2.1 and 6.2.2). Numerous experimental data show that, in many cases, Eq. 2.6.18 is valid in changing viscosity by 6-8 decimal orders. However, corrections to Eq. 2.6.18 are sometimes needed. The more general equation for (MM, MMD) looks as follows[69]

$$\eta_0 = K\overline{M}_w^\alpha \left(\frac{\overline{M}_w}{\overline{M}_n}\right)^b \left(\frac{\overline{M}_z}{\overline{M}_w}\right)^c \qquad [2.6.19]$$

where $b \approx 0.24$; $c \approx 0.44$.

Steady-state compliance, J_s^0, depends primarily on higher average MM. The following equation is commonly used:[70]

$$J_s^0 = k\frac{\overline{M}_{z+1}\overline{M}_z}{\overline{M}_w\overline{M}_n}$$

[2.6.20]

where k is an empirical constant.

Many publications support the existence of strong dependence of steady-state compliance on higher average values of MMD. These two examples (for viscosity and steady-state compliance) demonstrate that different rheological parameters depend mainly on various average values of MM. It is rather difficult to expect that some very general laws of mixing can be easily written, and formulation of the rule of mixing is still a serious challenge to molecular theories of the polydisperse polymers.

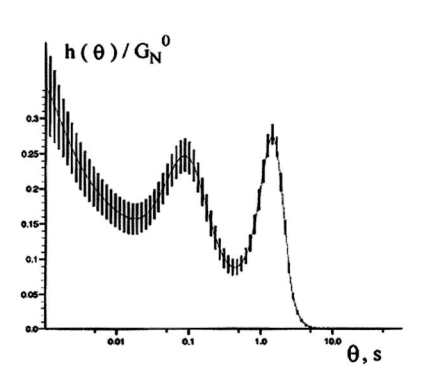

It is also worth mentioning that the experimental curves of monodisperse polymers are quite smooth. The same is true for the majority of industrial polydisperse polymers. However, from the experience with the mixtures of two monodisperse polymers as a model of polydisperse sample, it appears that both fractions behave – to some extent – as independent, and sometimes it is possible to suspect that they are not completely miscible. Then, the relaxation properties of both fractions are separated as in Fig. 2.6.7 and give clear separate peaks. The short relaxation side of the spectrum seems the same for both fractions. Then, it is reasonable to think that the smooth spectrum of polydisperse polymers is a consequence of the superposition of numerous peaks responsible for each fraction.

Figure 2.6.7. A logarithmic relaxation spectrum (normalized by the plateau modulus, G_N^0) of the mixture of two monodisperse polystyrene samples: 80% of polymer with M=177*10^3 and 20% of a polymer with M=60*10^3. Bars show possible errors of calculations. [Adapted, by permission, from W. Thimm et al., *J. Rheol.*, **44**, 429 (2000)].

The mixing rules for polydisperse polymers discussed in modern literature[71] are based on some molecular model arguments. They are usually obtained empirically and confirmed in experiments made with a rather limited number of polymers. Though there is some degree of independence in the rheological behavior of fractions in a mixture, it is commonly assumed that the relaxation properties of fractions entering a mixture are modified as a result of intermolecular interaction. A rather general form of the mixing rule, which reflects this phenomenon, was formulated for a relaxation function, G(t), in the following form:[72]

$$\frac{G(t)}{G_N^0} = \left(\int_0^\infty F^{1/\beta}(t, m)\frac{w(m)}{m}dm\right)^\beta$$

[2.6.21]

where $m = M/M_0$ is reduced MM, M_0 is MM of a monomer unit in a polymer chain, $m_e = M/M_e$ and M_e is average MM between two neighboring entanglements. This value is found as:

$$M_e = \frac{\rho RT}{G_N^0} \qquad [2.6.22]$$

and G_N^0 is the plateau modulus value, ρ is the density, R is the universal gas constant, and T is the absolute temperature.

The function F(t,m) in Eq. 2.6.21 reflects the effect of mixing and the scaling factor originates from a molecular model of mixing: for the reptation model $\beta = 1$ ("linear mixing rule") and for a model of double reptation $\beta = 2$ ("quadratic mixing rule"). In some publications, a model combining linear and quadratic mixing rules was also discussed.[73]

The central point of the formulation of the mixing rule is the form of the kernel F(t,m) reflecting the mutual influence of different fractions in a polydisperse polymer.[74] Based on very accurate experimental data for polystyrene and statistical comparison of different kernels it was found[71] that the best fit can be achieved with $\beta = 3.84 \pm 0.1$ and the kernel represented by:

$$F(t, m) = e^{-t/\theta_0(m)} \qquad [2.6.23]$$

where the maximum (terminal) relaxation time θ_0 is related to MM by the standard scaling law:

$$\theta_0 = km^\alpha \qquad [2.6.24]$$

and the scaling exponent $\alpha \approx 3.5$ (more exactly, for polystyrene $\alpha = 3.67$).

This mixing rule (with β close to the standard value of 3.5) is more realistic than the theoretical values of β (1 or 2) because it directly connects with a natural rule of mixing in calculating viscosity through $\overline{M}_w$ as in Eq. 2.6.18. Viscosity is calculated from

$$\eta_0 = [w(M_i)M_i^{1/\alpha}]^\alpha \qquad [2.6.25]$$

It is worth mentioning that the final results of calculations, at least in cases of some polymers with moderately wide MMD, show that viscosity is not very sensitive to the choice of the scaling factor (3.5 or another). It is also difficult to verify which coefficient is correct.[75]

The high value of the scaling factor (~3.5) in comparison with the lower theoretical value (1 or 2) may be explained by difficulties in estimation of the relaxation spectrum responsible for slow relaxation processes.[76] Eq. 2.6.21 includes relaxation modes for fractions with $m > m_e$ only. A measured relaxation spectrum includes all types of molecular movements and can be treated as consisting of two parts: for $m < m_e$ and $m > m_e$. Then, the whole spectrum, $F(\theta)$, is a sum:

$$F(\theta) = F_{entangl}(\theta) + F_{rapid}(\theta) \qquad [2.6.26]$$

Only the first part (entanglement effect) is used in Eq. 2.6.21, but not the second term that reflects the rapid molecular movements occurring between neighboring links. It was shown[74] that the scaling factor in Eq. 2.6.21 appears close to 2 if rapid molecular movements are included, as predicted by the double reptation (entanglement) model.

The concept of the relationship between MMD and relaxation properties of the polymer is viewed from two perspectives. First is a mixing rule, which permits the calculation of viscoelastic properties of polydisperse polymers. The solution is hidden in Eq. 2.6.21 with an appropriate choice of exponent β and kernel F(t,m). The second is the determination of MMD based on the results of rheological measurements made in the range of linear viscoelastic behavior. It is not important whether the relaxation curve or frequency dependence of dynamic moduli are used for this purpose.

It is also necessary to find function w(m) included in Eq. 2.6.21, i.e., from a mathematical point of view to solve the first kind of the Fredholm integral equation.

Some necessary preliminary steps are required in this method. First of all, a mixing rule must be known, i.e., the kernel should be written in an analytical form and the scaling factor β selected. Different types of kernels are used. From a practical point of view, it may not be important which kernel is selected. Usually, the simpler it is, the more convenient it is to make calculations, but even complicated forms can be handled by modern computer techniques). Then, it is necessary to formulate how MMD will be determined.

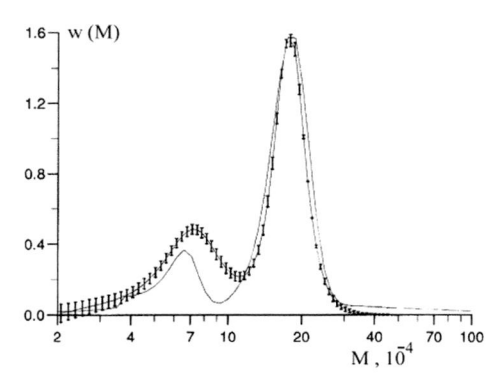

The first task involves the determination of the MMD parameters, assuming that its shape is known *a priori*. The function w(m), is substituted into Eq. 2.6.21 and the results of calculations are compared with experimental data. The MMD parameters are fitting factors found by standard procedure. In industrial synthesis, the shape of MMD is determined by the process chemistry. Studies usually follow changes in MMD caused by variation of technological conditions. Finding MMD of an unknown sample is more complex. More reliable results are obtained from unimodal MMD, i.e., MMD has a single maximum. However, some multi-modal materials MMD are prepared and used for technological applications and laboratory experiments, and these materials are difficult to study.

Figure 2.6.8. Results of calculation of MMD (bars show the limits of errors of calculation) in comparison with experimental data (solid line found by chromatographic method). The same sample as in Fig. 2.6.6. [Adapted, by permission, from W. Thimm et al., *J. Rheol.*, **44**, 429 (2000)].

Based on Eq. 2.6.21 the following analytical equation was obtained,[71,75] which helps to determine MMD using measured relaxation spectrum:

$$w(M) = \frac{1}{\beta}\left(\frac{\alpha}{G_N^0}\right)^{1/\beta} h_{entangl}(m) \left[\int_{m_e}^{\infty} \frac{h_{entangl}(m')}{m'} dm'\right]^{\frac{1}{\beta}-1} \qquad [2.6.27]$$

where $h_{entangl}$ is a logarithmic relaxation spectrum related to the entanglement relaxation modes only, i.e., obtained as written in Eq. 2.6.25 after extraction of part related to rapid relaxation processes, m' is a variable of integration, and M is the molecular weight. The spectrum $h_{entangl}$, as a function of m, is determined in the following way

$$h_{en\,tangl}(m) = h_{en\,tangl}[\theta(m)] \qquad [2.6.28]$$

in accordance with Eq. 2.6.24. All other constants entering this equation are the same as above. The relaxation spectrum used for calculations related to the molecular mass is in the range of $m > m_e$.

An example of the final results of MMD calculations is shown in Fig. 2.6.8. The rheological method gives realistic behavior of MMD including positions of peaks for bimodal samples. The correspondence between MMD and relaxation properties of a sample does exist.

2.7 TIME-TEMPERATURE SUPERPOSITION. REDUCED ("MASTER") VISCOELASTIC CURVES

2.7.1 SUPERPOSITION OF EXPERIMENTAL CURVES

In real experimental or technological practice, the frequency dependence of elastic modulus can be measured for a limited frequency range because of limitations of experimental techniques. Let the frequency be from ω_1 to ω_2. This range is sometimes called an experimental window. Measurements can be carried out at different temperatures. The experimental data for a linear high-molecular-mass polymer are presented by a set of curves, as shown in the vertical column on the left side of Fig. 2.7.1. The curves 1-2, 3-4,..., 15-16 relate to different temperatures decreasing from lower to upper curves. The curves in the column might have some practical interest, but they do not show a broader picture. They do not represent the general picture of frequency dependence on the modulus, because the frequency range from ω_1 to ω_2 is narrow. An interesting question arises as to what happens with a modulus beyond the experimental window.

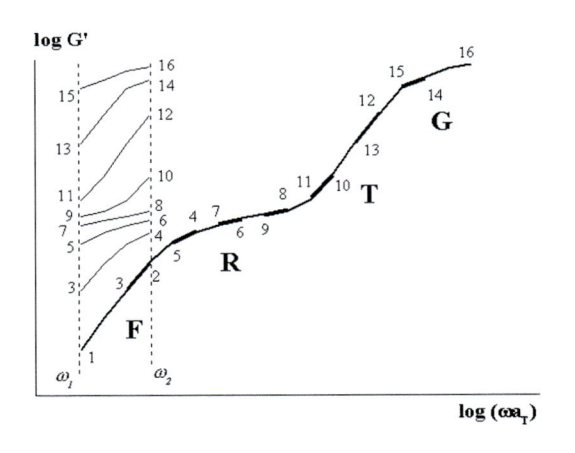

Figure 2.7.1 An example of experimental dependencies of G'(ω) measured at different temperatures (vertical column at the left part of the figure) and construction of the reduced (master) curve of dependence G'(ω) in the wide frequency range. Qualitative illustration.

It was noticed that the time (or frequency) dependencies of viscoelastic properties, measured at different temperatures, are similar in their shape though shifted along the time (frequency) axis.[79] In practice, it is sufficient to see that the boundary parts of the curve are similar, and then it is assumed that the whole curve should be similar too. Then, it is reasonable to suspect that each curve can be extended beyond the range of its real measured values with the conservation of the shape which is general for all curves. In this approach, it is supposed that each curve is shifted along the frequency scale and the distance between the curves, a_T, depends on temperature.

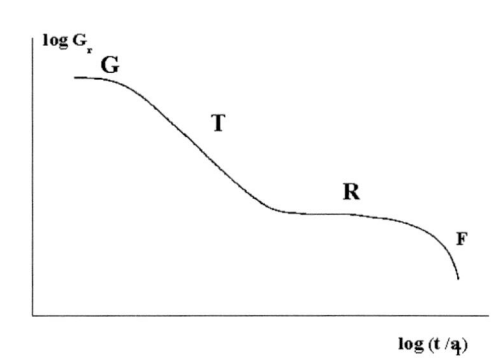

Figure 2.7.2. Relaxation modulus as a function of a reduced time. Qualitative illustration.

Based on this concept the initial (experimental) curves can be *superimposed* by shifting all curves to one arbitrarily chosen as the base curve. The resulting *reduced* (or *master*) *curve*[80] is shown in Fig. 2.7.1 by a solid line and the position of initial curves are marked along the master curves by the same numbers as those used at the ends of the initial curves. For an illustration, the overlapping portions of neighboring curves are shown by doubled lines (they would not superimpose completely, but, in fact, the points of both curves lie on the same curve).

An analogous curve for relaxation modulus is shown in Fig. 2.7.2. The characteristic zones (domains) of viscoelastic behavior are marked in these curves. Their explanation can be found below.

It is assumed that the reduced curve has the same shape in the whole temperature range, though changing the temperature of reduction, or *reference temperature*, leads to a shift of the curve along the $\log \omega$ scale. The temperature dependence of this shift is expressed by the function $\log a_T(T)$.

This master curve is related to the reduction temperature (which is the largest of the examined intervals in the example under discussion). The method permits us to obtain the viscoelastic curves in the frequency range which is much wider than the experimental window of initial experimental curves. The reduced curve contains more information than any initial curve obtained at a single temperature. It is now a standard method of treating experimental data obtained at different temperatures but with narrow frequency (time) windows.

The method of reduced variables helps to exceed the direct experimental capabilities of any measuring device. The frequency (or time) dependence of any viscoelastic function can be obtained for a broader range.

From the beginning of the application of the method of reduced variables, it was recognized that shifting viscoelastic curves is based on the concept of the same temperature dependence of all relaxation times of material. This can be illustrated by an example of the relaxation spectrum for an individual macromolecular chain. Let the frequency dependencies of $G'(\omega)$ and $G''(\omega)$ be described by Eq. 2.3.10 and the distribution of relaxation times as per Eq. 2.6.6. Then, after the substitution, the following equations can be obtained:

$$G'(\omega) = \sum_{n=1}^{M} G_n \frac{(\omega\theta_n)^2}{1+(\omega\theta_n)^2} = \sum_{n=1}^{M} G_n \frac{(\omega\theta_{max})^2}{n^2+(\omega\theta_{max})^2} \qquad [2.7.1a]$$

$$G''(\omega) = \sum_{n=1}^{M} G_n \frac{(\omega\theta_n)}{1+(\omega\theta_n)^2} = \sum_{n=1}^{M} G_n \frac{n(\omega\theta_{max})}{n^2+(\omega\theta_{max})^2} \qquad [2.7.1b]$$

(In these equations the dummy index, i, used in Eq. 2.3.10, is changed to n, in order not to confuse the index with the imaginary unit).

It is immediately seen that G' and G" are the functions of the single dimensionless argument, $\omega\theta_{max}$, and temperature dependence of modulus is completely described by the temperature dependence of the relaxation time θ_{max}. Then, the frequency dependencies of G' and G" measured in a wide temperature range as a function of the dimensionless argument ($\omega\theta_{max}$) should form a single master curve.

Frequency can be replaced by time with the same argumentation and the single dimensionless argument, in this case, is (t/θ_{max}). The same conclusion concerning the construction of a general master curve from the experiments performed at different temperatures is true for any viscoelastic curve – creep, relaxation, etc. It is also true of a relaxation spectrum itself, which can be treated as a function of the single reduced parameter (θ/θ_{max}).

Above, it was tacitly assumed that the coefficient G_n does not depend on temperature. This is not correct. This parameter is proportional to the factor ρT, where ρ is density and T is the absolute temperature. It is necessary to introduce a factor of vertical shift equal to $\rho T/\rho_0 T_0$. The values with zero-index are related to a reference temperature. The origin of this correction factor is explained in the theory of rubbery elasticity because the value G_n has a meaning of rubbery modulus of the material in extension (compare with Eq. 2.6.22 rearranged for G_N^0). However, this correction factor is small and becomes important only in the domain of rubbery behavior of polymeric material where the frequency dependence of modulus is rather weak.

The construction of the master curve is based on the following general assumptions: all relaxation times are characterized by the same temperature dependence (possibly due to the common mechanism of material relaxation) and it is possible to neglect the temperature correction factor for G_n. Then, it is reasonable to suppose that the opposite conclusion is also valid: temperature superposition becomes impossible if different temperature dependencies exist for various relaxation times, or there are different relaxation mechanisms characterized by various temperature dependencies. The latter conclusion is the general principal limitation of the application of the method of frequency (time) superposition.

Numerous experimental evidence confirm that the time superposition method is valid for linear polymers and their solutions but the method is not valid in the following cases:

- mixtures of different polymers characterized by their own relaxation spectra with their own temperature dependences of relaxation times
- branched polymers: relaxation processes in long branches can be different than in the backbone chain
- materials with inherent supermolecular structure, which can change with changing temperature
- block copolymers
- materials with properties changing with time (during an experiment), for example, because of slow chemical reactions.

One must be careful in applying the superposition method to unknown materials because doubt is always there as to whether the master curve reflects the physical reality in the very wide frequency range. New relaxation mechanisms may appear at very low or very high frequencies.

The temperature shift factor, a_T, is a function of current temperature, T, and the reference temperature, T_0. Using the definition of viscosity as the integral measure of a relaxation spectrum, it is easy to show that a_T can be calculated as the ratio of viscosities at temperatures T and T_0:

$$a_T = \eta(T)/\eta(T_0) \qquad [2.7.2]$$

Temperature dependence of viscosity can be expressed by any standard method, in particular by the Arrhenius equation:[81]

$$\frac{\eta(T)}{\eta(T_0)} = \exp\left[\frac{E_a}{R}\left(\frac{1}{T} - \frac{1}{T_0}\right)\right] \qquad [2.7.3]$$

where E_a is activation energy of the relaxation processes (and viscous flow), T_0 is some arbitrarily chosen reference temperature, and R is the universal gas constant.

The Williams-Landel-Ferry (WLF) equation is another popular expression of the temperature dependence of relaxation times:[82]

$$\log a_T = \frac{-C_1^0(T - T_0)}{C_2^0 + (T - T_0)} \qquad [2.7.4]$$

It is interesting to notice that $T_0 \approx T_g + 50$ (T_g is the glass transition temperature), then, for many polymers, C_1^0 and C_2^0 are universal constants having the following values: $C_1^0 \approx 7.60$ and $C_2^0 \approx 227.3$.

The classical illustration of the log a_T dependence on temperature decrement reproduced in many editions, is presented in Fig. 2.7.3. The average line that is drawn through these points is described by the WLF equation with the universal values of the constant. The relationship is true for an enormous range of 16 decimal orders of a_T values.

Using any of these equations, it is easy to see that by changing the temperature it is possible to vary a_T, and consequently the reduced time (frequency) by many decimal orders. It allows one to broaden the range of the argument up to 10-15 decimal orders. Such results cannot be obtained by direct measurements but only by the method of reduced variables. Direct experiments in a narrower frequency range, carried out for different polymer materials by many authors, confirmed that the superposition gives realistic data coinciding with those obtained by direct measurements.

The dependence of relaxation modulus in reduced variables (in this case, it is reduced time scale) is similar to the frequency dependence of a dynamic modulus, but it is a mirror reflection of the latter (compare the curves in Figs 2.7.1 and 2.7.2). Curves of both types contain the same physical information on the viscoelastic properties of materials under investigation.

Experimental data for "almost" monodisperse (polymers with very narrow MMD) polybutadiene sample, presented in form of master curves, are given in Fig. 2.7.4 for $G'(\omega)$ and $G''(\omega)$. This figure shows the relative position of the components of dynamic modulus as well as the existence of a "cross-over" point – the frequency at which $G' = G''$. The second important example based on real experimental data is presented in

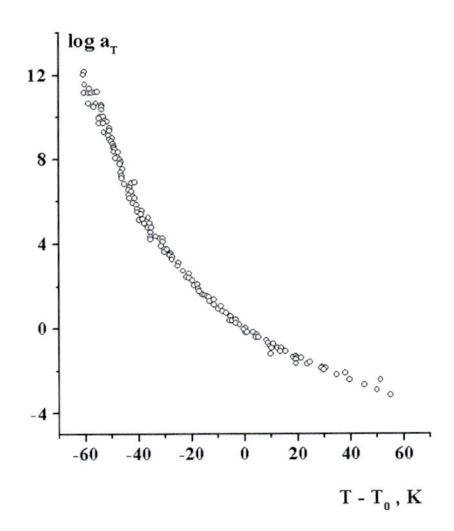

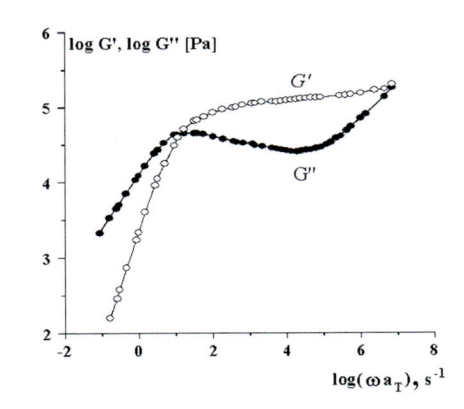

Figure 2.7.4. Frequency dependencies of the dynamic storage and loss moduli reduced to 28°C. Initial experimental data were obtained in the temperature range from -70 to 28°C. Sample is a monodisperse polybutadiene, M = $9.7*10^4$. [Adapted, by permission, from M. Baumgaertel et al., *Rheol. Acta*, **31**, 75 (1992)].

Figure 2.7.3. Temperature dependence of relaxation times, expressed as the ratio of relaxation times as a function of temperature difference according to the Williams-Landel-Ferry. Points present experimental data for 17 different polymeric systems.

Fig. 2.7.5 for the dependencies G'(ω) and G"(ω) related to the homologous series of polymers — materials having identical structure but with different chain lengths. These curves demonstrate the influence of MM on the range of frequencies where relaxation phenomena are observed.

Analogous experimental data were observed and published for various polymers and they form the basis for general conclusions concerning the physical properties of polymeric materials. These conclusions are summarized in the following section.

2.7.2 MASTER CURVES AND RELAXATION STATES

Qualitative curves, as well as the results of experimental observations, show that there are several typical and very general features of the viscoelastic curves.

First of all, there are four characteristic domains (or zones) on the viscoelastic curves, designated by the letters F, R, T, and G on the curves of Figs. 2.7.1 and 2.7.2. Real experimental data presented in Figs 2.7.4 and 2.7.5 show that only F and R zones and the transition to the T zone can be observed, even in the frequency range covering 9 decimal orders. However, other experimental data for different polymeric samples are shown in Fig. 2.7.6. These data can be treated as supplementary to Figs 2.7.4 and 2.7.5 because they start from the same level of modulus as in Figs 2.7.4 and 2.7.5. The curve in Fig. 2.7.6 covers the R, T, and initial portion of G zones. These domains spread over 12 decimal orders of frequency.

Combining experimental data of these figures, one can see the full range of relaxation covering about 20 decimal orders of frequency and a range of modulus from 10^2 to 10^{10} Pa. Based on real numerical values of the storage modulus, the physical sense of different domains, marked in Figs 2.7.1 and 2.7.2 as F, R, T, and G is given below.

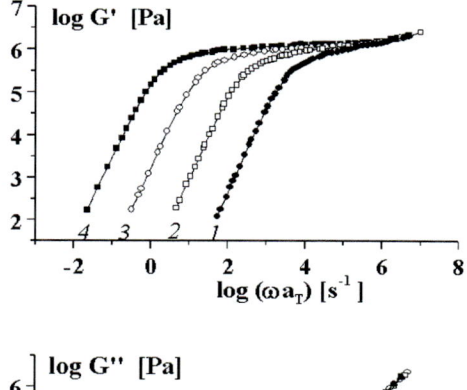

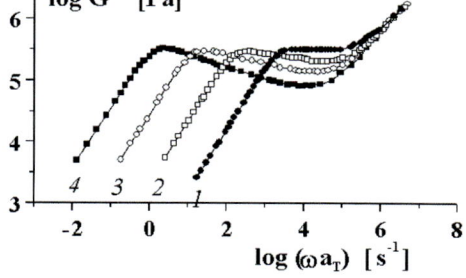

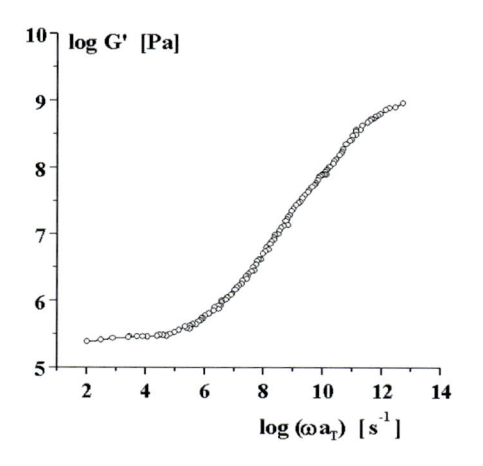

Figure 2.7.6. Frequency dependence of dynamic storage modulus in the transition zone (from rubbery to glassy state) for NBS polyisobutylene at a reference temperature of 25°C. [Adapted, by permission, from E. Catsiff, A.V. Tobolsky, *J. Colloid Sci.*, **10**, 375 (1955)].

Figure 2.7.5. Frequency dependencies of the dynamic storage and loss moduli reduced to 28°C for mono-disperse polybutadienes of different MM: $2.07*10^5$ (1); $4.41*10^5$ (2); $9.7*10^5$ (3); $20.1*10^5$ (4). Initial experimental data were obtained in the temperature range from -70 to 28°C. [Adapted, by permission, from M. Baumgaertel et al., *Rheol. Acta*, **31**, 75 (1992)].

F − is a flow zone. In the frequency range corresponding to this domain, material is treated as liquid. Storage modulus is considerably small ($G' \ll G''$). In the limiting case (at $\omega \to 0$) $G' \propto \omega^2$ and $G'' \propto \omega$, and it corresponds to the relaxation process determined by the maximum relaxation time, θ_{max}. The relaxation processes in this zone are governed by the chain movement as a whole as described by the reptation model and the model of entanglements, as discussed in section 2.6.

R − is a rubbery zone. In the frequency range corresponding to this domain, the material behaves like rubber. For rubbers and rubber-like materials, storage modulus is constant and its plateau value lies between 10^5-10^7 Pa, depending on the chain rigidity. The transition from F to R zone is arbitrary and can be attributed to the maximum of G'' or the crossover frequency. For real rubbers (cured polymers with the permanent network of chemical bonds) the F zone disappears and the R zone, corresponding to equilibrium storage (rubbery) modulus, continues to unlimited low frequencies, formally to steady or equilibrium) state of deformations. Adding solvent to a rubbery material (plasticization) decreases the plateau values of the modulus, and adding a solid filler increases its value.

The relaxation process in the R zone cannot be observed, even in the wide time interval (as seen from the plateau of relaxation modulus in Fig. 2.7.2). It is interesting to mention that the influence of the length of the molecular chain is important in the F zone only. The transition from F to R zone also depends on MM. The position of the G''(ω) dependence in the F zone is determined by viscosity and therefore it is proportional to $M^{3.5}$. It is seen (Fig. 2.7.5) that the plateau end in the high-frequency region does not depend on MM. It is easily proven that the plateau width is proportional to $M^{3.5}$.

T – is a transition zone. In the frequency range corresponding to this domain, properties of the material, determined by its modulus, resemble leather materials and therefore this zone is sometimes called the leather-like zone. It is a wide domain (Fig. 2.7.6) and relaxation properties here are determined by small-scale molecular motions of the portion of chains between entanglements. It is a different relaxation mechanism than in the F and R zones and, possibly, their temperature dependencies may be different. Therefore, it is necessary to be careful in applying the superposition method over all relaxation states of a material.

G – is a glassy zone, and the behavior of the material at very high frequencies resembles the deformation of glass. It is reflected by the high values of the modulus (about 10^9 – 10^{10} Pa) characteristic for inorganic and organic glasses and independent of the modulus values on frequency, i.e., it is possible to think that frequencies in this domain are so high that all relaxation processes are absent.[83]

The main and principal conclusion obtained from the application of the time (frequency)-temperature superposition method consists of a statement that the same values of modulus can be reached by changing either temperature or frequency. It means that the question: what is the state of this material? must be substituted by another, more accurate question: how does material behave under these conditions (frequency)?, i.e., at the fixed temperature the same material can behave in a very different manner either like liquid (at very prolonged intervals of observation or at very low frequencies), like a rubber (if the duration of observation or frequency corresponds to the rubbery plateau), like a leather (if the frequency is positioned in the glass transition zone), or like a glass (if frequencies are very high).

This conclusion is very important for numerous applications of viscoelastic polymeric materials because their properties can be estimated in one frequency domain and the application can be related to quite different conditions. Then, one may encounter such effects as glassy behavior of rubber articles – they break as fragile material at high frequencies, a behavior not expected from rubbers (for example, aircraft tires at high speed of landing in winter may break due to high frequencies of glassy state deformation, though they are safe at the same temperature based on the result of a static or low-frequency experiment). Polymer melts at high shear rates cannot flow, due to forced transition into the rubbery state (see section 3.6).

The success of the application of the time-temperature superposition method in numerous examples urged investigators to propose an analogous method for various phenomena. For example, the time of aging depends on temperature. Then, it was proposed to accelerate the experiments of aging by temperature increase and extrapolating results of such experiments to lower temperatures as a basis for predicting material behavior for long periods of time at lower temperatures. The acceleration of relaxation processes can be reached by adding a solvent; then the results of experiments with polymer-solvent systems are used for predicting the behavior of materials at other concentrations of a solvent or without it (this is a method of time-concentration superposition), and so on. Sometimes, this approach gives quite positive and useful results. However, this is not a universal method and one ought to be sure that the method of superposition used reflects the physical effects caused by relaxation phenomena.

Generally speaking, it is a dangerous practice to extend any of these methods for predicting beyond more than, say, one decade of time.

2.7.3 "UNIVERSAL" RELAXATION SPECTRA

There were a number of attempts to propose some "universal' functions for describing viscoelastic properties (primarily, a relaxation spectrum). The main difficulty in an approximation of real experimental data is caused by the polydisperse nature of real polymers. In reality, then, the relaxation spectrum of polydisperse material is achieved by superposition of relaxation properties of different fractions. This means that no "universal" spectrum exists for a polymer with arbitrary MMD because polymers with different MMD are different materials.

Indeed, one meets with two separate problems: describing relaxation properties of individual fractions and combining these properties in their mixing. This problem is discussed in section 2.6 based on molecular models of relaxation properties of a polymer chain.

It is necessary to divide the general problem of searching a relaxation spectrum into two parts. First, these are relaxation properties in the F and R zones and second in the T zone.

All these problems are actively discussed in modern rheological literature, because their solution gives a chance to find general methods for calculating viscoelastic properties of real polymeric materials. Existing molecular models do not give a realistic (close to an experiment) representation of the relaxation spectrum, even for a monodisperse polymer. That is why it is important to propose at least an empirical approximation to experimental data.

Concerning monodisperse polymers, the general solution was proposed for relaxation behavior in the F and R zones.[84] This "universal" relaxation spectrum is described by the function:

$$G(\theta) = G_0\left[\left(\frac{\theta}{\theta_0}\right)^{-n} + \left(\frac{\theta}{\theta_{max}}\right)^{m}\right] \qquad [2.7.5]$$

where G_0 is constant, m and n are empirical parameters of a spectrum, and θ_0 and θ_{max} are minimum and maximum boundary values of relaxation spectrum for F and R zones.

The relaxation spectrum in the T zone is regarded as caused by movements of short portions of a macromolecular chain. In the first approximation, these movements can be described by the spring-and-bead model (see section 2.6).

The problem of summation of inputs of relaxation properties of different fractions into a relaxation spectrum of polydisperse polymers is also discussed in section 2.6. However, it is worth mentioning that at the moment this problem is far from a clear understanding and quantitative description.

In conclusion to this section, it is worth mentioning that many relaxation properties of polymers are successfully treated in the framework of some "universal" equations. For example, it is true for the temperature dependence of relaxation times (Eq. 2.7.4), "universal" form of a relaxation spectrum (Eq. 2.7.5), regardless of the chemical nature of a polymer chain. It is reasonable to suspect that it is not coincidental but it reflects the basic peculiarity of relaxation properties of polymers: movement of long flexible chains interacting with their surrounding; in this concept, the primary role is played by geometrical restrictions to the Brownian movement, and in the first approximation, these restrictions do not depend on details of the chemical structure of a chain.

2.8 NON-LINEAR EFFECTS IN VISCOELASTICITY

2.8.1 EXPERIMENTAL EVIDENCE

There is ample experimental evidence, which clearly demonstrates that all concepts discussed in the previous section are related to the limiting cases of low deformations (or low deformation rates). The increase of deformation or deformation rate immediately leads to non-linear effects which are characteristics of all real technological materials. Thus, the linear viscoelastic behavior should be considered as the limit of the generalization of viscoelastic properties.

The principle difference between the linear and non-linear behavior of a matter is that the linear deformation (either oscillation or any other mode of deformation) deals with the initial (undisturbed) structure of a sample under study, while the transition to the non-linear domain of the deformation causes changes in the structure at the super-molecular level. Then, the results of measurement under the linear deformations characterize the static state of a material, while the results of the non-linear studies allow us to follow the transformations taking place under the action of external forces.

Below, the principal experimental evidence of the non-linear behavior of viscoelastic materials are discussed.

2.8.1.1 Non-Newtonian viscosity

Viscosity appears in the linear theory of viscoelasticity as the first moment of a relaxation spectrum, according to Eq. 2.5.12. By this definition, it is expected that viscosity is constant, i.e., does not depend on shear rate or stress. This regards Newtonian viscosity according to the definition of Eq. 2.1.1. However, it is well known (and this is one of the basic effects of rheology) that the viscosity of numerous liquids is not constant but depends on conditions of flow such as shear rates, $\eta(\dot{\gamma})$, or shear stresses, $\eta(\sigma)$. This effect is known as the non-Newtonian behavior of liquids. Non-Newtonian flow is a non-linear effect not described within the framework of the linear theory of viscoelasticity.

The theoretical and applied meaning of the non-Newtonian behavior of liquids is complex. Therefore, a separate chapter of this book is devoted to non-Newtonian flow, and these effects are further discussed in Chapter 3.

The same is valid for the coefficient of normal stresses, determined in the theory of viscoelasticity as the second moment of a relaxation spectrum. According to Eq. 2.5.14, the coefficient of normal stresses is also a material constant. However, this is not true for all real viscoelastic liquids. In fact, the coefficient of normal stresses depends on shear rate, $\Psi_1(\dot{\gamma})$. This non-linear phenomenon will be discussed in more detail in Chapter 3, along with viscosity.

2.8.1.2 Non-Hookean behavior of solids

According to the linear theory of viscoelasticity, the elastic modulus depends on time (or frequency), but not on stress or deformation. In fact, the elastic modulus of many real materials depends on deformation in static conditions (regardless of time effects). This non-Hookean behavior is a main non-linear phenomenon characteristic of solids. This is of fundamental significance for rheology and the non-linear elastic behavior of solids is discussed in Chapter 4.

2.8.1.3 Non-linear creep

The creep function in the linear viscoelasticity limit is, by definition, independent of stress. But the real behavior of mechanical properties of the material is more complex. A typical and general example of mechanical behavior of viscoelastic behavior of the material at different constant stresses is shown in Fig. 2.8.1 by creep functions (ratios of deformation to stress applied), ψ (t). At low stresses, σ_1 in this example, the limit of ψ (t) at t $\rightarrow \infty$ exists. The same creep function (in the limits of experimental errors) is observed for the range of low stresses, at least at $\sigma < \sigma_1$. This is the linear limit of the viscoelastic behavior of the material.

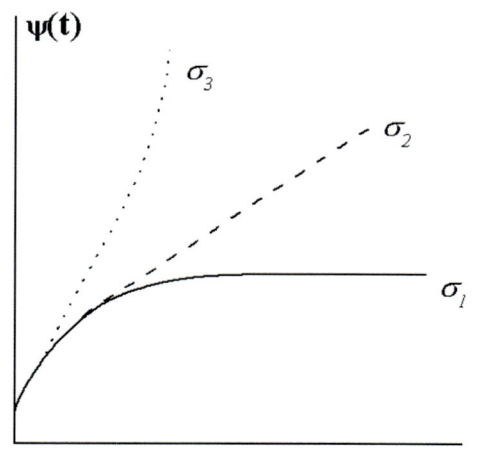

The increase of stress results in a deviation from the linear limit, and at stress σ_2 the creep function grows approximately linearly in time, i.e. the *stationary* regime is observed. At higher stresses, σ_3, deformation increases with acceleration; this is an *unsteady* regime of creep, which ends with rupture of the sample.

It is interesting to mention that in the initial part of deformations all curves coincide, i.e., at low deformations (or at times shorter than some "critical" value t*) the behavior of the material at any stress is linear.

Figure 2.8.1. Creep function at different stresses ($\sigma_1 < \sigma_2 < \sigma_3$). Qualitative picture.

It is seen from the scheme in Fig. 2.8.1, that this critical time t* depends on the applied stress. Real experimental data illustrating this dependence are shown in Fig. 2.8.2 for the extension of solid plastic. It is seen that, in the range of stresses increasing by about 2.5 times, the limit of linearity, t*, decreases by 2 decimal orders, i.e., by 100 times. It demonstrates that the non-linearity limit is a very strong function of stresses. In Fig. 2.8.2, t* is exponentially dependent on stress. Such strong dependence of the limit of linearity on stress is valid for the whole domain of non-linear behavior of viscoelastic materials.

Experimental data of the type shown in Fig. 2.8.1 can be treated in the following manner: up to t/t* $\approx$ 1, the behavior of material obeys the laws of linear viscoelasticity and at t > t* a correction ought to be introduced, which is a function of stress and the ratio t/t*(σ). Even this simple example demonstrates that the description of non-linearity of mechanical properties of the material is complex because the stress factor enters in two different manners – limiting the boundary of linearity and determining the form of the deformation vs. the stress curve. This becomes even more complex in the transition to

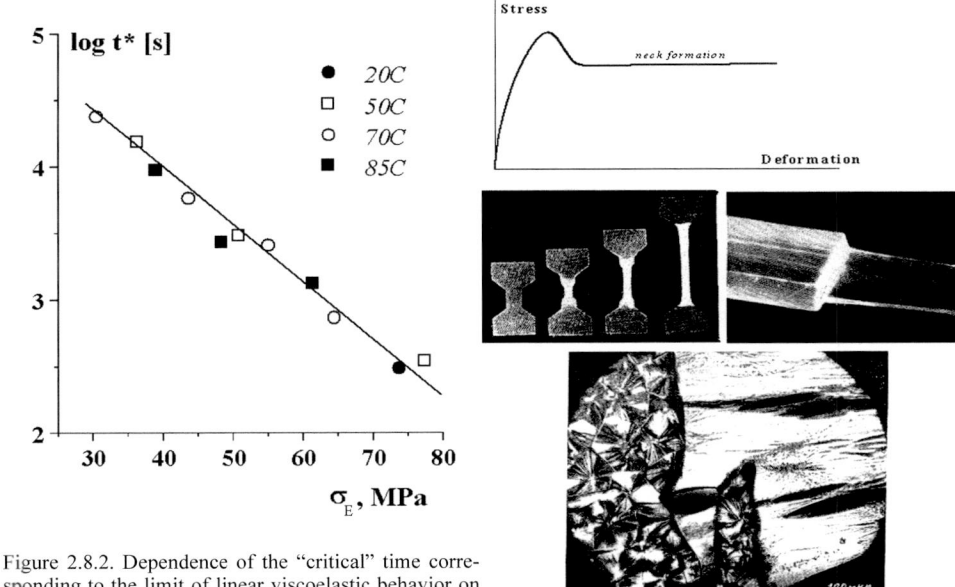

Figure 2.8.2. Dependence of the "critical" time corresponding to the limit of linear viscoelastic behavior on stress in extension of polycarbonate at different temperatures. [Adapted, by permission, from A.Ya. Malkin, A.E. Teishev, M.A. Kutsenko, *J. Appl. Polym. Sci.*, **45**, 237 (1992)].

Figure 2.8.3. Neck formation at the transition from isotropic to oriented state of a polymeric material (lower photograph was made by G.P. Andrianova).

three-dimensional deformations because invariants must be used instead of unidimensional deformations and stresses.

However, this problem can be solved on the basis of fitting experimental data with a suitable analytical expression giving the formal (phenomenological) representation of non-linearity. This is true for homogeneous deformations of material. In reality, creep (for example in the extension of solid viscoelastic materials) is even more complex, due to the effects of instability of deformations.

The most well-known example of instability is the formation of a neck (the effect of *necking*). This effect is illustrated in Fig. 2.8.3 and it consists of a sudden transition from homogeneous extension to a jump-like transit from the wide to the narrow section of a sample. This figure shows a graphical representation of stress vs. deformation, necking development (the middle part of a sample with a bright stripe is a neck), and a microscopic picture of changes (lower part of the picture clearly shows that necking is the transition from an isotropic structure to an oriented state of material).

The extension appears here not as the homogeneous decrease of the cross-section of the sample but as the co-existence of two different parts of the sample and the continuous decrease of the length of the wide part and increase of sample length because of increase in the length of the narrow part. This transition is typical of any polymer, either crystalline or amorphous. In fact, this is a kind of phase transition, from homogeneous to oriented. It is also possible that macromolecules and their assemblies exist in two distinct states.

In the formulation of laws of the non-linear behavior of viscoelastic materials, it is necessary to include some critical states and effects of instability at some specific conditions. These effects may be treated as phase non-linearity, and these phenomena are not restricted to solid materials only.

2.8.1.4 Non-linear relaxation

According to the principles of the linear theory of viscoelasticity, a relaxation function (σ relaxation modulus) does not depend on deformation and time at which dependence of stresses is measured. The same is true for relaxation after sudden cessation of steady flow at different shear rates. Eq. 2.4.13 shows that the ratio of relaxing stress, $\sigma(t)$, to deformation rate, $\dot{\gamma}$, or to its initial value, σ_0, which equals $\eta\dot{\gamma}_0$, does not depend on shear rate but on time only. Then, one would expect that the experimental points obtained for relaxation after flow at different shear rates, presented in coordinates σ/σ_0–t, lie on the same curve.

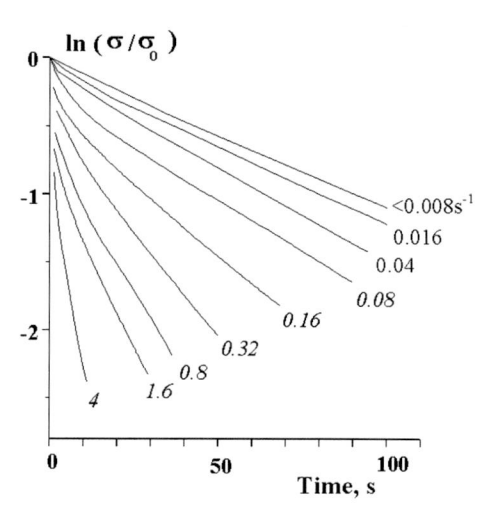

The reality appears much more complex, as shown in Fig. 2.8.4 for the sudden cessation of steady shear flow. The linear domain is only a limiting case of a very low shear rate, and the higher the shear rate before relaxation, the faster the relaxation. The viscoelastic behavior of material in the wide range of deformation conditions is strongly non-linear. It is more a characteristic of viscoelastic properties of material than the limiting linear case. The increase in stress (that is equivalent to an increase in

Figure 2.8.4. Shear stress relaxation after sudden cessation of steady flow at different shear rates for polyisobutylene with MM = $1*10^5$ at 20°C.

shear rate) leads to acceleration of relaxation as a general rule for non-linear behavior.

The analogous situation is observed with normal stresses (Fig. 2.8.5): relaxation of shear stresses is much faster than relaxation of normal stresses. An increase in shear rate, which precedes relaxation, accelerates the relaxation of shear and normal stresses.

Based on these (and many other analogous) experimental data it is reasonable to suppose that the increase of deformation rate results in suppressing slow relaxation processes.[85] In other words, at high deformation rates slow relaxation modes do not have enough time to be realized. Formally, this phenomenon can be discussed in terms of the dimensionless parameter, $\gamma\theta$: viscoelastic effects can be observed when values of this parameter are of the order of 1. The increase of deformation rates results in a shift of the boundary of relaxation times, which are still active in a relaxation spectrum of material, to lower relaxation times (i.e., faster relaxation processes).

Fig. 2.8.6 shows that the increase in deformation rate leads to a shift of high relaxation times of a relaxation spectrum. This effect should be taken into account in the formu-

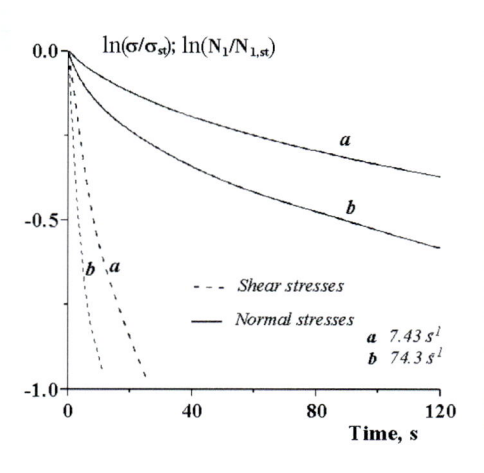

Figure 2.8.5. Relaxation of stresses after cessation of steady shear flow for butyl rubber at 25°C.

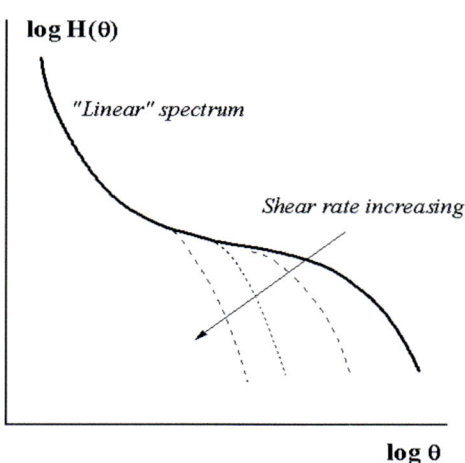

Figure 2.8.6. Changes in a relaxation spectrum induced by increase in shear rate. Qualitative picture.

ation of the rheological equation of state, describing a non-linear behavior of real viscoelastic materials.

The relaxation spectrum shortening can be expressed by simple equations relating to its integral characteristics, such as viscosity and coefficient of normal stresses. The influence of deformation rate on shear flow depends on the relationship between shear rate and relaxation time. Then, it is possible to assume that the following equations are the generalization of Eqs. 2.5.12 and 2.5.14, respectively:

$$\eta(\dot{\gamma}) = \int_0^\infty \theta G(\theta, \dot{\gamma}\theta)d\theta \qquad [2.8.1]$$

$$\Psi_1(\dot{\gamma}) = 2\int_0^\infty \theta^2 G(\theta, \dot{\gamma}\theta)d\theta \qquad [2.8.2]$$

These equations are valid for the steady-state flow, demonstrating that non-Newtonian behavior is due to a change in the relaxation spectrum of a flowing liquid induced by intensive shearing.

In light of this experimentally established fact, the question arises how the initial relaxation spectrum $H(\theta)$ at low shear rates corresponding to the Newtonian behavior shifts with an increase of the shear rate corresponding to the non-Newtonian behavior. This question is closely related to the generalized (invariant) representation methodology of non-Newtonian viscosity that was developed in many papers.[87-90] In particular, it was proposed[87,88] that the shear rate dependence of the apparent viscosity of concentrated polymer solutions can be represented in an invariant form by plotting the ratio of η/η_{in}, where η is the apparent viscosity and η_{in} is the initial Newtonian viscosity, as a function of the product $\theta_{in}\dot{\gamma}$, where θ_{in} is the initial (characteristic) relaxation time at low shear rates. The following invariant representation was obtained:

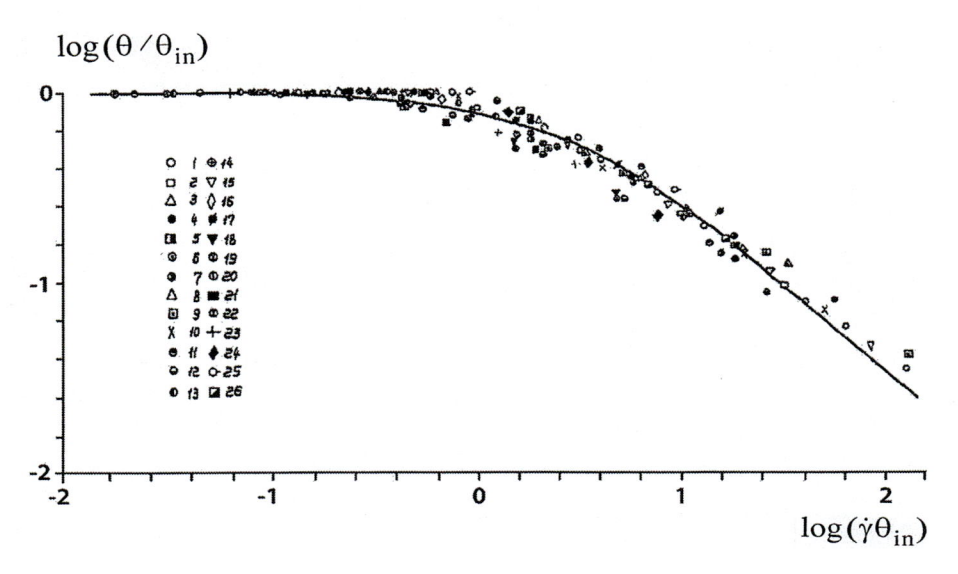

Figure 2.8.7. Master curve for apparent relaxation time for polymer melts of different molecular weights and molecular weight distributions and their solutions (symbols and numbers next to them indicate various polymers). Initial relaxation time and materials: (1) 400 s, polyisobutylene (PIB), 22°C; (2) 100 s, PIB, 40°C; (3) 20 s, PIB, 60°C; (4) 5.7 s, PIB, 80°C; (5) 2.2 s, PIB, 100°C; (6) 0.04 s, polystyrene in decalin (PS-D), 25°C, 18.4 vol%; (7) 0.58 s, PS-D, 29 vol%; (8) 3.4 s, PS-D, 38 vol%; (9) 41.7 s, PS-D, 46.6 vol%; (10) 400 s, PS-D, 57.3 vol%; (11) 52.5 s, polybutadiene, M_w=1.52x10^5, M_w/M_n=1.1, 22°C; (12) 0.026 s, polybutadiene (M_w=2.4x10^5, M_w/M_n=1.1) in α-methylnaphthalene (PBD-N), 22°C, 10 vol%; (13) 1.32 s, PBD-N, 30 vol%; (14) 81.2 s, PBD-N, 50 vol%; (15) 2.75 s, poly(methylmethacrylate) in diethyl phthalate, 30°C, 5 wt%; (16) 2.14 s, PIB in tetralin, 25°C, 12 wt%; (17) 0.155 s, PS in toluene (PS-T), 9 wt%; (18) 0.048 s, PS-T, 15 wt%; (19) 4.8 s, PS-T, 20.1 wt%; (20) 0.076 s, PS in chlorinated diphenyl (PS-CD), 27°C, 13 wt%; (21) 0.224 s, PS-CD, 16 wt%; (22) 0.32 s, polydimethylsiloxane (PDMS), M_w=4.68x10^5, M_w/M_n=1.29, 20°C; (23) 0.045 s, PDMS, M_w=9.71x10^4, M_w/M_n=1.18; (24) 1.82 s, polyethylene (PE), 150°C, 1.4x10^3 Pas; (25) 9.55 s, PE, 8x10^3 Pas; (26) 226 s, PE, 2.5x10^5 Pas. [Adapted, by permission, from A. I. Isayev, *J. Polym. Sci., Polym. Phys.*, **11**, 2123 (1973)].

$$\eta = \frac{\eta_{in}}{1 + (\theta_{in}\dot{\gamma})^{0.75}} \qquad [2.8.3]$$

Later,[89] it was found that the shear rate dependence of the apparent viscosity for a wide range of bulk industrial polymer melts can be represented by plots of η/η_{in} as a function of $\eta_{in}\dot{\gamma}$. This function was found to be independent of a temperature. It was fitted to the following equation:

$$\eta = \frac{\eta_{in}}{1 + A_1(\eta_{in}\dot{\gamma})^{0.355} + A_2(\eta_{in}\dot{\gamma})^{0.72}} \qquad [2.8.4]$$

Plots of η/η_{in} versus $\eta_{in}\dot{\gamma}$ or $\theta_{in}\dot{\gamma}$ were found to strongly depend on the polydispersity of polymers.

Subsequently, it was established[90] that the generalized invariant form can only be established for the shear rate dependence of the apparent relaxation time, $\theta = N_1/\sigma_{12}$

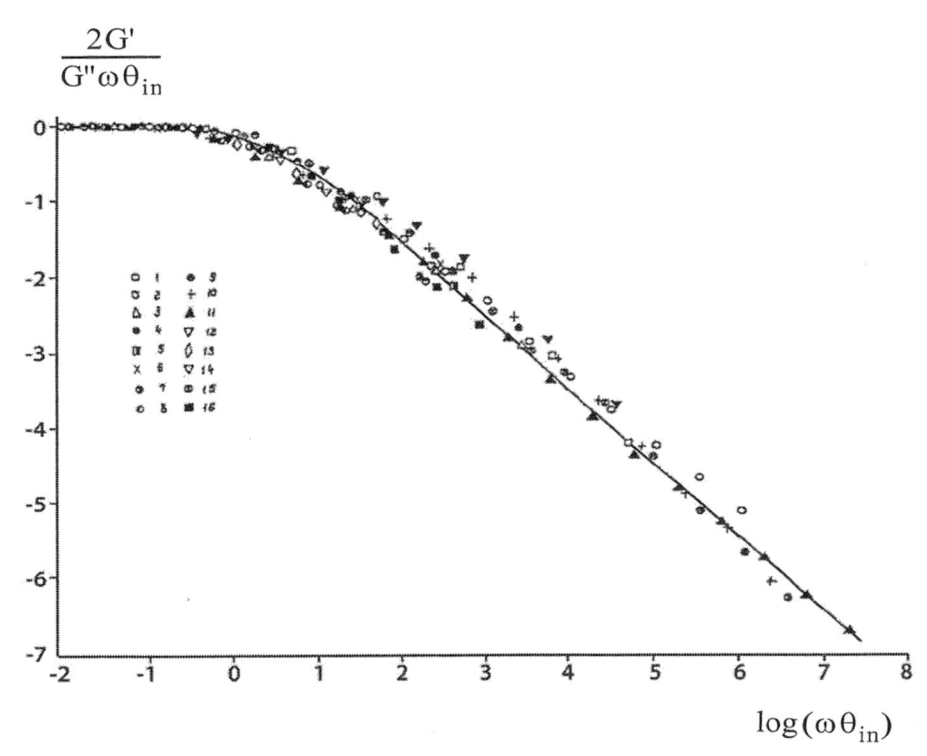

$$\frac{2G'}{G''\omega\theta_{in}}$$

$$\log(\omega\theta_{in})$$

Figure 2.8.8. Master curve for apparent relaxation time of polymer melts of different molecular weights and molecular weight distributions and their concentrated solutions (symbols and numbers next to them indicates various polymers). Initial relaxation time and materials: (1) 107 s, PIB NBS in decalin, 25°C, 20 wt%; (2) 0.55 s, PS S-111, 190°C; (3) 27.6 s, PIB in cetane, 2 wt%; (4) 27 s, PMMA, M_w=1.91x10^5, M_w/M_n=1.26; (5) 0.04 s, PMMA, M_w=5.21x10^4, M_w/M_n=1.39; (6) 3.24 s, PMMA mixture, M_w=1.43x10^5, M_w/M_n=1.23 and M_w=8.01x10^4, M_w/M_n=1.59, 220°C; (7) 0.0178 s, PIB (M_w=8.4x10^5) in mineral oil, 25°C, 2 wt%; (8) 0.0025 s, PS in Arochlor (PS-A), M_v=2.67x10^5, 25°C, 10 wt%; (9) 0.02 s, PS-A, 20 wt%; (10) 24 s, PS, M_v=5.18x10^6, 200°C; (11) 400 s, PIB, 2°C; (12) 0.316 s, PIB in cetane (PIB-C), M_w=10^6, 25°C, 8.7 wt%; (13) 0.069 s, PIB-C, 5.4 wt%; (14) 6170 s, PS, M_w=5.81x10^5, M_w/M_n=1.06, 160°C; (15) 3720 s, mixture of PS, M_w=5.81x10^5, M_w/M_n=1.06, M_w=5.87x10^4, M_w/M_n=0.94 and M_w=8.9x10^3, M_w/M_n=1.01; (16) 0.276 s, PIB in mineral oil, 25°C, 4 wt%. [Adapted, by permission, from A. I. Isayev, *J. Polym. Sci., Polym. Phys.*, **11**, 2123 (1973)].

when plotted in coordinates of θ/θ_{in} versus $\theta_{in}\dot{\gamma}$. This generalized plot for various polymer melts and their concentrated solution is shown in Fig. 2.8.7. These data were fitted to

$$\theta = \frac{N_1}{\sigma_{12}\dot{\gamma}} = \frac{\theta_{in}}{1+0.3(\theta_{in}\dot{\gamma})} \qquad [2.8.5]$$

as indicated in Fig. 2.8.7 by the solid line. Clearly, this quantitative representation collaborates with schematics shown in Fig. 2.8.6 indicating that an increase of the shear rate leads to a shift of the relaxation spectrum to lower relaxation times with this process being universal for polymer melts and their concentrated solutions. However, it should be noted that a deviation from this universal behavior was found to occur in the case of low molecular weight polymer melts and dilute polymer solutions.[88]

Such an invariant representation can also be obtained for the frequency dependence of dynamic properties of polymer melts and solutions in the linear range of their behavior. In particular, Fig. 2.8.8 shows a master curve of the ratio of relaxation times determined by a value of $2G'/(G''\omega\theta_{in})$ as a function of the product of a frequency and initial relaxation time, $\theta_{in}\omega$. The solid line in this figure corresponds to the following equation:

$$\theta = \frac{2G'}{G''\omega} = \frac{\theta_{in}}{1 + 0.3(\theta_{in}\omega)} \qquad [2.8.6]$$

It is seen that the right-hand side of this equation coincides with that of Eq. 2.8.5 at $\omega = \dot{\gamma}$. Therefore, the dependence of relaxation time in the steady-state flow at various shear rates corresponding to the nonlinear region of behavior is similar to that of the oscillatory flow at small amplitudes corresponding to the linear region of behavior.

2.8.1.5 Non-linear periodic measurements

Within the framework of the linear theory of viscoelasticity, dynamic modulus (its both components G' and G'') does not depend on the amplitude of deformations. However, if the amplitude of deformation is increased, two possible situations occur:
- the response is still harmonic but the stress/deformation ratio becomes amplitude-dependent
- harmonic deformation leads to a periodic but inharmonic response in stress and *vice versa*.

These two cases are characteristic of non-linear behavior of the material at large deformations. First of all, it is necessary to make some comments concerning the term "large" deformations. In continuum mechanics, deformations become large when they approach 1 (100%). This leads to the effects of *geometrical* non-linearity. However, the structure of the material can be destroyed at much lower deformation. This is characteristic of materials containing particles of filler. Displacement comparable with the size of solid particles may lead to the destruction of the initial structure and change the mechanical properties of the material. This is the case of *physical* non-linearity, and in this case, deformations can be very small, for example, a few percent.

If the response to harmonic deformations is also a harmonic stress evolution, then the definitions of the components of dynamic modulus are the same as in the linear theory of viscoelasticity.

However, if the response is non-harmonic, it appears that it is necessary to introduce some different definitions for the components of (complex) elastic modulus. These "new" generalized definitions are based on the physical meaning of both G' and G'' from Eq. 2.2.34. According to this equation, G'' is the measure of energy dissipation during the cycle of oscillations.

Let us come back to Fig. 2.2.2. The area of the hysteresis loop corresponds to the dissipation energy. In the non-linear domain, the hysteresis loop can also be measured, though it can be non-elliptic in shape (Fig. 2.8.9 is an example of real experimental data analogous to Fig. 2.2.3). Its surface area corresponds to energy losses and it gives the ground for calculating loss modulus, G''. So, "conventional" or apparent loss modulus is calculated from Eq. 2.2.34. The next step is a calculation of loss angle, δ, based on the

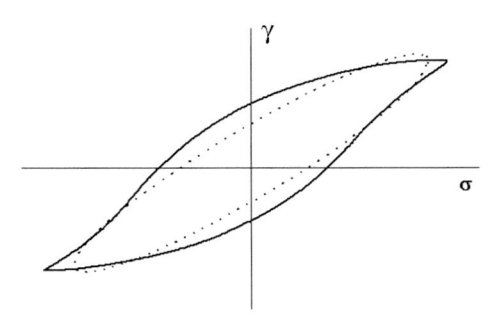

Figure 2.8.9. Hysteresis loops for a linear viscoelastic material (dots) and non-linear viscoelastic behavior (solid curve).

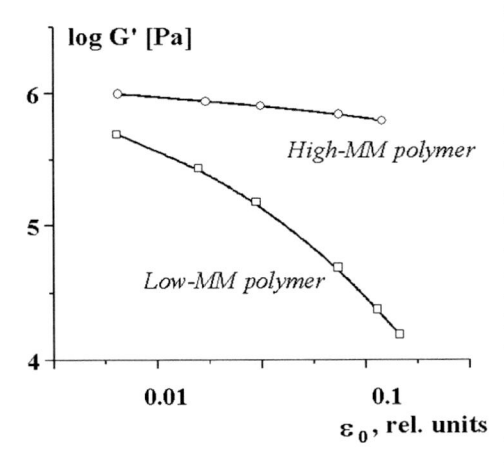

Figure 2.8.10. Amplitude dependence of the storage modulus (measured in oscillatory shear deformations) for carbon black filled polyisobutylene at frequency of $63 \, s^{-1}$; ε_0 – amplitude of deformation.

hysteresis loop using Eq. 2.2.30 or Eq. 2.2.31. Apparent storage modulus, G', is calculated by means of Eq. 2.2.24.

The above-mentioned calculations of the apparent G' and G" values in the non-linear domain are not the only method. Other ways can also be proposed. It is possible to decompose the non-harmonic wave to a series of harmonics and to take the main harmonic. Then, G' and G" are calculated in a similar way as in linear theory, but based on the main harmonic only.

It is also possible to decompose the response non-linear signal into several harmonics at different frequencies and to calculate modulus for these higher harmonics. The Fourier analysis of non-harmonic signals is well known and can also be applied.[90] In these different ways, it is possible to find G' in the range of non-linear viscoelastic behavior of material and evaluate its amplitude dependence of G' (see Fig. 2.8.10). Data are given for two polymer samples having different MM but with the same content of solid filler. Analogous curves can be easily drawn for G" as a function of the amplitude of deformations, though in the latter case the influence of the amplitude on modulus is not so extensive.

The method of large amplitude (non-linear) oscillation (so-called LAOS method) became very popular because this allows one to determine constants of non-linear rheological models in a rather simple experiment.[92-94]

The meaning of the term "large" deformations varies with polymer properties. The elasticity of the matrix in the dispersion of high-MM polymer makes the decrease of modulus a slower process, while in the case of the inelastic binder the transition to a non-linear region is rapid. In the latter case, it is difficult to establish the boundaries of linearity because even small deformations lead to non-linear effects (the decrease of apparent modulus). This is due to changes in the filler-matrix interphase even at very small deformations. This is typical behavior of a low viscosity matrix. It is difficult to measure their real linear viscoelastic properties because they easily manifest physical non-linear effects at small deformations.

The effect of non-linearity at large periodic deformations was often observed for multi-component systems such as emulsions or suspensions. It was repeatedly demon-

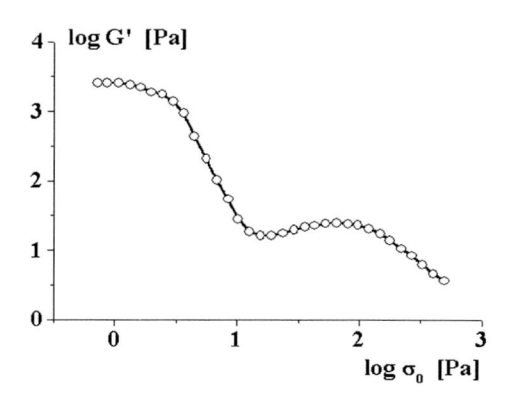

Figure 2.8.11. Amplitude dependence of the elastic modulus in oscillatory amplitude sweep test of a suspension of PMMA spheres in low-molecular-mass PDMS at a frequency of 6.28 s⁻¹. Average diameter of the particles is 5 µm and concentration is 0.45. [Adapted, by permission, from L. Heymann, S. Peukert, N. Aksel, *J. Rheol.*, **46**, 93 (2002)].

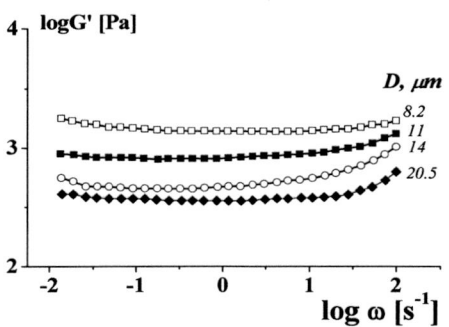

Figure 2.8.12. Frequency dependencies of the storage modulus for highly concentrated emulsions with varying droplet size at constant concentration.

strated that the increase of deformation (or stress) amplitude results in several decades of modulus drop (see Fig. 2.8.11). In the low-amplitude range, suspensions behave as Hookean solids, whereas at high shear stress amplitude they show a Newtonian fluid behavior.[95] This effect is explained by the degradation of their structure induced by high stresses.

For highly concentrated emulsions the linear-to-non-linear transition at large deformations looks rather expressive.[96] At low deformations these emulsions behave like a quasi-solid substance. The reflection of this behavior is independence of elastic modulus on frequency in a wide frequency range. An example presented in Fig. 2.8.12 demonstrates that this is true in the frequency range exceeding 4 decimal orders.

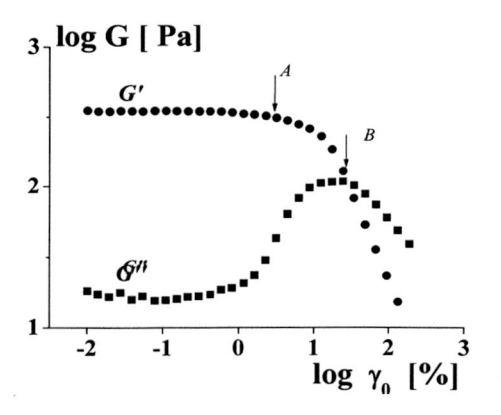

Fig. 2.8.13 Amplitude dependencies of storage and loss moduli for highly concentrated emulsion. Concentration of an internal phase 92%. Frequency 1 Hz.

However, an increase in the deformation amplitude leads to a decrease of the storage modulus and the growth of the loss modulus (Fig.2.8.13). Point A in this Figure is the limit of linearity defined by the elastic modulus, and point B is the deformation amplitude at which G' = G". At higher deformation amplitudes emulsion behaves in a liquid-like manner that is opposite to the domain of low amplitudes.

The transition from the solid-like to liquid-like behavior is explained by the transformation of the initial (related to the static state) emulsion structure. Then it is possible to see the analogy between the behavior of a solid-like medium at low deformations and properties measured at low shear rates with the similar transition observed at large deformations and large shear rates. In both cases, the linear-to-non-linear transition was

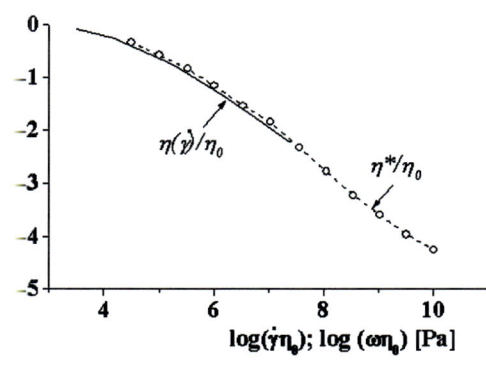

Figure 2.8.14. Correlation between dynamic viscosity and apparent viscosity for polyisobutylene (experimental data were obtained at different temperatures and reduced to 22°C). [Adapted, by permission, from G.V. Vinogradov, A.Ya. Malkin, Yu.G. Yanovsky, L.A. Dzyura, V.F. Shumsky, V.G. Kulichikhin, *Rheol. Acta*, **8**, 490 (1969)]

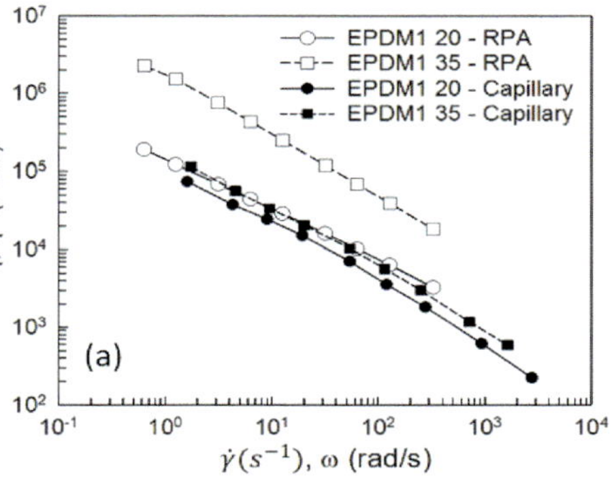

Figure 2.8.15. Apparent viscosity and complex dynamic viscosity versus shear rate and frequency, respectively, for EPDM compounds at carbon black concentrations of 20 and 35 vol%. EPDM is Vistalon 8731 (ExxonMobil), carbon black is N660.

observed. However, the quantitative criterion of the transition in both cases can be different due to the difference in the mechanics of shearing.[97]

2.8.2 LINEAR – NON-LINEAR CORRELATIONS

An interesting phenomenon was elucidated from a comparison of linear and non-linear viscoelastic properties of polymeric materials. It was found that functions $|\eta^*|(\omega)$ and $\eta(\dot{\gamma})$ practically coincide, i.e., $|\eta^*| = \eta$ if $\omega = \dot{\gamma}$. Here $|\eta^*|$ is the absolute value of the complex dynamic viscosity. The first of these functions is characteristic of linear viscoelastic properties of the material, whereas the second one is a non-linear function (apparent viscosity dependence on shear rate). This correlation is called the *Cox-Merz rule*[98] and an example of such correlation is presented in Fig. 2.8.14. The physical meaning and different aspects of the Cox-Merz rule have been discussed elsewhere.[99]

It should be noted that Cox-Merz rule applies only to pure polymer solutions and polymer melts. This rule fails for melts of particle-filled polymers, thermoplastic elastomers, thermoplastic vulcanizates, polymer blends, and concentrated suspensions.[100-103] This is indicated in Figure 2.8.15 showing the apparent viscosity vs. shear rate obtained using a capillary rheometer and the complex viscosity vs. frequency obtained using the rubber process analyzer (RPA) for EPDM rubber compounds filled with carbon black at concentrations of 20 and 35 vol%. For both compounds, the complex viscosity is significantly higher than the apparent viscosity, which is a typical trend observed for filled polymer melts. In fact, this difference increases with an increase in the concentration of carbon black. Note, deformations in the dynamic rheology tests were in the linear region, whereas deformations in the steady shear tests were in the nonlinear region. The failure of the Cox-Merz rule for the filled compounds was due to a significantly greater loss of structure compared to the unfilled rubber in the nonlinear region.[103] In the transition from linear to nonlinear deformations, the EPDM gum experienced a loss of entanglements. The filled compounds, on the other hand, experienced a loss of entanglements as well as a loss of filler-filler and rubber-filler interactions, causing a much greater loss of structure. It

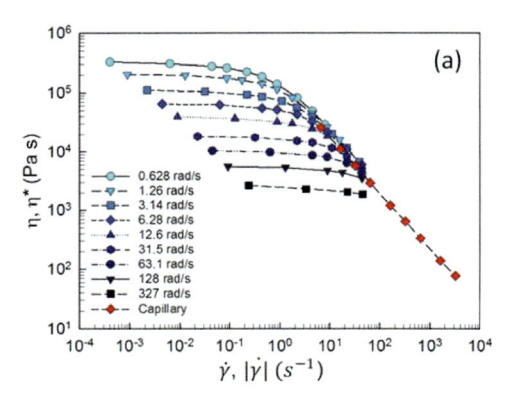

Figure 2.8.16. Apparent viscosity and complex dynamic viscosity versus the shear rate at the wall and amplitude of shear rate for EPDM compound filled with 20 vol% of carbon black. EPDM is Vistalon 8731 (ExxonMobil), carbon black is N660. [Taken from Edward Norton, MSc Thesis, University of Akron, 2019].

should be noted that recent theoretical calculations[107] supports this experimental observation.

An alternative approach for correlation of the apparent viscosity and complex dynamic viscosity was proposed for dilute polymer solutions a while ago by Philippoff.[104] Specifically, his study showed that the dependence of the out-of-phase component of shear stress amplitude as a function of the shear rate amplitude at different frequencies in large amplitude oscillatory shear (LAOS) flow coincides with the shear stress as a function of the shear rate in steady shear flow. Since then, a similar approach has been applied to concentrated suspensions.[105] It was shown that the complex dynamic viscosity as a function of the shear rate amplitude at different frequencies obtained in LAOS coincides with the apparent viscosity as a function of shear rate in a steady shear flow. Other researches have had success in correlating the steady rotational shear viscosity to the complex dynamic viscosity obtained for rubber compounds by applying LAOS.[106] Extensive experimental and theoretical studies of correlations in rheological behavior between LAOS flow and steady shear flow of filled polymer melts and rubber compounds were carried out recently.[107,108] Figure 2.8.16 shows the apparent viscosity vs. shear rate, obtained using the capillary rheometer, and the complex dynamic viscosity vs. shear rate amplitude at various frequencies, obtained using the rubber process analyzer (RPA) for EPDM compound containing 20 vol% of carbon black. The complex viscosity data was obtained in a shear rate amplitude range of 3×10^{-4} to 10^2 s^{-1}, and the steady shear viscosity data was obtained in a shear rate range of 6 to 3×10^3 s^{-1}. The complex viscosity and the steady shear viscosity coincide at all frequencies in the high shear rate amplitude region, forming an "envelope." The presence of such an envelope was also predicted by theoretical calculations reported recently for silica-filled SBR compounds,[107] and carbon nanotube-filled PP melts.[108] This established correlation is very important for rapid characterization of processability of filled polymer melts since LAOS data can be obtained much faster than steady shear data using a capillary rheometer.

An analogous correlation exists between the dependencies $2G'(\omega)/\sin\delta$ and $N_1(\dot{\gamma})$ at the same condition of comparison ($\omega = \dot{\gamma}$).[109]

The existence of a very close correlation between linear viscoelastic and non-linear viscous (apparent viscosity) and elastic (normal stresses) properties is clearly seen if one presents experimental data using dimensionless variables − tanδ and the ratio $2\sigma/N_1$ as functions of the same arguments $\omega = \dot{\gamma}$ (Fig. 2.8.17).

Another correlation was established between transient shear and normal stresses (see section 3.5.1) and shear rate dependence of apparent viscosity and coefficient of normal stresses (known as the *Gleissle mirror relations*).[110]

Based on these correlations, it appears possible to widen the experimental window beyond the limits of direct measurements.

2.8.3 RHEOLOGICAL EQUATIONS OF STATE FOR NON-LINEAR VISCOELASTIC BEHAVIOR

Non-linear viscoelastic behavior means the absence of linear proportionality between input and response, for example, between stresses and deformations. From the physical point of view, the non-linearity means:
- absence of superposition of consequent actions
- energetic exchange among different relaxation modes.

This is directly opposite to linear behavior, for which all relaxation modes are independent and the superposition of different actions is true.

One can distinguish three basic reasons for non-linearity:[111]
- *geometrical* (or *weak*) non-linearity is caused by large deformations; in this case, a relaxation spectrum is not changed
- *physical* (or *strong*) non-linearity is caused by forced-induced structure rearrangements; in this case, a relaxation spectrum is continuously changed because of an increase in deformation (deformation rates or stresses)
- *thermodynamic* (or *rupture*) non-linearity is caused by deformation-induced phase or relaxation transitions; in this case, a relaxation spectrum is changed rapidly at the point of transition.

It is noteworthy that, due to very different reasons for the non-linear behavior of real materials, it is *a priori* impossible to expect to build a universal rheological equation of state for various materials. That is why it is reasonable to emphasize here an introduction to the non-linear theory as opposed to the theory of non-linear viscoelasticity. A quantitative description of non-linear effects, i.e., the building block of the rheological equation of state (or constitutive equation), is expected to be different due to a variety of mechanisms of non-linearity, which depend on the material under deformation, its structure, and a method of sample loading. It is unlikely that a constitutive equation will be derived based on some general concepts, but the description of non-linear behavior must be based on experimental (empirical) arguments.

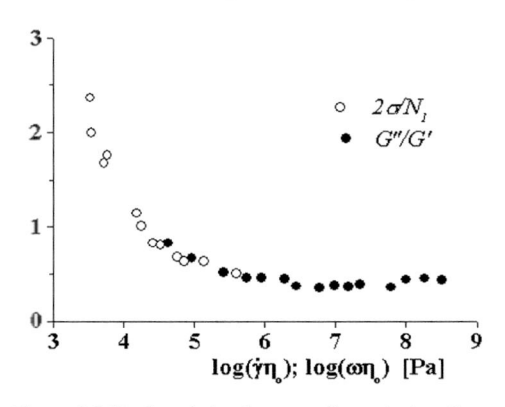

Figure 2.8.17. Correlation between dimensionless linear and non-linear characteristics of a polymer melt (polyisobutylene at 22 and 60°C reduced to 22°C).

However, several very general principles can be formulated:[112]
- stress state at time, t, is completely determined by the pre-history of deformation, i.e., it is a function of current time, t, and previous times, t'
- stress state at a point is determined by deformations in the infinitesimal surroundings of this point, but not in an entire body
- a rheological equation of state must not depend on the choice of the coordinate system

- there is no preferable configuration of the body structure, i.e., a rheological equation of state is equivalent for all points inside the material
- memory of previous deformations is fading.

A great number of liquids, for which these rules are satisfied are called *simple liquids*.

These principles must be taken into account to formulate the rheological equations of state, i.e., relationships between stresses and deformations. Different modes of deformation should give a consistent solution of rheological equations of state.[113] This establishes the limitations in the formulation of rheological equations of state but does not show a way of constructing suitable constitutive equations within the framework of these limitations. Every general invariant formulation is possible. But such an approach is too broad, and it is difficult to construct an equation that satisfies several practical rules:

- material constants and/or functions entering an equation of state should be determined using a set of simple experiments and it should be unambiguous within the framework of these experiments
- these constants (and/or functions) should be sufficient for the prediction of the behavior of the material in any arbitrary mode of deformation
- it is desirable that the structure of a constitutive equation be convenient for applied calculations (this is not a severe limitation with the availability of modern computer techniques)
- it is desirable that constants (or functions) entering a rheological equation of state are related to the molecular (structure) parameter of material; it is not necessary to limit the task by a phenomenological approach and apply a constitutive equation to a particular dynamic problem only.

In constructing rheological equations of state for viscoelastic materials, it is reasonable to pay attention to the inherent analogy between mechanical properties of rubbery solids (which cannot flow due to the existence of a network of chemical bonds) and elastic liquids (which can flow because of the limited lifetime of fluctuating entanglements – physical bonds). The nature of elasticity is the same in both cases and it is primarily the conformational rearrangement of segments of long flexible chains. It means that the same measures of large (finite) deformations should be used for a flowing liquid in formulating the phenomenological constitutive equation, as for rubbers. This is the Cauchy-Green tensor, C_{ij}, and the Finger tensor, C_{ij}^{-1}, which is inverse to the previous one, as was discussed in section 1.2.

Elastic effects in the range of large deformations are expressed *via* an elastic potential function, W, depending on the invariants of deformations: $W(I_1, I_2)$. For an incompressible material, the third invariant of the deformation tensor equals 1 (compare with Eq. 1.2.16) and it is not included as an argument for the elastic potential. The physical meaning of the function W is "energy stored in a material as a result of deformations". Chapter 4 of this book is devoted to a much more detailed discussion of this function because it determines the behavior of solid materials. In this place, only one remark is important: it is very likely that this function is the same for rubbery solids and elastic liquids. This idea is widely used in the formulation of constitutive equations for elastic materials of any type. Among many others, the following phenomenological rheological equation of state (or models) became popular in modern rheological literature.

2.8.3.1 The K-BKZ model

The following constitutive equation can be written in a general form:[114]

$$\sigma_{ij} = 2\int_{-\infty}^{t} \left(\frac{\partial \hat{W}}{\partial I_1} C_{ij}^{-1}(t, t') - \frac{\partial \hat{W}}{\partial I_2} C_{ij}(t, t') \right) dt' \qquad [2.8.7]$$

where $\hat{W}(I_2, I_2)$ is a function of invariants I_1 and I_2 of the tensors C_{ij} and C_{ij}^{-1}.

The elastic potential is assumed to be time-dependent not only *via* time dependency of deformations but additionally through the memory function:

$$\hat{W}(I_1, I_2) = m(t - t')W(I_1, I_2) \qquad [2.8.8]$$

where the memory function $m(t - t')$ reflects the concept of fading (decreasing) influence of the previous deformation; $W(I_1, I_2)$ is an elastic potential function which is supposed to be the same for elastic deformations of a material of any kind (i.e., liquids and rubbers).

Combining Eqs. 2.8.7 and 2.8.8 leads to a *factorable K-BKZ model,* where memory and finite deformation effects are separated. The structure of this equation supposes that the relaxation spectrum (represented by the memory function) is determined in the linear viscoelastic domain and it is unaffected by deformations. The non-linearity appears as a result of large deformations.

It was shown that the K-BKZ model correctly describes many special rheological effects in various modes of deformation.[115] It was noticed that "in the choice of $W(I_1, I_2)$ lies the art of fitting rheological data".[116] For adequate fitting of real experimental results complicated forms of the function $W(I_1, I_2)$ are needed. Simple potential functions, for example, the Mooney-Rivlin function (cf. Chapter 4), are not satisfactory in this case.

For fitting results of experiments carried out not only in shear but also in elongation deformation, the following form of the potential function was used:[117]

$$W = \frac{3}{2\xi} \ln\left[1 + \frac{(I - 3)\xi}{3} \right] \qquad [2.8.9]$$

where:

$$\xi = C_0 + C_2 \arctan\left[\frac{C_1(I_2 - I_1)^3}{1 + (I_2 - I_1)^2} \right]$$

$$I = 2 + (1 - \alpha)(I_1 - 3) + [1 + 2\alpha(I_2 - 3)]^{1/2}$$

and C_0, C_1, C_2, and α are empirical numerical parameters.

Complicated equations are unlikely to be used in practical applications, though they present a rather general view of constitutive equations for elastic liquids.

2.8.3.2 The Wagner models

The direct analogy between elasticity of rubbers and flow of polymeric systems is the ground for several versions of the W-models.[118] In its initial form of a W-I model,[119] it was suggested that large deformations influence a relaxation spectrum and a constitutive equation of state has the following general form:

$$\sigma_{ij} = \int_{-\infty}^{t} M(t-t';I_1, I_2)C_{ij}^{-1}\,dt' \qquad\qquad [2.8.10]$$

where M is a non-linear memory function dependent on invariants of the deformation tensor. The comparison of Eq. 2.8.10 with the K-BKZ model shows that the W-I model does not contain the Cauchy-Green tensor of deformations. The next simplification of dividing the memory function into the product of a linear member and the "function of influence":

$$M(t-t';I_1, I_2) = m(t-t')h(I_1, I_2) \qquad\qquad [2.8.11]$$

where the memory function m(t - t') is determined in the range of linear viscoelasticity, and the function $h(I_1, I_2)$ is called the damping function, and the latter can be treated as the consequence of the influence of deformations on the viscoelastic behavior of the material.

The final equation of the W-I model

$$\sigma_{ij} = \int_{-\infty}^{t} m(t-t')h(I_1, I_2)C_{ij}^{-1}\,dt' \qquad\qquad [2.8.12]$$

can be considered as a very special case of a factorable K-BKZ model, where

$$\frac{\partial W}{\partial I_2} = h(I_1, I_2) \text{ and } \frac{\partial W}{\partial I_1} = 0$$

Another more general the W-model (called the W-II model) contains two measures of finite deformations:[120]

$$\sigma_{ij} = \int_{-\infty}^{t} m(t-t')h(I_1, I_2)[(1-\beta)C_{ij}(t, t') - \beta C_{ij}^{-1}(t, t')]\,dt' \qquad\qquad [2.8.13]$$

This equation can also be treated as a simplified version of the K-BKZ equation, and again the damping function, $h(I_1, I_2)$, plays a central role in the W-model.

Several experimental investigations were carried out in order to find the analytical expression for a damping function. It is an empirical function and it is preferable to choose the simplest possible form of it. Based on some experimental data, the following expression for $h(I_1, I_2)$ was found:[121]

$$h(I_1, I_2) = [1 + \alpha(I-3)]^{-1}, \text{ where } I = \beta I_1 + (1-\beta)I_2$$

The values α and β in these equations are empirical numerical parameters.

The W-models can be explained (at least on a semi-quantitative level) by some molecular model arguments.[122] This leads to a "universal" form of the calculated damping function, which is shown in Fig. 2.8.18 in comparison with the most reliable experimental data of several authors. It is seen that the "theoretical" damping function correctly represents the main peculiarities of experimental data.

This search for "universality" of viscoelastic properties of polymers reflects the fundamental idea of the analogy of behavior of any long-chain flexible polymer macromolecules. It is supposed that the presence of such a molecular structure by itself explains the main features of viscoelastic behavior, whereas details of the molecular (chemical) structure are of secondary value only. This concept can be satisfactory for qualitative explana-

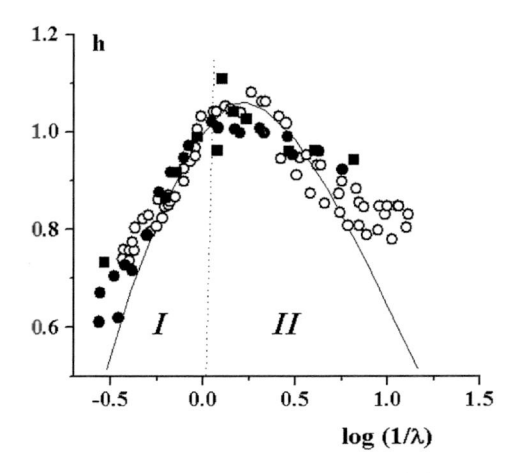

Figure 2.8.18. Experimental (various symbols correspond to data of three groups of authors obtained for different rubbers) and theoretical (line) damping factor, h. I – domain of elongation; II – domain of compressions. [Adapted, by permission, from M.H. Wagner, J. Schaefer, *J. Rheol.*, **37**, 643 (1993)]

tions, but it is rather difficult to use "universal" viscoelastic functions to compare different polymers, having differences in chemical structure important for technological applications.

2.8.3.2 The Leonov model[123]

The Leonov model belongs to the state variable theories that use irreversible thermodynamics as the basis for its development.[124] Presently, it plays an important role in rheological literature. It has been used for some realistic modeling of polymer processing. The Leonov model is based on a hypothesis that the rubbery state (where equilibrium elastic deformations have been stored) is the internal thermodynamic equilibrium state in the flow of viscoelastic fluids. Any deviation from this state causes non-equilibrium. In contrast to the KBZ and Wagner models, this model is of differential type. Its original derivation was based on irreversible thermodynamics and the classical potential function of the network theory of elasticity (see section 4.4). In recent years, other potential functions were also proposed. Irreversible thermodynamics supplies the necessary relationship between the dissipative part of the strain rate and the dissipative part of the stress. This constitutive model is derived from the thermodynamic idea that the stress in flowing polymers is related to the stored elastic energy. The theory operates with multi-relaxation modes for Maxwell fluid, each having relaxation time, θ_k, modulus, G_k, and the elastic strain tensor (the Finger measure) in each mode, C_k, that is dependent on the strain rate tensor, D, according to the following equation:

$$C_k^{\nabla} - C_k(D - D_k^p)C_k = 0 \qquad [2.8.14]$$

with

$$C_k^{\nabla} = C_k + \omega\dot{C}_k - \dot{C}_k\omega$$

being a Jaumann tensor derivative with respect to time, where ω is the vorticity tensor (see section 1.3). In Eq. 2.8.14, the quantity D_k^p is the irreversible strain rate tensor that is determined from the following equation:

$$D_k^p = \frac{1}{2G_k\theta_k}\left[\left(C_k - \frac{1}{3}I_{k,1}\delta\right)W_{k,1}^s - \left(C_k^{-1} - \frac{1}{3}I_{k,2}\delta\right)W_{k,2}^s\right] \qquad [2.8.15]$$

with

$$2W_k^s = W_k(I_{k,1}, I_{k,2}) + W_k(I_{k,2}, I_{k,1})$$

$$W_{k,j} = \frac{\partial W_k}{\partial I_{k,j}}, I_{k,1} = trC_k, I_{k,2} = trC_k^{-1} \qquad [2.8.16]$$

In Eqs. 2.8.15 and 2.8.16, $I_{k,1}$, $I_{k,2}$ are the first and second invariants of C_k, and W_k is the elastic potential. In the simplest case, W_k is taken according to the classical potential function of the network theory of elasticity (Eq. 4.4.10). In the model under discussion

$$W_k = G_k(I_{k,1} - 3)$$

The stress tensor, σ, is determined from the following equation:

$$\sigma = 2\eta_0 sD + 2\sum_{k=1}^{N}(W_{k,1}C_k + C_k^{-1}W_{k,2}) \qquad [2.8.17]$$

where η_0 is Newtonian viscosity of the fluid, s is a rheological parameter lying between 0 and 1. Models such as Eq. 2.8.17 were developed by Mooney and later used by Philipoff.[113] The elastic strain tensor is determined from the kinematics of the flow that can be shear, elongation, biaxial stretching, or other viscometric or non-viscometric flows. By solving the governing equation, Eq. 2.8.14, one can find values of the elastic strain tensor components and then various components of the stress tensor. In a simple shear flow, the stress tensor for an N-mode of the Leonov model is given by

$$\sigma = \eta_0 s\dot{\gamma}\begin{bmatrix}0 & 1 & 0\\1 & 0 & 0\\0 & 0 & 0\end{bmatrix} + 2\sum_{k=1}^{N}G_k\begin{bmatrix}C_{11,k} & C_{12,k} & 0\\C_{12,k} & C_{22,k} & 0\\0 & 0 & 1\end{bmatrix} \qquad [2.8.18]$$

The tensor components of the elastic strain tensor are governed by the following system of equations:

$$\frac{dC_{11,k}}{dt} - 2C_{12,k}\dot{\gamma} + \frac{1}{2\theta_k}(C_{11,k}^2 + C_{12,k}^2 - 1) = 0$$

$$\frac{dC_{12,k}}{dt} - 2C_{22,k}\dot{\gamma} + \frac{1}{2\theta_k}(C_{11,k} + C_{12,k})C_{12,k} = 0 \qquad [2.8.19]$$

$$C_{11,k}C_{22,k} - C_{12,k}^2 = 1$$

In this system of equations, the last equation is obtained based on the incompressibility condition that **det**C_k = 1. This system of equations can be solved for a given value of the shear rate, $\dot{\gamma}$, and then the shear stress, σ_{12}, the first, N_1, and second, N_2, differences of normal stresses can be found from Eq. 2.8.18.

The model parameters required for fluid under consideration can be obtained from the measurements of G' and G" for oscillatory shear deformations in the linear range of behavior in the wide frequency range or by measuring flow curve (shear stress vs. shear rate) in a wide range of shear rates.[125] These data can then be fitted to equations for the discrete relaxation spectra using known methods of nonlinear regression (see section

2.5.3). Then, these parameters can be utilized to describe various time-dependent rheological experiments for viscometric and non-viscometric flow[126] or applied to polymer processing operation to calculate dynamics and kinematics of the process.[127] The available scientific literature contains various examples of successful utilization of this model. The advantage of the Leonov model is that to determine model parameters it requires measurements on the viscoelastic fluid in the linear range of its behavior only. Then these parameters are used to describe the fluid nonlinear behavior.

2.8.3.4 The Marrucci models

Marrucci proposed a number of molecular models for the description of the rheological behavior of polymer melts and solutions of flexible chain and liquid crystalline polymers. For the flexible chain polymers and solutions, Marrucci used a multi-mode Maxwell model with the convected derivative for the evolution of stress tensor dependent on the first Rivlin-Erickson strain rate tensor.[128] The moduli and relaxation times of each mode of the model were assumed to be related to structural evolution parameters governed by the first-order differential equation containing an adjustable parameter. The model provides a realistic prediction of many fundamental rheological experiments in many flow situations. Predictions in step shear are not in agreement with experimental observations since the structure does not evolve from equilibrium before beginning stress relaxation. Marrucci has also made important contributions to the rheology of polymer melts by incorporating chain stretch and convective constraint release (CCR) in constitutive models.[129] In the field of liquid crystals and liquid crystalline polymers, Marrucci explained the source of the negative first normal stress difference that appears in shearing flows of nematic polymers as being due to director tumbling.[130] Also, the viscosity of liquid crystalline polymers in the nematic phase at low shear rates is known to depend on shear rates and does not approach a Newtonian plateau. In this regard, Marrucci has proposed a theory for the structure of defects in liquid crystals that can be applied to their flow.[131] His theory explains this non-linear effect as the result of large distortions of the LCP polydomain (the defect structure). The defects play a leading role in LCP behavior at low shear rates by acting as pseudo-walls for director anchoring.

2.8.4 COMMENTS – CONSTRUCTING NON-LINEAR CONSTITUTIVE EQUATIONS AND EXPERIMENT

Initial information necessary for constructing a non-linear constitutive equation is the correct description of the linear viscoelastic behavior of the material, which is the evident limit of the non-linear range. After that, it is necessary to move beyond this limit and know something about the non-linear properties of the material.

The non-linearity might be introduced in two different ways.

First, it is possible to measure the *elastic properties* of a sample and then to propose an elastic potential function. This determines the limit of the non-linear domain corresponding to stationary states of stored elastic deformations at different stresses.

Second, it is possible to measure the *flow curve*, i.e., the dependence between shear stresses and shear rates at different stationary states of flow.

Both characteristics of material correspond to the non-linear domain additive to a linear relaxation spectrum.

However, a great field of possibilities lies between these experimental borders, and different versions of constitutive equations can be written which degenerate to these boundary conditions – linear relaxation spectrum in the small deformations limit and steady states at different stresses at $t \to \infty$.

In search of the equation for this field, two general problems need to be solved:
- time-dependent (transient) behavior of the material
- three-dimensional generalization of experimental data.

The simplest way to take into account time effects is to use the factorable non-linear equations, as was discussed above. In this approach, a linear memory function is assumed to apply (without any changes) to the domain of the non-linear viscoelastic behavior. If it appears to be insufficient for fitting experimental data, a more complicated concept must be used. One of them is introducing some arbitrary dependence of a relaxation spectrum on stress or deformation rate. For example, it can be the mathematical description of the picture illustrated in Fig. 2.8.6. If one supposes that non-linearity is the consequence of some structural process, it is reasonable to introduce a kinetic equation describing this process. In this approach, the structure is characterized by a generalized parameter, X. The kinetic equation for X is written as

$$\frac{dX}{dt} = -k_-X^n + k_+X^m\sigma^\varepsilon \qquad\qquad [2.8.20]$$

where k_- and k_+ are kinetic constants, m, n, and ε are empirical parameters, and σ is stress.

It is also assumed that various rheological properties (for example, apparent viscosity) depend on the structure parameter, X. Many experimental data can be fitted on the basis of this equation by choice of the appropriate values of constants.

Three-dimensional generalization of experimental data obtained in a unidimensional experiment is achieved by introducing invariants of stress and deformation tensors instead of the components of these tensors.

The value of any approach to the generalization of initial experimental data is to be checked against independent experiments because no universal recipe exists on how to do it.

The experimental possibilities are usually restricted by limitations of use of standard or slightly modified experimental devices (see Chapter 5). According to modern practice and traditions, the following schemes of experiments are regularly used for the construction and verification of non-linear constitutive equations.

Shear
- Flow in simple shear. Shear stresses in steady-state characterize apparent viscosity; however, pre-stationary stress evolution gives additional information about the transient behavior of the material
- Shear stress relaxation; different versions of the relaxation process are possible to observe: relaxation after steady shear flow, relaxation after different moments at the pre-stationary stage of deformation

- Superposition of small-amplitude harmonic deformations on steady flow; the results of measurements characterize the changes in a relaxation spectrum induced by shearing. The direction of oscillations can be parallel or normal to the direction of flow
- Two-step deformations. In the time interval $0 - t_1$, deformation is γ_1, and then deformation increases in a jump-like manner to $\gamma_2 > \gamma_1$; multi-step changes of deformation can be also used
- Large-amplitude periodic deformations
- Up-and-down sweeping of shear rate with measurement of a hysteresis loop
- Normal stress measurements including the pre-stationary part of the transient curve, determination of the steady-state values of the coefficient of normal stresses and normal stress relaxation
- Non-linear creep and elastic recoil measurements; elastic recovery can be free or constrained by hard walls of an apparatus
- Flow through different channels, primarily with a simple cross-section (capillary flow)

Extension
- Uniaxial extension; the evolution of stresses using different modes of deformation characterizes the combination of flow and elastic properties in this geometry of deformation, which is quite different than the shear
- Two-dimensional (biaxial) extension. The rates of deformation in two normal directions can be the same (equibiaxial extension) or different

A more complicated experimental technique can also be used and sometimes is used for constructing constitutive equations and search of material functions and constants of an experimental material (sample).

In performing the experimental study of material and use of experimental data for constructing a constitutive equation of this material, it is useful to have in mind the answer to the following question: What is the final goal of this study?

When considering the simplest linear materials (Newtonian liquid and Hookean solid), two answers to these questions exist. First, the constants (viscosity and Young's modulus) are the material parameters and different materials can be compared by values of these parameters. Second, these constants are used in solving different boundary static and/or dynamic problems.

The first answer is invalid for non-linear materials, because constitutive equations of state are different for different materials, even if their structures are very similar. Therefore, it is uncertain which constants or material functions must be used for comparison. Besides, it is not necessary to know the complete complicated form of a constitutive equation: in order to compare two materials, it is enough to compare some constants measured under well-defined conditions.

However, the second answer is valid – a constitutive equation is necessary for solving different boundary problems. The former discussion shows (though the conclusion is disappointing) that no universal rheological equation of state for non-linear viscoelastic materials exists. It is *possible* to make a vast set of experiments and as a result to build a possible constitutive equation for this material. It takes a lot of time and effort. If it is an important material, widely used in different applications. However, in many cases (judging from numerous publications) it is not so, and constructing a constitutive equation appears to be the final goal of research.

In constructing rheological equations of state, it is reasonable to have in mind the goal of this operation. Actually, two approaches can be pursued. First, based on some very general principles, one may try to formulate a global concept, which describes numerous experimental facts. Such an "absolute" theory might appear to be too complicated and require a lot of independent experiments for the determination of material constants and functions entering the equation of state. As a result, such a theory appears inconvenient and in many cases, a lot of effort is spent to investigate the validity of a theory itself.

Second, it is possible to propose a rather simple model which is convenient in application. But one has to remember that such a model is limited by its origin only and can be applied within the framework of the basic assumptions used for its construction. So, in many applications, it is possible to neglect elastic effects and to solve different technological problems, e.g., in predicting transportation characteristics of the material, through tubes using the model of non-Newtonian inelastic liquid. Even the simplest power-law model of viscous properties is not the worst in this case.

REFERENCES

1   Sir Isaac Newton (1642-1727) – great English scientist and philosopher, one of the founders of modern physics and mathematics. He formulated the principal laws of mechanics, gravitation and optics, discovered many new optical effects. He is the founder (simultaneous with Leibnitz but independent from him) of differential and integral calculus. His main publication is **"Philosophiae Naturalis Principia Mathematica"** (1687), where he formulated the famous basic laws of mechanics.
2   R. Hooke (1635-1703) – outstanding English experimentalist, physicist and architect. He invented or improved different scientific devices and formulated several basic concepts of modern mechanics. However his publications are very scarce and this is the reason why his achievements are not as well known as they are worthy.
3   The phenomenon of creep was first observed and quantitatively described by W. Weber who experimented on silk threads: W. Weber, *Ann. Phys. Chem.*, **34**, 247 (1835) and **54**, 1 (1841). The complete description of this phenomenon and realization of its general importance, in particular to deformation of Earth, is attributed to W. Thomson, *Proc. Roy Soc.*, **14**, 289 (1865); **"Elasticity"** – a paper in Encyclopedia Britannica, 9th Ed. (1875).
     W. Thomson (Lord Kelvin) (1824-1907) – British physicist, one of the pioneers in the field of thermodynamics, an author of numerous studies on the theory of electricity and magnetism. He also introduced the idea of "viscosity of metals" supposing that even such typical solids as metals can have inherent friction (viscosity).
4   Relaxation came from Latin "retaxätiõ" what means "abatement of relief".
5   The idea of relaxation was introduced into modern scientific literature by Maxwell.
     J.C. Maxwell (1831-1879) – English physicist, a founder of classical electrodynamics, author of fundamental studies in the field of statistical physics, organizer of the famous Cavendish laboratory in Cambridge. He proposed the simplest model combining the effects of viscosity and elasticity, and introduced the concept of relaxation in his studies on dynamic theory of gases: J.C. Maxwell, *Phil. Trans. Roy. Soc. London*, **157**, 49 (1867); *Phil. Mag.* **35**, 129 and 185 (1868).
6   F.W. Kohlrausch, *Pogg. Ann. Physik*, **119**, 337 (1863); **128**, 1, 207 and 399 (1866); **158**, 337 (1876).
7   The first who paid attention to the effect of viscoelasticity was C.A. Coulomb, *Mém. Acad. Sci.*, 1784. He studied torsional stiffness of wires and damping in torsional oscillations. He proved that the damping in oscillations are due not to resistance of ambient air but are caused by inherent properties of material. C.A. Coulomb (1736-1806) – French engineer and physicist, one of the founders of electrostatics, invented the precise torsional balance and established the law of friction of solids.
8   Deformations in uniaxial shear and extension are usually denoted as γ and ε, respectively; shear and normal stresses are marked as σ and σ_E, respectively, moduli of shear and extension are G and E, and compliance in shear and extension are J and D, respectively. In this and the following chapters of the book both types of symbols will be used. In this chapter, the notation related to shear will be used primarily.
9   P.A.M. Dirac (1902-1984) – British physicist, one of the pioneers in the field of quantum mechanics and

quantum statistics. Nobel prize 1933.

10 See reference to Kelvin's publications. The same model was proposed in: W. Voigt, *Abh. Ges. Wiss. Göttingen*, **36**, 1(1889); *Ann. Physik*, **47**, 671 (1892).

11 See reference 5.

12 The first application of dynamic method for measurement of viscoelastic properties: R. Eisenschitz, W. Philippoff, *Naturwiss.*, **21**, 527 (1933).

13 J.M. Burgers, **First Report on Viscosity and Plasticity**, *Nordemann Publisher*, Amsterdam, 1935.

14 Current-voltage-time dependencies in electrical circuits and dielectrics were analyzed in literature and the effects of delay and slow current decay were described by formal mathematical equations independently and, possibly, earlier than the analogous mechanical phenomena. The theory of electrical circuits, though using different terms, formally is a complete analogue of the theory of viscoelasticity.

15 The ideas of linear superposition were formulated in: L. Boltzmann, *Sitz. Kgl. Akad. Wiss. Wien*, **70**, 275 (1874) and **76**, 815 (1877); *Pogg. Ann. Physik*, **7**, 624 (1876); *Wied. Ann.*, **5**, 430 (1878). L Boltzmann (1844-1906) – Austrian physicist, one of the founders of modern concepts of statistical physics and physical kinetics, author of fundamental studies on thermodynamics and optics.

16 The formulation of the superposition principle in the framework of general mathematical concept was given in: V. Volterra, **The Theory of Functional and of Integrals and Integro-differential Equations**, 1931; V. Volterra, J. Pérèz, **Théorie Générale de Function**, *Gauthier-Villars*, Paris, 1936.

17 A.Ya. Malkin, *Rheol. Acta*, **7**, 335 (1968).

18 In older publications symbol ζ is used instead of Ψ for the coefficient of normal stresses. The relationship between them is $\Psi = 2\zeta$.

19 This problem was studied for viscoelastic models of macromolecules in: M.C. Williams, *J. Chem. Phys.*, **42**, 2988 (1965); L.C. Akers, M.C. Williams, *J. Chem. Phys.*, **51**, 3834 (1969). The obtained results are of general validity for viscoelastic continuum.

20 The term "quasi-" here means that the viscoelastic behavior is linear in respect to one-dimensional deformation (simple shear) but the non-linear effect of second order, i.e., normal stresses, exists.

21 H. Kajiura, H. Endo, M. Nagasawa, *J. Polymer Sci.: Polymer Phys. Ed.*, **11**, 2371 (1971); G.V. Vinogradov, A.I. Isayev, D.A. Mustafayev, Yu.Ya. Podolsky, *J. Appl. Polym. Sci.*, **22**, 665 (1978).

22 H.C. Booij, *Rheol. Acta*, **7**, 202 (1968); A. N. Prokunin, A. I. Isayev, E. Kh. Lipkina, *Polym. Mech.*, **13**, 589 (1977); A. I. Isayev, C. M. Wong, *J. Polym. Sci., Polym. Phys.*, **26**, 2303 (1988).

23 The main ideas of the theory of viscoelasticity were developed over a century, starting from the classical publications already cited. The great interest in the theory of viscoelasticity and related practical phenomena developed with industrial production of synthetic polymers and rubbers. The beginning of modern studies in this field is connected with the publication: H. Leaderman, **Elastic and Creep Properties of Filamentous Materials and Other High Polymers,** Washington, D.C., 1943, where all concepts of the theory of viscoelasticity can be found. The complete review of the theory of linear viscoelasticity was summarized by B. Gross, **Mathematical Structure of the Theories of Viscoelasticity**, *Hermann*, Paris, 1953. He called his approach "dogmatic" because it was based on some fundamental principles and it appeared possible to present the theory in a compact and clear form. Later, several important and complete monographs devoted to this subject were published: D.R. Bland, **The Theory of Linear Viscoelasticity**, *Oxford*, 1960; R.M. Christensen, **Theory of Viscoelasticity. An Introduction**, *Acad. Press*, N.-Y., London, 1971; N.W. Tschoegl, **The Phenomenological Theory of Linear Viscoelasticity**, *Springer*, Berlin, 1989. These monographs contain the mathematical background of the theory and all necessary justifications of the statements cited in this book.

24 J.B.J. Fourier (1768-1830) – French mathematician and physicist, author of fundamental studies in analysis, algebra and the theory of heat exchange.

25 The relationships were obtained by R.L. Kronig, *J. Optic Soc. Amer.*, **12**, 547 (1926); H.A. Kramers, *Atti Cong. Dei Fizici*, p. 545 (1957). The first publication was devoted not to mechanical phenomena but, due to the analogous concept, the results reported in these publications can be applied to viscoelasticity. See also: H.C. Booij, G.P.J.M. Thoone, *Rheol. Acta*, **21**, 15 (1982).

26 This chapter discusses shear deformations. If extension or volume deformations are discussed, only the symbols should change.

27 Sometimes this value is also called *strain recoil*.

28 For more details and experimental data see section 3.4.

29 The discussion in this section is devoted to a relaxation spectrum. The same words and analogous equations can be used for a retardation spectrum.

30 P.S. Laplace (1749-1847) – French astronomer and mathematician, author of classical studies on mathematical statistics, dynamics of the solar system, theory of differential equations of mathematical

physics.
31   N.W. Tschoegl, I. Emri, *Rheol. Acta*, **32**, 322 (1993).
32   H.H. Winter, *J. Non-Newtonian Fluid Mech.*, **68**, 225 (1997).
33   A.Ya. Malkin, I. Masalova, *Rheol. Acta*, **40**, 261 (2001).
34   A lot of studies related to this approach were done and published in 50s-70s by J.D. Ferry, K. Ninomiya,
      M.L. Williams, N.W Tschoegl, F. Schwarzl, A. Staverman, H. Fujita, T.L. Smith and others. One can find a
      review of various relationships among viscoelastic functions based on kernel approximation in a
      monograph by J.D. Ferry, **Viscoelastic Properties of Polymers**, *Wiley*, New York, published in three
      editions (the last in 1980). This book summarized the state of knowledge in the field of viscoelasticity and
      was very important for the education of several generations of rheologists.
35   This approach was developed in R.K. Upadhyay, A.I. Isayev, S.F. Shen, *Rheol. Acta*, **20**, 443 (1981);
      M. Baumgartel, H. H. Winter, *Rheol. Acta*, **28**, 511 (1989); *J. Non-Newt. Fluid Mech.*, **44**, 15 (1992);
      M. Baumgartel, A. Schausberger, H.H. Winter, *Rheol. Acta*, **29**, 400 (1990); M. Baumgartel,
      M.E. De Rosa, J. Machado, M. Masse, H.H. Winter, *Rheol. Acta*, **31**, 75 (1992); H. H. Winter,
      M. Baumgartel, S. Soskey, in A.A. Collier (Ed.) **Techniques for Rheological Measurements**, *Chapman
      and Hall*, London, 1993; J Jackson, M. DeRosa, H.H. Winter, *Macromolecules*, **27**, 2426 (1994).
      The complete account of the method including the analysis of possible errors and limitations was published
      by H.H. Winter, *J. Non-Newt. Fluid Mech.*, **68**, 225 (1997). The computer program (called the IRIS-
      program) is at http://128.119.70.193/Lab/IRIS.html. Also, commercially available software, for example
      Sigmaplot^{TM} can be used.
36   A.Ya. Malkin, V.V. Kuznetsov, *Rheol. Acta*, **39**, 379 (2000).
37   This method is the direct consequence of some molecular models (section 2.6): according to any of them
      the *a priori* distribution of the relaxation modes is the direct consequence of molecular movements.
      However, this method can be applied regardless of any molecular model and the form of relaxation time
      distribution can be given arbitrarily. The latter idea was widely used in the following publications: I. Emri,
      N.W. Tschoegl, *Rheol. Acta*, **32**, 311 (1993) and **36**, 303 (1997); N.W. Tschoegl, I. Emri, *Rheol. Acta*, **32**,
      322 (1993).
38   This approach to the problem under discussion was proposed, discussed, and widely used in many
      publications: J. Honerkamp, J. Weese, *Macromolecules*, **22**, 4372 (1989); *Continuum Mech. Thermodyn.* **2**,
      17 (1990); *Rheol. Acta*, **28**, 65 (1993); C. Elster, J. Honerkamp, J. Weese, *Rheol. Acta*, **31**, 161 (1992);
      C. Elster, J. Honerkamp, *J. Rheol.*, **36**, 911 (1992); T. Roth, D. Maier, Ch. Friedrich, M. Marth,
      J. Honerkamp, *Rheol. Acta*, **39**, 163 (2000).
39   C.W. Groetsch, **The Theory of Tikhonov Regularization for Fredholm Equations of the First Kind**,
      Pitman, London, 1984.
40   The modern state of molecular-level concepts of rheological properties of individual flexible-chain
      macromolecules (in dilute solutions) is discussed in the comprehensive review: R.G. Larson, *J. Rheol.*, **49**,
      1 (2005).
41   A spring-and-bead model was first introduced by V.A. Kargin, G.L. Slonimskii, *Doklady Akad. Nauk SSSR*
      (Reports of the USSR Academy of Sciences), **62**, 239 (1948); *Zh. Fiz. Khim.* (J. Phys. Chem.), **23**, 563
      (1949). Then this model was analyzed in details in Yu.Ya. Gotlib, M.V. Vol'kenshtein, *Zh. Tech. Fiz.*
      (J. Tech. Phys.), **23**, 1963 (1953). The spring-and-dashpot model became popular after the publication:
      P. Rouse, *J. Chem. Phys.*, **21**, 1272 (1953) and is frequently called the "Rouse model". It continues to be
      widely used in different theoretical models and applications.
42   It can be proven that the set of retardation and relaxation times in the spring-and-bead model is expressed in
      a similar form, only the values of the constants are different.
43   The principal ideas of this model were proposed by J. Kirkwood and J. Riseman, the final form of the
      model was formulated in: B. Zimm, *J. Chem. Phys.*, **24**, 269 (1956); B. Zimm, G. Roe, L. Epstein, *J. Chem.
      Phys.*, **24**, 279 (1956).
44   See, e.g., N.S. Tschoegl, *J. Chem. Phys.*, **39**, 139 and 40, 473 (1963); N.S. Tschoegl, J.D. Ferry, *J. Phys.
      Chem.*, **68**, 867 (1964); A. Peterlin, *J. Polym. Sci., A-2*, **5**, 179 (1967); A.S. Lodge, Y.-J. Wu, *Rheol. Acta*,
      10, 539 (1971).
45   Among these studies the following are of special importance: J.C. Harrison, L. Lamb, A.J. Matheson,
      J. Phys. Chem., **68**, 1072 (1964); N.W. Tschoegl, J.D. Ferry, *Koll.-Z.*, **189**, 37 (1963); A review of
      numerous investigations was published primarily by J. D. Ferry, J. L. Schrag *et al.* in the beginning of the
      70th; see in J.D. Ferry, *Accounts of Chem. Res.*, **6**, 60 (1973). The last comprehensive review on this subject
      is: J. D. Ferry, *Pure and Appl. Chem.*, **50**, 299 (1978). Later the interest in this subject diminished, possibly,
      because the problem in whole became exhausted and rather clear and no new results were expected.
46   R.W. Rosser, J.L. Schrag, J.D. Ferry, *Macromolecules*, **11**, 1060 (1978).
47   F. Bueche, *J. Chem. Phys.*, **22**, 603 (1954).
48   This equivalency is well known from numerous experimental data obtained for polymer solutions and

melts and usually called the empirical Cox-Merz rule: W.P. Cox, E.H. Merz, *J. Polym. Sci.*, **28**, 619 (1958).

49 F. Bueche, *J. Chem. Phys.*, **20**, 1959 (1952); F. Bueche, **Physical Properties of Polymers**, *Interscience Publisher*, New York, 1962.

50 A.Ya. Malkin, A.V. Semakov, V.G. Kulichihkin, *Rheol. Acta*, **50**, 485 (2011); *Appl. Rheol.*, **22**, 32575 (2012).

51 J.D. Ferry, R.F. Landel, M.C. Williams, *J. Appl. Phys.*, **26**, 359 (1955).

52 A.J. Chompff, J.A. Duiser, *J. Chem. Phys.*, **45**, 1505 (1966); A.J. Chompff, W. Prince, *J Chem. Phys.*, **48**, 235 (1968).

53 G.V. Vinogradov, V.N. Pokrovsky, Yu.G. Yanovsky, *Rheol. Acta*, **11**, 258 (1972).

54 D.R. Hansen, M.C. Williams, M. Shen, *Macromolecules*, **9**, 345 (1976); S. Hayashi, *J. Phys. Soc. Japan*, **18**, 131 and 249 (1963); **19**, 101 (1964).

55 W.W. Graessley, *J. Chem. Phys.*, **54**, 5143 (1971).

56 R.S. Marvin, H. Oser, *J. Res. NBS, 66B*, **4**, 171 (1962); H. Oser, R.S. Marvin, *J. Res. NBS, 66B*, 87 (1963).

57 S.D. Hong, D.R. Hansen, M.C. Williams, M. Shen, *J. Polym. Sci., Polym. Phys. Ed.*, **15**, 1869 (1977); S.D. Hong, D. Soong, M. Chen, *J. Appl. Phys.*, **48**, 4019 (1977).

58 A.S. Lodge, *Rheol. Acta*, **7**, 379 (1968); **10**, 539 (1971).

59 S.F. Edwards, *Proc. Phys. Soc.*, **92**, 9 (1967); M. Doi, S.F. Edwards, *J. Chem. Soc., Faraday Trans.*, *II*, **74**, 1789, 1802 and 1818 (1978); **75**, 39 (1979); M. Doi, S.F. Edwards, **The Theory of Polymer Dynamics**, *Oxford University Press*, Oxford, 1986.

60 The term "reptation" was introduced by P.-G. de Gennes, *J. Chem. Phys.*, **55**, 572 (1971). The physical foundations of the theory, its main ideas, methods and principal results are set forth in the fundamental monograph: P.-G. de Gennes, **Scaling Concepts in Polymer Physics**, *Cornell University Press*, Ithaca, 1979.

61 M. Roland, L. A. Archer, P. H. Mott, J. Sanchez-Reyes, *J. Rheol.*, **48**, 395 (2004).

62 W.W. Graessley, *J. Polym. Sci., Polym. Phys. Ed.*, **18**, 27 (1980).

63 J. Des Croizeau, *Eur. Letters*, **5**, 437 (1988); *Macromol.*, **23**, 4678 (1990) and **25**, 835 (1992). This model can be very useful for calculating MMD from viscoelastic properties of polymer melts, see J.D. Guzmán, J.D. Schieber, R. Pollard, *Rheol. Acta*, **44**, 342 (2005).

64 C. Tsenoglou, *ACS Polym. Preprints*, **28**, 185 (1987); *Macromol.*, **24**, 1762 (1987).

65 G. Bishko, T.C.B. McLeish, O.G. Harlen, R.G. Larson, *Phys. Rev. Lett.*, **79**, 2352 (1997); T.C.B. McLeish, R.G. Larson, *J. Rheol.*, **42**, 81 (1998); N.J. Inkson, T.C.B. McLeish, O.G. Harlen, D.J. Groves, *J. Rheol.*, **43**, 873 (1999).

66 T.C.B. McLeish, S.T. Milner, *Adv. Polym. Sci.*, **143**, 195 (1999); R.J. Blackwell, T.C.B. McLeish, O.G. Harlen, *J. Rheol.*, **44**, 121 (2000).

67 M. Doi, W.W. Graessley, E. Helfand, D.S. Pearson, *Macromol.*, **20**, 1900 (1987).

68 T.G. Fox, *J. Polym. Sci., Ser. C*, **N9**, 35 (1965). The complete review of viscosity vs. MM relationship: G.C. Berry, *Adv. Polym. Sci.*, **21**, 261 (1968).

69 P.A.M. Steeman, *Rheol. Acta*, **37**, 583 (1998).

70 P.K. Agarwal, *Macromol.*, **12**, 342 (1979).

71 The results of earlier investigations in this field can be found in: G. Eder, H. Janeschitz-Kriegl, S. Liedauer, A. Schausberger, W. Schindlauer, *J. Rheol.*, **33**, 805 (1989).

72 D. Maier, A. Eckstein, Cr. Friedrich, J. Honerkamp, *J. Rheol.*, **42**, 1153 (1998); W. Thimm, Cr. Friedrich, M. Marth, J. Honerkamp, *J. Rheol.*, **43**, 1663 (1999).

73 R.S. Anderson, D.W. Mead, J.J. Driskoll IV, *J. Non-Newton. Fluid Mech.*, **68**, 291 (1997); R.S. Anderson, D.W. Mead, *J. Non-Newton. Fluid Mech.*, **76**, 299 (1998).

74 Five different types of kernels are usually discussed in literature, some of them based on molecular model arguments and others are purely phenomenological. See a review: H.S. Wasserman, W.W. Graessley, *J. Rheol.*, **36**, 543 (1992).

75 Y. Lee, M.T. Shaw, *J. Rheol.*, **42**, 267 (1998).

76 W. Thimm, Cr. Friedrich, M. March, J. Honerkamp, *J. Rheol.*, **44**, 429 (2000). This problem was later discussed in J.M. Dealy, *J. Rheol.*, 45, 603 (2001); see also the authors' reply on the same issue: *J. Rheol.*, **45**, 604 (2001).

77 This inverse problem is, in essence, the same as will be discussed in Chapter 3, section 3.3.5, in relation to the analysis of the correspondence between the flow curve and MMD of a polymer.

78 J. Honerhamp, J. Weese, *Rheol. Acta*, **32**, 65 (1993).

79 Possibly, this experimental fact was first noticed in: A.P. Aleksandrov, Yu.S. Lazurkin, *Acta Physico-chim. USSR*, **12**, 647 (1940).

80 The method of reduced time (or frequency) scale was introduced in modern rheological literature, proven experimentally for many polymers, and became very popular due to numerous and fundamental

publications of J.D. Ferry and A.V. Tobolsky and their co-authors in 40-50th.

81 S.A. Arrhenius (1859-1927) – Swedish chemist, one of the founders of the chemical kinetics, the theory of electrolytic dissociation. An author of publications on astronomy and biology. Nobel Prize 1903.

82 M.L. Williams, *J. Phys. Chem.*, **59**, 95 (1955); M.L. Williams, R.F. Landel, J.D. Ferry, *J. Amer. Chem. Soc.*, **77**, 3701 (1955).

83 Relaxation phenomena can be detected even at very high frequencies and extremely low temperatures (down to the liquid helium temperatures) but these effects are due to some different nature not directly related to the viscoelastic behavior of a material and will not be considered here.

84 M. Baumgaertel, M.E. De Rosa, J. Machado, M. Masse, H.H. Winter, *Rheol. Acta*, **31**, 75 (1992); J. Jackson, M.E. De Rosa, H.H. Winter, *Macromolecules*, **27**, 2426 (1994); H.H. Winter, *J. Non-Newton. Fluid Mech.*, **68**, 225 (1997).

85 M. Baumgaertel, M.E. De Rosa, J. Machado, M. Masse, H.H. Winter, *Rheol. Acta*, **31**, 75 (1992); J. Jackson, M.E. De Rosa, H.H. Winter, *Macromolecules*, **27**, 2426 (1994); H.H. Winter, *J. Non-Newton. Fluid Mech.*, **68**, 225 (1997). Another version of a power-type "universal" spectrum for polymer melts was proposed in: A.Ya. Malkin, *Intern. J. Appl. Mech. Eng.*, 2005, in press.

86 This concept was first introduced in: A.I. Leonov, G.V. Vinogradov, *Doklady Akademii Nauk* (Reports of the USSR Acad. Sci. – in Russian), **155** (1964); A.I Leonov, *Prikl. Mekh. Tekhn, Fiz.* (Applied Mech. and Techn. Phys. – in Russian) No. 4 (1964).

87 F. Bueche and S. W. Harding, *J. Polym. Sci.*, **32**, 177 (1958).

88 W.W. Graessley and J.S. Prentice, *J. Polym. Sci.*, **A-2**, 6, 1887 (1968).

89 G.V. Vinogradov and A.Ya. Malkin, *J. Polym. Sci.*, **A-2**, 4, 135 (1965).

90 A.I. Isayev, *J. Polym. Sci., Polym. Phys.*, **11**, 2123 (1973).

91 B. Debbaut, H. Burhin, *J. Rheol.*, **46**, 1155 (2002); A.I. Isayev, C.A. Hieber, *J. Polym. Sci., Polym. Phys.*, **20**, 423 (1982).

92 K. Hyun, M. Wilhelm, *Macromolecules*, **42**, 411-422 (2009); K. Hyun, W. Kim, S.J. Park, M. Wilhelm, *J. Rheol.*, **57**, 1-25 (2013); Klein C., Spiess H.W., Calin A., Balan C,. Wilhelm M. *Macromolecules*, **40**, 4250 (2007).

93 R.H. Ewoldt, Clasen C., A.E. Hosoi, G.H. McKinley, *Soft Matter.*, **3**, 634 (2007); R.H. Ewoldt, A.E. Hosoi, G.H. McKinley, *J. Rheol.*, **52**, 1427 (2008); R.H. Ewoldt, P. Winter, J. Maxey, G.H. McKinley, *Rheol. Acta*, **49**, 191 (2010); R.H. Ewoldt, P. Winter, J. Maxey, G.H. McKinley, *Rheol. Acta*, **49**, 191 (2010); K. Hyun, M. Wilhelm, Ch. O. Klein, K. S. Choc, J. G, Nam, K. H. Ahn, S. J. Lee, R.H. Ewoldt, G.H. McKinley, *Prog. Polym. Sci.*, **36**, 1697 (2011).

94 S. Ilyin, V. Kulichikhin, A. Malkin, *Appl. Rheol.*, **24** 13653 (2014); S.O. Ilyin, *Polym. Sci.*, **57**, 910 (2015).

95 L. Heymann, S. Peukert, N. Aksel, *J. Rheol.*, **46**, 93 (2002). See also A.I. Isayev, Yu.G. Yanovsky, G.V. Vinogradov, L.A. Gordievsky, *J. Eng. Phys.*, **18**, 675 (1971).

96 Highly concentrated emulsions are mixtures of two immiscible fluids, in which the concentration of an internal phase exceeds the concentration of the closest packing of spherical particles. So, higher concentrations above this limit can be created by compression of droplets with the transformation of spheres into polygonal shape. Figs 2.8.12 and 2.8.13 are related to the rheology of highly concentrated emulsions [After I. Masalova, A.Ya. Malkin, *Colloid J.*, **69**, 185 (2007)].

97 R. Foudazi, I. Masalova, A.Ya. Malkin, *Appl. Rheol.*, **18**, 4, 44709 (2008).

98 W.P. Cox, E.H. Merz, *J. Polym. Sci.*, **28**, 619 (1958).

99 H.H. Winter, *Rheol Acta*, **49**, (2009) 241.

100 A.I. Isayev, Yu.G. Yanovsky, G.V. Vinogradov, L.A. Gordievsky, *J. Eng. Phys.*, **18**, 675-78 (1970).

101 T. Kitano, T. Nishimura, T. Kataoka, T. Sakai, *Rheol. Acta*, **19**, 671-3 (1980).

102 N. Nakajima, H.H. Bowerman, E.A. Collins, *J. Appl. Polym. Sci.*. **21**, 3063-75 (1977).

103 T.S.R. Al-Hadithi, H.A. Barnes, K. Walters, *Coll. Polym. Sci.*, **270**, 40-46 (1992).

104 W. Philippoff, *Trans. Soc. Rheol.*, **10**, 317-334 (1966).

105 D. Doraiswamy, A.N. Mujumdar, I. Tsao, A.N. Beris, S.C. Danforth, A.B. Metzner, *J. Rheol.*, **35**, 647-8 (1991).

106 A.M. Randall, C.G. Robertson, *J. Appl. Polym.Sci.*, **131**, 41818 (2014).

107 S. Pole, A. I. Isayev, *J. Appl. Polym. Sci.*, **138** (28), e50660 (2021).

108 S. Pole, A. I. Isayev, J. Zhong, *Intern. Polym. Process.*, **36** (3), 233-254 (2021).

109 G.V. Vinogradov, A.Ya. Malkin, Yu.G. Yanovsky, V.F. Shumsky, E.A. Dzyura, *Mekh. Polym.* (Polymer Mechanics - in Russian), 1, 164 (1969); G.V. Vinogradov, A.Ya. Malkin, Yu.G. Yanovsky, E.A. Dzyura V.F. Shumsky, V.G. Kulichikhin, *Rheol. Acta*, **8**, 490 (1969).

110 W. Gleissle, in **Rheology**, Eds G. Astarita, G. Marrucci, G. Nicolais, (Proc. 8th Intern. Congr. Rheol. Italy, Naples), *Plenum Press*, v. 2, 457 (1980).

111 A.Ya. Malkin, *Rheol. Acta*, **34**, 27 (1995).

112 W. Noll, *J. Rat. Mech. Anal.*, **4**, 3 (1955) and *Arch. Rat. Mech. Anal.*, **2**, 197 (1958); B.D. Coleman,

W. Noll, *Arch. Rat. Mech. Anal.*, **3**, 289 (1959) and **6**, 355 (1960); *Rev. Mod. Phys.*, **33**, 239 (1961).

113   A.I. Leonov, *J. Non-Newt. Fluid Mech.*, **42**, 323 (1992); Y. Kwon, A.I. Leonov, *J. Non-Newt. Fluid Mech.*, **47**, 77 (1993).

114   This concept was first proposed in: A. Kaye, Note No 134 of the College of Aeronautics, Cranford, England, 1962, and became widely discussed after the publication: B. Bernstein, E.A. Kearsley, L.J. Zapas, *J. Nat. Bur. Stand.*, **68B**, 103 (1964). The constitutive equation is called by the names of its authors: K-BKZ model. They follow earlier developments in linear elasticity by G. Green, finite elasticity by Finger, incompressible elasticity by Rivlin, and works by Mooney and Philippoff who used the Rivlin constitutive equation for elasticity for viscoelastic fluids with recoverable strain.

115   R.J. Tanner, *J. Rheol.*, **32**, 673 (1988).

116   A. Kaye, *Rheol. Acta,* **31**, 3 (1992).

117   R.G. Larson, K. Monroe, *Rheol. Acta,* **26**, 206 (1987).

118   This term will be used for different versions of constitutive equations proposed by M.H. Wagner.

119   M.H. Wagner, *Rheol. Acta*, **15**, 136 (1976).

120   M.H. Wagner, A. Demarmels, *J. Rheol.*, **34**, 943 (1990).

121   M.H. Wagner, *Rheol. Acta*, **18**, 33 (1979); A.C. Papanastasiou, L.E. Scriven, C.W. Macosko, *J. Rheol.*, **27**, 387 (1983).

122   M.H. Wagner, J. Schaeffer, *J. Rheol.*, **37**, 641 (1993); M.H. Wagner, *J. Non-Newt. Fluid Mech.*, **68**, 169 (1997). The first of these references contains detailed calculations of the damping function, based on molecular arguments.

123   A.I. Leonov, *Rheol. Acta*, **15**, 85 (1976).

124   The application of thermodynamic arguments as the basis for rheological equation of state of rheologically complex (non-linear viscoelastic) materials is one of the "hot spots" in modern rheology. A collection of papers presented at the last (3rd) International Workshop on Nonequilibrium Thermodynamics and Complex Fluids (Princeton, New Jersey, USA, August 2003) one can find in a special issue of *J. Non-Newton. Fluid Mech.*, **120**, issues 1-3 (2004).

125   R.K. Upadhyay, A.I. Isayev, S.F. Shen, *Rheol. Acta*, **20**, 443 (1981).

126   R.K. Upadhyay, A.I. Isayev and S.F. Shen, *J. Rheol.*, **27**, 155 (1983); R.K. Upadhyay and A.I. Isayev, *Rheol. Acta*, **22**, 557 (1983); R.K. Upadhyay and A.I. Isayev, *J. Rheol.*, **28**, 581 (1984); A.I. Isayev, *J. Rheol.*, **28**, 411 (1984); R K. Upadhyay and A.I. Isayev, *Rheol. Acta*, **25**, 80 (1986); A.I. Isayev and A.D. Azari, *Rubber Chem. Technol.*, **59**, 868 (1986).

127   A.I. Isayev and C.A. Hieber, *Rheol. Acta*, **19**, 168 (1980); A.I. Isayev, *Polym. Eng. Sci.*, **23**, 271 (1983); M. Sobhanie and A.I. Isayev, *Rubber Chem. Technol.*, **62**, 939 (1989); X. Guo and A.I. Isayev, *Intern. Polym. Process.*, **14**, 377 and 387 (1999); G.D. Shyu, A.I. Isayev and H.S. Lee, *Korea-Australia Rheol. J.*, **15**, 159-166 (2003).

128   D. Acierno, F.P. La Mantia, G. Marrucci and G. Titomanlio, *J. Non-Newt. Fluid Mech.*, **1**, 125 (1976); D. Acierno, F P. La Mantia, G. Marrucci, G. Rizzio and G. Titomalio, *J. Non-Newt. Fluid Mech.*, **1**, 147 (1976); D. Acierno, F.P. La Mantia, G. Marrucci and G. Titomalio, *J. Non-Newt. Fluid Mech.*, **2**, 271 (1977).

129   Ianniruberto G. and Marrucci G., *J. Rheol.*, **45**, 1305 (2001); Ianniruberto G. and Marrucci G., *J. Non-Newt. Fluid Mech.*, **102**, 383 (2002).

130   P.L. Maffettone and G. Marrucci, *J. Non-Newt. Fluid Mech.*, **38**, 273 (1991).

131   G. Marrucci, and F. Greco, *J. Non-Newt. Fluid Mech.*, **44**, 1 (1992).

QUESTIONS FOR CHAPTER 2

QUESTION 2-1

For Maxwellian liquid with a relaxation time θ, what is the residual stress (in comparison with the initial stress σ_0), if the process of stress relaxation continues for the duration of time $t = 2\theta$?

QUESTION 2-2

For a solid material with rheological properties described by the Kelvin-Voigt model, with a retardation time λ, what is the time necessary to reach 95% of its equilibrium (limiting) value?

QUESTION 2-3

Viscoelastic properties of the liquid are described by two relaxation modes: the first with modulus G_1 and a relaxation time $\theta_1 = 1$ s and the second with modulus G_2 and a relaxation time $\theta_2 = 100$ s. Describe the evolution of stress in time. How do relaxation curves look if a linear time-scale and a logarithmic stress scale are used?

Additional question

What is the value of equilibrium stress in this case?

QUESTION 2-4

Explain why the value θ_K, entering the Kohlrausch function, Eq. 2.1.6 is not a relaxation time. How do you find relaxation times for this relaxation function?

QUESTION 2-5

Analyze the evolution of deformations in the following loading history: stress σ_0 was applied at the time $t = 0$; then additional stress σ_1 was added at the time t_1 and finally at the time t^* both stresses were taken away. The material is a linear viscoelastic solid.

Additional question

What will be the final deformation at $t \to \infty$?

QUESTION 2-6

What is the shape of the frequency dependencies of the components of dynamic modulus for Maxwellian liquid?

QUESTION 2-7

An experimental relaxation curve was approximated with the sum of three exponential functions with the following parameters:

$G_1 = 2*10^3$ Pa, $\theta_1 = 100$ s; $G_2 = 10^4$ Pa, $\theta_2 = 20$ s; $G_3 = 10^5$ Pa, $\theta_3 = 6$ s.

What is the viscosity of this liquid?

QUESTION 2-8

Eq. 2.3.11 and its solution show that the Burgers model describes the behavior of a material with two relaxation times. The same behavior is represented by two parallel Maxwell elements with their relaxation times $\theta_1 = \eta_1/G_1$ and $\theta_2 = \eta_2/G_2$ where η_1 and G_1 are the viscosity and elastic modulus of the first and η_2 and G_2 of the second Maxwell element joined in parallel. Calculate the values of the constants of the Burgers model expressed via constants of the two Maxwell elements.

QUESTION 2-9

Is it possible to measure dynamic modulus using non-harmonic periodic oscillations? How this is done?

QUESTION 2-10

In measuring a relaxation curve, it is assumed that the initial deformation is set instantaneously. In fact, it is impossible, and a transient period always exists. Estimate the role of this period for a single-relaxation mode ("Maxwellian") liquid.

QUESTION 2-11

Application of the theory of large deformations to a linear viscoelastic body leads to the following equation for the time evolution of the first normal stress difference, N_1^+ (t), at a constant shear rate, $\dot{\gamma}$ = const:

$$N_1^+(t) = 2\dot{\gamma}^2 \int_0^t t'\varphi(t')dt' \quad \text{(Can you prove this equation?)}$$

Calculate the function N_1^+ (t) for an arbitrary relaxation spectrum, $G(\theta)$.

Additional question 1

Find the $N_1(t)$ dependence for stress relaxation after a sudden cessation of steady flow. Compare the rates of relaxation of shear and normal stresses.

Additional question 2

For a single-mode viscoelastic liquid with relaxation time, θ, calculate the relative residual shear and normal stresses after relaxation continuing for time 4θ.

QUESTION 2-12

Explain the procedure of transition from a discrete to a continuous relaxation spectrum (from Eq. 2.6.6 to Eqs. 2.6.7 and 2.2.8).

QUESTION 2-13

Let a small solid dead-weight of mass m be attached to a rod at its end and the rod is fixed at the other end. Some initial displacement from the equilibrium position of the weight (deforming the rod) was created by an applied force, and then the force was ceased.

Analyze the movement of the weight after the force is ceased. Is it possible to find the components of the dynamic modulus of a rod material following the movement of a weight?

Comment

A rod can be of different lengths and cross-sections. Not specifying the sizes and the geometrical form of a rod, the latter is characterized by the value of a "form-factor" k.

Answers can be found in a special section entitled Solutions.

3

LIQUIDS

3.1 NEWTONIAN AND NON-NEWTONIAN LIQUIDS. DEFINITIONS

Rheology deals with materials in liquid and solid states. However, as can be derived from its name, liquids (or flowing media) attract the main attention of rheologists. This is why the information in this chapter is the backbone of rheology.

The quantitative approach to the flow properties of liquids began with "*Theorem XXXIX*" of the classical monograph of Newton, where he introduced such terms as "*defectus lubricitatus*" and "*attritus*", which are equivalent to the modern terms *"internal friction"* and *"viscosity"*. He proposed that resistance appearing due to internal friction is proportional to the relative velocity of fluid particles. This is the fundamental hypothesis. However, he argued about the circular movement. Later, the analogous supposition was formulated by Navier[1] and then by Stokes,[2] who gave the modern form to the Newton hypothesis.

Liquid with flow properties obeying the Newton hypothesis is called *Newton-Stokes liquid*, or *Newtonian liquid*. These properties can be formulated in a standard form as:

$$\sigma = \eta\dot{\gamma} \qquad\qquad [3.1.1]$$

where σ is the shear stress, $\dot{\gamma}$ is the shear rate, and the constant coefficient of proportionality, η, is called viscosity (or shear viscosity).[3]

This equation suggests that the shear stress is proportional to the deformation rate (the shear rate, in this case), and such liquid is a *linear* rheological medium.

If the shear stress is not proportional to the shear rate, such liquid is called *non-Newtonian* and the ratio $\sigma/\dot{\gamma}$ is called *apparent viscosity*, which is not necessarily constant.

The three-dimensional formulation of the basic rheological law for Newtonian liquid is

$$\sigma_{ij} = 2\eta D_{ij} \qquad\qquad [3.1.2]$$

where σ_{ij} is the stress tensor and D_{ij} is the deformation rate tensor.

As a rule, it is assumed that Newtonian liquid is incompressible. Therefore, the spherical component of the rate of the deformation tensor is absent. The spherical component of the stress tensor (which represents the hydrostatic pressure) can exist but it is immaterial for the rheological behavior of liquid.

In a simple shear

$$\dot{\gamma} = \frac{1}{2}D_{ij} = \frac{\partial u_1}{\partial x_2}$$

and therefore Eq. 3.1.2 refers to the standard definition of Eq. 3.1.1.

The basic Eq. 3.1.2 is the *rheological equation of state* (or a *constitutive equation*) of Newtonian, incompressible liquid. It is used for formulating the dynamic equilibrium equations, known as the *Navier-Stokes equations*, describing the mechanics of such liquids in any situation. These equations are the basis for formulating and solving any applied boundary problems.

The definition of Newtonian liquid can be easily extended to the uniaxial extension. It is evident that in this case there is only one component of the stress tensor – normal stress, σ_E. This stress tensor can be decomposed into spherical and deviatoric parts in the following manner:

$$\sigma = \begin{bmatrix} \sigma_E & 0 & 0 \\ 0 & 0 & 0 \\ 0 & 0 & 0 \end{bmatrix} = \frac{\sigma_E}{3}\delta_{ij} + \frac{\sigma_E}{3}\begin{bmatrix} 2 & 0 & 0 \\ 0 & -1 & 0 \\ 0 & 0 & -1 \end{bmatrix} \tag{3.1.3}$$

where the first term is a spherical part (i.e., negative hydrostatic pressure) and the second term is a deviator of the stress tensor.

The rate of deformation tensor for a uniaxial extension of an incompressible medium is given by:

$$\dot{\gamma} = \frac{1}{2}\dot{\gamma}_{11}\begin{bmatrix} 2 & 0 & 0 \\ 0 & -1 & 0 \\ 0 & 0 & -1 \end{bmatrix} \tag{3.1.4}$$

If a general definition of a Newtonian liquid is applied and compared with the deviatoric parts of both tensors, the following equation is obtained:

$$\sigma_E = 3\eta\dot{\gamma}_{11} \tag{3.1.5}$$

The coefficient in this equation, η_E,

$$\eta_E = 3\eta \tag{3.1.6}$$

is known as the coefficient of *extensional viscosity* or the *Trouton viscosity*, and Eq. 3.1.6 is the *Trouton law*.[4]

The same calculations can be done for a two-dimensional or biaxial extension. In this case, the coefficient of biaxial extensional viscosity, η_B, equals 6:

$$\eta_B = 6\eta \tag{3.1.7}$$

These two examples (uniaxial and biaxial extension) demonstrate that the results of different experiments can be treated as the consequences of the same basic rheological equation of state establishing the general, tensor relationship between components of the stress and deformation rate tensors. Therefore the elongational viscosity appears not to be an independent constant (property) of material but only another image of the same Newtonian viscosity.

It is also useful to formulate the rheological equation of the state of Newtonian liquid in an invariant form. The liquid is defined here as a medium for which all work done by deformation dissipates. The intensity of dissipation, A, is expressed as

$$A = \sum_{i,j} \sigma_{ij} \dot{\gamma}_{ij} \qquad [3.1.8]$$

Then it is easy to show that the rheological equation of state Eq. 3.1.2 is equivalent to the following relationship:

$$A = -4\eta D_2 = -4\eta \sum_{i,j} \dot{\gamma}_{ij} \dot{\gamma}_{ij} \qquad [3.1.9]$$

where D_2 is the second invariant of the deformation rate tensor.

Eq. 3.1.9 can be considered as the invariant definition of the Newton-Stokes liquid. Simple relationships between the stresses and the deformation rates for different deformation modes, such as Eq. 3.1.1 or Eq. 3.1.5, are obtained as consequences of this definition.

The coefficient of Newtonian viscosity, η, is a single characteristic of flow properties of liquid if its rheological behavior is determined by Eq. 3.1.2 or Eq. 3.1.9. Here are examples of ranges of viscosities of some liquids:

	viscosity range in Pa*s
Gases	0.00001
Water (at 20°C)	0.001
Sulfuric acid	0.03
Lubricating oils	0.1-3
Glycerin	2
Oligomers	0.010-10
Glues, paints	1-200
Melts of thermoplastics	100-100,000
Rubber compounds	10,000-10,000,000
Bitumens	100,000-100,000,000
Melted inorganic glasses	1,000,000-100,000,000,000
Glassy liquids	> 100,000,000,000

It is apparent that the viscosity of liquids varies in a wide range of values exceeding 15 decimal orders.

If an experiment shows that apparent viscosity is not constant, then Eq. 3.1.2 becomes invalid and it is necessary to formulate some rheological equation of state for non-Newtonian liquid. In this case, it is necessary to distinguish between two cases of non-Newtonian purely viscous liquid and viscoelastic liquid. In the first case, the work of deformation dissipates completely. In the second case, a part of deformation work is stored in the form of elastic energy and returned as elastic deformation.

The rheological equation of state for a purely viscous liquid can be written in the following general tensor form:

$$\sigma_{ij} = 2\eta(D_2)\dot{\gamma}_{ij} \qquad [3.1.10]$$

or in the invariant form:

$$A = -4\eta(D_2)D_2 \qquad [3.1.11]$$

The apparent viscosity is some function of the second invariant of the deformation rate tensor. The function $\eta(D_2)$ can be found experimentally in simple flow experiments, for example in simple shear or uniaxial extension, when only one non-zero component of the deformation rate tensor exists. However, it is necessary to confirm that the experiments performed in different geometries give the same function $\eta(D_2)$. If this principal condition is not fulfilled, it means that the rheological equation of state, taken for fitting one set of experimental data, is chosen incorrectly. Thus, it should not be used for any arbitrary flow geometry.

Example
For any purely viscous liquid it is possible to prove that the Trouton law is always valid if one compares the shear viscosity, η, and the elongational viscosity, η_E, at the following condition:

$$\dot{\gamma}_{11} = \frac{1}{\sqrt{3}}\dot{\gamma}_{12}$$

In this case, D_2 from Eq. 3.1.10 is the same for both simple shear flow and elongational flow. It is thus expected that if η is a decreasing function of the shear rate, then the elongational viscosity, η_E, should also be a decreasing function of the deformation rate. The first is frequently observed for various materials, but experiments demonstrate that the second statement is not true for many such liquids. It means that a model of a purely viscous liquid is not applicable to experiments in extension, and then it follows that such a model is of no general meaning for formulating an invariant rheological equation of state.

In the discussion of non-Newtonian behavior in shear flow and, in fact, any rheological problem, there is no unique interrelation between stresses and deformation rates, but stresses are functions of both deformation rates *and* time (or deformations) simultaneously. Hence it is reasonable to distinguish between two cases:

1. If deformations continue for a sufficiently long time such that the *stationary (or steady) regimes* of flow have been reached, there is an unambiguous relationship between the shear stresses and shear rates. This relationship is called a *flow curve*.
2. If this stationary deformation regime has not yet been reached, the state of the material continuously changes with time, and it is the function of deformation pre-history. In such a *transient deformation regime*, stresses are functions of deformation rate and deformation experienced by the material occurs from the beginning of the flow.

It is usually assumed in the *non-Newtonian shear flow* that the set of stationary deformation states is reached after the material is subjected to deformations for a sufficiently long time to pass through all transient states.

Experimental results of viscosity measurement in stationary deformation regimes may be presented as the relationship between shear stress, σ, and shear rate, $\dot{\gamma}$, or the opposite, as the relationship between shear rate and shear stress. For Newtonian liquid, this dependence is always linear. The *apparent viscosity* of non-Newtonian liquid, defined as the ratio, $\sigma/\dot{\gamma}$, can be evaluated as a function of either σ or $\dot{\gamma}$. This ratio is constant for Newtonian liquid but not for non-Newtonian liquid.

Non-Newtonian behavior may appear at different ranges of stresses. Therefore, it is useful to measure the $\sigma(\dot{\gamma})$ dependence in a wide range of variables. This dependence, graphically presented in log-log coordinates, reveals peculiarities of a flow curve in different ranges of the shear stress.

If the shear stress is proportional to the shear rate (Newtonian liquid), its graphic representation on the log-log scale is a straight line equally inclined to both coordinate axes. If the angle of a line is not 45° or experimental points do not lie on a straight line, it means that the material under study is non-Newtonian liquid.

3.2 NON-NEWTONIAN SHEAR FLOW

3.2.1 NON-NEWTONIAN BEHAVIOR OF VISCOELASTIC POLYMERIC MATERIALS

Fig. 3.2.1 gives the results of viscosity measurements of solutions of monodisperse polymer in a wide concentration range. The left upper line is the representation of the flow behavior of a low-molecular-weight solvent. This behavior is expressed by a straight line inclined at 45° in log-log coordinates. It means that the solvent is a Newtonian liquid.

Adding any amount of high-molecular-weight, long-chain, flexible polymer radically changes the rheological behavior of the resultant liquid. Even 1% solution of the polymer causes shear rate dependence of the *apparent viscosity*, η, ($\eta = \sigma/\dot{\gamma}$). However, the apparent viscosity is constant at very low shear rates. This is the so-called initial, or *zero-shear-rate Newtonian viscosity* (or *limiting viscosity at zero shear rate*), η_0. Then the apparent viscosity decreases with an increase of shear rate, and at higher shear rates the viscosity reaches its minimum value, which is called the upper Newtonian viscosity (or *limiting viscosity at the infinite shear rate*), η_∞.

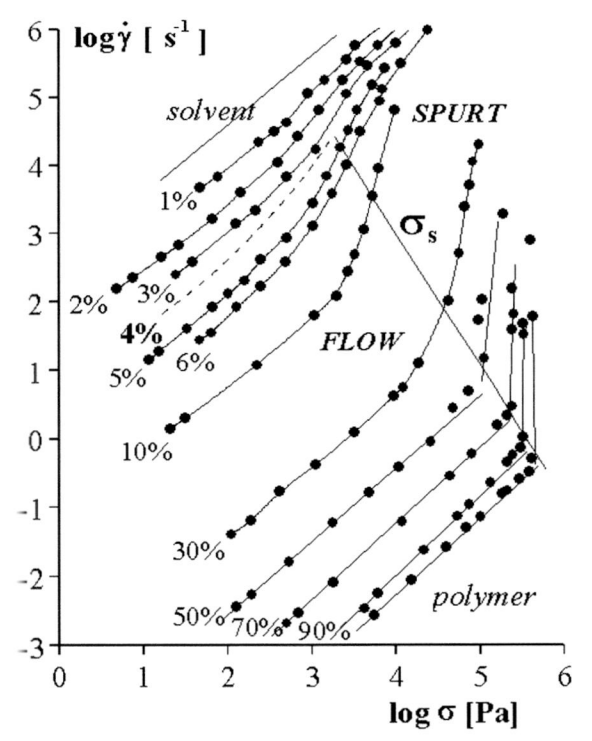

Figure 3.2.1. Flow curves of polybutadiene solutions in the full concentration range. MM = 2.4×10^5; $\overline{M}_w/\overline{M}_n$ = 1.1. Solvent: methyl naphthalene. T = 25°C. [Adapted, with permission, from G.V. Vinogradov, A.Ya. Malkin, N.K. Blinova, S.I. Sergeyenkov, M.P. Zabugina, L.V. Titkova, Yu.G. Yanovsky, V.G. Shalganova, *Europ. Polymer J.*, **9**, 1231 (1973)].

It is supposed that macro-molecules in dilute solution are independent in their molecular movements and do not have permanent contact with each other. Hence, the non-Newtonian behavior of the dilute polymer solution is explained by the deformation of the individual molecules in the stream. Macromolecules change their shape, resulting in a decrease in their resistance to flow due to streamlining at higher shear rates.

The increase of polymer concentration in solution (shift from the upper left corner of Fig. 3.2.1 downwards) results in more frequent contacts of macromolecules and the formation of a network of entanglements. Chains at the sites of contact may slide over each other and that is why the network is quasi-permanent. The contacts may have some "*lifetime*". The intermolecular contacts causing the formation of the entanglement network are believed to occur at some critical concentration, c*, depending on the length of the chain. According to Debye,[5] a criterion for "dilute" solution can be written as follows:

$$c[\eta] < 1$$

where $[\eta]$ is the intrinsic viscosity which can be treated as a measure of the size of a macromolecule.[6]

Non-Newtonian behavior becomes more and more pronounced with the increase of polymer concentration in solution. If one analyzes these graphs beginning from the lower right corner of the graph, the viscosity of the polymer is constant until some critical shear stress, σ_s, is reached. At higher stresses flow becomes impossible: attempts to increase shear stresses result in loss of material fluidity and material begins to slide along the walls of the measuring device. This effect is called *spurt*.[7] At high shear rates, the limit of flow is reached. This phenomenon is of principal importance for the flow of viscoelastic liquids, though the outward appearance of this limiting state can be very different (see Section 3.6 specially devoted to this topic). The addition of solvent to melt does not change general behavior because material can only flow until some critical shear stress is reached and then it begins to slide when shear is further increased. The critical shear stress, σ_s, decreases with an increase in the content of the solvent in the solution. This is shown by a straight line in Fig. 3.2.1, which presents the limiting region of flow and corresponds to the transition from flow to spurt behavior of the fluid. It is also worth mentioning that for a dilute solution the region of decreasing apparent viscosity appears between the lower Newtonian branch of a flow curve and the spurt depicted by the vertical line on the graph. Such a region is absent in monodisperse polymer melts.

Experimental data in Fig. 3.2.1 are for a monodisperse polymer and its solutions. However, practically all real polymers are polydisperse, i.e., contain fractions of chains of different lengths. Hence they can be treated as mixtures of different substances. A typical flow curve of a polydisperse polymer is shown in Fig. 3.2.2 for polyethylene melt. The data points were obtained using various experimental techniques (geometric flow setups). This data shows that the apparent viscosity "at a point" is a real physical effect acting regardless of the geometry of flow or other peculiarities of the experimental methods used.

Fig. 3.2.2 shows exemplary non-Newtonian behavior of melt, the apparent viscosity of which decreases more than 1000 times. At low shear rates, the apparent viscosity reaches its initial Newtonian limit, η_0. The upper Newtonian viscosity limit for concentrated solutions and melts is not reached due to the instability of the stream at high flow rates (see Section 3.6).

The non-Newtonian flow of melts of polydisperse polymers and their concentrated solutions depends on the superposition of deformation of long flexible chains and mutual sliding of macromolecules in network entanglements. In dilute solutions (interchain inter-

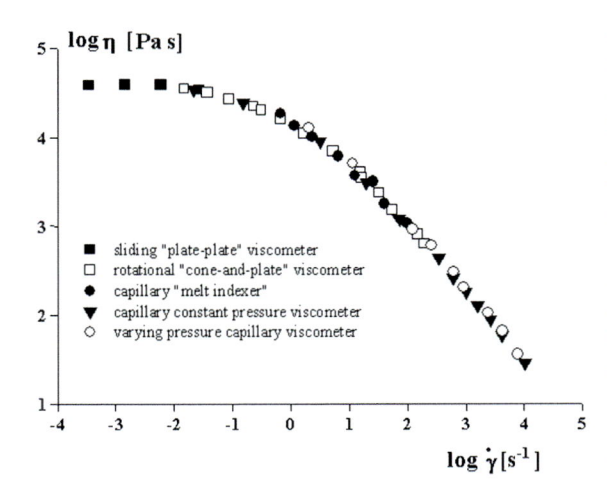

Figure 3.2.2. Flow curve obtained using different experimental techniques. LDPE with MI = 2 (ICI). 150°C. [Adapted, with permission, from G.V. Vinogradov, M.P Zabugina, A.A. Konstantinov, I.V. Konyukh, A.Ya. Malkin, N.V. Prozorovskaya, *Vysokomol. Soedin.* (Polymers - in Russian), **6**, 1646 (1964)].

actions are practically absent), the mechanism of flow is controlled by the deformation of individual macromolecular chains. The intermolecular contacts may be considered as the dominating cause of non-Newtonian behavior.

A qualitative explanation of non-Newtonian flow is based on analysis of the interrelation between the lifetime of entanglements, T_{ent}, and characteristic time of deformation, T_{def}, (the latter can be defined as the reciprocal shear rate, $T_{def} = \dot{\gamma}^{-1}$). If $T_{def} \ll T_{ent}$, the entanglements behave as stable joining points. The polymer melt behavior is analogous to the deformation of the three-dimensional network in cured rubbers (which cannot flow). Then it is possible to think that there is a distribution of the lifetimes of entanglements and at some shear rate, some junctions behave as a stable (permanent) network, but some polymer chains may slide in the case of some entanglements. The shares of quasi-stable and disintegrating junction points depend on T_{def}. It means that resistance to flow and, consequently, apparent viscosity depends on shear rate. The range of deformation rates at initial Newtonian viscosity corresponds to a very slow flow when for all T_{ent} the condition $T_{def} \ll T_{ent}$ is fulfilled. Therefore, the value of η_0 is a parameter of material directly related to its molecular structure, which is not affected by deformation. Indeed, η_0 is a measure of the molecular mass of polymer, which can be used for comparative purposes with different samples.

The concept based on the influence of shear rates (or stress) on the stability of macromolecular entanglements explains the non-Newtonian behavior of polymer systems. However, a quantitative description of non-Newtonian flow can be constructed only within the framework of viscoelastic behavior because the above-mentioned physical events are related to different time effects.

3.2.2 NON-NEWTONIAN BEHAVIOR OF STRUCTURED SYSTEMS – PLASTICITY OF LIQUIDS[8]

Fig. 3.2.3 shows non-Newtonian behavior of water suspension of bentonite. The relationship is composed of three parts:
1. a flow with very high viscosity at low shear stresses
2. an abrupt drop in the viscosity at some critical shear stress, σ_Y, (or at least in a very narrow stress range)
3. non-Newtonian flow at stresses exceeding σ_Y.

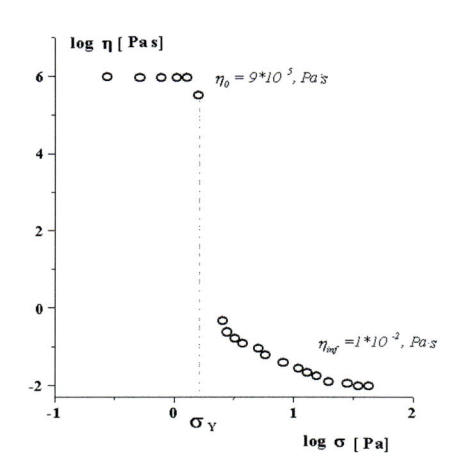

Figure 3.2.3. Complete flow curve, including the yield stress, of 5% bentonite dispersion in water. [Reconstructed from experimental data of L.A. Abduraghimova, P.A. Rehbinder, and N.N. Serb-Serbina (1955).]

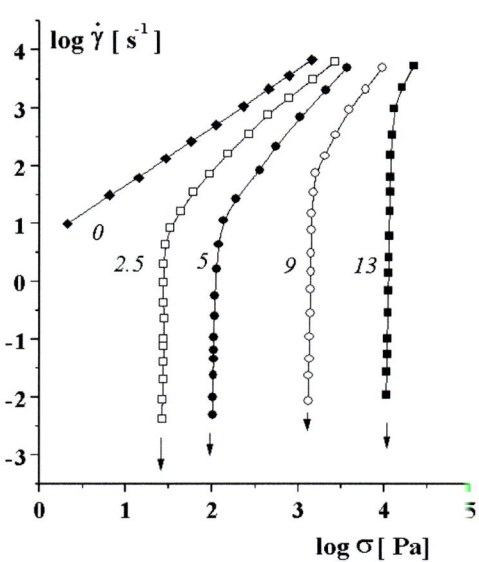

Figure 3.2.4. Flow curves of a filled low-MM polyisobutylene. Filler: active carbon black. Concentration of a filler in vol% is given on the curves. Arrows show σ_Y values.

Here, there is no gradual decrease in viscosity but a rapid drop in viscosity by several decimal orders. This is a case of a solid dispersed phase organized in a continuous three-dimensional structure (*coagulated structure*) having some *strength*. At low shear stresses, flow takes place because of the sliding of thin (possibly molecular size) layers between elements of a solid phase. The resistance to slicing is high and therefore the viscosity at low stresses is also high.

The strength of the coagulated structure is characterized by stress σ_Y called the *yield stress*. After destroying the coagulated structure of a solid phase (at stresses exceeding σ_Y) the dispersion flows like a low viscosity liquid, which does not contain residues of a solid phase. The viscosity of such a system is close (within the order of values) to the viscosity of a dispersion medium. This viscosity is several decimal orders lower than the viscosity of a system with the undestroyed coagulated structure.

This type of rheological behavior is called *plasticity*, and the behavior is called *viscoplastic flow*. Plasticity is typical of many liquid-like media, such as greases, coal suspensions, concrete mixes, crude oil, mud, different pastes (toothpaste, cosmetic pastes, shoe polish, ice cream, and many other products), biological liquids, and so on.[9]

Superposition of viscoelastic and viscoplastic effects in non-Newtonian flow is often observed if a dispersion medium is a viscoelastic liquid. This is observed in filled polymers with an active (structure forming) filler. Fig. 3.2.4 shows flow curves of polyisobutylene filled with varying amounts of carbon black. The similarity of shapes of curves in Figs. 3.2.3 and 3.2.4 is pertinent.

Fig. 3.2.5 shows rheological behaviors of two polyisobutylenes having different molecular masses (viscosities differ with molecular mass) and the concentration of carbon black. The viscosity of materials with non-destroyed structure (very low shear stresses) is

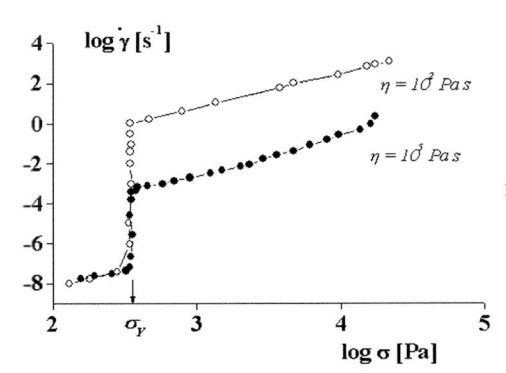

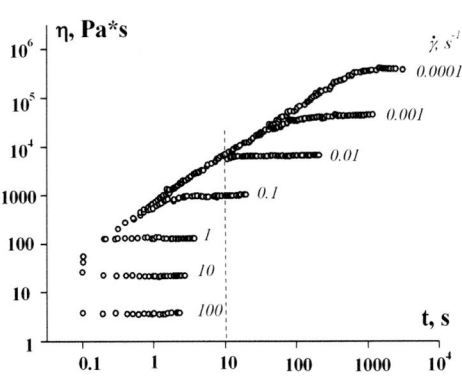

Figure 3.2.5. Complete flow curves of two filled polymers of different MM but with the same active filler concentration. Polymer: polyisobutylene. Filler: carbon black. The viscosity at $\sigma < \sigma_Y$ is 10^9-10^{10} Pa*s.

Figure 3.2.6. Observed values of apparent "viscosity" at different given shear rates for the highly concentrated emulsion.

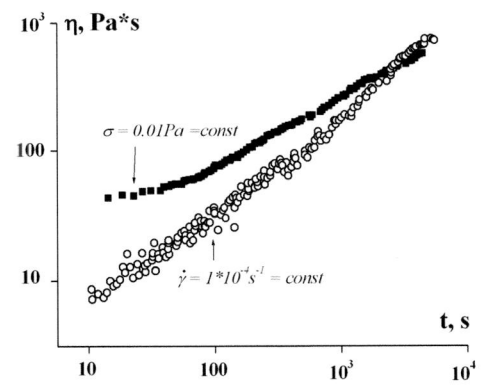

Figure 3.2.7. Time dependencies of apparent viscosity gel-forming supramolecular polyacrylonitrile solutions in dimethylsulfoxide.

the same for both systems because the flow in this stress range does not depend on the viscosity of the binder but is determined by molecular forces and interaction between solid particles. The yield stress in both cases is the same because the strength of the structure formed by carbon black does not depend on the nature of the liquid binder.

The upper Newtonian viscosity (at stresses below the yield stress), as in Fig. 3.2.3, was "found" and presented in numerous publications. But, most likely, this is an artifact and no flow is possible at $\sigma < \sigma_Y$, and the apparent existence of this viscosity is related to the incorrect treatment of the transient regime of deformation.[10]

Indeed, let us consider the results of measurements of the apparent viscosity (determined as $\sigma(t)/\dot{\gamma}$) as a function of time (Fig. 3.2.6) at different shear rates for the superconcentrated emulsion of a visco-plastic material.[11]

Then, if to take, e.g., 10 s as the base (time) for measurement of the values of the apparent viscosity for all shear rates lower than 0.01 s^{-1}, the apparent viscosity seems to be constant and one can erroneously think that it is the upper Newtonian viscosity because all experimental data will fall at the same point on the left branch as shown by the dashed line. However, the increase of the time of deformation for the lower shear rates clearly demonstrates the growth of the apparent viscosity. So, there is a typical transient regime of deformation without any steady value of the viscosity and with a sufficiently long transient period of shearing. The apparent viscosity increases without limit, approaching the

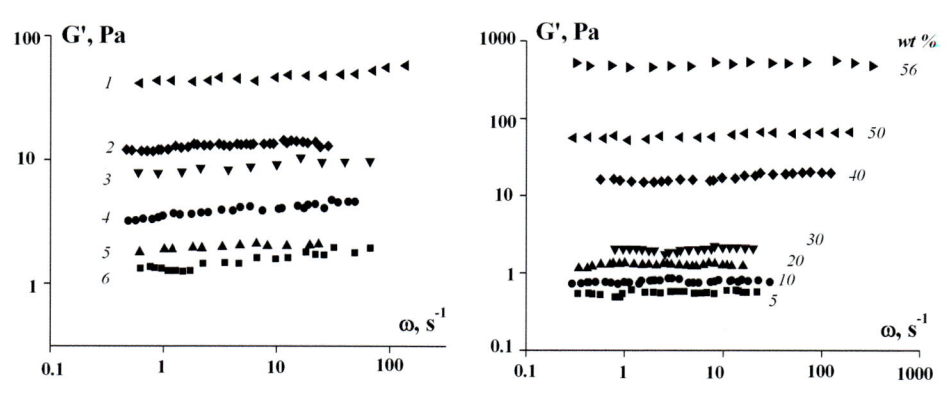

Figure 3.2.8. The storage modulus in a wide frequency range for cysteine/Ag gels with different composition of electrolyte (see details elsewhere[13]) and suspensions of Geothite with different content of a solid phase, shown at the curves. [Adapted, by permission, from Ilyin S.O., Malkin A.Ya., Kulichikhin V.G., *Colloid J.*, **74**, 492-502, (2012)].

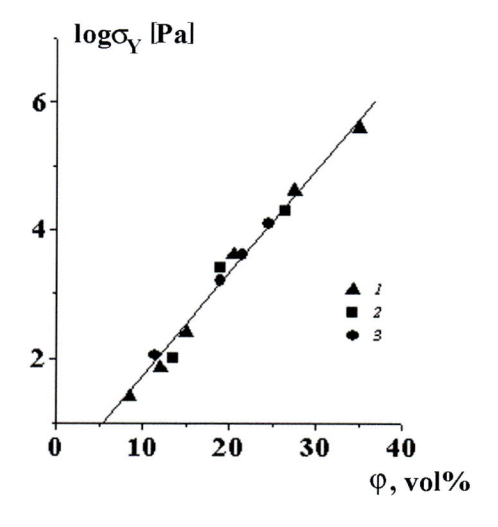

Figure 3.2.9. Concentration dependence of the yield stress for the same filler and different binders: Polybutadiene MM = $1.35*10^5$ (1); Polybutadiene MM = $1*10^4$ (2); Silicone oil - low viscosity liquid (3).

yield stress.[12] This statement is supported by the direct experimental data, presented in Fig. 3.2.7.[13]

The state of the material below the yield stress is solid-like. This is confirmed by measuring viscoelastic properties of a visco-plastic material at low-stress amplitude (in the domain of linear viscoelasticity) where deformation does not change the structure of a material. In this case, the storage modulus appears independent of frequency and the loss modulus is much lower than the storage modulus. Typical experimental data for the storage modulus measured in a wide frequency range are shown in Fig. 3.2.8. These viscoelastic properties are the characteristic behavior of solid materials and in particular of gels.

For an observer, the transition from the low shear stress ($\sigma < \sigma_Y$) to the higher stress ($\sigma > \sigma_Y$) looks like a transition from solid-like to liquid-like state or from gel to sol. At low stresses, such material is a non-flowing solid and at higher stresses, it is an "ordinary" liquid that can flow as any other liquid.

The value of the yield stress mainly depends on the strength of the structure formed by solid filler. In particular, σ_Y depends on the concentration of filler but not on the nature of a liquid binder. Fig. 3.2.9 shows the concentration dependence of σ_Y for different liquid binders containing the same filler. The concentration dependence of $\sigma_Y(\varphi)$ is close to exponential (φ is concentration of filler in vol%). However, it is not always true, because

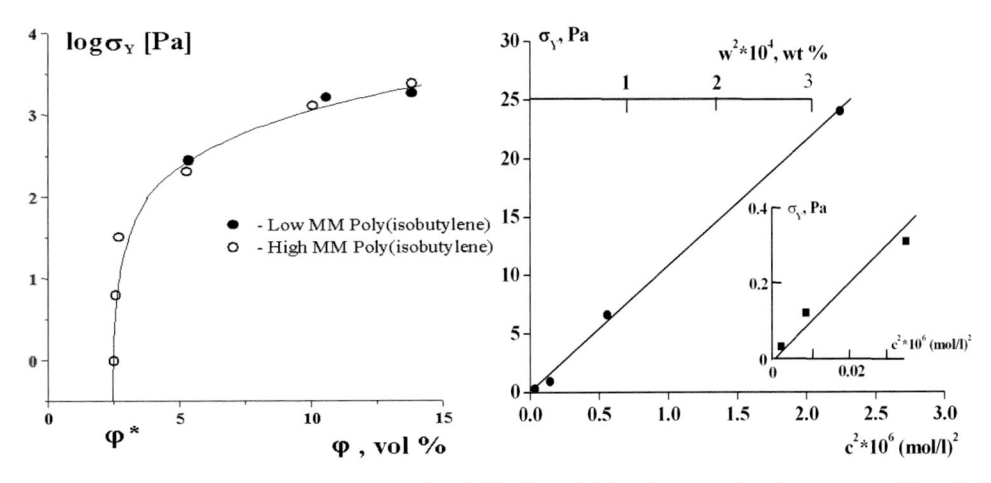

Figure 3.2.10. Concentration dependence of the yield stress showing the existence of a "critical' concentration of structure formation, $\varphi *$. Filler: active carbon black. [Adapted, by permission, from G.V. Vinogradov, A.Ya. Malkin, E.P. Plotnikova, O.Yu. Sabsai, N.E. Nikolaeva, *Intern. J. Polym. Mater.*, **2**, 1 (1972)].

Figure 3.2.11. Concentration dependence of the yield stress for gels of cysteine/Ag-based dilute colloid systems.

plasticity occurs only when the concentration of filler is sufficiently high to form a continuous structure. This is demonstrated in Fig. 3.2.10, where a linear scale instead of a logarithmic scale is used for filler concentration. Plasticity of filled systems occurs when $\varphi > \varphi *$.

The yield stress values depend on the nature of the bond-forming the structure of a material and consequently, its strength. These values can be very low for supramolecular networks as shown in Fig. 3.2.11.[15] It appears to be possible to measure the yield stress on the level of 3×10^{-3} Pa. The experimental data are presented in Fig. 3.2.9 show the yield stress of the order 10^6 Pa. So, Y can change in the range exceeding 8 orders of magnitude.

The transition from the solid-like to liquid-like behavior with increasing stress may not be necessarily catastrophic but may proceed in a wide stress range. This happens when the structure of the material is not "rigid" and there is a distribution of bond strength. In this sense, it is not always correct to consider "the yield stress" or "yield point" as a "point" on the flow curve, but this term is only used as an approximate description of the real rheological behavior of the material.

In this case, start of the flow begins not at the threshold stress but after some duration of stressing, as shown in Fig. 3.2.12.

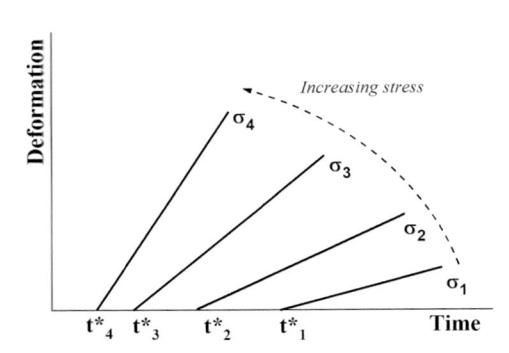

Figure 3.2.12. Time-dependent start of the flow.

The time dependence of the start of the flow characterized the durability of the structure depending of the applied stress, t*(σ). The increasing slope of the deformation vs. time lines shows at the increasing shear rate that corresponds to the increasing stress in the domain of the flow. So, the t*(σ) dependence is generalization of the concept of viscoplasticity or yielding for so-called "soft" matters.[16]

The phenomenon of durability is typical for numerous engineering applications and determines the guaranteed service life of various products, e.g. transportation tubes in gas and oil industry.

Structured systems are *non-Newtonian viscoplastic liquids*. Their viscosity below the yield stress and yielding are affected by the same factors – structure and concentration of filler. At stresses exceeding σ_Y, the viscosity of systems depends on the viscosity of the liquid medium.

Some authors consider the "structure" (though this term is being treated as a generalized but not rigorously defined phenomenon) as a unique cause of non-Newtonian behavior. The apparent viscosity, depending on the shear rate, is then called the "structural viscosity" (e.g., *strukturviscosität* in German),[17] and a relative decrease of the apparent viscosity in comparison with the initial Newtonian viscosity is treated as a measure of structure breakdown.

In reality, there is no strict separation of flow of materials into two different regimes exhibiting a low shear stress limit of Newtonian flow and viscoplastic flow with yield stress. However, it may be useful to define some limiting cases:
- yielding may develop in a wide stress range but not as an abrupt change in a flow curve. A continuous decrease in the apparent viscosity, as in Fig. 3.2.2, is sometimes called *pseudo-plasticity*
- one can always suspect that the observed low shear stress of a Newtonian branch is a pseudo-Newtonian region and it consists of part of a complete flow curve. It could be envisioned that at much lower stresses a yield point is reached and flow becomes impossible. It is possible to find real materials which are close in behavior to the described extreme cases of non-Newtonian flow, as well as materials whose behavior is intermediate and approaching the extreme cases.

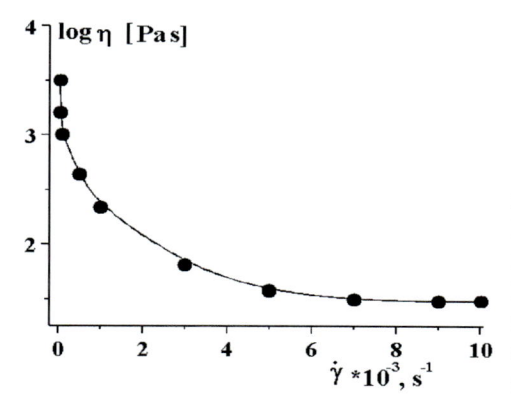

The method of presentation of experimental data ("points") can be illusive and lead to false conclusions. For example, Fig. 3.2.2 clearly demonstrates that there is a low-shear-rate Newtonian limit on the flow curve of polyethylene melt. However, if one presents the same experimental data in a semi-log scale (as is done in Fig. 3.2.13, where several points at low shear rates are taken from Fig. 3.2.2), then the relationship

Figure 3.2.13. Flow curve of polyethylene as in Fig. 3.2.2, but presented in semi-log scale. Several points at low shear rates are arbitrarily taken from Fig. 3.2.2.

looks different, showing unlimited growth of the apparent viscosity on approaching the yield point.

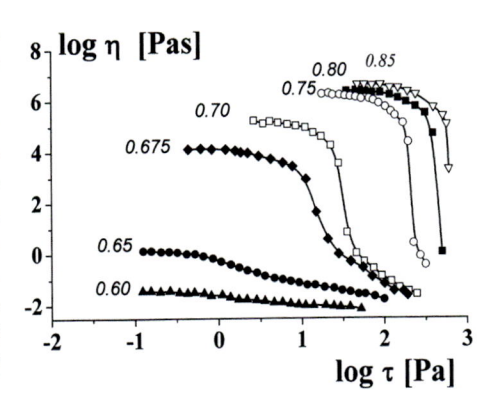

The answer to the question concerning such physical reality as the yield stress is the following: for some materials, there is a more or less wide stress range in which their apparent viscosity decreases in a sudden manner; this stress range is treated as the yield stress. This is similar to the *strength* of solid materials. The majority of engineers believe that the values of strength given in handbooks are real physical characteristics of materials. At the same time, these reference values come from measurements made under some standardized conditions. The same material may break at different stress, depending on numerous factors, including duration of stress application.

Figure 3.2.14. Flow curves of emulsions at different concentration of an inner phase (shown at the curves) [After R. Lapasin, A. Trevisan, A. Semenzato, G. Baratto, Presented at AERC, Portugal, Sept. 2003).

The difference between two extreme cases of rheological behavior is reflected in the difference of definitions between "*solution*" and "*gel*". The term "solution" is related to systems exhibiting the Newtonian branch of a flow curve at low shear stress, i.e., formally, by definition, it is a material that flows at any stress regardless of how low this stress is. At the same time, the viscosity of the Newtonian region may be very high. Gels, on the contrary, are materials which, by definition, do not flow at low stresses because of the existence of a yield point, though their viscosity at higher stresses may be low. Moreover, the term gel is often used to characterize low-modulus (very soft) rubbery materials (stable gels) with a permanent molecular or supermolecular (crystalline) network. Such materials cannot flow at all because the lifetime of bonds in these materials is infinite, whereas the lifetime of network entanglements in flowing gels is limited. Then, the yield stress in stable gels equals their macroscopic strength.

The transition between solutions and gels can be continuous and sometimes it is impossible to give a self-evident definition of a matter. An example illustrating the last statement is presented in Fig. 3.2.14 where flow curves of emulsions with different concentration of an inner phase are presented.[17]

It is seen that the flow curves of emulsions at a concentration range of 0.6-0.675 exhibits typical non-Newtonian behavior. However, the change in apparent viscosity as a function of shear stress is strongly affected by a slight increase of concentration from 0.65 to 0.675. A further increase of concentration till 0.70 creates a medium with clearly expressed yielding behavior. At concentrations of 0.75-0.85 a medium becomes gel-like which does not flow after the transition through the yield stress. So, even small changes in the concentration of an emulsion lead to radical changes in rheological behavior. The latter is explained as a result of the approach to the state of the closest packing of spherical liquid droplets and their compression after the transition through this threshold.

The term "plasticity" (and the concept of plastic behavior) is also widely used in the mechanics of solids: there are materials that can be treated as purely elastic up to the yield point and they can be deformed unlimitedly at stresses exceeding the yield stress.[18] In other words, such solids flow at high stresses. The term "flow" has a rather peculiar meaning for materials such as metals, but in reality, these materials also flow, i.e., deform irreversibly at high stresses. Plastic flow of many "solid" materials is essential in different technological operations, such as punching of silver or golden articles (coins, jewelry), rolling of steel, pulling wire through dies, and so on.

The difference between viscoplastic and solid plastic materials is hidden in their behavior at $\sigma < \sigma_Y$. In a formal approach, describing the mechanical behavior of the material, this difference is not principal. In real technological practice, the yield stress value for viscoplastic liquid is much lower than the yield stress of solid plastic materials. It should be remembered that both concepts, viscoplastic liquid and plastic solid, are no more than a model presentation of rheological properties of real technological materials.

3.2.3 VISCOSITY OF ANISOTROPIC LIQUIDS

In many cases, viscosity is treated as a scalar value and it means that the viscosity does not depend on the flow direction (or the direction of its measurement). Meanwhile, there are liquid substances that have their inherent or stress-induced structure, and not all directions in such liquids are equivalent They are known as *anisotropic liquids*, which are analogous to anisotropic solids (see Eq. 4.3.1 with the change of deformation for the rate of deformation and tensor of elastic modulus for the tensor of viscosity). It means that the simplest Newtonian approximation does not have a general meaning. In a general case, viscosity must be considered as a parameter of a tensor nature.

The quantitative approach to the description of anisotropic viscosities was first formulated by Miesowicz.[19] He proposed to distinguish three coefficients of viscosity related, correspondingly, to the direction of flow, the direction of the velocity gradient, and direction perpendicular to both these directions. Actually, his three coefficients are components of the complete viscosity tensor.

There are two main causes of anisotropy. First, there are liquid crystals,[20] which initially have a structure. Their viscosity depends on the direction of measurement in relation to the orientation of crystallographic axes. Indeed there are a lot of investigations of this phenomenon frequently treated in terms of the three Miesowicz coefficients.[21] The main body of investigations in this field is devoted to viscosity anisotropy in nematic liquid crystals.[22]

Further development of the rheology of anisotropic nematic crystals was proposed by Leslie,[23] Ericksen,[24] and Parodi.[25] They introduced a conception of the *director*, which is a vector characterizing the average molecular orientation. The degree of order in structure orientation is described by the average deviation of long molecular axes from the director. The existence of a specified direction results in the anisotropy of all physical properties of a liquid including its viscosity and viscoelastic properties.

Leslie considers five independent viscosity values, which characterize the behavior of the director (see also reviews).[26] There are connections between three Miesowicz coefficients and Leslie coefficients of viscosity.

The second case is that of induced anisotropy. Deformations may lead to the dominating effect of orientation of structural elements in the matter. This is typical of poly-

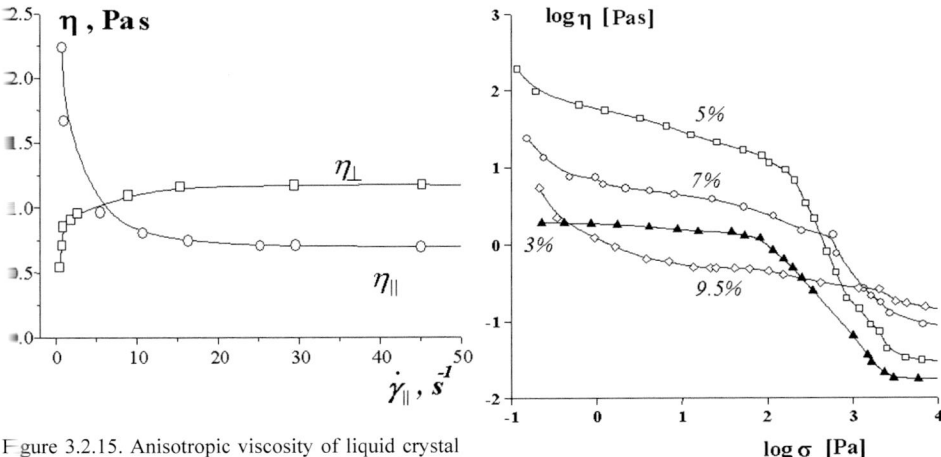

Figure 3.2.15. Anisotropic viscosity of liquid crystal solution of poly-p-benzamide in dimethyl acetamide: $\eta_\parallel$ – viscosity measured in the direction parallel to shear; $\eta_\perp$ – viscosity measured in the direction perpendicular to shear. [Adapted, by permission from A.Ya. Malkin, N.V. Vasil'eva, T.A. Belousova, V.G. Kulichikhin, *Kolloid. Zh.* (Colloid. J.— in Russian), **41**, 200 (1979)].

Figure 3.2.16. Flow curves of poly-p-benzamide solutions in dimethyl formamide. (1) – 3%; (2-4) – 5-9.5%. [Adapted, with permission, from S.P. Papkov, V.G. Kulichikhin, V.D. Kalmykova, A.Ya. Malkin, *J. Polym. Sci., Polym. Phys. Ed.*, **12**, 1753 (1974)].

meric substances having macromolecules oriented in flow. Orientation also influences the results of measurement of elongation viscosity (see Section 3.7).

Methods of measuring different coefficients of anisotropic viscosity are described in the review.[27] The fluid dynamics of nematic crystals includes the equation for orientation movement of the director in addition to all other equations of equilibrium.

First measurements of anisotropic viscosity of low molecular weight liquids were performed by Miesowicz and Tsvetkov and Mikhailov.[28]

The viscosity of anisotropic liquids depends on the direction of flow. Moreover, the effect of anisotropy depends on flow conditions. Therefore components of anisotropic viscosity are also shear rate dependent and contribute to the non-Newtonian behavior of such systems. Anisotropic viscosity of polymer liquid crystal is given in Fig. 3.2.15. For one-dimensional shear flow, the apparent viscosity was measured along with two directions: parallel to flow lines and in perpendicular direction. Non-Newtonian behavior of the liquid crystal solution manifests itself in two ways:

- the initial viscosity values (at very low shear rates) are different, due to the anisotropic structure of liquid
- a shear rate dependence is observed for both components of the viscosity and it is due to the structure rearrangement induced by shear deformation: the viscosity decreases in the parallel direction due to sliding of aligned crystal layers, and it increases in the perpendicular direction, reflecting an increase in the resistance to penetration through regular layers of macromolecules.

The non-Newtonian behavior of liquid crystals combines different mechanisms of the phenomenon (Fig. 3.2.16). A flow curve of solution at low polymer concentration (3%), below the concentration threshold of phase transition, is typical of ordinary (isotro-

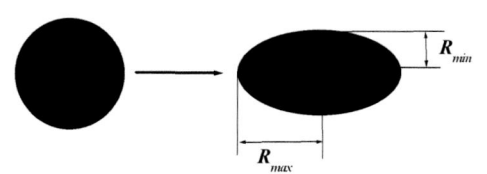

Figure 3.2.17. Transformation of a spherical particle to ellipsoid.

pic) polymer solutions. The solutions containing sufficiently high polymer fractions have an inherent structure. So, it is reasonable to expect that they might have yield stress. It is proven by an increase of apparent viscosity at very low shear stresses for solutions with a high concentration of the polymer (curves at 5 to 9.5%). In the domain of higher shear stresses, the part of the flow curves is similar to that observed in isotropic polymeric materials. In this region, the regular shear rate dependence of the apparent viscosity is observed.

The concentration dependence of viscosity of LC polymers will be discussed in more detail in Section 3.3 (see Fig. 3.3.7). This dependence convincingly demonstrates the transition from isotropic to LC state.

The shear-induced anisotropy of viscosity in initially isotropic polymer solutions was observed in Ref.[29] It was found that the velocity of falling ball in a gap between two coaxial cylinders depends on the rate of rotation of one cylinder i.e., on the shear rate in flow of 10% solution of polyisobutylene in tetralin. The transverse apparent viscosity decreased in the increase of the primary shear rate. It is reasonable to think that this is caused by the orientation of macromolecules in flow resulting in induced anisotropy of the solution.

The important case of deformation-induced anisotropy is changing shape of spherical deformable particles in the flow. This is the case of the flow of emulsions. What happens with liquid droplets in the flow of emulsions is shown in Fig. 3.2.17, Here, the emulsion becomes anisotropic.

The degree of asymmetry is characterized by a dimensionless factor D expressed as

$$D = \frac{R_{max} - R_{min}}{R_{max} + R_{min}}$$

The meanings of the values R_{max} and R_{min} are given in Fig. 3.2.17.

According to the classical calculations by Taylor,[30] the degree of anisotropy is expressed by

$$D = \frac{16 + 19\lambda}{16(\lambda + 1)}Ca$$

After cessation of deformation and return to the stationary state, the shape of the droplet comes back to the initial state. The rate of this process is described by the Maxwellian exponential law:

$$D(t) = D_0 e^{-t/\theta}$$

where D_0 is the degree of asymmetry in the deformed state and θ is the characteristic relaxation time expressed as[31]

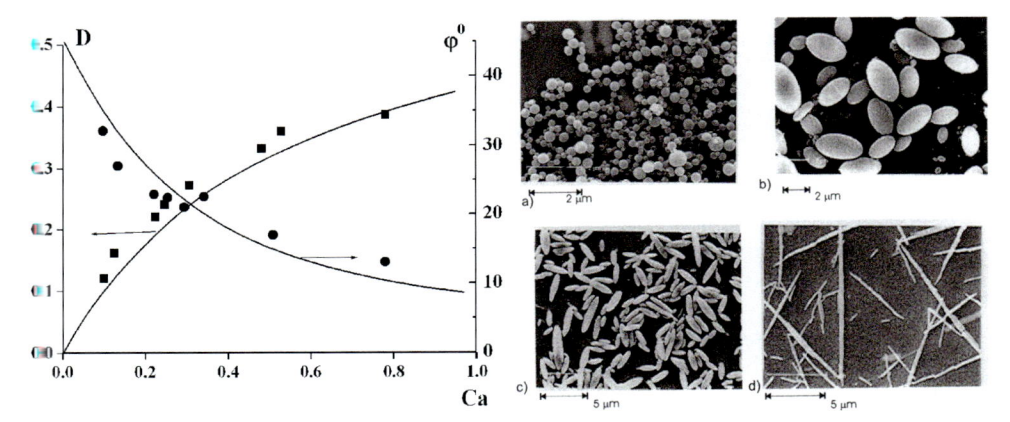

Figure 3.2.18 Comparison of the theoretical predictions for the liquid droplet deformation (D) and orientation (φ) in a viscous liquid flow (λ=3.6) with experimental data (points). [Adapted, by permission, from N.E. Jackson, C.L. Tucker III, *J. Rheol.*, **47**, 659 (2003)].

Figure 3.2.19. Deformation of PMMA droplets in PS matrix PMMA/PS ratio = 15/85. a – Initial state of non-deformed particle, b – rate of extension 0.2 s^{-1}, deformation H = 0.5; c – rate of extension 0.02 s^{-1}, deformation H = 1.5; d – rate of extension 0.02 s^{-1}, deformation H = 2.9. [Adapted, by permission, from Oosterlinck, M. Mours, H.M. Laun, P. Moldenaers, *J. Rheol.*, 49, 897 (2005)].

$$\theta = \frac{\eta_0 R}{\Gamma} \frac{(3 + 2\lambda)(16 + 19\lambda)}{40(\lambda + 1)}$$

where R is the radius of a spherical droplet, η_0 is the viscosity of a continuous medium, Γ is the surface tension, and λ is the ratio of the droplet viscosity to that of the medium.

This equation clearly demonstrates that the cause of elastic recovery is surface tension.

It is necessary to add that a droplet orients in shear flow and this effect is characterized by the angle of inclination, φ, of the principle axis in relation to the direction of flow. This phenomenon is determined by the dimensional ratio of the driving force (shear stress σ) to the resistance force (surface tension, Γ). This ratio is called the Capillary Number, Ca, and is expressed as

$$Ca = \frac{\eta_0 \dot{\gamma}}{\Gamma / R}$$

The dependencies of D and φ on the capillary number are shown in Fig. 3.2.18.

Polymer molecules change their conformation in a flow and, generally speaking, they pass from coiled to the extended shape. This also leads to the formation of the structure with different properties in elongation and transverse directions.

Melts of polymer blends can be considered as the particular case of emulsions (the deformation of a dispersed droplet of one polymer in the matrix of the other). Deformation of the blend (in extension) leads to the strong anisotropy of dispersed polymer as shown in Fig. 3.2.19 and consequently, to the blend in whole.

The strong orientation of macromolecules of the dispersed phase in a polymeric matrix may result in effects of self-reinforcing.[33]

Mechanical field-induced structure anisotropy can appear even in a dispersion of spherical particles. As was first observed,[34] initially random distribution of rigid spherical particles in a viscoelastic liquid tends to rearrange into the regular necklace structure.

The tendency to the alignment of disperse particles in flow has been confirmed by computer simulation.[35] The same effect was observed experimentally by the small-angle light scattering method.[36]

It is quite possible that this effect is related not to the inherent rheological properties of a continuous medium but to the influence of solid walls and the migration of particles because the structure formation definitely depends on the gap in flow between two parallel plates.[37]

One can soundly expect that this effect might be expressed much stronger in dispersions of anisotropic particles. Indeed, it was found that viscosity anisotropy in discontinuous fibers suspended in a viscous fluid expressed as the ratio of the axial elongational viscosity to the transverse elongational viscosity and both axial and transverse shear viscosities appears to be as high as 10^4-10^6.[38]

Anisotropy of structure leading to the viscosity anisotropy can be easily created in magnetic liquids by application of a magnetic field.[39]

The wide interest in viscosity anisotropy is also related to the problem of geodynamics because it is thought that strong anisotropy of the rheology exists in rocks and other earth substances.[40]

The combination of various possible mechanisms of non-Newtonian behavior is typical of many real materials. Specific mechanisms are developed for model systems to describe particular functions of the model.

3.3 EQUATIONS FOR VISCOSITY AND FLOW CURVES

3.3.1 INTRODUCTION – THE MEANING OF VISCOSITY MEASUREMENT

Viscosity, or in a more general case, a flow curve, is a fundamental characteristic of the mechanical properties of the liquid. Hence, it is widely used in different applications. There are two principal ways of using viscometric (as well as other rheological) data in practice:

- As a physical method to characterize material; in this case, it is necessary to choose one or several well-defined points on a flow curve or some constants in an equation used for fitting experimental data. The numerical values for these points or constants are compared with objective parameters of material (its chemical structure, contents of components in the multi-component system, concentration, molecular mass, and so on). It is a physical method of material control and a method of control of its standardized parameters relevant in technological processes. In this case, the discussion of viscometric data includes a correlation of the experimental constants with product specifications.
- As objective characteristics of mechanical properties of the material; a stress-deformation rate relationship is used as the basis for solving various dynamic problems, such as transportation of liquid through pipelines, movement of solid bodies through liquid media, and so on. Experimental points can be approximated by different fitting equations, and the choice among them is a matter of personal preference or convenience in practical calculations.

In any case, it is useful to describe a set of experimental points with an appropriate equation that best fits a flow curve. If a flow curve is utilized as a method of material characterization, the most important point is to standardize the methods of finding typical data suitable for this purpose. It could be a "zero-shear-rate" (Newtonian) viscosity, yield stress, degree of non-Newtonian behavior, apparent viscosity at the specified (strictly defined) stress, and so on. In this case, the *standardized procedure* for measuring and treatment of experimental data is developed. The flow curve does not need to be measured in a wide shear rate range.

If the main interest concentrates on solving dynamic problems, it is necessary to measure and describe the flow (viscous) properties in the shear rate range covering the range of shear rates (or stresses) used in a specific application. It is frequently dangerous to use a fitting equation beyond the limits of direct viscometric measurements. Moreover, it is necessary to confirm that the material, for which a dynamic problem is analyzed, is purely viscous, i.e., a flow curve is an adequate presentation of its rheological properties. The equation describing a flow curve is the necessary part of the formulation of any dynamic problem related to non-Newtonian liquids.

There are several general and commonly used approaches for fitting experimental data obtained in viscometric measurements. They are based on some typical models representing flow curves. As shown in Section 3.2, there are two characteristic types of flow curves dependent on the behavior of the material at low shear stresses – curves with the initial ("zero-shear-rate") Newtonian viscosity limit and curves with the yield stress limit. Neglecting intermediate situations, both types of flow curves require different approaches for their quantitative characterization.

It is possible to find dozens of analytical formulas proposed in the literature for fitting flow curves. It is reasonable to distinguish a molecular approach intended *to explain* real flow curves of materials having non-Newtonian behavior and various fitting methods useful *to describe* flow curves only. The first approach usually leads to complicated analytical expressions, which are not convenient in applications to fluid mechanics, though they could be useful in the physics of matter. An example of this approach is discussed in Section 3.3.5. In the next section, the usefulness of fitting equations is discussed for solving applied problems.

3.3.2 POWER-LAW EQUATIONS

Fitting experimental data to construct an equation for a flow curve requires that two conditions are satisfied:

- a wide range of stresses and shear rates data gives linear $\sigma(\dot{\gamma})$ dependence in log-log coordinates
- a low-stress range exists in which the apparent viscosity is constant (range of zero-shear-rate or initial Newtonian viscosity).

The four-constant Carreau-Yasuda[41,42] and the Cross[43] equations, respectively, are used to fit experimental data:

$$\eta = \eta_0[1 + (\lambda\dot{\gamma})^m]^{(n-1)/m} \qquad [3.3.1a]$$

$$\eta = \frac{\eta_0}{1 + (\eta_0\dot{\gamma}/\sigma^*)^{1-n}} \qquad [3.3.1b]$$

where η_0 is the initial Newtonian viscosity, λ is a characteristic constant with the dimension of time, σ^* is the characteristic shear stress, m and n are empirical factors.

The mathematical peculiarities of these equations include: the dependence $\eta(\dot{\gamma})$ has the limit at low shear rates equal to η_0 and at high shear rates this equation transforms to:

$$\eta = \eta_0(\lambda\dot{\gamma})^{n-1} \qquad [3.3.2a]$$

$$\eta = \eta_0(\eta_0\dot{\gamma}/\sigma^*)^{n-1} \qquad [3.3.2b]$$

Eqs. 3.3.1 satisfy the main conditions, which are necessary for fitting experimental data of non-Newtonian liquids. They can be generalized if there is a need to include the upper Newtonian viscosity, $\eta_\infty(\eta_\infty \ll \eta_0)$. This generalization is:

$$\frac{\eta - \eta_\infty}{\eta_0 - \eta_\infty} = [1 + (\lambda\dot{\gamma})^m]^{(n-1)/m} \qquad [3.3.3]$$

Simple analysis shows that at $\dot{\gamma} \to 0$, $\eta \approx \eta_0$ and at $\dot{\gamma} \to \infty$, $\eta \approx \eta_\infty$. In the intermediate shear rate range, the power-law dependence of $\sigma(\dot{\gamma})$ is predicted.

The intermediate-range of stresses and shear rates is the most important range for practical applications, where Eq. (3.3.2) is valid. This equation is usually used in a form known as the "*power law*" model or the Ostwald-De Waele equation:[44]

$$\sigma = k\dot{\gamma}^n \qquad [3.3.4]$$

where k and n are empirical constants and n < 1.

In this case, the apparent viscosity is a decreasing function of shear rate expressed as

$$\eta = \frac{\sigma}{\dot{\gamma}} = k\dot{\gamma}^{n-1} \qquad [3.3.5]$$

The value of the exponent, n, in Eqs 3.3.4 and 3.3.5 lies between 0 and 1. This provides a decrease in the apparent viscosity with an increase in the shear rate or stress.

The power law, like all other above-discussed equations, is obtained in the experiments carried out in simple shear flows. However, for fluid mechanics applications it is necessary to formulate a three-dimensional generalization of the fitting one-dimensional equation because the majority of real dynamic problems deal with two- or three-dimensional flows. This task does not appear important for Newtonian liquid because the viscosity of such liquid is its unique material constant, and from a mathematical point of view the viscosity is a scalar value.

A rheological equation for the viscosity of a non-Newtonian liquid is formulated in invariant terms. The power law for three-dimensional flows is written as:

$$\eta = k\left(\frac{D_2}{2}\right)^{\frac{n-1}{2}} \qquad [3.3.6]$$

where D_2 is the second invariant of the deformation rate tensor.

Such generalization is no more than an assumption. It is correct for various shear flows, but it is not proper for extensional flow: Eq. 3.3.6 predicts that the apparent viscosity decreases with an increase of D_2, but it is not so in extension (see Section 3.7).

3.3.3 EQUATIONS WITH YIELD STRESS

Section 3.2 shows real materials which do not flow at low shear stresses, or their flow is negligible because their viscosity is so high at low shear stress. However, at higher shear stresses, these materials flow and can be transported like any other liquid. Commercial materials of this kind include concrete and other construction products, greases, food products (yogurt, ketchup, tomato puree, ice cream, chocolate), pharmaceutical pastes, etc.

The yield stress, σ_Y, is one of their fundamental parameters. The most popular and simple equations reflecting viscoplastic behavior are as follows:

the Bingham equation[45]

$$\sigma = \sigma_Y + \eta_p \dot{\gamma} \qquad [3.3.7]$$

the Casson equation[46]

$$\sigma^{1/2} = \sigma_Y^{1/2} + (\eta_p \dot{\gamma})^{1/2} \qquad [3.3.8]$$

the Hershel-Bulkley equation[47]

$$\sigma = \sigma_Y + K\dot{\gamma}^n \qquad [3.3.9]$$

In these equations, σ_Y is the yield stress, η_p, is the "plastic" viscosity, and K and n are the experimental parameters.

A more general rheological equation with the yield stress was proposed to describe shear rate dependence of viscosity of filled elastomers:[48]

$$\sigma = \sigma_Y + \frac{A\dot{\gamma}}{1 + B\dot{\gamma}^{1-n}} \qquad [3.3.10]$$

where A and B are parameters. This equation is capable to describe rheological data for the carbon black filled NR, SBR, CR, and EPDM compounds.[48] Also, this equation was further developed to incorporate the temperature and state of cure dependence of the apparent viscosity of rubber compounds during the vulcanization process.[49]

The "plastic" viscosity η_p is not equivalent to the apparent viscosity, η. By using a standard definition: $\eta = \sigma/\dot{\gamma}$, the apparent viscosity for the Bingham viscoplastic medium is:

$$\eta = \frac{\sigma_Y + \eta_p \dot{\gamma}}{\dot{\gamma}} = \eta_p + \frac{\sigma_Y}{\dot{\gamma}} \qquad [3.3.11]$$

The apparent viscosity of the Bingham medium is a decreasing function of the shear rate ($\eta \to \infty$) at very low shear rates and it approaches the constant limit η_p at high shear rates. All such equations are valid, by definition, at stresses $\sigma > \sigma_Y$ only.

These equations predict that just above σ_Y, the apparent viscosity is high. But this does not exactly correspond to experimental data (see, e.g., Fig. 3.1.3). Hence, it is preferable to modify the Hershel-Bulkley equation in the following manner. Viscosity is assumed to be unlimitedly high at $\sigma < \sigma_Y$, and in the stress range $\sigma > \sigma_Y$, the flow curve is described by the power law, Eq. 3.3.5, without σ_Y value in formula for the apparent viscosity, but the power law equation is correct only when $\sigma > \sigma_Y$. This model can be written as:

$$\eta = \begin{cases} \infty & \text{at} \quad \sigma < \sigma_Y \\ k\dot{\gamma}^{n-1} & \text{at} \quad \sigma > \sigma_Y \end{cases} \qquad [3.3.12]$$

A three-dimensional generalization of any of the equations proposed for viscoplastic liquids is based on mechanical arguments representing a "critical" point σ_Y in the invariant form. The *Von Mises criterion of plasticity*[51] is the most popular and widely used for real materials. The state of plasticity is reached at the certain "critical" value of the second invariant of the stress tensor. This criterion can be written through the principal stresses as

$$K_{cr} = \left[\frac{(\sigma_1 - \sigma_2)^2 + (\sigma_1 - \sigma_3)^2 + (\sigma_2 - \sigma_3)^2}{6} \right]^{1/2} \qquad [3.3.13]$$

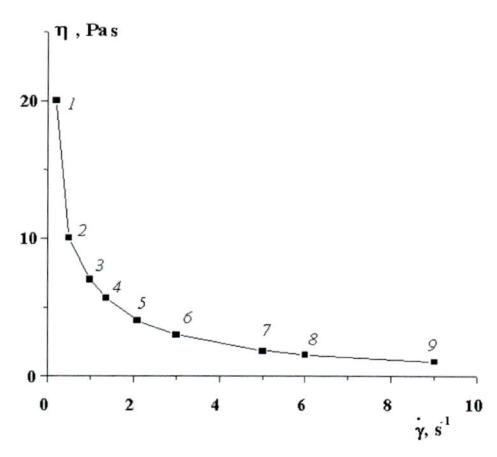

Figure 3.3.1. Illustration of the ambiguity in determination of yield stress.

where σ_1, σ_2, and σ_3 are the principal stresses, and K_{cr} is some critical value of the second invariant of the stress tensor, corresponding to the yield point. In a simple shear flow, the value K_{cr} can be easily expressed through σ_Y measures. In a simple shear flow, $\sigma_1 = -\sigma_2 = \sigma$, and then

$$K_{cr} = \frac{1}{\sqrt{3}} \sigma_Y \qquad [3.3.14]$$

The equations proposed may be applied for the approximation of the rheological properties of viscoplastic bodies in discussing three-dimensional deformation (flow) problems.

The determination of the yield stress in two (or three-) dimensional deformations is not trivial, especially for Hershel-Bulkley liquid, due to a non-linearity of viscous properties of viscoplastic liquids at $\sigma > \sigma_Y$.[52] In order to solve three-dimensional dynamic problems it is necessary to formulate an invariant critical condition of yielding and use the generalized form of this equation.[53]

As noted above, the equations for viscoplastic bodies (Eq. 3.3.7 to Eq. 3.3.11, as well as many others proposed by different authors) are based on the following assumptions:
- there are no deformations at stresses below σ_Y
- the transition from solid to liquid occurs just at one point, namely yield stress.

Both assumptions are no more than model approximations of the observed rheological behavior of real materials made by the "best" fitting of experimental points. The yield stress is found by the procedure of extrapolation of experimental points based on the specified fitting equation.

It is necessary to stress again that all equations discussed in Sections 3.3.2 and 3.3.3 are no more than approximations applicable to a part of full flow curves of real materials. As was illustrated in Section 3.2, the general picture of non-Newtonian behavior can be richer. Indeed, measuring the rheology of different media demonstrated that their non-Newtonian properties are caused by the existence of two limiting values of Newtonian viscosity and the existence of an intermediate region of quasi-power law viscosity with a more or less abrupt transition from the upper Newtonian viscosity to decreasing apparent viscosity. What an experimentalist see and how one approximates experimental data depends on the range of explored shear rates or stresses and the analytical fitting of experimental points measured with some limited accuracy.

Example
Is it reasonable to say that it is possible to *measure the real yield point*?
The following numerical example analyzes the situation. Fig. 3.3.1 gives an example of experimental data: nine marked points are measured values of the apparent viscosity at different shear rates. The flow curve indicates that the material under investigation has yield stress. Let us try to find the σ_Y value by approximation of the experimental data at the low shear rate region using linear equation (3.3.7). The set of three points will be used for linearization. First, these are points 1, 2, and 3. Then, it is supposed that only 8 points, starting from point 2 are measured, and so on. The results are as follows:

Points used for approximation	1, 2, 3	2, 3, 4	3, 4, 5	4, 5, 6
Found σ_Y value, Pa	3.5	3.8	6.2	7.12

It is seen that the σ_Y value is no more than the result of approximation strongly dependent on the range of shear rates used for measuring and fitting. Then, if σ_Y values are used as a measure of material quality and to compare different media it is necessary to use a reproducible standardized method of searching for σ_Y value.

3.3.4 BASIC DEPENDENCIES OF VISCOSITY

In general, the viscosity of materials depends on their properties. This section is limited to dependencies of viscosity on the molecular mass in a homologous series and concentrations of fillers in dispersions.

3.3.4.1 Viscosity of polymer melts

It is commonly assumed[54] that the dependence of Newtonian viscosity on the molecular mass, $\eta_0(M)$, consists of two parts: close to linear in the low molecular weight domain, expressed by a power law with the universal exponent, close to 3.4 in the high molecular weight domain. It is expressed as follows:

$$\eta_0 = \begin{cases} K_1 M^a & \text{at} \quad M < M_c \\ K_2 M^b & \text{at} \quad M > M_c \end{cases} \qquad [3.3.15]$$

where K_1, K_2 are empirical constants, the exponent a is close to 1 and $b \approx 3.4$.

The boundary-value of molecular mass, M_c, separating low and high molecular mass domains is called "critical molecular mass". In the physics of polymers, it is a boundary

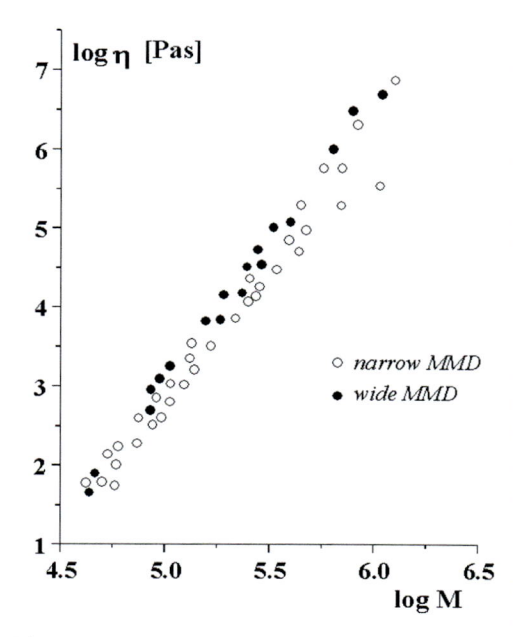

Figure 3.3.2. Zero-shear-rate viscosity of polystyrene melt as a function of molecular mass. T = 200°C. This figure in the original publication summarizes the experimental data of 18 studies. Here, only a part of experimental points is reproduced. [Adapted, with permission, from A. Casale, R.S. Porter, J.F. Johnson, *J. Macromol. Sci., Rev. Macromol. Chem.*, C, **5**, 387 (1971)].

between low molecular weight products and polymers. For example, non-Newtonian flow begins at $M > M_c$. The activation energy of viscous flow becomes independent of the length of a polymeric chain at $M > M_c$, and so on. The M_c value can be related to sufficiently long chains such that their parts behave independently and the spatial entanglement network is present. The "independent" part of the chain is called a *chain segment*. The idea of a chain consisting of many independently moving segments is one of the fundamental suggestions in polymer physics. The characteristic length of a chain is a statistical value, and it is meaningless to consider that individual segments may have some well-defined "boundaries". It is reasonable to treat M_c as a measure and reflection of the chain rigidity.

Examples

	M_c
Polyethylene	4,000
Polybutadiene	5,600
Polyisobutylene	17,000
Polymethylmethacrylate	27,500
Polystyrene	35,000

The origin of a segment can be attributed to the existence of the flexibility of a long macromolecule. This allows one part of the molecule to move independently from the other part(s) of a chain. Macromolecules in molten state or in a concentrated solution contact each other and form a *network of entanglements*. Then it is reasonable to introduce the average length of chain between two neighboring physical junctions (nodes). The part of a chain between two neighboring junctions consists of several segments (if the chain is flexible enough). The entanglements (junctions) can be characterized by some lifetime and the measure of this lifetime is a characteristic relaxation time. If deformation proceeds slower than this relaxation time, the chains move without "noticing" these junctions, they can slide over each other, which means that the polymer can flow. If the deformation is sufficiently rapid, these junctions do not have sufficient time to disintegrate. In this case, they form stable (quasi-chemical) entities and polymer *behaves* similar to a cured rubber: it cannot flow and it assumes a rubber-like physical (or relaxation) state. It is necessary to add that the lifetime is not a single constant of such a system, but a distribution of lifetimes of junctions (and consequently of relaxation times) always exists.

Chemical bonds between chains decrease the average length of chain between neighboring entanglements. This occurs in the process of curing (or crosslinking) rubbers. It is possible to treat chemical bonds as having an infinite lifetime (even though it is not absolutely true: some cases are known when bonds break under stress at a sufficiently long time). These permanent chemical bonds prevent flow because chains cannot slide over each other. The shorter the distance between crosslinks, the more rigid a material becomes.

An example illustrating the dependence of $\eta_0(M)$ in the high molecular mass domain is shown in Fig. 3.3.2. This dependence is represented by a power law with an exponent close to 3.4 (the averaged value of the slope for the points in Fig. 3.3.2 is 3.3 ± 0.1). This dependence is also valid for polydisperse polymers if one assumes that M in Eq. 3.3.15 means the *weight averaged molecular mass*, $\overline{M}_w$, i.e., the dependence $\eta_0(M)$ for high molecular mass polymers is given by:

$$\eta_0 = K_2 \overline{M}_w^b \qquad [3.3.16]$$

where constants K_2 and b have the same values as in Eq. 3.3.15.

One can see that the viscosity of polystyrene, given in Fig. 3.3.2, as an example, changes in a very wide range exceeding 5 decimal orders. The same is true for any other polymer. Therefore, in contrast to low molecular weight substances, it is meaningless to put the question: which polymer is more or less viscous? The answers depend on the chain length.

3.3.4.2 Viscosity of polymer solutions

An example of the viscous properties of polymer solutions in the whole concentration range is given in Fig. 3.2.1. The viscosity of solution, η, is higher than the viscosity of the solvent, η_s. In very dilute solutions, macromolecules move independently. One may postulate that there is a linear dependence of solution viscosity on concentration, c.[55] This linear dependence is written as

$$\eta = \eta_s(1 + c[\eta]) \qquad [3.3.17]$$

where the constant $[\eta]$ is introduced as a coefficient of the first-order term. A more rigorous definition of this constant is:

$$[\eta] = \lim \frac{\eta - \eta_s}{c\eta_s} \quad \text{at } c \to 0 \qquad [3.3.18]$$

The constant $[\eta]$ is called the *intrinsic viscosity* and its dimension is reciprocal to concentration. $[\eta]$ is a measure of the influence of polymer dissolved in a solvent on the viscosity of a polymer-solvent system. In various molecular theories, it was proven that $[\eta]$ depends on the size of the macromolecule. That is why $[\eta]$ directly correlates with the molecular mass of polymer and can be used as a simple measure of molecular mass. As a general rule, this correlation is expressed by the standard *Mark-Kuhn-Houwink equation*:

$$[\eta] = k\overline{M}_\eta^\alpha \qquad [3.3.19]$$

where the parameters k and α are constants that are characteristics of polymer and solvent used. In many cases, at least for flexible-chain polymers, α lies between 0.5 and 0.8. Molecular mass related to $[\eta]$ is known as a viscometric-averaged molecular mass which is intermediate between number-averaged, $\overline{M}_n$, and weight-averaged, $\overline{M}_w$, molecular mass.

The product c[η] is a measure of the volume occupied by macromolecules in solution. It is assumed that c[η] < 1 is typical of *dilute solutions*. In this case, one may neglect interactions between macromolecular chains. At c[η] > 1, intermolecular contacts influence viscous properties of solution.

The dependence η(c) is also represented by the *Huggins equation*,[56] which can be written as:

$$\frac{\eta - \eta_s}{\eta_s} = c[\eta] + K_H(c[\eta])^2 \qquad [3.3.20]$$

where the constant K_H is called the *Huggins constant*. The Huggins constant is a measure of interaction between polymer and solvent.

Many other approximations of η(c), dependence can be formally constructed as the sum of higher-order terms of concentration. Many fitting equations for η(c) dependence were proposed. The *Martin equation*[57] is one of them:

$$\eta = \eta_s \exp(K_M c[\eta]) \qquad [3.3.21]$$

where K_M is the *Martin constant*.

The *Kraemer equation*[58] is widely used in analytical practice:

$$\ln(\eta/\eta_s) = c[\eta] - K_K(c[\eta])^2 \qquad [3.3.22]$$

where K_K is the *Kraemer constant*.

In the range of low concentrations, any non-linear equation is approximated by the simplest linear Eq. 3.3.17. The method of determination of [η] depends on the equation used for approximation. If the Huggins equation is used for fitting experimental data, [η] is found by the presentation of the experimental data in coordinates $(\eta - \eta_s)/\eta_s c$ vs. c and extrapolation of this dependence to c = 0. The intercept gives the value of [η] and the slope of the straight line is a measure of K_H. This dependence is linear.

The structure of equations commonly used for η(c) dependence shows that they can be presented in a dimensionless form. The argument is expressed as c[η], and the dimensionless viscosity as $\eta = (\eta - \eta_s)/\eta_s[c]$. These values are used in the analysis of concentration dependencies of the viscosity of different poly-

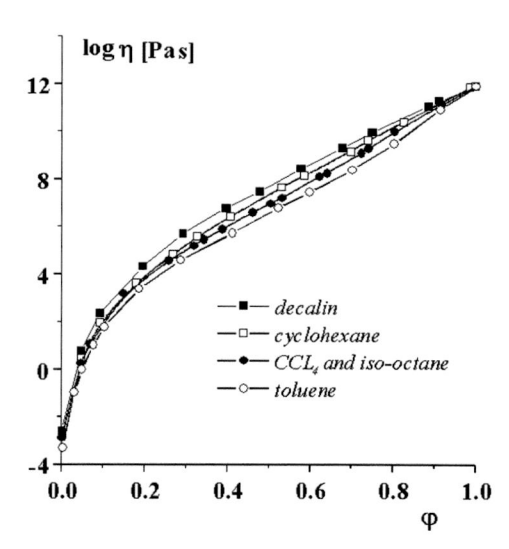

Figure 3.3.3. Concentration dependencies of viscosity of polyisobutylene solutions in different solvents in the whole concentration range. T = 20°C. [Adapted, with permission, from V.E. Dreval, A.Ya. Malkin, G.V. Vinogradov, A.A. Tager, *Europ. Polym. J.*, **9**, 85 (1973)].

mers. In the presentation of viscometric data with these dimensionless variables for different polymer-solvent systems, the common initial reference point exists: $\hat{\eta} = 0$ at

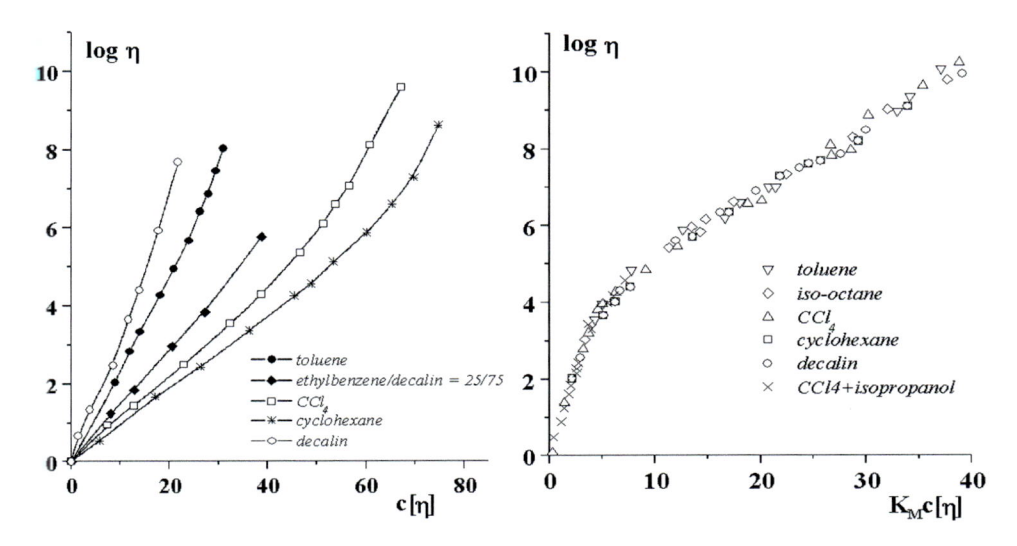

Figure 3.3.4. Concentration dependencies of viscosity of polystyrene in five different solvents. T = 25°C. [Adapted, with permission, from V.E. Dreval, A.Ya. Malkin, V.O. Botvinnik, *J. Polym. Sci.: Polym. Phys. Ed.*, **11**, 1055 (1973)].

Figure 3.3.5. Concentration dependence of polyisobutylene solutions in different solvents using the reduced coordinates. [Adapted, with permission, from V.E. Dreval, A.Ya. Malkin, V.O. Botvinnik, *J. Polym. Sci.: Polym. Phys. Ed.*, **11**, 1055 (1973)].

$c[\eta] = 0$. This permits the comparison of concentration dependencies of different polymer-solvent systems in reference to a common point.

Two polymers are further discussed: polyisobutylene (Fig. 3.3.3) and polystyrene (Fig. 3.3.4) solutions in different solvents. Polyisobutylene melts at the measurement temperature. This permits us to measure the viscosity of its solutions in the whole concentration range. Polystyrene is in the glassy state at the measurement temperature. Thus, its solution at a certain concentration will form a glassy state. In the first case, there is a single final point corresponding to the polymer melt. The difference in viscosity of equi-concentrated solutions is not large in the whole concentration range.

Experimental data in Fig. 3.3.4 are represented by *reduced* (dimensionless) variables in order to have a common initial point. The viscosity grows unlimitedly when concentration approaches a certain value (different for various solvents). This corresponds to the glass transition of solution. Depending on the nature of the solvent, the difference in viscosity of equi-concentrated solutions can reach several decimal orders.

The difference in the concentration dependence of viscosity for different polymers is determined by the relaxation state of the polymer, i.e., whether the polymer is in a melt or glass state at the temperature of viscosity measurements.

The influence of solvent on solution viscosity in the low concentration range is determined by the coefficients of the high-order term of the $\eta(c)$ dependence. The initial parts of these dependencies coincide if the reduced argument $K_M c[\eta]$ is used. This argument permits the generalization of $\eta(c)$ dependence in the whole concentration range. This is illustrated in Fig. 3.3.5. The reduced coordinates are used to construct a unique viscosity vs. concentration dependence for all solvents in the *whole* concentration range. It means that the parameters $[\eta]$ and K_M representing the $\eta(c)$ dependence in the low concentration range are responsible for the viscous properties of solutions up to the limit of very high concentrations.

The last statement is valid for many polymer-solvent systems (see Fig. 3.3.6).[59] There is an initial universal part of the reduced dependence, $\eta/\hat{c}$, common to all polymer-solvent systems. A universal character of this dependence for any polymer and different solvents in the whole concentration range exists with a continuous shift in the position of the curve of flexible and rigid polymer chains.

The $\eta(c)$ dependence in the high polymer concentration range can be represented by simple analytical equations. In many cases, the power-law equation is useful:

$$\eta = k_c c^m \qquad\qquad\qquad [3.3.23]$$

The values of the exponent are m = 5-7.

Combination of Eqs. 3.3.16 and 3.3.23 gives:

$$\eta = k_{M,c} M^b c^m = k_{M,c}(M^{b/m}c)^m \qquad\qquad [3.3.24]$$

Using ordinary values of constants (b = 3.5 and m = 5-7), the exponents ratio b/m is 0.5-0.7. The value of the ratio b/m coincides with exponent α in Eq. 3.3.19. Then, combining Eqs. 3.3.19 and 3.3.24, the following equation for solution viscosity is obtained:

$$\eta = K_c(c[\eta])^m \qquad\qquad\qquad [3.3.25]$$

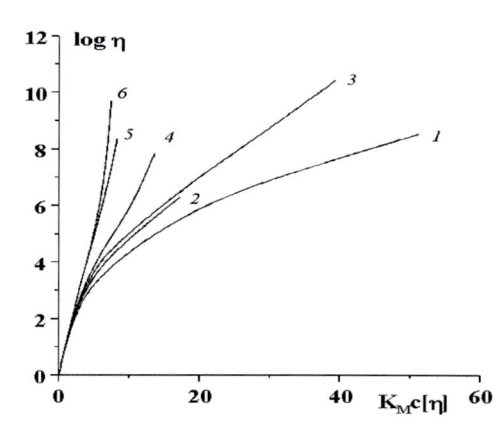

Figure 3.3.6. Concentration dependencies of viscosity for different polymers and solvents presented in reduced variables: 1 – polybutadienes of different molecular masses, φ = 0-1.0; 2 – polydimethylsiloxanes, φ = 0-1.0; 3 – polyisobutylenes of different molecular masses, φ = 0-1.0; 4 – acetyl cellulose, φ = 0-0.35; 5 - polyarylate, φ = 0-0.3; 6 – polyvinyl acetate, φ = 0-0.7; 7 – polystyrene, φ = 0-0.7. Each curve contains data obtained for several different solvents. [Adapted, by permission, from V.E. Dreval, A.Ya. Malkin, V.O. Botvinnik, *J. Polym. Sci.: Polym. Phys. Ed.*, **11**, 1055 (1973)].

and viscosity of highly concentrated solutions is a function of the product $c[\eta]$ as in the low concentration range. The same factor also determines the viscosity of concentrated solutions. There is no proof, but circumstantial evidence, that the structure of solvent does not change in the whole concentration range and that the macromolecular conformation (conformation of a statistical coil) continues to exist in the whole range of concentrations from dilute to concentrated solutions until the molten state. It means that solutions are homogeneous mixtures of polymer and solvent. It is not necessarily a general case. The opposite situation is met when polymer molecules in the solution are arranged in a regular structure. It is especially typical of rigid-chain (rod-like) polymers, which transfer into a *liquid crystalline* (LC) state.[60] This transition may take place at a certain temperature (for *thermotropic* polymer solutions) and in isothermal conditions

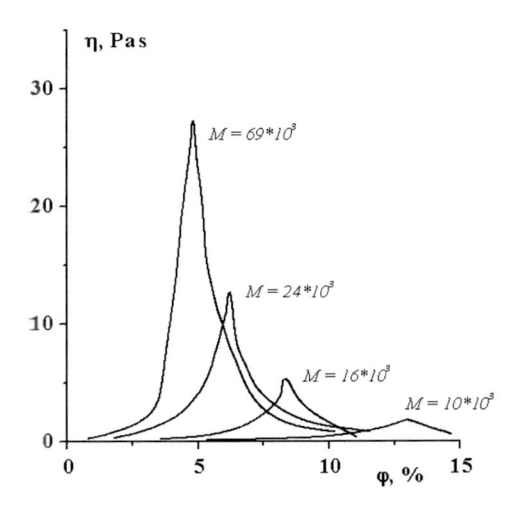

Figure 3.3.7. Concentration dependencies of viscosity of poly-p-benzamide solutions in dimethyl acetamide – transition to the LC state. Molecular masses of a polymer are shown on the curves. [Adapted, with permission, from S.P. Papkov, V.G. Kulichikhin, V.D. Kalmykova, A.Ya. Malkin, *J. Polym. Sci.: Polym. Phys. Ed.*, **12**, 1753 (1974)].

at a certain concentration (*lyotropic* solutions).[62] The latter case is the most interesting for discussion of the concentration dependence of viscosity. It is illustrated in Fig. 3.3.7.[61] The unexpected sudden decrease of viscosity at some critical concentration, c^*, is observed instead of continuous growth as in Figs. 3.3.3 to 3.3.6. It corresponds to a phase transition from a homogeneous solution to the LC state. Flory[62] has provided physical reasons for these transitions. He showed that a regular, parallel positioning of the rod-like molecules becomes thermodynamically preferable at some critical concentration. This critical concentration of LC phase transition corresponds to sharp maxima of viscosity in Fig. 3.3.7.

The monotonous viscosity growth is given by the factor of the volume filling $c[\eta]$ in the whole concentration range, and the rapid phase transition into the LC state with a drop of viscosity at $c > c^*$ are extreme cases (though the first one is rather typical of many polymer-solvent systems). In practice, it is possible to observe various intermediate cases of intermolecular interaction (for example, hydrogen bonding, the formation of colloid-like micelle clusters, and so on) leading to the appearance of various, more or less regular, structure aggregates of different sizes with various lifetimes. It results in different, sometimes very unusual anomalies of viscosity vs. concentration dependencies of real polymer solutions.

3.3.4.3 Viscosity of suspensions and emulsions

Suspensions and emulsions are multicomponent media consisting of the continuous phase and dispersed particles. So, in the discussion of the viscosity of suspensions, it is important to answer the following question: should a dispersion of solid particles (in suspension) or liquid droplets (in emulsions) in a flowing medium be treated as a homogeneous system neglecting its concentration and structure distribution? The answer to this question is not always positive. It depends on the ratio of particle size and the dimensions of the flow channel. It is reasonable to neglect the inner inhomogeneity of the medium and to consider a flowing system as a continuum to calculate some of its averaged characteristics, such as viscosity. In this case, viscosity represents a measure of energy dissipated in the flow of such a multi-component system.

The same is also true for mixtures: the size of component particles may be large, but in many cases (with some caution), it is reasonable to calculate the viscosity of such a multi-component system as a measure of energy dissipation during flow.

It is well known that the concentration dependencies of the viscosity of suspensions and emulsions in the range of dilute and semi-dilute dispersions are analogous.[63] This is correct till approaching the close-packing limit of particles of disperse phase in suspersions. However, in highly concentrated emulsions the concentration of the inner phase can exceed this limit. The latter case will be discussed at the end of this section.

The viscosity of liquid typically increases when the concentration of dispersed particles in the flowing medium increases. This is due to the additional energy loss for the liquid to flow around solid particles. Einstein[64] was the first who examined the problem of viscosity calculations of liquids containing small amounts of dispersed particles. He obtained the following relationship, which is a ground rule for all future studies in this field:

$$\eta = \eta_s(1 + 2.5\varphi) \qquad [3.3.26]$$

Here η is the viscosity of the suspension, η_s is the viscosity of a liquid phase, and φ is the volume concentration of solid spherical particles.

This equation is the first, or *linear* approximation, valid for non-interacting particles. It means that the dynamics of flow around one particle does not influence the velocity field around any other particle.

Many publications were devoted to the generalization of this equation for a range of higher concentrations. The complete form of expression for viscosity at higher approximations is a sum of concentrations with increasing power:

$$\eta = \eta_s(1 + 2.5\varphi + b\varphi^2 + c\varphi^3 + ...) \qquad [3.3.27]$$

The values of coefficients are different in various theories. The values cited for the coefficient b vary from 4.4 to 14.1. However, it is worth mentioning that the essential part in the estimation of b is played by the number of terms in Eq. 3.3.27.

A rather simple Quemada equation provides a good correlation with experimental data in many cases:[65]

$$\eta = \eta_0\left(1 - \frac{\varphi}{\varphi*}\right)^{-0.2} \qquad [3.3.28]$$

where $\varphi*$ is the dense random packing volume fraction that varies in the range of 0.48-0.57.

The other popular and convenient semi-empirical equation for the dependence $\eta(\varphi)$ was proposed by Mooney:[66]

$$\eta = \eta_s\exp\left[\frac{2.5\varphi}{1 - (\varphi/\varphi*)}\right] \qquad [3.3.29]$$

This equation is converted to the Einstein equation at low concentrations. It satisfies numerous experimental data in the intermediate concentration range, and it describes an important effect of unlimited viscosity increase when approaching some critical concentration, $\varphi*$. The latter has the meaning of limiting the possible degree of filling a space with filler particles. For example, if a solid filler is composed of spherical particles arranged in a hexagonal or cubic-centered manner $\varphi* = 0.74$.

The typical shape of the concentration dependence of viscosity of dispersion of solid particles is shown in Fig. 3.3.8.

We can find several dozen equations for concentration dependencies of the suspension viscosity in publications. The most general of them covering different versions of this dependence looks as given elsewhere[67] and it is formulated as

$$\eta_r \equiv \frac{\eta(\varphi)}{\eta_0} = \left[\frac{\varphi^* - \varphi}{\varphi^*(1 - \varphi)}\right]^{\frac{2.5\varphi^*}{1 - \varphi^*}} \qquad [3.3.30]$$

Formulas of the power or exponential type, as discussed above, treat the viscosity of suspensions as unique functions of concentration. The following effects can lead to more complicated results:[68]

- non-spherical shape of particles
- presence of particles of different sizes and shapes
- non-Newtonian properties of a liquid medium
- deformability of filler particles that are not necessarily solid (e.g., liquid or gas)
- physical interaction between liquid and solid particles, leading to the formation of stable surface layers
- interactions of various types between solid particles (for example, solid particles may mechanically touch each other, be charged, and so on).

Each of the effects on the list may have special importance for a particular system. For example, the deformability of solid non-spherical particles is especially important for the flow of blood. Physical interactions between solid particles are of special importance in processing and applied properties of rubbers, and so on.

Taking into account any of these effects requires modification of equations for the concentration dependence of viscosity. The modification has to include information about the content and structure of a flowing system. This is a problem that requires advanced and special investigations in the field of multi-component systems. Further discussion on the rheology of emulsions can be found in the reviews.[69,70]

As was mentioned above, there is a close packing limit of spherical particles in the continuum, which for monodisperse particles is close to $\varphi^* = 0.74$. The degree of filling can be higher for the polydisperse particles because small particles can occupy empty spaces between larger ones. Meanwhile, the threshold of filling of solid particles does not much exceed this limit.

If the concentration of a filler approaches φ^*, a suspension lost a possibility to flow due to the absence of sufficient free volume for displacements of hard particles. This is the case of "jamming" or "dynamic (mechanical) glass transition", i.e., the transition from a fluid to a non-fluids state.[71-73] This effect is well known to those, who tried to push a dispersed, in particular, granulated, material, such as sand, clay, any form of waste, etc., through a

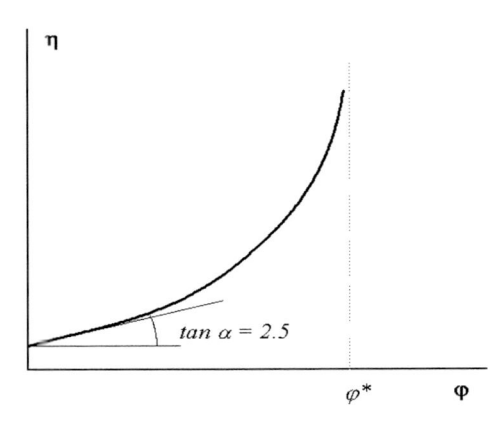

Figure 3.3.8. Typical concentration dependence of the viscosity of suspension of solid particles.

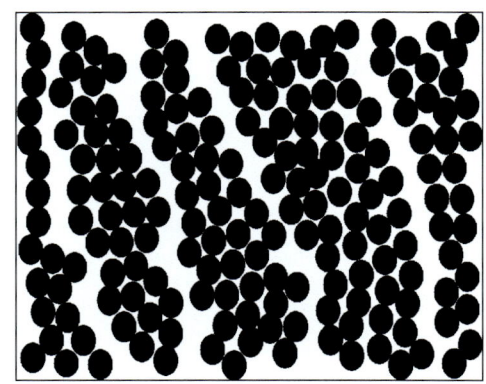

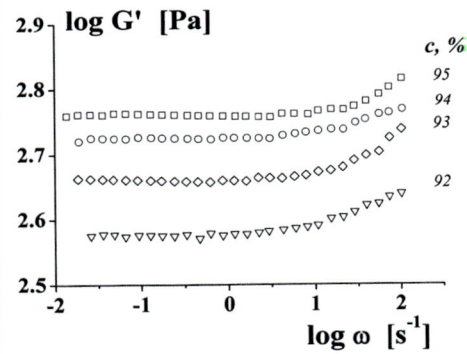

Figure 3.3.9. Scheme illustrating the character of deformation in a highly concentrated suspension (displacement along the slipping surfaces).

Fig. 3.3.10. Storage modulus in a wide frequency range for highly concentrated emulsions with different concentration of an inner phase (shown at the curves). After I. Masalova, A.Ya. Malkin, *Colloid J.*, **69**, 185 (2007).

pipe; as the pressure is increased, the pipe is "plugged" at some moment and displacement of the material either becomes impossible or requires a disproportionate increase in pressure. Different aspects of this phenomenon are discussed elsewhere.[74]

However, such highly concentrated suspension of solid particles (also applies to dry solid particles) can move in a heterogeneous manner by blocks along boundary lines as shown in Fig. 3.3.9.

Relative displacement of large clusters in highly concentrated suspensions was described in Ref. 75. Ruptures in movement of granular materials were also found in Ref 76. This kind of material displacement, which unlikely to be treated as "flow," is accompanied by strong and irregular stress oscillations.[77] Surely. it would be incorrect to try to estimate the "viscosity" of this movement.

The situation with emulsion is different. They can be compressed leading to filling some space due to the transformation of spheres to polygonal objects separated by thin interface films. Then, the volume degree of filling can be as high as 0.93. The transition to the highly concentrated emulsions significantly affects their rheological properties. First of all, they become gel-like or pastes. It means that they exhibit a solid-like behavior at low deformations or low stresses. This is illustrated in Fig. 3.3.10[78] (see also Fig. 2.8.12) where this effect is indicated by the absence of frequency dependence of the elastic (storage) modulus. It should be noted that in this frequency range the loss modulus exhibits low value.

The physical explanation of this kind of elasticity was proposed by Princen[79] and later developed in Ref.,[80] where the elasticity of compressed droplets was treated as a consequence of the increases in the droplet surface under the osmotic pressure. However, the latest understanding of this phenomenon includes the input of interface interaction between compressed droplets.[81]

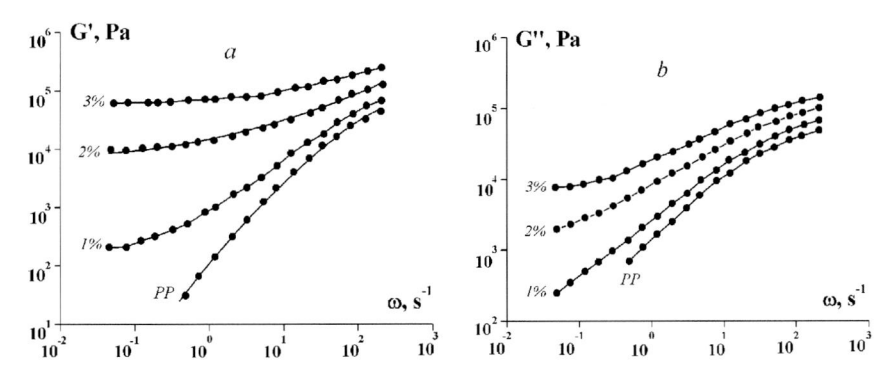

Figure 3.3.11. The storage modulus (a) and loss modulus (b) as a function of frequency in linear region for metallocene polypropylene (PP3825) melt containing various weight concentrations of carbon nanotubes (CNTs, NC7000) at 160°C.

3.3.4.4 Viscosity and viscoelastic behavior of nanocomposites

The rheological behavior of nanocomposites, i.e. polymers reinforced by nanometer-scale particles, such as carbon black (CB), carbon nanotubes (CNTs), nanoclays, etc. is extensively studied. The most frequently used technique in these studies is a small amplitude oscillatory shear (SAOS) flow. It is used to characterize the structure of polymers and filled polymer systems in their original state. SAOS has a solid theoretical foundation based on the linear theory of viscoelasticity. The measured dynamic properties in SAOS are independent of the strain amplitude and strain rate amplitude. Rheological behavior of nanocomposites is attributed to the dispersion and breakdown of aggregates and agglomerates in the case of CB and CNT nanofillers and the intercalation and exfoliation in the case of nanoclays. The most extensive review on the effect of interactive fillers such as black and white fillers in elastomeric systems on their dynamic properties was published by Donnet.[82] Four factors are identified to contribute to the storage modulus behavior of nanocomposites. These are matrix properties, hydrodynamic effects, nanofiller-matrix, and nanofiller-nanofiller interactions. Contributions of these factors to the overall dynamic properties in the most simplistic way can be considered to be additive. Although direct methods to fully evaluate each of these contributions are unavailable, the strain amplitude dependence of the storage modulus of nanocomposites is attributed to the nanofiller-nanofiller interactions and at high strain amplitudes, to the matrix-nanofiller interactions and the matrix properties. However, the dynamic properties alone would not lead to a full understanding of the problem. Applications of direct spectroscopic techniques that may reveal these interactions are beneficial. In contrast to microcomposites, containing micron-sized filler particles and exhibiting a solid-like behavior at a high filler concentration, nanocomposites exhibit such behavior at a low concentration of nanofillers. As an example of such behavior of nanocomposites, Figure 3.3.11 shows the storage modulus (a) and loss modulus (b) of polypropylene (PP) melt as a function of the frequency at various CNT (Nanocyl NC7000) concentrations in the linear region of their behavior. A typical terminal behavior at low frequencies with the scaling of $G' \sim \omega^2$ is observed for pure PP melt in Figure 3.3.11 (a) with incorporation of CNTs at a concentration of only 1 wt%, indicating that the long-range motion of polymer chains is restrained by the presence of

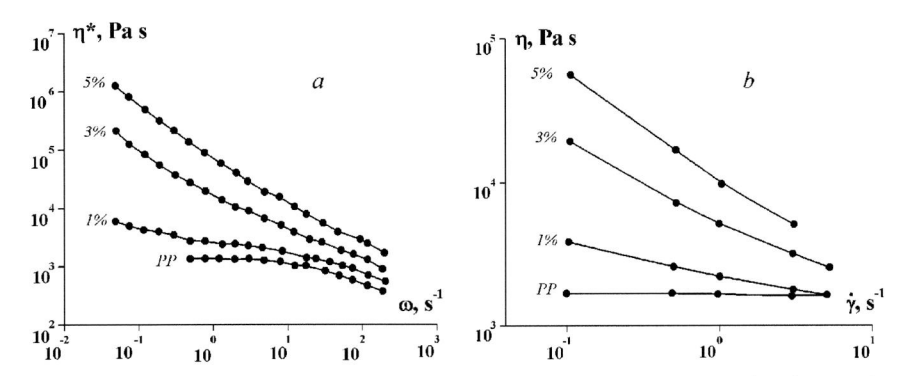

Figure 3.3.12 Complex dynamic viscosity as a function of frequency (a) and apparent viscosity as a function shear rate (b) for PP and PP/CNT composites of various concentrations at 160°C.

CNTs. With further increase of CNTs concentration, the storage modulus is increased by several orders of the magnitude accompanied by a significant decrease of the slope of G' vs. ω at low frequencies, eventually leading to the appearance of the low-frequency plateau of G' at 3 wt% and 5 wt% concentrations of CNTs. This is solid evidence of the formation of the CNTs network at such a low concentration of CNTs and increasing the interaction between the polymer and CNTs. At high frequencies, CNTs showed less effect on the storage modulus increase, suggesting that CNTs do not significantly influence the short-range dynamic motion of polymer chains. In addition, as seen from Fig. 3.3.11 (b), a significant increase of the loss modulus, G", with increasing CNTs concentration is also observed at low frequencies, which is also an indication of the effect of CNTs on the relaxation behavior of polymer chains leading to a solid-like behavior of nanocomposites at low concentrations of CNTs.

Figure 3.3.12 shows the complex dynamic viscosity as a function of frequency (a) and the apparent viscosity as a function of the shear rate (b) for PP and PP/CNT composites of various concentrations at 160°C. The complex viscosity of virgin PP shows a Newtonian behavior at low frequencies. However, with the incorporation of only 1 wt% CNTs, the Newtonian behavior disappears and a strong shear thinning behavior is observed at low frequencies. With the incorporation of 3 wt% and 5 wt% CNTs the complex viscosity shows a much stronger non-Newtonian behavior with the slope of the complex viscosity vs. frequency being 1, indicating the unbound rise of the complex viscosity with the decrease of the frequency, i.e., nanocomposites show a solid-like behavior. This is due to the formation of the CNT network leading to an increase of the largest relaxation time of nanocomposites. Similar behavior is also seen for the dependences of the apparent viscosity on the shear rate in the steady-state flow (Figure 3.3.12 b). However, a comparison of the complex viscosity with the apparent viscosity at frequencies being equal to shear rates indicates that these viscosities are equal for the pure PP melt only, indicating the validity of the Cox-Merz rule. However, a strong deviation from this rule appears for nanocomposites with the complex viscosity being much higher than the apparent viscosity. This deviation increases with an increase of CNT concentration in nanocomposites.

It should be noted that during processing, the strain and strain rate are high. Therefore, they affect the structure of materials. Accordingly, it is plausible to look at the strain

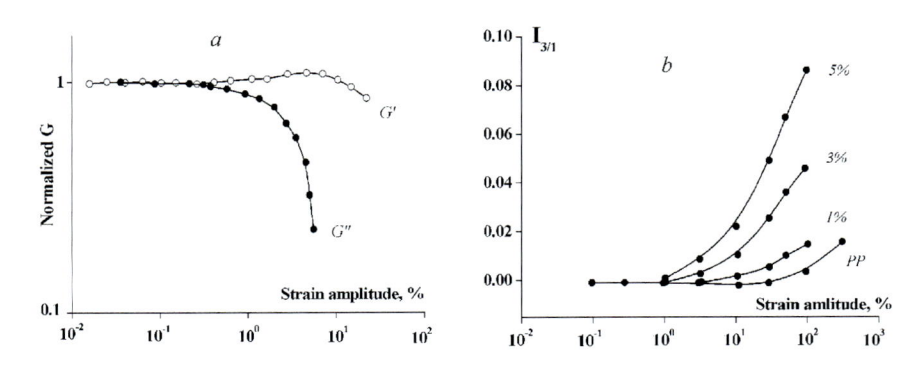

Figure 3.3.13. Normalized primary values of G' and G" for PP/5 wt% CNT composites (a) and ratio of the 3rd to a 1st harmonic of shear stress (b) as a function of strain amplitude at a frequency of 3 s^{-1} for various PP/CNT composites at 160°C.

amplitude dependence of dynamic properties of PP/CNT nanocomposites. In particular, the large amplitude oscillatory shear (LAOS) flow is used to provide additional information about structural changes in polymer melts.[83] However, understanding and analysis of the dynamic properties of the LAOS flow are more complicated and less developed. In the LAOS flow, a sinusoidal oscillatory strain/stress with a large amplitude is imposed and the corresponding stress/strain response is not sinusoidal. The non-sinusoidal response signal is processed using the Fourier transform to calculate contributions from higher-order harmonics. In the LAOS flow, the imposed strain is sinusoidal and the shear stress response is not sinusoidal. It contains odd harmonics, as given in the following Fourier series:[84]

$$\sigma(t) = \sum_{n = odd} \sigma_n \sin(n\omega t + \delta_n) \qquad [3.3.31a]$$

where σ_n is the nth harmonic of the shear stress and δ_n is the phase angle between nth harmonic of shear stress and imposed strain. One can define a ratio of $I_{n/1} = \sigma_n/\sigma_1$ to measure the intensity of the nth harmonic defining the level of the nonlinearity of viscoelastic behavior. Eq. [3.3.31a] can be re-written as[84]

$$\sigma(t) = \gamma_0 \sum_{n = odd} [G'_n(\omega, \gamma_0)\sin(n\omega t) + G''_n(\omega\gamma_0)\cos(n\omega t)] \qquad [3.3.31b]$$

where G'_1 and G''_1 are the primary storage modulus and loss modulus corresponding to the first harmonic of the shear stress, respectively.

Figure 3.3.13 shows the normalized primary values of G' and G" (a) and the value of $I_{3/1}$ (b) as a function of the strain amplitude for PP/5wt%CNT nanocomposites. The normalized G' and G" are defined as G'/G'_0 and G''/G''_0, where G'_0 and G''_0 are values of G' and G" in the linear region corresponding to low strain amplitudes corresponding to the SAOS flow. It is seen that the normalized value of G' for the nanocomposite decreases with the strain amplitude, but the normalized value of G" exhibits a maximum with increasing strain amplitude (overshoot) and then decreases. The explanations for the overshoot of G" are very diverse.[85-87] It is claimed that the overshoot of G' is due to the rearrangement of the loose clusters in the system,[85] increase of the ineffective network density,[86] and destruction and reformation of the microstructure in the system[87] with the strain amplitude. The diversity of explanations means that the first harmonic of the shear stress is not sufficient in describing the nonlinearity. Therefore, the inclusion of the higher harmonic behavior is required to describe the nonlinearity. A variation of the values of the third harmonic, $I_{3/1} = \sigma_3/\sigma_1$, with the strain amplitude, is shown in Figure 3.3.13 (b). Theoretically, the value of $I_{3/1}$ is zero for the linear rheological behavior (SAOS) and is non-zero for the nonlinear behavior (LAOS). For the pure PP melt, the value of $I_{3/1}$ is seen to be close to zero at the strain amplitudes lower than 100%, and it increases to a value of 0.02 only at a strain amplitude of 300%. The value of $I_{3/1}$ increases monotonically with the strain amplitude and concentration of CNT. It is believed that in polymer nanocomposites when a nanofiller network is formed, the structure is analogous to that of a gel and exhibits a solid-like behavior. In these cases, the contribution of the higher harmonics of shear stress becomes large.

Significant differences between the behavior of nanocomposites and microcomposites are seen in Figure 3.3.14 showing the complex dynamic viscosity in the linear viscoelastic region of polystyrene (PS) melt filled with various silica fillers versus the volume concentration at a low frequency of 0.01 s⁻¹ (a) and a high frequency of 100 s⁻¹ (b). The PS melt containing the precipitated porous spherical silica (FK140) and fumed smooth spherical silica (Aerosil 90) with nano-sized particles having high surface areas forms nanocomposites, while the PS melt containing irregular shape silica (Minusil 5) and platelet shape silica (Novacite L337) of micron-sized particles, having low surface areas, form microcomposites. Large differences are clearly seen in the concentration dependences of the complex dynamic viscosity for mixtures containing these various fillers obtained at frequencies of 0.01 s⁻¹ (Fig. 3.3.14 a) and at 100 s⁻¹ (Fig. 3.3.14 b). Differences between the viscosity of microcomposites and nanocomposites cannot be attributed to a hydrodynamic effect alone, since the significant differences in the complex dynamic viscosity behavior are seen at a low volume concentration of silica particles. The overall behavior at low concentrations is affected by a large relaxation time mechanism. The relaxation spectrum of the matrix is modified by the presence of strong filler-filler and matrix-filler interactions, as was seen earlier from changes in the behavior of the storage modulus at low

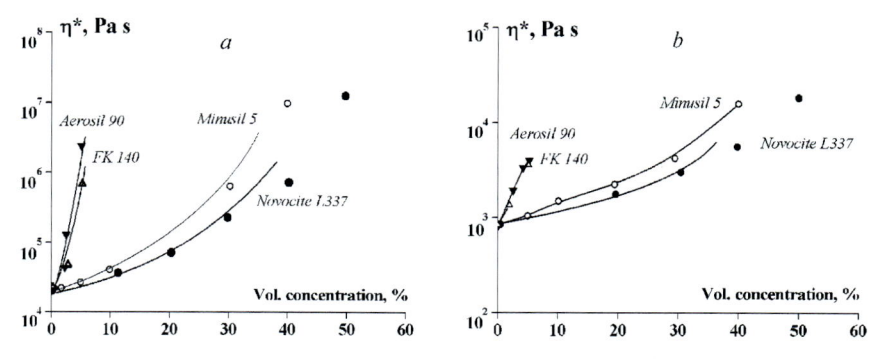

Figure 3.3.14. Dependences of the complex dynamic viscosity of various silica-filled polystyrene (Styrene 663) as a function of volume concentration of silica at frequencies of 0.01 s⁻¹ (a) and 100 s⁻¹ (b) at 200°C. Lines are drawn based on fitting to Maron and Pierce equation (Eq. 3.3.32).

frequencies in Figure 3.3.14 (a) for PP/CNTs mixtures. The presence of strong filler-filler and matrix-filler interactions is directly observed in the SEM micrographs shown in Ref. 88. Also, these large differences seen in the dependences of the complex dynamic viscosity on the filler volume concentration shown in Figure 3.3.14 are analyzed using Maron and Pierce empirical model[89]

$$\eta_r = (1 - \varphi/\varphi_M)^{-2} \qquad\qquad [3.3.32]$$

which is equivalent to the Quemada equation [3.3.28], where η_r is the relative value of the complex dynamic viscosity of the mixture with respect to that of the matrix and φ_M is the maximum volume concentration related to rheological percolation concentration. Due to its simplicity and ease of fitting, this model is considered the best empirical expression.[90] The results of this fitting are shown by lines in Figure 3.3.14. The maximum volume concentration, φ_M, obtained at a frequency of 0.01 s⁻1 and 100 s⁻¹ are 5.51 and 9.22 vol%, 6.07 and 9.98 vol%, 36.8 and 51.9 vol%, 43.2 and 63.6 vol% for Aerosil 90, FK140, Minusil 5, and Novacite L337, respectively. A large increase of the parameter φ_M is found at a high frequency in comparison with that at a low frequency. Therefore, the rheological behavior of these mixtures cannot be attributed to the particle-particle interactions alone, which shall follow frictional relationships being independent of the frequency. In addition, it is rather caused by the polymer-filler interactions. Obviously, one can argue that the PS/silica mixture containing high surface silica at a concentration of 5 vol% is a dilute suspension. However, based on the above obtained low values of φ_M, nanocomposites can actually be considered as highly filled suspensions, even at this low concentration level. A question arises what is the relationship between the rheological percolation threshold and the maximum volume concentration, φ_M, in polymer/particle mixtures. An answer to this question in a qualitative sense is very clear. The mixtures exhibiting low maximum concentrations would show low rheological percolation thresholds, such as occurs in nanocomposites. However, microcomposites exhibit high values of φ_M and therefore, high values of the rheological percolation thresholds.

3.3.4.5 Effect of filler surface area on the viscoelastic behavior of polymer compounds

The effect of filler surface area on the viscoelastic behavior of polymers is demonstrated using silica-filled, tin-coupled star-shaped SBR (Duradene 739) compounds. Duradene 739 is a solution polymerized 3- and 4-arm star SBR (Firestone Polymers). Its weight averaged molecular mass is 334,000. The rubber is comprised of 20% bound styrene, 60% vinyl, and an antioxidant stabilizer system.[91] This rubber is a tin-coupled SBR that is commonly used in tire treads for its ability to improve various tire performance characteristics. Four different precipitated amorphous silica fillers (PPG Industries) were 532EP, 210, 320G, and 190G. Its contents in SBR was 60 phr. The table below indicates the surface areas of the four silica grades.[92] It is expected that a silica grade with a higher surface area will create stronger rubber-filler and filler-filler interactions, impacting both the morphology and rheology of the compounds. These interactions are crucial to understanding the structure and behavior of a filled-polymer system. Forming these interactions consists of a series of complex processes, which are not yet fully understood. As a filler mixed with a polymer, the polymer chains interact with the filler, wetting filler particles. Once the particles are wetted, dispersion takes place. Furthermore, in the case of rubber compounds, once the initial wetting and dispersion occur, the chains can form stronger bonds with the filler particles creating so-called "bound rubber." The wetting, dispersion, and bound rubber formation processes are all kinetically controlled processes that occur concomitantly.[93,94] The adsorption-desorption kinetics are influenced by a number of aspects related to the filler nature (e.g., loading, surface chemistry, surface area, etc.) and the polymer nature.[93-96] Bound rubber is one of the most important metrics associated with reinforcement of filled elastomers. It is defined as the content of a filled rubber compound that cannot be dissolved in a good solvent. The reason a portion of the rubber cannot be dissolved is that it is strongly bonded to the filler particles.[93-97] Thus, a high bound rubber content is indicative of a highly reinforced elastomer. Bound rubber content is considered as a representation of the equilibrium state of the adsorption-desorption process. The bound rubber content of a filled elastomer typically ranges from 20-40%,[93] although it can be higher in some cases. Several factors have been shown to impact the bound rubber content. Among the most important factors are the nature of the filler, the nature of the polymer, the filler surface area, and the mixing time. Generally, the higher surface area and higher mixing time lead to a higher bound rubber content, as seen in the table below. Additionally, the bound rubber content increases with storage time. The fraction of bound rubber usually plateaus within 3-4 weeks for many filled rubber systems.[97]

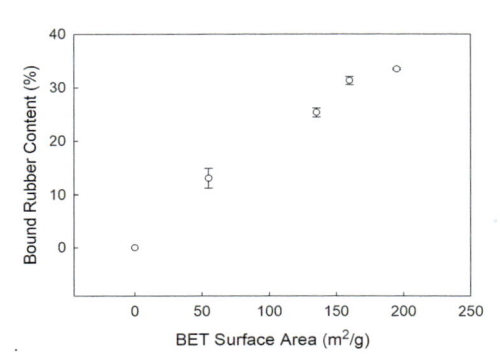

Figure 3.3.15. Bound rubber content as a function of filler surface area in SSBR/silica compounds. [Adapted, by permission, from Sandeep S. Pole, PhD Thesis, University of Akron, 2019].

The surface area of silica and bound rubber content in compounds

Compound	Surface area, m^2/g	Bound rubber, %
SBR/532EP	55	13.1 ± 1.9
SBR/210	135	25.3 ± 0.8
SBR/320G	165	31.3 ± 0.8
SBR/190G	195	33.5 ± 0.2

In addition to the surface area of silica, the table above shows the measured bound rubber content in compounds. It is seen that SBR/190G exhibits the highest bound rubber content, followed by the SBR/320G, SBR/210, and SBR/532EP. Additionally, Figure 3.3.15 depicts the bound rubber content as a function of filler surface area. Clearly, there was a direct influence of the filler surface area on the bound rubber content. The figure shows a monotonic increase in bound rubber content with filler surface area. This trend is due to the fact that the higher filler surface area promotes the greater rubber-filler interactions, resulting in a higher bound rubber content.

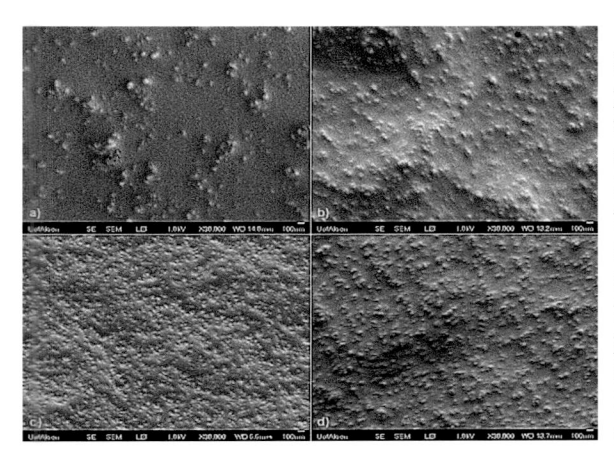

Figure 3.3.16. SEM images of SBR/532EP (a), SBR/210 (b), SBR/320G (c), and SBR/190G (d) at 30,000x magnification. [Adapted, by permission, from S. S. Pole, A. I. Isayev, *Rheol. Acta*, **56** (12), 983-993 (2017)].

Figure 3.3.16 shows the SEM images for the four compounds at 30,000X magnifications.[99] From the images, it is apparent that SBR/532EP had the least rubber-filler network penetration and also had the largest agglomerates. SBR/210 also showed strong agglomeration, but the agglomerates were better dispersed in comparison to those of SBR/532EP. SBR/320G and SBR/190G both showed good dispersion, with sizes of the agglomerates being smaller. A comparison of SBR/320G and SBR/190G indicates that agglomerates were smaller in SBR/320G. Evidently, this was due to the original forms of supplied silica fillers. Hi-Sil HDP-320G was supplied in the form of micro-granules, while Hi-Sil 190G was supplied in the form of granules.

Figure 3.3.17 depicts the storage (a) and loss (b) moduli, respectively, as functions of strain amplitude.[100] As can be seen from the storage modulus behavior, the limit of the linear viscoelastic region was a strain amplitude of ~0.01 for all the filled compounds and ~0.40 for the SBR gum. In addition, several important observations can be made. At low strain amplitudes, corresponding to the linear region of behavior, the storage modulus significantly increases with the silica surface area due to high filler-filler interactions.

A similar trend was also seen at different frequencies. Among various compounds, SBR/532EP had the weakest filler-filler network. At large strain amplitudes, the storage modulus also increases with the silica surface area due to higher polymer-filler interactions and an increase in bound rubber content. Furthermore, a maximum was observed in

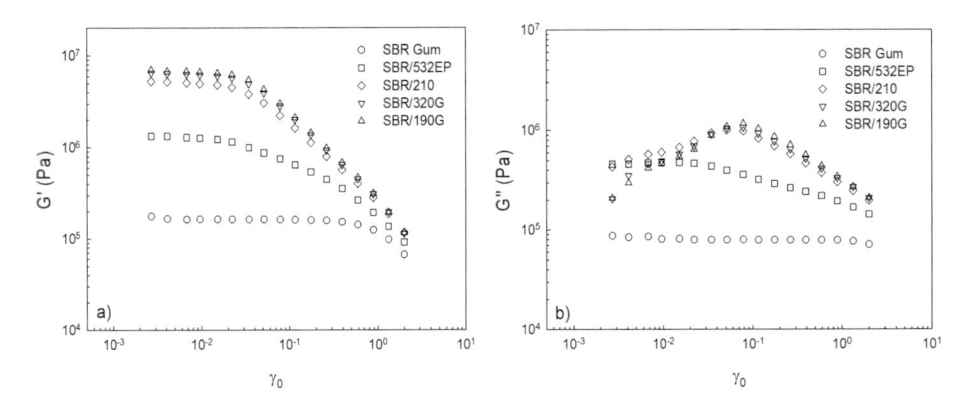

Figure 3.3.17. Storage (a) and loss (b) moduli as functions of strain amplitude at 1 rad/s and 90°C. [Adapted, by permission, from S. S. Pole, A. I. Isayev, *J. Appl. Polym. Sci.*, **138** (28), Article Number e50660 (2021)].

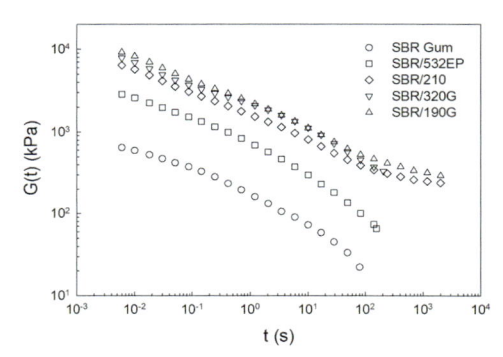

the loss modulus at a strain amplitude of ~0.10 for SBR/190G, SBR/320G, and SBR/210. This occurred due to significant viscous dissipation arising from the breakup of the filler-filler network. SBR/532EP showed a very slight maximum at a strain amplitude of ~0.015. The maximum was not as pronounced as it was for the other compounds since the filler-filler network is very weak. Once the filler-filler network was broken in the filled compounds, differ-

Figure 3.3.18. Relaxation modulus in the linear region as a function of time for SSBR and various compounds. [Adapted, by permission, from S. S. Pole, A. I. Isayev, *Rheol. Acta*, **56** (12), 983-993 (2017)].

ences in rheological behavior arose from rubber-filler interactions, hydrodynamic effects, and the polymer network. At higher strain amplitudes ($\gamma_0 > 0.50$), the differences in the storage moduli between SBR/190G, SBR/320G, and SBR/210 became insignificant. This occurred because the additional rubber-filler interactions caused by a higher filler surface area were destroyed in this high-strain-amplitude region, and there was no considerable difference in hydrodynamic effects or the polymer network. Also, at small amplitudes, compounds containing silica with high surface area exhibit a lower loss modulus. The latter means the compound has a lower loss tangent, indicating a more elastic material.

Figure 3.3.18 shows the time evolution of the linear relaxation modulus of SBR compounds containing silica of various surface areas.[101] Similar to the observation in the storage modulus, the SBR gum exhibited the lowest relaxation modulus, followed by SBR/532EP, SBR/210, SBR/320G, and SBR/190G. Again, the differences in the stress relaxation behavior among the samples are attributed to the surface area of the silica filler. It is clear that the relaxation modulus increased with the filler surface area due to an increase in rubber-filler and filler-filler interactions. Additionally, the development of a plateau in the relaxation modulus was observed at large times for SBR/210, SBR/320G, and SBR/190G. This occurred due to the high rubber-filler interactions hindering the relaxation behavior of the compounds. SBR/532EP did not exhibit a plateau since its rub-

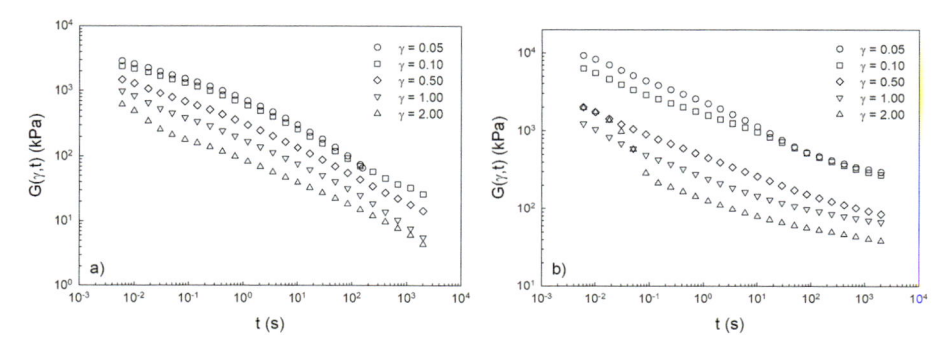

Figure 3.3.19. Relaxation modulus as a function of time at various strain levels for the SBR/532EP (a) and SBR/190G (b) at 90°C. [Adapted, by permission, from S. S. Pole, A. I. Isayev, *J. Appl. Polym. Sci.*, **138** (12), Article Number e50080 (2021)]

ber-filler network was not strong enough to significantly hinder chain relaxation. It is also apparent from the limit of the linear viscoelastic regime that the mere presence of a filler significantly increased nonlinearity in relaxation behavior.

The silica surface area also affects the behavior of relaxation modulus in the nonlinear region. This influence is seen from Figure 3.3.19.[101] The figure indicates relaxation modulus as a function of time at different strain levels for SBR/532EP (a) containing the lowest surface area silica and for SBR/190G (b) containing the highest surface area silica. It is obvious that the relaxation modulus decreased with both time and strain levels for both samples. Also, the development of a plateau in the relaxation modulus at long times of relaxation was observed for SBR/190G compound at all strain levels due to the strong rubber-filler interactions hindering relaxation behavior even in the nonlinear region. However, this plateau was not observed for SBR/532EP since the rubber-filler network was not strong enough to significantly hinder chain relaxation. For this reason, this compound did not exhibit a plateau at long times of relaxation. The relaxation behavior of SBR/532EP is similar to that of a rubber compound with a low filler content, which also does not exhibit a plateau at long times due to weak rubber-filler interactions.[102] Additionally, it was seen from the extent of the linear viscoelastic region that the mere incorporation of filler significantly increased the nonlinearity in the relaxation behavior of compounds.

It is also seen that the relaxation modulus values at strain levels of 0.05 and 0.10 were nearly identical for the SBR/532EP, indicating that the limit of the linear viscoelastic region was a strain level of 0.10. For SBR/190G, the relaxation modulus at strain levels of 0.05 and 0.10 deviated significantly, especially at short times, suggesting that a strain level of 0.05 was the limit of the linear viscoelastic region. Also, similar to the strain amplitude dependence of the storage modulus, shown earlier, the relaxation modulus of the compound containing the high surface area silica is more strongly affected by strain. Therefore, the nonlinearity of the relaxation modulus increases with the increase in silica surface area.

One important aspect of the stress relaxation behavior of polymer solutions and melts is the concept of time-strain separability. According to the principle of time-strain separability, the relaxation modulus can be factorized into two independent functions of time and strain as follows:

$$G(\gamma, t) = G_0(t) \times h(\gamma) \qquad\qquad [3.3.33]$$

where $G(\gamma,t)$ is the strain and time-dependent relaxation modulus, $G_0(t)$ is the time-dependent linear relaxation modulus, and $h(\gamma)$ is the strain-dependent damping function.[103,104] According to Eq. (3.3.33), by dividing the relaxation modulus, $G(\gamma,t)$, by the damping function, $h(\gamma)$, a superimposable relaxation modulus master curve can be obtained. The applicability of the principle of time-strain separability was also analyzed for the different compounds. In order to determine the damping function, the relaxation modulus at the various strain levels was divided by the linear relaxation modulus at long times of relaxation, where the damping function is typically assumed to be time independent. This time was taken as ~1 s for the SBR gum and ~10 s for the filled compounds because the damping function of the SBR gum is expected to become time independent at shorter times of relaxation than that of the filled compounds. Figure 3.3.20 shows the reduced relaxation modulus as a function of time for SBR/532EP (a) and SBR/190G (b). A deviation at short

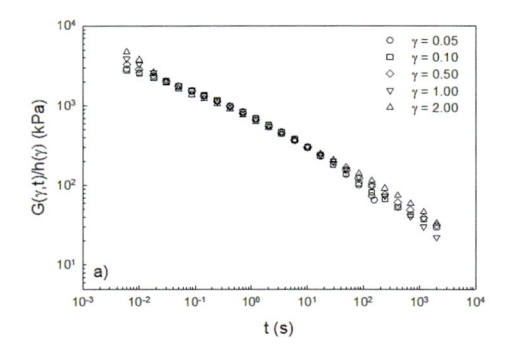

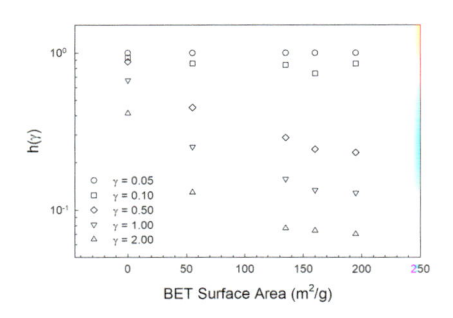

Figure 3.3.21. Damping function as a function of filler surface area at various strain levels and 90°C. [Adapted, by permission, from S. S. Pole, A. I. Isayev, *J. Appl. Polym. Sci.*, **138** (12), Article Number e50080 (2021)].

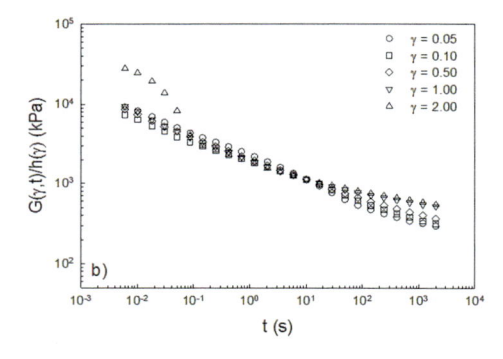

Figure 3.3.20. Reduced relaxation modulus as a function of time for SBR/532EP (a) and SBR/190G (b) at 90°C. [Adapted, by permission, from S. Pole, A. I. Isayev, *J. Appl. Polym. Sci.*, **138** (12), Article Number e50080 (2021)].

times of relaxation (t < 0.1 s) was observed for all the samples due to a transient response of the instrument due to a non-instantaneous imposition of strain. After ~0.1 s, the SBR gum showed good super-position, following the principle of time-strain separability. As seen from Figure 3.3.20 (a), the SBR/532EP compound containing silica with the lowest surface area, also showed good superposition after ~0.1 s, but a visible deviation was observed at longer times of relaxation (t > 10 s). At the same time, for the SBR/190G containing the highest surface area silica, the principle of time-strain separability was valid only in a very narrow range of times of relaxation, generally indicating the inapplicability of this principle for filled compounds.

The damping function is depicted as a function of silica surface area for all samples in Figure 3.3.21. The damping function is clearly a decreasing function of strain level. The sharpest drop in the damping function was observed for SBR/190G, followed by SBR/320G, SBR/210, SBR/532EP, and the SBR gum. This trend for filled compounds is directly correlated to the surface area of the filler. The higher surface area caused a steeper decrease in the damping function, indicating significantly greater nonlinearity. Evidently, the increased rubber-filler and filler-filler interactions, due to a higher filler surface area, caused the increased viscoelastic nonlinearity.

3.3.4.6 Effect of type of filler on the viscoelastic behavior of polymer compounds

In the previous section, the effect of filler surface area on the rheological properties of polymer melts was discussed. Rheological properties are also affected by the type of filler, surface treatment, concentration, the polarity of polymer, and filler. There are a number of widely used commercial fillers. Carbon black (CB), silica, clay, kaolin, quartz, talc, and graphite are the classic examples. Many investigations have been carried out to study the effects of CB,[105] silica,[106] clay,[107] and white rice husk ash[108] on the rheological properties of elastomers. However, the effect of the type of filler having the same surface area is seldom studied. Therefore, dynamic properties of silica-, nanoclay- and CB-filled EPDM rubber compounds are discussed below. It should be noted that the compounds were prepared utilizing the same mixing procedures, equipment, and mixing conditions. EPDM (Keltan 5636A, DSM Elastomers) with 5-ethylidene-2-norbornene (ENB) as a termonomer with 70% ethylene content, 4.5% ENB unsaturation was used. Three different kinds of fillers were utilized. These included: 1) amorphous precipitated silica, Hi-Sil 132 with BET surface area of 200 m^2/g (PPG Industries); 2) an organoclay, Closite$^®$ 15A (Southern Clay Products); 3) carbon black, V1391 with a surface area of 202 m^2/g (Cabot Corporation). It is seen that silica and CB have the same surface area. The effect of filler type and filler loading on dynamic properties of the mixtures of filled EPDM compounds is shown in Figure 3.3.22 depicting a comparison of the storage (a) and loss (b) moduli, complex dynamic viscosity (c) and Tan Delta (d) of EPDM and various filled EPDM mixtures versus frequency at 100°C obtained in SAOS.[109] Obviously, the storage and loss moduli and complex dynamic viscosity of the silica-, nanoclay- and CB-filled EPDM mixtures are higher than those of the unfilled EPDM over the frequency measurement range. These properties continuously increased with increased filler loading with different fillers showing their different level. In general, the storage and loss moduli and complex dynamic viscosity of the silica-filled EPDM rubber show much higher values than that of the CB-filled EPDM. There was no significant difference in the storage modulus between 10 phr (parts per hundred of rubber) silica- and nanoclay-filled EPDM mixtures at low-frequency values. However, the storage modulus of the 10 phr silica-filled EPDM mixture was slightly higher than that of the 10 phr nanoclay-filled EPDM mixture in the high frequency region. The storage modulus for the 10 phr nanoclay-filled EPDM mixture was slightly higher than that of the 10 phr CB-filled mixture in the low-frequency region. However, the storage modulus showed an opposite trend at high frequency. Similar observation can be seen on the loss modulus and complex viscosity behaviors.

The dependences of Tan Delta on frequency for EPDM and various EPDM/filler mixtures shown in Fig. 3.3.22 (d) indicate that Tan Delta decreases with frequency, but its frequency dependence decreases with increased filler loading. The reduction of Tan Delta with frequency was due to the fact that EPDM and EPDM/filler mixtures were in the transition region from the fluid state to the rubbery state within the frequency measurement range.[110] The Tan Delta of the various EPDM/filler mixtures decreased with increased filler loading. The samples at 10, 20, and 30 phr silica-filled EPDM mixtures showed lower Tan Delta values than the CB-filled samples at the same filler loading. This indicates that the silica-filled EPDM mixtures are more elastic than the CB-filled ones. In addition, Tan Delta of the 10 phr silica-filled EPDM mixture was higher than that of the 10 phr nanoclay-filled one in the low frequency region, whereas the Tan Delta of the 10 phr silica-filled EPDM mixture was lower than that of the 10 phr nanoclay-filled one at high-frequency region. This indicated that the 10 phr silica-filled EPDM mixture was less elastic than the 10 phr nanoclay-filled sample in the low-frequency region. At the same time, the 10 phr silica-filled EPDM mixture was more elastic than the nanoclay-filled sample. It is also seen that the effect of various fillers on Tan Delta diminishes in the high-frequency

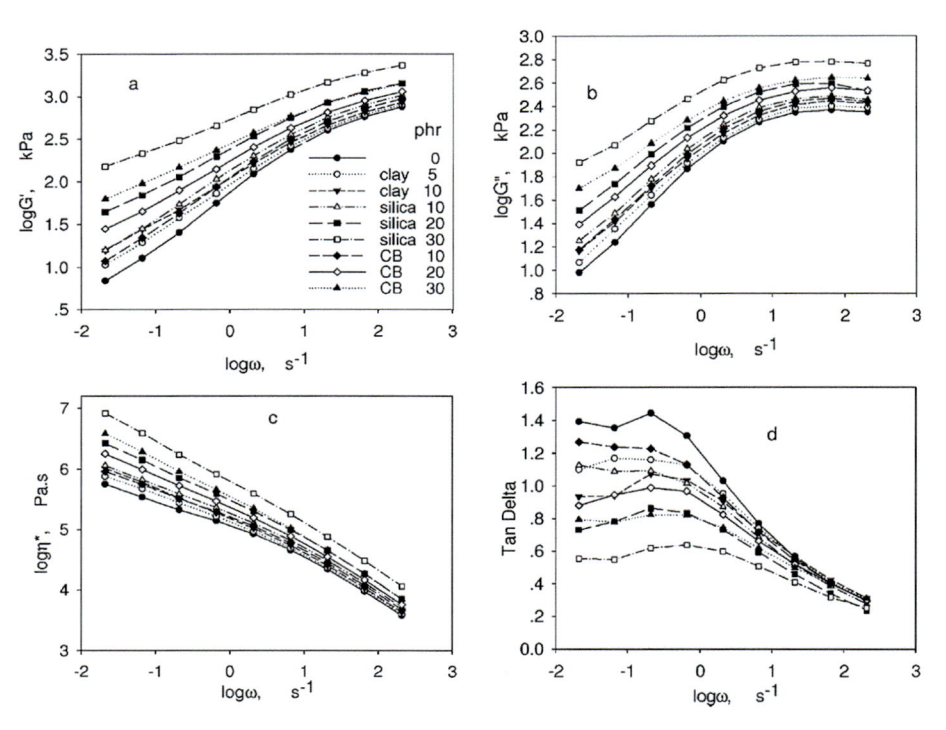

Figure 3.3.22. The storage (a) and loss (b) moduli, complex dynamic viscosity (c) and Tan Delta (d) as a function of a frequency for EPDM and EPDM/filler mixtures at 100°C.[Adapted, by permission, from H. Tan, A. I. Isayev, *J. Appl. Polym. Sci.*, **109**, 767-774 (2008)].

region. Finally, it is clearly established that in addition to the effect of filler surface area on rheological properties, different fillers with the same surface area affect the dynamic properties of compounds.

3.3.5 EFFECT OF MOLECULAR WEIGHT DISTRIBUTION ON NON-NEWTONIAN FLOW

Various attempts to construct the flow curve equations are based on molecular models. For this approach, the knowledge of the mechanism of non-Newtonian behavior is required. Polymeric materials are characterized by their polydispersity, i.e., they are *mixtures of fractions* having different molecular mass. As demonstrated in Fig. 3.3.23, the viscosity of the *monodisperse polymer* is constant, $\eta_0 = $ const, until critical stress, σ_s, after which no flow is possible. Formally, it can be written as:

$$\eta = \begin{cases} \eta_0 & \text{at} \quad \sigma \leq \sigma_s \text{ or } \dot{\gamma} \leq \dot{\gamma}_s \\ \sigma_s/\dot{\gamma} & \text{at} \quad \sigma > \sigma_s \text{ or } \dot{\gamma} \geq \dot{\gamma}_s \end{cases} \qquad [3.3.34]$$

Let us mix two fractions of different molecular masses, each behaving like a Newtonian liquid until shear stress is σ_s. The results are shown in Fig. 3.3.23. The blend of two quasi-"Newtonian" liquids has non-Newtonian behavior. The deviation from the low shear

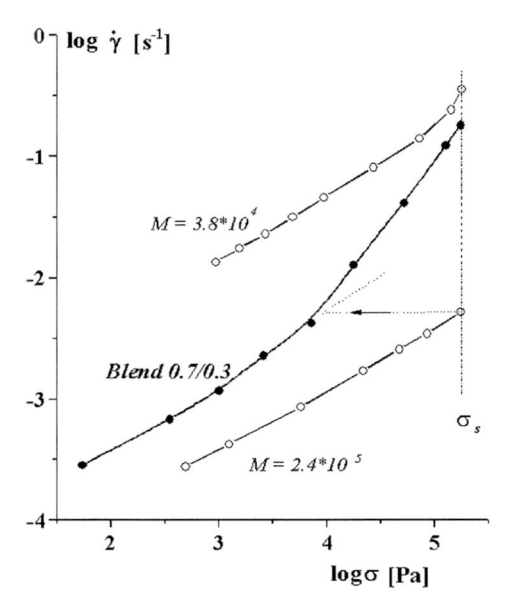

Figure 3.3.23. Non-Newtonian behavior of a polydisperse polymer sample. Flow curve of a blend (0.7/0.3) of two monodisperse polymers. Polyisoprenes. MM are shown at the curves. 25°C. Horizontal arrow shows the spurt shear rate for a high molecular weight fraction. [Adapted, by permission, from A.Ya. Malkin, N.K. Blinova, G.V. Vinogradov, M.P. Zabugina, O.Yu. Sabsai, V.C. Shalganova, I.Yu. Kirchevskaya, V.P. Shatalov, *Europ. Polym. J.*, **10**, 445 (1974)].

rate Newtonian branch begins at a shear rate corresponding to a critical point, $\dot{\gamma}_s$, of the high molecular fraction, as shown by a horizontal arrow in Fig. 3.3.23. This means that there are two domains in the flow curve of the blend and each of them is characterized by different rheological properties. In the low shear rate range, the blend, similar to both components, is a Newtonian liquid. In the shear rate range above the critical shear rate of the high molecular mass component, this blend is a non-Newtonian liquid.

The following model of the non-Newtonian behavior of a polydisperse polymer can be proposed.[111] Low shear rate viscosity of a polydisperse polymer is determined by some rule of mixing; non-Newtonian viscosity at the high shear rate range is calculated as the sum of inputs, first, low molecular mass components flowing as Newtonian liquids, and second, high molecular mass components flowing at shear rates exceeding their (different for different fractions) critical values $\dot{\gamma}_s$ and behaving as non-fluid "filler".

The rule of mixing may have different forms based on empirical or molecular arguments. In particular, it can be assumed that the Newtonian viscosity of a polydisperse polymer, η_0, is proportional to the weight-averaged molecular mass of a blend in the power b, as was discussed above (Eq. 3.3.16)

Based on the above-formulated suggestions, it is possible to write the following equation for a flow curve:

$$\eta(\dot{\gamma}) = \left[\int_0^{M(\dot{\gamma})} (K_2 M^\alpha)^{1/\alpha} f(M) dM + \left(\frac{\sigma_s}{\dot{\gamma}} \right)^{1/\alpha} \int_{M(\dot{\gamma})}^\infty f(M) dM \right]^\alpha \qquad [3.3.35]$$

Here f(M) is a molecular mass distribution function.

The structure of Eq. 3.3.35 demonstrates that non-Newtonian behavior results from polymer polydispersity. The first term in this equation is the input of "flowing" fractions, and the second term represents the input of high molecular mass fractions which do not flow at high shear rates. The boundary between these two terms is the function of shear rate. This boundary (the upper limit of the first integral and the lower limit of the second integral) is expressed as:

$$M(\dot{\gamma}) = \left(\frac{\sigma_s}{K_2\dot{\gamma}}\right)^{1/\alpha}$$ [3.3.36]

The validity of this approach is confirmed by the results of calculations of the non-Newtonian branch of the flow curve of a blend (the solid symbols in Fig. 3.3.23). This method attracted attention of many researchers because it permits solving the inverse problem – calculation of molecular weight distribution, function f(M), based on rheological measurements. Though it is an incorrect (ill-posed) problem,[112] various methods for solving it are widely discussed in modern rheological literature.[113]

The non-Newtonian flow of various fluids can originate from two different mechanisms. First, this can be stress-induced structure rearrangement which is typical of multi-component materials (emulsions, suspensions, and so on) or polymer with strong intermolecular interaction. Second, this can be the result of polymer polydispersity because, in this case, we encounter mixtures of several substances with different viscoelastic properties.

3.4 ELASTICITY IN SHEAR FLOWS

3.4.1 RUBBERY SHEAR DEFORMATIONS – ELASTIC RECOIL

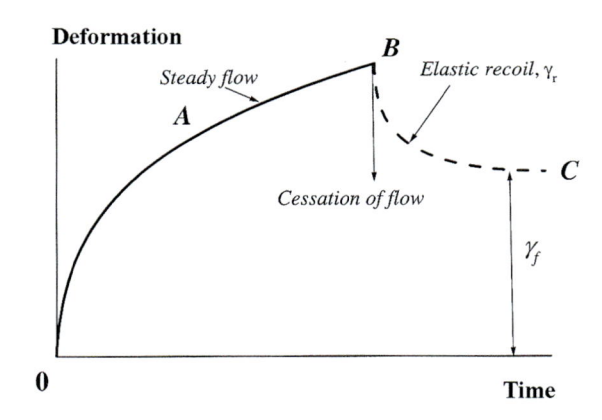

Figure 3.4.1. Deformation and elastic recoil.

Results of continuous development of shear deformation after application of constant shear stress are schematically shown in Fig. 3.4.1 by the 0AB part of the curve. Section 0A is an unsteady deformation range and the straight line AB represents the deformations corresponding to a steady flow. At point B stress is removed, flow ceases and elastic recoil takes place. The value of this *elastic recoil* is a measure of the *elastic* (or *rubbery*) *deformation*, γ_r, stored before the regime of steady flow was reached. This value is also called the *recoverable strain*, or *recoil strain*. The limits of elastic recoil are elastic deformations stored during the steady flow called *ultimate recoil*, γ_∞. The total shear deformation, γ, is the sum of γ_r and the irreversible (plastic) deformation of flow, γ_f:

$$\gamma = \gamma_r + \gamma_f$$ [3.4.1]

This sum reflects the superposition principle, which is valid in the linear viscoelastic behavior domain. However, it is not a universal principle. It is possible to carry out the same measurements at high shear stresses (the non-linear domain). In this non-linear case, the measured elastic recoil does not correspond to the elastic deformation stored at steady flow: if in the linear region there is no flow during elastic recoil, in the non-linear case, some flow (accompanied by partial dissipation of the stored elastic energy) can take place during elastic recoil, and the measured value of γ_r is lower than expected. Therefore, the values of γ_r measured in the non-linear viscoelastic domain must be treated as relative (though might be useful) characteristic of elasticity only but not as the absolute values of rubbery deformations in the state of steady flow.

Rubbery elasticity is characterized by the *rubbery modulus*, G_e, or by the reciprocal value − *equilibrium shear compliance*, J_s, or (in the non-linear domain of viscoelastic behavior) *ultimate recoil function*. These parameters are determined as:

$$G_e = J_s^{-1} = \frac{\sigma}{\gamma_\infty} \qquad\qquad [3.4.2]$$

By definition, G_e and J_s, are constants in the linear viscoelastic domain and they become dependent on the shear stress in the transition to the non-linear behavior domain.

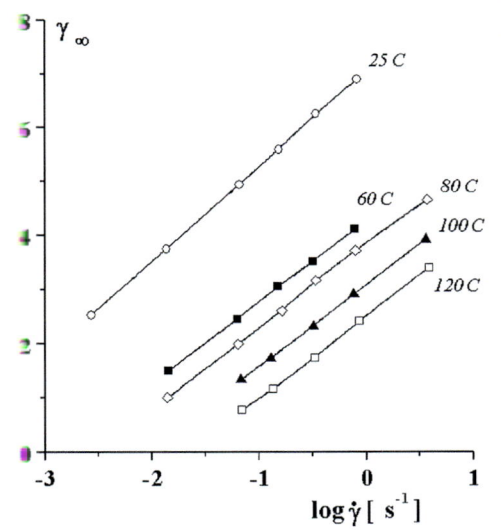

Rubbery elasticity is typical for all liquid polymeric materials − melts, solutions, dispersions, some colloid systems forming polymer-like structures, such as aluminum naphthenate colloid solutions, and some other materials.

A typical example of the $\gamma_\infty(\dot\gamma)$ dependence is shown in Fig. 3.4.2. The real measured values of recoverable strains may reach several hundred percent and they increase with the increase in shear rate (or shear stress). Changes in the rubbery modulus (or shear compliance) as a function of shear rate do not exceed one decimal order, even in the range of shear rates exceeding 6 decimal orders.[114] Measured values of the rubbery modulus in the non-linear domain have qualitative meaning only due to the noticed absence of superposition of elastic and flow deformations.

Figure 3.4.2. Elastic recoil as a function of shear rate for butyl rubber at various temperatures. [Adapted, by permission, from G.V. Vinogradov, A.Ya. Malkin, M.P. Zabugina, V.F. Shumsky, *Vysokomol. Soedin.* (Polymers − in Russian), **11A**, 1221 (1969)].

The initial ("linear") value of shear compliance, J_s^0, is an important rheological parameter (see Chapter 2). It is a factor related to the molecular structure of matter.

Many authors noticed that J_s^0 of the so-called "monodisperse" polymers does not depend on the molecular mass (MM) of a polymer.

The following values of J_s^0 were obtained experimentally for "monodisperse" polymers (or at least polymers with very narrow molecular-mass distribution):[115,116]

Polydimethylsiloxanes	$1.5*10^{-3}$ Pa^{-1}
Polybutadienes	$1.6*10^{-4}$ Pa^{-1}
Polyisoprenes	$1.4*10^{-5}$ Pa^{-1}
Polystyrenes	$7.1*10^{-6}$ Pa^{-1}

These values are only estimates because of difficulties in the preparation of "monodisperse" polymers.

The elasticity of polymer melts depends on the width of molecular-mass distribution (MMD). Regularly, MMD is characterized by the $\overline{M}_w/\overline{M}_n$ ratio. However, the $\overline{M}_w/\overline{M}_n$ ratio is not a representative characteristic of MMD in its correlation with J_s^0, because J_s^0 is determined primarily by higher averaged values of MMD, i.e., the presence of even very small amounts of species with very high MM dramatically increases rubbery elasticity (quantitatively presented by the steady-state compliance) of melt.

The J_s^0 values are determined by ratios of different average molecular weights, including the higher moment of molecular weight distribution. Two of these relations are the most popular. As was mentioned above, many authors used the ratio $\overline{M}_z/\overline{M}_w$ as the measure of molecular weight distribution determining J_s^0. In this case, the following equation is usually used:[117]

$$J_s^0 = A\left(\frac{\overline{M}_z}{\overline{M}_w}\right)^{3.7}$$

The other equation is as follows:[118]

$$J_s^0 = A\left(\frac{\overline{M}_z\overline{M}_{z+1}}{\overline{M}_n\overline{M}_w}\right)^{3.7}$$

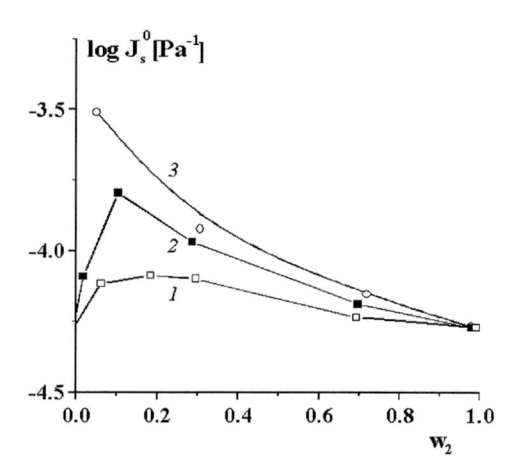

Actually, it is rather difficult to make the choice between these equations. Both equations can successfully describe the same experimental data, as was demonstrated for high-density polyethylenes.[119] At the same time the ratio $\overline{M}_w/\overline{M}_n$ can be used for the rough correlation with J_s^0, as was shown for polypropylenes.[120]

The role of small fractions of high MM fractions is illustrated in Fig 3.4.3, where J_s^0 is presented as a function of the relative concentration in a mixture of two "monodisperse" polymers with different MM. The J_s^0 values of mixed polymers are the same, but the blend has a very different value of J_s^0. Also, a very strong effect of small amounts of high MM fractions is observed. Analogous experimental evidence is known for various polymers. The J_s^0 values are also sensitive to details of chain architecture and particularly to branching. Fig. 3.4.3 also shows that compliance of solutions increases when solvent

Figure 3.4.3. Shear compliance of mixtures of "monodisperse" polyisoprenes – influence of MMD. w_2 – concentration of a component with higher MM. 1 – MMs of components: $1.6*10^5/5.75*10^5$; 2 – MMs of components $6.5*10^4/8.3*10^5$; 3 – Solution of a polymer (MM = $5.75*10^5$) in low MM oligomer (MM = $7.3*10^3$). [Adapted, by permission, from G.Zh. Zhangereyeva. M.P. Zabugina, A.Ya. Malkin, in Rheology of polymers and disperse systems and rheo-physics, *Inst. of Heat and Mass Transfer, Belorussian Academy of Sciences,* Minsk, **1**, 161 (1975)]

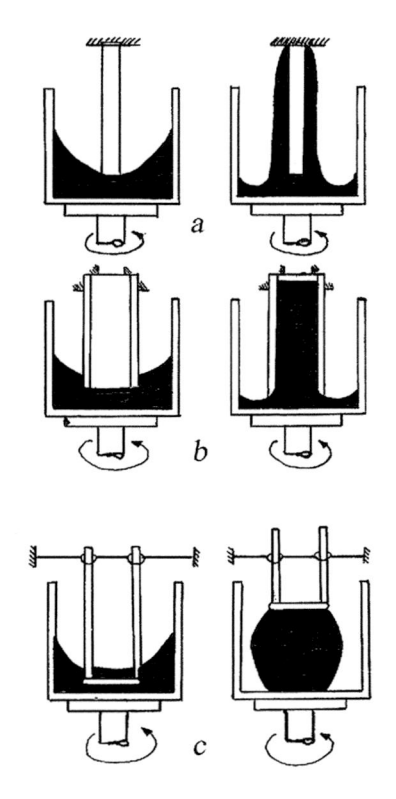

Figure 3.4.4. The Weissenberg effects (see text for details).

content increases and becomes much higher than J_s^0 of polymer melt.

The nature of recoverable deformations (elastic recoil) in polymer melts and solutions is the same as in the cured rubbers: conformational changes in chain configurations in space. The difference between linear and cured polymers is in the lifetime of junctions: in cured rubbers junctions can be treated as existing infinitely long; junction (entanglement) lifetime in flowing solutions and melts is limited and therefore the elastic recoil is accompanied by sliding in junctions, i.e., by flow.

3.4.2 NORMAL STRESSES IN SHEAR FLOW

3.4.2.1 The Weissenberg effect

Weissenberg described several abnormal observations from the point of view of classical fluid dynamics.[121] Saponified hydrocarbon colloid gels and polymer systems (solutions and gels) become the main materials for demonstrating the Weissenberg effect.[122]

Some examples of the phenomena called the *Weissenberg effect* are shown in Fig. 3.4.4, where the behavior of traditional viscous fluid and special "rheological" liquid are compared. The left pictures demonstrate what happens with an ordinary viscous liquid, whereas the pictures on the right side show the special phenomena observed in the deformation of "rheological" (or elastic) liquids. These are called the *Weissenberg effect*.

If a rod is rotating inside a "*rheological*" liquid, such liquid, instead of being displaced out of the rotor by the centrifugal forces towards the walls of a vessel, begins to climb around the rotor (Fig. 3.4.4, case a). In the case of two coaxial cylinders (inner hollow), the rotation of the outer cylinder forces liquid into the inner cylinder (Fig. 3.4.4, case b). Another characteristic example concerns the flow of liquid between two parallel discs. When the outer disc is rotated around a common axis, the inner disc is lifted up by the normal force generated due to rotation (Fig. 3.4.4, case c), and if a hole were made in the center of one of the discs, this "rheological" liquid, instead of being removed from the space between the discs to the periphery of the discs, is pressed through the hole. Some other related observations were also made which demonstrate similar unusual behavior of "rheological" liquid.

The immediate impression appears that in the flow of "rheological" liquid some forces exist which compress liquid from outside normal to its surface, acting like a stretching elastic ribbon twining around a sample and forcing it to move to the center. Such *normal forces* act not only normal to the surfaces of the discs but also in the radial direction to the central axis as well.

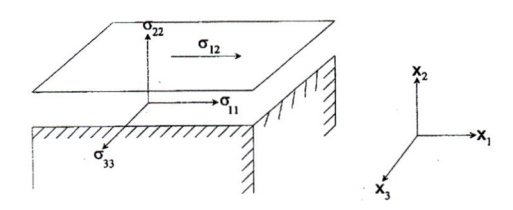

Figure 3.4.5. Stress tensor in shear between two parallel planes.

Phenomena associated with the Weissenberg effect are very common in technological applications: reactors are supplied with rotating mixing elements (mixers) of different geometry and type. Such equipment is used in various branches of the chemical, food, and pharmaceutical industries. Moreover, the last example of the Weissenberg effect shown in Fig. 3.4.4, case c was used to design a special type of machine for polymer processing, the so-called *screwless extruder*. In this design, a material is mixed between two discs and pressed through a nozzle by the action of forces originated from the Weissenberg effect. Normal forces caused by rotation of parallel discs also give additional support in slide bearings if the "rheological" (elastic) liquid is used as a lubricant. Some investigators think that the Weissenberg effect provides the ability of articulations in biological organisms and that deficiency in this phenomenon leads to illness ("squeak in joints").

It is interesting to understand the rheological origin of this effect, i.e., what happens at the reference point. The common feature of all phenomena is the appearance of forces acting in the direction of shear and in the perpendicular direction. In the language of continuum mechanics it is equivalent to the appearance of the normal stresses in shear.

It can be anticipated that normal components of the stress tensor are related to the appearance of a diagonal component in the tensor of large elastic deformations. So, there is a common agreement that the Weissenberg effect is related to the existence of *normal stresses* in shear, and the latter can be explained by large elastic deformations developed in the flow. (It may be argued whether this explanation is general, but undoubtedly the large elastic deformations lead to normal stresses).

3.4.2.2 First normal stress difference – quantitative approach

Stress combination in a simple shear is expressed by the stress tensor shown in Fig. 3.4.5. The tensor has the following structure:

$$\sigma_{ij} = \begin{bmatrix} \sigma_{11} & \sigma_{12} & 0 \\ \sigma_{21} & \sigma_{22} & 0 \\ 0 & 0 & \sigma_{33} \end{bmatrix} \qquad [3.4.3]$$

The stress tensor can be presented as the sum of hydrostatic pressure and the deviatoric components. If the pressure is not high, it is possible to neglect the compressibility of liquid relevant in some real situations. Therefore, only the deviatoric part of the stress tensor is important for the deformation of fluids. It means that if the hydrostatic pressure is superimposed, it will change all normal components of the stress tensor but would not influence flow. The direct consequence of this approach is that in order to characterize the effect of normal stresses in shear flow, it is not the absolute values of normal stresses but their *differences* that are important.

The differences of normal stresses are defined as follows:

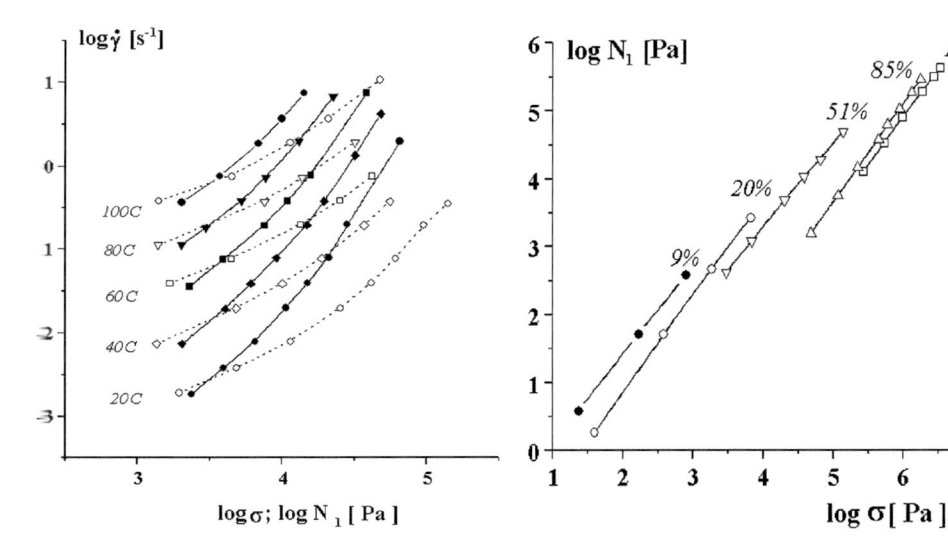

Figure 3.4.6. Shear and the first normal stress dependencies on shear rate in a steady flow at different temperatures. Polyisobutylene. M = 1*10⁵. Dotted lines – shear stresses; solid lines – normal stresses. [Adapted, by permission, from G.V. Vinogradov, A.Ya. Malkin, V.F. Shumsky, *Rheol. Acta*, **9**, 155 (1970)].

Figure 3.4.7. Relationship between the first difference of normal stresses and the shear stresses in a steady flow of butyl rubber solutions in cetane. T = 20°C. Concentrations are labelled on the curves. [Adapted, by permission, from G.V. Vinogradov, A.Ya. Malkin, G.V. Berezhnaya, *Vysokomol. Soedin.* (Polymers – in Russian), **12**, 2797 (1971)].

the first (or primary) difference of normal stresses: $\qquad N_1 = \sigma_{11} - \sigma_{22}$

the second (or secondary) difference of normal stresses: $\quad N_2 = \sigma_{22} - \sigma_{33}$

In rheological literature, N_1 and N_2 are not often used but their coefficients (analogous to the use of viscosity coefficient instead of the shear stress) are introduced.

The *first* and the *second normal stress coefficients* are defined as

$$\Psi_1 = \frac{\sigma_{11} - \sigma_{22}}{\dot{\gamma}^2} = \frac{N_1}{\dot{\gamma}^2} \qquad [3.4.4]$$

and

$$\Psi_2 = \frac{\sigma_{22} - \sigma_{33}}{\dot{\gamma}^2} = \frac{N_2}{\dot{\gamma}^2} \qquad [3.4.5]$$

It is interesting to compare the shear and normal stresses in a steady shear flow in order to understand the meaning and value of the normal stress effect. An example of experimental results is shown in Fig. 3.4.6 for the shear stress and the first normal stress difference. The normal stresses are smaller than the shear stresses at low shear rates but they are growing along with an increase of the shear rate much faster than shear stresses and they exceed the latter at high shear rates. The ratio between stress tensor components depends not only on the shear rate but on the nature of the liquid, temperature, and so on.

The direct relationship between the first normal stress difference, N_1, and the shear stress, σ, can be obtained after excluding the shear rate (see Fig. 3.4.7). N_1 is proportional to the square of the shear stress:

$$N_1 \propto \sigma^2$$

It means that the appearance of the first difference of normal stresses is the *second-order effect* in relation to shear stresses.

The coefficients of normal stress differences (and apparent viscosity) depend on shear rate. This is caused by a strong non-linearity of the rheological properties of the liquid. The initial values of the coefficients of normal stresses, Ψ_0, as their limits at low shear rates, can be defined analogously to Newtonian viscosity.

The first normal stress difference is the second-order effect in the whole shear rate range, including the range of non-Newtonian flow. The following relationship is valid:

$$\Psi(\dot{\gamma}) = \left(\frac{\Psi_0}{\eta_0^2}\right)\eta^2(\dot{\gamma}) \qquad [3.4.6]$$

The first normal stress difference is the effect of the second order in comparison with the viscous flow and this directly leads to the relationship between Ψ_0 and molecular mass of the polymer (based on Eq. 3.3.15):

$$\Psi_0 = K_\Psi M^{2b} \qquad [3.4.7]$$

where K_Ψ and b are constants.

However, normal stresses — contrary to shear stresses — are more dependent on higher moments of molecular mass distribution than viscosity, i.e., the M value in Eq.3.4.7 for polydisperse polymers must be substituted not by $\overline{M}_w$ but by higher averaged values of molecular mass.

3.4.2.3 Second normal stress difference

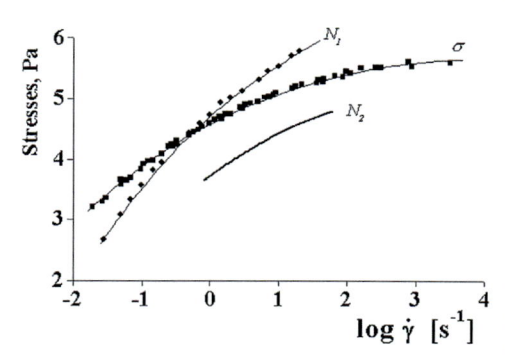

The *second normal stress difference* seems less important for practical applications because $N_2 \ll N_1$. However, the measurements of N_2 are important for constructing different rheological models and theories. The published experimental results of N_2 are controversial: in early publications, zero, negative, and positive values of N_2 were found. However, careful analysis of possible experimental errors leads to the general conclusion that N_2 is negative ($N_2 < 0$) and its absolute value is close to $(0.1-0.3)N_1$.[123] Some shear rate dependencies of all components of the stress tensor are presented in Fig. 3.4.8.

Figure 3.4.8. Shear and normal stresses in stationary flow of polyisobutylene [Adapted, by permission, from: N. Ohl and W. Gleissle, *Rheol. Acta*, **31**, 294 (1992)].

3.4.3 NORMAL STRESSES AND ELASTICITY

The common agreement that the normal stresses are related to the elasticity of liquid is quantitatively expressed by the Lodge equations valid for "linear" elastic liquids (see also Section 2.5.2):

$$J_s^0 = \frac{N_1}{2\sigma^2}$$

[3.4.8]

or

$$\gamma_\infty = \frac{N_1}{2\sigma}$$

[3.4.8a]

These equations can also be written as a relationship between coefficients:

$$G_e = J_s^{-1} = \frac{2\eta_0^2}{\Psi_1}$$

[3.4.9]

These equations demonstrate that the normal stresses are second-order effects with respect to the shear stresses.

Eq. 3.4.9 is for Newtonian liquid capable of exhibiting elastic deformations. Therefore this equation contains coefficients for Newtonian viscosity, η_0, and

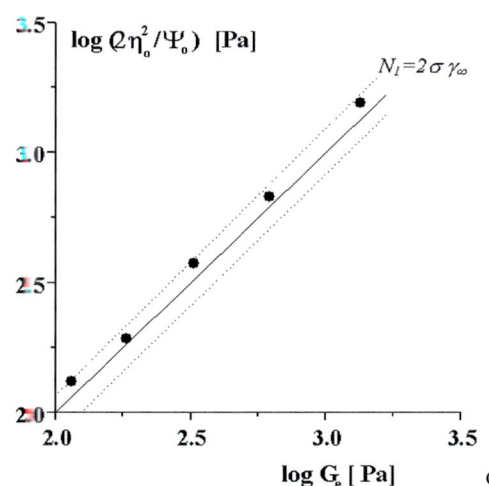

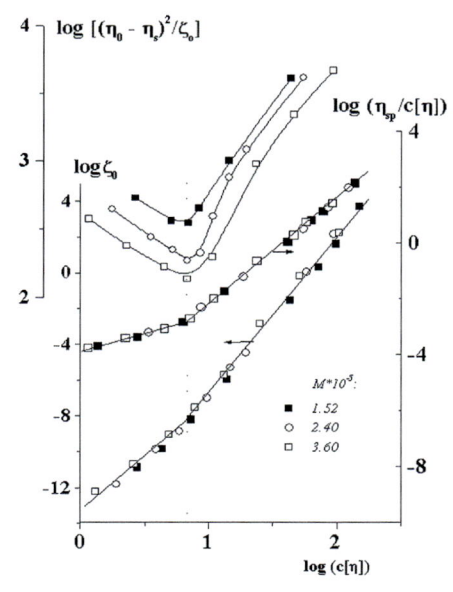

Figure 3.4.10. Comparison of concentration dependencies of Newtonian viscosity, η_0, coefficient of the first difference of normal stresses, ζ_0, and elastic modulus, G_e, calculated as $(\eta_0 - \eta_s)^2/\zeta_0$. In this Figure: $\eta_{sp} = (\eta_0 - \eta_s)/\eta$ and η_s is the viscosity of a solvent. Coefficient of normal stresses, ζ_0, used in the original publications equals $\zeta_0 = \Psi_0/2$. Polybutadiene solutions in methyl naphthalene. T = 25°C. [Adapted, by permission, from A.Ya. Malkin, G.V. Berezhnaya, G.V. Vinogradov, *J. Polym. Sci.: Symposia*, **42**, 1111 (1973)].

Figure 3.4.9. Comparison of independently measured values of G_e and the ratio (η_0^2/Ψ_0). Polystyrene solutions in decalin. T = 25°C. Dotted lines – limits of 25% experimental error. [Adapted, by permission, from A.Ya. Malkin, G.V. Berezhnaya, G.V. Vinogradov, *J. Polym. Sci., Symposia*, **42**, 1111 (1973)].

zero-shear-rate coefficient of the first normal stress difference, Ψ_1. The applicability of Eq. 3.4.9 is illustrated by experimental data in Fig. 3.4.9. The elastic modulus, G_e, and the coefficients η_0 and Ψ_0 were measured independently. The values of G_e and $(2\eta_0^2/\Psi_0)$ are equal.

Eq. 3.4.8 is not valid in the nonlinear range of viscoelastic flow, primarily because the directly measured elastic recoil does not correspond to the elastic energy stored at the steady flow regime (because of the absence of superposition of elastic and flow deformations in the nonlinear domain). However, the normal stresses can be used as a direct measure of the stored elastic energy in flow. The expression for the stored elastic energy, W, per unit volume can be obtained from the Lodge equation for an elastic liquid:

$$W = \frac{N_1}{4} \qquad\qquad [3.4.10]$$

This equation permits the estimation of stored elastic energy in non-linear regimes of steady flow.

Fig. 3.4.10 compares concentration dependencies of rheological characteristics of solutions, such as viscosity, normal stresses, and elastic modulus using the reduced concentration, $c[\eta]$, as an argument.

The Debye criterion, $c[\eta] = 1$ helps to establish a boundary of dilute solutions. This concentration range is on the left side of Fig. 3.4.10. There are two separate concentration ranges. In the low concentration range, at $c[\eta] < (c[\eta])_{cr}$, where $(c[\eta])_{cr} \approx 5\text{-}7$, η_0 is a weak function of concentration and ζ_0 is proportional to the square of viscosity (the second-order effect). In the high concentration range, i.e., at $c[\eta] > 7$, η_0 is proportional to c^5 and ζ_0 is proportional to c^{10} (again, the second-order effect). It is worth mentioning that concentration dependencies of the "apparent" modulus pass through a minimum at the same concentration $(c[\eta])_{cr}$.

Thus, there are *three* concentration domains in polymer solutions:
- domain of *dilute* solutions at $c[\eta] < 1$; in this concentration range, macromolecules deform independently of one another
- domain of *semi-dilute* or *semi-concentrated* solutions at $1 < c[\eta] < 7$; in this concentration range contacts between different macromolecules are possible but a continuous three-dimensional entanglement network is still absent
- domain of *concentrated solutions* at $c[\eta] > 7$; in this concentration range, a special entanglement network exists throughout the whole volume of solution.

Rubbery (large) deformations in solution happen in the range of concentrations in which macromolecular chains form a temporary network, i.e., at $(c[\eta]) > (c[\eta])_{cr}$. Fig. 3.4.10 shows that normal stresses and "apparent" modulus can be measured in the low concentration range, $(c[\eta]) < 7$. This means that the elastic energy can be stored not only by the molecular network but also by individual macromolecules, changing their conformation under shear deformation in solution. It becomes increasingly more "difficult" (modulus increases) to deform a macromolecule by diluting its solution.

It can be predicted that normal stresses always appear if the shear deformation results in a three-dimensional (3D) reaction of structural elements in a flowing medium. The deformation of polymer liquid (either melt, solution, or cured rubber) inevitably results in changes in the macromolecular conformations. The deformation or relaxation −

due to statistical reasons – always proceeds in 3D space. Macro- one-dimensional shearing leads to 3D micro-reaction and this is the reason for the appearance of normal stresses in a simple shear flow. This is also true for dilute polymer solutions where the 3D network is absent, but individual macromolecules deform in 3D space in any geometry of macrodeformations. The same effect, not related to the elasticity of medium, can be expected in the shear flow of suspensions of non-spherical particles (e.g., the fiber in a Newtonian matrix). The micro-reaction is 3D and it results in a relaxation process restoring statistical distribution of particle orientation in space and normal stresses in shear flow.

3.4.4 DIE SWELL

Figure 3.4.11. Extrudate swell in post extrusion of polyethylene melt. (Original photo of the authors).

When a stream of viscoelastic liquid leaves a die, the diameter of extrudate increases, and the ratio of the extrudate diameter to the die diameter is called the *swell ratio*, S_R. Calculations based on the kinetic equations of classical fluid dynamics predict that the jet diameter changes for Newtonian liquid do not exceed 13% and extrudate contracts at high Reynolds numbers. However, Barus[124] demonstrated that there are liquids (e.g., marine glue) for which a large swelling of a stream was observed. Later, analogous experimental results were published by Merrington[125] who experimented with rubber solutions and soap-thickened mineral oils and directly explained this effect by the elastic nature of materials used.

The typical example of *die swell* (or post-extrusion swelling) is shown in Fig. 3.4.11 for a polymer melt. The values of S_R for real technological materials can reach several units and they provide a measure of the elasticity of these materials. The observed values of S_R depend on the preceding kinematics of deformation – the shear rate at the capillary and the length of the channel. Also, S_R depends on the nature of the material, reflecting its capability to store elastic deformations.[126]

Elastic deformations in the post-extrusion stage of polymer processing are caused by the release of elastic energy stored during deformations inside a channel. However, the die swell occurs in the nonlinear viscoelastic behavior domain and the elastic deformations of a free stream are accompanied by a partial dissipation of the stored elastic energy. This is why it is not easy to establish direct quantitative relationships between S_R and the rheological parameters of the material. However, S_R can be used as an important and useful estimation of the elasticity of the material.

The extrudate swell distorts the shape of extruded articles in comparison with the sizes of calibrating devices (of a die). It gives an especially undesirable effect in the extrusion of complicated profiles because the elastic swell appears differently in different parts of a cross-section. The production of calibrating dies in the extrusion of such profiles is expensive, and it gives a good chance to develop numerical methods for calculation of S_R in the flow-through channels of various cross-sections.

3.5 STRUCTURE REARRANGEMENTS INDUCED BY SHEAR FLOW

3.5.1 TRANSIENT DEFORMATION REGIMES

In steady flows of liquids, there is a unique relationship between shear stress, σ, and shear rate, $\dot{\gamma}$. In transient deformation regimes, such an unambiguous relationship is absent. It means that at $\sigma = $ const, the shear rate continuously changes until it reaches its steady level and *vice versa*.

A transient deformation regime also exists for Newtonian liquid, however, it is due to inertial effects alone. In contrast, inertial effects are negligible for highly viscous liquids. The reasons for the transient behavior of non-Newtonian viscous liquids are:
- development and storage of elastic (rubbery) deformation, superimposed on flow
- structure changes induced by shearing.

Both reasons may exist simultaneously.

The transient behavior is observed for $\dot{\gamma} = $ const. In this case, a pre-stationary evolution of shear stress is observed. Typical relationships are shown in Fig. 3.5.1 for an elastic polymer melt and in Fig. 3.5.2 for an inelastic grease. Formally both sets of curves are similar: at low shear rates, a monotonic growth of shear stress, σ, is observed; at higher shear rates, a maximum, σ_m, on a stress-vs.-time curve, $\sigma^+(t)$, appears. This phenomenon is usually called an *overshoot*. In both cases, the steady-state stress, σ_{st}, is reached at $t \to \infty$. There are some principal peculiarities in both sets of experimental data, and the transient behavior in both cases reflects two different mechanisms of this effect.

Analysis of the $\sigma^+(t)$ curves for polymer solutions and melts shows that both σ_{st}

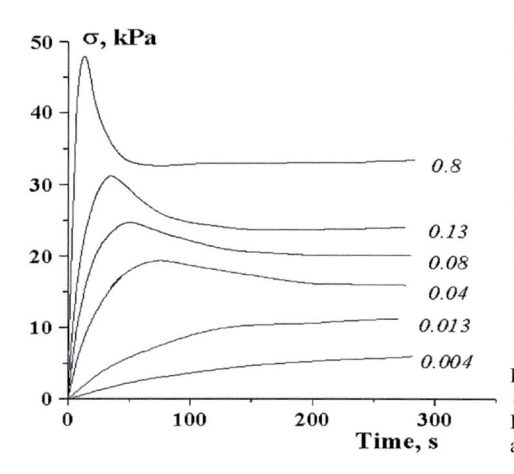

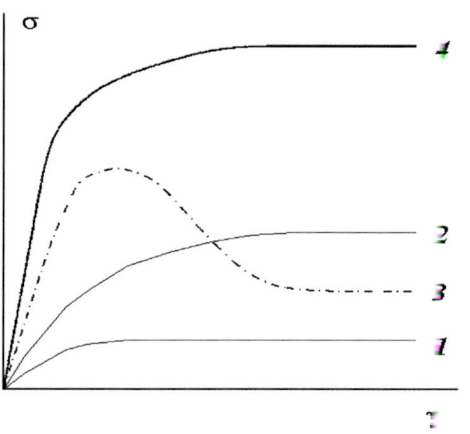

Figure 3.5.1. Transient shear stress evolution at startup of shear flow at different shear rates. Values of shear rate are shown at the curves (in s⁻¹). Polyisobutylene. MM = $1*10^5$. T = 0°C. [Adapted, by permission, from E. Mustafayev, A.Ya. Malkin, E.P. Plotnikova, G.V. Vinogradov, *Vysokomol. Soedin.* (Polymers – in Russian), **6**, 1515 (1964)].

Figure 3.5.2. Transient shear stress evolution at startup of shear flow of a grease at different shear rates. Increasing shear rates: $\dot{\gamma}_1 < \dot{\gamma}_2 < \dot{\gamma}_3 < \dot{\gamma}_4$. Curves 1 and 2 – creeping flow of a material with undestroyed structure; curve 3 – deformation in the yielding range (transition through the maximum of the structure strength); curve 4 – flow of material with destroyed structure. [Adapted, by permission, from V.P. Pavlov, G.V. Vinogradov, *Dokl. AN SSSR* (Reports of the Academy of Sciences of the USSR – in Russian), **122** 646 (1958); *Kolloid. Zh.* (Colloid J. – in Russian), **23**, #3 (1966)].

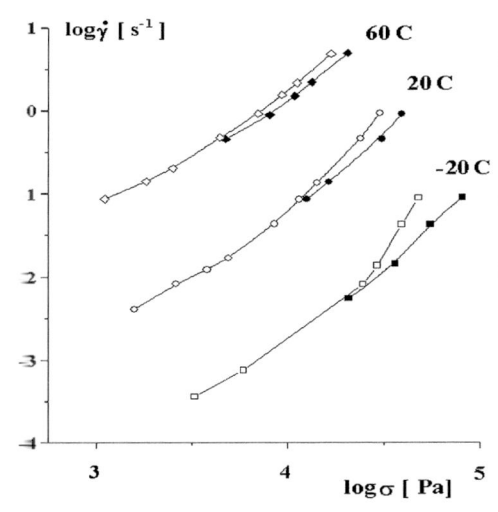

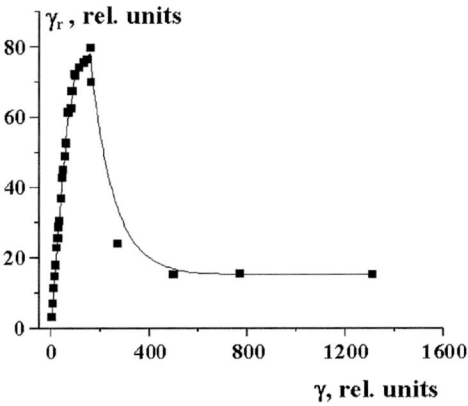

Figure 3.5.4. Evolution of elastic (rubbery) deformations for a 5% solution of polyisobutylene in vaseline. $\dot{\gamma} = 160$ s^{-1}; T = 20°C. [Adapted, by permission, from A.A. Trapeznikov, *Kolloid Zh.* (Colloid J. – in Russian), **28**, 666 (1966)].

Figure 3.5.3. Shear rate dependence of stresses at steady-state flow, σ_{st} (open symbols) and at maxima of the transient stage of deformations, σ_m (solid marks), P-lyisobutylene. MM = $1*10^5$. [Adapted, by permission, from G.V. Vinogradov, A.Ya., Malkin, E.P. Plotnikova, V.A. Kargin, *Dokl. AN SSSR* (Reports of the Academy of Sciences of the USSR – in Russian), **154**, 1-21 (1964)].

and σ_m are shear rate dependent and they increase monotonously with an increase in $\dot{\gamma}$. Fig. 3.5.3 shows the $\sigma_{st}(\dot{\gamma})$ dependencies and the $\sigma_m(\dot{\gamma})$ dependencies. Analysis of the transient deformation regimes of polymer systems compares $\sigma^+(t)$ curves with the development of elastic (recoverable) deformations, γ_r, stored during the *transient deformation range*. The dependence $\gamma_r(t)$ at $\dot{\gamma}$ = const is shown in Fig 3.5.4 for 5% solution of polyisobutylene in vaseline. Analogous dependencies are observed for many polymer solutions and melts as well as for some colloid systems; for example, for aluminum naphthenate dispersions.[127]

The following observations were made. At the beginning of shearing, the total deformation is mainly elastic. Hence, the material in this deformation range is primarily "stretched" but does not flow. The elastic deformation can be as large as 60-80 units (6,000-8,000%). These are very large values. They are typically observed in materials such as aluminum naphthenate, some gels, egg-white, and others. Typical elastic deformations of synthetic polymer melts do not exceed several units. Similar to the $\sigma^+(t)$ dependence, the $\gamma_r(t)$ dependence passes through a maximum. It is caused by the elastic deformation of the entanglement network, the knots of which are temporary bonds between macromolecules or polymer-like colloid particles. If the time of deformation is shorter than the characteristic lifetime of these junctions, they behave as permanent bonds and the material mechanical behavior is similar to that of rubber. At longer times, macromolecular chains slip in the junctions, partly destroying the network. It results at the beginning of flow and a decrease in the stored elastic deformations. Part of the entanglements (with a lifetime longer than $\dot{\gamma}^{-1}$) continue to exist even in a steady flow and therefore some elasticity can also be found in the flowing liquid.

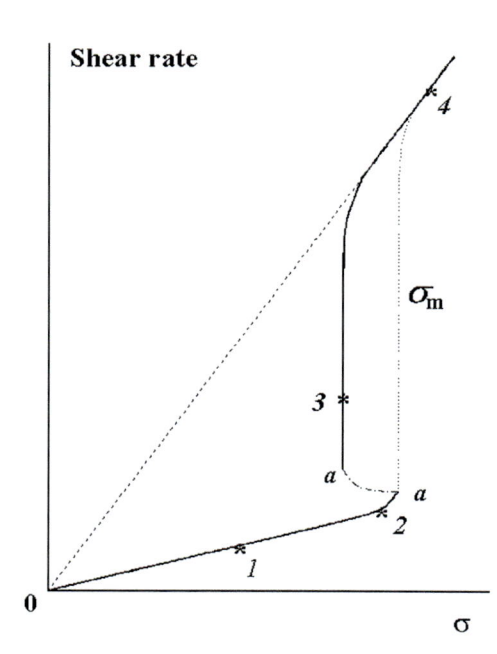

Figure 3.5.5. Complete rheological curve of a viscoplastic liquid (grease). 1-2 – creeping flow of material with undestroyed structure; aa – range of unsteady deformation regimes; 3 – range of yielding; σ_m – stress limit (strength) of structure; 4 – flow of liquid with destroyed structure. [Adapted, by permission, from G.V. Vinogradov and V.P. Pavlov (see Refs. in Fig. 3.5.2)].

The existence of a very sharp maximum on the $\gamma_r(t)$ curves suggests that the maximum on the $\sigma^+(t)$ curves reflects elastic deformation in shear: at short times material deforms predominantly as a rubber-like, and, only after partial disintegration of entanglements, the transition to viscous flow occurs. This physical reason explains the shapes of curves in Figs. 3.5.1.

The physical effects responsible for the transient behavior of greases (as well as numerous other technological materials) are different. At very low shear rates (curves 1 and 2 in Fig. 3.5.2), the transient $\sigma^+(t)$ curves are analogous to curves depicted in Fig. 3.5.1. These curves correspond to very high viscosity values, typical of structured materials with the yield value. An increase in the shear rate (curve 3) leads to the appearance of a maximum on the $\sigma^+(t)$ curves. However, the steady-state stress value is lower than that of curves 1 and 2. At very high shear rates (curve 4), the maximum on the $\sigma^+(t)$ curves disappears, the apparent steady value of the stress, σ_{st}, increases again, and the viscosity at these shear rates is low, corresponding to the viscosity of a dispersion phase (oil). It is also important to mention that the steady-state stress, σ_{st}, in the flow of plastic greases is constant in a wide range of shear rates and therefore it must be regarded as the yield stress, σ_Y, typical of any viscoplastic materials. The stress values at the maximum of the $\sigma^+(t)$ curves, σ_m are also constant in the same shear rate range.

The rheological behavior of this material is generalized in Fig. 3.5.5, where the dependencies of σ_{st} and σ_m on shear rate are shown, and the numbers at the points 1, 2, 3, and 4 correspond to the number of the curves in Fig. 3.5.2. It is evident that the material is viscoplastic, and the following features are characteristic of such materials in addition to the existence of the yield stress.

First, if one follows the $\sigma_{st}(\dot{\gamma})$ dependence, a multi-valued branch of the flow curve appears in the transition region from the range of very low shear rates to the central part of the flow curve. This branch is shown in Fig. 3.5.5 by the dashed line aa. Possibly, this part of the flow curve is not reached experimentally and the attempts to "measure" the apparent viscosity of this branch of a flow curve results in a sudden increase in flow rate.

Second, the difference in viscosity values between the lower part (straight line 012 in Fig. 3.5.5) and the upper part (straight line 04 in Fig. 3.5.5) of the flow curve are very large and reach many decimal orders (the graphs in this figure are not presented in the same scale). The initial low shear rate branch of the flow curve is characterized by high viscosity values, which are typical of viscoplastic materials with solid filler (see Section 3.2.2). Viscosity at the final high shear rate branch of the flow curve is low and close to the viscosity of a dispersion medium.

Third, the difference between σ_m and σ_{st} values shows that the properties of a material depend on its pre-history. The yield stress varies depending on whether the shear rate increases or decreases in the cycle of measurements: if the shear rate is decreased and measurements started, after the initial structure of the material is destroyed, it is possible not to "notice" the high yield stress value corresponding to σ_m. It is a reflection of the thixotropic behavior of the material, which will be discussed in more detail in the next section.

Systems such as greases, toothpaste, and other analogous viscoplastic materials are not rubbery at all. Rheological properties, as presented in Fig. 3.5.5, are attributable to structure effects and partial damage to their inherent structure caused by deformation. In this sense, the stress at the maximum σ_m of the $\sigma^+(t)$ curve can be treated as the strength of this structure.

The transient time dependence of stresses and the transitions through the stress maximum in materials of any type is generally explained by the "structure effects", though this structure is different. For polymer and polymer-like colloid systems, it is the network of rubbery chains, which can store large elastic deformations and slip in temporary entanglements at low deformation rates. For plastic materials, it is a rigid coagulated structure that is formed by solid particles. This structure is characterized by a certain strength and it is destroyed at the yield stress. Then, Figs. 3.5.1 and 3.5.2 show the difference in flow curves of elastic and plastic materials as discussed in Section 2.1. It is possible to find intermediate situations since the structure of the real natural or synthetic materials can be formed in various ways.

3.5.2 THIXOTROPY AND RHEOPEXY

The term *"thixotropic"*[128] means "sensitive to touching" and the effect of thixotropy is commonly regarded as slow changes in viscosity, or in a more common sense, any rheological properties induced by deformation and rest after deformation. The difference between "non-Newtonian" and "thixotropic" behavior is in the time-scale: by definition, it is supposed that the apparent viscosity of non-Newtonian liquid changes immediately along with the change in the shear rate, whereas the viscosity of the thixotropic medium changes slowly. This difference is evident in some cases, but it may not be easy to establish the time-scale of changes and in some cases, it may be reasonable to treat thixotropy as a cause of non-Newtonian behavior and *vice versa*. Deformation destroys the inherent structure of structured liquids, leading to non-Newtonian behavior, and recovery of the structure requires time, resulting in thixotropic effects. The difference between thixotropy and other rheological properties is not always clear for viscoelastic materials, because time effects due to elasticity are superimposed on structure rearrangements and can proceed within the same time-scale. Moreover, the question of whether the relaxation phenomenon (which is also caused by structure rearrangements, e.g., molecular chain movements and/or disappearance of fluctuation) can be treated as "thixotropy" or not, can be discussed.

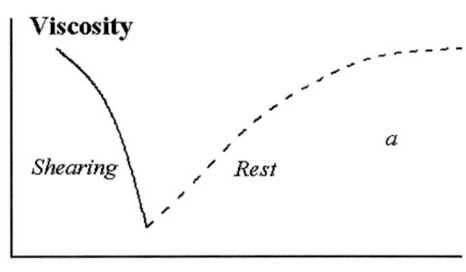

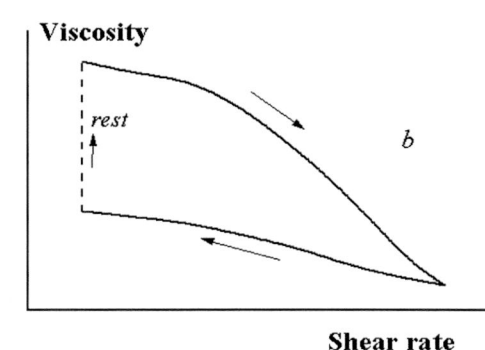

Figure 3.5.6. Typical thixotropic effects. a: viscosity decrease in deformation at constant shear rate and viscosity increase at rest; b: viscosity decrease as a function of ascending shear rate at steady flow and delayed restoration of viscosity at descending shear rate.

Structure processes in some cases may be thought to be similar to chemical reactions leading to the formation of chemical bonds with unlimited (or at least very long) lifetime. In this sense thixotropy is a part of a more general field of the rheokinetic effects, i.e. rheological transformations taking place in the synthesis and reactions of different materials, primarily oligomers and polymers (see Section 6.3).

Thixotropy is understood in two ways (Fig. 3.5.6): first, as a slow viscosity decrease due to deformations (at the constant shear rate), together with a reverse process of complete or partial restoration of the initial viscosity at rest after deformation ceases (Fig. 3.5.6a); second, as a decrease of steady-state values of the apparent viscosity at different shear rates accompanied by delayed restoration of the viscosity values with decreasing shear rates (Fig. 3.5.6b). Both cases are treated as the kinetic reactions of structural rearrangements induced by deformations and it is commonly accepted that thixotropic materials are those that have a *microstructure*, though the latter term can be understood in various ways.

There is a large number of commercial materials which are thixotropic in nature. They include:[129]

- paints, inks, and coatings
- sealants and adhesives
- detergent systems
- clay suspensions
- oils and lubricants
- coal suspensions
- metal slushes
- rubber solutions, especially filled with different fillers
- food and biological systems (including blood)
- creams and pharmaceutical products.

Viscosity decrease of many food products (yogurt, cream, pastes, gels), pharmaceuticals, or concrete as a result of their mixing encountered every day are typical and technologically important cases of thixotropic behavior. A typical example is shown in Fig. 3.5.7 for a body lotion: a decrease in viscosity on shearing, followed by a slow increase at rest. It is pertinent that the quality (or performance properties) of products are determined by a

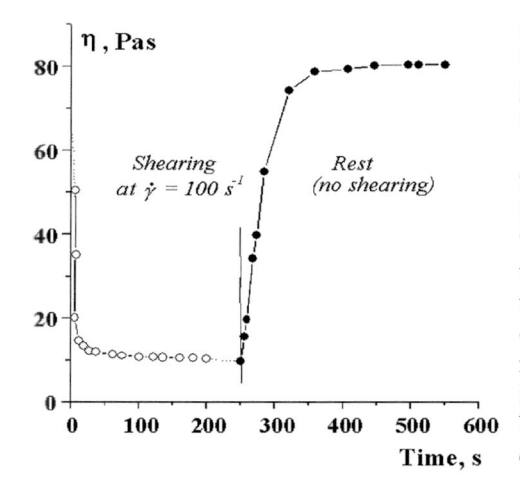

Figure 3.5.7. Thixotropic properties of a body lotion: decrease of viscosity in shearing and slow growth at rest. [Adapted, by permission, from G. Schramm, "A practical approach to rheology and rheometry", Haake, 1994].

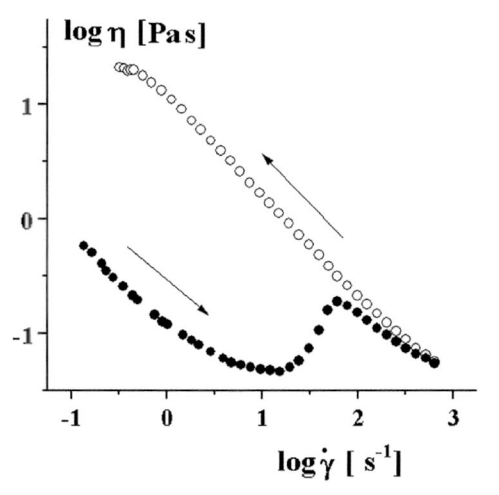

Figure 3.5.8. Apparent viscosity evolution in up-and-down changes of shear rate for a ternary system of alkali dimethyl oxide/alcohol/water. The period of shear rate scanning was 20 min. T = 25°C. [Adapted, by permission, from J. I. Escalante, H. Hoffman, *Rheol. Acta*, **39**, 209 (2000)].

degree of the viscosity decrease and the rate of its increase, though the quality of products in everyday observations are preferably judged as "good" or "bad".

The deformation process in some systems may lead, not to a decrease but to an increase in viscosity. This is called "*antithixotropy*" or "*rheopexy*". Sometimes, these effects are distinguished. Both terms are related to a continuous increase of viscosity with time at a constant shear rate, and/or thickening of material with the increase in shear rate, and the difference between them lies in the time scale. Both effects are much rarer than thixotropy, but they occur in some materials, especially in systems containing active chemicals, e.g., ionic or polar groups. Deformation may enhance the intermolecular interaction, resulting in a viscosity increase.

The thixotropic behavior of the material is well demonstrated in cyclic experiments. The shear rate is continuously increased up to the upper level and then decreased to the initial value. The structure of the sample in such a deformation regime is first destroyed and then gradually rebuilt during the entire $\dot{\gamma}(t)$ dependence. A thixotropic hysteresis loop is observed. The last point (at the lowest shear rate) may be different than the initial one if the characteristic time of scan is shorter than the time of structure formation, i.e., the time is not long enough for the initial structure to recover during the time of measurement.

Thixotropy, as well as rheopexy, is often observed for surfactant-water systems because intermolecular interactions in such systems are strongly pronounced. Shearing leads to structural transformation among colloid particles. Two examples from modern rheological literature illustrate the above-mentioned phenomena. First, the shear-rate dependence of the apparent viscosity measured in cyclic up-and-down deformation mode is shown in Fig. 3.5.8. The unusual effect of viscosity growth at about 10 s^{-1} is explained by the transition of the bilayer structure to more viscous vesicle particles.[130]

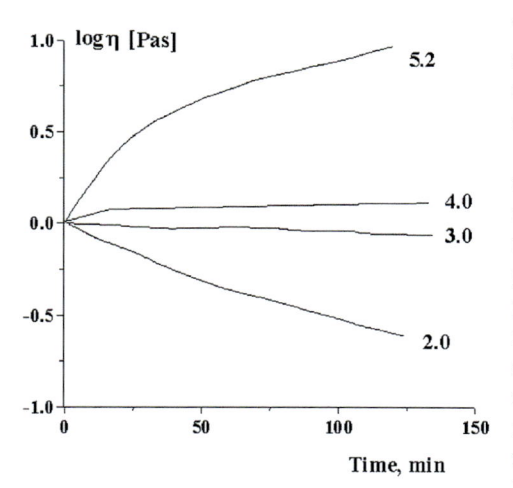

Figure 3.5.9. Steady-state viscosity evolution in time, measured at $\dot{\gamma}$ = 2.25 s⁻¹ for organogels containing 250 mg/mL lecithin. The ratio of water molecules to lecithin molecules is shown at the curves. [Adapted, by permission, from Yu.A. Shchipunov, H. Hoffman, *Rheol. Acta,* **39**, 542 (2000)].

Then the positions of ascending and descending branches of the flow curve are different because they represent different structures of the flowing colloid system.

The same material may exhibit either thixotropic or antithixotropic effect, and the observed rheological behavior sometimes depends on minor details of the composition. Fig. 3.5.9 illustrates this phenomenon for lecithin solutions in water, forming polymer-like micelles.[131] By changing the molar ratio of water to lecithin it is possible to cause a transformation from shear-thinning (thixotropic) to shear-thickening (antithixotropic) rheological behavior in this colloid system. The effects observed are explained by the restructuring of the polymer-like micelles and their network. Kinetics of restructuring is slow in contrast to quick viscoelastic effects observed in colloid systems.

A quantitative description of thixotropic phenomena depends on the approach selected. Some authors consider it as a particular case of time-dependent effects in the deformation of materials, analogous to the viscoelastic effect. They prefer to treat it in terms of constitutive equations, i.e., in the framework of a mechanical approach only.[132] However, it seems that this approach is not adequate, especially in practical applications, because the evident changes in rheological properties (in particular, viscosity) are observed even when no stresses can be measured. The situation is the same as with phase transitions or chemical reactions, which are not mechanical phenomena. It is, therefore, more fruitful to apply a quasi-chemical approach to thixotropy-antithixotropy phenomena, introducing a kinetic equation for some parameter of a system (it can be a number of bonds, or concentration of structure aggregates, or some arbitrary not rigorously defined "structure" parameter, and so on) determining viscosity or other rheological properties of the material.

Slow changes in rheological properties, definitely related to structural effects, are well demonstrated in examining the shear stress evolution measured at constant shear rates, $\dot{\gamma}$=const. The $\sigma(t)$ dependence has a maximum characterizing the "strength" of the structure destroyed by deformations. If one repeats this experiment just after cessation of the previous shearing, the maximum disappears and a monotonous shear stress growth is observed. However, if one lets the material rest, the maximum in the curve appears again and its height depends on the time of rest. This is shown in Fig. 3.5.10, where the subsequent curves correspond to successively longer times of rest. These observations can be easily explained by the structure rupture caused by deformations and slow structure restoration during rest.

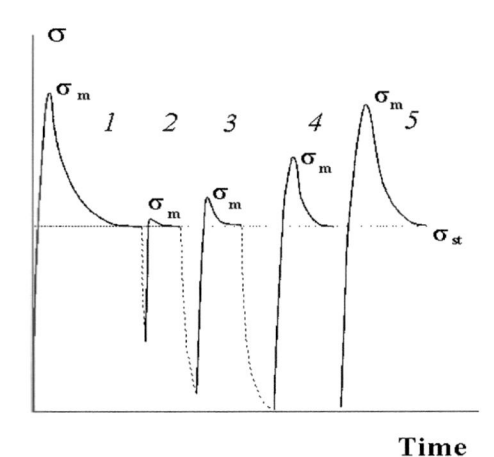

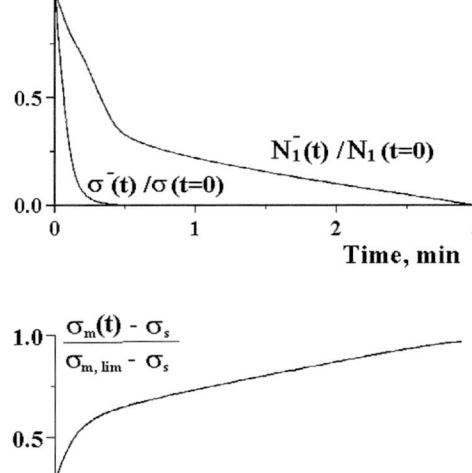

Figure 3.5.10. Thixotropic structure restoration of polymer material in consecutive cycles of shearing and rest. 1 – initial deformation; 2, 3, 4 – deformation after rest during different (increasing) times; 5 – deformation after very prolonged rest – complete restoration of the initial structure. Dotted lines – relaxation at rest.

Figure 3.5.11. Comparison of the relaxation rates of shear, s(t), and normal, N_1, stresses after sudden cessation of steady shear flow (upper figure) along with the kinetics of structure restoration at rest as observed by the relative increase of maximal stress in transient flow (lower figure). Polyisobutylene. T = 20°C. [Adapted, by permission, from G.V. Vinogradov, A.Ya., Malkin, E.P. Plotnikova, V.A. Kargin, *Dokl. AN SSSR* (Reports of the Academy of Sciences of the USSR – in Russian), **154**, 1421 (1964)].

It is worth mentioning that this phenomenon of structuring is not directly related to stress relaxation. Indeed, Fig. 3.5.11 clearly demonstrates that both are developing in absolutely different time scales: stresses completely disappear during several minutes while the approach to the initial value of the maximum, σ_m, on the $\sigma(t)$ dependence, continues for several hours.

Examination of the transient behavior in shearing gives even more pronounced effects if one measure, instead of shear stresses, the first normal stress difference, $N_1^+(t)$.[133] First of all, instead of a single maximum, two maxima are observed for normal stresses. Then, if the position of the first maximum is close (in time-scale) to the shear stress maximum, the second normal stress maximum appears after prolonged deformations reaching several hundred units. So, this effect must be related to a cooperative movement of macromolecules but not to the relaxation of the individual chains. It is definitely a structure formation effect due to intermolecular interactions induced by shearing. Structures formed during prolonged shearing in some cases can be found by direct optical observations. These structures are thixotropic by their nature: they are destroyed and again restored at slow deformation and long rest time.

It is interesting to notice that in measuring transient $N_1^+(t)$ dependencies, oscillations of stress are observed. Possibly this effect is also related to the formation and disintegration of larger structures in the flow, which influence the rheological properties of the system as a whole.[134]

Structure evolution in transient deformation regimes was well documented in the experiments performed with model immiscible polymer blends,[135] as well as with textured polymeric liquid crystals.[136] In particular, it was demonstrated that the yield stress appears to be different depending on whether the shear rate increases or decreases in the cycle of measurement: if the shear rate is decreased and measurements are started when the initial structure of the material is destroyed, it is possible not to "notice" the higher yield stress value corresponding to σ_m.

The results of numerous observations confirm that the overshoot on the $\sigma^+(t)$ and $N_1^+(t)$ curves and evolution of rheological properties at rest are accompanied by large-scale structure effects related to thixotropic properties of multi-component or textured systems.

Thixotropy is usually attributed to the time-dependent shear-induced phenomenon. Peculiarities in rheological behavior can also be related to volume effects. Effects of this kind are called *dilatancy*.[137] This phenomenon was first described by Reynolds,[138] who noticed that traces of footsteps on wet sand stay when deformed. It means that volume changed under shear deformation. Dilatancy is important for humid soils, such as wet sands and clays: shearing leads to change in water content and in an increase of viscosity. This phenomenon is called *shear thickening*.

Volume effects (dilatancy) caused by shear are also possible in elastic bodies, as well as in viscoelastic liquids. This phenomenon for solids was described by Kelvin,[139] who treated it as a second-order effect, i.e., volume changes are expected to be proportional to the square of shear deformations. In solids, it is known as the *Poynting effect* (see Chapter 4).

Many industrial materials are thixotropic. In technological practice, transportation of these materials through long pipes is an important and frequently complex problem. Therefore methods of designing pipelines for the transportation of thixotropic inelastic liquids were developed.[1407] The simulation procedures are based on the formulation of the constitutive equation for time-dependent (and non-Newtonian) viscosity and solving the dynamic problems. However, the complexity of the governing equations always calls for extensive numerical methods for finding the final results.

The effect of *rheopexy* also belongs to the group of structure rearrangement induced by deformation. According to the definition, rheopexy "is a solidification of a thixotropic system by the gentle and regular movement".[141] Fig. 3.2.6 demonstrates the effect of rheopexy for super-concentrated water-in-oil emulsion. It is seen that deformation at a low shear rate is accompanied by the slow growth of apparent viscosity covering several decades. Actually, the apparent viscosity increases unlimitedly until the yield stress is reached.

It is also important that rheopectic effects can be observed at low shear rates only (i.e., at "gentle movement"). The experimental points in Fig. 3.5.12 were obtained at increasing and decreasing shear rates. The experimental points coincide in the high shear rate range, but the curves are different in the range of slow deformations and the difference increases with the shear rate decreasing. It is also worth mentioning that viscosity drops from the upper to the lower curve almost immediately after the cessation of shear-

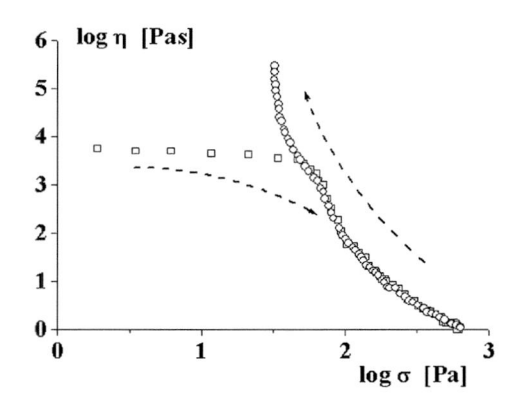

Figure 3.5.12. Flow curves of a water-in-oil super-concentrated emulsion measured at upward and downward changes of shear rate. [Adapted, by permission, from A.Ya. Malkin, I. Masalova, P. Slatter, K. Wilson, *Rheol. Acta*, **43** (2004)].

ing. Structure rearrangement at rest, after shearing proceeds very quickly, can be explained by elastic deformation of dispersed droplets that causes a rheopectic effect.

Deformation can lead to self-organization in a flowing medium. Some examples of this effect have been already described in Sections 3.2.3 and 3.6.3. Self-organization and induced anisotropy can also be created in dense suspensions of rigid particles.[142] Here, it is reasonable to add that self-arrangement can also lead to the formation of anisotropic micellar structures in the flow of block-copolymers.[143] It is interesting to report that structure effects in shearing can occur in self-oscillation (periodic) manner with corresponding periodic (in time) variation of apparent viscosity.[144]

Structure formation leads to the anisotropy of the rheological properties of a matter.

In discussing the rheology of fluids with their structure dependent on deformation, it is necessary to introduce additional structural (or internal) parameters that should be coupled with dynamic equations. An example of such an approach is the model incorporating the flow-induced anisotropic structural changes. The induced structure was described by the tensor, and its evolution was governed by a relaxation-type differential equation.[145]

Finally, it is interesting to report that structure effects in shearing occur in a self-oscillation (periodic) manner with corresponding periodic (in time) variation of apparent viscosity.[146]

3.5.3 SHEAR INDUCED SEPARATION

3.5.3.1 General approach

Shear flow can lead to a transformation of the material structure, and this is the main inherent cause of the non-linearity of the rheological properties. The structure evolution can be so fundamental that it results in the shear-induced phase transitions.

The rigorous concept of phase transitions treats this phenomenon as an equilibrium coexistence of thermodynamic equilibrium states of a matter. Any phase transition is characterized by the intermittent change of macroscopic properties.

The interrelation between the temperature of phase separation is well known in classical thermodynamics. The following dependence was established:

$$\frac{dT}{dp} = \frac{T\Delta V}{\Delta H} \qquad [3.5.1]$$

where dT/dp is a change of the phase transition temperature due to changes in pressure, T is temperature, ΔV is a change of molar volume at the phase transition point, and ΔH is the molar heat of transition.

This equation, known as the *Clausius-Clapeyron equation*,[147] relates the shift in the equilibrium phase transition temperature to hydrostatic pressure, i.e., to the volume changes of material.

It is evident that the concept of time, as well as the rate of the process, do not enter any thermodynamic relationships. However, rheology is interested in shifts in transitions induced by shear flow (proceeding mainly without volume changes).

The first observation related to the subject was the effect of extension-induced crystallization of rubbers.[148] The full theory of this effect as an analogue of the pressure-induced shift of the crystallization temperature is well known.[149] Later, it was shown that the shearing induces phase transitions in amorphous systems too.[150]

It was stressed by many authors that shearing may result in two opposite effects: homogenizing of multi-component systems and flow-induced phase separation.[151] Shear deformation always favors homogenizing of a multi-component system due to pure geometrical reasons, and this is not necessarily related to the phase transition from a two- to a one-phase system. The shear-induced phase separation is caused by the same thermodynamic reasons as the phase transition due to static pressure. It is caused by changes in the free energy of a system due to its deformation.[152] It becomes possible in the shear flow because the flow in some systems (primarily polymer solutions) is accompanied by the storage of rubbery deformations, i.e., elastic energy (see Section 3.4).[153]

There are three aspects of flow-induced transition.

First, the *thermodynamic effect*: deformations can lead to a shift of the equilibrium phase transition temperatures.

Second, the *kinetic effect*: deformations change the kinetics of a phase transition at the fixed temperature.

Third, the *morphological effect*: deformation-induced phase transition leads to the formation of different molecular and supermolecular structures, which determine the properties of the end product.

These three aspects are connected: a shift in the equilibrium phase transition temperature results in a shift of the distance from a given temperature to an equilibrium one, and this is the reason for changes in the process kinetics and so on.

The shear-induced structural transformations related to the chain alignment can result in the formation of an LC-state. Surely, this is possible only in the presence of rod-like elements (molecules or anisotropic supramolecular or colloidal particles). The LC-state formation of micellar solutions depends on boundary conditions, such as the nature of the boundary solid surface and the width of the gap between boundary surfaces (i.e., the width of a liquid layer).

The influence of deformation on the transition temperature was observed for both amorphous phase separation and crystallization. Fig. 3.5.13 shows the dependence of the shift in the equilibrium transition temperature, ΔT, as a function of shear rate for two polymer solutions of different concentrations. The dominating factor is shear stress. Fig. 3.5.14 demonstrates that the dependence of ΔT on the shear stress is similar for solutions of different polymers. The influence of shear flow on the equilibrium transition temperature can be as high as 30-40K.

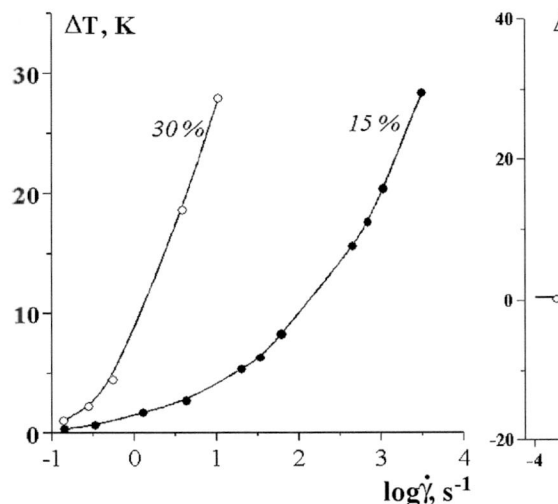

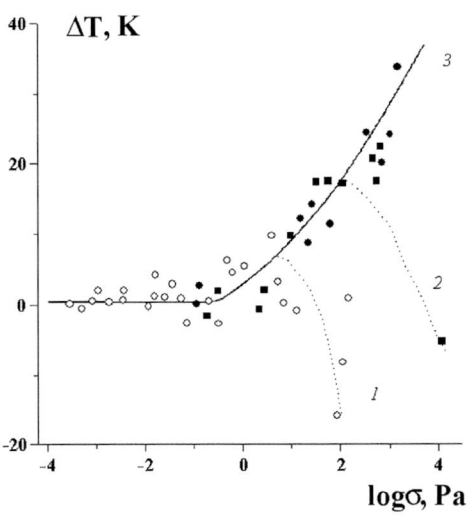

Figure 3.5.13. Shift in the phase separation temperature as a function of shear rate. Polyvinylacetate solutions in ethanol. Concentrations are shown on the curves. T = 25°C. [Adapted, by permission, from A.Ya. Malkin, S.G. Kulichikhin, G.K. Shambilova, *Vysokomol. Soedin.*, (Polymers – in Russian), **33B**, 228 (1991)].

Figure 3.5.14. General dependence of the shift of separation temperature on shear stress for: 1 – polystyrenes of different molecular masses; 2 – polybutadiene solutions in toluene/ethanol mixture; 3 – polyvinyl acetate solutions in ethanol. [Adapted, by permission, from A.Ya. Malkin, S.G. Kulichikhin, V.A. Kozhina, *Vysokomol. Soedin.* (Polymers – in Russian), **38A**, 1403 (1996)].

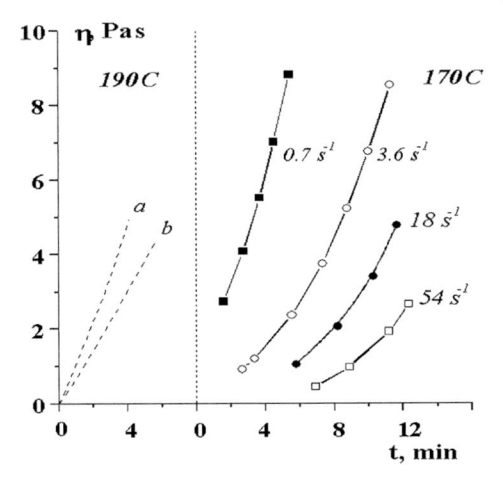

Figure 3.5.15. Viscosity evolution in polymerization of caprolactam at different temperatures and at different shear rates (shown on the curves). The field between the curves 0a and 0b covers points obtained in the shear rate domain from 0.1 to 100 s⁻¹. [Adapted, by permission, from A.Ya. Malkin, S.G. Kulichikhin, *Makromol. Chem., Macromol. Symposia.*, **68**, 301 (1993)].

Two dotted lines are drawn in Fig. 3.5.14. They show that the apparent transition temperature begins to decrease at very high shear rates. This observation is explained by the homogenizing effect at high shear rates.

The influence of shearing on the phase separation temperature is not only of purely academic interest. The following example illustrates its technological significance. Fig. 3.5.15 shows the increase of viscosity in a reacting medium during polymerization of caprolactam in bulk.[154,155] The viscosity increase parallels an increase in concentration and/or molecular mass of polymer formed, i.e., the kinetics of polymerization. At high temperatures, shearing does not influence the kinetics of polymerization. However, at lower temperatures, an increase in the shear rate decreases the rate

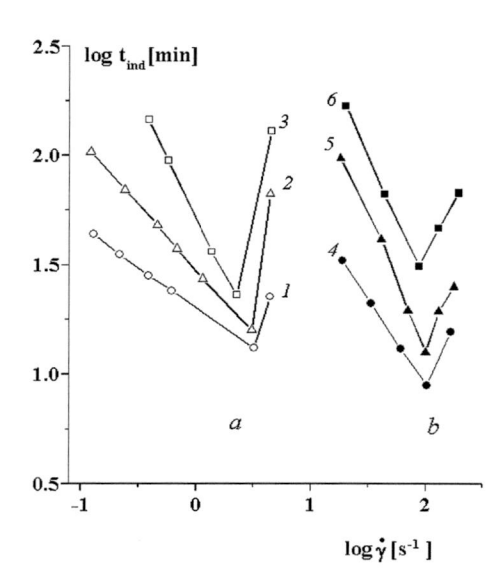

Figure 3.5.16. Influence of shearing at different shear rates on the induction period, t_{ind}, of crystallization for polyamide-6 melt (a) and 30% polyamide-6 solutions in caprolactam (b). Temperatures, °C: a: 1 – 230.0; 2 – 232.5; 3 – 235.0; b: 4 – 160.0; 5 – 162.5; 6 – 165.0. [Adapted, by permission, from A.Ya. Malkin, S.G. Kulichikhin, *Kolloid Zh.* (Colloid J. – in Russian), **41**, 141 (1979)].

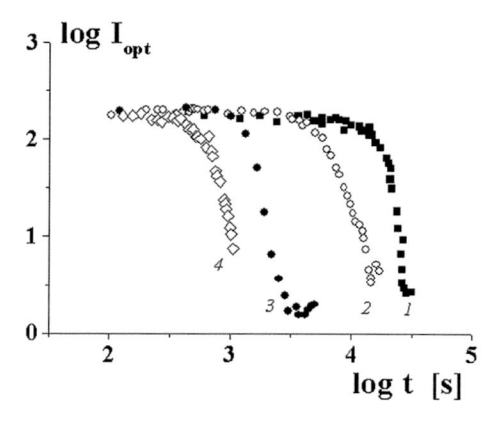

Figure 3.5.17. The intensity of the transmitted light as a function of time in crystallization of isotactic poly(1-butane) at 103°C under different shear rates, s⁻¹: 0 (1), 0.1 (2), 1 (3) and 10 (4). [Adapted, by permission from S. Acierno, B. Palomba, H.H. Winter, N. Grizzuti, *Rheol. Acta*, **42**, 243 (2003)].

of polymerization until almost complete cessation of the process. It is a peculiar behavior because it can be expected that an increase in the shear rate (increase in the intensity of mixing) should accelerate the reaction. But the observed effect is the opposite, and this is because of the phase separation of the polymer from the reactive solution.

The flow-induced phase separation has some special features in non-homogeneous polydisperse systems. In these cases, two effects take place. The forces forming curvilinear shear fields are different, especially if one bears in mind that all real polymers undergo separation differently since they contain species of various molecular masses. Therefore, the separation by the chain length (*shear-induced fractionation*) can proceed due to a difference in the migration speed.[156] The effect is weak and it becomes important for very long chains, primarily for biopolymers.

The influence of shear flow on the kinetics of crystallization is shown in Fig. 3.5.16 where the kinetics of crystallization is characterized by an induction period, t_{ind}. An increase in the shear rate shortens the induction period, i.e., accelerates the crystallization process. However, after passing through a minimum, the induction period begins to increase again. This can be explained by the destruction of crystallization nuclei at high stresses, which is the reason for slower crystallization.

In other experiments, the formation of the crystalline phase was followed by monitoring the transparency of polymer melt by the intensity of light transmission, I_{opt}, through a sample as a measure of crystalline content.[158] Fig. 3.5.17 demonstrates the strong influence of shearing on the kinetics of crystallization. The analogous effect is shown in Fig. 3.5.18 for isotactic polypropylene. Use of the Avrami kinetic

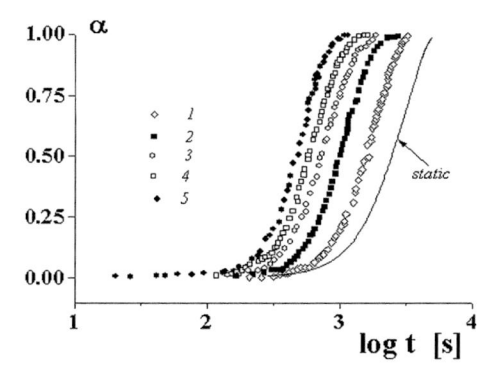

Figure 3.5.18. Kinetics of crystallization of isotactic polypropylene in static state and under shearing at different shear rates, s^{-1}: 0.15 (1); 0.3 (2); 0.5 (3); 1 (4) and 1.4 (5). T = 136°C. [Adapted, by permission, from N. Devaus, B. Monasse, J.-P. Haudin, P. Moldenaers, J. Vermant, *Rheol. Acta*, **43**, 210 (2004)].

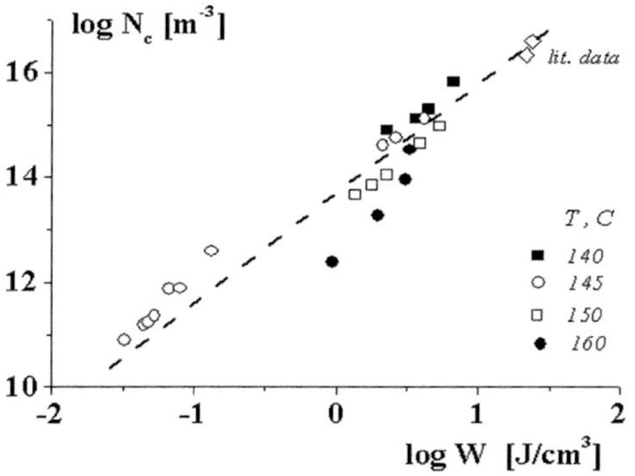

Figure 3.5.19. Density of nuclei formation as a function of work applied to polypropylene melt. Dashed "averaging" line is drawn by us. [Adapted, by permission, from H. Janeschitz-Kriegl, E. Ratajski, M. Stadbauer, *Rheol. Acta*, **42**, 355 (2003)].

Figure 3.5.20. Formation of ultra-fine fibers in the capillary flow of a two-component blend. Scanning electron microscope photograph of an extrudate transversal section of the polyoxymethylene/copolyamide blend. [Adapted, by permission, from M.V. Tsebrenko, G.P. Danilova, A.Ya. Malkin, *J. Non-Newton. Fluid. Mech.*, **31**, 1 (1989)].

equation shows that the rate of crystallization is enhanced as $\dot{\gamma}^{1.22}$, i.e., the influence of shear rate is strong.

The mechanism of influence of deformation on the kinetics of crystallization of polymers is related to nucleation promoted by the flow. The crystallization process relies on two components: nucleation and growth of crystals. Deformation influences the rate of nucleation due to the orientation of macromolecular chains. It was experimentally proven that the density of nuclei is directly related to the mechanical work applied to melt (Fig. 3.5.19).[159]

The morphological aspects of the deformation-induced phase transition are worth mentioning. Shearing may lead to phase separation and phase inversion in multi-component systems. The latter is especially frequent in blends of immiscible polymers. Direct microscopic observations confirm that morphological phase inversion is caused by shearing.[160] The phase separation of the components in polymer blends during the shearing of a blend may result in the appearance of various structures. One of the interesting examples is the formation of a system of ultra-fine fibers of one polymer in the matrix of the other (Fig. 3.5.20).[161] It is possible that it is related to the viscosity ratio of both components. The criterion of phase inversion in shearing is the equality of viscosities of both components in a blend.[162] Generally speaking, the observed picture of structure formation is multi-faceted and depends on numerous rheological and technological factors.

The conformational transition gives us a possibility to obtain a material with fully extended chains. It promises to produce material with outstanding mechanical properties. The theoretical possibility was proven experimentally.[163] This approach was realized in some technological processes, and polymeric materials were obtained with strength and Young's modulus approaching their theoretical (extremely high) limits.

The following consequences of shearing, related to the structure transitions, take place:

- increase in the equilibrium phase transition temperature
- decrease in the induction period of the isothermal crystallization
- increase in the degree of orientation of macromolecular chains
- decrease in the nuclei size of a new phase
- increase in the formation rate of the new phase nuclei
- phase separation and/or phase inversion in multi-component systems
- formation of oriented structures.

All these effects are more or less pronounced for different deformation modes (for example, for extension or shear), though they are always present. The scale of these effects may reach several decimal orders and deformations may induce the above-listed processes, which in turn may lead to their influence on the rheological properties of matter.

It is also interesting to mention that shearing may result in a post-flow effect of crystallization. One reason for this effect is the acceleration of nucleation.[164]

3.5.3.2 Shear banding

There are two types of shear-induced transitions – homogeneous in the bulk of a material (crystallization, creation of supramolecular structures, disorder-to LC transition) and appearance of spacial separated structures.

The phenomenon of shear banding has two aspects. First, this is a typical shear-induced transition and this is why consideration of shear banding is placed in this section. Second, this phenomenon is a manifestation of the instability in the flow of elastic liquids and might be considered in the next section devoted to instabilities in flow of elastic liquids. However, this is not the limit of flow but only its peculiar mode.

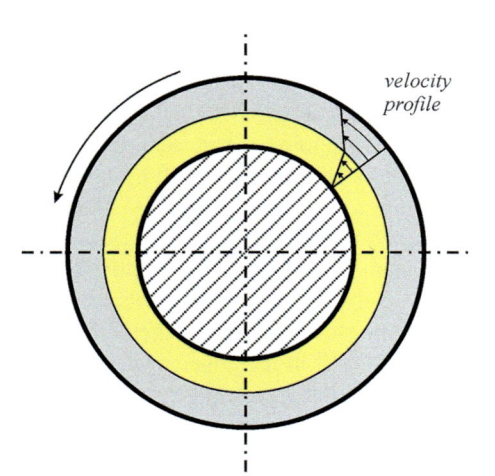
velocity profile

A picture illustrating this phenomenon is shown in Fig. 3.5.21.

Two circles in this Figure correspond to two different states of a material that was homogeneous before shearing but disintegrated into two liquids with different rheological properties.

Physically, this effect is due to a separation of the initially homogeneous multi-component fluid into two parts with differ-

Figure 3.5.21. The formation of two layers ("bands") in rotational flow in a narrow gap between two coaxial cylinders.

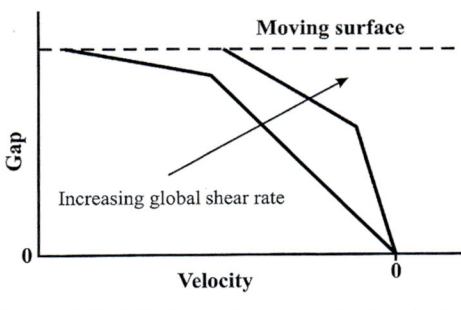
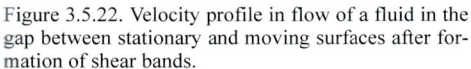

Figure 3.5.22. Velocity profile in flow of a fluid in the gap between stationary and moving surfaces after formation of shear bands.

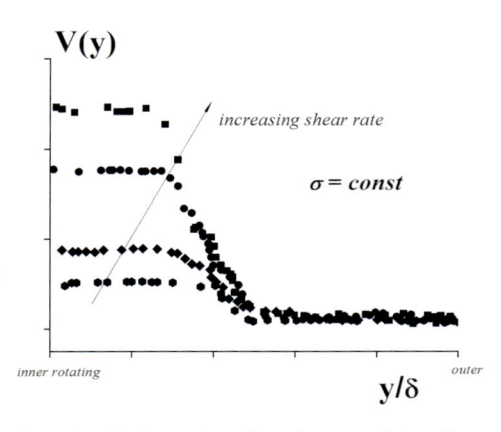

Figure 3.5.23. Separation of a polymer melt into fractions with different molecular weights; the highly viscous layer formed by a higher molecular weight fraction is almost motionless.

ent rheological properties. The phenomenon of gradient banding is the most widely observed and studied for worm-like micellar (colloid) systems (see e.g., ref.[165] discussing rheologically (and structurally) similar to polymer solutions). Moreover, micelles can be formed by polymeric substances, e.g., block copolymers.[166] The bands can contain different concentrations of a dispersed phase or can have a different order of structure organization. Accordingly, they exhibit different flow properties leading to different velocity gradients, as shown in Fig. 3.5.22.

One of the layers with very high viscosity can be almost motionless (or even solid-like at stress below the yield stress). Fig. 3.5.23 shows an example of this (appearing due to the separation of a polydisperse polymer to fractions with different molecular weight.[167] Shear banding consists of the formation of a high molecular weight layer with viscosity much higher than the viscosity of the low molecular-weight component). The highly viscous layer formed by a higher molecular weight fraction is almost motionless. Therefore, the velocity of this layer is close to zero and really, only the low-molecular-weight fraction flows.

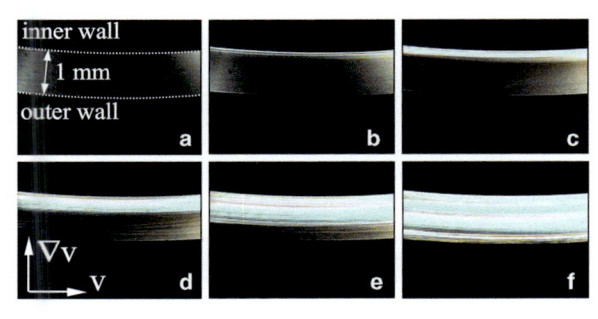

The physical state of a material in different bands can be in various physical states. This is shown in Fig. 3.5.24 which presents not only shear banding but also the formation of the LC-phase of one band and widening this state in increasing the shear rate.[168]

Figure 3.5.24. Shear banding with the phase transition into the LC-phase. The flow occurs in a narrow gap between inner and outer cylinders. Screen shots from a to f correspond to an increase in shear rate. [Adapted, by permission, from J.-F. Barret, **Rheology of wormlike micelles: equilibrium properties and shear-banding transition. Molecular gels**, Elsevier, 2005.]

Shear banding can be observed in a stationary mode and also can take place in an oscillatory mode, as shown in Fig. 3.5.25.[169] Similar periodic effects were observed by following the

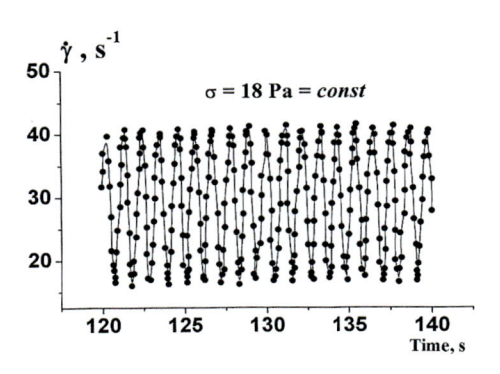

Figure 3.5.25. Self oscillations in shear banding formation (aqueous solution of cationic surfactant and Na-salicylate). According to Herle H., Fischer P., Windhab E.J., Langmuir, 21, 9051 (2005).

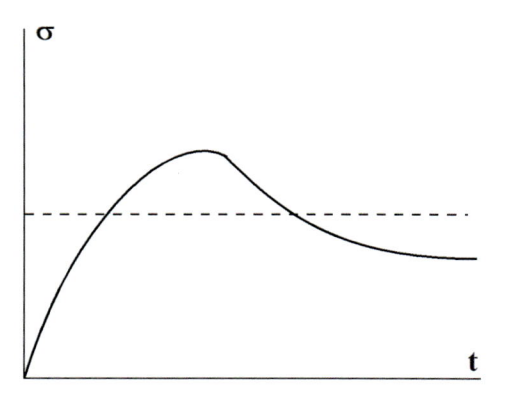

Figure 3.5.26. Overshoot at the deformation curve t a given shear rate.

structure of shear bands along a hysteresis loop.[170] Periodic formation of shear banding can be related to an isotropic-nematic phase transition induced during shearing and after-shearing rest.[171]

Usually, the phenomenon of shear banding is related to multi-valued flow curves of the type shown in Fig. 3.6.13. Then each band corresponds to lower or higher branches of a flow curve and consequently, its properties are characterized by different viscosity. The shear banding phenomenon can also happen in flow of thixotropic materials.

The hysteresis double branch flow curves reflect some delayed structural changes in increasing and/or decreasing stress. The up and down viscosity vs. shear rate dependences do not coincide. A role of thixotropy in the rheology of complex liquids and methods for its modeling were discussed in the review.[172] The double-valued flow curves of thixotropic fluids can be a source of shear banding because different points on a hysteresis curve correspond to various morphological states of a multi-component liquid.

The start-up of shearing at high enough shear rates at the stage of the pre-stationary flow can initiate another type of two-valued stress (shown in Fig. 3.5.26 by a horizontal line).

The correlation between the behavior of viscoelastic liquids at the start-up (at rate-control deformations) and its steady-state properties can be absent because of different relaxation scales of the transient and steady state of shearing. The only requirement is a possibility of the coexistence of a material in two different states (at any stage of deformation) can result in shear banding.[173] So, shear banding can happen for a monotonic stress vs. shear rate function for steady flows which are expected not to form shear bands. Banding starts immediately at the decreasing branch of the $\sigma(t)$ curve.

The stable shear banding observed in the entangled polymer solution is not the "true" effect but may be "trapped" transient banding. Experimental studies of the transient regime of shearing of worm-like micelles showed that following a start-up, large temporal velocity fluctuations take place. They reflect the appearance of bands with different concentrations, although the wall slip can at least play a partial role.[174]

The above discussion concerning shear-induced phase separation in amorphous mixtures (polymer solutions) was based on the thermodynamic arguments. The Clausius-Clapeyron equation was the basis for the analysis of the role of shear flow on structural transformations in the early stages of studying this phenomenon.

This phenomenon happens in the strong non-equilibrium conditions and it means that arguments of equilibrium thermodynamics are not adequate. The another approach to understanding the deformation-induced phase separation connects this effect to the stress-diffusion coupling. he mechanism of this effect is considered as the displaying of instability: large concentration fluctuations are enhanced due to an increase in the coefficient of diffusion under stress.

Helfand and Fredrickson in their pioneering works proposed a model explaining the physics of deformation-induced banding based on the concept of great deformation-enhanced concentration fluctuations in a bi-component system, which are not treated as the true phase transition.[175] As a result of such fluctuations and concentration inhomogeneity, the fundamental rheological parameters (apparent shear viscosity and normal stresses) being concentration-dependent become different through volume. This scenario leads to the movement of molecules along with the concentration gradient. This results in the growth of fluctuations, and finally in the phase separation (demixing) of a bi-component system. This approach was applied for shear-banding.[176]

The same model was successfully used for extension of polymer solutions in the extension of solvents used in the fiber spinning where phase separation is forced by mechanical stresses.[177]

An initial simplified picture of the demixing process used only hydrodynamic arguments, but the theoretical analysis also later included the thermodynamic factors (according to the Flory-Huggins concept).[178]

The direct experiment has proven the reality of the basic assumption of concentration fluctuations and the shear-induced shift in the concentration profile which confirms the basic concept of the model discussed above.[179]

Therefore, the dominating modern understanding of the driving mechanism for shear-induced phase separation is the coupling of the stress distribution to an inhomogeneous concentration profile.

This basic concept is included in conservation dynamic equations and their analysis explains the observed effects of instability, phase separation (demixing), and self-assembly in the flow of multicomponent complex fluids and the prediction of possible effects. The concrete results of calculations depend on the choice of a rheological constitutive model.

Two of the most popular physical methods for modeling the rheological behavior of complex liquids are also used in different versions for predicting shear-induced effects in their flow. This is a tube model[180] and a slip-spring model.[181] Any of these rheological models including the concept of stress-diffusion coupling allows for the main macro-effects discussed above to be described: demixing and self-assembly in multicomponent complex fluids. This principle is possibly a consequence of the instability predicted by non-linear rheological models which lead to the coexistence of different states of a fluid under external forces.

In the latest theoretical consideration, the shear induced demixing in chemically identical linear polymer blends with different molecular weight distributions was discussed (the generalized "multifluid" model).[182] There is the difference in the alignment in the flow direction of short and long macromolecules and this is enough reason for their separation at high Weissenberg numbers where elasticity begins to play a dominant role. Then elastic forces promote a migration of short chains resulting in enhancing mixing.

The modern state of the understanding the shear-induced phase transitions is discussed in.[183]

3.6 LIMITS OF SHEAR FLOW – INSTABILITIES

3.6.1 INERTIAL TURBULENCE

The term "*limits of shear flow*" in the title of this section should be understood as the limit of steady shearing before flow instability of any type appears. At higher velocities (or shear rates) shear flow continues but in an unstable mode.

Instability is the phenomenon of large and uncontrolled disturbances induced by a slight and accidental cause. The most general type of instability, common to all liquids, is known as the *Reynolds* or *inertial instability*. The origin of this phenomenon, first found by Reynolds,[184] is the inability in suppressing occasional flow fluctuations by viscous dissipation. It means that at sufficiently high velocities inertial forces become large and exceed viscous forces. At low velocities, viscous forces damp any stream fluctuations, therefore streamlines are smooth and parallel layers of liquid slip over each other. This type of flow is called *laminar*. At higher velocity, fluctuations increase due to inertial forces and become more pronounced. Above some critical (threshold) velocity, the fluctuations cannot be damped by viscous resistance. As a result, streamlines become irregular and the pathway of any individual particle in the stream appears chaotic. Such type of flow is called *turbulent*, and the transition to turbulent flow results in much higher energy dissipation than in smooth laminar flow.

The criterion of the appearance of inertial turbulence defines a threshold value of the dimensionless ratio of measures of inertial and viscous forces. The *Reynolds number*, Re, is such a criterion and it is expressed by:

$$Re = \frac{VD\rho}{\eta} \qquad\qquad [3.6.1]$$

where V is the velocity, D is the characteristic geometrical size (e.g., the diameter of a capillary as in Eq. 3.6.1; the radius of a round tube or the distance between two parallel plates or another geometrical factor can be used in the definition of the Reynolds number), ρ is density, and η is viscosity.

The critical value of the Reynolds number, corresponding to the transition from steady (*laminar*) to chaotic (*turbulent*) flow, is close to 2,300.

Inertial instability is very important for the flows of numerous liquids, such as water, low viscous oil products, and so on. Turbulent flow is a great field of study, and a lot of original publications and monographs are devoted to this subject. However, inertial (or Reynolds) turbulence is not primarily a rheological phenomenon, though it can also be observed for non-Newtonian liquids. This phenomenon is important for weak (dilute) solutions and suspensions (for example, blood flow). In all these cases it is necessary to re-define the value of the Reynolds number, taking into account the shear rate dependence of the apparent viscosity entering the expression for Re.

3.6.2 THE TOMS EFFECT

For liquids of all types, the dependence of flow resistance (or energy dissipation) on flow rate in a laminar flow is commonly expressed by the Hagen-Poiseuille law:[185]

$$Q = \frac{1}{\eta} \frac{\Delta p \pi R^4}{8L} \qquad [3.6.2]$$

where Q is the flow rate, Δp is the pressure drop, η is viscosity, R and L are the radius and the length of a capillary, respectively.

In publications and applied work on fluid dynamics, some other form of the Hagen-Poiseuille law is preferred. It is based on the dimensional variables. The dependence of the *coefficient of friction*, λ_R, on the Reynolds number is defined as:

$$\lambda_R = \frac{\Delta p}{0.5 \rho V^2} \frac{R}{L} \qquad [3.6.3]$$

Then, the Hagen-Poiseuille law is presented as

$$\lambda_R = \frac{16}{Re_R} \qquad [3.6.4a]$$

where the characteristic geometrical size in the Reynolds number is R but not D, unlike in Eq. 3.6.1. The physical meaning of both dimensionless variables, Re_R and λ_R is not changed if the diameter of a tube, D, is used as a characteristic geometrical factor, but not the radius. However, the numerical coefficient in the dependence $\lambda(Re)$ appears different, and Eq. 3.6.4a is changed to the following formula (if D but not R is used as a geometrical factor for the coefficient of friction and the Reynolds number):

$$\lambda = \frac{64}{Re} \qquad [3.6.4b]$$

The dependence $\lambda(Re)$, as expressed by Eq. 3.6.4b, is shown in Fig. 3.6.1 and marked by the letters H-P. The physical reality of this dependence was repeatedly confirmed by numerous experiments carried out for different liquids. It is illustrated by points that are related to both Newtonian (open circles) and non-Newtonian (solid circles) liquids. It is worth mentioning again that this dependence is the same for liquids of different types. At the Reynolds numbers exceeding the critical threshold, $Re^* \approx 2,300$, the coefficient of friction begins to increase; this is the transition zone (from laminar to turbulent regime). At the second decreasing branch of the $\lambda(Re)$ dependence, marked by the letter B, a well-developed turbulent regime of flow exists. The $\lambda(Re)$ dependence in turbulent flow is described by a very common equation, known as the *Blasius rule*:[186]

$$\lambda = \frac{0.3164}{Re^{0.25}} \qquad [3.6.5]$$

The Blasius rule is an empirical generalization of numerous experimental data obtained by different researchers for various liquids. It is an equivalent of the Hagen-Poiseuille law for a turbulent regime of flow. The coefficients of dynamic resistance in turbulent flow are much higher than in laminar flow. This is explained by more extensive dissipative losses in the chaotic movement of liquid particles in comparison with a regular displacement of layers in the laminar regime of flow.

Experimental data for non-Newtonian liquids in transient and turbulent regimes are very close to the generalized regularities found for Newtonian liquids if the convenient definitions for λ and Re are chosen.[187]

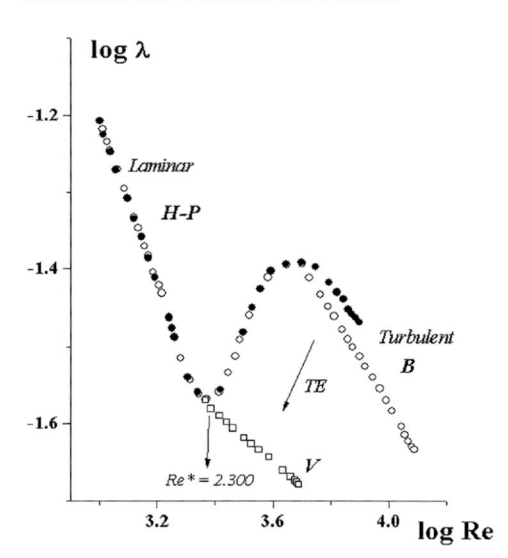

Figure 3.6.1. Dynamic resistance as a function of velocity (in dimensionless variables vs. Re) in flow through tubes. Open circles – Newtonian liquids; dark circles – non-Newtonian liquids; Squares – dynamic resistance due to the Toms effect. H-P – the Hagen-Poiseuille law; B – the Blasius rule; V – the Virk asymptote; TE – the Toms effect – decrease in dynamic resistance.

Toms[188] showed that very small amounts of additives, dissolved in liquid, can suppress turbulent flow, or at least decrease turbulent losses to a great degree. This effect, known as the *Toms effect*, sometimes is called *drag reduction*. These additives are polymers, and their amount does not exceed 100 ppm.[189]

There are two main classes of liquids to which the Toms effect applies: water (and therefore water-soluble polymers, such as polyoxymethylene, polyacrylamide, some natural polymeric substances are used) and oil products and some synthetic polyolefins.

The decreasing extent of flow resistance depends on many factors, including the concentration of additive, its chemical nature, temperature, and so on.[190] However, there is a limit of drag reduction by the Toms effect. This limit is shown in Fig. 3.6.1 and marked by the symbol V. The Toms effect causes the disappearance of the transition zone on λ vs. Re diagram and the shift of the λ(Re) dependence in the turbulent zone from B to V line in Fig. 3.6.1. This limiting curve corresponding to the maximum of drag reduction (line V) is called the *Virk asymptote* and it described by the equation:[191]

$$\lambda = \frac{2.36}{Re^{0.58}}$$

[3.6.6

The origin of the Toms effect is still the subject of intensive discussion. However, the common opinion is that the Toms effect is related to the elasticity of liquid caused by adding polymer to water (or an oil product) and, possibly, the crucial factor is the extensional (or elongational) viscosity of such solutions. It is thought that the turbulent losses are related to the existence of multi-frequency oscillations superimposed on the main stream. Every mode gives its own input to dissipative losses. Then, it can be suggested that the decrease in turbulent losses due to the Toms effect can be explained by oscillations at some modes that do not dissipate but continue because these oscillations become elastic.

The Toms effect leads to a decrease in the flow resistance, which can be as high as 75%. This is a considerable change, and it was reasonable to search for its different applications. The most likely applications include an increase in the pumping rate of water and oil products in transportation through pipelines, or surface coating of sport boats and torpedoes by polymer additives in order to decrease the resistance of their movement in the water and consequently to increase their speed. The Toms effect favors the stability of liquid streams, and thus the presence of a small amount of polymer additive in water permits an increase in the length of a jet from fire-pumps. The Toms effect was also applied to the investigation of chemical reactions and physical transformations of dilute solutions.[192]

3.6.3 INSTABILITIES IN THE FLOW OF ELASTIC LIQUIDS

3.6.3.1 Dynamic structure formation and secondary flows in inelastic fluids

Instability in the flow of fluids does not necessarily lead to the formation of chaotic patterns. In many cases, self-organization, appearance of shear-induced structure, and/or secondary flows can take place. The effect of the flow-induced crystallization (i.e., the appearance of three-dimensional order) has been described in Section 3.5.3. Below, we will concentrate mainly on a dynamic phenomenon not related to phase transitions, though there is no well-defined boundary between different types of structurization in the flow.[193]

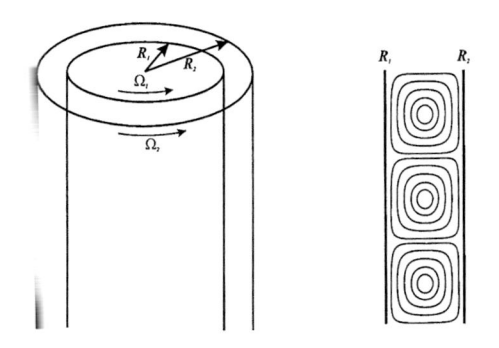

Figure 3.6.2. Taylor-Couette instability – formation of cell.

The formation of the Taylor cells in flow between two rotating coaxial cylinders is a well-known effect of dynamic shear-induced structure. One can expect fluid elements placed between coaxial cylinders move along circumferences when cylinders rotate at different speeds. It is so-called Couette flow. This suggestion is the base of rotational viscometry (see Section 5.3.2). Meanwhile at some relationship between speeds of rotation, streamlines form closed figures – cells, as shown in Fig. 3.6.2. These are so-called *secondary flows*, which organize streamlines in the planes perpendicular to the direction of the main flow.

This phenomenon is due to the centrifugal (inertial) forces which push a liquid from the inner cylinder towards the outer. This type of flow pattern and its theory were proposed first by Taylor[194] and then developed by other authors who discussed different kinds of rotational motions. When a stability threshold is overcome, inertial toroidal roll cells appear. Their height, h, is roughly equal to the gap clearance between the cylinders, i.e., the difference $(R_2 - R_1)$. The term "instability" must be accepted here with certain accuracy; it only means that there are some secondary effects superimposed on the main stream. Meanwhile, these secondary streams can be quite stable in time and space.

Since the Taylor-Couette instability is the consequence of inertial forces, it is determined by a definite relationship between centrifugal (as a measure of inertial effects) and viscous forces. This relationship is expressed by the dimensionless value called the Taylor Number, Ta, which is an analogue of the Reynolds Number by its physical meaning. The Taylor Number is expressed as

$$Ta = \frac{4\Omega^2 R^4}{v^2}$$
[3.6.7]

where Ω is the frequency of rotation, R is the average radius of coaxial cylinders, and v is the kinematic viscosity (Newtonian viscosity divided by density). The vortex structure (Taylor-Couette cells) becomes possible if the Taylor number exceeds some critical value Ta*, which increases with the increase of the Ω_1/Ω_2 ratio,[195] while the minimum value of Ta* is close to 1700.

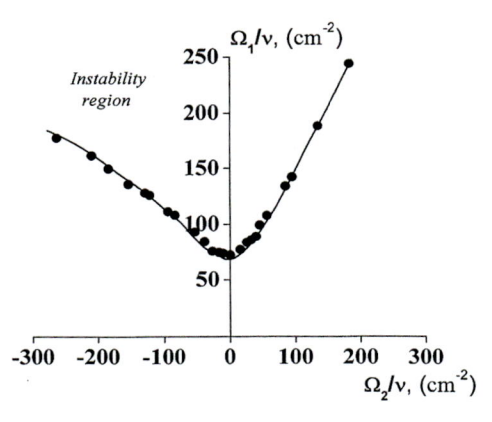

Figure 3.6.3. Limit of stability in formation of Taylor-Coutte cells.

The general "curve of stability" is shown in Fig. 3.6.3. This curve separates regions of stable and secondary flows. One can see that the experimental points perfectly correspond to the theoretical threshold curve.

The Taylor-Couette cells are also formed in a Bingham viscoplastic media[196] as well as in the flow of two immiscible viscous fluids.[197]

It is interesting that cells are multiplied with the increase of the Reynolds Number and finally the laminar flow, though complicated with enhancing secondary flows, transforms to the turbulent regime of flow.

3.6.3.2 Secondary flows in the flow of elastic fluids

Cells of the Taylor-Couette type can appear in the flow of elastic fluids. Systematic studies of the Couette flow of elastic liquids showed that cells are also formed in these liquids. It is possible to observe a continuous transition from inertial to elastic instability by controlling the rheological properties of a fluid.[198] Different types of coexisting cells of various sizes were observed in the flow of elastic liquids.[199]

Sometimes, the appearance of secondary streams in the flow of an elastic liquid is considered as the consequence of the second normal stress difference. Though these stresses are not large, they are thought to be responsible for

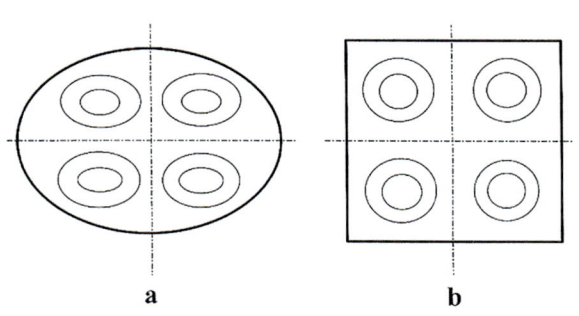

a   b

Figure 3.6.4. Secondary flows in channels of elliptical (a) and square (b) cross-sections. Main stream is directed perpendicularly to the plane of a drawing.

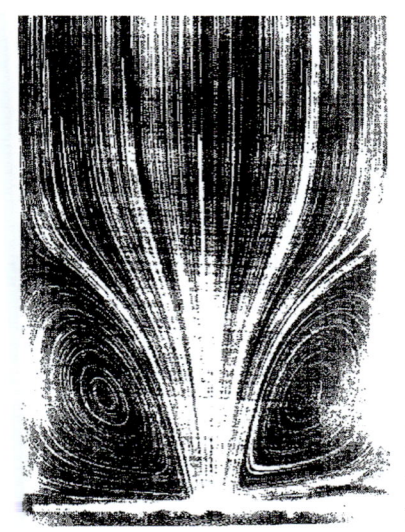

Figure 3.6.5. Secondary flows in sudden contraction – transition from a cylinder to a small diameter capillary.

the development of elastic-type secondary flows during the flow of a liquid through non-round channels.[200] Such secondary flows look as shown in Fig. 3.6.4 for a channel of an elliptical[201] (case – a) and square[202] (case – b) cross-sections.

Elastic instability can occur in different geometries of flow. So, effects of this type were investigated in the cross-channel flow,[203] which can result in a steady asymmetric state even when flow takes place in perfectly symmetric geometry.[204]

Secondary flows are also observed in the flows through channels with a sudden contraction – for example in the transition region from a round cylinder to a small diameter capillary (Fig. 3.6.5). Also, it was proven (by both, numerical and experimental methods) that the second normal stress difference is the driving mechanism of secondary flows in tapered channels.[206] Besides, the second normal stress difference can be responsible for the edge fracture of viscoelastic liquids and suspensions (such as lubricating grease and toothpaste) in the circular flow of a liquid between a cone and plate or two parallel plates.[207]

It was shown that a regular dynamic structure can be observed in the flow of different elastic liquids. The quantitative measure of the critical conditions is defined by the Deborah number, De, which is the ratio of characteristic internal (relaxation) and external (rate of deformation) times. The Deborah number in rotational flows of an elastic fluid characterized by a single relaxation time is determined as

$$De = \frac{2\Omega_1 \lambda (1 + \varepsilon)^2}{(1 + \varepsilon)^2 - 1}$$ [3.6.8]

where Ω_1 is the frequency of rotation of the inner cylinder ($\Omega_2 = 0$) in the Couette flow, ε is the dimensionless gap defined as

$$\varepsilon = \frac{R_2 - R_1}{R_1}$$ [3.6.9]

Definition of R_1 and R_2 is shown in Fig. 3.6.2 and λ is the relaxation time.

The Reynolds number in all studies devoted to instability in elastic fluids was very low, less than $7*10^{-3}$; this is four orders of magnitude less than the critical value corresponding to the onset of the centrifugal instability. So, the Taylor numbers in all cases were several orders less than the above-mentioned critical value of Ta*. It means that this effect has nothing in common with the inertial mechanism, but is completely governed by the elasticity of a fluid.

It was found that there is a critical Deborah number De_{cr} (or, equivalently, the critical Weissenberg number, Wi_{cr}) equal to 35.5, which is responsible for the onset of instability.

The critical Weissenberg number, Wi_{cr} depends on the gap clearance between coaxial cylinders, and according to Larson,[208]

$$Wi_{cr} = 5.9\varepsilon^{1/2} \qquad\qquad\qquad\qquad [3.6.10]$$

Secondary flows of the elastic nature can be observed not only in polymeric systems but also in so-called worm-like micellar colloids,[211] though the term Taylor-Couette instability is frequently used for these structures. Meanwhile, the origin of these cells is definitely elasticity but not inertia of a medium and can be connected with the inhomogeneity of normal stresses along the direction of the veloity gradient in the Couette flow.[208]

In concentrated solutions and polymer melts, other types of instabilities occur much before the onset of the inertial turbulence is reached. Indeed, their viscosity is very high and the denominator in the expression for the Reynolds Number is large. As a result, it is impossible to reach high values of Reynolds Number, close to the above-mentioned critical value, at any real velocity.

Hence, the appearance of unstable regimes of deformation for non-Newtonian polymer melts and concentrated solutions at high shear rates is not related to the inertial turbulence but has other physical reasons. In fact, different kinds of irregular flow, or instabilities, in the flow of polymer substances have been described, and generally speaking, they are related to the viscoelastic nature of deformation of these materials.[209]

Several examples of the ordinary sequence of irregularities developing with the increase of flow rate of industrial-grade polydisperse polymers pushed through a cylindrical channel (capillary, die) are shown in Fig. 3.6.6 (flow rate is increasing from left to right).

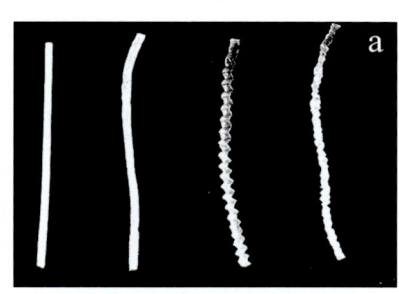
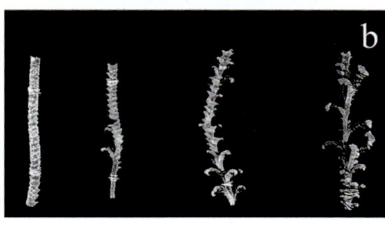

Instabilities appear initially as the surface defects of an extrudate. The unsteady flow is noticed firstly in the form of small-scale surface defects and as a result, the extrudate surface becomes matte. According to its look, this effect sometimes is called "sharkskin". The most severe form of the sharkskin effect is the appearance of small-scale regular (periodical) thread-like or screw-type defects on the surface of a jet (Fig. 3.6.7).

In many cases, instability of the sharkskin type in the flow of polymer melts appears at stresses close to 0.1MPa, though the exact limit and severity of the effect depend on the nature of a material and (to some degree) on the smoothness of a channel surface and the material which it is made from.[210]

Figure 3.6.6. Typical polymer extrudates obtained at different shear rates (increasing from left to right) – transition from smooth stream to unsteady regimes of deformations. a: polyethylene; b: polyvinyl chloride. (Photographs were made in laboratory of Polymer Rheology, Institute of Petrochemical Synthesis of Academy of Sciences.)

The appearance of these small-scale defects has negative consequences for the quality of industrial products such as films, which lose their gloss and brilliancy due to surface defects, wire insulation, and others.

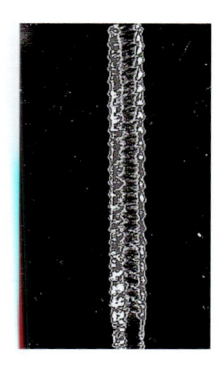

Figure 3.6.7. Small-scale screw-like surface defects – start of development of the unsteady flow of a polymer melt. Low density polyethylene.

The sharkskin effect is believed to initiate at the die exit. The origin of sharkskin is definitely related to a special dynamic situation near the exit section of a channel. Indeed, the sharp edge of a channel near its exit is a singularity and transition from a parabolic velocity profile inside a channel to a flat profile, which occurs right after the exit from the channel. The rearrangement of the velocity profile near the exit is quite evident and was well documented by visualization of a stream. Optical observations confirmed that small pulsations are visible in the exit zone with the same frequency as surface defects of the extrudate,[211] whereas flow along the whole length of a capillary stays stable and is not influenced by small-scale instabilities of the sharkskin type.

The stability of a stream along the whole channel in spite of the appearance of sharkskin (related to the channel exit) allows one to measure the apparent viscosity in the range of stresses corresponding to this type of instability.

Sometimes the origin of periodic ("sharkskin-like") defects on the surface stream is related to the cyclic generation of wall slip at the tube exit.[212] High-speed video microscopy technique demonstrates that cohesive failure downstream of the exit accompanies each sharkskin cycle while surface treatment can initiate a strong slip and suppress sharkskin defects.[213]

The stress distribution near the exit section of a channel was examined by means of the technique of flow birefringence, and it was confirmed that not only shear but also tensile stresses appear close to the point of singularity.[214] The following qualitative picture can clear up the origin of small-scale surface defects of an extrudate. When a melt is pushed out of a channel, the stream is hindered at the outlet section of the die. Meanwhile, the central part of a stream continues to move forward and pulls the other outer layers of a liquid. When this pulling force becomes high enough the adhesive contact between a melt and a wall is broken, and the material is detached from the die. This process is repeated periodically and it leads to periodic pulsations or oscillations of the surface defects. The appearance of the small scale defects is definitely related to stress conditions at the exit edge of a die: some extra energy is necessary for tearing a flowing matter off a solid surface, this energy is stored in the form of elastic deformations, After leaving a tube this energy is released and it results in the periodical swelling of a jet.

As a result of numerous visual observations and calculations, it was suggested that the mechanism of sharkskin is the *surface rupture under tensile stress* acting at the edge of a channel: periodic oscillations appear as the result of material rupture when this stress reaches some critical level.[215] Indeed, it was shown that small traction zones appear on the surface of an extrudate, and the dimensions of these zones are of the order of magnitude of sharkskin amplitude. So, we can think that the main reason for the sharkskin effect is the rubber-like behavior of a melt near the die exit and cracks which appear perpendicular to the flow direction. They are due to elongational (normal) stresses created at the die exit and stored elastic energy is responsible for crack creation.

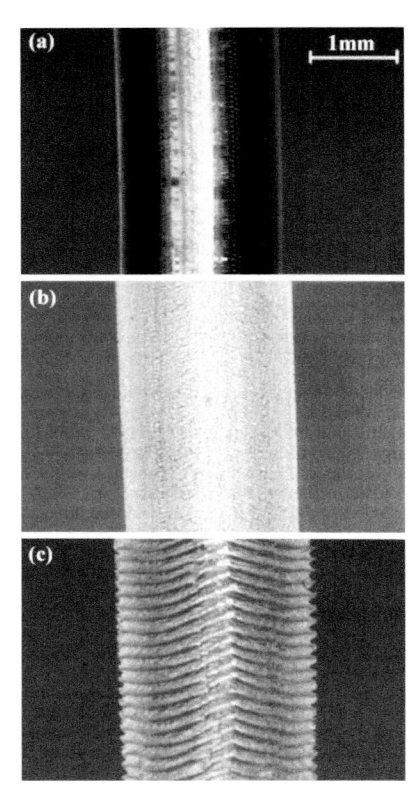

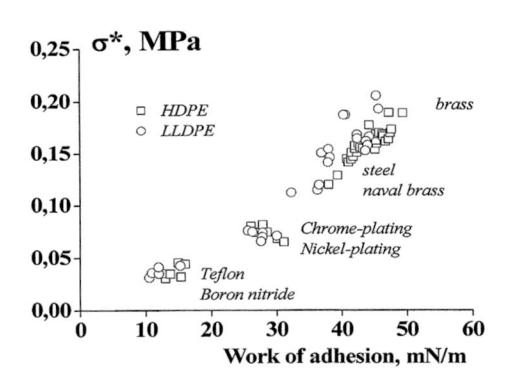

Figure 3.6.9. Influence of surface properties on the critical shear stress corresponding to the sharkskin effect. [Adapted, by permission, from H.J. Larrazabal, A.N. Hrymak, J. Vlachopoulos, Rheol. Acta, 45, 705 (2006)].

Figure 3.6.8 Developing of surface defects in extrusion. Velocity of flow increases from top to bottom. a - smooth surface; b -loss of gloss; c - sharkskin. [Adapted, by permission, from E. Miller, J.P. Rothstein, *Rheol. Acta*, **44**, 160 (2004)].

The appearance of the sharkskin effect is not a threshold effect but develops continuously. Fig. 3.6.8 illustrates the main stages of the formation of regular surface defects.[216] It was noted that the waviness length and amplitude of teeth is a monotonous function of the shear stress,[217] though it is possible to estimate the critical stress corresponding to the appearance of slight visual effects. It is reasonable to think that the loss of gloss means that the order of magnitude of the defects becomes of the order of the wavelength of visible light, i.e. 0.4 - 0.8 μm. For linear polyethylene, this stress is of the order 0.1 - 0.2 MPa.

A promising technological method of excluding sharkskin defects is related to adding some additives to polymer melt which create coating at the die wall and induce slip. This method was proposed, for example, for the extrusion of low-density polyethylene. Fluoropolymer processing additives were effectively used to eliminate sharkskin surface defects. It was proven that these additives adsorb first in the entrance region of the die and then migrate towards the capillary exit where they suppress sharkskin defects.[218]

It was noted that any tendency to slip at the die wall for the melt near the exit will reduce the acceleration of the surface layer and so reduce the stress level in that layer and so the severity of any rupture.[219]

Bearing in mind the slip conception, the role of surface properties of the die becomes important. This is illustrated by Fig. 3.6.9, which clearly shows that the critical shear stress strongly depends on the surface properties of a material used for preparing a die.[220]

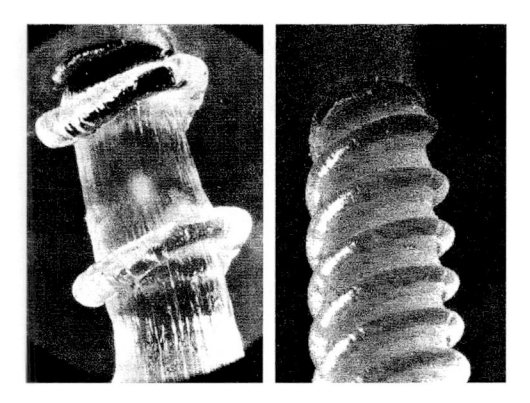

Figure 3.6.10. Large-scale periodic defects of the extrudate. [Adapted, by permission, from N. El Kissi, J.-M. Piau, T. Toussaint, *J. Non-Newton. Fluid. Mech.*, **68**, 271 (1997)].

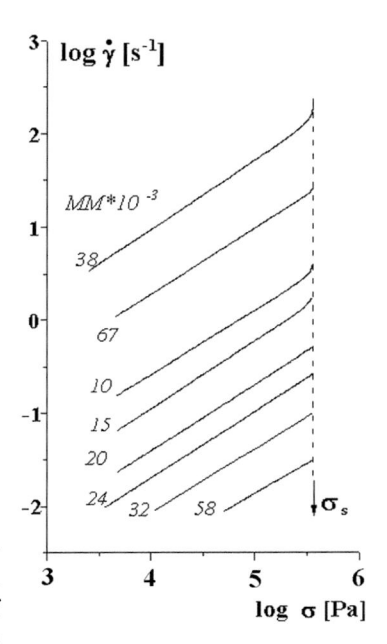

Figure 3.6.11. Flow curves and spurt stress for polybutadienes of different molecular mass (shown on the graph). [Adapted, by permission, from G.V. Vinogradov, A.Ya. Malkin, Yu. G. Yanovsky, E.K. Borisenkova, BV. Yarlykov, G.V. Berezhnaya, V.P. Shatalov, V.G. Shalganova, V.P. Yudin, *Vysokomol. Soedin.* (Polymers in Russian), **14A**, 2425 (1972)].

An increase in the flow velocity and corresponding stresses leads to the transition from sharkskin defects to the large-scale periodicity of surface defects. They can have a very impressive effect (Fig. 3.6.10).

Different terms were used to describe this phenomenon: fluid wraps, fringes, rings, collars, surface cracks, and so on. In all these cases, it is possible to point out the exit section of a die as the origin of these defects.

An attempt to increase the speed of flow of monodisperse polymer melts leads to the transition from regular gross scale to irregular defects. They may have different appearances. The origin of these gross scale defects is attributed to peculiarities of deformations *inside a die* or even at *the inlet section* of the die.

When a monodisperse polymer melt is extruded through a die, the sliding of material along the solid wall of a channel begins at certain shear stress, σ_s. This effect was mentioned in Section 3.2.1 and it is known as *spurt*. The origin of the spurt is attributed to *liquid-to-rubber transition* such as observed at high frequencies of oscillations (see Section 2.7.2), i.e., to the quasi-solidification (hardening) of a melt in a near-wall layer at high deformation rates.[221] It means that the lifetime of intermolecular contacts becomes longer than the characteristic deformation time (reciprocal shear rate) and the polymer melt behaves as a quasi-stable rubbery network. As a result, highly stressed boundary layers of material lose fluidity and adhesive contacts between polymer and wall is broken and the material begins to slide along the wall. Alternatively, cohesive breaks can happen because the strength of both is similar.

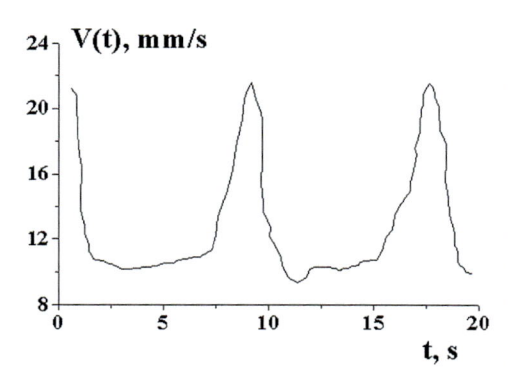

Figure 3.6.12. Velocity fluctuation (at the axis of channel) during stick-slip regime of capillary flow. [Adapted, by permission, from L. Roberts, Y. Demay, B. Vergnes, *Rheol. Acta*, **43**, 89 (2004)].

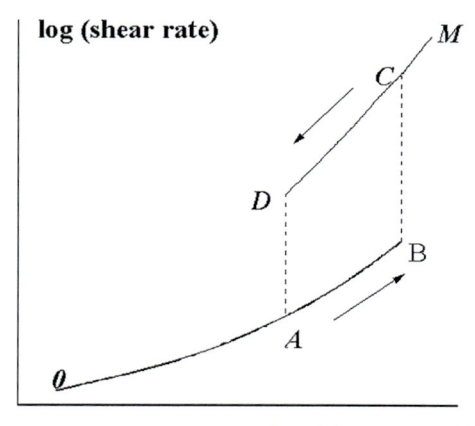

Figure 3.6.13. Hysteresis loop in measuring viscous properties of some polymer melts – multi-valued flow curve with spurt effect.

The existence of wall slip is accompanied by strong surface charge formation ("*tribological effect*") that is characteristic for the friction of a dielectric material moving along metal; the electrical charge is a strong function of apparent shear stress.[222]

It is interesting to note that critical shear stress, σ_s, corresponding to the transition from flow to spurt does not practically depend on the molecular mass of the polymer (Fig. 3.6.11) and/or temperature, i.e., i does not depend on the viscosity of the melt.

For many flexible-chain polymers (such as polybutadiene or polyethylene the spurt shear stress, σ_s, is close to 0.3MPa. However, σ_s is not the universal constant. More detailed analysis showed[223] that there is a correlation between the rigidity of the molecular chain and the σ_s values. It was found that for rigid-chain polyphosphazenes σ_s can be as low as 0.009 MPa.

Spurt is not typical of melts of industrial polydisperse polymers, though the phenomenon of transition from flow to sliding or periodic sliding is very general for all polydisperse polymers. This periodic sliding, known as the stick-slip phenomenon, leads to the appearance of periodical regular sequences of distorted and smooth parts on the surface of the extrudate and it is accompanied by periodic fluctuations of velocity (Fig. 3.6.12) and instant output.

This effect is attributed to the behavior of flowing material on solid surfaces. Indeed, direct measurements of velocity profiles (made by the Doppler velocimetry method) confirmed that a classical velocity profile in the smooth part of the flow curve of high-density polyethylene is observed while strong macroscopic wall slip is documented for the upper part of the flow curve.[224] It is worth adding that the existence of the surface slip is now a well-documented phenomenon.[225]

The typical level of shear stresses responsible for the appearance of stick-slip instabilities in capillary flow is the same as the spurt stress for flexible-chain polymer melts, i.e., it is about 0.3 MPa (compare with data in Fig. 3.6.9 where it is seen that sharkskin effect appears at lower stresses). The frequency of oscillations can be very different varying from 2-3 till 20-30 per second.

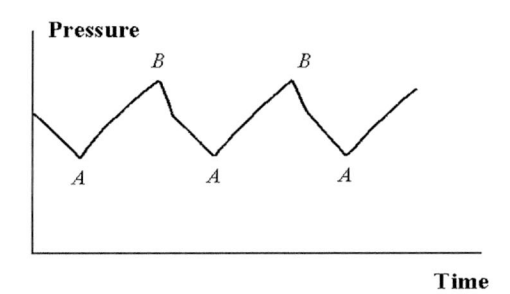

Figure 3.6.14. Regular pressure oscillations along the hysteresis loop. Points A and B correspond to the same points on the flow curve in Fig. 3.6.7.

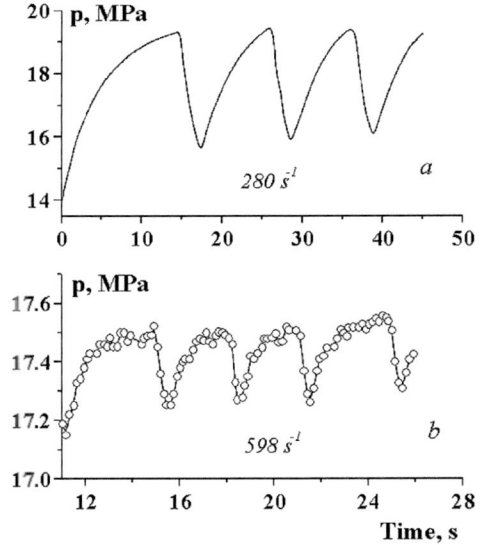

Figure 3.6.15. Double oscillation loops – pressure oscillations at two different shear rates in the capillary flow of linear polyethylene. a: main double-value zone on the flow curve (experimental points are omitted); b: secondary (high-shear rate) double value zone on the flow curve. [Adapted, by permission, from L. Robert, B. Vergnes, Y. Demay, *J. Rheol.*, **44**, 1183 (2000)].

The stick-slip phenomenon is directly related to the existence of *multi-valued flow curves*, or hysteresis loops as shown in Fig. 3.6.13. This phenomenon was observed in various polymers, including high-density polyethylene, polybutadiene, polyisoprene, etc. The area of a hysteresis loop depends on the molecular mass and the polydispersity of the polymer. When apparent viscosity, or flow curve, is measured with a capillary viscometer, two principal modes of the experiment are possible. First, a pressure-controlled regime can be set. In this case, the pressure is constant and the flow output (or calculated shear rate) is measured. When pressure is increased, the measured output follows curve branch 0AB. At point B spurt happens, accompanied by a jump to point C. Then, the points of the apparent flow curve lie on the CM branch. If the pressure is decreased beginning from point M, the points on the apparent flow curve move along the upper branch MCD, and rapid change to the lower branch happens from point D to point A, followed by a decrease of the shear rate along the curve branch A0.

Periodic oscillations of instant output also take place in a rate-controlled situation, i.e., when a total (averaged) flow rate is constant. For example, this occurs when a piston pushes material through a large radius cylinder and then through a die at the end of a cylinder. The velocity and output oscillate between upper and lower branches of a flow curve, as shown in Fig. 3.6.13. The oscillations of pressure in the extrusion of monodisperse polymers may happen in regular intervals, as shown in Fig. 3.6.14. There is a correlation between the points along the hysteresis loop and the measured pressure values.

In some cases, periodic oscillations look rather whimsical. For example, pressure oscillations appear at different shear rate ranges in a capillary flow and look like double oscillation loops (see Fig. 3.6.15).

In many cases, the two branches in a flow curve correspond to two different physical states of a matter, and oscillations reflect periodic transitions between these states.

The hysteresis effect accompanied by regular periodic oscillations of the instant output rate is related primarily to melt compressibility of material in a large volume cylinder before entering into a capillary ("capacity model"):[226] the elastic energy of volume deformations is stored in a large volume and periodically released. The second condition necessary for the appearance of a hysteresis loop in an apparent flow curve is the possibility of the stick-slip phenomenon and/or spurt at certain critical shear stress.

The position of points on a hysteresis loop is directly related to the types of extrudate distortions. The stream is smooth in the 0A zone, small periodic distortions (or a sharkskin effect) appear in the AB zone. Gross extrudate distortions are typical of higher pressures. However, in some cases a smooth extrudate surface again appears at the upper branch of the "flow curve"; this is called *super-extrusion* and can be useful for increasing the output in the processing of some polymers (for example, polytetrafluoroethylene).

A multi-valued flow curve can be also observed in a rather special case under a constant pressure condition if slip is built along half of the die wall. In this rather artificial situation, cyclic self-oscillations in capillary flow, accompanied with periodic changes in the appearance of an extrudate, are observed.[227]

As shown in Fig. 3.6.6 (the last samples on the right), the final stage of instability is the appearance of highly visible gross defects of different types. In extreme cases a stream can even disintegrate into separate pieces. This stage of instability is called the *melt fracture*.

It can be thought that such gross effects are the consequence of rubbery deformations and elastic recovery after material leaves the channel and restrictions (applied by the solid walls of the channel) to the elastic recoil are removed. Ruptures of the jet are caused by the large amount of elastic energy stored during deformations inside the channel.

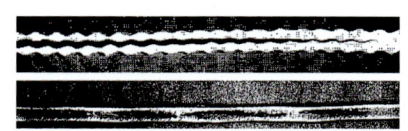

Figure 3.6.16. Deformations of stream inside a channel: oscillations of a stream correlated with periodical extrudate distortions; ruptures of a stream along the extrudate axis. [Adapted, by permission, from P.L. Clegg, in **Rheology of Elastomers**, *Pergamon Press*, 1958].

At higher shear stresses or velocities, distortions become irregular and in the limiting cases they result in the breakup of the stream. Ruptures happen even inside a tube (channel) as was proven by direct observations.[228] In this experiment, black pieces were introduced into a white melt flowing through the capillary. Inner oscillations at low flow rates and discontinuous streamlines at high flow rates were clearly noticeable (Fig. 3.6.16).

It is interesting to mention that breaks occur along the channel axis, where shear stresses are absent. It suggests that these effects are caused by tensile stresses present at the die inlet. At high flow rates, the deformation of material inside the channel becomes irregular. Possibly, the discontinuous flow lines first appear in the inlet sections of a channel at the corners near the entrance to the die, where the rearrangement of stream takes place and additional high shear and tensile stresses appear.

The transition from the steady flow regime to instability is clearly seen by direct optical observations and/or the birefringence method (Fig. 3.6.17).

The left picture demonstrates the regular stress distribution, whereas the chaotic flow pattern with an irregular variation of stresses is typical of the instability regime (the right picture). Very strong stress concentration is evident in the inlet zone of the channel (espe-

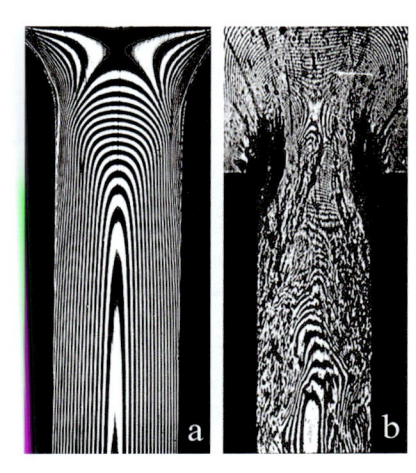

Figure 3.6.17. Birefringence pictures for steady (a) and chaotic (b) movement of polymer melt through a capillary. Polybutadiene. [Adapted, by permission, from G.V. Vinogradov, N.I. Insarova, V.V. Boiko, E.K. Borisenkova, *Polym. Eng. Sci.*, **12**, 323 (1972)].

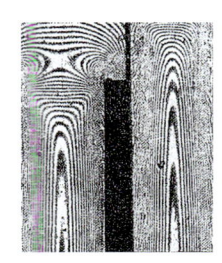

Figure 3.6.18. Optical fringe patterns during flow of polybutadiene in a rectangular duct under shear stress slightly above the spurt conditions. [Adapted, by permission, from G. V. Vinogradov, *Rheol. Acta*, **12**, 357 (1973)].

cially near the corners). The shear stress in inlet and outlet zones may achieve the critical value, whereas inside the die the shear stress at the wall may be substantially lower than its critical value. If the critical stress is reached at the sharp corners, the polymer passes into the rubbery state, its adhesion to the wall diminishes, and a local spurt occurs. As a result of the detachment of polymer from the walls, the acting stress decreases, the polymer relaxes, and then it begins to flow in the layer adjacent to the wall.

Then, in contrast to flow with the sharkskin effect, it is not useful to measure apparent viscosity or to apply the concept of steady flow (as well as to analyze dynamic problems of flow) in such situations, because in this case instability means that the real flow is absent.

It is possible to propose the following rough scheme of development of instabilities in the flow of viscoelastic polymer materials. First, *outlet instability* appears due to stress concentration at the outlet section of a die and its corners. It results in periodic small-scale defects. Then, with an increase in flow rate, oscillations spread inside a tube, leading to *surface (land) instability*, possibly in the form of spurt or stick-slip or other surface distortions. At last, inlet instability occurs at the entrance of the tube. Then it develops along the whole length of the die and demonstrates itself as the melt fracture.

As was discussed above, small-scale disturbances (sharkskin) are dependent on the quality and material of the tube (or die), while the onset of gross defects is not influenced by the material of a die surface at all.[229]

Obviously, instabilities of different types are related to the special surface effects, and, at least in some cases, they are connected with wall slip of material. Three possible mechanisms of near-wall effects are discussed: an adhesive rupture inside a polymer melt, a cohesive failure and the appearance of a lubricating layer at the wall.[230] It should be mentioned that the adhesive and cohesive strengths of polymeric materials are rather similar. Hence, whether it is an external or internal rupture it always depends on the nature of the surface material and the geometry of flow. The appearance of a lubricating layer is possible in the flow of polymer solutions or polydisperse polymers: a low viscous solvent or low molecular mass fractions can migrate to the solid wall and stay there in the form of a very thin coating, sliding along this lubricated layer looks like wall slip.

It is also interesting to examine the flow regimes corresponding to a pressure gradient that is a little larger than the critical value. This is depicted in Fig. 3.6.18, where the die entrance is shown in the left part; and the die proper in the right part; the lower edge of the photograph corresponds to the middle part of the die along its length. With pressure gradients that create the flow pattern presented here, the ruptures in the flowing polymer extend over a limited zone adjoining the edges, and the random movement of the polymer takes place only in layers adjacent to the wall. As polymer lumps advance along the die walls, their internal stress is relaxed, and the distribution of interference bands of ever-improving regularity appears near the walls.

The change in the slippage regime of the polymer – the movement of lumps of the polymer, which are randomly displaced relative to one another – explains the increase in the pressure gradient with the increasing rate of flow under above-critical regimes of polymer movement in the die.

It should be emphasized that at sufficiently high-pressure gradients and flow rates, a break in the continuity of the polymer occurs with a smooth entrance, as well. The sharp edges at the die entrance only facilitate this process but its origin is determined by the polymer in the rubbery state undergoing a rupture when the deformation rates and stresses exceed certain critical values.

The transition from true flow to wall instability (regardless of its detailed mechanism) is evident from the observation of capillary flow at high flow rates. However, there are two additional factors that prevent one from making definitive conclusions about wall slip: continuous input of the fresh material into a channel and the variation of stress conditions along the channel due to the hydrostatic pressure decrease from inlet to outlet.

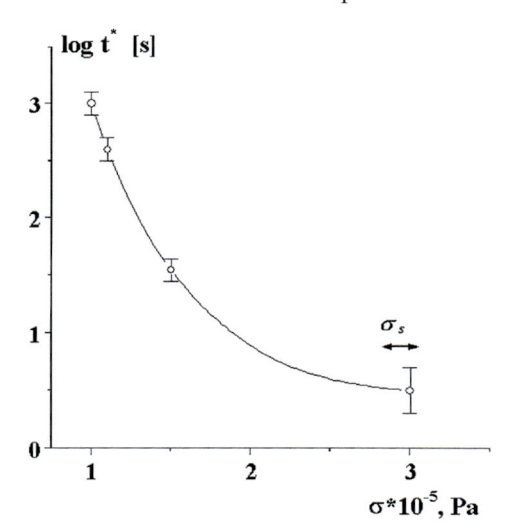

It is possible to exclude both factors and to follow a "pure" transition from flow to sliding. It is observed by imposing a continuous shear on a sample placed between a cone and a plate, where homogeneous stress exists (see Part 5, considering theory for these viscometers). This instrumentation permits observation of slow dewetting from the solid surface. A sample does not slide along the solid surface at low shear stresses (or rates). An increase in the shear stress leads to the separation of the sample from a solid wall, but it happens not just as the stress is applied but after some deformation or time of shear. This critical time, t^*, depends on the shear stress applied, as shown in Fig. 3.6.19. Hence, the *adhesive strength is*

Figure 3.6.19. Stress dependence of time corresponding to the break of the adhesion contacts between the polymer melt and a solid wall. Polyisobutylene. MM=$1*10^5$. T = 25°C. [Adapted, by permission, from A.Ya. Malkin, B.V. Yarlykov, *Mekh. Polym.* (Polymer Mech. – in Russian), #5, 930 (1978)].

time-dependent and at stresses corresponding to melt fracture in capillary flow (marked by the horizontal bar in Figure 3.6.19) this critical time becomes instantaneous for an observer.

Regular surface defects or regular variations of the form of a stream formally do not correspond to the rigorous definition of instability as large and uncontrolled disturbances. However, physical reasons for both effects – superposition of elastic and plastic deformations at high stresses – are similar and there is a continuous development of increasing extrudate distortions. That is why all extrudate defects are considered together in this Chapter.

Indeed, all above-discussed types of instabilities in the flow of polymer solutions and melts are related to their viscoelasticity. Therefore, it is reasonable to introduce a dimensionless criterion, characterizing the relationship between inherent relaxation properties of material and rate of deformation. This is the so-called *Weissenberg number*, Wi,[231] determined as

$$\mathrm{Wi} = \dot{\gamma}\theta \qquad\qquad [3.6.11]$$

where $\dot{\gamma}$ is the rate of deformation, and θ is a characteristic relaxation time.

It can be suggested that the instability begins at some critical value of the Weissenberg number.[232] The exact numerical value of this threshold Weissenberg number depends on the method of determination of the relaxation time.

The rubbery deformation, $\dot{\gamma}_r$ stored in the flow is one of the possible alternatives of the Weissenberg number, Wi. If one defines $\dot{\gamma}$ as σ/η and θ as η/G, where G is elastic modulus, then the Weissenberg number can be presented by

$$\mathrm{Wi} = \frac{\sigma}{G} \qquad\qquad [3.6.12]$$

The ratio σ/G is elastic (rubbery) deformation. In this approximation, the critical value Wi is equal to some critical value of rubbery (stored) deformation, $\dot{\gamma}_r$, responsible for the melt fracture effect. Different authors proposed that this critical value of $\dot{\gamma}_r$ lies between 3 and 7, and on average it is close to 5 units (500%). Possibly, this estimation is valid for polydisperse polymers and solutions, but $\dot{\gamma}_r$ (at the spurt stress) of monodisperse polymers is much less due to their high rigidity (high values of G), and for polymers of this type critical value of the Weissenberg number $\mathrm{Wi} \approx 1$.

Using the Weissenberg number as the criterion of melt fracture reflects the general physical possibility of the event. However, the description of the real situation in treating the instability needs to apply a mathematical model of the rheological properties of the material and to develop a rigorous analysis of the specific dynamic situation.

The theoretical analysis of the cell formation in elastic fluids is ordinarily limited by low Weissenberg numbers. One can expect that cells can duplicate themselves with an increase of Wi values, as occurs during transition to the Feigenbaum fractal chaotic structures. Indeed fractal approach was successfully applied to analyze the sharkskin effect.[233] Such chaotic structures were observed in strong capillary flows, as shown, for example, in Fig. 3.6.17b.

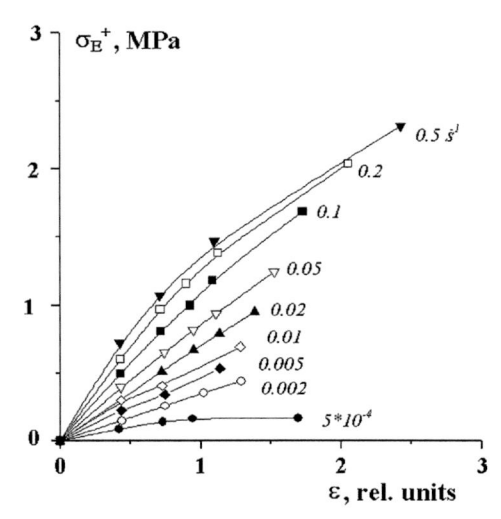

Figure 3.7.1. Stresses vs. deformations in uniaxial extension at different constant deformation rates. Polyisoprene. MM = 5.75*10⁵. 0°C. Deformation rates are shown on the curves. [Adapted, by permission, from G.V. Vinogradov, A.Ya. Malkin, V.V. Volosevitch, V.P. Shatalov, V.P. Yudin, *J. Polym. Sci.: Polym. Phys Ed.*, **13**, 1721 (1975)].

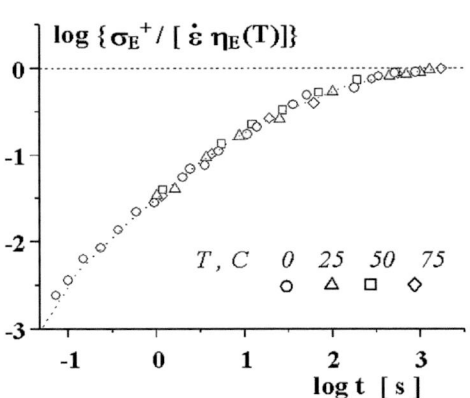

Figure 3.7.2. Development of normalized stresses in extension of polymer melt. Polyisoprene, MM=5.75*10⁵. Initial experimental data obtained in the temperature range from 0 to 75°C are "reduced" to 25°C. The stress scale is normalized by the final value of steady-state viscosity. [Adapted, by permission, from G.V. Vinogradov, A.Ya. Malkin, V.V. Volosevitch, V.P. Shatalov, V.P. Yudin, *J. Polym. Sci.: Polym. Phys Ed.*, **13**, 1721 (1975)].

3.7 EXTENSIONAL FLOW

3.7.1 MODEL EXPERIMENTS – UNIAXIAL FLOW

The uniaxial extension is not easy to induce in some liquids. It is difficult to maintain the shape of a stream of low viscous liquid and stretch it while measuring its properties. Elongational flow can be observed for substances having high viscosity, such as honey or resin. Molten glass threads are convenient objects for creating elongational flow. However, the most popular and important materials for studying elongational flows are polymer melts or concentrated solutions. The ability to stretch and to form fine fibers or thin films is a very special rheological property of polymers, which is the basis of many technologies in the textile industry and in polymer processing. It is easy to imagine why polymeric substances can be stretched: uniaxial extension of polymeric substances leads to alignment of macromolecules and creates a dominating orientation of matter. It results in increased resistance to further deformation of this stress-induced macromolecular structure.

All industrial polymers are polydisperse, but the rheological behavior of monodisperse polymers is easier to study.[234] This section is devoted to experimental results obtained in studies of uniaxial extension of monodisperse samples.

Stresses increase in uniaxial extension (see Fig. 3.7.1). Deformations in Fig. 3.7.1 are understood in the Hencky sense, i.e., $\varepsilon=3$ corresponds to about a 20-fold draw ratio.

The net tensile stress, σ_E, divided by the rate of deformation, $\dot{\varepsilon}$, gives the ratio which is the *elongational viscosity*, η_E:

$$\eta_E = \frac{\sigma_E}{\dot\varepsilon} \qquad\qquad [3.7.1]$$

It is an elongational viscosity of "pure" liquids (which deform without storing rubbery deformations). The theory predicts (see Section 3.1.1) that the *Trouton law* (Eq. 3.1.6) is valid for such inelastic liquids.

In experiments carried out at constant deformation rates ($\dot\varepsilon$ = const), one could expect that the stress is also constant. However, the real case is quite different. Fig. 3.7.2 shows the dependence of ratio ($\sigma_E/\dot\varepsilon$) on time at different deformation rates. The values of the ratio ($\sigma_E/\dot\varepsilon$) in Fig. 3.7.2 are normalized by their limiting value obtained at high deformations, and that is the elongational viscosity, η_E, at a steady flow. The argument can be also treated as deformation, ε, because at $\dot\varepsilon$ =const, the deformation is proportional to the time of elongation. The data presented in this figure were obtained at different temperatures and *reduced* to 25°C.[235]

At first glance, the experimental results presented in Fig. 3.7.2 seem to contradict the Trouton equation because the ratio ($\sigma_E/\dot\varepsilon$) is not constant but continuously increasing with the increase in deformation. However, one must be accurate in discussing this situation. Resistance to deformation increases, but it does not necessarily result in an increase of viscosity. Eq. 3.1.6 is related to a "pure" flow but the stretching of polymer melts consists of a superposition of the elastic (recoverable or rubbery) deformations and flow (or plastic deformation). Then, the stress evolution with deformation of a polymer melt under extension must be treated as a behavior of the viscoelastic body.

The theory of viscoelasticity is discussed in Chapter 2. The experimental results presented in Fig. 3.7.2 must be modeled to understand their significance.

For a linear viscoelastic material, the stress evolution in a uniaxial extension is expressed by

$$\sigma^+(t) = 3\dot\varepsilon \int_0^\infty \theta G(\theta)(1 + e^{-t/\theta})d\theta \qquad\qquad [3.7.2]$$

where $G(\theta)$ is the relaxation spectrum, measured, for example, at small-amplitude harmonic oscillation in a wide frequency range. This is the same relaxation spectrum used in treating experimental data in shear deformation. Similar to the rheological behavior in shear, the extension is expressed by the same form of the equation for stresses with the change of a shear rate, $\dot\gamma$, to an elongational deformation rate, $\dot\varepsilon$, and the difference is in the coefficient 3, only. In particular, the limiting value of the expression for stress, σ, at $\to \infty$ gives the Trouton viscosity, η_E, which equals 3η.

Dividing *tensile growth stress function*, $\sigma_E^+(t)$, by $\dot\varepsilon$, according to Eq. 3.7.2, one can expect the universal dependence of $\sigma_E^+(t)/\dot\varepsilon\,\eta_E$ vs. time. If this ratio is normalized with respect to its limiting value, which is the elongational viscosity, then the limit of the universal dependence of $\sigma_E^+(t)/\dot\varepsilon\,\eta_E$ and time should be equal to 1. This is true for a model polymer liquid, as shown in Fig. 3.7.2. It means that this material can be treated as a linear viscoelastic liquid with constant viscosity.

Fig. 3.7.3 presents results of direct measurements of the elongational viscosity at different deformation rates. It is seen that the elongational viscosity does not depend on the rate of deformation and it equals 3η.

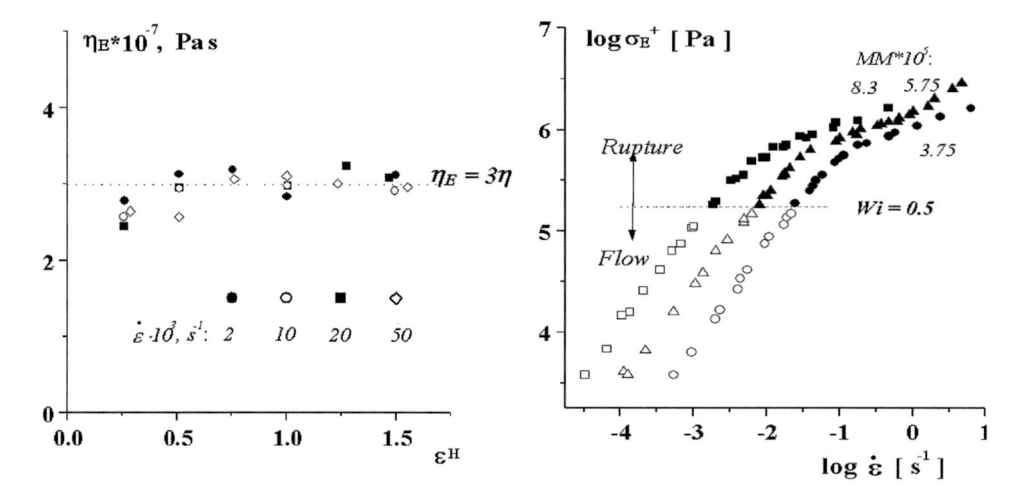

Figure 3.7.3. Measured elongational viscosity at steady-state flow at various deformation rates. Polyisoprene. M=5.75*10^5. T=25°C. [Adapted, by permission, from G.V. Vinogradov, A.Ya. Malkin, V.V. Volosevitch, V.P. Shatalov, V.P. Yudin, *J. Polym. Sci.: Polym. Phys Ed.*, **13**, 1721 (1975)].

Figure 3.7.4. Limiting stresses (at steady flow – open symbols, or at rupture – solid symbols) as a function of deformation rate for polyisoprenes of different molecular mass (shown on the curves). Experimental points were obtained in the temperature range from -25 to 75°C. [Adapted, by permission, from G.V. Vinogradov, A.Ya. Malkin, V.V. Volosevitch, V.P. Shatalov, V.P. Yudin, *J. Polym. Sci.: Polym. Phys Ed.*, **13**, 1721 (1975)].

The increase in the elongational viscosity at high deformation rates is sometimes considered as a necessary condition for the stability of a stream being stretched. However, this is not so as seen from Fig. 3.7.3. High draw ratios can be reached without a noticeable increase in viscosity. A many-fold increase in the deformation rate does not lead to the necessary growth of the apparent elongational viscosity.

The experimental results presented in Figs. 3.7.2 and 3.7.3 for monodisperse polymers as well as the constancy of their Newtonian viscosity in shear are not universal for real materials, such as commercial polydisperse polymer materials.

3.7.2 MODEL EXPERIMENTS – RUPTURE

Steady flow is possible up to some critical stress or shear rate level. What happens then on further extension? The material is ruptured above some critical stress or deformation rate level. This is predicted by a simple model of an *elastic liquid*.[236] If the rheological behavior of liquid is characterized not only by viscosity but also by a relaxation time, θ, then due to the effect of large deformations the elongational viscosity, η_E, as a function of deformation rate, $\dot{\varepsilon}$, is expressed as

$$\eta_E = \frac{\sigma_{11}}{\dot{\varepsilon}_{11}} = \frac{3\eta}{(1+Wi)(1-2Wi)} \qquad [3.7.3]$$

where

$$Wi = \theta \dot{\varepsilon} \qquad [3.7.4]$$

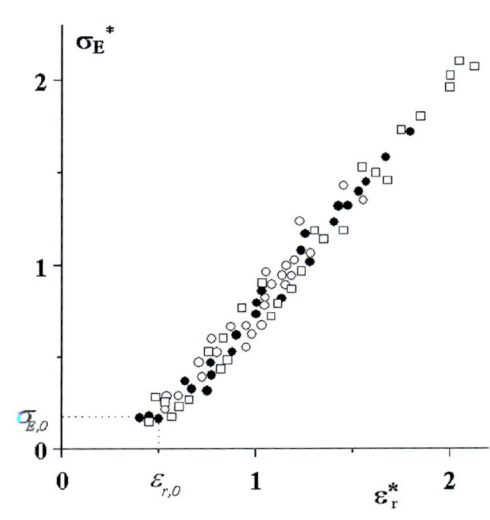

Figure 3.7.5. Relationship between stress and rubbery deformations at the moments of rupture reached at different rates of deformation. Solid circles – polybutadienes of different MM at different temperatures; Open circles – polyisoprenes of different MM at different temperatures; Squares – polybutadienes at 25°C in different liquid media (water and alcohols). [Adapted, by permission, from A.Ya. Malkin, C.J.S. Petrie, *J. Rheol.*, **41**, 1 (1997)].

is the Weissenberg number. It is the dimensionless ratio of the rate of deformation and rate of relaxation.[237]

At large deformation rates, namely at $Wi \to 0.5$, the apparent elongational viscosity and consequently the normal stresses increase unlimitedly. This means that the rupture occurs before a steady-state flow regime is reached. There is a critical value of the Weissenberg number separating domains of flow and stretching, which leads to rupture.

Fig. 3.7.4 shows dependencies of limiting stress values reached in elongation at different deformation rates. In the domain of low deformation rates, the limiting stresses are reached when the steady elongational flow regime is reached. These points correspond to constant values of elongational viscosity (that is why these limiting stresses are proportional to deformation rates). Regimes of steady flow are marked by open symbols in Fig. 3.7.4.

If the deformation rate is increased, the steady flow regime cannot be reached because the sample breaks before this state is attained. Then, the maximum (limiting) stress corresponds to the moment of rupture. The points corresponding to rupture are solid symbols in Fig. 3.7.4. There is a point of transition from deformations regimes, at which stretching leads to a steady elongational flow, to regimes when stretching ends by rupture before steady flow occurs.

The location of transition from flow to rupture is important for understanding the rheological behavior of viscoelastic polymer materials in a uniaxial extension.[238] The most important experimental result concerning this transition is that it happens at a certain value of the Weissenberg number, Wi, as predicted by theory, Eq. 3.7.4.

This criterion can be presented in different forms. In particular, the following rearrangements are pertinent:

$$Wi = \frac{\sigma_E \eta_E}{\eta_E E} = \frac{\sigma_E}{E} = \varepsilon_r \qquad [3.7.5]$$

where ε_r is elastic (rubbery) deformation stored in extension or tensile recoil.

According to the theory for linear elastic liquids, there is a critical value of the Weissenberg number corresponding to unlimited stress growth in extension, indicating that steady flow is impossible. The experimental confirmation that the transition from open to solid symbols in Fig. 3.7.4 (the transition to unsteady extension and rupture) takes place at $Wi = 0.5$.

However, the extension cannot be carried any further. An increase in the deformation rate leads to the growth of limiting (maximum attainable) stress, as shown in Fig. 3.7.4. There is universal dependence between this limiting (breaking) stress, σ_E^*, and the stored (elastic) deformation at the moment of break, ε_r^*. The latter, according to Eq. 3.7.5, is one of the possible interpretations of the Weissenberg number. This dependence is presented in Fig. 3.7.5 and it is expressed by the following linear relationship:

$$\sigma_E^* = \sigma_{E,0} + K(\varepsilon_r^* - \varepsilon_{r,0}) \quad \text{at} \quad \sigma_E^* > \sigma_{E,0}, \varepsilon_r^* > \varepsilon_{r,o} \qquad [3.7.6]$$

where $K = 1.2$ MPa is an empirical constant; $\sigma_{E,0} = 0.18$ MPa and $\varepsilon_{r,0} = 0.5$ are the parameters of the "transition point", where the flow becomes impossible and stretching causes breakup of a sample.

Fig. 3.7.2 demonstrates that the development of stresses in the extension of polymer liquids is the consequence of their viscoelastic properties. Eqs. 3.7.3 to 3.7.6 confirm that elasticity plays the dominating role in the extension of these liquids. The understanding of the rheological behavior of polymeric substances in the uniaxial extension must be based on the treatment of stress development as a result of their viscoelastic properties − superposition of elastic (rubbery) deformations and plastic flow.

The total deformation, ε_t, can be separated[239] into a recoverable (elastic), ε_r, and flow (plastic), ε_f, part to follow the development of both components as a function of the deformation rate. The results of such separation in the linear region of the viscoelastic behavior are shown in Fig. 3.7.6, where limiting values of deformations that can be reached at different deformation rates are presented.

There are several regions in which the character of the deformation evolution is different. At low deformation rates, the total deformation may grow unlimitedly due to a steady flow (both upper parts of Fig. 3.7.6). Rubbery deformation, ε_r, increases with an increase in deformation

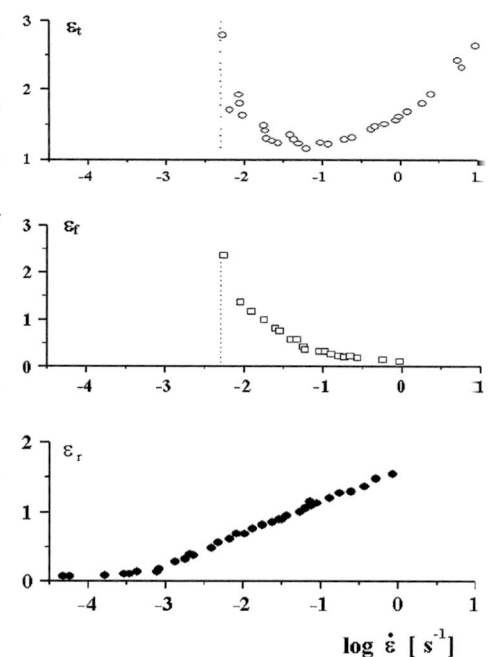

Figure 3.7.6. Separation of the total deformation into plastic and elastic parts: limiting total, t, plastic, f, and elastic, r, deformations at the moment of rupture. Experimental points are "reduced" to 25°C. [Adapted, by permission, from G.V. Vinogradov, A.Ya. Malkin, V.V. Volosevitch, V.P. Shatalov, V.P. Yudin, *J. Polym. Sci.: Polym. Phys Ed.*, **13**, 1721 (1975)].

rate (and, accordingly, stress). The values of ε_r, presented as a function of $\dot{\varepsilon}$, are the equilibrium values corresponding to the steady flow limit. The vertical dotted lines in Fig.

3.7.6 have the same meaning as the horizontal line in Fig. 3.7.4 – they correspond to Wi = 0.5 and divide the whole deformation rate scale into two domains – slow (left of the line), where the steady flow regime is acceptable and fast (right of the line), where the steady flow is not possible and stretching ends with sample failure. Beyond this transition line, the flow becomes less and less noticeable but rubbery deformations (at the moment of rupture) continue to increase together with the growth of stress. Combination of the decreasing ε_f and increasing ε_r results in the appearance of the minimum on the total $\varepsilon_t(\dot\varepsilon)$ dependence. At a sufficiently high deformation rate, the flow becomes negligible and the total deformation of stretching liquid appears to have a rubbery character.

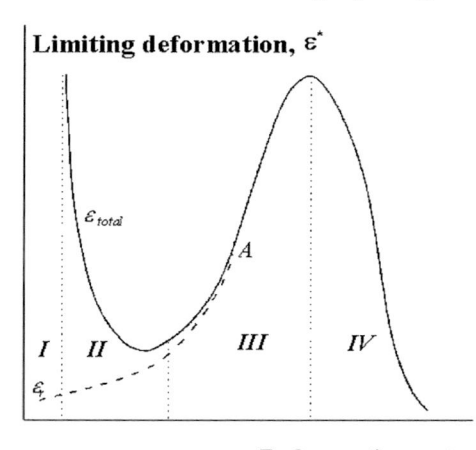

The experimental data presented in Fig. 3.7.6 constitute only a part of the characteristics of the limiting deformation evolution as a function of the deformation rate because they do not show what happens in the domain of very high deformation rates. The complete sequence of events taking place during extension as a function of increasing deformation rate is shown in Fig. 3.7.7. The domains I, II, and the left part of domain III are the same as in Fig. 3.7.6, but in comparison with Fig. 3.7.6 additional domains (part of III and IV) exist. In a single experiment, it is very difficult to move through the whole deformation range covering all domains, as in Fig. 3.7.7, because, in fact, the $\dot\varepsilon$ scale spreads through 8-10 decimal orders. That is why the generalized picture, presented in Fig. 3.7.7 is obtained by "reducing" experimental data to one reference temperature (see Chapter 2).

Figure 3.7.7. General picture illustrating deformation rate dependencies of limiting total (solid line) and elastic (dashed line) deformations. Roman numbers show principal regions of viscoelastic behavior observed as a function of deformation rate. Vertical dotted lines are approximate boundaries of these regions. [Adapted, by permission, from A.Ya. Malkin, G.V. Vinogradov, *Vysokomol. Soedin.* (Polymers – in Russian), **27**, 245 (1985)].

The solid line in Fig. 3.7.7 is for breaking (total) deformations and the dotted line, converging to the solid line at high deformation rates, is for elastic deformations. The physical meaning of the four domains in Fig. 3.7.7 is as follows:

- Domain I is a *flow* region, where extension causes a steady flow and the total deformation is unlimited. The boundary between domains I and II corresponds to Wi = 0.5
- Domain II is a *viscoelastic* region characterized by the superposition of elastic and plastic deformations, the latter equals the difference between the solid and dotted lines
- Domain III is a rubbery or pure *elastic* region, where flow is negligible and large deformations realized in the extension are completely elastic (rubbery)

- Domain IV corresponds to a decrease in possible deformation due to high-deformation-rate-induced solidification of material and the end of this domain could be called a *glassy* region.

The transition through the domains $I \rightarrow II \rightarrow III \rightarrow IV$ corresponds in polymer physics to well-known transitions through different relaxation states of amorphous polymer induced by increasing deformation rates.[240]

There is a special point in Fig. 3.7.7 – that is, a maximum located in the transition from the rubbery to elastic region. It was suggested[241] that the maximum value of the elastic deformation in the extension of the polymer melt is close to 2 (it corresponds to the draw ratio of about 7.3). It seems that this value is in agreement with the experimental data in Fig 3.7.5. However, a more general idea was advanced:[242] it was proposed that the maximum elastic deformation (the maximum on the curve in Fig. 3.7.7) equals the maximum extensibility of a macromolecular chain, which forms a statistical coil in an undeformed state. Then this maximum ε_e value depends on the molecular mass of the polymer.

Figs. 3.7.5 and 3.7.7 present results of model experiments carried out for monodisperse polymers and related primarily to the linear regime of viscoelastic behavior of these materials. They represent a general tendency in deformation evolution during the extension of polymer melts and solutions. The real relationships observed for industrial (polydisperse) polymers may be more complex and more difficult to understand. The common features of the superposition of flow and rubbery deformations and the dependence of their relative values on the deformation rate and transitions through different relaxation states are the same for any polymeric material. Analogous qualitative results confirming the applicability of the general concept of Fig. 3.7.7 were obtained for industrial-grade polydisperse polyethylene and polyisobutylene.[243]

The rupture of liquids described in this section is, by its physical mechanism, equivalent to the rupture of crosslinked rubbers. Temporary molecular junctions in the melt, present due to macromolecular physical interactions (for example, chain entanglements), are analogous to chemical crosslinking in rubbers. The difference is in the lifetimes of these junctions. However, if the characteristic time of deformation is short (or the deformation rate is high), physical junctions behave like permanent chemical junctions (at Wi > 0.5). Therefore, the mechanism of rupture during the extension at high deformation rates is similar to that of rupture of rubbers that can be treated as elastic.

3.7.3 EXTENSION OF INDUSTRIAL POLYMERS

Uniaxial extension of polymeric materials is considered as a model for different technological processes, primarily fiber formation and film orientation. This is also a physical method that can be useful for material structure characterization. These are the reasons why this method is widely used in investigations of real industrial polymers.[244] The experiments in uniaxial extension of industrial polymers were initially based on a technique that utilized measurements in the regime of the constant deformation rate.[245] Then many experimental investigations were made along the same lines. Their goal was to develop a variety of different polymers and to compare the results of these experiments with the molecular structure of the material (molecular mass distribution, branching of the chain, the chemical structure of the polymer, or the content of blend).[246]

The most obvious difference in the results of experimental investigations of many industrial polymers and the model data, as presented in Fig. 3.7.2, is the potential influence of *strain hardening*. It is often thought that the effect of strain hardening is of primary importance for polymer processing in fiber formation and in some other technologies. A comparison of the stress evolution in uniaxial extension of several polymers is shown in Fig. 3.7.8,[247] where both possibilities are presented: a smooth increase of the stress up to the regime of steady flow is reached at low deformation rates (a) and an increase of stresses – "*strain hardening*" observed as the unlimited growth of stresses until the break at a high deformation rate (b). The situation presented in Fig. 3.7.8a is quite equivalent to the model case shown in Fig. 3.7.2, whereas what is seen in Fig. 3.7.8b is different. The unlimited growth of stresses until breakup can be expected, as predicted by Eq. 3.7.4, and it happens earlier in time with an increase in the deformation rate.

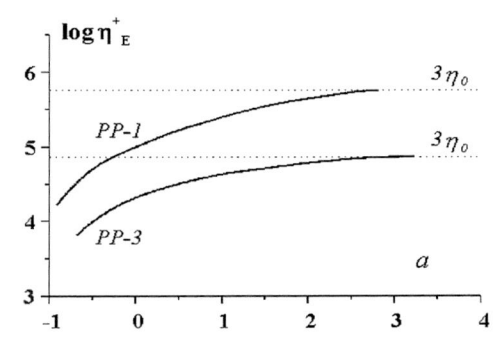

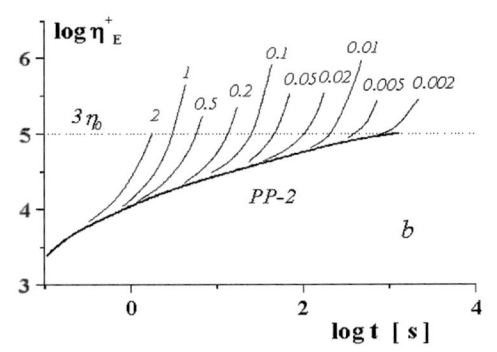

The increase in stress was discussed by the authors of original publications in terms of *strain hardening*, though the mechanism of this effect can be treated as the transition to domain III in Fig. 3.7.7 when elastic deformations become dominating and finally lead to rupture. Quantitative conditions of the breakup were not specified in the experiments under discussion, though it is possible to think that the final stress increase at higher deformation rates is similar to experimental results obtained for model systems, as shown in Fig. 3.7.5.

The difference in the evolution of normal stresses in Fig. 3.7.8 is remarkable and this is definitely the reflection of differences in the molecular structure of materials under investigation. As the closest version, a smooth increase in stresses is expected from melts of linear polymers (PP-1 and PP-3), whereas the strain harden-ing effect reflects the existence of branch-

Figure 3.7.8. Stress growth in elongation of polypropyl-ene samples. Experimental data points, presented by the authors, are omitted in reproducing this figure. PP 1, 2, and 3 are different PP samples. [Adapted, by permis-sion, from S. Kurzbeck, F. Oster, H. Münstedt, T.Q. Nguyen, R Gensler, *J. Rheol.*, **43**, 358 (1999)].

ing in the macromolecular structure (PP-2). The next reason for differences in rheological behavior can be connected with the molecular mass distribution of commercial polymers. Also, it can be suggested that, at least in some cases, an intensive stress growth in exten-sion can be caused by the stress-induced phase transition, possibly, even the crystallization of oriented chains.

Some cases are known when the steady-state regime during extensional flow cannot be reached at all, even at very low deformation rates. This is typical, for example, for ther-motropic liquid crystalline polymers (*Vectra A950*),[248] for which the regime of constant elongational viscosity does not exist but only strain hardening is observed. Possibly it corresponds to the well-known fact of the absence of a Newtonian flow range for liquid crystals because at low stresses a tendency of viscosity growth with a decrease of stresses (apparent approach to the yield stress) is observed.

3.7.3.1 Multiaxial elongation

Multiaxial elongation is a special case of elongational flow. There are various possible combinations of deformation modes. The reasonable classification of these cases is based on the analysis of the deformation rate tensor written as[249]

$$d(t) = \dot{\varepsilon}_0 \begin{bmatrix} 1 & 0 & 0 \\ 0 & m & 0 \\ 0 & 0 & -(1+m) \end{bmatrix}$$

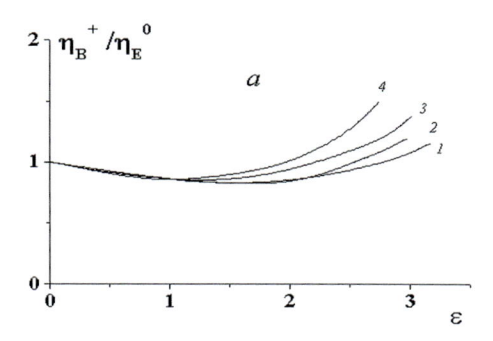

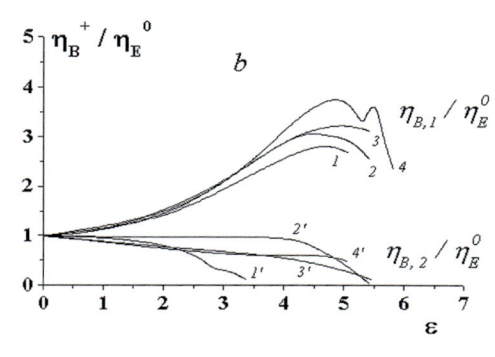

In the special case (discussed above) $m = -0.5$. This is *simple elongation*. The case of $m = 0$ is treated as *planar elongation* and the case of $m = 1$ is called *equibiaxial elongation*. An example of apparent elongational viscosity evolution as a function of strain is presented in Fig. 3.7.9.[250] Here elongational viscosity is normalized by the initial Trouton viscosity value (i.e., $\eta_E^0 = 3\eta_0$) in order to begin all curves from the same initial point. For equibiaxial elongation, only one elongational viscosity, η_B^+, (called the *biaxial stress growth coefficient*) exists, while in planar elongation two viscosity values can be measured along with two directions of elongation. In this Figure, η_B^+ is the *biaxial stress growth coefficient*, which is defined as the ratio of normal stress to the deformation rate. This value coincides with elongational viscosity in the absence of elastic deformations only (see Eq. 3.7.1 and its discussion).

Figure 3.7.9. Elongational viscosity (normalized by the initial Trouton value) in equibiaxial (a) and planar (b) elongation of high density polyethylene as a function of strain (Hencky measure). 150°C. Deformation rates, s^{-1}, a: 0.003 (1); 0.010 (2); 0.030 (3); 0,099 (4); b: 0.003 (1, 1'); 0.009 (2, 2'); 0.028 (3, 3'); 0.093 (4' 4'). [Adapted, by permission, from: P. Hachmann, J. Meissner, *J. Rheol.*, **47**, 989 (2003)].

The elongational viscosity varies in a very different manner, depending on the mode of deformation. In equibiaxial stretching, apparent viscosity is almost constant, i.e., the behavior of melt is close to the linear viscoelastic, though a weak effect of strain-softening also exists. In planar stretching, one apparent viscosity demonstrates the remarkable effect of strain hardening (like in simple elongation) and the second (normal to the first) viscosity value decreases at high strains. It is also worth mentioning that the type of elongational viscosity behavior depends on the nature of the material.

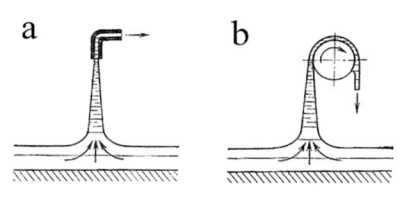

a   b

Figure 3.7.10. The tubeless siphon effect.

3.7.4 THE TUBELESS SIPHON EFFECT

Uniaxial extension of elastic (rubbery elastic) materials, capable of storing very large deformations, produces unusual effects. One of them is the so-called *tubeless siphon*.[251] This effect is observed during extension of dilute solutions of polymers (synthetic or biological) of ultra-high molecular mass. Such solutions are not very viscous but they are capable of storing large elastic deformations. In Fig. 3.7.7, the behavior of such solutions corresponds to domain III.

The tubeless siphoning effect is schematically presented in Fig. 3.7.10a. The fluid from a large free surface is drawn up through a nozzle by a vacuum. Then the nozzle is raised above the surface of the liquid. The liquid stream preserves its form and stability and it continues to flow upwards into the nozzle, even if a distance from the free surface to the die reaches several centimeters.

Another example of the siphoning effect is shown in Fig. 3.7.10b: in this case, a stream taken from a free surface is winding on rotating roller.

The driving force for this effect is undoubtedly the elasticity: initially created elastic deformations pull liquid upwards. In this regard, the behavior of liquid in a tubeless siphon is rubbery-like rather than fluid-like. It is interesting to note that it is impossible to create the siphoning effect for Newtonian liquid even if its viscosity is high.[252]

The siphoning effect produces high deformation rates in the extension of low viscosity (but highly elastic) liquid. This can be the basis of the original experimental method for measuring rheological properties of such liquids,[253] based on direct observation of the stream profile. However, the interpretation of the experimental results is not easy and it is intimately related to the rheological model used for treating the experimental data.

3.7.5 INSTABILITIES IN EXTENSION

3.7.5.1 Phase separation in extension

The influence of deformation (either in extension or in shear, though much more pronounced in extension) on the phase separation is related to the partial or complete uncoiling of a polymer chain. This phenomenon can be directly observed, and it was demonstrated that the transition from a coil to an extended conformation of a polymer chain is a phase transition. It was proven by birefringence measurements in very dilute solutions, where individual macromolecules are deformed separately.[254] The ratio of current birefringence to its limiting value corresponding to a completely extended chain, $\Delta n / \Delta n_{\infty}$, is a measure of chain conformation, or extension of a polymer statistical coil. It is useful to apply the dimensionless deformation rate as a measure of the intensity of loading. This dimensionless value (which was named above as the Weissenberg number) in the publications related to the subject under discussion, is usually named the *Deborah number*, De, which is (similar to Eq. 3.7.4) defined as

$$De = \theta \dot{\varepsilon} \qquad\qquad [3.7.7]$$

where θ is the characteristic relaxation time and $\dot{\varepsilon}$ is the deformation rate.[255] According to the molecular theory the θ value is determined as

$$\theta = \frac{aM[\eta]\eta_0}{RT}$$ [3.7.8]

where a is a constant of the order of one, M is the molecular mass of macromolecule, $[\eta]$ is the intrinsic viscosity, η_0 is the viscosity of solution, R is the universal gas constant, and T is the absolute temperature.

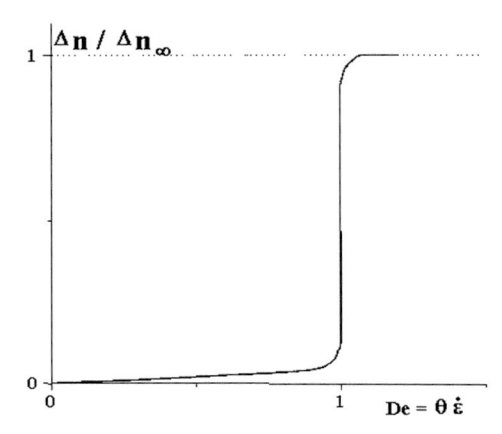

Figure 3.7.11. Increase in the degree of orientation (characterized by $(\Delta n)/(\Delta n_\infty)$) of individual macromolecules as a function of the dimensionless shear rate ($De = \theta\dot\epsilon$). Phase transition takes place at $De \to 1$.

Numerous experiments showed[256] that the dependence of the degree of uncoiling on the deformation rate is similar to that shown in Fig. 3.7.11. It means that change in macromolecular conformation occurs in a rapid manner as a typical phase transition. It also happens at the critical Deborah number of the order of 1, i.e., this phenomenon is determined by the ratio of deformation and relaxation rate measures.

It is also well known that aligning macromolecules (and their forced ordering) favors crystallization. Therefore, this phenomenon must be taken into account in constructing models of technological processes (kinetics of stress-induced crystallization is a necessary part of simulating the melt spinning process).[257]

The review of various types of instability phenomena of polymers in extension is discussed elsewhere.[258]

At high enough deformation rates, the phase separation takes place consisting in the appearance of solvent droplets on the surface of oriented macromolecules. The mechanism of this phenomenon was discussed above in Section 3.5.3. Shortly, this is stress-

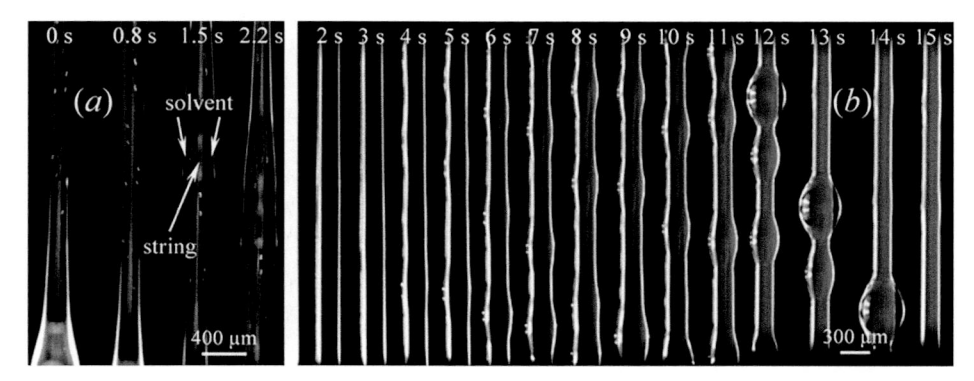

Figure 3.7.12. A sequence of photographs showing the formation of droplets on the surface of the jet. Photos by I.Yu. Skvortsov – Lab of rheology, Institute of Petrochemical Synthesis RAS. a – Short section of a jet. b – Full set of the stages of stretching. [Adapted, by permission, from I.A.V. Semakov, V.G. Kulichikhin, A.K. Tereshin, S.V. Antonov, A.Ya. Malkin, *J. Polym. Sci.: Part B: Polym. Phys.*, **53**, (8), 559 (2015).]

induced phenomenon creating giant fluctuations which migrate from the volume to a surface of a stream.

The development of the phase separation is illustrated in Fig. 3.7.12.[259]

The formation of the individual droplets is well seen in Fig. 3.7.13.[260]

The above presented pictures demonstrate the peculiarities of instability in the extension of viscoelastic liquids that lead to the phase separation as the mechanism especially related to the elasticity of polymer solutions.

3.7.5.2 Rayleigh instability[261]

Experimental data presented in Figs 3.7.6 and 3.7.7 show that in domain I, the total deformation is unlimited. But the extensional deformation cannot be unlimited because unlimited jet thinning is not possible. In this sense, extension, in contrast to shear, cannot continue unlimitedly. A rupture of a stream in domain I occurs due to different causes than the elastic rupture in other domains in Fig. 3.7.7. The breakup of the stream, in this case, is the consequence of the *Rayleigh instability*.

Figure 3.7.13. Images of solvent droplets on fiber surfaces for aromatic polyamide solution in DMAc (left) and polyacrylamide solution is DMSO (right).

It is well known and observed every day that when a liquid stream leaves a die (or a capillary) it disintegrates into separate drops. The mechanism of this phenomenon is as follows. If the speed of stream movement is not high, the breakup of liquid into drops is caused by the action of surface (capillary) forces. The thermodynamic cause of drop formation is a tendency for the minimization of the surface energy of the fluid. Surface energy is not at minimum for a stream but the optimum shape is a spherical drop.

The surface of a stream is disturbed by occasional factors. These small disturbances create surface waves which intensify until they disrupt the stream. The final result of the theoretical investigation is an expression for length, L, of the stream before disintegration:

$$L = 8.46u_0\left(\frac{\rho R^3}{\sigma}\right)^{1/2} \qquad [3.7.9]$$

where u_0 is the stream speed, R is stream radius, ρ is density, and σ is the surface tension.

Viscosity does not enter this expression. However, detailed analysis[262] showed that if the viscous forces are taken into consideration, another expression for the limiting length of a stream, L, is valid:

$$L = \frac{5u_0R\eta}{\sigma} \qquad [3.7.10]$$

where η is viscosity. This equation is correct for fluid moving through a medium with much lower viscosity than that of a viscous liquid stream flowing in the air.

The development of a theory of fluid stream disintegration leads to the field of spraying or atomization of streams leaving a nozzle with a very high speed. This problem is principal for the operation of a carburetor in car engines and many other engineering devices. However, an analysis of this problem goes beyond the field of rheology.

In extension of a viscous jet (during free falling), a regime of stable jet extension can exist and the profile of this jet can be exactly determined by solving the balance equations.[263] The length of a free falling jet before its rupture is very long,[264] greatly exceeding the limit predicted by the Plateau-Rayleigh limit. This delayed capillary breakup of falling viscous jets was treated as "paradoxical."[265] The general explanation of this effect was attributed to the transition from the capillary dominating regime to the viscous one.[266]

3.7.5.3 Instabilities in the extension of a viscoelastic thread

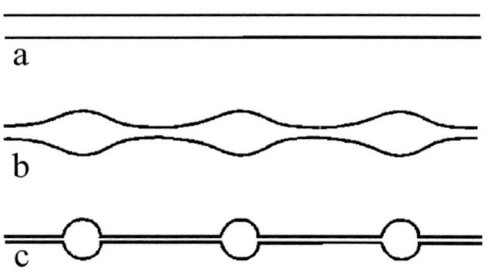

a

b

c

Figure 3.7.14. A view of viscoelastic jet leaving a nozzle: regular extrudate (a), draw resonance – periodic changes in thread diameter (b) and formation of a series of droplets joined by thin threads (c).

If the liquid is not purely viscous and elastic (rubbery) deformations superimpose on flow (some polymeric solutions, biological liquids, such as saliva), a stream's breakup mechanism is changed dramatically.

There are two main types of regular instabilities appearing in the extension of a thread. This is a draw resonance and the formation of small droplets along the thread (Fig. 3.7.14).

The draw resonance is the effect of large-scale periodic smooth variation in thread diameter (Fig. 3.7.14b) when this thread is taken up at constant speed.[267] This phenomenon is of special interest for the technology of fiber formation because periodic oscillations in thread diameter lead to a decrease in fiber quality.

A principal possibility of oscillations exists even for Newtonian liquid.[268] It was proven that the draw resonance appears at a high draw ratio,[269] and the critical value of the draw ratio, λ_{cr}, is close to 20.[270] The theoretical analysis of the stability problem in an extension of viscoelastic liquid showed[271] that the transition from stable to the unstable zone is determined by the dimensionless Deborah number, De, which is similar to that in Eq. 3.7.7, and defined as

$$De = \frac{\theta V_0}{L}$$

where θ is the characteristic relaxation time, V_0, is the extrusion velocity (at the outlet of a fiber-forming nozzle) and L is the length of thread.

The theory predicts that elasticity stabilizes thread diameter. It was shown that for any Deborah number there are two limits of the stability zone:
- the lower limit, which is practically constant for any De and corresponds to the $\lambda_{cr} \approx 20$

- the upper limit, which appears due to elasticity, and therefore the upper limit depends on De. The position of this upper limit of stability depends on the choice of model, which incorporates the elasticity into consideration.

The stability in relation to the draw resonance phenomenon depends on whether fluid has extension thinning or extension thickening properties; the latter stabilize the form of the thread. Thickening might also be understood as the transition from the flow regime of deformation to the regime dominated by elastic deformations in extension.

The draw resonance effect can be avoided by quick thread cooling,[272] which is an ordinary procedure in real technological processes.

In some cases, which are typical for low viscosity but highly elastic liquids, it was observed,[273] that instead of the breakup of a stream into separate drops, a thread looks like that are shown in Fig. 3.7.14c, i.e., the stream rearranges into a series of small droplets joined by a thin thread. The elasticity of liquid stabilizes the stream and prevents it from the formation of separate drops.[274] This effect can be explained by the transition of viscoelastic fluid connecting threads to domain III type behavior (in Fig. 3.7.7), in which they deform like an extendable rubber cord. This effect can be treated as the "extension thickening" property of liquid, though it is related primarily not to viscosity growth but to the rubbery elasticity of the medium.

Theoretical and direct visual observations confirm the difference in the mechanisms of the stream and drop breakup for viscous Newtonian and viscoelastic liquids.[275]

3.8 CONCLUSIONS – REAL LIQUID IS A COMPLEX LIQUID

The concept of *liquid* seems commonly known. Nobody doubts that water and gasoline at room temperature are liquids. But toothpaste or adhesive are not commonly considered liquids. This shows that liquid is not necessarily a material that flows. A more rigorous definition is needed to properly assign the behavior of real materials.

It may be suggested that liquid is material that undergoes *unrecoverable (irreversible) deformations*, i.e., the changes in shape remain after the action of external forces is removed. It should be noted, however, that this definition is too wide because it covers all real materials. For example, one would need to consider metals as liquids because during some technological operations, e.g., punching of golden articles, wire drawing of silver, or rolling steel ingots, unrecoverable deformations are undoubtedly created. These examples show that the above definition covers two different types of behavior: *viscous* and *plastic*. In the first case, unrecoverable deformations (or *flow*) can be detected at *any* stress, regardless of how small it may be. In the second case, unrecoverable deformations appear only when stress overcomes a certain level, which is called the *yield stress* or *yield point*. This means that a plastic medium can be called "liquid" only with some precautions. Nevertheless, treating deformations of plastic materials at stresses exceeding the yield stress as flow seems reasonable.

Liquids can also be defined as materials that can flow (or are able to accumulate unrecoverable deformations) under the action of infinitesimal stress. The possibility to flow under infinitesimal stress means that liquid at rest cannot store any stress. Formally, this definition is acceptable. But two questions arise, as follows:

- if stress is decreased by one (two, three, etc.) orders-of-magnitude, is it possible to reach the yield point at which flow, at very low stress, does not occur? Or is it reasonable to think that yield stress exists in the case of any liquid but can be so small that it cannot be observed under ordinary experimental conditions but can be attained if conditions are changed? Indeed, it is never certain that the level of stress attained in an experiment is *sufficiently* low to assume that material is a liquid *from a rigorous point of view*.

- perhaps during the period of observation (or an experiment) unrecoverable deformations are so small that they cannot be detected by ordinary methods, even though they presumably exist. If an experiment is continued, should the flow of material under investigation occur? It was demonstrated in Section 3.2 by several examples for viscoplastic materials, which can flow below the yield stress, as high viscosity materials, that a very long time is needed to detect flow.

Rheologists like to cite the famous exclamation from the Bible's Deborah: "*The mountains melted from before the Lord*",[276] bearing in mind that in the scale of eternity, the Lord can observe the flow of rocks (mountains). That is true, and the general answer to the problem raised by the theoretician lies in the comparison of the inherent time scale of material, t_{inh}, and time of observation, t_{obs}. This characteristic inherent time, t_{inh}, can be treated as the time of relaxation, i.e., the time necessary for the recovery of a stable structure after the removal of external forces. Then it is reasonable to introduce the dimensionless criterion t_{obs}/t_{inh}, which is called the *Deborah number*, De. The general form of this number is:

$$De = t_{obs}/t_{inh}$$

If De $\gg$ 1, the material behaves like a liquid. This happens when t_{inh} is small and relaxation occurs very quickly (in comparison with the time of observation). On the opposite side of the time-scale (when De $\ll$ 1), unrecoverable deformations cannot be detected, flow cannot be detected and material must not be treated as a liquid.

Another citation also illustrates this idea: "*Measuring – measure time in thou thyself*".[277] Indeed, estimation has real meaning in the scale of human existence.

Finally, a very general definition of liquid might be constructed on the basis of an energy concept. Any action is connected with energy consumption. Then two types of the post-effects can be established:

- the energy can be stored in the material and the stored energy returns after the removal of external forces. Storage of energy is characteristic of an elastic medium (elastic behavior in rheological media is discussed in Chapter 4.).
- the energy of deformations can be dissipated by its conversion into heat, and this type of behavior is characteristic of viscous liquids because the viscous resistance to movement means heat dissipation of the work produced by the forces applied.

A viscous liquid, then, can be defined as a medium deformed in such a manner that the *energy needed for deforming completely dissipates* in the process of deformation. In essence, it means that no energy source for further deformation exists in the material after the action of external forces ceases and that is why deformation cannot be recovered (no driving force exists for the process).

Intermediate cases may exist when the energy of deformation is partly stored by material and only part of this energy can be dissipated. Such is the case of viscoelastic bodies, and in particular, of viscoelastic liquid.

The energy concept dividing materials according to their reaction to the work of deformation is the most general approach of characterization of the type of material behavior. Such a concept is not related to any considerations of local values of stresses and deformations and does not need to be related to the observation of material behavior in the coordinate axis. In this sense, the energy approach is invariant with respect to coordinate transformations and satisfies the general requirements of rheological equations of state.

All real liquids are *complex liquids* and the following principal effects are more or less pronounced in their deformations:
- non-Newtonian viscosity in a steady flow
- elasticity (rubbery deformations)
- time-dependent slow structure transformations to self-organization and/or phase separation
- existence of limits of flow (instabilities of various types at high velocities or deformation rates).

The fundamental reasons for these phenomena are:
- superposition of large elastic deformations on viscous flow, this leads to numerous second-order effects
- changes in the intensity of energy dissipation depending on the deformation rate due to changes in streamline conditions at deformations or orientation of disperse phase particles in multi-component systems (stress-induced deformation of macromolecules in solutions, rotation of solid particles, and formation of necklace structures in suspensions)
- increase in the share of solid-like modes in molecular movements and decrease of energy dissipation in the flow of viscoelastic media (this is the most typical for concentrated polymer solutions and melts)
- destruction of the molecular or supermolecular structure of matter, which exists due to physical interactions (polar forces or other secondary interactions) or mechanical contacts between particles in multi-component systems (this is typical for liquids loaded with an active solid filler, highly concentrated suspensions, polar macromolecules in solutions, liquid crystals).

The observed behavior of various media can be different and very strongly dependent on the peculiarities of conditions of deformation and observation. Sometimes, effects characteristic of complex liquids are evident in the behavior of real liquids; in other cases, these effects appear under very special deformation conditions. This is the reason for a rheologist to say: do not ask what type of a material is, either liquid or solid, but try to understand how this material behaves – like a liquid or a solid.

REFERENCES

1   H. Navier (1785-1836) – French engineer, physicist, and mathematician. The author of basic works on civil engineering, mechanics, the strength of materials, the theory of elasticity, and fluid dynamics. The publication on the subject under discussion is: H. Navier, *Mém. de l'Acad. des Sciences*, **6**, 389 (1823).
2   G.G. Stokes (1819-1903) – English mathematician and physicist, the author of fundamental works on optics, spectroscopy, gravitation, fluid dynamics, and vector analysis. The publications related to the subject under discussion are: J.G. Stokes, *Trans. Cambridge Phyl. Soc.*, **8**, 287 (1845); *Math. Phys. Papers*, **1**, 76, Cambridge, 1880.
3   Sometimes this value is called a dynamic viscosity in order to distinguish it from kinematic viscosity, defined as the ratio of η to the density of the liquid.

4   F.T. Trouton, *Proc. Roy. Soc.*, **A77**, 426 (1906). Trouton experimented with pitch and similar very viscous materials. He proposed Eq. 3.1.6 based on theoretical arguments.

5   P.J.W. Debye (1884-1966) – an outstanding Dutch physicist, graduated and worked in Germany and in the US from 1940. He is an author of numerous works on quantum mechanics of solids at low temperatures, X-ray analysis of polycrystals, molecular physics, and application of physical methods to chemistry. Nobel prize in chemistry (1936).

6   See the definition of this value and some other details concerning [η] in Section 3.3.4.

7   G.V. Vinogradov, *Vysokomol. Soedin.*, (Polymers – in Russian), **A13**, 294 (1971); G.V. Vinogradov, A.Ya. Malkin, Yu.G. Yanovskii, E.K. Borisenkova, B.V. Yarlykov, G.V. Berezhnaya, *J. Polymer Sci.*, **10**, part A-2, p. 1061-1084 (1972); G.V. Vinogradov, A.Ya. Malkin, Yu.G. Yanovsky, E.K. Borosenkova, B.V. Yarlykov, G.V. Berezhnaya, V.P. Shatalov, V.G. Shalganova, V.P. Yudin, *Vysokomol. Soedin.*, (Polymers in Russian) **14A**, 2425 (1972); G.V. Vinogradov, *Pure Appl. Chem., Macromol. Chem.*, p. 417 (1973).; *Polymer*, **18**, 1275 (1977), *Polymer Eng. Sci.*, **21**, 339 (1981).
      G.V. Vinogradov (1910-1988) – a well-known USSR (Russian) expert on friction, wear, and rheology of greases and polymeric materials, a founder of the modern Russian school of rheology.

8   See also a special issue of Rheological Acta devoted to the Centennial of the first publication of E. Bingham introducing the concept of visco-plasticity and the yield stress – *Rheol. Acta*, **56**, 1 (2017). This issue contains a historical aspect of these phenomena and several original papers on the subject.

9   A comprehensive review specially devoted to the yielding in liquids was published by H.A. Barnes, *J. Non-Newton. Fluid Mech.*, **81**, p. 133 (1999). This review also contains the historical perspective of the subject; S.O. Il'yin, V.M. Spiridonova, V.S Savel'eva, M.M. Ovchinnikov, S.D. Khizhnyak, E.I. Frenkin, P.M. Pakhomov, A.Ya. Malkin, *Colloid J.*, **73**, 5 (2011).

10   P.C.F. Moller, A. Fall. D. Bonn, *Europhys. Lett.*, **87**, 38004 (2009).

11   I. Masalova, M. Taylor, E. Kharatiyan, A.Ya. Malkin, *J. Rheol.*, **49**, 839-849 (2005).

12   M.M. Denn, D. Bonn, *Rheol. Acta*, **50**, 307-315 (2011).

13   A.Ya. Malkin, S.O. Ilyin, T.B. Roumyantseva, V.G. Kulichikhin, *Macromolecules*, **46**, 1, 257-266 (2013)

14   Ilyin S.O., Malkin A.Ya., Kulichikhin V.G., *Colloid J.*, **74**, 492-502, (2012).

15   S. Ilyin, T. Roumyantseva, V. Spiridonova, E. Frenkin, A. Malkin, V. Kulichikhin, *Soft Matter.*, 7, 19, 9090-9103, (2011).

16   A. Malkin, V. Kulichikhin, S. Ilyin, *Rheol. Acta*, **56**, 177 (2017).

17   This term was very popular in early rheological literature. It was introduced by W. Ostwald, *Kolloid-Z.*, **5**, 99 (1925) and widely used for colloid systems and polymer solutions, see for example a monograph of W. Philippoff, "**Viscosität der Kolloid**", *Steinkopff*, Leipzig, 1942, which contains a complete survey of works published before Word War II; R. Lapasin, A. Trevisan, A. Semenzato, G. Baratto, Presented at AERC, Portugal, Sept. 2003.

18   Theory of plasticity is a special branch of the mechanics of solids. The formulation of principles of the theory of plasticity and solving of the main problems was given by: A. Nadai, **Plasticity**, *McGraw-Hill*, N.-Y, 1931; R. Hill, **Plasticity**, *Clarendon Press*, Oxford, 1950; W. Prager, P.G. Hodge, **Theory of Perfectly Plastic Solids,** *Wiley*, N.-Y., 1951.

19   M. Miesowicz, Nature, 136, 261 (1935) and 158, 27 (1946)

20   Liquid crystals are a special form of structure organization with two-dimensional order. There are numerous publications devoted to liquid crystals, for example, see a fundamental monograph: P.C. deGennes, J. Prost, *The theory of liquid crystals*, 2nd ed., *Clarendon Press*, Oxford, 1993. Rheology of liquid crystals of different types is also extensively studied. See, for example L. Walker, N. Wagner, *J. Rheol.*, **38**, 1525 (1994); S. Guido, N. Grizzuti, *Rheol. Acta*, **34**, 137 (1995); W.R. Burghardt, *Macromol. Chem.Phys.*, **199**, 471 (1998).

21   L. Léger, A. Martinet, *J. Phys. Colloques*, **37**, 89 (1976); R. Meyer, F. Nbergy, I. Aratutsae, T.H. Fraden S.-D. Lee, A. J. Hurd, *Faraday Discuss. Chem. Soc.*, **79**, 125 (1985); V.V. Belyaev, **Viscosity of nematic liquid crystals** (in Russian). *Fizmatlit,* 2002.

22   S. Hess, J.F. Schwarzl, and D. Baalss, *J. Phys.: Condens. Matter.*, **2**, 279 (1990); E.E. Pashkovsky, T.G. Litvina, *J. Phys. II France*, **2**, 521 (1992); A.M. Smondyrev, G.B. Loriot, R.A. Pelcovits, *Phys. Rev. Lett.*, **75**, 2340 (1995); R. Meyer, F. Nbergy, I. Aratutsae, T.H. Fraden, S.-D. Lee, A.J. Hurd, *Faraday Discuss. Chem. Soc.*, **79**, 125 (1985); M. Inoue, K. Yoshino, H. Moritake, K. Toda, *Jpn. J. Appl. Phys.*, **40**, 3528 (2001); V.V. Belyaev, **Viscosity of nematic liquid crystals,** (in Russian). *Fizmatlit,* 2002.

23   F.M. Leslie, *Arch. Ration.Mech. Anal.*, **28**, 265 (1979); *Adv. Liq. Cryst.*, **4**, 1 (1979).

24   J.L. Ericksen, *Trans. Soc. Rheol.*, **5**, 23 (1961).

25   O. Parodi, *J. Phys. (Paris)*, **31** 581 (1970).

26   J. Marrucci, *Pure & Appl Chem.*, **57**, 1545 (1985); A.D. Rey, M.M. Denn, *Annu. Rev. Fluid Mech.*, **34**, 233 (2002).

27   V.V. Belyaev, *Uspekhi Fiz Nauk* (in Russian), **171**, 267 (2001).

28   V.N.Tsvetkov, G.M. Mikhailov, *Acta Physicochim. URSS*, **8**, 77 (1938).
29   D.J. Highgate, R.W. Whorlow, *Brit. J. Appl. Phys.*, **18**, 1019 (1967).
30   G.I. Taylor, *Proc. Roy. Soc. A*, **146**, 501 (1934).
31   J.G. Oldroyd, *Proc. Roy. Soc. A*, **232**, 567 (1955).
32   Jackson N.E., Tucker III C.L., *J. Rheol.*, **47**, 659, 2003.
33   Oosterlinck, M. Mours, H.M. Laun, P. Moldenaers, J. Rheol., v. 49, p/ 897 (2005)
34   H. Giesekus, *Rheol. Acta*, **8**, 411 (1969); M.R. Lyon, D.W. Mead, R.E. Elliot, L.G. Leal, *J. Rheol.*, **45**, 881 (2001)
35   J.F. Brady, *Chem. Eng. Sci.*, **56**, 2921 (2001).
36   R. Scirocco, J. Vermant, J. Mewis, *J. Non-Newton. Fluid Mech.*, **117**, 183 (2004).
37   R. Pasquino, F. Snijkers, N. Grizzutti, J. Vermant, *Rheol. Acta*, **49**, 993 (2010).
38   R.B. Pipes, J. W. S. Hearle, A. J. Beaussart, A. M. Sastry, R. K. Okine, *J. Compos. Mater.*, **25**, 1204 (1991).
39   R. E. Rosensweig, **Ferrohydrodynamic**, *Cambridge University Press*, Cambridge, England, 1985; P. Ilg, M. Kröger, S.Hess, *Phys. Rev.*, *E*, **71**, 051201 (2005).
40   U.R. Christensen, *Geophys. J. Royal Astronom. Soc.*, **91**, 711 (1987); S.H. Treagus, Th. Kocher, S.M. Schmalholz, N.S. Mancktelow, *Tectonophysics*, **421**, 77 (2006); Z.Y. Yin, M. Karstunen, 12 Intern. Conf. Intern. Ass. Computer Methods and Advances in Geomechanics (IACMAG), Goa, India, Oct. 2008; Y. Takei, *J. Geophys. Research*, **115**, B03204, 19 pages (2010).
41   P.J. Carreau, *Trans. Soc. Rheol.*, **16**, 99 (1972).
42   K. Yasuda, R.C. Armstrong, R.E. Cohen, *Rheol. Acta*, **20**, 163 (1981).
43   M.M. Cross, *J. Colloid Sci.*, **20**, 417 (1965).
44   A. De Waele, *Oil Color Chem. Ass. J.*, **6**, 23 (1923) was the first who proposed a power-type function to describe non-Newtonian flow curves. W. Ostwald, Kolloid-Z., 36, 99 (1925); W. Ostwald, R. Averbach, *Kolloid-Z.*, **38**, 261 (1926).
45   This equation was first proposed by T. Schwedoff, *J. de Phys.*, **9**, 34 (1880) based on experiments with gelatin solutions (gels). It became popular after publications of E.C. Bingham, *Bull. US Bur. Stand.*, **13**, 309 (1916) and it is usually called after him.
      E.C. Bingham (1878-1945) – an American scientist, one of the pioneers in the field of modern rheology. He experimented with oil paints and, describing the distinction of their properties from the properties of the base (oil), introduced the concept of viscoplastic materials. His studies were summarized in the monograph **"Fluidity and Plasticity"**, *McGraw-Hill*, N.Y., 1922.
46   C.C. Mills (Ed.), **"Rheology of Disperse Systems"**, *Pergamon Press*, Oxford, 1959.
47   W.H. Hershel, R. Bulkley, *Proc. Amer. Assoc. Test Materials*, **26**, 621 (1926); *Kolloid Z.*, **39**, 291 (1926). See also G.V. Scott Blair, **An Introduction to Industrial Rheology**, *J.& A. Churchill Ltd.*, 1938.
48   J.L. White, Y. Wang, A. I. Isayev, N. Nakajima, F.C. Weissert and K. Min, *Rubber Chem. Technol.*, **60**, 387 (1987).
49   J.L. White, **Rubber Processing Technology Materials Principles**, *Hanser*, Munich, 1995.
50   A.I. Isayev and M. Wan, *Rubber Chem. Technol.*, **69**, 277 (1996).
51   R. Von Mises, *Göttinger Nachrichten. Math.-Phys. Klasse*, S. 582 (1913).
52   This problem was examined in several classical publications: W. Prager, P.G. Hodge, **Theory of Perfectly Plastic Solids**, *Wiley, Chapman, & Hall*, N.Y. 1951; J.D. Oldroyd, *Proc. Cambridge Philos. Soc.*, **43**, 100 (1947). Modern state is included in: G.R. Burgos, A.N. Alexandrou, V. Entov, *J. Rheology*, **43**, 463 (1999).
53   There are many publications devoted to boundary problems in deformation of viscoplastic media. See, e.g., G.R. Burgos, A.N, Alexandrou, *J. Rheology*, **43**, 485 (1999).
54   These are results of numerous experimental evidences related to different polymer materials. The first publications presenting this two part "power law" were: T.G. Fox, P.G. Flory, *J. Amer. Chem. Soc.*, **70**, 2384 (1948); *J. Appl. Phys.*, **21**, 581 (1950); *J. Polymer Sci.*, **14**, 315 (1954). For a review of earlier publications see T.G. Fox, S. Gratch, S. Loshaek, in **Rheology**, ed. F.R. Eirich, *Acad. Press*, N.-Y.- London, 1, 1956; G.S. Berry, T.G. Fox, *Adv. Polymer Sci.*, **5**, 261 (1968).
55   Concentration of polymer in solution can be expressed in a different way. Below, concentration, c, is expressed in grams of a polymer dissolved in 1 cm^3 of solution. The dimensionless measure, φ – volume share of a polymer in solution, will be also used.
56   M. Huggins, *J. Amer. Chem. Soc.*, **64**, 2716 (1942).
57   A.F. Martin, Amer. Chem. Soc. Meeting, Memphis, April 1962.
58   E.O. Kraemer, *Industr. Engng Chem.*, **30**, 1200 (1938).
59   Use of the concept of the "reduced concentration" c[η] was proposed by R. Simha, L. Utracki, *J. Polymer Sci.*, A-2, **5**, 853 (1967). See also: S. Onogi et al., *J. Polymer Sci.*, **C15**, 381 (1966) and *J. Polymer Sci.*, *A-2*, **5**, 899 (1967). Later, this approach was developed in many publications and the most general results were obtained in: V.E. Dreval, A.Ya. Malkin, G.O. Botvinnik, *J. Polymer Sci., Polymer Phys. Ed.*, **11**, 1055 (1973); V.E. Dreval, A.Ya. Malkin, G.V. Vinogradov, A.A. Tager, *Europ. Polymer J.*, **9**, 85 (1973);

A.Ya. Malkin, *Rheol. Acta*, **12**, 486 (1973).

60   LC state of polymer solutions is an acute problem extensively discussed during the last 20 years. See Ref. 11.

61   S.P. Papkov, V.G. Kulichikhin, V.D. Kalmykova, A.Ya. Malkin, *J. Polymer Sci., Polymer Phys. Ed.*, **12**, 1753 (1974).

62   The criterion of the transition of lyotropic solutions into LC state was discussed in P.J. Flory, *Proc. Royal Soc.*, **A234**, 73 (1956).
P.J. Flory (1910-1985) – an outstanding American physicist, one of the pioneers in the field of statistical physics of polymers. Author of many results concerning relationships between molecular structure and various physical properties of polymer materials. Nobel prize 1974.

63   S.R. Derkach, *Adv. Colloid Interface Sci.*, **151**, 1 (2009).

64   A. Einstein (1879-1955) – one of the greatest physicists of the XX Century, the author of special and general theories of relativity and many other fundamental works on theoretical physics. Nobel prize 1921. However, his new ideas seemed too new and he had to present the investigations on the flow of suspensions, carried out by the method of classical fluid mechanics, as his dissertation: A. Einstein, *Ann. Phys.*, **19**, 289 (1906); **34**, 591 (1911).

65   D. Quemada,. *Rheol. Acta*, **16**, 82 (1977).

66   M. Mooney, *J. Colloid Sci.*, **6**, 162 (1951).

67   S.A. Faroughi, Ch. Huber, *Rheol. Acta*, **54**, 85, 2015.

68   This problem was discussed and continues to be discussed in numerous publications. The review of earlier works, including classical studies of main models, was given in H.L. Goldsmith, S.G. Mason, in **Rheology**, ed. F.R. Eirich, *Acad. Press*, N-Y - London, 4, 86 (1967).

69   S.R. Derkach, *Adv. Colloid Interface Sci.*, **151**, 1-2,1-23, 2009.

70   R. Foudazi, S. Qavi, I. Masalova, A.Ya. Malkin, *Adv. Coll. Interface Sci.*, **220**, 78-91 (2015).

71   L. Berthier, G. Biroli, *Rev. Mod. Phys.*, **83**, 587, 2011.

72   J. Kaldasch, B. Senge, J. Laven, *J. Thermodyn.*, Art. ID 153854, 2015.

73   P. Holmqvist, M.P. Lettinga, J. Buitenhuis, J.K.G. Dhont, *Langmuir*, **21**, 10976, 2005.

74   A.Ya. Malkin, V.G. Kulichikhin, *Colloid J.*, **78**, 3, 2016.

75   X. Bian S. Litvinov, M. Ellero, N. J. Wagner, *J. Non-Newton. Fluid Mech.*, **213**, 39 (2014)

76   H. Kawabata, D. Nishiura, H. Sakaguchi, Y. Tatsumi, *Rheol. Acta*, **52**, 1 (2013)

77   V. Rathee, D. L. Blair, J. S. Urbach, *J. Rheol.*, **64**, 299 (2020)

78   A.Ya. Malkin, I. Masalova, **Advances in rheology and its applications**, Eds. Y. Luo, Q. Rao, Y. Xu., *Sci. Press*, N.-Y., USA, pp. 5-12 (2005); I. Masalova, A.Ya. Malkin, *Colloid J.*, **69**, 185 (2007).

79   H.M. Princen, *J. Colloid Interface Sci.*, **91**,160 (1983); **105**, 150 (1985); **128**, 176 (1989); H.M. Princen, A.D. Kiss, *J Colloid Interface Sci.*, **112**, 427 (1986).

80   M.-D. Lacasse, G.S. Grest, D. Levine, T.G. Mason, D.A. Weitz, *Phys. Rev. Lett.*, **76**, 3448 (1996); T.G. Mason, *Current Opinion Coll. Interface. Sci.*, **4**, 231 (1999).

81   R. Foudazi. I. Masalova, A.Ya. Malkin, *Appl. Rheol.*, **20**, 45096 (2010); *Colloid J.*, **72**, 74 (2010).

82   J. B. Donnet, Black and white fillers and tire compound. *Rubber Chem. Technol.*, **71**, 323-341 (1998).

83   G. V. Vinogradov, YU. Yanovsky, and A. I. Isayev, Viscoelastic behavior of an amorphous polymer under oscillations of large amplitude. *J. Polym. Sci: Part A-2*, **8**, 1239-1259 (1970).

84   A. I. Isayev and C. A. Hieber, Oscillatory shear flow of polymeric systems. *J. Polym. Sci. Polym. Phys.*, **20**, 423-440 (1982)

85   M. Parthasarathy and D. J. Klingenberg, D. J., Large amplitude oscillatory shear of ER suspensions. *J. Non-Newton. Fluid.*, **81** (1-2), 83-104 (1999).

86   V. Tirtaatmadja, K. C. Tam and R. D. Jenkins, Superposition of oscillations on steady shear flow as a technique for investigating the structure of associative polymers. *Macromolecules*, **30** (5), 1426-1433 (1997).

87   K. Hyun, M. Wilhelm, C. O. Klein, K. S. Cho, J. G. Nam, K. H. Ahn, S. J. Lee, R. H. Ewoldt and G. H. McKinley, A review of nonlinear oscillatory shear tests: Analysis and application of large amplitude oscillatory shear (LAOS). *Prog. Polym. Sci.*, **36** (12), 1697-1753 (2011).

88   G. Havet and A. I. Isayev, Rheology of silica-filled polystyrene: From microcomposites to nanocomposites. *Polym. Sci., Ser. A*, **54** (6), 476-492 (2012).

89   S. H. Maron and P. E. Pierce, Application of Ree-Eyring generalized flow theory to suspensions of spherical particles. *J. Colloid. Sci.*, **11**, 80-95 (1956).

90   A. B. Metzner, Rheology of suspensions in polymeric liquids. *J. Rheol.*, **29**, 739-775 (1985).

91   Firestone Polymers (2016) Datasheet of Duradene 739

92   PPG Industries (2011) Datasheets of Hi-Sil precipitated silica products

93   J. L. Leblanc, *Progress Polym. Sci.*, **27**, 627-687 (2002).

94   J. L. Leblanc, *J. Appl. Polym. Sci.*, **78**, 1541-1550 (2000).

95 J. L. Leblanc, *J. Appl. Polym. Sci.*, **66**, 2257-2268 (1997).

96 J. Fröhlich, W. Niedermeier, H-D. Luginsland, *Compos. Part A: Appl. Sci. Manufacturing*, **36**, 449-460 (2005).

97 S. Wolff, M-J. Wang, E-H. Tan, *Rubber Chem. Technol.*, **66**, 163-177 (1993).

98 J. L. Leblanc, A. Staelraeve, *J. Appl. Polym. Sci.*, **53**, 1025-1035 (1994).

99 S. S. Pole, A. I. Isayev, *Rheol. Acta*, **56** (12), 983-993 (2017).

100 S. S. Pole, A. I. Isayev, *J. Appl. Polym. Sci.*, **138** (28), Article Number e50660 (2021).

101 S. S. Pole, A. I. Isayev, *J. Appl. Polym. Sci.*, **138** (12), Article Number e50080 (2021).

102 S. Montes, J. L. White, N. Nakajima, *J. Non-Newt. Fluid Mech.*, **28**, 183-212 (1988).

103 K. Osaki, M. Kurata, *Macromolecules*, **13**, 671-676 (1980).

104 R. B. Bird, R. C. Armstrong, O. Hassager, **Dynamics of Polymeric Liquids: Fluid Mechanics**. *Wiley-Interscience*, New York, NY, vol. 1, p. 451 (1987).

105 J. Yun, A. I. Isayev., *J. Appl.Polym. Sci.*, **92**(1), 132-138 (2004).

106 S. Bazgir, A. A. Katbab, H. Nazockdast, *J. Appl. Polym. Sci.*, **92** (3), 2000-2007 (2004).

107 D. Kang, D. Kim, S. H. Yoon, D. Kim, C. Barry, J. Mead, *Macromol. Mater. Eng.*, **292** (3), 329-338 (2007).

108 S. Siriwardena, H. Ismail, U. S. Ishiaku , *Plast. Rubber Compos. Process. Appl.*, **31** (4), 167-176 (2002).

109 H. Tan, A. I. Isayev, *J. Appl. Polym. Sci.*, **109**, 767-774 (2008).

110 W. Feng, A. I. Isayev, *J. Polym. Sci. B Polym. Phys.*, **43**, 334-344 (2005).

111 This model was proposed in A.Ya. Malkin. N.K. Blinova, G.V. Vinogradov, M.P. Zabugina, O.Yu. Sabsai, V.G. Shalganova, I.Yu. Kirchevskaya, V.P. Shatalov, *Europ. Polymer J.*, **10**, 445 (1974) and later quite independently in B.H. Bersted, *J. Appl. Polymer Sci.*, **19**, 2167 (1975); B.H. Bersted, J.D. Slee, *J. Appl. Polymer Sci.*, **21**, 2631 (1977). This model was cast into the final quantitative form in: A.Ya. Malkin. A.E. Teishev, *Vysokomol. Soedin.* (Polymers - in Russian), **29A**, 2230 (1987); *Polymer Eng. Sci.*, **31**, 1590 (1991). The latter publications initiated the discussion on the possibility to use this model for solving the inverse problem of MMW calculation (see ref. 37).

112 A.Ya. Malkin, *Rheol Acta*, **29**, 512 (1990).

113 There were numerous publications devoted to this problem. See, for example W.H. Tuminello, N. Cudré-Mauroux, *Polymer Engng Sci.*, **31**, 1496 91991); M.T. Shaw, W.H. Tuminello, *Polymer Engng Sci.*, **34**, 159 (1994); Y.-M. Liu, M.T. Shaw, W.H. Tuminello, *J. Rheology*, **42**, 453 (1998); D.W. Mead, *J. Rheology*, **38**, 1797 (1994); M.R. Nobile, F. Cocchini, J.V. Lawler, *J. Rheology*, **40**, 363 (1996); P.M. Wood-Adams, J.M. Dealy, *J. Rheology*, **40**, 761 (1996); D. Nichetti, I. Manas-Zloczower, *J. Rheology*, **42**, 951 (1998).

114 See experimental data obtained for polyisobutylene: G.V. Vinogradov, A.Ya. Malkin, V.F. Shumsky, *Rheol Acta*, **9**, 155 (1970) and for low-density polyethylene melt: H.M. Laun, *J. Rheology*, **30**, 459 (1986).

115 M.B. Peirotti, J.A. Deiber, J.A. Ressia, M.A. Villar, E.M. Vallés, *Rheol. Acta*, **37**, 449 (1998).

116 K. Oda, J.L. White, E.S. Clark, *Polymer Engng Sci.*, **18**, 25 (1978).

117 This equation was first proposed in H.J.M.A. Mieras, C.F.H. Rijn, *Nature*, **218**, 865 (1968). Later it was successfully used for different polymers: N.J. Mills, *Nature*, **219**, 1249 (1968) and *Eur. Polym. J.*, **5**, 675 (1969).

118 This equation was proposed in P.K. Agarwal, *Macromolecules*, **12**, 342 (1979) and used by many authors.

119 M. Ansari, S.G. Hatzikiriakos, A.M.Sukhadia, D.C. Rohlfing, *Rheol. Acta*, **50**, 17 (2011).

120 J.A. Resch, J. Kaschta, F. Wolff. H. Münstedt, *Rheol. Acta*, **50**, 53 (2011).

121 K. Weissenberg, *Nature*, **159**, 310 (1947); Proc, 1st Inter., Congress on Rheology, v. 1, 29, v. 2, 114 (1949). K. Weissenberg (1893-1976) – one of the pioneers in the field of rheology and biorheology; also, the author of original publications on crystallography. His name is connected, in particular with introducing the Weissenberg number. He was the first who described the action of normal stresses in shear flow.

122 Also, see the early description of the phenomena related to this effect and demonstrated for polymer materials in the paper: F.H. Garner, A.H. Nissan. G.F. Wood, *Phys. Trans. Royal Soc. London*, **A243**, 37 (1950).

123 A. Kaye, A.S. Lodge, D.G. Vale, *Rheol. Acta*, **7**, 368 (1968); M.J. Miller, E.B. Christiansen, *AIChE J.*, **18**, 600 (1972); O. Olabishi, M.C. Williams, *Trans. Soc. Rheol.*, **16**, 727 (1972).

124 C. Barus, *Amer. J. Sci., Ser. 3*, **45**, 87 (1893).

125 A.C. Merrington, *Nature*, **152**, 663 (1943); *Nature*, **155**, 669 (1945).

126 A comprehensive analysis of extrudate die swell is presented in the paper: Z. Zhu and A.-Q. Wang, *J. Rheol.*, **48**, 571 (2004), which contains numerous references to previous experimental and theoretical studies of this effect.

127 Aluminum naphthenate dispersed in hydrocarbonates is a rubber-like matter widely used as a thickener in lubricants and firing systems, napalm in particular.

128 A thixotropic effect was first described for aqueous dispersions of iron oxide in E. Schalek, A. Szegvari, *Kolloid Z.*, **32**, 318 and **33**, 326 (1923). The term "thixotropy" appears in the paper T. Peterfi,

Arch. Entwicklungsmech.Organ., **112**, 680 (1927). Then a special monograph devoted to this phenomenon was published: H. Freundlich **"Thixotropy"**, *Hermann and Co.*, Paris, 1935. Later and until now this term is widely used in rheological literature. The comprehensive review on thixotropic effects was presented by J. Mewis, *J. Non-Newton. Fluid Mech.*, **1**, 6 (1979).

129 This list of thixotropic materials with numerous references to original publications devoted to rheological investigations of these materials is cited in review: H.A. Barnes, *J. Non-Newton. Fluid Mech.*, **70**, 1 (1997).

130 J.I. Escalante, H. Hoffman, *Rheol Acta*, **39**, 209 (2000).

131 Yu.A. Shchipunov, H. Hoffman, *Rheol. Acta*, **39**, 542 (2000).

132 This point of view was expressed in the following manner: "Fortunately today we hear less and less about 'thixotropy' and more and more about constitutive equations", C.A. Truesdell, VIII Intern. Congress on Rheology, Naples, Italy, 1980. A. Slibar, P. R. Pasley, *J. Appl. Mech.*, **29**, 107 (1964).

133 A. Peterlin, D.T. Turner, W. Philippoff, Koll. *Z. Z. Polymere*, **204**, 21 (1965); Z. Laufer, H.L. Jalink, J. Staverman, *J. Polymer Sci. Polymer Chem. Ed.*, **11**, 3005 (1973); S. Mani, M.F. Malone, H.H. Winter, *Macromolecules*, **25**, 5671 (1992); J.J. Magda, C.S. Lee, S.J. Muller, R.G. Larson, *Macromolecules*, **26**, 1696 (1993).

134 Some modern instruments have special computer-aided systems of acquisition of experimental data. In some cases, it is dangerous to trust the data obtained in such systems because they can smooth out physically meaningful fluctuations of the original data.

135 M. Minale, P. Moldenaers, J. Mewis, *J. Rheology*, **43**, 815 (1999); P. Van Puyvelde, H.Yang, J. Mewis, P. Moldenaers, *J. Rheology*, **44**, 1401 (2000).

136 P. Moldenaers, G.G. Fuller, J. Mewis, *Macromolecules*, **22**, 960 (1989); P.L. Maffettone, G. Marrucci, M. Mortier, P. Moldenaers, J. Mewis, *J. Chem. Phys.*, **100**, 7736 (1994).

137 The term dilatancy came from Latin *dilato* – expansion. Cf. dilatometer – an instrument for measuring changes in volume.

138 O. Reynolds, *Phil. Mag.*, **20**, 469 (1885).

139 Lord Kelvin (Sir William Thompson) (1824-1907) – an outstanding English physicist and mathematician, the author of fundamental results in the field of equilibrium thermodynamics and electricity. His ideas on continuum mechanics were summarized in Lord Kelvin **"Elasticity"**, in **Encyclopedia Britannica**, London, 1890.

140 See, for example, Z. Kemblowski, J. Petera, *Rheol. Acta*, **20**, 311 (1981); J. Billingham, J.W.J. Ferguson, *J. Non-Newton. Fluid Mech.*, **47**, 21 (1993).

141 M. Reiner, D.W. Scott Blair in **Rheology. Theory and Applications** (1967) Ed. F.R Eirich, *Acad Press, NY-London*, vol 4, Ch. 9. Interesting examples of shear-thickening of viscosity (or "negative thixotropy") in prolonged deformations can be found in: A. Peterlin, D.T. Turner, *Nature*, **197**, 488 (1963) and *J. Polymer Sci., B*, **3**, 517 (1965); T. Matsuo, A. Pavan, A. Peterlin, D.T. Turner, *Colloid Interface Sci.*, **24**, 273 91967); S.T.J. Peng, R.F. Landel, *J. Appl. Phys.*, **52**, 5988 (1981); P. Branda, O. Quadrat, *Colloid & Polymer Sci.*, **262**, 189 (1984).

142 G.G. Fuller, K. Smith, W.R. Burghardt, *J. Statistic. Phys.*, **62**, 1025 (1991).

143 H. Watanabe, T. Kanaya, Y. Takahashi, *Macromolecules*, **34**, 662 (2001); K. Mortensen, E. Theunissen, R. Kleppinger, K. Almdal, H. Reynaers, *Macromolecules*, **35**, 7773 (2002).

144 F. Bagusat, B. Böhme, P. Schiller, H.-J. Mögel, *Rheol. Acta*, **44**, 313 (2005).

145 H. Zhu, D. De Kee, K. Frederic, *J. Non-Newton. Fluid Mech.*, **157**, 108 (2009).

146 F. Bagusat, B. Böhme, P. Schiller, H.-J. Mögel, *Rheol. Acta*, **44**, 313 (2005).

147 R.J.E. Clausius (1822-1888) – German mathematical physicist, one of the pioneers of classical thermodynamic and kinetic theory of heat introduced the terms "entropy" and "ideal gas"; B.P.E. Clapeyron (1799-1864) – French physicist and engineer.

148 J. Gough, *Proc. Lit. Phys. Soc. Manchester*, **1**, 288 (1805).

149 P.J. Flory, *J. Chem. Phys.*, **15**, 387 (1947); W.R. Krigbaum, R.J. Roe, *J. Polymer Sci., Ser. A.*, **2**, 4391 (1964).

150 A. Silberberg, W. Kuhn, *Nature*, **170**, 450 (1952); *J. Polymer Sci.*, **13**, 21 (1954); F. Eliassaf, A. Silberberg, A.Katchalsky, *Nature*, **176**, 1119 (1955).

151 R.G. Larson, *Rheol. Acta*, **31**, 497 (1992). This is a comprehensive review touching all aspects of the problem of flow-induced phase mixing, separation, and phase transitions in polymer systems.

152 B.A. Wolf, *Macromol. Chem Rapid Commun.*, **1**, 231 (1980); *Macromolecules*, **17**, 615 (1984).

153 This approach is formulated by C. Rangel-Nafaile, A.B. Metzner, K.F. Wissbrun, *Macromolecules*, **17**, 1187 (1984) and also discussed in detail in the review A.Ya. Malkin, S.G. Kulichikhin, *Vysokomol. Soedin,*
Ser B (in Russian) **38**, 362 (1996).

154 Polymerization of caprolactam is the process of synthesis of polyamide-6. In the experiments cited the process of anionic polymerization of caprolactam was studied.

155 A.Ya. Malkin, S.G. Kulichikhin, *Makromol. Chem.: Macromol Symposia*, **68**, 301 (1993).
156 This effect was discussed in several theoretical and experimental works: R. Shafer, N. Laiken, B. Zimm, *Biophys. Chem.*, **2**, 180 and 185 (1974); K.A. Dill, *Biophys. Chem.*, **10**, 327 (1979); K.A. Dill, B. Zimm, *Nucleic Acid Res.*, **7**, 735 (1979).
157 Brochard, P.-G. de Gennes, *Macromolecules*, **10**, 1157-1161 (1977); Helfand E., Fredrickson G.H., *Phys. Rev. Lett.*, **62**, 2468 (1989); M. Doi, A. Onuki, *J. Phys. II France*, **2**, 1631-1656 (1992); S. T. Milner, *Phys. Rev. E*, **48**, 3674-3691 (1993); H. Ji, E. Helfand, *Macromolecules*, **28**, 3869-3880 (1995); Cromer M, Villet M.C., Fredrickson G.H., Leal L.G., *Phys. Fluids*, **25**, 051703 (2013); M. Cromer, M. C. Villet, G. H. Fredrickson, L. G. Leal, R. Stepanyan, and M. J. H. Bulters, *J. Rheol.*, **57**, 1211 (2013)
158 S. Acierno, B. Palomba, H.H. Winter, N.Grizzui, *Rheol. Acta*, **42**, 243 (2003).
159 H. Janeschitz-Kriegl, E. Ratajski, M. Stadbauer, *Rheol. Acta*, **42**, 355 (2003; M. Stadbauer, H. Janeschitz-Kriegl, G. Eder, E. Ratajski, *J. Rheol.*, **48**, 631 (2004).
160 There is a number of publications demonstrating this effect. One of the most current is M. Astrus, P. Navard, *J. Rheology*, **44**, 693 (2000).
161 M.V. Tsebrenko, A.V. Yudin, T.I. Ablazova, G.V. Vinogradov, *Polymer*, **17**, 831 (1976); A. I. Isayev, M. Modic, *Polymer Composites*, **8**, 269 (1987); M.V. Tsebrenko, G.P. Danolova, A.Ya. Malkin, *J. Non-Newton. Fluid Mech.*, **31**, 1 (1989).
152 L.A. Utracki, *J. Rheology*, **35**, 1615 (1991).
163 The first publications devoted to producing polymer samples with fully extended chains were: A.J. Pennings, M.J.F. Pijpers, *Macromolecules*, **3**, 261 (1970); A.J. Pennings, A.Zwijnenburg, R. Lageveen, *Koll. Z. u. Z. Polymere*, **251**, 500 (1973).
164 R. Zheng, P.K. Kennedy, *J. Rheol.*, **48**, 823 (2004).
X. Hu Y. Th., *J. Rheol.*, **54**, 1307 (2010).
1. S. Varchanis., G. Makrigiorgos, P. Moschopoulos, Y. Dimakopoulos, J. Tsamopoulos, *J. Rheol.*, **63**, 609 (2019).
R. G. Larson, Y. Wei, *J. Rheol.*, **63**, 477 (2019).
Y A. Jain, R. Singh, L. Kushwaha, V. Shankar Y. M. Joshi, *J. Rheol.*, **62**, 1001 (2018).
R. L. Moorcroft, S. M. Fielding, *Phys. Rev. Lett.*, **110**, 086001 (2013) and *J. Rheol.*, **58**, 103 (2014).
Z R. N. Al-Kaby, J. S. Jayaratne, T. I. Brox, S. I. Codd, J. D. Seymour, J. R. Brown, *J. Rheol.*, **62**, 1125 (2018)
R. N. Al-Kaby, S. L. Codd, J. D. Seymour, J. R. Brown, *Appl. Rheol.*, **30**, 1 (2020)
165 J.-F. Berret, G. Porte, J.-P. Decruppe, Phys. Rev. E, **55**, 1668 (1997); M.M. Britten, T.P. Callaghan, *Phys. Rev. Lett.*, **78**, 4930 (1997); J.-B. Salmon, S. Manneville, A. Colin, *Phys. Rev., E*, **68**, 051503 (2003); A. Raudsepp, P. Callaghan, Y. Hemar, *J. Rheol.*, **52**, 1113 (2008); M.E. Helgeson, P.A. Vasquez, E.W. Kaler, N.J. Wagner, *J. Rheol.*, **53**, 727 (2009).
166 A.G.Dankova, E. Mendes, M.-O. Coppens, *J. Rheol.*, **53**, 1087 (2009).
167 Y.T. Hu, C. Palla, A. Lips, *J. Rheol.*, **52**, 379 (2008)
169 H. Herle, P. Fischer, E.J. Windhab, *Langmuir*, **21**, 9051 (2005).
170 J. Delgado, A. Castillo, *J. Coll. Interface Sci.*, **312**, 481 (2007).
171 K. Kang, M.P. Lettinga, J.K.G. Dhont, *Rheol. Acta*, **47**, 499 (2008).
172 S. Varchanis, G. Makrigiorgos, P Moschopoulos. Y., Dimakopoulos. *J. Rheol.*, **63**, 609 (2019); R G. Larson. Y. Wei, *J. Rheol.*, **63**, 477 (2019).
173 R.L.Moorcroft, S.M. Fielding. *Phys. Rev. Lett.*, **110**, 086001 (2013) and *J. Rheol.*, **58**, 103 (2014). A. Jain, A., R. Singh, L. Kushwaha, V. Shankar Y. M. Joshi, *J. Rheol.*, **62**, 1001 (2018).
174 R.N. Al-Kaby, J.S. Jayaratne, T.I. Brox, S.I. Codd, J.D. Seymour, J. R. Brown, *J. Rheol.*, **62**, 1125 (2018) and R. N. Al-Kaby, S. L. Codd, J. D. Seymour, J. R. Brown, *Appl. Rheol.*, **30**, 19 (2020).
175 E. Helgand, H. Fredrickson, *Phys. Rev. Lett.*, **62**, 2468 (1989); *Phys. Fluids*, **25**, 051703 (2013); M. Cromer, M. C. Villet, G. H. Fredrickson, L. G.. Leal, R. Stepanyan, M. H. Bulters, *J. Rheol.*, **57**, 1211 (2013).
176 M. l. Cromer, G. H. Fredrickson, L. G. Leal, *Phys, Fluids*, **26**, 063101 (2014).
177 M. Cromer, M. C. Villet, G. H. Fredrickson, L. G.. Leal, R. Stepanyan, M. H. Bulters, *J. Rheol.*, **57**, 1211 (2013).
178 A. Onuki,. R. Yamamoto, T. Taniguchi, *Prog. Colloid Polym. Sci.*, **106**, 150 (1997).
179 M. E.Helgeson, L. Porcar, C. Lopez-Barron, N. J. Wagner, *Phys. Rev. Lett.*, **105**, 084501 (2010).
180 A. H Boudara, J. D. Peterson, L. G. Leal, D. J. Read, *J. Rheol.*, **63**, 71 (2019); X. Xu, F. Tang, W. Tian, W. Chen, L. Li, *J. Rheol.*, 64, 941 (2020); E. Narimissa, M. H. Wagner, *J. Rheol.*, **63**, 361 (2019).
181 A.E. Likhtman, *Macromolecules*, 38, 6128 (2005); T. Sato, S. Moghadam, G. Tan, R.G. Larson, *J. Rheol.*, **64**, 1045 (2020).

182   J. D. Peterson, G. H. Fredrickson, L. G. Leal, *J. Rheol.*, **64**, 1391 (2020).
183   A.Ya. Malkin, *Adv. Colloid Interface Sci.*, **290**, 102381 (2021).
184   O. Reynolds (1842-1912) – English physicist and engineer. He carried out pioneering works (1883) devoted to the transition from steady (laminar) flows to turbulent regimes of flow. Also, he is the author of the first publications on lubrication theory and flows in thin liquid layers.
185   J.L.M. Poiseuille (1799-1869) – French physiologist and physician who, while experimenting with blood, found that flow rate is proportional to the fourth power of the radius of a capillary, though blood is a non-Newtonian liquid. He proposed to use a mercury manometer to measure the blood pressure. The historical publications formulating the Hagen-Poiseuille law are G. Hagen, *Ann. d. Phys.*, **46**, 423 (1839); J.L.M. Poiseuille, *Mém. Savants Étrangers*, **9**, 423 (1846).
186   H. Blasius, *Z. de Ver. deutscher Ing.*, 639 (1912).
187   D.W. Dodge, A.B. Metzner, *AIChE J.*, **5**, 189 (1959).
188   B.A. Toms, Proc. 1st Intern. Congress on Rheology, 2, 135, North-Holland, Amsterdam, 1949.
189   ppm means "parts per million", i.e. the concentration of an additive equal to 0.01 percent.
190   Several hundred original publications and comprehensive reviews were devoted to studies of the Toms effect. As an example of existing approaches the following publications are quite typical: J.L. Lumley, *Ann. Rev. Fluid Mech.*, **1**, 367 (1969); *Macromolecules*, **7**, 263 (1969); *Phys. Fluids*, **20**, 564 (1977); F. Durst, R. Haas, W. Interthal, *Rheol. Acta*, **21**, 572 (1982); M. Renardy, *J. Non-Newton. Fluid Mech.*, **59**, 93 (1995); M.P. Escudier, F. Presi, S. Smith, *J. Non-Newton. Fluid Mech.*, **81**, 197 (1999).
191   P.S. Virk, *J. Fluid Mech.*, **45**, 225 (1971); *AIChE J.*, **21**, 625 (1975).
192   A.Ya. Malkin, G.V. Nesyn, A.V. Ilyushnikov, V.N. Manzhai, *J. Rheol.*, **44**, 371 (2000); *J. Non-Newton. Fluid Mech.*, 97, 195 (2001).
193   Two reviews devoted to experimental evidence and their theoretical understanding related to the structure formation in the flow of colloid and polymer substances were recently published: A.Ya. Malkin, A.V. Semakov, V.G. Kulichikhin, *Adv. Colloid Interface Sci.*, **157**, 75 (2010) and A.V. Subbotin, A, Ya. Malkin, V.G. Kulichikhin, *Adv. Colloid Interface Sci.*, **162**, 29 (2011).
194   G.I. Taylor, *Phil. Trans. Royal Soc. London*, **A223**, 289 (1923).
195   R.J.Donnely, D. Fritz, *Proc. Roy. Soc. A (London)*, **258**, 101 (1960). The general theory covering different types of instabilities in a flow of viscous fluids as well as magnetic liquids has been presented in a classical monograph: S. Chandrasekhar. Hydrodynamic and hydromagnetic stability. Cambridge: *Cambridge University Press* (1981).
196   M.P. Landry, I.A., Frigaard, D.M. Matrinez, *J. Fluid Mech.*, **560**, 321 (2006).
197   G.Baier, M.D. Graham, *Phys. Fluids*, **10**, 3045 (1998).
198   V. Steinberg, A.Groisman, *Phys. Mag., Part B*, **78**, 2563 (1998).
199   See original publications S.J. Muller, R.G. Larson, E.S.G. Shaqfeh, *Rheol. Acta*, **28**, 499 (1989); E.S.G. Shaqfeh, S.J. Muller, R.G. Larsen, *J. Fluid Mech.*, **235**, 285 (1992) and reviews: R.G. Larson, *Rheol. Acta*, 31, **213** (1992) and E.S.G. Shaqfeh, *Ann. Rev. Fluid Mech.*, **28**, 129 (1996).
200   J.G. Oldroyd, *Proc. Royal Soc. London*, **A283**, 115 (1965); B. Gervang, P.S. Larsen, *J. Non-Newton. Fluid Mech.*, **39**, 217 (1991); S.-C. Xue, N. Phan-Thien, R.I. Tanner, *J. Non-Newton. Fluid. Mech.*, **59**, 191 (1995).
201   H. Giesikus, *Rheol Acta*, **4**, 85, 65 (1965); *Rheol. Acta*, **5**, 239 (1966); *J. Non-Newton. Mech.*, **11**, 69 (1982).
202   A.G. Dodson, P. Townsend, K. Walters, *Comput. Fluids*, **2**, 317 (1974).
203   P.E. Arratia, C.C Thomas, J. Diorio, J.P. Gollub, *Phys. Rev. Lett.*, **96**, 144502 (2006).
204   R.J. Poole, M.A. Alves, P.J Oliveira, *Phys. Rev. Lett.*, **99**, 164503 (2007).
205   B. Debbaut, J. Dooley, *J. Rheology*, **43**, 1525 (1999).
206   M. Keentok, S.-C. Xue, *Rheol. Acta*, **38**, 321 (1999).
207   J.G Dhont, W.J. Briels, *Rheol. Acta*, **7**, 257 (2008).
208   E. Fischer, P.T. Callaghan, *Phys. Rev. E*, **64**, 011501. (2001).
209   The first description of unstable flow in polymer melt extrusion were made in H.K. Nason, *J. Appl. Phys.*, **16**, 338 (1945); M. Mooney, *J. Coll. Sci.*, **2**, 69 (1947). The first criterion of instability based on the systematic study of this effect was proposed in R.S. Spenser, R.E. Dillon, *J. Coll. Sci.*, 4, 241 (1949). Several comprehensive reviews on the problem were published and they cover an absolute majority of original publications that appeared before those reviews. The most important are the following: J.P Tordella, in "**Rheology. Theory and Applications**", ed. F.R. Eirich, **5**, 57 (1969); C.J.S. Petrie, M.M. Denn, *AIChE J.*, **22**, 209 (1976); R.G. Larson, *Rheol. Acta*, **31**, 213 (1992).
210   It was shown that sharkskin-type instability can be smoothened or may completely disappear if a die is made from brass: A.V. Rumamurthy, *J. Rheol.*, **30**, 337 (1986); V.G. Ghanta, B.L. Rüse, M.M. Denn, *J. Rheol.*, **43**, 435 (1999). See also N. El Kissi, J.-M. Piau, *J. Non-Newton. Fluid Mech.*, **37**, 55 (1990).
211   Such observations were published by many authors; most remarkable results were documented in the

papers: J.-M. Piau, N. El Kissi, F. Mezghani, *J. Non-Newton. Fluid Mech.*, **59**, 11 (1995); N. El Kissi, J.-M. Piau, F. Toussaint, *J. Non-Newton. Fluid Mech.*, **68**, 271 (1997).

212 H. Mizunuma, H. Takagi, *J. Rheol.*, **47**, 735 (2003).

213 K.B. Migler, Y. Son, F. Qiao, K. Flynn, *J. Rheol.*, **46**, 382 (2002).

214 V.I. Brizitsky, G.V. Vinogradov, A.I. Isayev, Yu.Ya. Podolsky, *J. Appl. Polymer Sci.*, **20**, 25, 1976.

215 This idea was expressed first by F.N. Cogswell, *J. Non-Newton. Fluid Mech.*, **2**, 37 (1977), and later supported by many authors. Rather evident numerical evidence supporting this mechanism was advanced by C. Venet, B. Vergnes, *J. Non-Newton. Fluid. Mech.*, **93**, 117 (2000).

216 E. Miller, J.P. Rothstein, *Rheol. Acta*, **44**, 160 (2004).

217 R. Rutgers, M. Mackley, *J. Rheol.*, **44**, 1319 (2000).

218 S.B. Kharchenko, P.M. McGuiggan, K.B. Migler, *J. Rheol.*, **47**, 1523 (2003).

219 F.N. Cogswell, *J. Rheol.*, **43**, 245 (1999).

220 H.J. Larrazabal, A.N. Hrymak, J. Vlachopoulos, *Rheol. Acta*, **45**, 705 (2006).

221 See Ref. 3 in Section 3.2.

222 V.E. Dreval, G.V. Vinogradov, V.P. Protasov, in Proc. IX Intern. Congr. Rheol., Eds. B. Mena, A. García-Rejón, C. Rangel-Nafaile, Mexico, 3, 185 (1984); E. Lee, J.L. White, *Polym. Eng. Sci.*, **39**, 327 (1999); J. Pérez-González, *J. Rheol.*, **45**, 845 (2001).

223 E.K. Borisenkova, G.B. Vasil'ev, V.G. Kulichikhin, S.A. Kuptsov, D.R. Tur, *Vysokomol. Soedin.* (in Russian), **40**, 1823 (1988); English translation: *Polymer Science USSR, Ser. A*, **40**, 1124 (1988).

224 L. Robert, Y. Demay, B. Vergnes, *Rheol. Acta*, **43**, 89 (2004).

225 F. Koran, J.M. Dealy, *J. Rheol.*, **43**, 1291 (1999); H. Münstedt, M. Schmidt, E. Wassner, *J. Rheol.*, **44**, 413 (2000); T. de J. Guadaramma-Medina, J. Pérez-Gonsáles, L. de Varges, *Rheol. Acta*, **44**, 278 (2005).

226 The multi-valued flow curves were first described by E.B. Bagley, J.M. Cabott, D.C. West, *J. Appl. Phys.*, **29**, 109 (1958) and then repeatedly observed and quantitatively explained by a "capacitance model" in publications of A.P. Metzger, C.W. Hamilton, *SPE Trans.*, **4**, 107 (1964); J.M. Lupton, J.W. Regester, *Polymer Eng. Sci.*, **5**, 235 (1965); Myerholtz, *J. Appl. Polymer Sci.*, **11**, 687 (1967). The later publications devoted to this subject are: J. Molenaar, R.J. Koopmans, *J. Rheol.*, **38**, 99 (1994); K.P. Adewale, A.I. Leonov, *Rheol Acta*, **36**, 110 (1997); M. Ranganathan, M.R. Mackley, P.H.J. Spitteler, *J. Rheol.*, **43**, 443 (1999).

227 See details in: S.Q. Wang, N. Plucktaveesak, *J. Rheol.*, **43**, 453 (1999).

228 P.L.Clegg, in "**The Rheology of Elastomers**", ed. P. Mason and N.Wookey, *Pergamon Press*, 1958.

229 A.V. Ramamurthy, *J. Rheology*, **30**, 337 (1986); S.G. Hatzikiriakos, J.M. Dealy, *J. Rheology*, **35**, 497 (1991).

230 A comprehensive review devoted to the problem of wall slip in the flow of polymers and its correspondence to instabilities of various types was published by M.M. Denn, in *Annual Review of Fluid Mech.*, **33**, 265 (2001). This author discussed numerous experimental evidences and arguments concerning possible mechanisms of instabilities and their interrelations with surface effects.

231 Sometimes the Weissenberg number is denoted as We. The symbol We is traditionally used as the dimensionless ratio of inertial forces to surface tension (the Weber number). Therefore the symbol Wi for the Weissenberg number is used here.

232 G.V. Vinogradov, A.Ya. Malkin, A.I. Leonov, *Kolloid-Z. Z. Polymere*, **191**, 25 (1963).

233 C. Tzoganakis, B.C Price, S.G. Hatzikiriakos, *J Rheol.*, **37**, 355.(1993).

234 Pioneering works on the uniaxial flow of monodisperse polymers were: G.V. Vinogradov, A.Ya. Malkin, V.V. Volosevitch, V.P. Shatalov, V.P Yudin, *J. Polymer Sci.: Polymer Phys. Ed.*, **13**, 1721 (1972); G.V. Vinogradov. A.Ya., Malkin, V.V. Volosevitch, *Appl. Polymer Symposia*, **27**, 47 (1975).

235 The meaning of this term and the operation of "reducing" the initial experimental data to a single temperature is discussed in detail in Chapter 2.

234 The complete theory of rheological behavior of linear elastic liquids was developed in the monograph A.S. Lodge, **Elastic Liquids**, *Acad. Press*, New York-London, 1964.

237 There are two different names for the dimensionless ratio of characteristic times of relaxation (inner time scale) and deformation (outer time scale), the Weissenberg number, Wi, or the Deborah number, De, Though it is possible to point out some subtle differences between Wi and De, they do not seem to be principal and we shall use them as being equivalent. The Deborah number is of more philosophical nuance.

238 The conditions of these transitions and criteria of rupture of viscoelastic materials were discussed in detail in the review: A.Ya. Malkin, C.J.S. Petrie, *J. Rheol.*, **41**, 1 (1997). Simple scaling model describing the conditions of rapture in elongational flows as a consequence of critical recoverable deformations was developed in: Y.M. Joshi, M.M. Denn, *J. Rheol.*, **47**, 291 (2003). Later this theory was extended to viscoelastic liquids for the case when dissipation is superimposed on elastic deformations: Y.M. Joshi, M.M. Denn, *J. Rheol.*, **48**, 591 (2004).

239 It is useful to remind that the equality $\varepsilon_t = \varepsilon_r + \varepsilon_f$, i.e., the separation of the deformation into its components

has unambiguous physical meaning if and only if the Hencky (logarithmic) measure of deformations is used. Besides, the superposition of both components of deformation is physically unambiguous in the linear viscoelastic behavior domain.

240   This phenomenon is described in details in many classical text books on polymer physics and mechanics, for example A.V. Tobolsky, **Properties and Structure of Polymers**, *Wiley*, New-York, 1960; J.D. Ferry, **Viscoelastic Properties of Polymers**, 3rd Ed., *Wiley*, New-York, 1980.

241   F.N. Cogswell, *Appl. Polymer Symp.*, **27**, 1 (1975).

242   V.E. Dreval, G.V. Vinogradov. B.V. Rudushkevich, *J. Polymer Sci.: Polymer Phys. Ed.*, **22**, 1155 (1984).

243   J.M. Maia, J.A.Covas, J.M. Nóbrega, T.F. Dias, *J. Non-Newton. Fluid Mech.*, **80**, 183 (1999).

244   Extensional flow of polymer materials was the subject of numerous investigations. One can find the review of early publications in the monograph: C. J. Petrie, **Elongational Flows**, *Pitman*, London, 1977. The review of results of the last decade of the XX Century in the field of uniaxial extension was presented by G.H. McKinley, Proc. XIII Intern. Congr. Rheology, Cambridge, August 2000. Useful discussion on rheological models related to uniaxial deformations of elastic liquids was published by V.M. Entov, *J. Non-Newton., Fluid Mech.*, **82**, 167 (1999).

245   The pioneering work in this field was carried out by J. Meissner, *Rheol Acta*, **8**, 78 (1969); *Trans. Soc. Rheol.*, **16**, 405 (1972); *Pure Appl. Chem.*, **42**, 553 (1975); G.V. Vinogradov, B.V. Radushkevich, V.D. Fikhman, *J. Polym. Sci., A-2*, **8**, 1 (1970).

246   There were number of publications based on this experimental technique. Some of the most typical are as follows: H. Münstedt, *Rheol. Acta*, **14**, 1077 (1975); *J. Rheol.*, **23**, 421 (1979); **24**, 847 (1980); H. Münstedt, H. Laun, *Rheol. Acta,* **18**, 492 (1979); **20**, 211 (1980); H. Münstedt, S. Kurzbeck, L. Egersdörfer, *Rheol. Acta*, **37**, 21(1998). The latest publications also contain theoretical description of extensional behavior of viscoelastic polymer melts: M. Wagner, H. Bastian, P. Hachmann, J. Meissner, S. Kurzbeck, H. Münstedt, F. Langouche, *Rheol. Acta*, **39**, 97 (2000); M.H. Wagner, P. Rubio, H. Bastian, *J. Rheol.*, **45**, 1387 (2001); M.H.Wagner, M. Yamaguchi, M. Takahashi, *J. Rheol.*, **47**, p.779 (2003); M.H. Wagner, S. Kheirandish, M. Yamaguchi, *Rheol.Acta*, **44**, 198 (2004); M.H. Wagner, S. Kheirandish, K. Koyama A. Nishioka, A. Mineigishi, T. Takahashi, *Rheol. Acta*, **44**, 235 (2005).

247   S. Kurzbeck, F. Oster, H. Münstedt, T.Q. Nguyen, R Gensler, *J. Rheol.*, **43**, 359 (1999).

248   A.D. Gotsis, M.A. Odriozola, *J. Rheology*, **44**, 1205 (2000).

249   J. Meissner, T. Raible, S.E. Stephenson, *J. Non-Newton. Fluid Mech.*, **11**, 221 (1982).

250   P. Hachmann, J. Meissner, *J. Rheol.*, **47**, 989 (2003).

251   The first description of this effect is in G. Funo, *Arch. Fisiol.*, **5**, 365 (1908).

252   A.I. Leonov, A.N. Prokunin, *Izv. AN SSSR: Mekh. Zhid. Gas* (Reports USSR Acad. Sci.: Mech. Fluids and Gases - in Russian), **5**, 25 (1973).

253   W.C. MacSporran, *J. Non-Newton. Fluid Mech.*, **8**, 119 (1981).

254   A special experimental method permits observation of an elongational (or extensional) flow of a low viscosity dilute solutions. This so-called four-roll method was first proposed by G.I. Taylor, *Proc. Royal Soc. London*, **145**, 501 (1934). This method was developed for polymer solutions: D.G. Crowly, F.C. Frank, M.R. Mackley, R.G. Stephenson, *J. Non-Newton. Fluid Mech.*, **14**, 1111 (1974); D.P. Pope, A. Keller, *Colloid Polymer Sci.*, **255**, 633 (1977). Also, a very effective method for creating high strain rate flow between two opposite nozzles was proposed in: M.R. Mackley, A.Keller, *Phil. Trans. Royal Soc.* (Lond), **278**, 29 (1975).

255   It is evident that definitions of the Weissenberg number, Wi, (Eq. (3.7.5)) and the Deborah number, De, (Eq. (3.7.8) are equivalent.

256   C.J. Farrel, A. Keller, M.J. Miles, D.P. Pope, *Polymer*, **21**, 1292 (1980); V.G. Baranov, D.Kh. Amribahshov, Yu.V. Brestkin, S.A. Agranova, I.S. Saddikov, *Vysokomol. Soedin.* (in Russian), **29**, 1190 (1987).

257   A.K. Doufas, A.J. McHugh, Ch. Miller, *J. Non-Newton. Fluid Mech.*, **92**, 27 (2000).

258   A.Ya. Malkin, A. Arinstein, V.G. Kulichikhin, *Progr. Polym. Sci.*, **39**, 959 (2014).

259   A.V. Semakov, V.G. Kulichikhin, A.K. Tereshin, S.V. Antonov, A.Ya. Malkin, *J. Polym. Sci.: Part B: Polym. Phys.*, **53**, (8), 559 (2015).

260   Kulichikhin V.G., Skvortsov I.Yu., Subbotin A.,V., Kotomin S.V., Malkin A.Ya., *Polymers*, **10** (8) 856 (2018)

261   Lord Rayleigh, *Proc. Lond. Math. Soc.*, **10**, 4 (1979); *Phyl. Mag.*, **34**, 177 (1982). J.W.S. Rayleigh (1842-1919) – English physicist, author of fundamental works in fluid dynamics, acoustics, light scattering in media; he also discovered argon. Nobel Prize 1904.

262   S. Tomotika, *Proc. Royal Soc.*, **150**, 322 (1935); **153**, 302 (1936).

263   N.S. Clarke, J. Fluid Mech., 31, 481 (1968); J.M. Montanero, M.A. Herrada, C. Ferrera, E.J. Vega, A.M. Gañán-Calvo, *Phys. Fluids*, **23**, 122103 (2011).

264   S. Senchenko, T. Bohr, *Phys., Rev., E*, **71**, 056301 (2005).

265   A. Javadi, J. Eggers, D. Bonn, M. Habibi, N.M. Ribe, *Phys. Rev. Lett.*, **110**, 144501 (2013).

266 A.V. Subbotin, I.Yu. Skvortsov, M.E. Kuzin, P.S. Gerasimenko, V. G. Kulichikhin, A.Ya. Malkin, The shape of a falling jet formed by concentrated polymer solutions, *Phys. Fluids*, **33**, 083108 (2021).

267 R.E. Christensen, *Soc. Plast. Engng J.*, **18**, 751 (1962).

268 M.A. Matovich, J.R.A. Pearson, *End. Engng Chem. Fund.*, **8**, 512 (1969).

269 Draw ratio is the ratio of the take-up velocity to the extrusion velocity, and this is the same as the ratio of the initial cross-section area of the fiber to the final cross-section area of the extended fiber.

270 C.J.S., Petrie, M.M. Denn, *AIChE J.*, **22**, 209 (1976).

271 R.J. Fisher, M.M. Denn, *AIChE J.*, **22**, 236 (1976); J.C. Chang, M.M. Denn, Rheology (Proc. VIII Intern. Congr. Rheol., Naples), v. 3, Eds. G. Astarita, G. Marrucci, L. Nicolais, Plenum Press, 1980.

272 J.R.A Pearson, Y.T. Shah, R.D. Mhaskar, *Ind. Engng Chem. Fund.*, **15**, 31 (1976). S. Kase, T. Matsuo, Y. Yoshimoto, *Seni Kikai Gakkaishi*, **19**, T63 (1966).

273 M. Goldin, H. Yerushalmi, R. Pfeffer, R. Shinnar, *J. Fluid Mech.*, **38**, 689 (1969).

274 J.W. Hoyt, J.J. Taylor, *Phys. Fluids*, **20**, 256 (1977).

275 The theory of instability of viscoelastic threads was published in: J. F. Palierne, F. Lequeux, *J. Non-Newton. Fluid Mech.*, **40**, 289 (1991). See also: H.B. Chin, C.D. Han, *J. Rheol.*, **23**, 557 (1979). The series of photographs demonstrating the unstable motion of viscoelastic drops in stretching accompanied by the formation of periodicity on a jet were presented by W.J. Miliken, L.G. Leal, *J. Non-Newton. Fluid Mech.*, **40**, 355 (1991). This problem was also investigated from both theoretical and experimental points of view in D.W. Bousfield, R. Keunings, G. Marrucci, M.M. Denn, *J. Non-Newton. Fluid Mech.*, **21**, 79 (1986); V.M. Entov, A.L. Yarin, *J. Fluid Mech.*, **140**, 91 (1984); *Fluid Dynamics*, **19**, 27 (1984).

276 Judges, 5:5. Sometimes this phrase is cited as "The mountains flowed before the Lord" – see, for example H.A.Barnes, J.F., K. Walters, **An Introduction to Rheology**, *Elsevier*, 1989, p.5; and possibly this translation is closer to the sense of the idea.
The introduction of the Deborah number as one of the key conceptions of rheology is by M. Reiner (1886-1976), who lived mainly in Palestine (then Israel). He was one of the founders of modern rheology, developing primarily fundamental aspects of rheometry and constitutive equations. His textbook on rheology: M. Reiner, **Twelve Lectures on Theoretical Rheology**, *North Holland Publ. Co*, Amsterdam, 1949) played an important role in teaching new generations of rheologists after the Second World War.

277 Ezra, Non-canonical 9:1.

QUESTIONS FOR CHAPTER 3

QUESTION 3-1

Can viscosity be negative? Explain the answer.

QUESTION 3-2

In measuring the viscous properties of the polymer solution, it appeared that the experimental data within the experimental range of shear rates can be fitted with the power-law equation (Eq. 3.3.4). Analyze the possibility of extrapolating this equation to the range of very high shear rates.

Additional question

Which kind of rheological behavior at high shear rates is expected in this case?

QUESTION 3-3

What is the difference in stress relaxation of viscous liquids and viscoplastic materials?

QUESTION 3-4

Can we expect that the values of the yield stress, σ_Y, found by treating a set of experimental data by means of Eqs. 3.3.7 to 3.3.9, are the same?

QUESTION 3-5

Calculate shear stresses in the flow of liquid through a straight tube if the flow is created by the pressure gradient $\Delta p/L$ (L is the length of a tube).

Additional question

Are the results valid for Newtonian liquid only?

QUESTION 3-6

Calculate the radial distribution of shear rates and flow velocity of Newtonian liquid (having viscosity η) through a straight tube with radius R.

Additional question 1

Calculate the volume output, Q, for the flow of Newtonian liquid.

Additional question 2

Express maximum shear rate, $\dot{\gamma}_R$, via volume output.

Additional question 3

Is the last expression valid for a liquid with arbitrary rheological properties?

QUESTION 3-7

Calculate the velocity profile in the flow of a power-law type liquid through a straight tube with a round cross-section. The radius of a tube is R.

Additional question

Calculate the volume output, Q, as a function of Δp for a power-law type liquid.

QUESTION 3-8

An experimenter obtained two pairs of data: at $\dot{\gamma}_1 = 1*10^{-3}$ s^{-1} $\sigma_1 = 100$ Pa and at $\dot{\gamma}_2 = 1*10^{-2}$ s^{-1} $\sigma_2 = 600$ Pa.

Assuming that the flow curve is described by a power-law type equation, find the constants of this equation for a liquid under study.

Additional question

How do you find the constants of the power-law type equation if an experimenter obtained three or four pairs of experimental points?

QUESTION 3-9

Analyze the flow of a viscoplastic ("Bingham-type") liquid through a straight tube of radius, R. Find radial stress and velocity distributions and calculate volume output as a function of the pressure gradient.

QUESTION 3-10

A ball with a radius R is falling in a Newtonian liquid having viscosity η. After some transient period, the velocity of ball movement becomes constant. Find the velocity of steady movement, V_∞.

QUESTION 3-11

An experimenter measured the viscous properties of the material at different shear rates and obtained a flow curve. What can he say concerning viscous properties of this material in the uniaxial extension? Explain the answer.

QUESTION 3-12

Prove the validity of Eq. 3.1.7 – the dependence between normal stress and deformation rate for Newtonian liquid in two-dimensional (biaxial) extension.

QUESTION 3-13

Normal stresses in shear appear as a second-order effect. However, at high shear rates, they exceed shear stresses. Estimate the condition under which it becomes possible.

QUESTION 3-14

Can normal stresses appear in the shear flow of suspension of solid particles? Explain the answer.

Additional question

Estimate the characteristic time ("relaxation time"), θ, of this process.

QUESTION 3-15

An experiment was carried out in shear at the constant shear rate, $\dot{\gamma} = \text{const}$, and the curve similar to shown in Fig. 3.5.1 or Fig. 3.5.2 was obtained. Can the ratio $\sigma(t)/\dot{\gamma}$ be treated as the evolution of viscosity of liquid? Explain the answer.

QUESTION 3-16

A liquid layer is intensively sheared at shear rate $\dot{\gamma} = 1*10^2$ s^{-1}. A liquid is Newtonian and its viscosity $\eta = 500$ Pa*s. Shearing continued for 10 s. Temperature dependence of viscosity is neglected; density is assumed to be 1 g/cm^3 and heat capacity is 0.5 J/(g*K).

What temperature rise is expected?

Additional question

If shearing proceeds for a longer time, what physical phenomena must be taken into consideration and what final thermal effect of shearing can be expected?

QUESTION 3-17

Analyze the Mooney equation (3.3.27) for the concentration dependence of viscosity for the limiting case and, in particular, calculate the intrinsic viscosity of dilute suspensions.

QUESTION 3-18

Newtonian viscosity of polymer with molecular mass $M_1 = 3*10^5$ is $\eta_1 = 5*10^5$ Pa*s. There is also another polymer of the same chemical structure with molecular mass $M_2 = 4*10^4$. How can one decrease the viscosity of the polymer by 10 times?

QUESTION 3-19

Experiments show that an electrical charge appears on the surface of the polymer stream leaving a capillary in an unstable or spurt regime. Explain the origin of the charge.

Answers can be found in a special section entitled Solutions.

4

SOLIDS

4.1 INTRODUCTION AND DEFINITIONS

The concept of *solid* is an idealization of the real behavior of numerous materials. Some of them are close to this model, for example, steel and stone. Other materials are far from this ideal model, but in some applications, they can also be treated as elastic solids, for example, wood, rubber, and concrete.

In the previous chapters, it was emphasized that any material can be treated as solid or liquid depending on the Deborah number, which is the ratio of inherent time-scale and characteristic time of loading. This is also true for the materials listed above. Therefore, the elastic solid is in fact a concept describing *behavior* rather than a particular *material*. The same material may behave as *solid-like* or *liquid-like*, depending on the time-scale of observation. Steel is a typical example in this respect. Nobody doubts that steel is solid in all its applications. However, in many technological operations (such as forging, rolling, drawing) steel *flows*, i.e., it is able to undergo irreversible deformation. This ability of solid material is characterized as *plasticity*.

In spite of departures from the model, it is important to discuss the concept of a solid, or *elastic solid*, because it is a limiting case in the rheological properties of real materials.

There is also another aspect of rheological interest in the elastic behavior of solid materials. There is a great number of *elastic liquids*, i.e., real liquids which also demonstrate elastic behavior. These are polymeric substances, such as melts, solutions, and emulsions. The qualitative description of mechanical properties of such liquids must also include the characterization of their elasticity. For these reasons, we consider the "pure" case of elastic behavior of a solid, which is a suitable model useful in the rheological equation of the state of elastic liquids.

Main concepts and experimental facts on the mechanical behavior of solids are included in the framework of rheology. At the same time, considering that rheology is mainly a science devoted to liquids or flowing media, then this chapter is a secondary subject, though necessary in the general structure of rheology.

The basic idea of a solid is its ability to experience elastic (reversible) deformation. These materials store work done by external forces. They store work in the form of elastic energy and return this energy when forces are removed.

Elastic solids are also treated as materials with a clearly defined relationship between stresses and deformations, i.e., if the stress field is known, then the spatial distribution of deformations is also known, and *vice versa*.

The main point in both concepts is the *absence of time or time-dependent effects*. However, if the deformation is time-dependent (in a permanent stress field), then a characteristic feature of the viscous (dissipative) behavior of matter should be combined with elasticity. Therefore, the material is viscoelastic, as was discussed in Chapter 2.

Formulating the rheological model of an elastic solid, i.e., writing its rheological equation of state (or a constitutive equation), the transformation from components of stress and deformation tensors observed in an experiment to an invariant formulation is required. The formulation must not depend on the choice of the coordinate system. It can be an expression in the form of *elastic potential* (stored energy), W, as a function of the invariants of stress or deformation tensor or both. Also, this invariant formulation can be a relationship between the invariants of stress and deformation tensors. Both approaches are suitable in formulating a rheological equation of state. The result of the formulation based on the energy concept can be reformulated into the components of stress vs. deformation tensors.

This transition is based on a fundamental expression for stored energy:

$$dW = \sum_{i,j=1}^{3} \sigma_{ij} d\varepsilon_{ij} \qquad [4.1.1]$$

Then, the following formula for stress components, in terms of an elastic potential, can be written:

$$\sigma_{ij} = \frac{\partial W}{\partial \varepsilon_{ij}} \qquad [4.1.2]$$

If W (a function of deformations) is known, the components of the stress tensor can be calculated from Eq. 4.1.2. A method of conversion from stress-deformation function to the elastic potential function is needed to be written using invariants.

In a general case, W can be written as a function of three invariants:

$$W = W(E_1, E_2, E_3) \qquad [4.1.3]$$

where E_1, E_2, and E_3 are invariants of the tensor of large deformations, as was discussed in section 1.2.

Using an ordinary rule of function differentiation, the following formula can be obtained:

$$\sigma_{ij} = \frac{\partial W}{\partial \varepsilon_{ij}} = \sum_{k=1}^{3} \frac{\partial W}{\partial E_k} \frac{\partial E_k}{\partial \varepsilon_{ij}} \qquad [4.1.4]$$

The last equation gives an answer to the problem formulated above: if a function $W(E_1, E_2, E_3)$ is known, the components of the stress tensor can be found from Eq. 4.1.4.

The above-written definitions and equations are applicable to any elastic material regardless of the form of the elastic potential function, Eq. 4.1.3.

4.2 LINEAR ELASTIC (HOOKEAN) MATERIALS

The basic concept and equation of Hookean material were frequently mentioned in the previous chapters (see Eq. 2.1.2) because it correctly represents numerous experimental data of real materials, but with two important limitations:
  - the equation is written for one-dimensional deformations (extension)
  - the equation is valid for small deformations.

It is thus necessary to formulate the general (invariant) form of Hooke's law. The formulation of Hooke's law for an extension is:

$$\sigma_E = E\varepsilon = \frac{1}{J}\varepsilon \qquad [4.2.1]$$

where E is *Young's modulus*, $J = E^{-1}$ is the elastic compliance, σ_E is the normal stress and ε is the deformation in extension.

It would be incorrect to write Hooke's law by adding indices in the following form:

$$\sigma_{ij} = E\varepsilon_{ij}$$

Such an equation is correct for uniaxial extension (ij = 11). But experiments show that in uniaxial extension, the cross-section of the sample is reduced, and this effect is described by *Poisson's ratio*, μ. This experimental fact means that the components of the deformation tensor ε_{22} and ε_{33} do not equal zero, but the external forces in the directions normal to 11 (22 and 33 directions) are absent, i.e., σ_{22} and σ_{33} equal zero. This is the direct proof that this generalization of Hooke's law is invalid.

An experiment in simple shear shows that a linear relationship between deformations and stresses exists:

$$\sigma = G\gamma \qquad [4.2.2]$$

and all-directional (hydrostatic) compression:

$$p = -B\varepsilon_v \qquad [4.2.3]$$

where σ is the shear stress, γ is the shear deformation, ε_v is the volume deformation (relative change in volume of material) and G and B are coefficients:

G – *shear modulus*
B – *bulk modulus of compressibility*

which are different than Young's modulus.

It is now necessary to find out whether these constants (E, G, μ, and B) are independent and whether the number of constants is sufficient to characterize the material properties. For other forms of deformation, e.g., biaxial extension, the question arises whether it will be necessary to introduce a "modulus" characterizing the linear behavior of the material in these types of deformation.

Then the main questions are:
  - how a general rheological equation of state should be written to reflect its linear elastic properties in different modes of deformation, which would include all these relationships.

- what is the minimal number of independent characteristic constants ("moduli')
  describing all types of deformation of material?

In uniaxial extension, not only the *shape* (form), but also the *bulk* (volume) of a body changes, and these two are *independent effects*. It is thus necessary to introduce *at least two independent constants* for the characterization of material resistance to volume and shape changes. This may suggest the use of previously proven methods to decompose stress and deformation tensors into spherical (isotropic) and deviatoric parts.

Let us assume that the linear relationships between spherical and deviatoric parts of both tensors exist separately. This assumption gives two independent invariant relationships:

$$I_1 = kE_1 \tag{4.2.4}$$

and

$$\sigma'_{ij} = 2G\varepsilon'_{ij} \tag{4.2.5}$$

where I_1 is the first invariant of the stress tensor (sum of normal stresses, which is the measure of *hydrostatic pressure*), E_1 is the first invariant of the deformation tensor (a measure of relative volume changes), σ'_{ij} are deviatoric components of the stress tensor, ε'_{ij} are deviatoric components of the deformation tensor, and k and G are material constants, the first of them characterizing resistance to volume changes and the second one to the material shape changes.

Hydrostatic pressure is expressed *via* the first invariant of the stress tensor (Eq. 1.1.15) as

$$p = -\frac{I_1}{3} \tag{4.2.6}$$

Then, Eq. 4.2.4 can be rewritten:

$$p = -BE_1 \tag{4.2.7}$$

where $B = -k/3$ is the bulk modulus of elasticity. The two fundamental coefficients, B and G, are sometimes called the *Lamé constants*. The value of G in Eq. 4.2.5 is the shear modulus, and coefficient 2 in this equation appears due to the formal definition of components of the deformation tensor.

The basic assumption of the generalized (three-dimensional) Hooke law is that Eqs. 4.2.4 and 4.2.5 are valid for *any* type of deformation and that they are the invariant *definition* of Hookean elastic material (body).

Then, it is important to relate the constants in Eqs. 4.2.4 and 4.2.5 with those, which are directly measured in a standard experiment of uniaxial extension, namely, Young's modulus and Poisson's ratio.

Let σ_E be the extensional (normal) stress. Then Eq. 4.2.1 written for all components of the stress and deformation tensors is

$$\varepsilon_{11} = \frac{\sigma_E}{E}; \ \varepsilon_{22} = \varepsilon_{33} = -\mu\varepsilon_{11}$$

and

$$E_1 = (1-2\mu)\frac{\sigma_E}{E}$$

where E is Young's modulus and μ is Poisson's ratio (coefficient).

Hydrostatic pressure, p, in uniaxial extension is

$$p = -\sum_{i=1}^{3} \sigma_{ij} = -\frac{\sigma_E}{3}$$

Then, based on Eq. 4.2.4, the following equality can be written

$$\frac{\sigma_E}{3} = B(1-2\mu)\frac{\sigma_E}{E}$$

This gives the desirable relationship between the constants:

$$E = 3(1-2\mu)B$$

Based on the same arguments, the following relationship can be easily obtained:

$$\sigma_E = 2G(1+2\mu)\frac{\sigma_E}{E}$$

and this gives the following relationship between the other constants:

$$E = 2G(1+2\mu)$$

The results obtained in these relationships permit the calculation of any pair of constants for any pair of variables, as summarized below.

- for known constants E and μ

$$B = \frac{E}{3(1-2\mu)}, \ G = \frac{E}{2(1+\mu)} \qquad [4.2.8]$$

- for known constants B and G

$$E = \frac{9BG}{3B+G}, \ \mu = \frac{3B-2G}{6B+2G} \qquad [4.2.9]$$

- for known G and μ

$$E = 2G(1+\mu), \ B = \frac{2G(1+\mu)}{3(1-2\mu)} \qquad [4.2.10]$$

- for known E and G

$$B = \frac{EG}{3(3G-E)}, \ \mu = \frac{E-2G}{2G} \qquad [4.2.11]$$

The calculations for any other pairs of constants can also be done.

This set of relationships permits finding any value of material constant from two other values measured experimentally. It is important to note that there are only *two independent constants* that must be measured.

There are several principal physical limitations. If a body is compressed, its volume cannot increase; it can only decrease or may not change at all if a body is incompressible. It means that there is a principal limitation:

$$B \geq 0$$

Then, it is evident that $E > 0$, and this inequality can be fulfilled only if

$$\mu \leq 0.5$$

Incompressible material is of special interest. For many solids

$$B \gg E$$

i.e., it is much easier to change the shape of a body than its relative volume (density).

In some cases, it is reasonable to accept that $B \to \infty$, i.e., to suppose that some materials are completely incompressible. It is a good model (or it is almost true) for all liquids and rubbers. This leads to two simple relations:

$$\mu = 0.5$$

and

$$E = 3G$$

This means that *for incompressible media there is only one independent or "free" material constant*.

Reformulating the rheological equation of state for a linear Hookean elastic material in terms of the elastic potential function, let us consider (for a sake of simplicity) an incompressible body. Then, $E_1 = 0$, and W can be a function of E_2 and E_3 only. The simplest is an assumption of a linear relationship between W and E_2, i.e.,

$$W = -BE_2 \qquad\qquad [4.2.12]$$

where B is the single independent constant of the material.

It is possible to calculate all items entering this equation and to find derivatives:
- with the same indices

$$\frac{\partial E_2}{\partial \varepsilon_{ii}} = -\varepsilon_{ii}$$

- with different indices

$$\frac{\partial E_2}{\partial \varepsilon_{ij}} = -\frac{1}{2}\varepsilon_{ij}$$

Then, a combination of these results with Eq. 4.1.4 gives Hooke's law, i.e., the linear relationship between deviatoric components of the stress and deformation tensors is obtained, as indicated by Eq. 4.2.5. This means that Eq. 4.2.12 is equivalent to the above-formulated concept of Hookean elastic material and this equation can be treated as an invariant definition of an incompressible linear elastic body in the limits of infinitesimal deformations.

Both definitions are equivalent and it seems that Eq. 4.2.12 does not offer any additional advantages in comparison with the standard definition, describing the relation between components of stress and deformation tensors. However, it is not completely true, and when finite (large) deformations of an elastic body will be discussed, it will be seen that the formulation of the rheological equation of state through an elastic potential function is preferable.

The real range of changes in modulus for some typical materials is as follows:

Material	Young's modulus
High modulus, oriented fibers	> 300 GPa
Steel	200 GPa
Copper, aluminum, and alloys	100 GPa
Stones	40 to 60 GPa
Engineering plastics	5 to 20 GPa
Ice	10 GPa
Wood	1 to 10 GPa
Leathers	1 to 100 MPa
Rubbers	0.1 to 5 MPa
Polymer and colloid solutions	1 to 100 Pa

Young's modulus may vary in the range of more than 11 decimal orders. Poisson's ratio for many solids ranges from 0.3 to 0.4, and for rubbers, it is close to 0.5.

The concept of a linear elastic (Hookean) material is the basis of many engineering disciplines, first of all, the theory of elasticity and strength of materials. The basic rheological equations combined with the equilibrium equations (see section 1.1.6) are widely used for solving numerous applied problems. The discussion of all these problems goes beyond the scope of the present book.[1]

4.3 LINEAR ANISOTROPIC SOLIDS

In formulating and discussing Hooke's law, it was tacitly assumed that material has the same properties along any arbitrary direction, i.e., that material is *isotropic*. The consequence of this assumption is that only two independent moduli, e.g., extension and shear are modeled. In fact, many real materials are *anisotropic*, i.e., their properties depend on a direction of measurement. The closest examples are oriented fibers, reinforced plastics, and wood – their rigidity and strength are very different along the fibers and normal to the fibers. The properties of monocrystals are different in different crystallographic directions due to different intermolecular interactions and different distances between atoms in the crystalline cell.

The behavior of all these materials can be approximated (at least at small deformations) by linear relationships between stresses and deformations. However, it is rather evident that Hooke's law in its standard formulation does not describe properties of anisotropic materials. Then, it is reasonable to make more general suggestions concerning a possible relationship between the components of stress and deformation tensors within the framework of linear approximation.

There are six independent components of the stress tensor, σ_{ij}, and six independent components of the deformation tensor, ε_{ij}. The general *linear* relationship among them is:

$$\sigma_{ij} = \sum_{m, n = 1}^{3} E_{ijmn}\varepsilon_{mn} \qquad [4.3.1]$$

where E_{ijmn} are the components of elastic modulus for an anisotropic material and summation takes place by the indices corresponding to the components of the deformation tensor, m and n.

Comment

In order to elucidate Eq. 4.3.1 one of the components of the stress tensor is written below in an expanded form:

$$\sigma_{12} = E_{1211}\varepsilon_{11} + E_{1222}\varepsilon_{22} + E_{1233}\varepsilon_{33} + E_{1212}\varepsilon_{12} + E_{1213}\varepsilon_{13} + E_{1223}\varepsilon_{23}$$

Eq. 4.3.1 can also be written in an inverse form to consider deformation as a function of the components of the stress tensor:

$$\varepsilon_{ij} = \sum_{m, n = 1}^{3} J_{ijmn}\sigma_{mn} \qquad [4.3.2]$$

It is evident that this complete formulation includes 36 values of "moduli", E_{ijmn}, which, in fact, are the components of the *modulus tensor*, or the same number of components of the *compliance tensor*, J_{ijmn}. The correspondence between both is established by the rules of matrix algebra.

Based on some theoretical arguments concerning the mathematical properties of tensors, it is possible to prove that there are not 36 but only 21 independent values of modulus (or compliance).

This is the maximum possible number of independent characteristics of mechanical properties for a linear elastic material (body). The decrease of the number of moduli depends on the type of symmetry of material.

It is necessary to use all 21 constants to describe the properties of a crystal of triclinic structure. The increase of the number of axes of symmetry results in a decrease in the number of independent constants. For a monoclinic crystal with the axis of symmetry of the second order, several moduli equal zero, and the number of independent moduli decreases to 13. For a rhombic crystal, only 9 independent moduli exist. The number of independent moduli for a cubic crystal with the axes of symmetry of the fourth-order decreases to 3, and coming back to an isotropic material with an infinite number of axes of symmetry, only 2 independent moduli can be determined. The latter relates to amorphous materials or to polycrystals. In polycrystals, the properties of individual crystals are averaged because of the coexistence of a large number of individual anisotropic crystals oriented statistically in space. Elastic properties of the material are measured using relatively large samples (at least, much larger than the sizes of individual crystals in a polycrystal body), and therefore the properties of such materials can be described by a model of an ideal Hookean body.

The elastic potential of an anisotropic elastic material is determined by the following sum of products of the components of stress and deformation tensors:

$$W = \sum_{i,j=1}^{3} \sum_{m,n=1}^{3} E_{ijmn}\sigma_{ij}\varepsilon_{mn} \qquad [4.3.3]$$

The tensor E_{ijlm} is symmetrical, and this reduces the number of independent constants characterizing elastic properties of the material.

The determination of the components of the modulus tensor is very complex from the experimental point of view. It can be done for real monocrystals, and the results of such kind are not used in rheology but are of interest in the physics of solids. The situation is somewhat different for anisotropic structures, such as glass-fiber reinforced plastics. These materials are of great technological interest in various applications and therefore quantitative description of their properties is necessary for designing articles made of these materials because many of them are used for the construction of vital importance. These anisotropic materials possess several axes of symmetry and the number of moduli is limited: as a rule, they are oriented along one or two axes and used for producing thin articles (covers, hulls, cases, roofs, and so on). Nevertheless measuring constants for such anisotropic materials requires special experimental techniques and is not a trivial task.

4.4 LARGE DEFORMATIONS IN SOLIDS AND NON-LINEARITY

4.4.1 A SINGLE-CONSTANT MODEL

Large elastic deformations result in non-linear dependence between stresses and deformations, and that is why both phenomena (large deformations and non-linearity) appear in the title of this section. Large deformations are not a single cause of non-linearity, as was discussed in section 2.8 for viscoelastic materials and will be illustrated here for elastic solids.

Discussion of large elastic deformations in solids is the most important for rubbers and rubber compounds (*elastomers*) because their main characteristics determine the quality and applicability of these materials and their ability to undergo large recoverable deformations.

The articles made out of rubbery materials work in various applications in which they are subjected to a three-dimensional stress state. The mechanical testing of these (and other) materials is carried out primarily in a unidimensional extension. Therefore, in formulating a constitutive equation for these materials it is necessary to solve the problem, frequently mentioned in this book, of generalization of unidimensional experiments for three-dimensional (invariant) form. And again, it is worth noticing that this problem does not have a unique solution but continues to be the subject of many different attempts, involving personal experience and luck, though some general principles must be fulfilled.[2]

Discussion of properties of materials in the domain of large deformations is based on some fundamental definitions advanced in section 1.2, where some results related to simple modes of deformations were also considered.

The invariants, E_i, of the large deformation tensor are written, as usual, *via* the principal values of this tensor, ε_i, and the following notation will be used below:

The first (linear) invariant:

$$E_1 = \varepsilon_1 + \varepsilon_2 + \varepsilon_3 \qquad [4.4.1]$$

the second (quadratic) invariant:

$$E_2 = \varepsilon_1\varepsilon_2 + \varepsilon_1\varepsilon_3 + \varepsilon_2\varepsilon_3 \qquad [4.4.2]$$

the third (cubic) invariant:

$$E_3 = \varepsilon_1\varepsilon_2\varepsilon_3 \qquad [4.4.3]$$

Rubbery materials are practically incompressible (at least, their bulk modulus is by several orders of magnitude higher than shear modulus), and Poisson's ratio, $\mu = 0.5$. Therefore, the third invariant of the deformation tensor (determining volume changes in deformations) equals 1 and can be excluded from further discussion.

The first invariant of the deformation tensor, in the limit of small deformations, has the meaning of volume changes (see Chapter 1.2), and that is why in this case $E_1 = 0$. In the range of large deformations, E_1 does not have such a simple meaning, but the condition of constant volume at the deformation of any type permits us to reduce the number of independent invariants to two because the equality

$$(1 + E)_1(1 + E_2)(1 + E_3) = 1 \qquad [4.4.4]$$

is always valid for incompressible materials.

Then, any invariant expressed as a function of two others can be excluded. For example, the third invariant, which is expressed by means of Eq. 4.4.4, can be written as

$$E_3 = \frac{1}{(1 + E_1)(1 + E_2)} - 1 \qquad [4.4.5]$$

These arguments reduce the dependence of W to two independent variables:

$$W = W(E_1, E_2) \qquad [4.4.6]$$

An invariant form of Hooke's law providing the linear relationship between stresses and deformations has been formulated above (Eq. 4.2.12) as a linear dependence of an elastic potential on the second invariant of the deformation tensor. In the limits of Hooke's law, E_2 is the second invariant of the tensor of infinitesimal deformations. As discussed in section 1.3, in the case of large deformations, it is necessary to utilize the theory of finite deformations and use some measures of large deformations. It is a natural way of generalization of Hooke's law, though the ambiguity of measures of large deformations may lead to different possibilities of representation of the relationship under discussion.

As the first approximation (or as the first reasonable simple idea), let us assume that an elastic potential is a linear function of the first invariant of the tensor of large deformations:

$$W = AE_1 \qquad [4.4.7]$$

Below, the main consequences of this approximation, i.e., Eq. 4.4.7, for different geometries of deformation are discussed. Eq. 4.4.7 can be written in an expanded form as

$$W = A(\varepsilon_1 + \varepsilon_2 + \varepsilon_3) = A\left(\frac{\lambda_1^2 - 1}{2} + \frac{\lambda_2^2 - 1}{2} + \frac{\lambda_3^2 - 1}{2}\right) \qquad [4.4.8]$$

or

$$W = \frac{A}{2}(\lambda_1^2 + \lambda_2^2 + \lambda_3^2 - 3) \qquad [4.4.9]$$

where λ_i are the principal extension ratios.

This formula is equivalent to the linear relationship between an elastic potential, W, and the first invariant, C_1, of the Cauchy-Green tensor of large deformations

$$W = \frac{A}{2}(C_1 - 3) \qquad [4.4.10]$$

The problem of three-dimensional elongation of a body "at a point" can be analyzed by calculating the elastic potential of deformations in the principal axes. From the definition of W, it is easy to show that the elastic potential is expressed *via* principal extension ratios as

$$dW = \sigma_1 \frac{d\lambda_1}{\lambda_1} + \sigma_2 \frac{d\lambda_2}{\lambda_2} + \sigma_3 \frac{d\lambda_3}{\lambda_3} \qquad [4.4.11]$$

The following equality is valid for an incompressible material

$$\lambda_1 \lambda_2 \lambda_3 = 1$$

and therefore

$$d(\lambda_1 \lambda_2 \lambda_3) = 0$$

Then, after some simple rearrangements, the following formula for dW can be written

$$dW = (\sigma_1 - \sigma_3)\frac{d\lambda_1}{\lambda_1} + (\sigma_2 - \sigma_3)\frac{d\lambda_2}{\lambda_2} \qquad [4.4.12]$$

Moreover, the following expression for dW can be obtained beginning from Eq. 4.4.11 and the condition of material incompressibility:

$$dW = A\left[(\lambda_1^2 - \lambda_3^2)\frac{d\lambda_1}{\lambda_1} + (\lambda_2^2 - \lambda_3^2)\frac{d\lambda_2}{\lambda_2}\right] \qquad [4.4.13]$$

A direct comparison of the last formulas for dW (Eqs. 4.4.12 and 4.4.13) gives the following system of equations widely used in the theory of rubber elasticity:

$$\begin{cases} \sigma_1 - \sigma_3 = A(\lambda_1^2 - \lambda_3^2) \\ \sigma_2 - \sigma_3 = A(\lambda_2^2 - \lambda_3^2) \end{cases} \qquad [4.4.14]$$

The last system of equations is a solution to the problem of calculation of normal (principal) stresses at known principal elongations.

This solution is not complete because the system of Eqs. 4.4.14 contains only two separate equations for three independent variables, σ_1, σ_2, and σ_3. This result is not unexpected because deformation in an incompressible medium is considered. The last limitation means that, in principle, stresses can be determined up to an uncertain constant: $\sigma_i + C$, where the constant C cannot be determined unambiguously. Superposition of arbitrary hydrostatic pressure changes the stress state of the medium but does not influence its deformation state. As a result, the system of Eq. 4.4.14 determines components of the stress tensor in relation to the constant C that is dependent on hydrostatic pressure.

Therefore, the general solution in determining principal stresses in a three-dimensional deformation state (i.e., when deformations are known or preset) can be written, in accordance to Eqs. 4.4.14, in the following form:

$$\begin{cases} \sigma_1 = A\lambda_1^2 + C \\ \sigma_2 = A\lambda_2^2 + C \\ \sigma_3 = A\lambda_3^2 + C \end{cases} \qquad [4.4.15]$$

where the constant C may be found if the hydrostatic pressure is known beforehand.

With this background, it is possible to analyze the main cases of deformations of an elastic solid body with rheological properties obeying Eq. 4.4.7. In uniaxial extension along the axis x_1, $\sigma_2 = \sigma_3 = 0$, if the extension ratio along the axis x_1 equals λ, the condition of the constant volume of the body under deformation results in the following relationship:

$$\lambda_2 = \lambda_3 = \lambda^{-1/2}$$

Then, any of the last two equations of the system Eq. 4.4.15 gives the value of the "free" constant C:

$$C = -A\lambda^{-1}$$

The formula for normal (principal) stress in uniaxial extension for material with rheological properties described by Eq. 4.4.7 is as follows:

$$\sigma_1 = A\left(\lambda^2 - \frac{1}{\lambda}\right) \qquad [4.4.16]$$

Comment

Eq. 4.4.16, within the limits of small deformations, degenerates to Hooke's law. The value of the principal elongation is

$$\lambda = \frac{l_0 + \Delta}{l_0}$$

where l_0 is the initial length of a sample and Δ is the increase of the length due to deformations; it is assumed that $\Delta \ll l_0$. Then, by direct substitution of the expression for λ in Eq. 4.4.16, after necessary calculations and neglecting higher-order terms of Δ, it is possible to demonstrate that Eq. 4.4.16 degenerates to the following linear relationship

$$\sigma_1 = 3A\frac{\Delta}{l_0} = 3A\varepsilon$$

which is, evidently, Hooke's law with Young's modulus equal to E = 3A.

Materials obeying Eq. 4.4.16 are sometimes called *neo-Hookean* because this equation is the most evident and simplest generalization of Hooke's law for a region of large deformations.

In engineering applications, it is frequently more convenient to use, not the true stress as in Eq. 4.4.16, but the *engineering stress*, f_E, that is the force divided by the initial cross-section of the sample. Simple geometrical arguments show that

$$f_E = \frac{\sigma_1}{\lambda}$$

and therefore the constitutive equation for uniaxial extension of rubbers is formulated as

$$f_E = \frac{E}{3}\left(\lambda - \frac{1}{\lambda^2}\right) \qquad [4.4.17]$$

The reason for changing A to E/3 was discussed in the **Comment** above.

Eq. 4.4.16 is one of the possible methods of representation of experimental data for large deformations of elastic materials. This equation is a consequence of the invariant Eq. 4.4.7. This equation is applied to the analysis of simple shear, which is relatively easy to study by experimental methods. For example, this mode of deformation can be realized by twisting a thin-walled cylinder.

The principal elongations for simple shear were calculated in section 1.2 (see Eq. 1.2.35, where ε_{ii} are the principal elongations). Elastic potential for this mode of deformation is

$$W = \frac{1}{2}A\gamma^2 \qquad [4.4.18]$$

Shear stress is found from:

$$\sigma = \frac{dW}{d\gamma} = A\gamma \qquad [4.4.19]$$

Rheological properties of the material, which are described by the invariant Eq. 4.4.7, give linear dependence of shear stress on deformation. Thus, the value of the constant A in this equation has the meaning of shear modulus. In the linear limit of elasticity, modulus in extension, $E = 3A$, but for the non-linear domain of large deformations, this simple relationship is not valid, as is clearly seen from Eq. 4.4.16. Therefore, the following conclusions can be made:

- Shear behavior of two different kinds of materials (Hookean and those described by the rheological equation of state, Eq. 4.4.7) can be the same, even though they are quite different rheological materials. This is confirmed by the difference in their behavior in extension. It is the direct proof of the thesis that investigation of stress deformation behavior in one mode of loading does not give enough information for estimating the type of rheological model of the material.

- Elastic potential (Eq. 4.4.7) predicts the non-linear behavior of the material in extension and this non-linearity is a direct consequence of large deformations by itself. This rheological equation of state contains only one material constant, which has the meaning of shear modulus and can be used for the prediction of deformation behavior of the material at any mode of loading.

Elastic potential in the form of Eq. 4.4.7 was formulated as a consequence of the molecular (kinetic) statistical theory of rubbery elasticity and is called the *Kuhn-Guth-James-Mark potential*.[3] The potential function, expressed by Eq. 4.4.7, was proposed for rubbers. It can be considered as the first approximation describing the deformation of rubbers at equilibrium conditions. The last limitation implies that possible time effects are not included in consideration, though effects of such kind are quite typical for rubbers and observed in measuring stress-deformation relationship (see Chapter 2).

4.4.2 MULTI-CONSTANT MODELS

4.4.2.1 Two-constant potential function

Comparison of experimental data obtained for a typical rubbery material with a curve calculated in accordance with Eq. 4.4.17 is presented in Fig. 4.4.1. The experimental curve consists of three sections:

- At small deformations, the stress-deformation relationship is close to linear

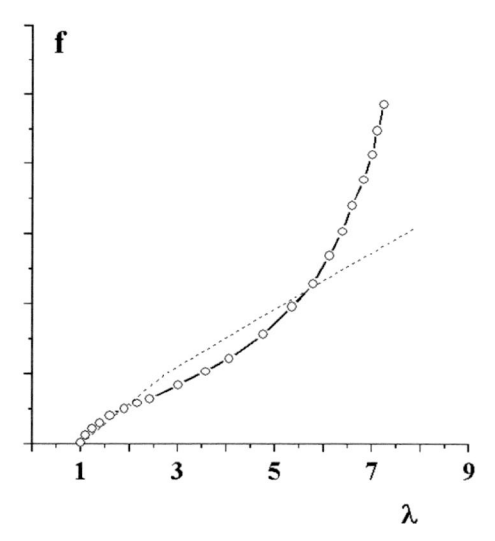

Figure 4.4.1. Typical dependence of engineering stress, f, on relative elongation, λ (experimental data points) for a soft rubber and its theoretical representation by Eq. 4.4.17 (dashed line).

- In the intermediate deformation region, the relationship is predicted by Eq. 4.4.17
- At very large deformations, stresses increase rapidly and this effect is not described by Eq. 4.4.17.

It is necessary to develop more comprehensive methods for the description of experimental data. For this purpose, rheological equations of state can be developed based on Eq. 4.4.7. Representation of rheological properties of solid materials *via* an elastic potential function $W(C_1, C_2)$ is equivalent (after some formal mathematical transformations) to represent it as a function $W(C_1, C_1^{-1})$, i.e., it is possible to use first invariants of the Cauchy-Green and the Finger tensors of large deformations. There is great freedom in varying any conceivable form of this function in an attempt to fit various experimental data.

Formally, no limitation in approximating function $W(C_1, C_1^{-1})$ exists because any approximation obeys the general principle of invariance. Certainly, in real practice, it is desirable to search for approximations having the simplest form, possibly a linear form.

The linear form is the simplest, and one of the examples of linear elastic potential is given by Eq. 4.4.7, but it contains only one argument, the first invariant E_1, (or C_1 as in

Eq. 4.4.10). Then the next possible approximation is a combination of linear functions in the form

$$W = AE_1 + BE_2 \qquad [4.4.20]$$

where A and B are material constants.

This elastic potential can be rewritten *via* the principal values of extension ratios. After mathematical transformation, the following formula is obtained:

$$W = G_1(\lambda_1^2 + \lambda_2^2 + \lambda_3^2) + G_2\left(\frac{1}{\lambda_1^2} + \frac{1}{\lambda_2^2} + \frac{1}{\lambda_3^2}\right) \qquad [4.4.21]$$

where "new" constants, G_1 and G_2, are expressed by "old" ones, A and B, as

$$G_1 = \frac{A - B}{2}, \; G_2 = \frac{B}{4}$$

and the final result of the rearrangements is the expression of W, as a function of the first invariants of both the Cauchy-Green and the Finger tensors given by

$$W = G_1(C_1 - 3) + G_2(C_1^{-1} - 3) \qquad [4.4.22]$$

This result confirms the equivalence of functions $W(C_1, C_2)$ and $W(C_1, C_1^{-1})$. An elastic potential in the form of Eq 4.4.22 was proposed by Mooney[4] and independently by Rivlin[5] on the basis of his general approach to the construction of elastic potential functions for large deformations. The function $W(C_1, C_1^{-1})$ in Eq. 4.4.22 is known as the *Mooney-Rivlin potential*.

The role of the second term in Eq. 4.4.22 is illustrated by stress-deformation dependencies for uniaxial extension and shear. In uniaxial extension

$$\lambda_1 = \lambda, \; \lambda_2 = \lambda_3 = \lambda^{-1/2}$$

where λ is an extension ratio.

Using formulae for components of stresses discussed above, it is possible to show that normal stress in the direction of stretching, σ_E, is

$$\sigma_E = 2\left[G_1\left(\lambda^2 - \frac{1}{\lambda}\right) + G_2\left(\lambda - \frac{1}{\lambda^2}\right)\right] \qquad [4.4.23]$$

with two other stress components equal to zero.

The acting force can be related to the initial cross-section of the stretched sample. This engineering stress, f_E, is written in the following manner

$$f_E = 2\left(\lambda - \frac{1}{\lambda^2}\right)\left(G_1 + G_2\frac{1}{\lambda}\right) \qquad [4.4.24]$$

The structure of Eq. 4.4.24 demonstrates that the addition of the second term in Eq. 4.4.22 leads to the appearance of the "correction" term in Eq. 4.4.24. If $G_2 = 0$, Eq. 4.4.24 becomes the well-known formula for the stress related to one member of elastic potential function (Eq. 4.4.7). The difference is because of $2G_2/\lambda$ in Eq. 4.4.24. The influence of this correction can be estimated, bearing in mind that according to the experimental data, $G_2 \approx 0.1 G_1$. However, the addition of the second term is of principal value, especially considering that other relationships between both constants are not excluded.

Treating experimental data in terms of Eq. 4.4.24 assumes that they ought to be presented in the coordinates f_M vs. λ^{-1}, where

$$f_M = \frac{f}{\lambda - \lambda^{-2}} \qquad [4.4.25]$$

It is expected that the experimental data points, presented in these coordinates, will lie on a straight line. It is also worth mentioning that the function f_M is, in fact,

$$f_M = \frac{\partial W}{\partial C_1} + \frac{1}{\lambda}\frac{\partial W}{\partial C_2} \qquad [4.4.26]$$

Fig. 4.4.2 is an illustration of this approach. The dependence under discussion gives a straight line at $\lambda^{-1} > 0.4$ (i.e. at $\lambda < 2.5$). At higher degrees of extension, strong divergence from the two-constant potential is evident. However, $\lambda = 2.5$ is already a large deformation, and it is important that at least in this deformation range Eq. 4.4.24 is valid.

Is it possible to be certain that this potential function is a correct image of real changes (at least at $\lambda < 2.5$)? In order to answer this question, another type of deformation other than uniaxial extension must be examined. Let it be shear deformation. It is easy to prove that Eq. 4.4.21 leads to the following dependence of shear stress, σ, on shear deformation, γ,

$$\sigma = 2(G_1 + G_2)\gamma \qquad [4.4.27]$$

where the sum $(G_1 + G_2)$ is the shear modulus.

The constitutive equation, expressed by Eq. 4.4.20, predicts linear behavior in shear, though it is non-linear in extension.

Repeatedly, this is proof that the same rheological behavior in deformations of one type (in this case, linear stress-deformation in shear) does not mean that the type of deformation under other conditions (in this case, stress-deformation in uniaxial extension) must be the same. The result demonstrates again that the data obtained in experiments of one type cannot be a criterion for the selection of rheological equation of state (constitutive equation) and cannot be used for unambiguous prediction of stress-deformation dependencies in different modes of deformations.

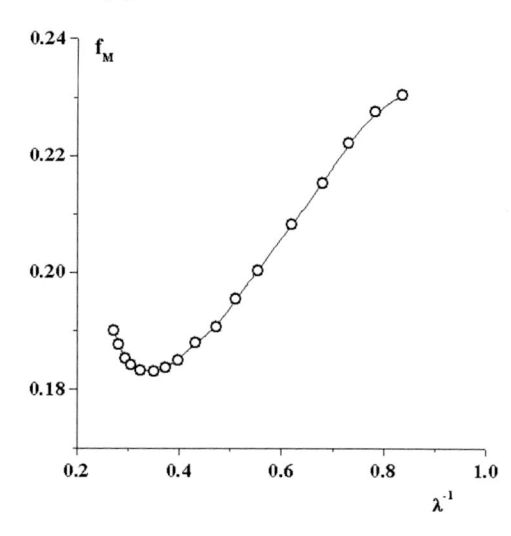

Figure 4.4.2. Linearization of experimental data in coordinates of Eq. 4.4.24. [Adapted, by permission, from R.S. Rivlin, D.W. Saunders, *Phil. Trans. Roy. Soc.*, **A243**, 251 (1951)].

4.4.2.2 Multi-member series

Eqs. 4.4.20 and 4.4.21 can be formally treated as the first approximation (the first-order terms of a series) of some non-linear functions $W(C_1, C_2)$ or $W(C_1, C_1^{-1})$. If one adds higher-order terms with their own material constants, one may expect to improve the correspondence between the theoretical predictions and the experimental data. In particular, it is important for shear studies because the linear relationship in Eq. 4.4.27 is not a realistic result.

If linear functions (Eq. 4.4.7 or Eq. 4.4.20) are not sufficient (and they are not for three-dimensional deformations), the quadratic term can be added and the expression for elastic potential is

$$W = AE_1 + BE_2 + ME_1^2 = G_1(C_1 - 3) + G_2(C_1^{-1} - 3) + G_3(C_1 - 3)^2 \quad [4.4.28]$$

where A, B, M, G_1, G_2, and G_3 are material constants, and the latter three are expressed *via* A, B, and M.

Based on Eq. 4.4.28, the following formula can be derived for the normal engineering stress in uniaxial extension:

$$f_E = 2\left(G_1 + G_2\frac{1}{\lambda} + G_3\lambda^2\right)\left(\lambda - \frac{1}{\lambda^2}\right) \quad [4.4.29]$$

It is evident that Eq. 4.4.29 contains a new quadratic term with its own material constant in a "correction" term, compared to Eq. 4.4.24. Certainly, three empirical constants allow us to fit experimental data points much better than one or even two "free" constants in Eq. 4.4.24, and that is why Eq. 4.4.29 gives a satisfactory approximation of different experimental data. Moreover, Eq. 4.4.29 predicts the nonlinear dependence of the stress-deformation function in simple shear.

The most general *phenomenological* form of an elastic potential function is

$$W = \sum_{\alpha, \beta, \gamma = 0} G_{\alpha, \beta, \gamma}(C_1 - 3)^{\alpha}(C_2 - 3)^{\beta}(C_3 - 3)^{\gamma} \quad [4.4.30]$$

where $G_{\alpha,\beta,\gamma}$ are empirical constants and $G_{000} = 0$ because elastic energy of undeformed body is assumed to be zero.

For an incompressible material, $C_3 = 0$, and a simpler general form of the elastic potential can be written:

$$W = \sum_{\alpha, \beta = 0} G_{\alpha, \beta}(C_1 - 3)^{\alpha}(C_2 - 3)^{\beta} \quad [4.4.31]$$

A very interesting and principal problem arises concerning the practical possibility of finding W as a function of invariants. The general answer to this problem is that it is necessary to compare the results of experiments carried out at *different geometrical schemes of loading*. For example, normal stress in uniaxial extension for an arbitrary function $W(C_1, C_2)$ can be expressed as

$$f_E = 2\left(\lambda - \frac{1}{\lambda^2}\right)\left(\frac{\partial W}{\partial C_1} + \frac{1}{\lambda}\frac{\partial W}{\partial C_2}\right) \qquad [4.4.32]$$

as is also seen from Eqs. 4.4.25 and 4.4.26.

The general form of the dependence of shear stress on deformation for simple shear (its measure is γ) can be formulated as

$$\sigma = 2\gamma\left(\frac{\partial W}{\partial C_1} + \frac{\partial W}{\partial C_2}\right) \qquad [4.4.33]$$

The dependencies of f_M and $\sigma/2\gamma$ on the elastic potential function are expressed in different manners. From the comparison of dependencies of f_M (see Eq. 4.4.25) and $\sigma/2\gamma$ on the right-hand sides of Eqs. 4.4.32 and 4.4.33, it is possible to find the dependence $W(C_2)$ as

$$\frac{\partial W}{\partial C_2} = \frac{\dfrac{\sigma}{2\gamma} - f_M}{1 - \lambda^{-1}}$$

Then the dependence $W(C_1)$ is found from Eq. 4.4.32 or Eq. 4.4.33.

Other types of fundamental experiments can also be used to find an elastic potential function $W(C_1, C_2)$ but it is essential that at least two different geometrical schemes of loading are used and compared.

The idea that a function $W(C_1, C_2)$ can be expanded into a power series, and that it is possible to use any desired number of terms of series, permits, by adding new arbitrary ("free") empirical constants, a reasonable degree of correspondence between a theoretical (phenomenological) curve and experimental data.

The same concept can be realized in a somewhat different way. From the very beginning, it was assumed that coefficients used, even in a very general formulation of an elastic potential, Eq. 4.4.30, are constant but in reality, the material properties are not constant and may depend on deformation. The coefficients must be expressed as dependencies of "modulus" A in Eq. 4.4.7 on invariants of the deformation tensor. It means that nonlinearity appears not only as a consequence of large deformations by itself but also as a function of some physical phenomenon (for example, structure transformation happening in the course, or as a consequence, of large deformations), i.e., the simplest quasi-linear potential (Eq. 4.4.7) is not sufficient for fitting experimental data when physical nonlinear effects are encountered.

4.4.2.3 General presentation

It was proven that the most general constitutive equation for elastic materials can be written as[6]

$$\sigma = \alpha_0\delta + \alpha_1 C + \alpha_2 C^2 \qquad [4.4.34]$$

In this equation, tensor values of stress, σ, unit tensor, δ (see comments in section 1.1.1), and the Cauchy-Green tensor, C, are used. The tensor C can be equivalently substituted by the tensor C^{-1}.

It is important that the coefficients, α_0, α_1, and α_2 in Eq. 4.4.34 are arbitrary scalar functions of the deformation invariants. For the particular case of incompressible material (such as rubbers) $\alpha_0 = -p$ (i.e., it is hydrostatic pressure). Also, it is easy to treat Eqs. 4.4.7 and 4.4.20 as particular cases of the more general Eq. 4.4.34.

Both approaches expand the function $W(C_1, C_2)$ into a power series or treat the material parameters α_1 and α_2 in Eq. 4.4.34 as functions of invariants of the deformation tensor. These are formal presentations of the elastic properties of solids. Eq. 4.4.7 can be based on some reasonable physical arguments (*"statistical theory of rubber elasticity"*).[7] Its generalization, in spite of numerous theoretical attempts, has no such universally accepted physical ground and must be treated as an empirical relationship invented for fitting the experimental data.

The last remark in this section regards time effects. *Time must not be mentioned* in this section at all, because this concept is not consistent with the idea of elastic (instantaneous) reaction of the material to the applied force. This fundamental idea already has been emphasized by stating that stress-deformation relationships discussed in this chapter are valid for *equilibrium* conditions. However, there is a great difference between instantaneous and equilibrium reactions and the gap between both is the field of time-dependent effects. Moreover, large deformations and rubbery elasticity are relevant to polymeric materials and various time-dependent effects (relaxation, etc.) are typical of these materials. That is why it is important (though in some cases difficult) to separate time-dependent effects and distinguish "pure" (equilibrium) stress-deformation dependence.

The formulation of the elastic potential function should be based on fundamental molecular arguments determining the structure of the elastic potential function. In the simplest case, such an approach was used in the single-constant Kuhn-Guth-James-Mark potential, which was previously discussed. This potential function is based on the affine transformation of a network created by macromolecular chains. Later a two-constant potential was proposed that incorporated a concept of limited extensibility of macromolecular chains:[8]

$$W = K_1 E_1 + K_2 \ln[(E_2 + 3)/3] \qquad [4.4.35]$$

where K_1 and K_2 are constants and the small deformation limit of elastic modulus (Hooke's modulus) E is expressed as

$$E = 6K_1 + 2K_2$$

This concept is useful in the explanation of the effect of hardening in the extension of polymer melts (see section 3.7.3). The construction of general non-linear models of viscoelastic liquids also requires such molecular-based models of non-linear elasticity (see section 2.8.2).

4.4.2.4 Elastic potential of the power-law type

A method of introducing a multi-component elastic potential function is based on the generalized (nonlinear) measures of deformations. The following assumption for the $W(\lambda_i)$ function was proposed:[9]

$$W = \frac{2G}{n} I_E + B I_E^m \qquad [4.4.36]$$

where G, m, n, and B are material constants, and I_E is the first invariant (the sum) of the generalized measure of deformations, E_α, which is expressed as

$$E_\alpha = \frac{1}{n}(\lambda_\alpha^n - 1)$$

This approach leads to non-linear dependencies in the deformation of any arbitrary type. The potential function, including extension and simple shear, is based on a comparison of invariant dependencies of stresses and deformations, including the geometry of deformations of different types. The calculations or two-dimensional model gives the following equation:

$$\sigma_\alpha - \sigma_\beta = (\lambda_\alpha^n - \lambda_\beta^n)\left(\frac{2G}{n} + mBI_E^{m-1}\right) \qquad [4.4.37]$$

where indices α and β define principal stresses and extension ratios defined in principal axes.

The left-hand side of this equality will be noted as X and the right-hand side as Y. Then the dependence of X on Y is expected to be a straight line inclined by 45° to both axes regardless of the geometry of deformation. Experimental results for two different materials and four geometries of deformation are presented in Fig. 4.4.3 in the coordinates X-Y. It is seen that the power-law type elastic potential fits the experimental data in the invariant form, and that is why this equation can be considered as acceptable.

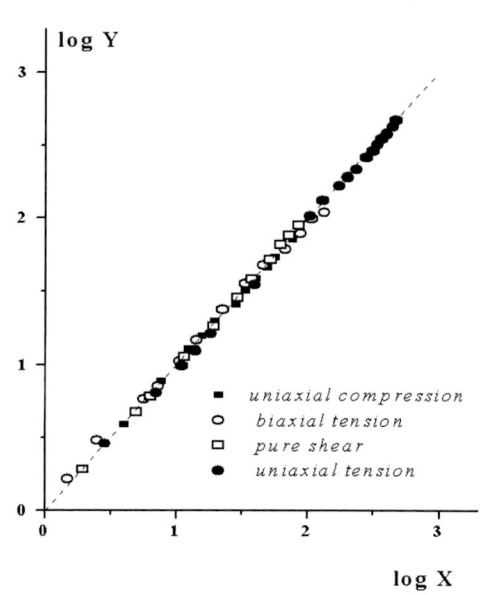

This approach is useful for fitting experimental data obtained from studies of different rubber-like materials.

4.4.3 THE POYNTING EFFECT

Large deformations, even uniaxial, lead to three-dimensional effects. In liquids, it s the Weissenberg effect (see section 3.4.2). In solids, a similar effect is called the *Poynting effect*.[10]

The experiments demonstrated that twisting thin wires affects their length. This effect becomes noticeable at relatively large angles of twisting, for example, after several turns of wire.

The characteristic (dimensionless) geometrical size of wire is the ratio of its radius to length, R/L. This ratio is smaller than 1. The values of relative deformations are also small, but they are comparable with values of $(R/L)^2$. The Poynting effect is a typical phenomenon of the second

Figure 4.4.3. Experimental data for two materials and four types of deformation presented in the coordinates of Eq. 4.4.36. [Adapted, by permission, from B.J. Blatz, S.C. Sharda, N.W. Tschoegl, *Trans. Soc. Rheol.*, **18**, 145 (1974)].

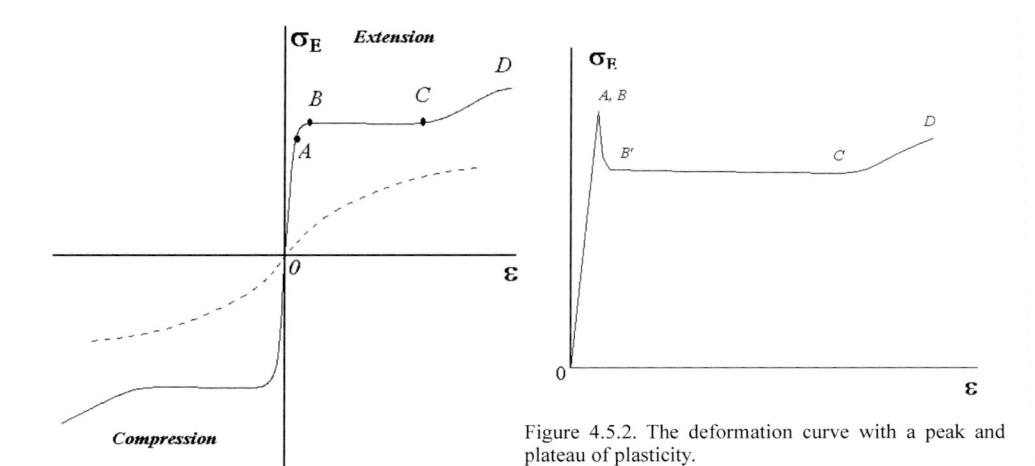

Figure 4.5.1. Typical deformation curves for mild steel (solid line) and non-ferrous metals (dashed line). Characteristic points are marked.

Figure 4.5.2. The deformation curve with a peak and plateau of plasticity.

order. The changes in length are proportional to the squared value of $R\Omega/L$, where Ω is an angle of twisting.

The Poynting effect applies to large deformations in shear. Shearing does not cause volume changes. However, this is not true in large deformations. That is why the other outcome of the Poynting effect is volume change. It is analogous to *dilatancy* (volume change in shear) in liquid-like materials.

4.5 LIMITS OF ELASTICITY

4.5.1 STANDARD EXPERIMENT – MAIN DEFINITIONS

The *limit of elasticity* is a result of standard experiments in uniaxial extension of solid samples. Two typical cases are shown in Figs. 4.5.1 and 4.5.2. Sometimes, the normal stress-deformation dependence, $\sigma_E(\varepsilon)$, is monotonous (e.g., for mild, low carbon, steel, or copper, as in Fig. 4.5.1). In other cases, a sharp maximum is observed on the $\sigma_E(\varepsilon)$ dependence (e.g., for many metallic materials and plastics, as in Fig. 4.5.2).

Point A in these figures corresponds to the *limit of proportionality*, and the OA part of the full curve is a domain of validity of Hooke's law. Elastic deformations continue up to point B, which is called a *limit of elasticity*. After this point, irreversible deformations are detected. Deformations of this kind increase greatly along the plateau in the range BC (or B'C) and this is known as a domain of *plasticity* or *plastic deformation*. The maximum stress (points A and B in Fig. 4.5.2) is called a *limit of plasticity*, or *yield stress*.[11] The $\sigma_E(\varepsilon)$ curve ends at point D, marking a failure of the sample, and this point is called *ultimate* or *tensile strength* of the material. Deformation, corresponding to this stress, is called the *ultimate elongation* or *elongation at break*.

There is a large number of *fragile* or *brittle* solid materials which break close to point B (e.g., reinforced plastics, many ceramics, inorganic glass, pig iron, monocrystals, etc.).

Two phenomena define the limits of elasticity:

- *elastic-to-plastic transition*
- *failure.*

It is noteworthy that both effects are observed not only in extension but also in deformation of any geometry, for example, in uniaxial compression, shear, a combination of twisting and extension, etc. Only all-dimensional (hydrostatic) compression occupies a special position. Quantitative measures of conditions, corresponding to the limit of elasticity, are different depending on the deformation mode.

The elastic-to-plastic transition, as well as fracture, can be considered as some critical event on the deformation-stress dependence. Therefore, both phenomena can be considered in analogous terms, especially when the three-dimensional effects are discussed and the general criteria are searched for.

4.5.2 CRITERIA OF PLASTICITY AND FAILURE

The limit of elasticity for elastic-to-plastic transition or failure is a critical point on the stress-deformation diagram if time-dependent effects are not taken into account (the time effects are typically assumed to be absent in the concept of solids). This critical point can be found in experiments of standard geometry. The principal question is how to construct the three-dimensional (invariant) criterion which can be used for an arbitrary stress state of the material. This is important in applications because stress states in real situations are three-dimensional.

The generalized criteria of plasticity and failure ought to be formulated in the analogous form because both of them are critical phenomena. For an isotropic material, the criterion of transition must be the symmetrical function of principal stresses. The role of the average normal stress (i.e., hydrostatic pressure) is negligible. Therefore, it is sufficient to use two invariants of the stress tensor. Then, the critical state criterion must depend on the difference of principal stresses. It leads to the formulation of two criteria mostly used for applied purposes.

The search for the criterion of transition (or failure) in an arbitrary three-dimensional stress state is based on the concept of the equivalency of this stress state to some unidimensional state determined by a single critical (or limiting) value of stress.

4.5.2.1 Maximum shear stress

Material transforms to the state of plasticity (flow) or breaks at a stressed state when the maximum shear stress exceeds some critical value, σ^*.[13] This condition is called the *Tresca-Saint-Venant criterion.*[14]

For uniaxial extension, at normal stress, σ_E, shear stress equals $\sigma_E/2$ (see section 1.1) and therefore the condition of plasticity in extension is

$$\sigma^* = 0.5\sigma_E^*$$ [4.5.1]

It means that the plasticity state (or failure) is reached when shear stress equals $0.5\sigma_E^*$ for σ_E^* measured in uniaxial extension.

For a three-dimensional stress state, three maximum ("invariant") shear stresses $\sigma_{1,max}$, $\sigma_{2,max}$, and $\sigma_{3,max}$, are calculated as

$$\begin{cases} \sigma_{1,\,max} = \left| \dfrac{\sigma_2 - \sigma_3}{2} \right| \\[2mm] \sigma_{2,\,max} = \left| \dfrac{\sigma_1 - \sigma_3}{2} \right| \\[2mm] \sigma_{3,\,max} = \left| \dfrac{\sigma_1 - \sigma_2}{2} \right| \end{cases} \qquad [4.5.2]$$

One of these differences is maximum (by its absolute value) and it determines the critical condition. According to this approach, σ^* equals the difference of two normal stresses, maximum and minimum. The middle value of principal normal stress has no influence on the limiting state.

As an example, let us calculate maximum shear stress in a superposition of shear stress, σ, and normal stress, σ_E. Then the condition of limiting state, according to the results of calculations of maximum shear stress, is:

$$\sigma^* = \sqrt{\sigma_E^2 + 4\sigma^2} \qquad [4.5.3]$$

The critical (limiting) state is reached when and if the right-hand side of this equation becomes equal to the critical number – the acceptable value of shear stress.

4.5.2.2 The intensity of shear stresses ("energetic" criterion)

This approach accounts for three differences of principal normal stresses or three principal shear stresses, defined by Eqs 4.5.2.[15] The criterion is formulated as

$$\sigma_E^* = \sqrt{\frac{1}{2}[(\sigma_1 - \sigma_2)^2 + (\sigma_1 - \sigma_3)^2 + (\sigma_2 - \sigma_3)^2]} \qquad [4.5.4]$$

This condition is called the *Huber-von Mises criterion*. The value T, is defined as

$$T = \sqrt{\frac{1}{6}[(\sigma_1 - \sigma_2)^2 + (\sigma_1 - \sigma_3)^2 + (\sigma_2 - \sigma_3)^2]} \qquad [4.5.5]$$

In the theory of stresses, it is sometimes called the *intensity of shear stresses*. Both values, σ_E and T, can be expressed by the "average" value of shear stress, σ_{av}, which is

$$\sigma_{av} = \frac{1}{3}\sqrt{(\sigma_1 - \sigma_2)^2 + (\sigma_1 - \sigma_3)^2 + (\sigma_2 - \sigma_3)^2}$$

Then, the criterion is written as

$$\sigma_E^* = \sqrt{3}\,T = \sqrt{\frac{9}{2}}\,\sigma_{av} \qquad [4.5.6]$$

If $T = \sigma$, the following relationship between σ^* and σ_E^* takes place:

$$\sigma^* = 0.577\sigma_E^* \qquad [4.5.7]$$

It is close, though not equivalent, to the criterion of maximum shear stress in Eq. 4.5.1.

In the deformation of elastic solids, the work done is stored in the form of elastic potential energy. The total energy of deformation can be separated into components related to shape and volume changes. Then, accurate calculations show that the part of the energy that is responsible for shape changes, W_{sh}, is expressed as:

$$W_{sh} = \frac{1+\mu}{3E}(\sigma_1^2 + \sigma_2^2 + \sigma_3^2 - \sigma_1\sigma_2 - \sigma_1\sigma_3 - \sigma_2\sigma_3) = \frac{1+\mu}{E}T^2 \qquad [4.5.8]$$

The last equation shows that the condition of a critical state is directly related to the stored energy, not to the total energy, but only to the part that is responsible for shape changes. This is the reason for calling the criterion an *energetic criterion*. Volume changes do not influence the state of the material.

4.5.2.3 Maximum normal stress

The above-discussed cases were based mainly on the estimation of shear stresses, which determine conditions of the limiting state. This is directly related to the concept of shape changes of solid materials. However, volume changes in extension of solids also take place, and sometimes they lead to transitions. Neck formation in uniaxial extension of polymers is an example of such an effect (see Fig. 2.8.3). Though different explanations of the mechanism of neck formation were proposed and discussed, the following seems adequate:[16] a neck forms when a relative volume of the body increases up to a certain level. This formation of an extra free volume provides the conditions for relaxation transition. This is a transition from an isotropic to an oriented state. The role of volume changes is dominating.

The following condition of the limiting state can be introduced. The criterion of the condition is a critical level of normal stress, σ^*. For a two-axis stress state, when shear, σ, and normal, σ_E, stresses act simultaneously, the criterion of the limiting state is written as

$$\sigma^* = \frac{1}{2}[\sigma_E + \sqrt{(\sigma_E^2 + 4\sigma^2)}] \qquad [4.5.9]$$

Normal and shear stresses influence the condition of the limiting state to a relatively equivalent extent.

4.5.2.4 Maximum deformation

The physical meaning of this approach is close to the previous one. However, the quantitative expression for the critical stress appears somewhat different. The detailed calculations show that for superposition of shear, σ, and normal, σ_E, stresses, the condition of the limiting state is written as

$$\sigma^* = \left[\frac{1}{2}(\sigma_E + \sqrt{\sigma_E^2 + 4\sigma^2}) - \frac{1}{2}\mu(\sigma_E - \sqrt{\sigma_E^2 + 4\sigma^2})\right] \qquad [4.5.10]$$

In this case, too, there is a complicated superimposing influence of normal and shear stresses on the critical state of the material.

4.5.2.5 Complex criteria

Summarizing the above-mentioned, one can see that two concepts of a critical state condition exist: the dominating role of either shear or normal stresses is assumed. In the first

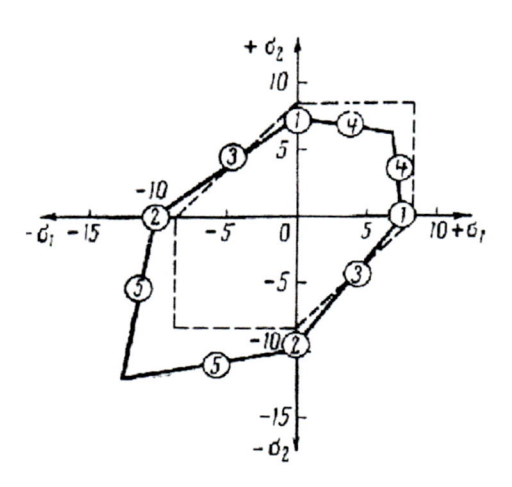

Figure 4.5.3. Limiting states of polystyrene as predicted by the maximum shear stress criterion (dashed line) and measured experimentally. [Adapted, by permission, from W. Whitney, R.D. Andrews, *J. Polymer Sci., C*, **16**, 2981(1967)].

case, shear sliding (shape changes without volume increase) determines the critical state, and in the second, volume effects are a dominating factor, and the critical state is reached due to the volume transition or separation of one part of the material from another.

It is reasonable to suggest that in reality both effects are superimposed and the dominating mechanism of the critical state depends on the nature of the material and/or the geometry of loading. Some situations are rather evident. For example, neck formation is a consequence of volume effects and thus of normal stresses. During the extension of many ferrous metals, the lines on the sample surface (so-called *Luders lines*) inclined by 45° to the axis of the extension are formed. This is a result of the sliding of structure elements caused by the action of shear stresses. In some cases, the angle of inclination of these lines is not exactly 45°, and this is the direct proof of the superposition of both mechanisms – shearing and extension.

A comparison of experimental results measured in a planar (two-dimensional) deformation field and predictions of the maximum shear stress criterion are shown in Fig. 4.5.3. The points marked are 1 – related to the uniaxial extension, 2 – compression, 3 – shear in twisting, 4 – biaxial extension, and 5 – biaxial compression. Points marked as 3 are the direct measures of the critical shear stress, σ^*. Starting from this value, it is possible to construct the complete contour of limiting states, Eq. 4.5.1. This contour is drawn by a dashed line. The solid line is the contour built using the experimental data points.

It is evident that the experimental data do not exactly correspond to the theoretical predictions. The main reason in the case of data presented in Fig. 4.5.3 is the difference of limiting states in extension and compression (i.e., in two different uniaxial stress states). This phenomenon is well known for many real materials, such as stones, pig iron, and others. This phenomenon is the reflection of the role of hydrostatic pressure on a limiting state. The correct formulation of a criterion requires the introduction of a normal stress factor in addition to the maximum shear stress. Then, the critical condition can be written as

$$\frac{\sigma_1 - \sigma_3}{2} = \sigma^* + \sigma_E \tan \varphi \qquad [4.5.11]$$

The introduction of the angle, φ, reflects the inclinations of the direction of the maximum resistance from the line of the action of maximum shear stresses in uniaxial extension. This direction is oriented by the angle $\theta = (\pi/4 + \varphi/2)$ to the axis of extension.

According to the experimental data for polystyrene (Fig. 4.5.3), the angle φ equals $13°$ and $\theta = 51°30'$.

A general criterion of limiting states must reflect the influence of both shear and normal stresses. Normal stress modifies the conditions of transition (or break) according to the following linear relationship obtained from modification of the Eq. 4.5.4:[17]

$$\sigma^* = \sigma_{av} + \frac{v}{3}I_1 \tag{4.5.12}$$

where I_1 is the first invariant of the stress tensor (see Eq. 1.1.7) and v is an empirical factor. The condition $v = 0$ or deformation mode with $I_1 = 0$ corresponds to the criterion of the limiting intensity of shear stresses. If $I_1 > 0$, i.e., the material is stretched and the sample can resist lower shear stresses. Hydrostatic compression increases the ability to withdraw at higher shear stresses.

A criterion accounting for different resistances to compression and extension is presented in the following way[18]

$$9\sigma_{av}^2 + 6(C - M)\sigma_E = 2CM \tag{4.5.13}$$

where C and M are limiting values of normal stress in compression and extension, respectively. If $C = M$, the material is isotropic and, again, the criterion of maximum intensity of shear stresses appears valid. However, if $C > M$ (as usual), the situation appears more complex.

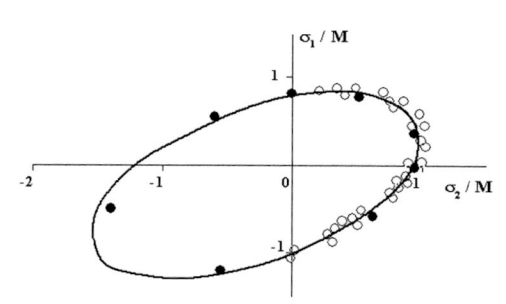

The applicability of Eq. 4.5.13 to the analysis of experimental data is illustrated in Fig. 4.5.4. The asymmetry of the curve (due to the difference of properties in compression and extension) is evident. Eq. 4.5.13 fits the experimental data very well. It is worth mentioning that Eq. 4.5.12 also describes these experimental points in a satisfactory manner. Both equations are empirical and it is not possible to say which one of them is better.

Figure 4.5.4. Comparison of theoretical predictions based on Eq. 4.5.13 (solid line) with experimental data for polystyrene (solid circles) and some other polymers (open circles). The ratio of C/M is chosen to be equal 1.3. Normal stresses σ_1 and σ_2 are normalized by limiting extension stress M. [Adapted by permission from S.S. Sternstein, L. Ongchin, *ACS Polymer Preprints*, **10**, 1117 (1969)].

The criteria discussed above are related to isotropic materials. Numerous materials of engineering interest are anisotropic in their structure, for example, reinforced plastics or monocrystals. Criteria of the limiting state also can be constructed for such materials, based on the same approaches that were used for isotropic materials. An analytical form of such criteria is complicated, though the physical principles are the same.

4.5.3 STRUCTURE EFFECTS

Departing from the limits of elasticity inevitably has an effect on structure and, thus, changes in mechanical (rheological) properties of matter. Such effects are, in many cases, irreversible. However, this irreversibility may be, at least, partly reversible on rest, though the rates of physical processes in solids are much slower than in liquids because of much longer relaxation times. The characteristic relaxation times in solids are at least in hours. In some cases, the increase of temperature and long storage at elevated temperature (*annealing*) may accelerate the process of structure restoration and make it observable. On the other hand, the damage of structure at large deformations may be so extensive that its complete restoration is not possible even with high-temperature treatment. This is why the initial state (shape) of material after plastic deformation cannot be reached by applying the same stresses but with the opposite sign.

Example
    The car body deformed by a single impact in an accident regrettably cannot be repaired by a single impact in the opposite direction.

The structure destruction and its (partial or complete) restoration in the case of a solid are similar to the *thixotropy* of liquid-like materials, though this term is rarely used for solids. The concept of *structure* is more definite and clear for solids than for liquids, because the structure of solids can easily be inspected by direct optical, electronic, or X-ray methods. The mechanisms of structure rearrangement are also easy to investigate in many cases, though these mechanisms may be very different. One may be caused by sliding along crystal defects (dislocations), another by destroying relatively large structure aggregates (e.g., spherulites), or by orientation, deformation-induced phase transitions, and so on.

The reversibility of structure rearrangements suggests that this process proceeds in time, and it is necessary to consider the kinetics of this phenomenon.

It is reasonable to distinguish between time-dependent effects of two types:
- a domain of viscoelastic effects consisting of a superposition of energy storage and dissipation (see Chapter 2)
- the structure effects caused by changes in the macrostructure of materials induced by deformations. Effects of such kind are observed primarily by irreproducibility of the results of material properties measurements in repeated experiments and their slow (complete or partial) restoration at rest.

Some typical structural phenomena leading to the changes of mechanical properties in the domain of nonlinear inelastic deformations are discussed below. It is noteworthy to remind one that these effects are *not viscoelastic* phenomena but a *structure time-dependent* effect. The situations cannot be well defined in various cases and it is not always possible to establish clear boundaries between the effects of these two different classes.

4.5.3.1 Strengthening

A crystalline material after additional loading, beyond the limit of elasticity (in a plastic zone), becomes more rigid on repeated loading. The limit of elasticity increases and the plasticity of material becomes suppressed. Also, impact strength and resistance to shock resistance decreases. This effect is called *strengthening* or *stress hardening*. Strengthening of the surface appears as a result of mechanical treatment in some technological operations and can be a purposeful technological procedure to improve properties of products.

Strengthening in plastic deformation decreases the resistance of the material to loading in the opposite direction (this is known as the *Bauschinger effect*). Strengthening partly or completely disappears after prolonged rest or annealing at elevated temperature, which is similar to thixotropic effects.

4.5.3.2 Thixotropy

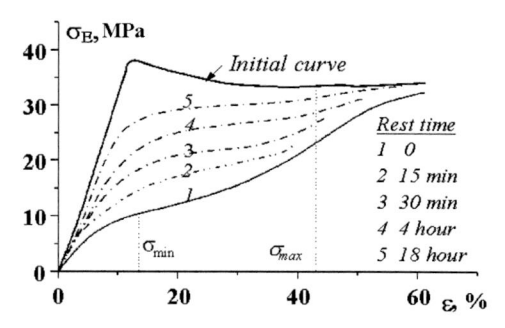

Figure 4.5.5. Thixotropy in deformation and rest of crystalline polypropylene – structure destruction and restoration. [Adopted, by permission, from G.P. Andrianova, Nguen Vin-Chii, *Vysokomol. Soedin.* (Polymers – in Russian), **14A**, 1545 (1972)].

The effect of reversible changes of material properties in the range of large deformations was first described for filled rubbers (known as the *Mullins effect*)[19] and later for deforming crystalline polymers (see Fig. 4.5.5). The plateau part of the deformation curve decreases from σ_{max} at an initial experiment to σ_{min} in the repeated experiments carried out just after one another. However, during prolonged rest (several hours) this plateau value of stress returns to its initial value.

Structure destruction and restoration is a typical thixotropic phenomenon. It is very difficult to establish, which state of the material is in equilibrium due to very slow structure transformations.

4.5.4 PLASTICITY (INELASTIC DEFORMATIONS)

In previous chapters, the term "plasticity" was mentioned many times. When rheological properties of liquids were discussed, viscoplasticity meant a material behavior as solid-like at stresses lower than the yield stress and plastic (i.e., it can flow) at stresses higher than this critical stress. In this Chapter devoted to solids, plasticity meant the transition through the critical point corresponding to the boundary of elasticity though it was not determined what happens at higher stresses. So, it is necessary to define what is understood as the plasticity of solids.

Before the critical point, visco-plastic liquids behave in a solid-like manner, and the deformations are elastic (reversible). The same is true for solids. The difference is in the level of the elastic modulus, which can be less by several orders of magnitude in liquids in comparison with ordinary solids.

However, above the critical stress (that is defined in the previous Section), the understanding of plasticity becomes different.

Intuitively, plasticity is a possibility of irreversible deformations. This is true in both cases. However, the mode of these deformations in liquids and solids appears different.

In liquids, these irreversible deformations correspond to flow, i.e., the unlimited increase in deformations under applied stress, and there is no correspondence between stress and deformations developing in time.

In solids, irreversible deformations also take place, but applied stress creates only limited deformation, and there is an unambiguous correspondence between stress and deformation. Sometimes, speaking about the plasticity of solids, the term "*inelastic* deformations" is used that seem to be more accurate.

The transition from elastic to plastic behavior of solids is a very important phenomenon from two points of view.

First, the articles made of solids must store their form in the application. It is possible only if deformations in loading an article do not exceed the limit of elasticity. If it is not so, an article becomes useless, and in engineering applications, this transition is dangerous and crucial for construction as a whole. The solid articles after plastic deformations cannot be easily restored just because the deformed part does not "remember" its initial state. This situation is well known to everybody who had a car accident: dented parts are easier to replace than to restore.

This is why the limit of elasticity (or "limiting stress") is determined in the standard experiments for all materials. It is the most important engineering property of a material used as a starting point in all design and applied calculations.

Second, many technological operations in preparing different articles are based on plastic deformations of a material that is forced to accept the pre-defined form via plastic deformations. The examples are quite evident: that is such operations as stamping, rolling, drawing through a die, and so on. So, technological operations require passing through the elastic-to-plastic transition.

A very important and independent field of application of the concept of plasticity of solids is geology and geophysics. It is widely accepted that many soils and rocks can be plastically deformed under the action of giant forces acting inside them, though the speed of their movement can be very slow. Displacement of glaciers is another important example of plastic deformations of solids. Quantitative description of plasticity in geology is not yet developed well (mainly due to experimental difficulties in the determination of material properties of these media), though its importance is quite evident.

So, the following differences and similarities between visco-plastic "liquids" and plastic solids can be distinguished.

- In the domain of small deformations preceding plasticity, elastic deformations of visco-plastic "liquids" are negligible and not important in applications, whereas small deformations in solids are critically important because they determine the applied properties of a material.
- The development of plastic deformations in solids is limited by the failure of a sample, whereas plastic deformations of liquids (i.e., their flow) can continue for an unlimitedly long time.
- values of elastic modulus in the domain of small deformations preceding to the plasticity are of the order 10^{10}-10^{11} Pa for typical solids, but not higher than 10^3 Pa for typical visco-plastic media;
- the values of stresses in the domain of plastic flow do not exceed 10^6 Pa (depending on viscosity and shear rate) for flowing liquids but of the order of 10^{10} Pa for metals and 10^7-10^8 Pa for plastics.

Besides metals (which are beyond the frames of this book), inelastic deformations are especially important for highly concentrated suspensions. In approach the limit of the closest packing of such suspensions, the apparent viscosity begins to increase along with the increase in shear rate (Fig. 4.5.6). This is the effect of shear thickening, sometimes treated as *dilatancy*.

Usually, this kind of rheological behavior is observed in the concentration range of 50-65 vol%.

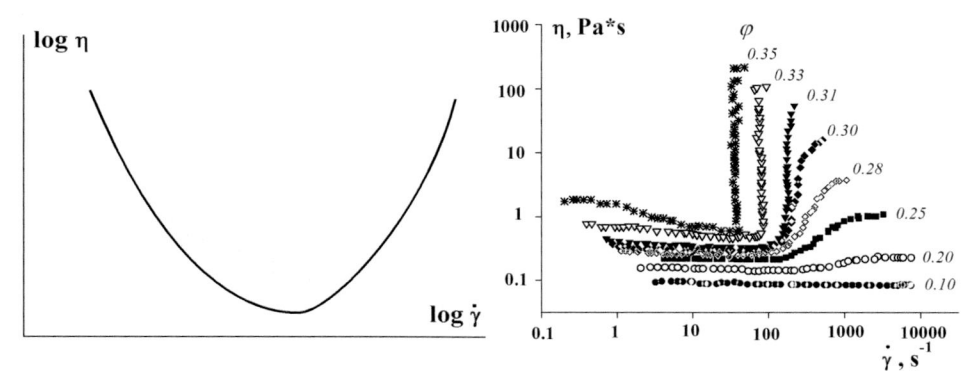

Figure 4.5.6 The transition from shear-thinning to shear thickening along with the increase of shear rate.

Figure 4.5.7. Abrupt transition to shear thickening. [Adapted, by permission, from R.G. Egres, N.J. Wagner, *J. Rheol.*, **49**, 719 (2005).]

Visual observations show that this effect is accompanied by easily detected periodic creation and destruction of jamming clusters that are concentration inhomogeneities.[20-22] These oscillations demonstrate the coexistence of several structures with periodic transi tions between them like it happens in the shear banding phenomenon (see Section 3.5.3.2) However, these fluctuations are strongly suppressed by the increase of the viscosity of a liquid continuous phase.[23]

Meanwhile, two important peculiarities of this effect related to highly concentrate d suspensions are noteworthy to mention.

First, the effect of thickening cannot be smooth like in *continuous shear thickening* (**CST**) but happens as the sudden unlimited growth of viscosity like in *discontinuous shear thickening* (**DST**), and this transition occurs at the minor increase in concentration (Fig. 4.5.7).[24]

Second, actually, the final stage of the shear thickening is the transition to the solic- state that is called *jamming* and can be treated as the glass transition. This transition excludes the possibility of the flow. It is necessary to stress that this effect has the other nature in comparison of the flow-to-rubbery state transition discussed in Section 3.6.3.2, which is related to the relaxation on the macromolecular level. The liquid-to-solid transi- tion is due to jamming obliged to macroscopic movement – friction of clusters in hetero- geneous structure initiated by the deformation.

The creation of a quasi-solid structure happens very quickly, as well as its decay after cessation of loading, so that one can neglect the relaxation phenomena. The detailed the- ory of jamming, taking into account the role of the interparticle friction, was developed in.[25]

The solid-like structure that appeared at a high concentration of solid particles after *DST* is not elastic because the packing of solid particles allows for retaining some free vol- ume that provides the possible movement of large structure clusters. This is reflected in strong and chaotic oscillations of stresses acting at the cluster boundaries and leading to their sliding in opposite directions.[26] This is really an uncertain movement that is different

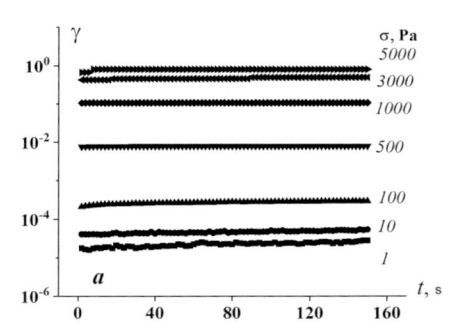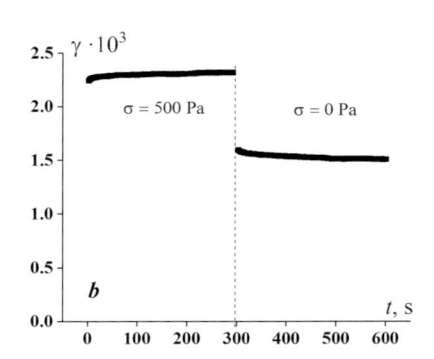

Figure 5.6.8. The behavior of elastic (a) and elasto-plastic (b) media - high concentration suspensions. [Adapted, by permission, from A.Ya. Malkin, A.V. Mityukov, S. V. Kotomin, A.A. Shabeko, V.G. Kulichikhin, *J. Rheol.*, **64**, 469 (2020); Ya. Malkin, V.G. Kulichikhin, A.V.Mityukov, S.V. Kotomin, *Polymers*, **12**, 1038 (2020).]

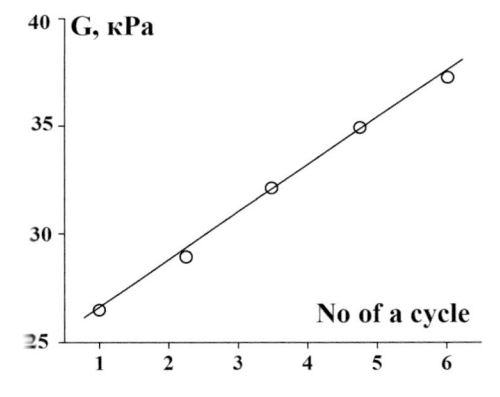

Figure 4.5.9 An increase in the elastic modulus for 60% suspensions in sequential loading-unloading cycles. The stress at loading was 500 Pa.

from the bifurcation leading to the formation of two coexisting structures like it takes place in shear banding.

So, highly concentrated suspensions retain some movability, but this does not allow them to flow. In fact, this is an elastic-plastic state, and the complete deformation at some given stress consists of two components − elastic and inelastic parts. The principal difference between elastic and elasto-plastic media are shown in Fig. 4.5.8.[27,28]

Fig. 4.5.8a illustrates the behavior of an elastic medium at different shear stresses: deformation γ appears instantaneously after the application of stress and does not change in time. After cessation of the stress, deformation comes back to zero, and there are no residual effects. Fig. 4.5.8b illustrates the behavior of an elasto-plastic medium. After the application of stress, deformation γ also appears instantaneously and does not increase, which corresponds to the absence of flow. However, a large portion of deformations retains after cessation of loading, and this is inelastic (not flow-induced!) deformation.

The ratio between elastic and inelastic deformations depends on the concentration of suspensions and the level of loading.

A possibility of inelastic (plastic) deformations of highly concentrated suspensions says about shear-induced structure rearrangements which accompany these deformations. It is possible to postulate that these deformations promote structure ordering in particles ensembles. As a consequence, material becomes more rigid that leads to an increase in the elastic modulus. This is seen in the repeated loading-unloading cycles (Fig. 4.5.9).

So, elasto-plastic domain of the rheological behavior exists inside the solid-state and can be important at some technological operations, e.g., powder molding.

REFERENCES

1   There are many fundamental and applied textbooks devoted to solving the problem of linear elastic bodies and engineering designs. See, for example, classical and standard books: A.E.H. Love, **A Treatise on the Mathematic Theory of Elasticity**, 4th Ed., *Dover*, NY, 1927; S. P. Timoshenko and G. H. MacCullough, **Elements of Strength of Materials**, *van Nostrand*, NY, 1949; S. P. Timoshenko and J. M. Gere, **Mechanics of Materials**, *van Nostrand Reinhold*, NY, 1972; B. Sandor, **Strength of Materials**, *Prentice Hall*, Englewood Cliffs, 1978; E. P. Popov and T. A. Balan, **Engineering Mechanics of Solids**, *Prentice Hall*, Upper Saddle River, 1999.
2   A general approach to solving this problem was advanced by R.S. Rivlin, *Proc. Roy. Soc.*, **A193**, 260 (1948); *Phil. Trans. Roy. Soc. Lond.*, **A240**, 459 (1948); *J. Rat. Mech. Anal.*, 4, 681 (1955) and 5, 179 (1956). R. S. Rivlin, *Phil. Trans. Roy. Soc.*, **A241**, 379 (1948) and **A242**, 173 (1949); *Proceed. Cambridge Phil Soc.*, **45**, 485 (1949); *Proceed. Royal Soc.*, **A195**, 563 (1949).The review of "**Forty Years of Non-linear Continuum Mechanics**" was presented in: R.S. Rivlin, **Advances in Rheology**, v. 1, 1, Eds. B. Mena, A. Garcia-Rejon, C. Rangel-Nafaile, IX Intern. Congress on Rheology, Mexico, 1984. The general theory of large deformations is also summarized in the monograph: B.W. Ogden, **Non-linear Elastic Deformations**, *Ellis Harwood*, Chichester, 1984 and the latest survey is G. Saccomandi, R.W. Ogden, **Mechanics and Thermodynamics of Rubberlike Solid**s, *Springer*, 2004.
3   This theory was proposed and developed in the following publications: W. Kuhn, *Koll. Zs.*, **68**, 2 (1934); E. Guth, H. Mark, *Monatsch. Chem.*, **65**, 93 (1934); E. Guth, *J. Appl. Phys.*, 10, 201 (1939); E. Guth, H.M. James, *J. Ind. Eng. Chem.*, **33**, 634 (1941); H.M. James, E. Guth, *J. Ind. Eng. Chem.*, **34**, 1365 (1942) and *J. Chem. Phys.*, 11, 455 (1943); E. Guth, H.M. James, H. Mark, *Adv, Colloid Sci.*, **2**, 253 (1946). One can find a very good review of the fundamental concepts in rubber elasticity in a classical monograph: L.R.G. Treloar, **The Physics of Rubber Elasticity**, *Oxford Univ. Press*, 3rd ed., 1975.
4   M. Mooney, *J. Appl. Phys.*, 11, 582 (1940).
5   See the publications by Rivlin of 1948 in Ref.2.
6   M. Reiner, *Am. J. Math.*, **70**, 433 (1948). R.S. Rivlin, Proc. 1st Symp. Naval Struct. Mech., 169 (1960), Pergamon. Oxford. The absence of higher terms of C are based on mathematical arguments, in particular on the Hamilton-Cayley theorem, proving that higher terms can degenerate to linear and quadratic terms of a tensor.
7   See Ref. 3.
8   M.H. Wagner, *J. Rheol.*, **38**, 655 (1994); E. Fried, *J. Mech. Phys. Solids*, **50**, 571 (2002). See also A.N. Gent, *J. Rheol.*, **49**, 271 (2005).
9   B.J. Blatz, S.C. Sharda, N.W. Tschoegl, *Trans. Soc. Rheol.*, **18**, 145 (1974).
10 H.H. Poynting, *Proc. Roy. Soc. London*, **A82**, 546 (1909); **86**, 534 (1912). It was the first experimental demonstration of the second-order effects in the mechanics of continuum.
11 Some authors treat the stress corresponding to the plateau B'C on the $\sigma(\varepsilon)$ curve as the yield stress.
12 The theory of plasticity of solids was first developed by A. Saint-Venant (1797-1866) – a French engineer and scientist – as a separate branch of mechanics. This theory is set forth in the following classical monographs: A. Nadai, **Plasticity**, 1931 and **A Theory of Flow and Fracture of Solids**, 1950, *McGraw Hill Co*, New York; R. Hill, **Plasticity**, 1950, *Clarendon Press*, Oxford; W. Prager, P.G.Hodge, **Theory of Perfectly Plastic Solids**, 1951, *Wiley & Sons*, New York.
13 French engineer H. Tresca (1864) was the first who introduced this idea. The complete mathematical formulation of this concept was developed by A. Saint-Venant.
14 Sometimes this approach is also called the Mohr criterion.
15 This criterion was first proposed by Huber (1904) and became popular due to publications of R. von Mises (1913).
16 G.P. Andrianova, V.A. Kargin, *Vysokomol. Soedin.* (Polymers - in Russian), **12A**, 3 (1970).
17 S.S. Sternstein, L. Ongchin, *ACS Polymer Preprints*, **10**, 1117 (1969).
18 R. Raghava, R.M. Caddel, G.S. Yeh, *J. Mater. Sci.*, **8**, 225 (1973).
19 L. Mullins, *Rubber Chem. Technol.*, **42**, 339 (1969).
20 V.T. O'Brien, M.E.Mackley, *J. Rheol.*, **46**, 557 (2002).
21 Y.S. Lee, N.J. Wagner, *Rheol. Acta*, **42**, 199 (2003).
22 F. Bagusat, B. Böhme, P. Schiller, H.-J. Mögel, *Rheol. Acta*, **44**, 313 (2005).
23 Q. Xi, A. Singh, H.M. Jaeger, *J. Rheol.*, **64**, 321 (2020).
24 R.G. Egres, N.J. Wagner, *J. Rheol.*, **49**, 719 (2005).

25   A. Singh, R. Mari,M.M. Denn, J.F. Morris, *J. Rheol.*, **62**, 457 (2018).
26   V. Rathee, D. L. Blair, J. S. Urbach, *J. Rheol.*, **64**, 299 (2020).
27   A.Ya. Malkin, A.V. Mityukov, S. V. Kotomin, A.A. Shabeko, V.G. Kulichikhin, *J. Rheol.*, **64**, 469 (2020).
28   Ya. Malkin, V.G. Kulichikhin, A.V.Mityukov, S.V. Kotomin, *Polymers*, **12**, 1038 (2020).

QUESTIONS FOR CHAPTER 4

QUESTION 4-1

Values of Young's modulus, E, and the bulk modulus of compressibility, B, are measured. Find shear modulus for a Hookean solid.

   Additional question

Show that for an incompressible material the last equation transfers to the relationship between extensional and shear modulus known for rubber-like materials.

QUESTION 4-2

A bar is placed between two rigid walls. Its temperature is 20°C. Then the bar is heated to 200°C. What are the stresses that appear in the bar?

   Additional question

Using standard values of parameters for steel, estimate the level of stresses. Ordinary values of the parameters of material are: $\alpha = 1.2*10^{-5}\,K^{-1}$, $E = 2.1*10^5$ MPa. It was assumed that $\Delta T = 180K$. Then direct calculation gives $\sigma \approx -450$ MPa.

QUESTION 4-3

Analyze the stress field in torsion of a cylindrical shaft caused by torque, T. This occurs in transmitting torque in a gearbox.

QUESTION 4-4

Compare the stress state in torsion of a solid cylindrical bar of radius R_o and a tube with the same outer radius and the inner radius equal R_i. What is the increase of the maximum shear stress produced by a decrease in cross-section of bar caused by changing solid cross-section to a tube?

QUESTION 4-5

Calculate the principal stresses and maximum shear stress, if torque, T, and the stretching force, F, act simultaneously on the shaft of radius R.

   Additional question

Are these results valid for shafts made out of rubber?

QUESTION 4-6

A shaft is twisted with a torque T, as in Question 4-3. However, the torque is high enough to produce stresses exceeding the yield stress, σ_Y, of material. Describe the stress situation along the shaft radius.

   Additional question

What will be the deformations after unloading the shaft?

QUESTION 4-7

Prove that at small deformation, Hooke's law is the limit of the rubber elasticity equation.

QUESTION 4-8

A rubber-like strip is stretched by the applied force F = 0.2N. The area of the cross-section of the strip is S = 1 mm². The elastic modulus, E, was measured at small deformations and it equals $3*10^5$ Pa. What is the elongation of the strip? What would be the estimated elongation if one would use Hooke's law for calculations?

Additional question 1

Why does the coefficient 1/3 appear in this equation?

Additional question 2

Why was Eq. 4.4.17 used for calculations but not Eq. 4.4.16?

QUESTION 4-9

According to Hooke's law, the use of compression instead of extension leads to the symmetrical change of normal stresses. Is it the same for a rubbery material with rheological properties characterized by Eq.4.4.17?

Additional question

Can the last result be treated as proof of anisotropy of material, i.e., the existence of different values of elastic modulus in extension and compression, as is known for some other engineering materials, for example, concrete?

QUESTION 4-10

How are time effects taken into account in the formulation of the constitutive equation for large deformations, e.g., Eqs. 4.4.7 or 4.4.20?

QUESTION 4-11

A cylindrical rod of radius R was studied in uniaxial extension. It was found that it can work below the critical force F*. Then, this rod was used as a shaft working at torsion deformation mode. What is the limiting value of the torque, T*, that can be applied to the shaft?

Answers can be found in a special section entitled Solutions.

RHEOMETRY
EXPERIMENTAL METHODS

5.1 INTRODUCTION – CLASSIFICATION OF EXPERIMENTAL METHODS

This part of the book is devoted to the selection and evaluation of modern experimental methods of rheology. The main attention is given to methods used in studies of the rheological properties of liquids.

To measure rheological properties, the numerical values of constants that are included in rheological equations of the state of various materials have to be found. Depending on the choice of the equation, experiments are carried out in order to establish the influence of these constants on results. However, irrelevant of the choice of a model of rheological behavior, the measurement of two fundamental characteristic functions of material always plays a central role – viscosity as a function of a shear rate, and viscoelastic properties, such as dynamic moduli, as a function of frequency.

Experimental methods of measurement of rheological properties are defined by the general term *rheometry*, while, a more narrowly defined term, *viscometry*, is typically used in measurements of viscosity.

Two approaches are possible to determine the rheological properties of materials – *absolute and relative* measurements. Both approaches are widely utilized in modern research and technological practice, with each playing its own role.

Absolute methods of viscosity measurement are based on direct utilization of the main equation, Eq. 3.1.1, that defines the concept of viscosity. In this case, the shear stress and shear rate are measured and viscosity is calculated as a ratio of these quantities. Both values are termed as *local* values, i.e., they are referred to as some point in space occupied by liquid. Thus, in determining the shear stress and shear rate, the solutions of the problem of hydrodynamics are utilized to provide a relationship between measured macroparameters and dynamic (stress) and kinematic (shear rate) characteristics of the stream at a point. In the experiment, force, pressure, torque, etc., may serve as values determining the dynamic macroparameters. A linear or angular velocity plays a role of kinematic macroparameters. Thus, the main goal of absolute methods of viscosity measurement is to establish a relationship between pairs of values dependent on flow geometry defined by a design of a measuring device:

- *force (torque) – stress*
- *flow velocity – strain rate.*

Relative methods of viscosity measurement are based on the comparison of properties of fluid under investigation with a model fluid of known properties. Sometimes it is sufficient to find some characteristics of one fluid in comparison with another that is considered to be standard even if its absolute viscosity values are unknown. Relative characteristics of viscosity can be assumed, such as the time required for fluid to empty a vessel through a nozzle (similar to hourglass). In this experiment, the main focus is on keeping the same flow conditions and vessel dimensions. In this case, the term *calibrated* (strictly reproducible) dimensions are used. In contrast to the absolute method of viscosity measurement, relative viscosity characteristics are determined, for example, flow time.

Classification of methods of viscosity measurement, absolute as well as relative, is based on the *geometry of flow*.

Three main cases of flow are possible:
- flow of fluid between solid surfaces or through a hole in a solid body
- flow of fluid around a solid body
- free stream flow, relevant only to the extension of a fluid stream.

Fluid flow between two solid surfaces can be realized in the following geometries:
- fluid flow through a capillary with a cross-section of capillary being usually, but not necessary, circular
- rotational flow in which fluid is subjected to a circular motion in a gap between rotating cylinders, in a gap created by cone and plate or two conical surfaces, in a gap between two spherical surfaces, or other combinations of circular bodies
- shear flow of fluid between two parallel plates
- squeezing flow of fluid layer between two parallel plates approaching each other
- indentation of a solid body into the material.

Viscosity measurement in fluid flowing around solid bodies is usually carried out according to the following schemes:
- flow around a spherical or other surface moving in fluid with its resistance to flow depending on fluid viscosity. The space occupied by fluid may be restricted by solid walls or be infinite
- indentation of a solid body (*indentor*) into the fluid layer with shapes of the indentor being different – conical, spherical, cylindrical, etc.

Experimental methods also differ because of kinematics or dynamics of deformation, namely: force or velocity may be maintained constant or varied according to a given protocol; in modern instruments variations and/or combinations thereof are frequently used

Properties of the final product may be defined and the technological process may be controlled either by removing samples from a process and studying them in the laboratory or by using in-line measurement techniques during processing. The design and construction of the utilized instruments may vary accordingly, but the principles of measurements remain similar.

As mentioned in Chapter 3, the shear flow of much liquid media is accompanied by the storage of elastic (recoverable) deformations and the development of normal stresses. Measurement of the normal stresses at various shear rates is a separate problem of rheometry that is important for the selection and evaluation of the adequacy of rheological models.

Measurement of viscoelastic characteristics of materials represents observation involving transient regimes of deformation (see Chapter 2). When one of its characteristics (kinematic or dynamic) is kept constant, the time dependence of other characteristics can be measured.

Thus measurement of viscoelastic characteristics is usually carried out in the following regimes:
- constant stress is imposed and variations of strain with time are measured, i.e., *creep* is studied at various stresses
- a constant strain is imposed and variations of stress with time are measured, i.e., a *relaxation* of stress at various deformations is studied.

A special place among the rheological measurements is occupied by periodic deformations when the frequency of oscillations of strain (or stress) is given and changes in stress (or strain) response are measured. This most important method in rheology is called the *dynamic or vibrational* method and is widely utilized for the study of viscoelastic properties of materials as well as viscosity.

Similar to viscosity, the measurement of viscoelastic characteristics of materials can be performed according to different geometrical schemes. Thus, it is not necessary to include a separate classification of methods of measurement of viscoelastic characteristics according to their geometrical schemes of deformation. In addition, many modern instruments permit a combination of viscometric measurements determining the relationship between the shear stress and the shear rate, as well as dynamic testing and/or creep and relaxation of stresses.

Rheological measurements can be combined with various physical methods that are especially important for the study of structural transformations caused by deformation. Here it is important to utilize optical methods in various ranges of frequency. These methods permit direct observation of characteristics of fluid flow by tracing particles in a stream to measure velocity fields. Of particular interest is a measurement of *double refraction* or *birefringence* during flow, since dynamic anisotropy of optical properties is directly related to the material stress state. The use of other physical methods such as, for example, X-ray analysis, neutron scattering, calorimetry, and others along with rheometry is also of special interest.

5.2 CAPILLARY VISCOMETRY

5.2.1 BASIC THEORY

Capillary viscometry is the oldest and most widely used method of qualitative estimation and viscosity measurement. Its ubiquity is due to the obviousness of experiment, simplicity of experimental units, relatively inexpensive, and its easy to standardize test procedure.

The essence of the method consists in measuring the resistance to the flow of liquid through a calibrated channel. The central task of capillary viscometry is establishing the correspondence between volumetric flux (output), Q, and pressure drop in capillary, Δp, which induces flow. The pressure at the entrance is usually much higher than at the exit. Therefore, it is possible to replace the pressure drop, Δp, with the pressure at the capillary entrance, P.

The term "capillary" usually means any tube (channel) with arbitrary length and cross-section, though, as a general rule, cylindrical tubes (capillaries) with large length-to-radius, L/R, (or diameter, D) ratio are used.

The basic theory of capillary viscometry uses the following assumptions:

- the Newton-Stokes law (proportionality of stresses and shear stresses) is valid at any point of a stream
- velocity is not very high and flow is laminar
- flow along the main part of a capillary is steady and rearranging the velocity profile along the length does not take place; this assumption is not valid near the ends of a capillary: at the entrance and near the exit, the radial component of velocity appears due to velocity rearrangements
- circular and radial fluxes are absent; it might be not true if channels with non-round cross-section are used; in the latter case, circular flux appears
- flow is isothermal in the whole volume of liquid[1]
- a capillary radius is constant along the full length and in all cases L/R >> 1; if so the flux can be considered as steady along the dominating part of the length
- velocity at wall equals zero ("*hypothesis of stick*"); this is the reason for velocity distribution along the radius of a channel; this assumption may not be valid in some cases, and, if so, the case requires special treatment.

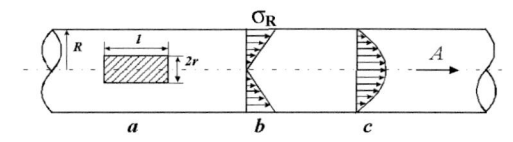

Figure 5.2.1. Basic scheme for viscosity calculation during liquid flow through capillary. Direction of flow is shown by arrow A. a: liquid element separated in stream; b: radial profile of shear stresses with σ_R being the shear stress at wall; c: radial velocity profile.

Quantitative analysis of flow through a channel is based on the formulation of equilibrium conditions for an axisymmetric, cylindrical element shown in Fig. 5.2.1a. The following forces act on this element: pressure drop, dp, along the length, dl, shear stresses, σ, applied on the surface $2\pi r dl$. The balance of forces for a stationary flow (i.e., in absence of acceleration) is written as

$$dp(\pi r^2) = 2\pi\sigma dl \qquad [5.2.1]$$

Then, the following expression for the distribution of shear stresses along the radius is obtained:

$$\sigma(r) = \frac{1}{2}\frac{dp}{dl}r \qquad [5.2.2]$$

In the stationary flow, dp/dl does not depend on the longitudinal coordinate. Therefore,

$$\frac{dp}{dl} = \frac{\Delta p}{L} = \frac{P}{L}$$

and, finally, the expression for the radial distribution of shear stress takes a form:

$$\sigma(r) = \frac{Pr}{2L} \qquad [5.2.3]$$

This expression is conveniently represented by shear stress, σ_R, which acts on a capillary wall, and it is calculated from Eq. 5.2.3:

$$\sigma_R = \sigma(R) = \frac{PR}{2L} \qquad [5.2.4]$$

Then the final formula for $\sigma(r)$ takes the following form:

$$\sigma(r) = \sigma_R \frac{r}{R} = \frac{PR}{2L} \frac{r}{R} \qquad [5.2.5]$$

Thus, the radial distribution of shear stress during flow in a capillary of a circular cross-section is always linear. This distribution is shown in Fig. 5.2.1b. It is significant that this result is not connected with the Newton-Stokes hypothesis, but directly ensues from the common formulation of equilibrium condition, i.e., Eq. 5.2.5 is valid for any liquid with arbitrary rheological properties, including Newtonian liquids.

Shear stresses are applied in the direction opposite to the flow direction, i.e., to arrow A in Fig. 5.2.1. Therefore (if the longitudinal axis is oriented along with the arrow A), the stresses should be assigned a negative sign.

It is now necessary to examine the kinematics of flow, i.e., to determine the distribution of velocities along a radius of the capillary, u(r), and to calculate shear rate:

$$\dot{\gamma} = \frac{\partial u}{\partial r} = \frac{du}{dr}$$

For Newtonian liquid, the calculation of velocity distribution is based on the Newton-Stokes equation, which is written as:

$$\dot{\gamma} = \frac{du}{dr} = \frac{1}{\eta}\sigma$$

Hence, the radial distribution of shear rate is easily found:

$$\dot{\gamma} = \frac{du}{dr} = \frac{1}{\eta}\frac{PR}{2L}\frac{r}{R} \qquad [5.2.6]$$

The distribution of deformation rates along the radius, as well as stresses, are linear.

Hence, taking into account boundary conditions ($u = 0$ at $r = R$), the following expression for the distribution of velocity along the capillary radius is obtained:

$$u(r) = \frac{1}{\eta}\frac{PR^2}{4L}\left[1 - \left(\frac{r}{R}\right)^2\right] \qquad [5.2.7]$$

Thus, the radial distribution of velocity during the flow of Newtonian liquid through a channel of a circular cross-section is expressed by a parabolic function (equation of the second-order).

Velocity is at maximum on the axis of the channel, i.e., at $r = 0$. This maximum velocity, V_{max}, is expressed as

$$V_{max} = u(R) = \frac{PR^2}{4\eta L} \qquad [5.2.8]$$

Then the radial distribution of velocities can be expressed through V_{max} in the following manner:

$$\frac{u(r)}{V_{max}} = 1 - \left(\frac{r}{R}\right)^2 \qquad [5.2.9]$$

The obtained radial distribution of axial velocity is shown in Fig. 5.2.1c. The formulas, obtained for the radial distribution of velocity, permit the calculation of the volumetric flow rate. For a Newtonian fluid, it is expressed as follows:

$$Q = \int_0^R 2\pi r u(r) dr = \frac{\pi P R^4}{8\eta L} \qquad [5.2.10]$$

Sometimes, instead of Q, it is convenient to use the average velocity, $\overline{V}$, which is expressed as

$$\overline{V} = \frac{Q}{\pi R^2}$$

i.e., the average velocity is the volumetric flow rate divided by the cross-sectional surface area of a channel.

Eq. 5.2.10 and following from it, the proportionality of the volumetric flow rate to the fourth power of the radius, is known as the *Hagen-Poiseuille law*.[2]

Thus, for Newtonian liquid, the volumetric flow rate is directly proportional to a pressure applied (the proportionality factor depends on the geometric dimensions of a channel) and inversely proportional to viscosity.

The method of determining the viscosity of a Newtonian liquid is based directly on Eq. 5.2.10. If the volumetric flow rate at the assigned pressure is measured, then the viscosity is calculated as

$$\eta = \frac{\pi P R^4}{8 Q L} \qquad [5.2.11]$$

Although Eq. 5.2.11 gives a completely obvious and single-valued method for enumerating the viscosity, it is an integral value, i.e., it does not directly use the basic determination of viscosity such as Eq. 3.1.1. This is not essential for Newtonian liquids, since the stresses and the shear rates at any point of flow are pre-determined, but it is important for liquids with arbitrary and *a priori* not specified rheological properties.

The convenient expression for $\dot{\gamma}_R$ is found directly from Eq. 5.2.10, which, after simple transformation, leads to the following formula:

$$\dot{\gamma}_R = \frac{4Q}{\pi R^3} = \frac{4\overline{V}}{R} \qquad [5.2.12]$$

This formula for the shear rate at a wall of the channel is valid only for the Newtonian liquid since it is derived from the Hagen-Poiseuille law also valid for Newtonian liquid. The integral definition of viscosity, i.e., Eq. 5.2.11, can be represented in a somewhat different form, namely:

$$\eta = K\frac{P}{Q} \qquad [5.2.13]$$

where $K = \pi R^4/8L$ is the shape factor (or form-factor), determined from known dimensions of the measuring device (capillary).

The last formula is easily generalized for capillaries of any cross-section. The exact analytical solutions are known for channels with simple geometrical forms. The values of shape factor for channels of an arbitrary cross-section can be calculated using modern computer technology with any desirable accuracy. An approximate method for calculating the flow of fluids in channels of noncircular cross-section, based on dimensional analysis, is also proposed.[3]

Another empirical method for determining the shape factor for the channel with an arbitrary cross-section is based on an experiment: Q and P are measured for liquid of known viscosity. Then, from Eq. 5.2.13, the geometric constant is calculated in an obvious manner.

The theory of measurements of the apparent viscosity of non-Newtonian liquids is based on the same prerequisites, which were formulated for Newtonian liquids. Here one fundamental exception is made that at any point of the flow field, the Newton-Stokes law is not satisfied, i.e., the assumption about the proportionality of shear stress to the shear rate does not hold.

If the experimentally observed dependence Q(P) is nonlinear, then liquid is non-Newtonian, and determination of apparent viscosity must be based on a general definition of the apparent viscosity, according to Eq. 3.1.1.

Eqs. 5.2.3-5.2.5 for shear stresses remain valid for liquids with any rheological properties. But the expressions for shear rates obtained above are not applicable to non-Newtonian liquids, since the linear relationship between shear rate and shear stress is not fulfilled.

The general solution of calculation of apparent viscosity of non-Newtonian liquid according to a method of capillary viscometry is based on enumeration of shear rate at one specific point, namely at a wall of the channel. The shear rate is a certain, *a priori* unknown, the function of shear stress:

$$\dot{\gamma} = f(\sigma)$$

Volumetric flow rate for liquid of any type is calculated as

$$Q = \int_0^R 2\pi r u(r)dr \qquad [5.2.14]$$

the integration of this expression leads to the following formula:

$$Q = \pi\int_0^R r^2 \left|\frac{du}{dr}\right| dr$$

In the last expression, the variable of integration, r, is substituted with Eq. 5.2.5 for σ. After performing corresponding operations, the following formula is obtained:

$$\dot{\gamma} = \sigma_R^{-3} \int_0^{\sigma_R} \sigma^2 f(\sigma) d\sigma \qquad\qquad [5.2.15]$$

where $\dot{\gamma} = Q/(\pi R^3)$ is the average shear rate.

The last expression is differentiated with respect to σ_R resulting in

$$\frac{d\dot{\gamma}_0}{d\sigma_R} = -\frac{3}{\sigma_R^4} \int_0^{\sigma_R} \sigma^2 f(\sigma) d\sigma + \frac{1}{\sigma_R^3} [\sigma_R^2 f(\sigma_R)] \qquad\qquad [5.2.16]$$

The function $f(\sigma_R)$ is the shear rate at the wall of the channel, i.e. $\dot{\gamma}_R$. Then, the comparison of Eqs. 5.2.15 and 5.2.16 makes it possible to obtain the following relationship:

$$\dot{\gamma}_R = 3\dot{\gamma}_0 + \sigma_R \frac{d\dot{\gamma}_0}{d\sigma_R} \qquad\qquad [5.2.17]$$

or equivalent to the latter, but sometimes the following expression is more convenient in applications

$$\dot{\gamma}_R = \dot{\gamma}_0 \left[3 + \frac{d\log\dot{\gamma}_0}{d\log\sigma_R} \right] \qquad\qquad [5.2.18]$$

Eqs. 5.2.17 and 5.2.18 are known as the *Rabinovitsch-Weissenberg equation*.[4] These formulas solve the stated problem, since they allow one to calculate the shear rate at a wall of the channel, $\dot{\gamma}_R$.

Instead of the average shear rate, $\dot{\gamma}_0 = Q/\pi R^3$, sometimes a quasi-Newtonian shear rate, $\dot{\gamma}_N = 4\dot{\gamma}_0$, i. e., $\dot{\gamma}_N = (4Q)/(\pi R^3)$ is used. However, in the general case, $\dot{\gamma}_N$ is not equal to the shear rate at wall $\dot{\gamma}_R$; $\dot{\gamma}_N = \dot{\gamma}_R$ only for Newtonian liquids.

The present method of calculation of flow curves applies to measurements using a circular capillary. In the practice of capillary viscometry, the slit capillaries, having thickness $H = 2h$ and width $B \gg h$, are also used.

The shear stress at the wall of the slit capillary, σ_H, and the corresponding shear rate, $\dot{\gamma}_H$, are calculated as:

$$\sigma_H = \frac{PH}{2L} \qquad\qquad [5.2.19]$$

and

$$\dot{\gamma}_H = 4\dot{\gamma}_0^H + 2\sigma_H \frac{d\dot{\gamma}_0^H}{d\sigma_H} = \dot{\gamma}_0^H \left(4 + 2\frac{d\log\dot{\gamma}_0^H}{d\log\sigma_H} \right) \qquad\qquad [5.2.20]$$

where the average shear rate $\dot{\gamma}_0^H$ during the flow in the slit capillary is calculated as $\dot{\gamma}_0^H = q/(BH^2)$, where B is the width of a slit channel.

If results of measurement are processed correctly, the dependence $\sigma(\dot{\gamma})$ that characterizes viscous properties of liquid being investigated must be independent of dimensions and geometric form of the capillary.

Thus, the dependence $\dot{\gamma}_N(\sigma_R)$ is the initial result obtained from experimental data. Further treatment of experimental data depends on the selection of a rheological model, i.e., function $\dot{\gamma}(\sigma)$. If the model is unknown *a priori*, then it is necessary to use Eq. 5.2.18. If the model is known, then the task is reduced to the determination of a limited number of constants, entering the appropriate equation. This can be done by a computer program suitable for the determination of constants of function by minimization of error − deviation of the calculated function from experimental data.

Such computation method of determining a flow curve is sometimes examined within the framework of s general approach to the solution of incorrectly posed inverse problems. The special feature of such tasks is that an unknown function is under the integral, while the measured function is related to it by means of an integral equation. It is significant that even a small (inevitable) measurement error can bring a large error in the calculation of the unknown function. In connection with capillary viscometry, the analysis of this task is based on somewhat modified Eq. 5.2.15, which is written in the following form:

$$Q = \frac{\pi \sigma_R}{R} \int_0^R \frac{r^3}{\eta(\dot{\gamma})} dr \qquad [5.2.21]$$

Here the function $Q(\sigma_R)$ is measured, and the function $\eta(\dot{\gamma})$ is to be determined. For this purpose, the function $\eta(\dot{\gamma})$ is represented by a certain analytical expression with a small number of constants to be determined. Minimizing the mean value of deviation of the calculated dependence from experimental data, the constants are determined and also the flow curve of material under study is known.[5-7]

5.2.2 CORRECTIONS

In reality, the measurements of viscosity require a number of corrections, which are intended to account for deviations of specific conditions of the experiment from idealized requirements, which were formulated at the beginning of Section 5.2.1. Even in measurements of the viscosity of the Newtonian liquid, deviations from linearity of the dependence of volumetric flow rate on imposed pressure can be observed. These deviations are caused by factors, which lead to the introduction of *corrections*. Corrections are of general importance in the practice of capillary viscometry.

5.2.2.1 Kinetic correction

The liquid being investigated typically enters a capillary from a large reservoir. Then, the flow velocity of the stream substantially accelerates as a result of a change in cross-section, i.e., the kinetic energy of flow increases. This change requires the additional expenditure of energy, which looks like a growth of viscosity since viscosity is a measure of expenditure of energy required to create flow.

If the total measured pressure drop is P, then the part of this pressure, P_k, is spent on an increase of the kinetic energy of the stream, and only the remaining part, P_v, is responsible for overcoming the resistance of flow through a capillary, i.e., for the measured viscosity.

Thus

$$P_v = P - P_k \qquad\qquad [5.2.22]$$

where the value of P_k is responsible for the *kinetic correction*.

The value of P_k can be calculated as follows:

$$P_k = \frac{\rho \overline{V}^2}{\alpha} = \frac{\rho Q^2}{\alpha \pi^2 R^4} \qquad\qquad [5.2.23]$$

where, as earlier, $\overline{V}$ is the mean (output-based) velocity, ρ is density, and α is the coefficient reflecting the influence of the velocity distributions on the value of kinetic correction. For Newtonian liquid with the parabolic velocity profile in the capillary, it is considered that $\alpha = 1$, although different theoretical estimates give the value of α in the range from 0.74 to 2.0.

The correction taking into account a change in the kinetic energy leads to the following expression for Newtonian viscosity

$$\eta = \frac{\pi P R^4}{8 Q L} - \frac{\rho Q}{8 \pi L \alpha} \qquad\qquad [5.2.24]$$

which is a modification of Eq. 5.2.11.

If we accept the standard value, $\alpha = 1$, then the introduction of kinetic correction leads to the measurement error, which can reach even 10%. For non-Newtonian liquids, it is difficult to determine the value of α *a priori*, but it is also of the order of 1.

The calculation of kinetic correction is important during measurements of viscosity of low-viscosity liquids, for example, dilute polymer solutions, where a high accuracy of measurements is required.

5.2.2.2 Entrance correction

An important role in the theory of capillary viscometry is played by the so-called *entrance corrections*, which combine different dynamic phenomena at the entrance to a capillary as a result of rearrangement of the inlet velocity profiles.

If we neglect the entrance corrections of different origins, then the results of measurements become dependent on the length of the capillary, due to the fact that the relative contribution of transient phenomena becomes greater when a shorter capillary is utilized for measurements. This effect can be seen in Fig. 5.2.2, where

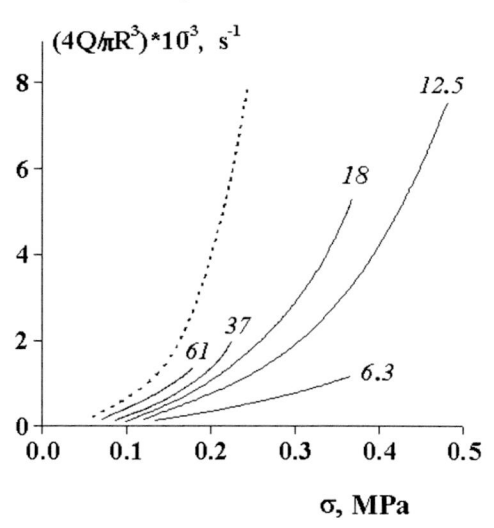

Figure 5.2.2. Dependence of quasi-Newtonian shear rate at a wall, $4Q/\pi R^3$, on the shear stress at a wall – role of L/D ratio. The values of L/D are shown on curves. Dashed line is a dependence of $4Q/\pi R^3$ on shear stress extrapolated to a capillary of infinite length. 60% solution of high molecular weight polyisobutylene in toluene. [Adapted, by permission, from E. Brenschade, J. Klein, *Rheol. Acta*, **9**, 130 (1970)]

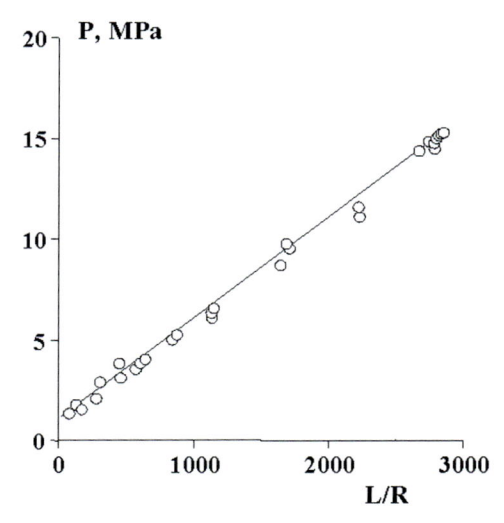

Figure 5.2.3. Dependence of pressure required to impose flow through capillary of different lengths at a constant value of quasi-Newtonian shear rate $\dot{\gamma}_N = 7 \times 10^4$ s^{-1}. Experimental data for 60% solution of high molecular weight polyisobutylene in toluene. [Adapted, by permission, from E. Brenschade, J. Klein, *Rheol. Acta*, **9**, 130 (1970)].

the results of measurements of dependencies of $\dot{\gamma}_N = (4Q)/(\pi R^3)$ on the shear stress at the wall of the capillary are shown. Even for sufficiently long capillaries with L/D = 61, the results of measurements are not invariant relative to the capillary length.

The same figure shows dependence on shear stress determined by extrapolation for an infinitely long capillary (dashed curve). Only in this case is the desired result achieved – the obtained dependence precisely reflects the viscous properties of the liquid.

For low-viscosity, inelastic liquids the so-called *Couette correction* plays the dominant role in the transient phenomena. It reflects effects that do not yield direct evaluation, such as additional expenditures of energy for rearrangement of the velocity profile, the formation of the entry cone by the liquid in a reservoir prior to capillary entrance, maintenance of vortices/eddies (secondary flows) around the entry cone, etc.

The influence of entrance effects becomes noticeable with the values of the Reynolds number, at least, the order of several tens, i.e., the corresponding effects become relatively significant for low-viscosity liquids.

In practice, a general method is developed for the introduction of entrance correction as a measure of the effective length of a capillary.[8] Let all additional losses of energy (with respect to losses caused by the flow of liquid through capillary) be determined by a certain fictitious additional pressure drop, P_k. Then, it is possible to imagine a certain fictitious capillary with the same radius, as utilized for measuring viscosity, but having a larger length. This additional length is selected in such a way that at this length the pressure drop, P_k, would occur exactly. This additional length of the capillary is expressed through a radius, R, as $m_k R$, where m_k is called the *entrance correction factor*.

Fig. 5.2.3 shows the experimental example of the dependence of pressure at the entrance into capillary, on its relative length, L/R. According to the basic theory of capillary viscometry (see Eq. 5.2.4), the dependence of pressure, P, on L/R at any constant stress, σ_R, must be linear and start from the origin of coordinates, i.e., for a capillary of zero length, pressure is zero. This does not happen and even at the zero-length of a capillary, the pressure is non-zero, however, as a result of entrance correction.

Data of such type, as shown in Fig. 5.2.3, can be represented in the form:

$$P - P_k = \frac{2\sigma_R L}{R} \qquad [5.2.25]$$

where P_k is the pressure loss that is responsible for the entrance correction (i.e., the value of the intercept of the straight line along the Y-axis in Fig. 5.2.3).

It should be noted that the entry pressure losses, ΔP_e, (defined as P_k in Eq. 5.2.25) measured by means of capillary viscometry for polymer melts are also important for various processing operations, including flow in spinnerets during fiber spinning, flow in the dies during extrusion, flow in the molds during injection molding, etc. Therefore, for application purposes, it would be useful to find empirical correlations between these losses and the structural, rheological and processing characteristics of polymers. For this reason, attempts were made to develop master curves that would provide such a correlation. For a given class of polymers, it has been found that the shear rate dependence of ΔP_e can be correlated by using the parameter $\eta_0 \dot{\gamma} / \beta$, where $\beta = M_w / M_n$ is the index of polydispersity.[9] These master curves are shown in Figure 5.2.4 for two homopolymers, polystyrenes (left) and HDPEs (right). In particular, the master curve for two polystyrenes from different manufacturers were tested at different temperatures with results shown in Figure 5.2.4 (left). It shows that the approach works reasonably well, since all the experimentally measured entry pressure losses are scattered around a unique master curve. Moreover, Figure 5.2.4 (right) shows the similar master curve for six HDPEs of different molecular weights at a temperature of 190°C. The original data used here to plot this master curve for various HDPEs were taken from earlier study.[10] Again, the proposed correlation is validated, since all the measured data collapse onto a single curve. This suggests that for any class of polymers there exists a unique curve independent of the molecular weight and the index of polydispersity. However, a further work is needed to obtain such correlations for other classes of polymers.

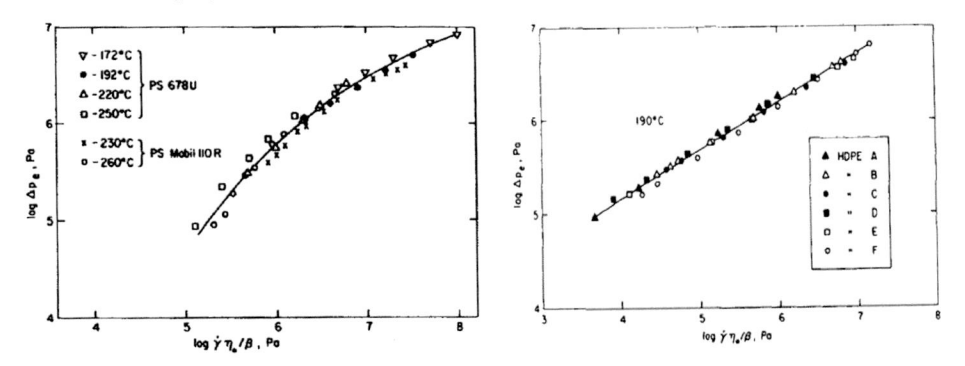

Figure 5.2.4. Master curves for two polystyrene melts at different temperatures (left) and for six HDPE melts at 190°C (right). [Adapted, by permission, from A. I. Isayev, B. Chung, *Polym. Eng. Sci.*, **25**, 264 (1985)].

Introducing m_k, the calculation of the shear stress at the wall of a capillary is performed as follows:

$$\sigma_R = \frac{PR}{2(L + m_k R)} \qquad [5.2.26]$$

with m_k varying with the flow rate or pressure.

For Newtonian liquids $m_k \approx 1$. The value can be much greater. In the practice of capillary viscometry a general, so-called differential, the method is used for correction of m_k.

The essence of this method is seen in Fig. 5.2.3: measurements on capillaries of different lengths are made and the results of measurements are extrapolated to the zero-length of the capillary, to determine P_k and thus m_k.

The use of a differential method and elimination of entrance corrections is required in the capillary viscometry of polymeric materials.

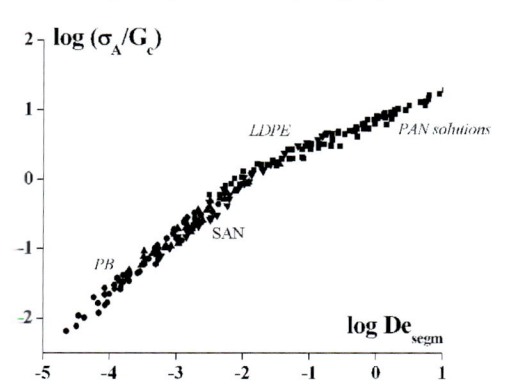

Figure 5.2.5. Master curve for four polymeric materials in dimensionless variables.

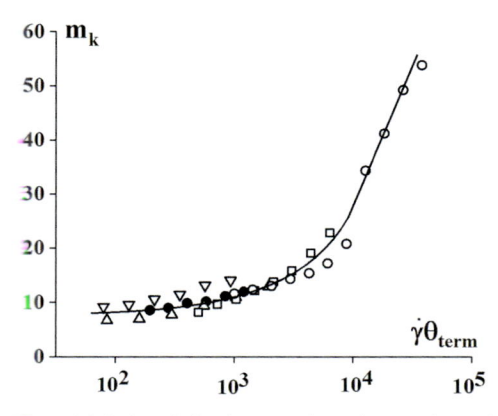

Figure 5.2.6. Correlation between the end correction and the Weissenberg Number for polyethylenes with different MWD. Various symbols correspond to different samples. [Adapted, by permission, from A.Ya. Malkin, V.G. Kulichikhin, I.V. Gumennyi, *Phys. Fluids*, **33**, 013105 (2021).]

The concept of end correction reflects the dependence of the experimental results on the relative length of a capillary. However, this approach is formal and does not carry clear physical meaning. It is reasonable to think that this effect is dependent on the elasticity of polymeric fluids. Then it is possible to develop an alternative method for calculating the resistance to the flow of viscoelastic polymeric fluids in capillaries.[11] The method is based on the assumption that pressure losses in short channels are determined by the average time of deformation in passing a fluid through a channel. Then it is possible to build a master curve (see Fig. 5.2.5) using dimensionless variables, which presents the dependence of the normalized shear stress σ_A on the Deborah Number for several polymers.

Here σ_A is the shear stress at the channel wall (see eq. 5.2.4) scaled by the elastic modulus G_c and the Deborah Number is the ratio of the characteristic relaxation time λ_c found from the frequency dependencies of the components of the dynamic modulus to the mean residence time T of material in a capillary: $De = \lambda_c/T$.

As said above, the physical nature of the end correction m_k is uncertain and, generally speaking, can be very different. However, the largest values of the end corrections are observed for colloidal and polymeric liquids. Their main rheological peculiarity is elasticity and therefore it is reasonable to presume that just viscoelastic properties of these matters determine large end corrections in the flow through capillaries. This point of view is supported by experimental data presented in Fig. 5.2.6 for set of different polyethylenes. Their viscoelasticity was varied due to the difference in molecular-weight distributions.

One can see that there is a direct correlation between the end correction and the Weissenberg Number that is determined by the product of the shear rate and the relaxation time. In this case the relaxation time corresponding to the terminal zone of the viscoelastic properties has been chosen. The similar results are obtained with using another characteristic relaxation time, e.g., the relaxation time corresponding to the crossover point on the frequency dependencies of the components of the complex modulus.

The physical connection of this result with the data presented in Fig. 5.2.4 is rather evident; in both cases the viscoelasticity of polymeric liquid plays the most important role.

5.2.2.3 Pressure losses in a reservoir of viscometer

The capillary viscometer is a combination of two cylindrical channels – a reservoir of a large radius, whence the material being investigated enters the measuring part of a capillary having a small radius. If the ratio of radii is very large, then losses of pressure during flow through the reservoir can be disregarded. However, this is not always correct, since if a constant pressure, P, is imposed in a reservoir, the pressure at the entrance of capillary, P_0, that is required for calculation of viscosity, increases as the reservoir is being emptied. This results in the growth of the volumetric flow rate with time during the duration of the experiment.

If a constant volumetric flow rate is imposed in an experiment, instead of constant pressure, then the pressure at the entrance into the measuring capillary decreases with time. This reduces pressure loss in the reservoir.

The above-indicated phenomena caused by flow in a reservoir prior to the entrance into the measuring capillary, indicate that a change in the apparent viscosity with time occurs. The reasons for these time-dependent effects, as explained above, are different.

It is possible to consider the pressure losses during the flow in the reservoir quantitatively if the rheological properties of the liquid being investigated are known. Then flow in the pair of cylinder-capillary channels is considered as flow in two sequential channels, and pressure losses as functions of shear rates are calculated for each of them.

5.2.2.4 Temperature correction

In the basic theory of viscometry, a flow is assumed to be isothermal, although it is well known that the viscous flow is accompanied by dissipation of energy leading to temperature increase. The rate of heat generation depends on shear rate, and as such, it is non-uniform along a radius. Furthermore, a part of the heat is transferred to the environment as a result of heat transfer. Therefore, temperature varies along both radius and length of the channel, which leads to appropriate changes in viscosity.

The temperature rise, ΔT, during the adiabatic flow of a Newtonian liquid with flow time, $\dot{t}$, is expressed as

$$\Delta T = \frac{\eta \dot{\gamma}^2 \dot{t}}{c\rho} \qquad [5.2.27]$$

where c is the heat capacity of liquid and ρ is its density. A convenient expression for calculating viscosity change caused by heat dissipation is obtained from the following form:[13]

$$\eta_0 = \frac{\pi R^4 P}{8QL}\left(1 + \frac{kQP}{16\pi\lambda L}\right) \qquad [5.2.28]$$

where λ is the heat transfer coefficient, and k is a coefficient of the temperature dependence of viscosity, which is expressed as

$$\eta = \eta_0 e^{k(T - T_0)} \qquad [5.2.29]$$

In this case, η_0 is the viscosity at temperature, T_0, imposed during the experiment, which in the course of experiment rises to temperature, T, causing viscosity reduction to a value of η.

The true value of viscosity at temperature, T_0, can be found using Eq. 5.2.28. For this purpose, the measurements should be conducted at several values of factor QP/L, entering this formula. Then, data are extrapolated to zero value of QP/L that corresponds to the absence of dissipative contribution. The values of viscosity, η_0, and the value of the coefficient, k, can be found.

5.2.2.5 Pressure correction

Viscosity measurement during the flow of liquid through a capillary may be affected by pressure. The higher the shear rate at which it is desirable to measure the apparent viscosity, the higher the imposed pressure. The pressure increase by itself may lead to viscosity increase, which sometimes leads to a very unique phenomenon. If a pressure necessary for viscosity measurement at high shear rate rises, it is expected that viscosity should decrease for shear-thinning liquids. Contrary to expectations, the measured viscosity increases. This occurs as a result of an increase in pressure, which prevails over viscosity decrease because shear rate also increases.

The dependence of viscosity on pressure is usually expressed by exponential function:

$$\eta = \eta^0 e^{bP} \qquad [5.2.30]$$

where η^0 is viscosity value P = 0, b is the baric (piezo) coefficient of viscosity.

The theoretical analysis, based on the solution of the dynamic equation of flow of Newtonian liquid, for which Eq. 5.2.30 is satisfied, gives the following expression for the volumetric flow rate:

$$Q = \frac{\pi R^4}{8L\eta^0 b}(1 - e^{-bP}) \qquad [5.2.31]$$

where P is the pressure at the entrance into a capillary.

At low pressures (where bP << 1), this formula degenerates into the standard Hagen-Poiseuille equation. However, at large pressures, due to the influence of this factor, the effect cannot be disregarded.

The influence of pressure on rheological properties is substantial for compressible (foamed) materials. However, no reliable methods of rheological studies of such materials have been developed.

5.2.2.6 Correction for a slip at a wall

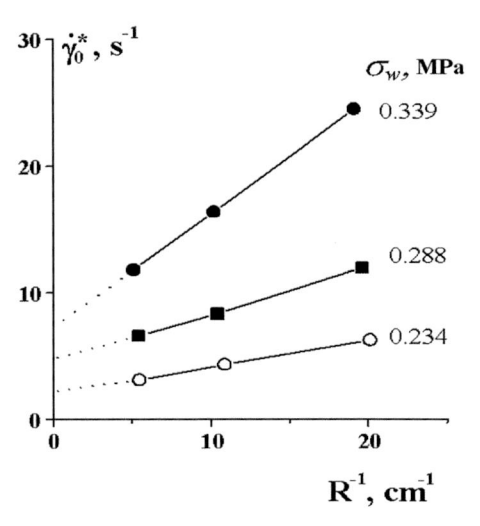

The hypothesis that the velocity of the liquid at a wall equals zero (*no-slip hypothesis*) is assumed in the analysis of liquid flow. There are situations when this hypothesis is not satisfied and slip occurs near a capillary wall. Usually, it is not essential whether the slip of liquid along the solid wall actually takes place or the liquid ruptures near a wall or a very thin layer of liquid appears on a wall, along which the remaining liquid slides.

Slip may not necessarily occur continuously, it may alternate with adhesion. This phenomenon, which is characteristic of the flow of melts and concentrated solutions at high shear stresses, is called the stick-slip phenomenon. In this case, the corresponding mechanism of motion in the channel proves to be unstable (see Section 3.6).

Figure 5.2.7. Determination of slip velocity according to the Mooney method of extrapolation to $R^{-1} = 0$. The example is given for butadiene rubber at 120°C.

The slip effect at a wall in viscometric measurements can be found by observing the dependence of results of viscosity measurements on capillary diameter. The procedure for calculation of slip velocity, V_s, and therefore viscosity in the shear flow based on these experimental results is called the *Mooney method*.[14]

Let the velocity of the liquid at $r = R$ be equal to V_s. Then the velocity profile is expressed as

$$u(r) = V_s + \int_0^r \dot{\gamma}(r)dr = V_s + \int_0^r f(\sigma)dr \qquad [5.2.32]$$

where, as for any liquid subjected to flow, $\dot{\gamma} = f(\sigma)$ is the flow curve.

It is possible, as usual in the theory of capillary viscometry, to introduce the value of an average shear rate $\dot{\gamma}_0^* = Q/(\pi R^3)$. However, in the case of flow with slip at a wall, the value of $\dot{\gamma}_0^*$ consists of two terms. The first term is the previously introduced value for flow without slip, $\dot{\gamma}_0$, and the second term is a contribution of slip at a wall and equal to V_s/R. Thus:

$$\dot{\gamma}_0^* = \dot{\gamma}_0 + \frac{V_s}{R} \qquad [5.2.33]$$

where the slip velocity at a wall, similar to the shear rate, can be a function of shear stress.

The true shear rate at a wall, $\dot{\gamma}_R$, (assuming that the thickness of the wall boundary layer, if this layer does exist, is very small) equals $\dot{\gamma}_0$. It is computed from the Rabinovitsch-Weissenberg formula, but not from $\dot{\gamma}_0^*$.

The method of determining the shear rate, $\dot{\gamma}_0$, and slip velocity directly follows from equation (5.2.33): the dependence of $\dot{\gamma}_0^*$ on R^{-1} should be plotted and extrapolated to $R^{-1} = 0$. An example of this plot is shown in Fig. 5.2.7.

The procedure assumes that the greater the radius of a capillary, the lesser the contribution of the slip effects such that it can be completely disregarded at $R^{-1} = 0$.

The phenomenon analogous to a slip at a wall can be observed during the flow of multicomponent materials. The diffusion of a low-viscosity component to a wall of a capillary may affect shearing to be preferentially accomplished with this low-viscosity wall boundary layer and change velocity gradient. The measured apparent viscosity is less than the mean viscosity in the volume of a multi-component material. A similar phenomenon can be observed during the flow of polydisperse polymers. As a result of diffusion of low-viscosity fractions to a channel surface, the enrichment of the wall boundary layer by low-viscosity fractions occurs leading to a decrease of apparent viscosity.

5.2.2.7 Adsorption on a channel surface

In the study of viscous properties of dilute polymer solutions, the effect of adsorption of macromolecules on the surface of a capillary was noted,[15] which is a reverse phenomenon to slippage at a wall. This phenomenon is especially noticeable during the use of capillaries of a very small radius. It leads to a decrease in surface area of effective flow, and, correspondingly, to an increase in apparent viscosity. Adsorption is strongly enhanced in studies of polymer solutions capable of interactions with the capillary surface.

Adsorption phenomena must be considered with capillaries having a diameter of up to several tens of microns (e.g., filtration through porous media or capillary flow of biological substances).

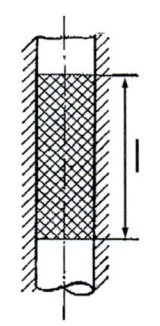

The effect of adsorption is usually expressed by a certain effective thickness of the adsorption layer on the surface of a capillary and is computed using volumetric flow rates, measured on capillaries of different radii.

5.2.3 FLOW IN INCOMPLETELY FILLED CAPILLARY

Observation of the boundary shift of a sample, which partially fills a capillary, is a unique version of the capillary method of viscosity measurement. Such experiments are carried out at low external pressures, or, more generally, in the absence of external pressure. The driving force in such an experiment may be the gravitational forces or the surface tension forces.

Figure 5.2.8. Partially filled capillary.

5.2.3.1 Motion under the action of gravitation forces

Let us consider that liquid is filled into a capillary up to a certain height and it is moving in the absence of external pressure under its own gravitational force, as shown in Fig. 5.2.8. The height of the sample, l, the density of the liquid, ρ, and viscosity, η, of liquid do not depend on shear rate. Then, the distribution of velocity along the radius is described by a parabolic law:

$$u(r) = \frac{\rho g R^2}{4\eta}\left[1 - \left(\frac{r}{R}\right)^2\right]$$

[5.2.34]

where g is the gravitational acceleration, and product ρg determines the stresses arising under the action of the weight of the column of liquid.

If we now measure the velocity of meniscus movement, U_0, (on the axis of the channel at $r = 0$), then viscosity is calculated from the formula:

$$\eta = \frac{\rho g R^2}{4 U_0} \qquad [5.2.35]$$

Velocity measurement of meniscus curving of an initially strictly cylindrical sample is a special case of the method under discussion. Viscosity is calculated from Eq. 5.2.35.

This procedure is especially convenient for viscosity measurements of high-viscosity materials at low shear rates and stresses. The range of shear rates in these measurements is 10^{-4} to 10^{-2} s^{-1}, and shear stresses are of an order of 10^2 Pa.

5.2.3.2 Motion caused by surface tension forces

Let the liquid in a capillary, placed at angle φ to the horizontal surface, rise under action of surface tension. Then, from the condition of force equilibrium, the following formula is obtained for calculating the length of capillary, l_0, filled with liquid:

$$l_0 = \frac{2\sigma\cos\beta}{\rho g R \sin\varphi} \qquad [5.2.36]$$

where σ is the coefficient of surface tension, β is the angle of contact formed by meniscus and surface of capillary, ρ is the liquid density.

The measurement of l_0 under equilibrium conditions makes it possible to determine $\sigma\cos\beta$, which defines the velocity of liquid in a capillary. The shear stress at a capillary wall is calculated as

$$\sigma_R = \frac{\sigma\cos\beta}{l} - \frac{R\rho g \sin\varphi}{2} \qquad [5.2.37]$$

where l is the variable length of a capillary filled with liquid under study.

Hence, viscosity is calculated from the formula:

$$\eta = \frac{R(2\sigma\cos\beta - Rl\rho g \sin\varphi)}{8l(dl/dt)} \qquad [5.2.38]$$

This procedure can be used for measurements of the viscosity of Newtonian liquids and yield stress of viscoplastic media.

5.2.4 LIMITS OF CAPILLARY VISCOMETRY

The capability of capillary viscometry is determined by fulfilling certain requirements that were formulated in Section 2.1.1. Therefore, the limitations of capillary viscometry are determined by the following effects:

- transition from laminar to turbulent flow conditions
- instability of flow as a result of fluid elasticity
- mechanical and thermal degradation of the test specimen
- strong thermal effects at the high deformation rates.

The upper boundary of shear rates in capillary viscometry is $\sim 10^6$ s^{-1}. In usual instruments, this limit is up to 10^4 s^{-1}. The upper range of 10^4–10^6 s^{-1} is only possible with special experimental techniques.

The lower boundary of shear rates is determined, to a considerable extent, by the patience of the experimenter, because the flow is very slow, and by some other considerations. First, the measurement of small movements of a fraction of a millimeter over a long time requires the use of a high-precision measurement technique, because of potential systematic instrument errors, caused by prolonged measurements. Second, the permissible duration of the measurement is limited by both the chemical and structural stability of the investigated material under the imposed experimental conditions. Therefore, as a general rule, during measurements by capillary viscometry, it is very difficult to go below shear rates of $\sim 10^{-2}$ s^{-1}.

The upper limit of shear stress is determined by the appearance of instability, excessive thermal effects, and mechanical and thermal degradation. The upper limit, in any case, does not exceed the value of 1 MPa.

The lower limit of shear stress is determined by the influence of parasitic resistance, which appears practically in any design of capillary viscometers and contributes to the error that is difficult to account for during measurements. In capillary viscometers, it is practically impossible to go below the shear stress of $\sim 10^2$ Pa.

The comparison of the above-indicated limitations on shear rates and shear stresses shows that capillary viscometers can measure viscosity in the range of 10^{-3} to 10^7 Pa*s. Different measurement techniques are used for different ranges of viscosity.

5.2.5 NON-VISCOMETRIC MEASUREMENTS USING CAPILLARY VISCOMETERS

In addition to viscosity, numerous attempts are made to use capillary instruments for evaluating other rheological properties of materials. These evaluations are based on general principles of rheology. In these cases, the estimates of various properties are obtained but not their absolute values.

The most obvious is the use of a shape of flow curve as characteristics related to the structural features or elastic properties of polymeric materials.

The elasticity of melts and solutions of polymers are evaluated according to two parameters – the capillary entrance correction and the extrudate swell at the exit from the capillary.

The problems concerning the determination of the entrance correction were previously discussed.

As far as the extrudate swell is concerned, evaluation of elasticity is based on the experimental fact that the diameter of extrudate at the capillary exit is larger than the diameter of the capillary. This effect is related to the elasticity of flowing material. The extrudate swell (coefficient of swelling), α, defined as the ratio of diameters of extrudate and capillary, serves as a measure of the elastic properties of the material.

The basic methodological difficulty in measuring the extrudate swell, α, is related to the fact that the measurements must be carried out under isothermal conditions on extrudate leaving a capillary since α depends on temperature. The relationship between α and other rheological properties, or characteristics of the molecular structure of the polymer, is established on the basis of corresponding rheological theories and concepts.

5.2.6 CAPILLARY VISCOMETERS

5.2.6.1 Classification of the basic types of instruments

From the general theory of capillary viscometry, it follows that for determining the viscosity, two parameters, pressure and volumetric flow rate, have to be measured. Typically, one of these parameters is *assigned*, and another is *measured*.

Pressure in the capillary tube viscometers is *imposed* by one of the following methods:

- by a load of specific weight (in the load viscometers)
- by pressure of compressed gas (in the gas viscometers)
- by a pre-compressed spring
- by a power drive, in which force with aid of the feedback control is maintained or is regulated according to a required program
- by the weight of liquid being investigated.

Imposition and regulation of *velocity* of piston stroke (or volumetric flow rate) are achieved with aid of a mechanical or hydraulic drive by a method of varying velocity.

The viscometers of the first group are usually simpler by construction and more frequently used for standardized measurements.

The viscometers of the second group are usually constructed on the basis of standard testing machines (used for measurements of mechanical characteristics of solids) since modern machines of such type are equipped with a high-precision drive with an adjustable speed of motion over a wide range.

Modern instruments usually contain measuring devices (sensors) with an output of the measured parameters acquired by a data acquisition system connected to a computer for automatic processing of measurement results using computer programs, supplied by the instrument manufacturer.

Below, the basic standard solutions will be examined, which have found wide applications in contemporary research practice.

5.2.6.2 Viscometers with the assigned load

5.2.6.2.1 Load viscometers

The most popular instrument of such type, utilized in many laboratories, is a *Melt Indexer*. This instrument measures the *melt flow index*, MFI, or the *melt flow rate*, MFR, of polymers. The MFI is the value of the volumetric flow during 10 min, measured under strictly standardized conditions at a specified temperature and load on an instrument with the specific dimensions of capillary and reservoir. According to the definition, MFI is expressed in g/10 min.

The schematic diagram of such an instrument is shown in Fig. 5.2.9. This instrument basically consists of a calibrated capillary with the following standardized dimensions: length, $L = 8.000 \pm 0.002$ mm; diameter, $D = 2.098 \pm 0.005$ mm.

According to ASTM D 1238, the length of the capillary must be 9.550 ± 0.007 mm.

The weight of the load, which creates pressure, is also fixed by appropriate standards. The basic weight is $2,160 \pm 10$ g. It is also permissible to use loads with weights of $5,000 \pm 10$ g, $10,000 \pm 15$ g, and $21,600 \pm 20$ g.

According to ASTM, different loads can be utilized, such that different combinations of temperature and load weight are possible, which determine measurement procedures.

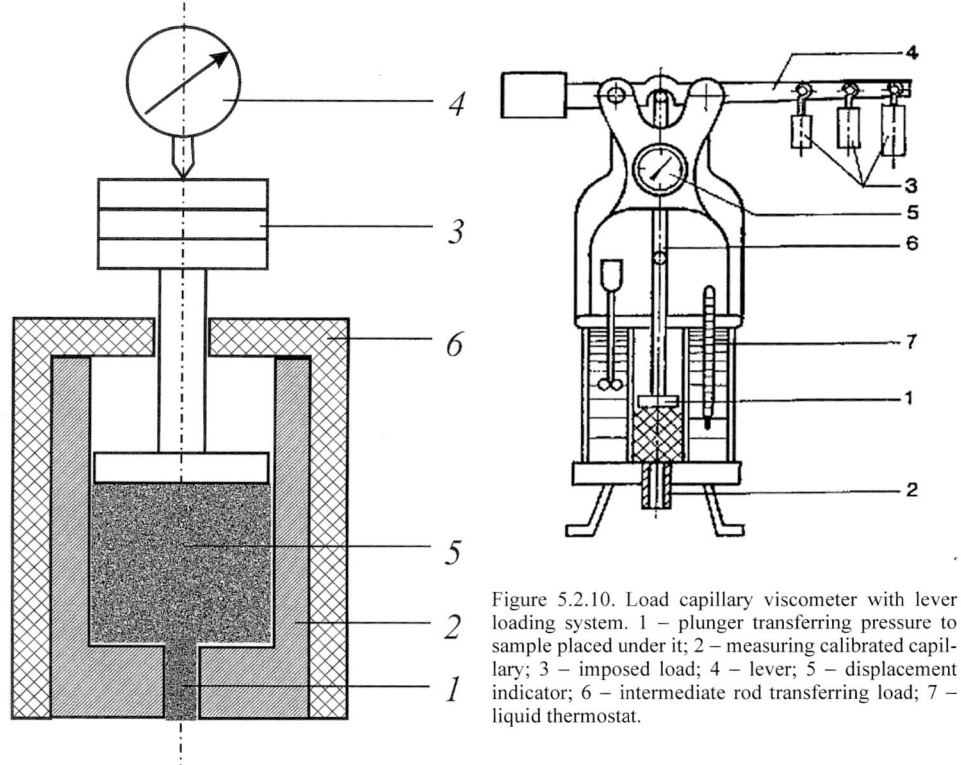

Figure 5.2.10. Load capillary viscometer with lever loading system. 1 – plunger transferring pressure to sample placed under it; 2 – measuring calibrated capillary; 3 – imposed load; 4 – lever; 5 – displacement indicator; 6 – intermediate rod transferring load; 7 – liquid thermostat.

Figure 5.2.9. Typical scheme of a load capillary viscometer. Instrument for measurement of melt flow index. 1 – measuring calibrated capillary; 2 – reservoir; 3 – load; 4 – displacement indicator; 5 – polymer melt; 6 – heating element and heating chamber.

Fig. 5.2.9 shows the simplest manual method of measurement of volumetric flow rate by measuring the speed of a lowering piston with the aid of a dial indicator and stopwatch. However, in the majority of modern instruments, the automated method of measurement of the piston speed with an aid of various electromagnetic sensors is used. The standardized instruments for the measurement of melt flow index are manufactured by a number of companies.

The pressure can simply be generated with the aid of a load as shown in Fig. 5.2.9, or using a lever system (Fig. 5.2.10). Here, the pressure is varied not only by a change of the load weight but also by varying the length of an arm of the lever, through which the load is transferred to the test specimen. This way, it is possible to somewhat widen the range of the imposed loads.

5.2.6.2.2 Gas viscometers

A tendency to increase a range of imposed loads, and, correspondingly, the shear stresses, led to an idea of using the pressure of compressed gas. This method may easily reach pressures of an order of 20 MPa (200 bar), and using a multiplier it is possible to increase this pressure to 60 MPa (600 bar). With such loads, maximum shear stresses suitable for conducting rheological measurements can be reached.

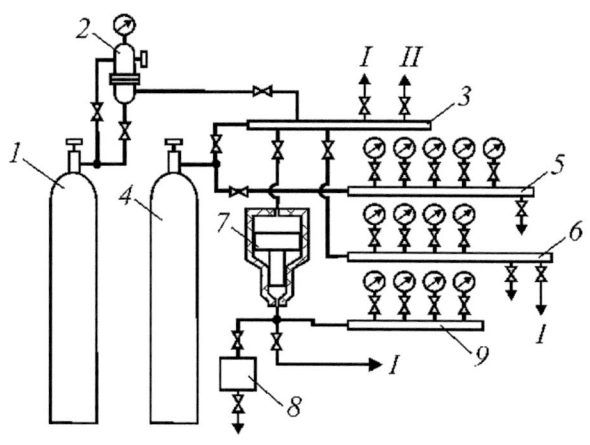

Figure 5.2.11. Scheme for imposing pressure in gas capillary viscometer. 1 – replaceable high pressure tank; 2 – pressure regulator; 3 – distributing manifold; 4 – intermediate tank maintaining imposed pressure; 5, 6, 9 – pressure gauge display including gauges of various ranges of pressure; 7 – hydraulic multiplier; 8 – liquid pump. I – line of pressure supply to reservoir of viscometer. II – line connected to vacuum system.

The system for generating and controlling the pressure in a typical gas viscometer is shown in Fig. 5.2.11. A standard pressurized tank of nitrogen gas is the primary source of pressure generation. The initial pressure is reduced and it is maintained at the required level using an intermediate tank of a large volume so that the gas flow during sample extrusion from the viscometer would not cause a noticeable pressure drop. The error of pressure measurement does not exceed 1%. The pressure ranges from 0.01 to 60 MPa (from 0.1 to 600 bar).

Gas viscometers are instruments typically used for research purposes. For the realization of their great possibilities, such viscometers are equipped with a series of capillaries of different lengths and diameters, made from various materials and supplied with inserts in order to vary the entrance geometry. These instruments are also used as components of different *rheo-optical* devices, in which the direct measurements of viscosity are supplemented by observation of the optical monitoring of flow.

Discharge velocity is usually measured by determining the weight of extrudate cut-off at assigned time intervals.

Gas viscometers measure in the range of shear rates from about 10^{-3} to 10^4 s^{-1} and shear stresses from about 10^2 to 10^6 Pa. Instruments are supplied with temperature controllers permitting measurements in the range of temperatures from -40 to 250°C using liquid for cooling/heating and electrical heating up to 400°C.

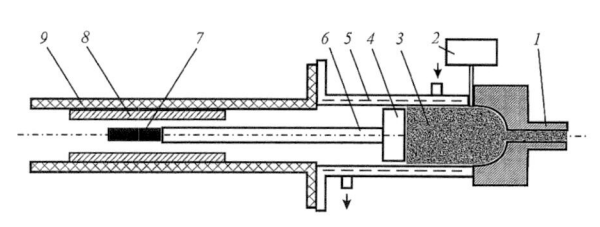

Figure 5.2.12. Scheme of gas capillary viscometer with automatic remote measurement of volumetric flow rate using a linear voltage differential transformer (LVDT). 1 – measuring calibrated capillary; 2 – pressure transducer; 3 – sample; 4 – plunger; 5 – thermostat; 6 – rod made of glass fiber filled plastics; 7 – LVDT rod; 8 – LVDT coil; 9 – main body of instrument. [Adapted, by permission, from J. E. Bujake Jr., *Rev. Sci. Instr.*, **36**, 1368 (1965)].

The method of measurement of volumetric flow rate is an essential drawback of an instrument of this type. In this instrument, the weight of extrudate (not its volume) is measured. The density of melt at a temperature of the experiment has to be measured in a separate experiment.

A possible scheme of non-contact (remote) automatic measurement of volumetric flow rate is realized in the instrument shown in Fig. 5.2.12. In essence, this is a typical gas viscometer. However, this instrument is equipped with a closed system of the automatic measurement of speed, and also with a pressure sensor, mounted directly before the entrance into the capillary. Experimental data are processed by computer.

A modern industrial version of a gas capillary tube viscometer (in accordance with the requirements of various standards, ISO 11443, ASTM D 3836 and DIN 54811) is manufactured by Goettfert (Germany) under the name *Rheograph 200*. In this instrument, measurements are accomplished using two cylindrical reservoirs installed in parallel (a version of this instrument with one cylinder is also possible). The instrument applies pressure in a broad range, from 0.1 to 210 bar. The heating chamber helps to obtain 400°C with deviations from the assigned temperature of not more than 0.1°C. The instrument can be used with the change of cylinders and capillaries.

5.2.6.2.3 Viscometers with varying load

Two types of capillary tube viscometers with varying loads are available. In the first case, the load is varied with the aid of various design solutions. This is based on a feedback control in which load changes according to a predetermined program, controlled by the pressure sensor. Another version is based on automatic load change.

Instruments with the spring load were very popular in the 1960-1970's, but at present, these instruments should be considered antiquated since the varying load (if necessary) is simpler to achieve using an automatic system for feedback control.

Instruments with automatic load control are traditional according to their design features. The control system in these instruments is their main attribute, although it is also solved by conventional methods of control.

Instruments are loaded by the weight of the column of liquid whose height changes during the experiment. Instruments of this type are used for a relative measurement of viscosity. The viscosity measurements are carried out by using some measures (for example, by measuring the duration of discharge of a calibrated volume of liquid), or by comparison with the standard liquids.

These instruments include viscometers of free discharge, which can be called "cup" and "glass" viscometers.

5.2.6.3 Cup viscometers

A characteristic example of a cup viscometer is shown in Fig. 5.2.13.[16] The liquid being investigated is filled into a container of a specific size, at the end of which the calibrated capillary is installed. Usually, the time of discharge of a specified (standardized) volume of liquid through a capillary is measured. After removal of the plug from the instrument shown in Fig. 5.2.13, liquid flows to a measuring flask having a volume of 60 cm³. The time required to fill this volume is used as a measure of viscosity.

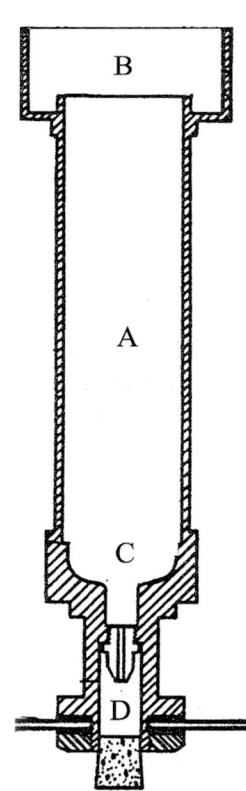

Figure 5.2.13. Measuring part of the Saybolt viscometer. A – internal reservoir; B – die of large diameter; C – lower die; D – calibrated capillary; E – cork.

The reliability and reproducibility of results are characterized by the variation of the time of discharge of a standard sample. If measured values of the standard sample are not within ±1 %, the viscometer is rejected as a measuring device.

There is a wide group of instruments and methods similar to the instrument in Fig. 5.2.13. The *Engler* and the *Redwood viscometers* are utilized in European countries. These instruments were initially intended for measuring the viscosity of petroleum products and lubricating oils, but they are now used for other liquids. Results of measurements are in relative units. Conversion tables of these values to the absolute viscosity are available.

These instruments seem primitive and obsolete, but they play an important role in the standardization of methods of measurements and they are successfully used in the industry.

5.2.6.4 Glass viscometers

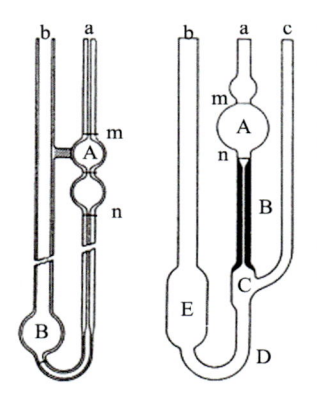

Figure 5.2.14. The Ostwald-Fen-ske (left) and the Ubbelohde (right) capillary viscometers. In Ostwald-Fenske viscometer: A and B bulbs; a and b – arms; capillary inserted in arm a; m and n – marks. In Ubbelohde viscometer: A – bulb with marks m and n; B – capillary; C – bulb; D – transition tube; E – reservoir bulb; a, b and c – arms.

Glass viscometers are included in a special group because they are made out of glass and utilized at comparatively small pressures, which cause flow. Pressure in such instruments is created by the weight of a column of liquid being investigated, although the application of additional pressure is not excluded. Instruments of such type are widely used in laboratories and industrial practice for evaluating the viscosity of dilute polymer solutions, which, in turn, is used as a measure of the molecular weight of the polymer. Although many original constructions of capillary tube viscometers are known, two basic versions of glass viscometers used in practice are shown in Fig. 5.2.14.

The *Ostwald-Fenske viscometer* is filled through arm b with investigated liquid, which is then sucked above marker m. The liquid is allowed to flow and the time of flow from marker m to marker n is measured. This time serves as a measure of viscosity. If dimensions of a capillary are known, then the absolute value of kinematic viscosity can be calculated by comparison of time for unknown sample with a time of discharge of a standard liquid of known viscosity. In reality, this instrument measures relative values of kinematic viscosity.

The *Ubbelohde viscometer* is an improved version of the Ostwald-Fenske viscometer. An important advantage of this instrument, which caused its wide acceptance, is the automatic maintenance of an identical level of liquid that is independent of their individual properties. This method provides high accuracy and excellent reproducibility of measurements. To achieve reliable results of measurements, the viscometer is placed in the thermostatic bath, which ensures the maintenance of the assigned temperature with high accuracy.

Glass viscometers can be used for measurement of viscosity from fractions of cSt to a few hundred St. Shear stress from 1 to about 100 Pa, and shear rates from 10^{-2} to 10^4 s^{-1} can be measured.

Pressure changes in the process of measurement. It is not always clear at what shear stress viscosity is measured. A certain smoothing of influence of pressure change in the process of measurement is achieved by using a monostat, that is, a device for maintaining constant pressure, in essence similar to that discussed with gas viscometers. The discharge of liquid through a capillary occurs under external pressure acting in addition to the weight of a column of liquid. Then, flow is accomplished in the opposite direction. The effect of the pressure of the column of liquid is excluded from the external pressure.

Glass viscometers are used for relative measurements. The calibration of the viscometer is performed on liquids of known viscosity. This permits the measurement of absolute values of kinematic viscosity using a ratio of discharge times. The main characteristic of the viscometer is the discharge time, t_0, for a standard liquid. The discharge time depends on geometric dimensions, especially the diameter of the capillary. It is desirable that the discharge time satisfies this condition: $t_0 > 100$ s. If so, it is possible to disregard corrections for kinetic energy. Instruments have acquired primary acceptance in the determination of *intrinsic viscosity* of highly dilute polymer solutions. Intrinsic viscosity is extrapolated from results of relative viscosity of polymer solutions at different concentrations to zero concentration of polymer.

Glass viscometers are manufactured with different sizes of measuring reservoirs and capillaries. This permits variation of mean pressure in sequential measurements. It is also possible to monitor if there is a dependence of viscosity and shear stress. The automation of measurement procedure is the basic tendency with capillary tube viscometers. Higher accuracy and reproducibility of automated equipment reduce the measurement error of time to 0.001 s.

In the most improved version of this instrument, all basic elements of instrument operation are automated, such as the measurement of discharge time using photoelectric transducers, the stepwise dilution of solution (for example, by 1.5, 2, 3, and 6 times), which is necessary for obtaining the series of points for extrapolation to zero concentration, and calculation of intrinsic viscosity according to one of the known procedures. In commercially manufactured glass viscometers, a measurement error is 1%, in the most precise viscometers, it can be lowered to 0.01%.

For viscosity measurements at low shear rates (less than 1 s^{-1}), the capillary is made in the form of a long spiral. In this case, the curvature of the channel can be disregarded because of the low flow rate. Several designs of glass viscometers were also proposed for chemically active, unstable, or highly volatile liquids.[17]

The number of versions of glass viscometers includes the combination of viscometer with a system of fractionation of polymers (achieved by a method of gel permeation chromatography). This makes it possible to fully automate the process of determining the molecular-mass distribution of polymers and/or evaluation of their degree of branching.[18]

5.2.7 VISCOMETERS WITH A CONTROLLED FLOW RATE

Various methods are known for the imposition of the constant or controlled flow rate of liquid through a capillary varying according to a predetermined program. In this case, the measured parameter is the pressure at the entrance into the capillary or the force necessary for maintaining a given rate.

The following methods of regulating the rate of flow are available: mechanical power drive, which creates the motion of the plunger, hydraulic drive, and extrusion of material through a capillary, installed in a discharge head.

5.2.7.1 Instruments with a power drive

These instruments are most frequently used for the imposition of a constant volumetric flow rate. The adjustable drive creates a constant velocity of displacement of the plunger, extruding material from the reservoir of the viscometer through a capillary. The popularity of such instruments is caused by wide acceptance in the research practice of mechanical testing machines for solid materials in extension or compression. These machines, produced by a number of companies, are well-developed, reliable devices, supplied with heavy-duty drives. The entire construction of the machine is sufficiently rigid to ensure strict maintenance of the assigned conditions of deformation. Contemporary testing machines of such type are supplied with reliable force transducers capable of measuring pressure over a wide range and with good accuracy.

All this makes such testing machines an excellent base for manufacturing viscometers of constant flow rates. For this purpose, the upper clamp, rigidly attached to the adjustable drive, is connected to a plunger of the viscometer. The viscometer itself is installed on the lower fixed base plate of the testing machine.

Viscometers of such type, built on the base of the *Instron* testing machine, were manufactured under the name of *Rheometer 3210* or *Rheometer 3211*. These instruments measure the viscosity of plastics in the range of shear rates from 10^{-2} to 10^5 s^{-1} and at practically unlimitedly high shear stresses. The lower range of reliably measured shear stresses is usually close to 100 Pa.

Control features of Instron viscometers permit conducting measurements not only at a constant speed but also at a variable plunger speed according to a predetermined program.

Another instrument, similar to Instron, is the *Monsanto Processibility Tester*, MPT, originally manufactured by *Monsanto Instrument Company* (now *Flexsys*) for measurement of viscosity of elastomers under conditions of imposing flow rate. This instrument operates in the range of shear rates from 1 to several thousand s^{-1}. The pressure transducer in the MPT is placed in a reservoir just before the capillary entrance. It is also supplied with a laser source to measure the extrudate swell. The instrument is supplied with capillaries of various lengths and diameters if the need arises to measure the slip effect during the flow of rubbers. A thermostat chamber is supplied to carry out measurements at different temperatures. This instrument is widely used in the rubber industry for the evaluation of the processibility of rubber compounds.

An instrument with all the basic capabilities of viscometers of this type is manufactured by *Rheometrics Scientific* (now *TA Instruments*, USA) under the name *Advanced Capillary Extrusion Rheometer* (ACER 2000). In this instrument, the drive makes it possible to regulate the speed of the plunger at a ratio of 1:200,000, providing the possibility to vary shear rates from 0.02 to about 2×10^5 s^{-1}. The replaceable force sensors make it possible to measure the maximal pressures up to 210 MPa, which, with the utilized sizes of operating units, ensures the possibility of reliable measurement of shear stresses on a wall of a capillary from 6×10^{-3} to 10 MPa. The temperature chamber is designed for studies in the temperature range from room to 400°C.

The instrument is supplied with an extensive library of application programs for the processing of experimental data. At the customer's will, the instrument can be fitted with additional devices, for example, for measuring extrudate swell after exiting from a capillary.

Instruments of this type have very high metrological characteristics: the stability of the assigned parameters, the high accuracy of measurement of force, the maintenance of baseline during a prolonged experiment, the compensation for parasitic loads, etc. All this makes these viscometers very valuable laboratory instruments for multi-purpose testing of different liquids.

5.2.7.2 Instruments with hydraulic drive

In such instruments, the constant velocity of the plunger is imposed and maintained using a hydraulic system, controlled by an adjustable hydraulic pump. This method was realized, for example, in instruments manufactured in the USA by Standard Oil.[19]

This instrument makes it possible to conduct viscosity measurements in the range from 2.5 to 10^4 Pa*s with shear stresses below 6.5×10^4 Pa. Since the output of the hydraulic pump is constant, the shear rate can be varied, only by a change of capillaries.

There is also a modification of this instrument, in which a continuous change of the output of the hydraulic pump is realized and, accordingly, shear rate can be changed during measurements.

5.2.7.3 Extrusion rheometers

An extruder is technological equipment, which creates a continuous flow of polymer melt. It suffices to install at the exit from the extruder a shaping die in a capillary form and to accurately measure pressure and temperature at the capillary inlet. Using these modifications a capillary tube viscometer of a constant flow rate is obtained. The viscosity of polymer melts can be estimated using a set of capillaries of different lengths.

This method of viscosity measurement is realized in some commercially produced instruments. The operating principle of these instruments differs by the construction of the pressure sensor. The uncertainty of sample prehistory, which may influence measurement results, is not considered in such instruments. With respect to their accuracy, extrusion viscometers are inferior to specially designed capillary tube viscometers.

5.2.7.4 Technological capillary tube viscometers

Capillary flow can be used as a method to control viscosity directly under production conditions. A device that has a measuring capillary is installed in such a way that there is a possibility of sampling from this installation. This can be a reactor, pipeline, etc. In essence, the viscometer itself, utilized for the purpose of continuous technological control, differs little from any other capillary tube viscometers; it is only essential that it is possible to install this instrument on the production line.

Some companies manufacture viscometers specially intended for use in the technological conditions of production. A characteristic example is a viscometer named the *Process Control Rheometer PCR-20*, manufactured by Rheometrics Scientific (TA Instruments, USA). In this instrument, sampling with a strictly controlled volumetric flow rate is achieved from a production line using a dosing pump. A sample being investigated is pushed through a slit capillary, at the ends of which pressure sensors are installed. This instrument is supplied with a set of capillaries of different sizes. Moreover, dosing pumps of different outputs can be used.

5.3 ROTATIONAL RHEOMETRY

5.3.1 TASKS AND CAPABILITIES OF THE METHOD

5.3.1.1 Viscometric and non-viscometric measurements

The use of rotational instruments makes it possible to measure various parameters characterizing the rheological properties of materials. Therefore, in the discussion of rotational viscometers, it is more appropriate to use a general term of *rheometry*.

Special features of application of rotational instruments for investigation of rheological properties of liquids are as follows. The use of rotational instruments makes it possible, firstly, to create within the sample the homogeneous regime of deformation with strictly controlled kinematic and dynamic characteristics, and, secondly, to maintain the assigned regime of flow for an unlimited period of time.

During material testing by rotational rheometry, different regimes of deformation are possible. The most important among them is the imposition of a constant rotational speed $\Omega = const$, or a constant torque, T = const. However, in many modern instruments, the method of scanning (or sweep) − the imposition of the controlled change of rotational speed or torque with time is realized.

Furthermore, in many rotational rheometers, the capabilities of imposing harmonic oscillations for measuring the viscoelastic properties of materials have been created.

In all cases of application of rotational rheometry, a strictly one-dimensional circumferential flow is assumed with secondary flows being absent.

Almost all modern rotational rheometers are supplied with the software that allows for carrying measurements in different preset automatic modes. For example, it can be shear rate or temperature scanning with certain duration of any step. This is very convenient and attractive for serial experiments but can lead to erroneous conclusions in studying new objects (a typical mistake in interpretation of the results of scanning measurements was presented and discussed in Fig. 3.2.6). So it is necessary to look at such data with caution and carefully differentiate transient and steady viscometric data.

5.3.1.2 The method of a constant frequency of rotation

The typical experimental results, obtained from tests by this method, are shown in Fig. 5.3.1 in the form of the time-dependent torque, T, which is related to the shear stress. In all cases, deformation at first leads to the appearance of a more or less extensive transient response. At the lowest speed (curve 1), the monotonic dependence of T(t) is observed until a steady-state flow process is reached. With an increase of speed (curve 2) during the transient stage, the shear stress maximum (stress overshoot) appears. With a further increase of speed (curve 3), the stress overshoot becomes more pronounced, and the region of steady flow, although it is observed, is followed by a drop in torque, which indicates that an unstable regime of deformation is approached. Finally, at very high speeds (curve 4) steady flow is generally impossible. As a general rule, a drop in torque is an indication of the appearance of ruptures in a sample or its detachment from the solid rotating or stationary surface (cohesive or adhesive rupture).

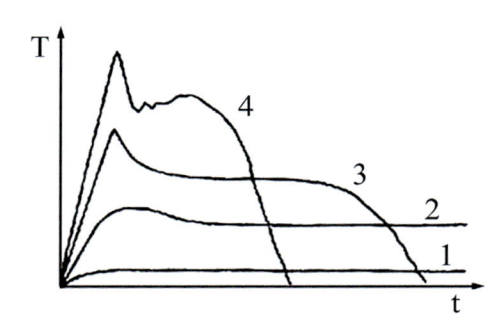

Figure 5.3.1. Typical relationships of torque, T, on time, t, at various rotational speeds. Rotational speed increases from curve 1 to 4.

Sometimes, a periodic stick-slip behavior occurs which is evident because of periodic oscillations of torque. A significant influence on this phenomenon renders the rigidity (deformability) of the torsional shaft-force transducer assembly since its deformation leads to deviation from the assigned regime of Ω = const, which contributes to the onset of oscillations.

During the imposition of a regime of deformation with Ω = const and use of a rigid force transducer, a change in measured torque with time is equivalent to shear stress, moreover, shear strains are easily calculated as

$$\gamma = \dot{\gamma}t \qquad [5.3.1]$$

such that at Ω = const or $\dot{\gamma}$ = const, the dependence of $\sigma(t)$ is equivalent to the dependence of $\sigma(\gamma)$.

From the dependence of $\sigma(\gamma)$, it is possible to determine the number of characteristic (non-viscometric) properties of the material being investigated.

The modulus of elasticity, G, in shear can be evaluated by different methods. The dependence of $\sigma(\gamma)$ at $t \to 0$ can be treated as a reflection of elastic deformations, because, at the initial stages of deformation, flow is absent. Then

$$G = \sigma/\gamma \qquad [5.3.2]$$

The initial section of the deformation curve is always measured with a significant error because of the inertia effects and deformation of a transducer itself. In some rotational instruments a possibility of direct measurements of the accumulated elastic deformations, γ_r, is provided. Then, the modulus of elasticity can be evaluated as

$$G = \sigma/\gamma_r \qquad [5.3.3]$$

where σ is shear stress.

Rubbery deformation, γ_r, can be measured in different stages of shear, in particular in a transient stage. Then it is possible to trace the evolution of elastic deformations, in particular, to find the maximum value of γ_r, which is attained in the transient stage of shear. This value of $\gamma_{r,max}$ may be related to the maximum deformability of individual macromolecules or their segments before the structural network is destroyed by deformation.

The limit of shear strength, σ_{max}, corresponding to its deformation, γ_m, at a given shear rate (i.e., the point of maximum on a shear stress-deformation curve) is frequently treated as a condition corresponding to the destruction of the structural physical network, which impedes development of flow. Both values, σ_{max} and γ_m, depend on the deformation rate.

5.3.1.3 The method of a constant torque

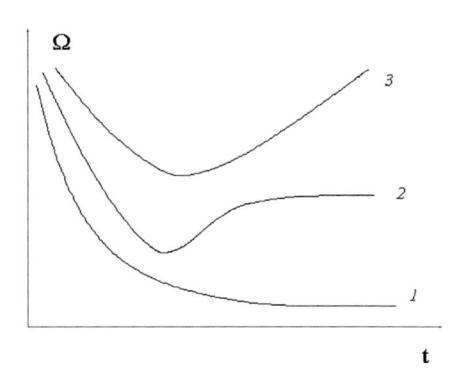

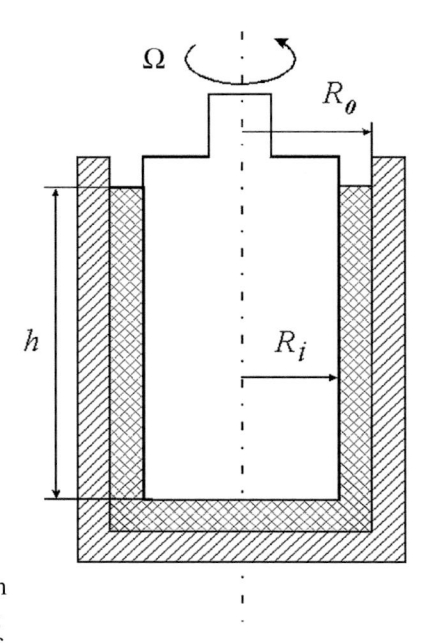

Figure 5.3.2. Typical dependencies of angular velocity on time at different levels of imposed constant torque: 1 – low torque (monotonic change of velocity before achieving steady state); 2 – medium torque (steady state flow is achieved after passing through a minimum of velocity); 3 – high torque (steady state flow is absent).

The typical character of an observed evolution with time of the rotation speed of measuring surface at different imposed constant values of torque is shown in Fig. 5.3.2. In all cases, in a region of small times, it is difficult to perform

Figure 5.3.3. Schematic representation of viscometer with working cell consisting of two coaxial cylinders.

reliable measurements because of the high initial velocity. Stress cannot be imposed instantly and, therefore, the initial section of the curve remains undetermined. This uncertainty is further aggravated by the presence of inertia effects.

At low torque values (and, respectively, shear stresses) slow monotonic transition to the steady viscous flow is observed (curve 1). At higher stresses, speed passes through a minimum, and only then the regime of steady-state flow is reached (curve 2). At very high shear stresses, after achieving the minimum of speed, a steady-state flow is generally impossible due to a gradual adhesive detachment of the sample from the measuring surface or a cohesive rupture of the sample.

In the practice of rotational rheometry, the application of a method of $T = const$ is limited. However, this method can be useful, at least, in the following cases:

- measurements of creep, which is one of the basic methods of determining the viscoelastic properties of the material
- scanning regimes of tests, when torque changes according to a predetermined program
- evaluation of lifetime of adhesive or cohesive joints
- evaluation of elastic deformations at different stresses.

5.3.2 BASIC THEORY OF ROTATIONAL INSTRUMENTS

5.3.2.1 Instruments with coaxial cylinders[20]

A schematic diagram of a rotational viscometer of the cylinder-cylinder type is shown in Fig. 5.3.3. Here R_o and R_i are the outer and inner radii of cylinders, respectively. Both cylinders have a common axis (i.e., they are coaxial). The ratio of radii R_o/R_i will subsequently be designated as ε. The height of the liquid in the clearance between the cylinders is h, and Ω is the rotational speed of the inner cylinder. It is assumed that the outer cylinder is fixed, although only the speed of rotation of cylinders relative to each other is important.

The liquid being investigated is also filled at the bottom of the cylinder end. Torque, T, for a layer of liquid, which is located at a distance r from the axis of cylinders is expressed as

$$T = 2\pi r^2 h\sigma \qquad [5.3.4]$$

where σ is shear stress, which acts over the area $2\pi rh$ at a distance r from the axis of the cylinders. At equilibrium conditions, torque does not depend on the radius. Therefore, the following expressions, which relate the shear stresses σ_o and σ_i acting on surfaces of the outer and inner cylinders, respectively, are valid:

$$T = 2\pi R_o^2 h\sigma_o = 2\pi R_i^2 h\sigma_i = const \qquad [5.3.5]$$

Hence, it follows that the ratio between the stresses acting on the surfaces of cylinders is expressed as

$$\frac{\sigma_i}{\sigma_o} = \left(\frac{R_o}{R_i}\right)^2 = \varepsilon^2 \qquad [5.3.6]$$

It follows from the given formulas that the distribution of shear stresses along a radius is expressed as:

$$\frac{\sigma(r)}{\sigma_o} = \left(\frac{R_o}{r}\right)^2 \qquad [5.3.7]$$

or

$$\frac{\sigma(r)}{\sigma_i} = \left(\frac{R_i}{r}\right)^2 \qquad [5.3.8]$$

Thus, the degree of heterogeneity of stresses in the liquid is determined by a value of ε, i.e., by the ratio of the radii of the cylinders. If ε is close to unity (which is typical of many rotational rheometers), then a practically uniform field of stresses in a clearance between the coaxial cylinders occurs. Specifically, the possibility of conducting an experiment in a practically uniform stress field is the major advantage of rotational viscometers.

Liquid resistance is determined by shear rate

$$\dot{\gamma} = r\frac{\partial\omega}{\partial r} \qquad [5.3.9]$$

where ω is the angular velocity, which depends on a radial coordinate.

For Newtonian liquid, a basic relation, Eq. 3.1.1, is fulfilled. Then, in a general case, obvious calculations give the following expression for velocity distribution, $u(r)$, in the gap between cylinders when the outer cylinder revolves with a frequency, Ω_o, and the inner cylinder with a frequency, Ω_i:

$$u(r) = \frac{(\Omega_o R_o^2 - \Omega_i R_i^2)r^2 - (\Omega_o - \Omega_i)R_i^2 R_o^2}{(R_o^2 - R_i^2)r} \qquad [5.3.10]$$

This formula contains, as special cases, two situations when the inner cylinder rotates and the outer cylinder is fixed ($\Omega_i \neq 0$ and $\Omega_o = 0$), and *vice versa*, when the inner cylinder is fixed and the outer cylinder rotates ($\Omega_i = 0$ and $\Omega_o \neq 0$).

According to the diagram in Fig. 5.3.3:

$$u(r) = \frac{\Omega R_i^2(R_o^2 - r^2)}{(R_o^2 - R_i^2)r} \qquad [5.3.11]$$

In this case, the final expression for viscosity takes the form:

$$\eta = \frac{T}{\Omega}\frac{R_o^2 - R_i^2}{4\pi R_o^2 R_i^2 h} \qquad [5.3.12]$$

where h is the height of the liquid sample between cylinders.

This formula is called the *Margules equation*, which can be rewritten in the form:[21]

$$\eta = K\frac{T}{\Omega} \qquad [5.3.13]$$

It means that the viscosity can be expressed through the ratio of T/Ω, and the geome try factor or *form-factor*, K:

$$K = \frac{R_o^2 - R_i^2}{4\pi R_o^2 R_i^2 h} \qquad [5.3.14]$$

Eq. 5.3.13, according to its structure and physical meaning, is identical to Eq. 5.2.13 in the theory of capillary viscometry.

Let us write down an expression for the distribution of shear rate along a radius:

$$\dot{\gamma} = r\frac{d\omega}{dr} = 2\Omega\frac{R_i^2 R_o^2}{R_o^2 - R_i^2}\frac{1}{r^2} \qquad [5.3.15]$$

i.e., the distribution of shear rates with an accuracy of a constant factor is equivalent to the distribution of shear stresses (see Eq. 5.3.8).

It is significant that for liquid of a constant viscosity, the torque is proportional to the frequency of rotation. If the ratio of T/Ω is constant, then this is direct proof that the medium being investigated exhibits properties of a Newtonian liquid.

The method of rotation of a cylinder in the infinite medium is frequently used for measurement of Newtonian liquids, i.e., conditions: $R \to \infty$ and $\Omega_o = 0$ are satisfied. The viscosity can be expressed as

$$\eta = \frac{T}{4\pi\Omega R_i^2 h} \qquad [5.3.16]$$

i.e., again the viscosity is proportional to the ratio of T/Ω with the multiplication factor being the form-factor, K.

For calculating the apparent viscosity of non-Newtonian liquids from a measured torque and frequency of rotation in the rotational instrument, the shear rate has to be estimated and correlated with the value of shear stress.

If the clearance between the cylinders, Δ, is small in comparison with the radii of cylinders, i.e.

$$\frac{\Delta}{R_i} = \frac{R_o - R_i}{R_i} \ll 1 \qquad [5.3.17]$$

the problem is solved simply.

Then, it is possible to determine the average shear stress, $\bar{\sigma}$, as

$$\bar{\sigma} = \frac{\sigma_o - \sigma_i}{2} \qquad [5.3.18]$$

The value of the average shear rate is found from Eq. 5.3.15. It is equal to:

$$\dot{\gamma} = \Omega\frac{R_i + R_o}{2(R_o - R_i)} \cong \frac{\Omega R}{\Delta} \qquad [5.3.19]$$

In this case, it is practically not important what value should be used for R in the last formula, the value of R_o or R_i.

However, the situation is more complicated if it is not possible to consider that the clearance is small, or if the dependence of $\sigma(\dot{\gamma})$ is very strong, such that even insignificant changes in the shear rate in the narrow clearance would bring substantial changes in the shear stress.

Then, in a general case, it is possible to write that

$$\dot{\gamma} = r\frac{d\omega}{dr} = f(\sigma) \qquad [5.3.20]$$

and since $T/2\pi h = const$, then $dr/r = d\sigma/2\sigma$ and $d\omega = f(\sigma)d\sigma/2\sigma$.

If the inner cylinder rotates, and the outer cylinder is fixed (i.e. $\omega = 0$ at $r = R_o$ and $\omega = \Omega$ at $r = R_i$), then:

$$\Omega = \frac{1}{2}\int_0^{\sigma_o} \frac{f(\sigma)}{\sigma}d\sigma \qquad [5.3.21]$$

When the outer cylinder rotates and the inner cylinder is fixed, Eq. 5.3.21 is also valid but the sign should be changed.

Two cases are possible: the cylinder rotates in an infinite medium (i.e., the clearance between cylinders is infinitely large) or inequality (5.3.17) is not fulfilled.

If $R_o \rightarrow \infty$, then Eq. 5.3.21 takes the form:

$$\Omega = \frac{1}{2}\int_{\sigma_i}^{0} \frac{f(\sigma)}{\sigma} d\sigma \qquad [5.3.22]$$

The last equation can be differentiated with respect to σ_i to get

$$\dot{\gamma} = f(\sigma) = -2\frac{d\Omega}{d\ln\sigma_i} \qquad [5.3.23]$$

Hence, the dependence of $\dot{\gamma}(\sigma)$ can be determined as follows. The dependence of Ω on σ_i is obtained and then the derivative of this dependence gives the function of $\dot{\gamma} = f(\sigma)$.

The most general (although inconvenient from the experimental point of view) situation arises, when the clearance between cylinders is arbitrary. This case is methodically undesirable. In the practice of rotational rheometry, such a situation is avoided. If this cannot be avoided, then the unknown function of $f(\sigma)$ can be found by analytical methods of regularization, using known algorithms for finding unknown functions on the array of experimental data, if the relation between them is established by an integral equation.

The above formulated classical theory of flow in the coaxial cylinders is valid for the homogeneous flow and can be applied to single-phase fluids. On page 202, the effect of shear banding, i.e., separation of the flux into layers with different properties was discussed. This effect is observed in rotational flows of multi-component fluids, in particular, the flow of emulsions and polydisperse polymer melts. In these cases, the jump on the shear rate appears and the standard theory of the flow between coaxial cylinders becomes unworkable because two fluids with different rheology co-exist in the gap between cylinders. The same is also true for the other rotational units described below.

The phenomenon of shear banding requires a special technique for measuring velocity profile in the flow between cylinders.

5.3.2.2 Instruments with conical surfaces

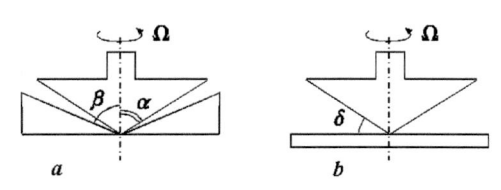

a   *b*

Figure 5.3.4. Schematic representations of rotational viscometers containing conical surfaces: a – cone-cone type viscometer; b – cone-plate type viscometer.

Instruments with conical surfaces are important in rheometry. They are especially valuable for the analysis of high viscosity fluids.

In this case, two versions of a device are possible, as shown in Fig. 5.3.4. A sample is placed into a clearance between two coaxial cones with a joint apex or into a clearance between the conical surface and the plate with the axis of the cone being normal to the flat surface. It is important that the angle between cones or the cone and the plate is small.

The theory of viscometers with conical surfaces is based on the analysis of the flow of liquid in spherical coordinates such that the angle, α, is counted from the vertical axis.

Torque with respect to the vertical axis generates shear stresses, σ. From the equilibrium condition for the volume element of a sample placed into a clearance, it follows that the dependence of $\sigma(\alpha)$ can be expressed as follows:

$$\sigma = \frac{C}{\sin^2\alpha} \qquad [5.3.24]$$

where C is the constant of integration, determined through torque, T, acting on the cone surface.

Thus, if the height of the outer cone in Fig. 5.3.4a equals T, the stress on its surface is σ_α, then the torque is expressed as:

$$T = 2\pi\int_0^H \sigma_\alpha h^3 \frac{\sin^2\alpha}{\cos^3\alpha}dh = \frac{2\pi H^3}{3}\frac{\sin^2\alpha}{\cos^3\alpha}\sigma_\alpha \qquad [5.3.25]$$

Hence, the shear stresses, σ_α, can be calculated from the measured torque values as:

$$\sigma_\alpha = \frac{3T}{2\pi H^3}\frac{\cos^3\alpha}{\sin^2\alpha} \qquad [5.3.26]$$

The angular distribution of shear stresses in a sample, located in a clearance between cones, is described by the following formula:

$$\sigma(\theta) = \sigma_\alpha\frac{\sin^2\alpha}{\sin^2\beta} \qquad [5.3.27]$$

The last formula estimates the degree of homogeneity of the stress state in a sample. It is expressed by the ratio $\sin^2\alpha/\sin^2\beta$. It also permits the calculation of shear stresses at any point of a sample being investigated at any measured (or assigned) torque.

For instruments of the *cone-plate* type, which are most frequently used in practice, the angle δ between the conical surface and the plate is typically made very small ($\delta < 5°$). Then, the degree of homogeneity of the stress state in a sample, that equals $\cos^2\delta$, is not less than 99%. Therefore, it is possible to consider that $\sigma = $ const in the entire volume of a sample.

Then the formula for calculating the shear stress in the cone-plate viscometer takes the following form:

$$\sigma = \frac{3T}{2\pi R^3} \qquad [5.3.28]$$

In the spherical coordinates a value of shear rate, which is a function of shear stress, is defined as

$$\dot{\gamma} = f(\sigma) = \sin\beta\frac{d\omega}{d\beta} \qquad [5.3.29]$$

where ω is the angular velocity, which depends on the angle β.

The instruments with small angles δ are of basic practical interest. The shear rate field, as well as the shear stress field, is practically uniform, and the shear rate with sufficiently high accuracy is calculated from the simplest formula:

$$\dot\gamma = \frac{\Omega}{\delta} \qquad\qquad [5.3.30]$$

Thus, the apparent viscosity is calculated as

$$\eta = \frac{T}{\Omega}\frac{3\delta}{2\pi R^3} \qquad\qquad [5.3.31]$$

where, as in the foregoing cases (see Eq. 5.3.13), viscosity is expressed as the ratio of T/Ω with multiplication factor being an instrument constant or form-factor K.

5.3.2.3 Bi-conical viscometers

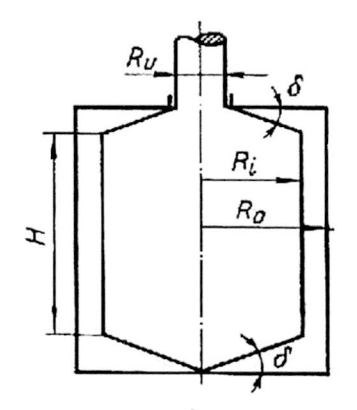

Figure 5.3.5. Schematic representation of rotational viscometer containing cylindrical and conical surfaces – bi-conical viscometer.

The combination of cylindrical and conical measuring surfaces serves as a convenient method for the elimination of the edge effect. The diagram of the rotational viscometer, called bi-conical, is shown in Fig. 5.3.5.

The use of a bi-conical viscometer makes it possible to maximally fill a liquid into a clearance between the rotating and the fixed surfaces of a working cell. This prevents viscoelastic material from escaping from a clearance during the action of normal stresses (the Weissenberg effect) and, therefore, makes it easier to conduct tests.

It is usually considered that the influence of coupling cylindrical and conical surfaces on the results of measurements is insignificant and the distribution of stresses and shear rates can be calculated separately for the cylindrical and conical parts of the working cell.

The optimum condition of operation of the viscometer of such type is realized when the equality of stresses and shear rates on both parts of the instrument is maintained. Then, disregarding possible local effects at the position of coupling of surfaces of different geometries, it is possible to formulate the following requirement for the geometrical dimensions of bi-conical viscometers:

$$\delta = \frac{R_o - R_i}{R_o} \qquad\qquad [5.3.32]$$

The relationship between shear stress and torque for the bi-conical viscometer is expressed by the following formula (notations according to Fig. 5.3.5):

$$\sigma = \frac{3T}{2\pi R_i^3}\left[2 + \frac{H}{R_i} - \left(\frac{R_o}{R_i}\right)^3\right] \qquad\qquad [5.3.33]$$

5.3.2.4 Disk viscometers

A disk viscometer can be presented as a viscometer in which both conical surfaces degenerate into the parallel plates (disk-disk) separated by a clearance of a height, h. In this case, the liquid being investigated is placed into a clearance between the disks, and one of the disks revolves relative to another around the common axis. This instrument is called the *disk viscometer*.

Torque arises as a result of the action of shear stresses, distributed over the surface of the disk. This torque is expressed as

$$T = 2\pi \int_0^R r\sigma(r)dr \qquad [5.3.34]$$

In this case, the shear rate varies along the radius (in contrast to flow between the cone and the plate, where the shear rate is constant). It is expressed as

$$\dot{\gamma}(r) = \frac{\Omega r}{h} \qquad [5.3.35]$$

where Ω is the angular velocity of disk rotation and h is the distance between disks. The variation of the shear rate throughout a volume of a sample being investigated is the main disadvantage of this measuring device. However, if the liquid being investigated possesses Newtonian properties, then torque is proportional to the frequency of rotation, and viscosity is calculated from a formula that is analogous in structure to that obtained for other rotational instruments, namely:

$$\eta = \frac{2Th}{\Omega\pi R^4} \qquad [5.3.36]$$

where the multiplication factor is the form-factor for this measuring device.

The apparent viscosity of non-Newtonian fluids at the assigned frequency of rotation can be expressed through the maximum shear rate, computed at the edge of disk $\dot{\gamma} = \Omega R/h$. In this case, the calculation of viscosity at this shear rate is reduced to a procedure, analogous to the Rabinowitch-Weissenberg equation in capillary viscometry, namely:[22]

$$\eta(\dot{\gamma}_m) = \bar{\eta}\left(1 + 0.25\frac{d\log\bar{\eta}}{d\log\dot{\gamma}_m}\right) \qquad [5.3.37]$$

where $\bar{\eta}$ is the average viscosity computed as

$$\bar{\eta} = \frac{2Th}{\pi R^4 \Omega}$$

Thus, by measuring the dependence of $\bar{\eta}(\dot{\gamma}_m)$ or $T(\Omega)$, it is possible to find the flow curve of non-Newtonian liquid, if, of course, one neglects the edge effects.

The following diagram is an interesting special case of disk viscometers: a thin disk is placed into the infinite medium filled with a viscous fluid. Theoretical calculations[23] give the following formula for torque, T, of a disk radius, R, as a function of the frequency of rotation, Ω:

$$T = 1.80R^4\sqrt{\eta\rho\Omega^3} \qquad [5.3.38]$$

where ρ is the density of the liquid, in which the disk rotates.

According to Eq. 5.3.38, the rotation of a thin disk in viscous fluid is a very special case: the viscosity is defined, not by the ratio of T/Ω, as in all other cases (see Eq. 5.3.13), but by the quantity T^2/Ω^3.

The *annular viscometer* is a version of the disk viscometer in which the disk is substituted by a ring with an outer radius of R_o and an inner radius of R_i. The apparent viscosity during the use of this geometry is also expressed by a formula of the type of Eq. 5.3.13, namely:

$$\eta = \frac{2Th}{\pi(R_o^4 - R_i^4)\Omega} \qquad [5.3.39]$$

The advantage of annular viscometers is the fact that the shear rate and stress field in the liquid being investigated can be considered as sufficiently uniform. In this case, the shear stress is calculated as

$$\sigma = \frac{T}{2\pi\bar{R}^2(R_o - R_i)} \qquad [5.3.40]$$

and the shear rate as

$$\dot{\gamma} = \frac{\Omega\bar{R}}{h} \qquad [5.3.41]$$

where $\bar{R} = (R_o + R_i)/2$ is the average value of radius and h is the distance between the ring and the plane.

Using Eqs. 5.3.40 and 5.3.41 it is not difficult to find the apparent viscosity and i s dependence on the shear rate.

5.3.2.5 Viscometers with spherical surfaces

Two types of viscometers, in which the deformation of liquid being investigated is accomplished by rotation of spherical surfaces, are possible:
- flow between two spheres (or hemispheres) with the same center
- rotation of one sphere in an infinite volume of liquid.

In this case, one of the spheres revolves at a constant angular frequency, Ω, and torque, T, is measured. According to theory, the viscosity of Newtonian liquid during flow is calculated as:

$$\eta = \frac{T(R_i^{-3} - R_o^{-3})}{8\pi\Omega} \qquad [5.3.42]$$

where R_o and R_i are radii of outer and inner spheres, respectively.

Calculation of apparent viscosity, measured in a spherical viscometer, is also possible for non-Newtonian liquids. But, in order to do so, it is necessary to introduce an assumption regarding the form of the expected flow curve.

The rotating sphere can be placed in a very large (theoretically infinite) volume of fluid. In this case, the wall effect of the vessel on the results of viscosity measurements can be disregarded, and the instrument for viscosity measurements is called a *mono-spherical viscometer*. Then, in the region of low speed of rotation (more precise, the region of Re <<1), the following formula can be obtained for the calculation of viscosity:

$$\eta = \frac{T}{8\pi\Omega R^3} \qquad\qquad [5.3.43]$$

where R is the radius of a rotating sphere.

If the rotational speed is not small, then secondary flows appear. In this case, the torque is expressed as follows:[24]

$$M = 8\pi R^3 \Omega\eta + \frac{\pi}{150}R^7\Omega^3\rho^2\eta \qquad\qquad [5.3.44]$$

where ρ is the density of the liquid being investigated.

The second term reflects a contribution of secondary flows to the value of the measured torque, which characterizes the viscous properties of the liquid. This correction becomes noticeable at sufficiently high values of the Reynolds number.

Eq. 5.3.44 permits finding values of viscosity by measuring torque at different rotational speeds with a subsequent extrapolation in the linear region, in which torque is proportional to the rotational speed.

Mono-spherical viscometers can be of interest as instruments for the control of production processes since it does not represent fundamental difficulties in placing the revolving sphere into an industrial reactor.

5.3.2.6 End (bottom) corrections in instruments with coaxial cylinders

As in the case of capillary viscometry, during the processing of experimental data obtained on the rotational instruments, it is necessary to introduce corrections of different kinds.

Although the geometric form of measuring surfaces in the rotational instruments influences the specific form of corrections, their nature remains one and the same for instruments of different geometry of deformation.

Flow near the bottom and in the bottom region of the coaxial cylinder, instrument differs significantly from the theoretical approach utilized for calculating viscosity. Consequently, it is necessary to consider this effect in some manner. Usually, this is done by one of the following experimental methods.

The method of two cylinders of different heights. The influence of the bottom effect is excluded by the fact that torque is measured at the same frequency of rotation, but using two different cylinders of the same diameter, but having different heights. In this case, the distance from the face of the cylinder to the bottom in both measurements should be identical. The calculation of viscosity is performed based on the difference in torques and heights of cylinders inserted into the calculation formula, being the difference in heights of two cylinders.

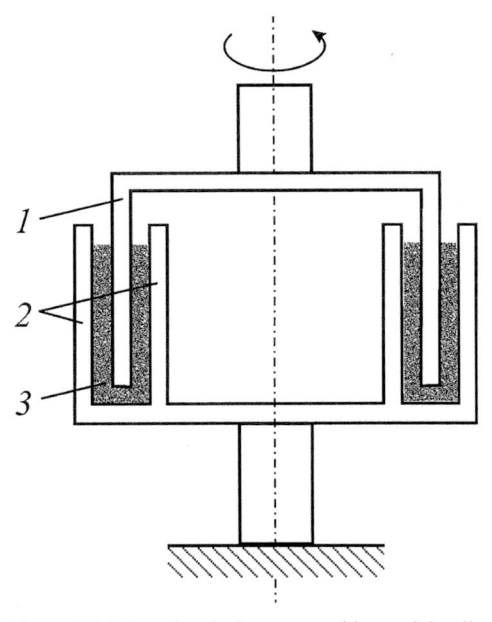

Figure 5.3.6. Rotational viscometer with coaxial cylinders of bell type. 1 – rotating intermediate cylinder; 2 – connected together outer cylinders; 3 – sample.

The device with guarding cylinders. In this device, the outer cylinder revolves. Above and below the inner cylinder are coaxially installed fixed cylinders of the same radii. These cylinders are not rigidly connected with the torque measuring device, but with the stationary casing of the instrument. Therefore, the faces of the measuring (inner) cylinder do not contribute to the measured torque.

The device with bell-type cylinders. In this device, shown in Fig. 5.3.6, a pair of hollow cylinders is made in the form of an inverted container, and between them, the intermediate revolving cylinder is installed. Due to the small thickness of the wall of this cylinder, the effect of the bottom of the cylinder on torque is negligible. The shear flow of fluid occurs simultaneously in two clearances. The latter must be taken into account in the viscosity calculations.

Arrangement of conical or spherical bottoms. In instruments of this type the bottom is made in the form of a cone or a hemisphere so that the combination of two rotational viscometers with different geometries of working surfaces is obtained.

5.3.2.7 On a role of the rigidity of dynamometer

Measurement of torque is usually accomplished with the aid of elastic elements (*torsional shafts, or dynamometers*) whose deformation serves as a measure of torque. The same scheme is also used in the compensating devices when the deformation of force transducer is compensated by action of external force and this force is measured. In any event, the dynamometer is deformed, and, thus, the conditions of liquid flow are changed. This factor plays no role during the steady-state flow. However, displacement of the dynamometer can have an essential effect on the results of measurements of torque in the transient regimes of deformation before a steady-state flow is achieved and during relaxation (after cessation of flow).

The theory of phenomenon under discussion shows that the greatest error in measurement appears at $t \to 0$.[25] Therefore, one should treat the initial stage of the transient regime of deformation with special care (see Section 5.3.1).

The general methodical recommendation is to use as rigid a dynamometer as possible. In the use of the compensating devices, the general requirement is that the response time of the controlling device should be much less than the time of measurement, for example, the characteristic time of a change in the deformation rate in the scanning mode.

5.3.2.8 Temperature effects

At sufficiently high deformation rates, the self-heating of liquid being investigated can be a reason for significant errors. This factor can play a more important role in rotational rheometers than in capillary tube viscometers because of the unlimited long stay of the same material under the action of intensive deformation.

During the viscous flow of fluid in a rotational instrument, different thermal effects are possible, including a rapid increase in temperature that is not compensated by heat transfer (so-called *heat explosion*). More frequent is an increase of temperature to a certain limit when heat generation is balanced by heat removal. If an increase in temperature is not very high, theory leads to the following formula for calculating apparent viscosity at an assigned temperature of measurement:[26]

$$\eta_0 = \frac{\sigma_r h}{u_r}\left(1 + \frac{0.083\,hk\sigma_R u_R}{\kappa}\right) \qquad\qquad [5.3.45]$$

where u_R and σ_R are, respectively, the velocity and shear stress at the boundary surface, κ is the coefficient of thermal conductivity of material being investigated, k is the coefficient characterizing temperature sensitivity of viscosity, h is the clearance between the bounding surfaces in viscometer.

The quantity in front of the brackets in Eq. 5.3.45 is the isothermal value of viscosity, and the second term in brackets is the temperature correction. Then, by measuring values of apparent viscosity at different values of u_R or σ_R and extrapolating the obtained data to zero, it is not difficult to obtain the unknown value of viscosity at an assigned temperature, and, at the same time, the coefficient characterizing temperature sensitivity of viscosity, k.

5.3.3 LIMITATIONS OF ROTATIONAL VISCOMETRY

Possible limitations of rotational viscometry at the low end are related to the design features of experimental techniques used. Thus, while conducting measurements under conditions of imposed strain rates, the required rotational speed is usually created by the power drive with speed control using a system of reducing gears. If such power transfers include many gears, then the natural mechanical imperfection of contacts in gears may lead to the inconstancy of the rotational speed, which is expressed by the appearance of shocks. The use of high-quality gears permits shear rates to be as low as 10^{-3} s^{-1}, although the construction of instruments with even lower deformation rates is known.

In using the method of measurement at imposed constant torque, it is necessary to eliminate or compensate resistance caused by friction in moving parts. By using the method of rotational viscometry, it is possible to measure the viscosity of any low-viscosity liquid and even gas.[27]

Basic limitations of rotational viscometry at the upper limit of the high shear rates (and high shear stresses) were indicated above. This, first of all, is an intensive heat generation and it is impossible to maintain a constant temperature of the sample, because of the dissipation of mechanical work at high shear rates and the escape of elastic liquid from a working cell as a result of the Weissenberg effect. The capabilities of rotational rheometry at high shear rates are limited as a result of the appearance of instabilities of the flow of different kinds. The physical causes of this phenomenon are the same as described for capillary tube viscometers, although the instability in rotational motion is manifested differently.

Thus, viscosity measurements constitute the right choice only in the region of laminar flow. The critical value of the Reynolds number, Re_c, during the circumferential flow of liquid between coaxial cylinders, for the case when motion is generated by the rotation of the inner cylinder, is calculated from the following formula:

$$Re_c = 41.3 \sqrt{\frac{R_o}{R_o - R_i}} \qquad\qquad [5.3.46]$$

For real dimensions of rotational instruments, this corresponds to a value of Re_c above 100.

If the motion is created by the rotation of the outer cylinder, then flow remains steady even at the higher values of the Reynolds number.

For viscoelastic liquids, the instability of flow can be caused by their elasticity. This effect may appear as a result of a secondary flow from the eddies arising in the clearance between cylinders, the oscillatory motions of liquid, the detachment of the medium from the walls of the viscometer, the escape of liquid from the working cell, and ruptures inside the sample. In all such cases, measurements are practically impossible. Different types of sample ruptures during the circumferential flows appear at the stresses 5-10 times smaller than the critical stress, σ^*, that corresponds to the appearance of elastic turbulence in the capillary flow. It is possible that the reason for this difference is the fact that the rupture is developed with the time of measurement, i.e., σ^* depends on the duration of deformation.

Sometimes to eliminate wall slip or ruptures (detachment from walls) in rotational instruments, hydrostatic pressure is applied. However, the described phenomena are related to the elasticity of the material, i.e., to its intrinsic properties, and not strictly to phenomena occurring near walls. Therefore, the imposition of additional pressure can be a useful method for warranting the initial adhesive contact of material (especially of high viscosity) with a solid wall. But the stress at which rupture occurs, as a result of deformation, hardly depends on the hydrostatic pressure.

Another recommendation is to corrugate the working surfaces (or increase their roughness). However, the strength of adhesive contact and the cohesive strength of flowing polymeric materials are usually close to each other. Therefore, corrugation of the surface, ensuring the best adhesive contact, does not remove slippage at apexes of reefs, because of the cohesive rupture that takes place on the top of reefs.

5.3.4 ROTATIONAL INSTRUMENTS

5.3.4.1 Introduction – general considerations

Rotational viscometry has attracted the attention of researchers and designers for many decades. During this time, hundreds of new constructions and many technical improvements of instruments of this type were proposed, described, patented, and found application in research practice. It is worthwhile to note a difference between individual samples of laboratory instruments and viscometers made in serial production. The latter always appear more attractively made and are supplied with a collection of standard programs for processing experimental results. They are sufficiently simple in operation so that they do not require a highly qualified experimenter. Instruments of such type are convenient for systematic and repeated measurements to carry out the same tasks each time. However, this does not prevent the possibility of using such instruments for scientific studies, since commercial instruments are frequently supplied with a wide collection of additional options of a high technical level.

At the same time, such instruments are not always suitable for studying new phenomena and refined effects that require additional consideration and understanding of the results of measurements and their possible errors. Certainly, the solution to such problems demands a qualified experimenter. Moreover, the data processing systems of some contemporary instruments are designed in such a way that they eliminate possible deviations from "smooth" dependences, considering them as random errors. The apparent (at first glance) random errors in the results of measurements can actually be hidden in new properties and important results for science.

In spite of a variety of designs and measuring schemes used, it is possible to isolate a number of similar fundamental solutions, common to all instruments. It is possible to identify the basic requirements for rotational viscometers. Thus, for any instrument it is required:

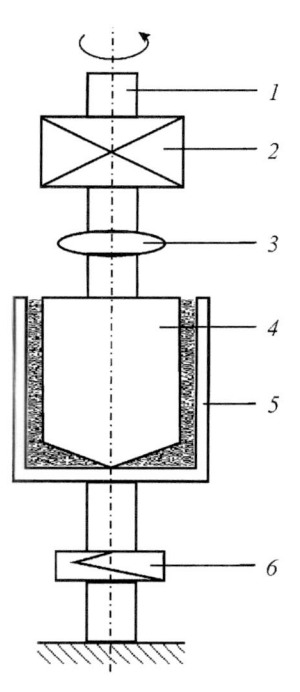

Figure 5.3.7. Schematic representation of rotational viscometer with its basic elements. 1 – rotation drive; 2 – reducer; 3 – angular velocity transducer; 4 – inner (rotating) cylinder; 5 – outer (fixed) cylinder; 6 – torque transducer.

- to have high-quality working surfaces
- to observe coaxiality (for the coaxial cylinders) or strict perpendicularity (for the instruments of the cone and plate type) during their assembly
- to obtain low parasitic friction in the bearing of the rotating part of a working cell (replacement of usual bearings by the gas-bearing and use of unsupported construction of the revolving shaft)
- to maintain assigned (constant or adjusted according to the required program) condition of deformation or torque without oscillations and jerks
- to quickly achieve a transient process while imposing the required regime of deformation or load (this requirement is not very important during viscometric measurement, but important during the study of transient processes such as relaxation and creep)
- to control the assigned temperature conditions of tests.

In accordance with these general requirements, the fundamental design of any rotational instrument can be represented in the form of a schematic diagram as depicted in Fig. 5.3.7. All basic elements of the construction of the viscometer are shown in this figure.

The basic element of the design is the working cell itself. In Fig. 5.3.7, this is the cylinder with a conical bottom, which revolves around a fixed cylinder that is coaxial with it. Other combinations of axially arranged symmetrical bodies, described above, can be used instead. The inner cylinder is subjected to rotation by means of a drive with the adjustable speed with the frequency of rotation during the experiment maintained constant or according to a predetermined program. The frequency of rotation is measured by an angular velocity transducer.

In Fig. 5.3.7, the torque transducer is installed on the shaft that connects the outer cylinder with the base of the instrument. This sensor can also be installed on the revolving shaft, measuring the torque experienced by it. Typically, an elastic member torsion shaft is utilized as the working (receiving torque) part of the sensor. The basic requirement for this member is to have a linearity of its performance characteristics, i.e., the dependence of displacement (or angular deflection) on torque. Finally, an important element of the construction of the viscometer (not shown in Fig. 5.3.7) is a temperature chamber since the entire working cell is usually thermostatically controlled. The temperature of the sample is monitored by a sensor, which is in direct contact with the material being investigated.

The schematic diagram that is shown in Fig. 5.3.7 can be realized in the existing rotational viscometers in various configurations. However, all basic elements of the construction of any viscometer are included in the drawing.

5.3.4.2 Rheogoniometers and elastoviscometers

A large group of instruments with diverse capabilities of rotational rheometry is united under these names. In these instruments, viscosity measurement is augmented by possibilities of determining different characteristics of the rheological properties of liquids.

A typical example and the prototype of many instruments of this type is the *Weisserberg Rheogoniometer*, which was manufactured by *Sangamo* (Great Britain).[28]

In essence, this instrument contains all the basic elements of the diagram, presented in Fig. 5.3.7. The working cell of this instrument is a system consisting of a cone and plate device. The drive (not shown in the figure) is connected with the driving member (lower plate) through a worm gear. The instrument operates in the regime of a constant frequency of rotation at a rotational speed of the main drive of 1,500 rpm. The construction of the gearbox makes it possible to vary shear rate in the range from 7.1×10^{-4} to 9×10^3 s^{-1} with steps of $10^{0.1} \approx 1.259$. Torque is received by a torsion shaft. The instrument is equipped with a collection of replaceable torsion shafts of different rigidity varying from 1 to 10^3 Nm/rad. By changing the torsion shaft, it is possible to measure shear stress in the range from 10^{-4} to 10^6 Pa.

Temperature control includes both liquid and electrical thermostats permitting tests in the range of temperatures from -50 to 300°C.

In addition to torque measurements, the instrument permits the determination of normal stresses using the axial force transducer. In the latest modification of the instrument, the drive was improved in such a way that it also permits the imposition of harmonic oscillations and the superposition of low-amplitude harmonic oscillations on steady rotation.

A similar instrument of high-level technical capabilities, according to the tasks performed, was produced by *Instron*[29] under the name of the *Rotary Rheometer 3250*. At present, several important instrument-manufacturing companies have the capability to produce such instruments. One of them is *Rheometrics Scientific* (now TA Instruments, USA), which makes several high-precision instruments. One of these instruments is the *ARES*.[30] This is a rotational rheometer with interchangeable working cells: cone-plate, disk-disk, coaxial cylinders. The drive system includes two direct current servomotors. The control system of the instrument allows one to conduct tests in different regimes, at a given shear rate or shear stress as well as periodic deformations. The instrument includes torque and axial force transducers (for measuring normal stresses). The instrument is supplied with a computer having an extensive library of programs.

The same company manufactures other high-precision elastoviscometers, known under the names *RDA III, RFS III,* and *RMS 800.*[31] These instruments allow one to conduct measurements in different regimes, including periodic oscillations and steady flow.

The recent achievement of the company in the area of instrument manufacture is the universal rheometer *SR5.*[32] This instrument permits the implementation of different regimes of tests including relaxation, creep, periodic oscillations, and superposition of different regimes. The instrument is also equipped with a normal stress transducer. The salient capability of this instrument is the wide range of the measured torque, from 10^{-4} to 50 mNm. As in other instruments produced by the company, the frequency can be varied in the range of 7 decimal orders. The temperature range is from -40 to 350°C.

Bohlin manufactures a number of rotational rheometers (general designation of a series C-VOR) with an adjustable frequency of a rotation and measured stresses. Several modifications of the instrument are intended for conducting tests in different ranges of torques. Another series of instruments, produced under the designation CVO, is intended for measurements at controlled shear stresses.

Paar Physica (Germany) manufactures a series of instruments. They possess a wide range of technical capabilities. The most complex instrument (produced under the abbreviation MCR) allows one to conduct measurements in different controlled regimes including a constant velocity of rotation, constant stress, varying frequency in the regime of relaxation, and so forth. It can also carry out the measurements in more complex regimes of tests such as the superposition of periodic oscillations on the steady flow.

A number of devices with similar characteristics are manufactured by *Haake* (Germany). In some modifications of the instrument, the rheological measurements are augmented by the possibilities of direct observation of structural transformations, caused by the deformation.

Instruments of the type under consideration possess many advantages. But they also have one essential disadvantage – they are expensive. This is due to the complexity of the design and technological solutions used for their creation. Therefore, these instruments are not made for mass production, but as measuring systems, intended, first of all, for studies conducted in specialized laboratories.

5.3.4.3 Viscometers with assigned rotational speed

5.3.4.3.1 Laboratory viscometers with adjustable rotational speed

Instruments of such type are simplified versions of rheogoniometers. They are intended exclusively for viscosity measurements. In these instruments, the frequency of rotation is assigned and the torque is measured. Accordingly, they are much simpler in construction in comparison with a rheogoniometer and substantially cheaper. All basic elements shown on the fundamental diagram in Fig. 5.3.7 are used in their design. Also, the possibilities of variation of shear rate in these instruments are sufficiently wide.

Some designs of these instruments, which are manufactured on an industrial scale, are examined below, as an example.

Fig. 5.3.8 shows the viscometer *Rotovisco* (*Haake*). Usually, this instrument is supplied with a cylinder-cylinder type working cell, although in some modifications of the

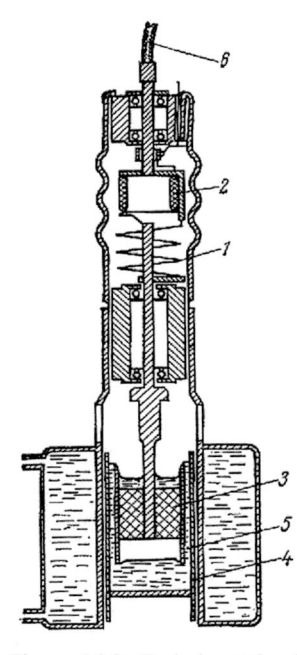

Figure 5.3.8. Typical rotational viscometer operating under imposed rotational speed of rotor. Viscometer *Rotovisco* of cylinder-cylinder type. 1 – torsion (elastic measuring) element; 2 – transducer for measurement of angle of torsional twist (potentiometer); 3 – rotating (inner) cylinder; 4 – fixed (outer) cylinder; 5 – sample; 6 – flexible connector from rotation control drive.

instrument the use of working cells of other geometry is possible (for example, cone and plate).

The torque in this instrument is measured on the drive shaft. The measured shear stress ranges from 1 to 10^5 Pa, the range of shear rate is from 10^{-2} to 10^4 s^{-1}.

A design similar to Rotovisco and accordingly close to its technical capabilities (although more limited) is a popular viscometer *Rheotest*, manufactured in Germany.

In the viscometer of *Ferranti-Shirley*, the working cell is made in a form of a pair of cone and plate with a very small angle between the generatrix of cone and plate. The special feature of this instrument is the stepless control of the frequency of rotation of drive shaft using a direct current electric motor operating according to a diagram of the generator-engine.

A known series of viscometers *Rheomat* are manufactured by *Contraves* (Switzerland). In one of the modifications of the instrument *Rheomat 15T*, a torque transducer allows one to measure the shear stresses in the range from 0.6 to 2×10^5 Pa with drive allowing one to vary speed (stepwise) to achieve a shear rate from 0.5 to 10^3 s^{-1}. In a more advanced modification of instrument *Rheomat 30*, the lower limit is substantially reduced to 4×10^{-3} s^{-1}. Moreover, in this instrument, it is also possible to attain continuous change of rotational speed with the recording of the dependence of torque on the frequency of rotation.

Of special interest is the version of the instrument known under the name of *Rheomat, Block DC50*. With this instrument, measurements can be carried out in vacuum or under pressure up to 50 bars in the temperature range from -50 to 300°C. Measurements can be carried out in the range of shear rates from 2.3×10^{-2} to 170 s^{-1} with a viscosity of fluid ranging from 0.1 to 8×10^3 Pa*s. The possibility of measurement under pressure usually is not realized in other serially produced rotational viscometers.

Simple and convenient rotational viscometers are also produced by *Rheometrics Scientific* (TA Instruments, USA). The instrument *RM 100* makes it possible to conduct tests at any of five fixed speeds of rotation[33] with a maximum rotation frequency of 600 rpm.

5.3.4.3.2 Viscometers with extension rotor (immersion type)

In these instruments the outer cylinder is absent, and the inner cylinder, fastened to the end of a console, is submerged in the liquid being investigated. The rotor can be made not only in the form of a cylinder but a disk or another body of the arbitrary geometric form. The rotor is driven by the electric motor through a reducer with an adjustable frequency of rotation. The measuring element (torsion shaft) is installed on the drive shaft. Such instruments are sufficiently simple in production and operation and are widely used for solving applied problems.

The most common instrument of such a type is the *Brookfield* viscometer. Different modifications of this instrument are produced. They are distinguished by a range of shear rates and by limits of measurements. The technical characteristics of the instrument are defined by the rigidity of the torsion shaft used, by the gearbox installed, and by the geometry of the utilized rotational member.

The most critical part of viscometers of this type is the twisted spring of the force transducer. This spring is made from beryllium bronze and it is calibrated by the manufacturer. Brookfield viscometers are mainly utilized in technological laboratories for quality control of produced materials with the purpose of comparing production samples with requirements of technical specifications. They are also used at different stages of the technological process of materials' manufacture.

Some modifications of viscometers of this type are intended, in particular, for the measurements of viscosity at high temperatures (up to 1,500°C), for measurements of the viscosity of corrosive media, and also for measurements of viscosity at a distance from the location of the technological process. The latter is achieved by the application of an adaptor, connected to a spindle of viscometers.

Instruments of the type of *Brookfield* viscometer are manufactured by various firms under different names. For example *Contraves* (Switzerland) manufactures a *Process Viscometer TO* and *Haake* (Germany) manufactures several modifications of the instrument, known under the names of *Viscotester VT 181/VT24* and *Viscotester VT 01*. *Rheometrics Scientific* manufactures the instrument with the extension rotor under the name *Rheomat RM 180*.

5.3.4.3.3 Constant torque viscometers

Load viscometers

The use of a measuring scheme in the regime of constant torque, T = const, has definite advantages in measurements not only of viscosity, but also of viscoelastic characteristics during deformation.

In practice, several different methods of the imposition of adjustable torque are used. The most simple method for imposing a constant torque is by use of descending loads, installed at a certain arm with respect to the rotational axis. The speed of rotation of the operating unit or the speed of motion of loads is measured. The instruments in which this method is used are simple in design and experimental procedure. They do not require complex measuring equipment. At the same time, they usually do not give high accuracy of the measurement. An increase in accuracy of measurements entails the need of using more advanced equipment, which is hardly compatible with measurements that are carried out manually. This is characteristic of the simplest viscometers of a fixed torque. At the same time, simplicity of construction and measuring scheme is justified when a fast comparative evaluation of material properties is necessary rather than obtaining precise absolute values of viscosity.

Spring viscometers

The simple measuring scheme for viscosity at an assigned torque is based on the use of energy of the initially twisted (loaded) spring, which is done by hand. In the process of measurement, after the release of the spring, torque decreases from a given maximum value to zero. The operating unit of the instrument can be made in the form of a rotational body of any configuration. The time required for releasing the spring is a measure of viscosity. The instrument is calibrated on the liquid of known viscosity. Then, viscosity is determined by comparing the time of spring release for the standard liquid and the liquid being investigated. This is a very simple measuring scheme that is applicable even under field conditions in the absence of electric power sources. Certainly, because of simplicity and universality, the accuracy of measurements is somewhat lost. Therefore, in this case, it is needless to speak about the variation of conditions of deformation and other special features if tests are conducted on non-Newtonian liquids. However, even this simplest measuring scheme can be useful under specific test conditions.

The automatic torque control systems

In contemporary instruments, the maintenance of the assigned torque, constant or variable according to a predetermined program, is achieved by methods of automatic control, based on the principle of feedback control.

Specifically, rheogoniometers described above and elastoviscometers operate in the stress-controlled regime. These can be both specific instruments and devices in which different conditions of the test are performed by adjustment of assigned torque and frequency of rotation.

5.3.4.4 Rotational viscometers for special purposes

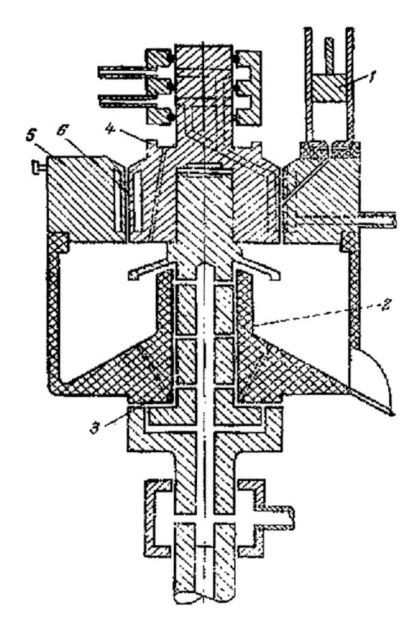

The principal capabilities of rotational instruments are frequently realized in viscometers for special purposes. They are intended to solve various scientific problems arising during the analysis of the physical properties of liquids. In this case, although the general rules of construction of rotational viscometers are retained, the need for essential modification and custom designs still exists. In the present section, some examples of the above-mentioned approach to the technology of rotational viscometry are given.

5.3.4.4.1 High-speed (thin-film) viscometers

High shear rates in rotational viscometers with coaxial cylinders require the solution of two basic problems: high accuracy in maintaining a very small clearance between the cylinders and temperature control under conditions of intensive heat dissipation.

Figure 5.3.9. High speed (thin film) viscometer with continuous flow of sample through working clearance between coaxial cylinders. 1 – ram supplying sample; 2 – radial air bearing; 3 – supporting air bearing; 4 – rotor; 5 – stator; 6 – working clearance between coaxial cylinders.

The technical tasks confronting the designer of this instrument were successfully solved during the development of *thin-film flowing viscometer*.[34] In this instrument (Fig. 5.3.9) the measurements are accomplished in the scanning regime by the continuous change of rotational speed. The clearance between coaxial cylinders is made open, such that the liquid being investigated, supplied with an aid of a press, can flow freely through the clearance in a vertical direction. Thus, a continuous change of volume of fluid being investigated is ensured and the effect of heat dissipation is reduced. The gap clearance between cylinders is 0.15 mm and the radius of the revolving cylinder (rotor) is 2.681 cm. Constant conditions of deformation are reliably maintained by the application of rigid radial and axial air bearings. The use of this instrument makes it possible to investigate liquids with viscosity from 0.1 to 100 Pa*s.

In one of the modifications of this instrument, the size of clearance between the cylinders is brought to several micrometers. In this case, the clearance between cylinders was calibrated using standard liquids, thus producing a geometric constant of the instrument. Control experiments showed that even in such thin layers, the effects of absorption or orientation of liquid being investigated are absent, so that one succeeds in obtaining reliable values of viscosity with error not higher than 2%.

5.3.4.4.2 Viscometers with the noncontact drive (with the very low torque)

For the solution of many applied and theoretical problems, it is necessary to know the properties of materials at very low shear stresses. Measurements must be conducted in rotational instruments at low torques. Such low shear stresses require the imposition of deformation for a long period of time and are important in the evaluation of creep of high-viscosity materials, speed determination of fluid runoff on vertical surfaces, etc. It is also needed for the evaluation of forces of intermolecular interaction, the strength of structural bonding, etc. In all these cases, there is interest in measurements in the shear stress range lower than 1 Pa.

From a methodology point of view, two problems must be solved: the creation of very low torque and elimination of friction in bearings.

Air bearings are used. This allows one to achieve torque due to the frictional forces as low as 10^{-11} Nm. Presently, air bearings are traditional elements in the construction of these instruments. In such bearings, the radial clearance does not exceed 0.05 mm with an end clearance of 0.1 mm. Thus, it is possible to decrease friction in bearings to values that correspond to a shear stress of the order of 10^{-5} Pa.

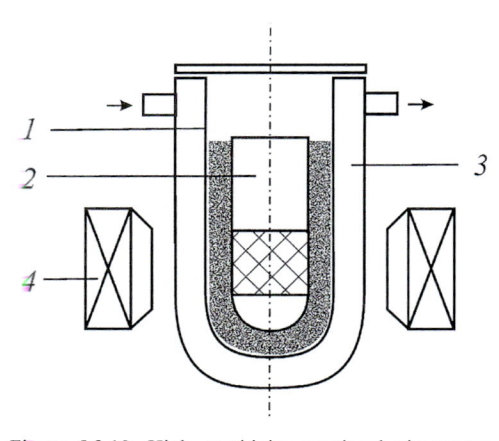

Figure 5.3.10. High sensitivity rotational viscometer having rotor without bearing used for measuring low torque values. 1 – test tube with sample; 2 – rotor with magnetic insert; 3 – thermostat; 4 – stator of electric motor.

A more radical method for further reduction of the measured shear stresses is by the elimination of friction using a noncontact method of the imposition of torque. The version of such an instrument in which this approach is realized is shown in Fig. 5.3.10. The torque here is created using electromagnetic forces such that it can be conveniently small. The revolving (inner) cylinder made of magnetic iron is weighed in the liquid being investigated, i.e., its weight is balanced by an Archimedean force. In a typical instrument of such type, the outer cylinder is made in the form of a glass test tube.

Using this viscometer, called the *Zimm-Crothers viscometer*,[35] it is possible to measure the viscosity of low-viscosity liquids and to observe the appearance of a yield point in the region of very low shear stresses. This is especially valuable for low-viscosity structured liquids, biopolymers, colloidal systems, and liquid-crystal polymer solutions.

The electromagnetic method of imposing torque for viscosity measurements in the region of low shear rates is quite innovative for testing solutions of electrolytes. In this case, it is generally possible to operate without rotating parts. Flow appears as a result of electromagnetic interaction of the external field with ions of electrolyte.[36]

Other variations of the method in question are possible. In particular, using a combination of layers of conducting liquid of high density (for example, a saturated solution of CsCl) and non-conducting liquid, it is possible to measure the viscosity of the latter by a differential method.

5.3.4.4.3 Viscometers for electro-rheological liquids

The so-called *magneto- and electro-rheological* liquids are of some interest. These are liquids having a structure that rapidly changes in a magnetic or electric field. Accordingly, their viscosity also changes. For viscosity measurements of such media and control of their change in electromagnetic fields, a number of rotational viscometers with the coaxial cylinders can be used. Their basic special feature consists of isolated outer and inner cylinders. This is achieved by modification of the usual rotational instrument by the installation of separating bushings made out of non-conducting materials. The instrument itself is an electrical capacitor because electro-rheological liquids are dielectrics. This design combines viscometric and electrical measurements.

Electrorheological measurements can be carried out also, using instruments with parallel plates (a modification of the rotational viscometer of *Rheometrics System IV*).[37] In this design oscillations were imposed on the disk, but it is also possible to rotate the disk with assigned frequency. The analogous scheme of shear between two parallel electrically isolated disks was realized by some modification of viscometer *RS -50 (Haake)*.[38]

5.3.4.5 Rotational instruments for technological purposes

The technological evaluation of polymeric materials is one of the basic tasks of rheological measurements. A number of specialized methods, which simulate real technological processes, were proposed. The procedures of measurement and basic geometric parameters of instruments are standardized. This makes it possible to conduct reliable comparative tests of real materials and to rely on previous experiences from processing these materials. Reliable recommendations can be made regarding their application and selection of a technological regime.

5.3.4.5.1 The Brabender Plasticorder

This instrument is a typical example of tools used in technological applications. The *Brabender Plasticorder* is manufactured in large quantities and widely used in the polymer industry. The instrument is basically a mixer with different mixing elements. Depending on the viscosity of the mixture, the shaft of the kneader arms experiences different torque. This torque is a measure of viscosity. By measuring torque, one can find some characteristics of the rheological properties of processed mixtures and trace the evolution of these properties during mixing. Thus, some characteristic values influencing the evaluation of material and selection of an optimum technological regime are determined.

Contemporary models of the Brabender Plasticorder are supplied with an adjustable drive allowing one to vary the frequency of rotation of rotors in the range from 1 to 600 rpm, although there are modifications of the instrument intended for use at a fixed speed of 31.5 or 63 rpm. The torque is measured using a torque transducer. The modifications of the instrument are produced for different maximum torque. In a working chamber, a vacuum or inert gas atmosphere can be created. The thermostatically controlled chamber makes it possible to conduct measurements up to 400°C.

5.3.4.5.2 The Mooney viscometer

This instrument, in various modifications, is manufactured in many countries, since the *Mooney test method*[39] appears in a number of standards for rubber compounds.[40]

The tested material undergoes shear strains in a closed chamber in clearances between a revolving rotor and the chamber walls. The rotor can be made in the form of a smooth or a serrated disk. During the deformation of the material in a chamber due to a complex configuration, the complex fields of velocities and stresses appear, so that the Mooney viscometer gives a conditional characteristic of rheological properties of the material being investigated.

The method of torque measurement is the original design feature of the Mooney viscometer. It is determined by measuring the force exerted on spring by the drive shaft.

The rotation of the rotor is achieved by means of a synchronous electric motor through the gear drive and the worm wheel. The material being investigated is located in the chamber under pressure, created during the closing of the upper and lower halves of the chamber. Typically, during the first compressing of a sample, the pressure created in the chamber is of the order of 200 Pa, and upon additional compression, with the aid of springs, the pressure is increased to 300-600 Pa. Heaters are installed in both upper and lower halves of the instrument chamber. Usually, tests begin 1 min after compressing the sample, without waiting to achieve steady-state temperature conditions.

Mooney viscometers are manufactured with standard dimensions of the working chamber and rotor. The usual frequency of rotation of the rotor in the Mooney viscometers is 2 rpm. In some modifications of the instrument, capabilities are augmented by setting the frequency of rotation to 4 and 8 rpm, and also by imposing oscillatory motions of the rotor.

The standardized requirements for tests according to Mooney are realized in a number of instruments. A typical example of a modern instrument of such type is the *Mooney-check Compact PC* (*Gibitre*, Italy).

A combination of possibilities of rotational and oscillatory motions of a rotor is realized in an instrument of the type of Mooney viscometer being manufactured by *Monsanto* (now *Alpha Technologies*). In an instrument known as *Rheometer-100*, capabilities exist to measure continuous changes of torque with time (according to ASTM D 2084) which makes it possible to follow the kinetics of vulcanization of rubber compounds. *Rheometer-100750* has a variable speed of rotation in the range of 1-150 rpm or a constant speed at 750 rpm. Additional options permit conducting automatic processing of test results, determining characteristic points on a vulcanization curve – initial and minimum viscosities, curing time, and maximum value of modulus of elasticity of vulcanized rubbers.

Further modifications of this instrument are in *Moving Die Rheometer MDR* (*Monsanto Instrument Company*, now *Alpha Technologies*, USA) and *Rotorless Curemeter* (*Goettfert*, Germany) in which one platen oscillates. These rheometers use thinner samples of rubber than Rheometer-100. Therefore, less effect of transient heat transfer during curing takes place. The most recent developments are two new instruments – *Rubber Pro-*

cess *Analyzer, RPA*, and *Advanced Polymer Analyzer, APA* (*Alpha Technologies*, USA), which permit tests over ranges of strain 0.7 to 1200%, frequencies of 0.1 to 2000 cycles per minute and at various temperatures. These instruments can measure the storage and loss moduli and $\tan\delta$ during the curing of rubber compounds.

5.3.4.5.3 The Goettfert instrument

Goettfert (Germany) developed a series of instruments intended to simulate real technological processes of processing plastics with the purpose of selecting parameters for an optimal regime of processing. One of these instruments is a rotational type viscometer.

5.3.5 MEASURING NORMAL STRESSES

Section 3.4.2 discussed normal stresses in shear flow (the Weissenberg effect). The general stress field produced in shear flow is determined by Eq. 3.4.3. It was also explained that from a rheological point of view, not normal stresses by themselves but two their differences are essential:

$$\text{the first difference} \qquad N_1 = \sigma_{11} - \sigma_{22}$$
$$\text{the second difference} \qquad N_2 = \sigma_{22} - \sigma_{33}$$

The necessity to operate with the differences of normal stresses but not with absolute values of normal stresses is based on the hypothesis of incompressibility of flowing liquid. Therefore, the addition of hydrostatic pressure (i.e., simultaneous change of all diagonal components of the stress tensor) would not influence the state of shear deformations and N_1 and N_2 values, consequently.

For this reason, an experimenter is interested to know how to measure N_1 and N_2 but not separate components of normal stresses.

5.3.5.1 Cone-and-plate technique

The basic approach to measurements of the first difference of normal stresses is related to the use of the cone and plate technique as in Fig. 5.3.4b.

The balance of force equation for a liquid element can be written (neglecting the mass forces) as

$$\frac{d\sigma_{33}}{dr} = \frac{\sigma_{11} + \sigma_{22} - 2\sigma_{33}}{r} \qquad [5.3.47]$$

It is reasonable to suppose that both normal stress differences $(\sigma_{11} - \sigma_{33})$ and $(\sigma_{22} - \sigma_{33})$ do not depend on the curvature of stream-lines and are determined by shear rate only. Then, integration of Eq. (3.5.47) gives the following equation for stress distribution:

$$\sigma_{22}(r) = \sigma_{22}(R) + [(\sigma_{11} - \sigma_{33}) + (\sigma_{22} - \sigma_{33})]\ln\frac{r}{R} \qquad [5.3.48]$$

The expression inside the square brackets does not depend on the coordinate r. It is reasonable to think (as many experimental works show) that $(\sigma_{22} - \sigma_{33}) \ll (\sigma_{11} - \sigma_{22})$ and that $\sigma_{22}(R) = 0$. The latter reflects the absence of stresses at the free surface. Then, it is seen that the stress distribution along the radius must be logarithmic.

Then, measuring the radial distribution of σ_{22}, one can find N_1 as a coefficient in the following dependence:

$$\sigma_{22}(r) = N_1 \ln \frac{r}{R}$$ [5.3.49]

Repeating these measurements at different shear stresses, one can obtain the dependence $N_1(\dot{\gamma})$.

The normal stress distribution along the radius was measured as shown in Fig. 5.3.11 by several piezo-tubes.[41] This *differential* method of measuring normal stresses was realized in the Roberts-Weissenberg rheogoniometer, model R8. Later this method was not used for practical purposes and normal stresses were measured by the *integral* method. The total force acting perpendicular to the radius of the rotating cone is measured. This force, F, is calculated by integrating Eq. 5.3.48. After some rearrangements, the following relationship is obtained:

$$F = \frac{1}{2}\pi R^2 N_1$$ [5.3.50]

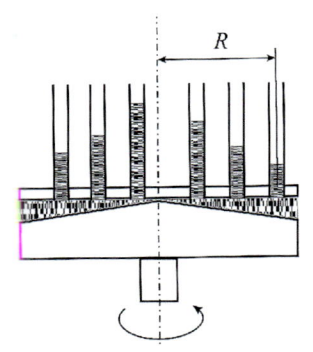

Figure 5.3.11. Measuring normal stress distribution along the radius of a cone-and-plate assembly.

This force, F, is the cause of the Weissenberg effect as described in Chapter 3. The last expression gives a direct method of determination of the first difference of normal stresses by measuring the total normal force.

This is the most popular method of measuring N_1, which is realized in several industrial rotational devices. The total force is measured by different transducers but the general rule is: a sensor must be as rigid as possible because the vertical shift (even very small) of a surface in a rotation device leads to distortion of the deformation field in the sample and to essential experimental errors. The maximum vertical shift in the best-known units does not exceed several microns and it permits measurement of normal stresses in the range from 0.1 to $1*10^6$ Pa at shear rates from 10^{-4} to 10^4 s^{-1}.

Modern rotational devices allow the experimenter to measure normal stress as a function of shear rate not only at steady-state flow regimes but also in transient deformation modes and in periodic oscillations as well.

The cone and plate technique gives an adequate and reliable method of measurement of the first normal stress difference and modern experimental devices realize this possibility. One example of such a technique is the *Rheometrics RAA rheometer*.[42]

Using micro-machining technology and miniature pressure sensors positioned along the radius allows one to extend the possibilities of a standard cone and plate rheometer (in particular *ARES* rheometer described above) and enables one to measure simultaneously the first, N_1, and the second, N_2, normal stress differences.[43]

5.3.5.2 Plate-and-plate technique

This method of measuring normal stresses is almost the same as the previous one, but the rotation of two parallel plates around their common axis is used instead of the cone and plate assembly. The main difference with the previous case is the variation of shear rate along the radius of the measuring device according to formula $\dot\gamma = (\omega r)/h$, where r is the current radius and h is the distance between the parallel plates.

Analysis of force balance of liquid element, and assuming that $(\sigma_{22} - \sigma_{33}) \ll (\sigma_{11} - \sigma_{22})$ results in the following relationship directly valid for measuring the first difference of normal stresses:

$$N_1 = \frac{d\sigma_{22}}{d\ln r}$$

[5.3.51]

Then, measuring the dependence of σ_{22} as a function of r at different shear rates, it is possible to find the dependence of N_1 on shear rate.

Simultaneous application of two geometries of flow (plate-plate and cone-plate) permits finding the second normal stress difference. The final equation is given by:[44]

$$N_2 = \dot\gamma \int_0^{\dot\gamma} \left[\left(r\frac{\partial\sigma_{22}}{\partial r} \right)_{p-p} - \left(r\frac{\partial\sigma_{22}}{\partial r} \right)_{c-p} \right] \frac{d\dot\gamma}{\dot\gamma^2}$$

[5.3.52]

where subscripts p-p and c-p relate to values measured in plate-plate and cone-plate geometries, respectively.

The alternative form of this equation, which is used in experimental practice, can be written as:[45]

$$N_2(\dot\gamma_R) = \frac{F_{c-p}}{\pi R_{c-p}^2} - \frac{F_{p-p}}{\pi R_{p-p}^2} \left(2 + \frac{d\ln F_{p-p}}{d\ln \dot\gamma_R} \right)$$

[5.3.53]

The value $N_2(\dot\gamma_R)$ is N_2 as measured at a shear rate corresponding to the outer radius of plates.

This is one of the possibilities of measuring N_2, which has a very low value (if not zero) in comparison with N_1. The application of Eq. 5.3.52 or 5.3.53 gives the best way to measure the second normal stress difference. However, though this method is applicable in principle it should be treated with caution due to uncertain experimental errors because both members of difference in these equations are comparable in value. Therefore, this difference is determined with the largest possible error.

Perhaps the most serious limitation in the application of rotational methods (either cone and plate or plate and plate) for measuring normal stresses is the edge effect including the edge fracture of the sample.[46]

5.3.5.3 Coaxial cylinders technique

The shear flow between rotating coaxial cylinders can be used to measure the second normal stress.[47] For this purpose, the difference of radial stresses $\Delta\sigma_{22}$ at the inner and outer cylinders should be measured:

$$\Delta\sigma_{22} \equiv \sigma_{22}(R_i) - \sigma_{22}(R_o)$$

Calculations based on the balanced equation show that if the difference between shear rates at inner and outer cylinders is large enough (i.e., the gap between cylinders is large) the following relationship is valid:

$$N_{2,i} = \frac{2}{1 + \dfrac{d\ln\eta}{d\ln\dot\gamma}} \frac{d\Delta\sigma_{22}}{d\ln\dot\gamma} \qquad [5.3.54]$$

The subscript i shows that N_2 determined by this equation is related to the values at the surface of the inner cylinder. The structure of this equation shows that N_2 can be calculated if the flow curve is known and $\Delta\sigma_{22}$ is measured as a function of shear rate.

Another way of measuring the second difference of normal stresses, N_2, can be realized on the base of the axial flow of liquid between two coaxial cylinders.[48] However, it is necessary to mention that all these methods are more exotic than these in everyday use.

5.3.5.4 Hole-pressure effect

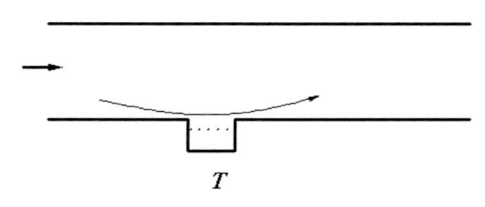

Figure 5.3.12. Measuring normal stresses by the hole-pressure method. Curvature of a streamline near a hole is shown. T – pressure transducer.

There have been several attempts to use flow through a capillary for measuring normal stresses.[49] A possibility to obtain accurate results by such a method is usually strongly dependent on the choice of rheological model and strict control of experimental details. The most widely discussed example is related to the measurement of pressure in a hole at the channel wall, as shown in Fig. 5.3.12. The channel should be flat (i.e., made as a slit) in order to avoid problems related to the wall curvature. The streamlines are distorted near a hole (as shown in Fig 5.3.12). Additional tension is related to the elasticity of liquid and its value might be used as a measure of normal stresses. This method is rather controversial: on one hand, a curvature of streamlines must be created, on the other hand, a diaphragm of a transducer should be flash-mounted to avoid uncertain experimental errors.

It was shown that the pressure gradient along a channel, as measured by the hole-pressure method is not only related to shear stresses but can be used as a method for estimating N_1.[50] Possible experimental errors of this method require special rigorous analysis.[51] The difficulties of correct estimation of details of flow near a hole lead to controversial results in attempts to apply this method to measuring N_2 (even the sign of N_2 appears different depending on not well documented experimental details).[52] That is why this approach is more of theoretical than applied value, and in modern laboratory practice, simpler (rotational) methods are preferred.

5.4 PLASTOMETERS

5.4.1. SHEAR FLOW PLASTOMETERS

Instruments of this type provide an ideal model of shear flow experiments, as shown in Fig. 5.4.1. This measuring scheme, most clearly corresponding to the definition of viscosity, seems to be optimal. However, in practice, the realization of this scheme of measurements encounters specific methodological difficulties.

If the surface area of the sample being investigated, placed between the plates, is S, and the distance between the plates equals h, then the shear rate is calculated from the

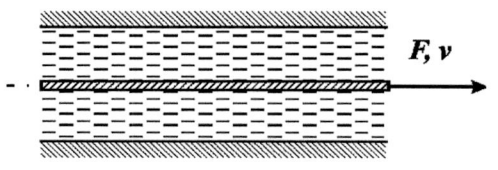

F, v

Figure 5.4.1. Sandwich shear flow plastometer.

measured velocity of movement of one of the plates relative to another, V, as $\dot\gamma = V/h$. Shear stress is determined through measured force, F, which must be created in order to accomplish the motion of the plate with a required velocity. This shear stress equals $\sigma = F/S$. Hence, according to the general formula, the apparent viscosity is found as a ratio of measured values $\eta = \sigma/\dot\gamma$.

Sometimes, shear plastometers are made in the form of a sandwich (Fig. 5.4.1): a moving plate is placed between two fixed ones, and the sample being investigated fills both clearances between the moving plate and each of the fixed plates. The velocity of the plate, V, and force, F, leading to its motion, are measured. In this case, the area entering the expression for calculating shear stress is doubled.

Shear plastometers are convenient for measurements of both steady and transient flows during creep under imposed constant shear stress value, σ, elastic recovery (elastic deformations) after removal of external load. The important advantage of shear plastometers is that they permit independent definition of shear rate and shear stress used then for calculation of apparent viscosity, i.e., these instruments can be considered as absolute.

The experiments carried out using shear plastometers can be performed in different regimes, either F = const or V = const; a possibility also exists (and it is actually used) to impose more complex conditions of deformation, in particular, the oscillatory conditions of deformation. Most frequently, these instruments are used with F = const, since it is easier to carry out experimental work by loading the moving plate with the falling weigh .

Shear plastometers are used as simple laboratory instruments, intended for the measurement of materials having a high viscosity (>10^5 Pa*s), at low shear stresses and shear rates. However, if we make the clearance between plates very small, in the order of a few fractions of cm, then these instruments can also be used for measurements of materials of low viscosity. Shear plastometers permit measurement of viscosity in the range from 10 to 10^{10} Pa*s. Shear plastometers are very convenient for measurements of yield points since they allow experimenters to conduct measurements at very low shear stresses.

The accuracy and reliability of results of measurements carried out on shear plastometers, depend on the quality of their manufacture and adjustment, especially because very frequently these are homemade instruments individually manufactured by the experimenter. Moving and fixed plates of these instruments must be parallel. The variance of thickness must not exceed 20-30 μm, and deviation from parallel alignment may reach several micrometers.

It is important to reduce friction in bearings to a minimum, especially during studies at low loads and speeds. The use of a scheme in which a load is unsupported is optimal. This is achieved, for example, by the installation of plates in a vertical position and by the direct suspension of the load on one of the plates. At a very low speed of movement of the moving plate, it is sometimes necessary to measure displacement using a microscope. The prolonged maintenance of a uniform temperature field in a sample is an additional problem, which requires a constructive solution.

Shear plastometers permit the determination of limiting values of shear rates and shear stresses, which is of interest for experimentalists. But it is usually difficult to obtain reliable results of measurements with an error of less than several percent.

5.4.2 SQUEEZING FLOW PLASTOMETERS

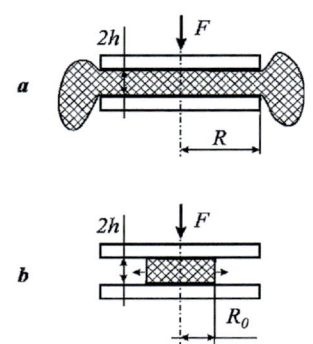

Figure 5.4.2. Squeezing flow plastometers. a – excess of material is squeezed out; b – squeezing sample of constant volume.

In these instruments, a sample, as in the preceding case, is placed between two parallel disks, but the motion of the upper disk occurs, not in parallel to the lower disk, but along the common axis. Thus, the sample being investigated is compressed and spreads along the radius of disks. As shown in Fig. 5.4.2, two cases are possible. In the first case (a), the investigated sample initially fills the entire space between disks. Therefore, during its squeezing, a surplus of the sample is extruded from the working volume. In the second case (b), the volume of the sample remains constant in the process of tests, and during squeezing it flows between disks, remaining completely between them.

The regime of squeezing can also be different. Tests can be carried out under a constant load, F = const, or under programmed change of force with time. In these cases, either speed of the upper disk is measured (in case a) or a change of radius of the sample is measured (in case b). It is also possible to impose speed of motion to the upper disk and measure an increase in the resulting force as a function of the time, F(t).

The calculation of viscosity according to the test results is simple if the liquid has Newtonian properties. The motion of Newtonian liquid between converging disks was described in detail in textbooks on hydrodynamics. Therefore, it is expedient here to provide the known solutions.

The basic calculation formula that relates squeezing force and speed of the vertical movement of disk takes the form:

$$F = \frac{3\pi\eta V R^4}{8h^3} \qquad [5.4.1]$$

where η is the measured viscosity. Notations of geometric dimensions are shown in Fig. 5.4.2. In this case, the value 2h is the variable distance between the disks. The rate of disks closure, V, equals dh/dt.

If the sample volume is $v = 2\pi R_0^2 h_0$ (where R_0 is the initial radius of the sample, and $2h_0$ is the initial distance between the disks), then from Eq. 5.4.1 it is possible to exclude R, since in the course of experiment the condition of the volume constancy, $R_0^2 h_0 = R^2 h$ is satisfied.

Then, it is possible to write the following equation:

$$F = \frac{3\pi\eta R_0^4 h_0^2}{8h^5}\frac{dh}{dt} \qquad [5.4.2]$$

This equation is integrated relative to h(t), resulting in the following equation:

$$\frac{1}{h^4} - \frac{1}{h_0^4} = \frac{32tF(t)}{3\pi\eta R_0^4 h_0^2}$$

[5.4.3]

Then, by plotting the dependence of $h^4(t)$ versus $F(t)$, it is not difficult to find viscosity from the slope angle of the obtained straight line.

It is usually more convenient to measure the dependence of $R(t)$ than $h(t)$. Then Eq. 5.4.3 can be converted after using substitution $h = h_0(R_0/R)^2$. This leads to the following relationship:

$$R^8 = R_0^8 + \frac{32tF(t)R_0^4 h_0^2}{3\pi\eta}$$

[5.4.4]

or

$$R^8 = R_0^8 + \frac{32tF(t)v}{3\pi^2\eta}$$

[5.4.5]

where V is the constant volume of the test specimen. Using these formulas it is convenien⁻ to determine viscosity by plotting dependence R^8 as a function of t at $F = $ const.

If a surplus of liquid being investigated is extruded from the working clearance between the disks, as shown in Fig. 5.4.2a, then integration of Eq. 5.4.1 under the condi tion $R = R_0 = $ const gives:

$$\frac{1}{h^2} - \frac{1}{h_0^2} = \frac{16tF(t)}{3\pi\eta R^4}$$

[5.4.6]

Then viscosity can be found from the dependence of h^{-2} and $tF(t)$. If the experiment is carried out under the condition of $F = $ const, then viscosity is determined by plotting the dependence of h^{-2} versus t.

The above-written relationships and calculation formulas permit the determination of viscosity based on experimental data, but with one fundamental limitation that viscosity must be constant. Analogous but more complex formulas can be obtained for other liquids, with rheological properties described by a certain model.

The solution to the hydrodynamic problem requires the assumption that there is no slip along the disk surface. This assumption is correct for Newtonian liquids. However, a squeezing plastometer is very frequently used for the rheological analysis of liquids, which can interact with a surface by complex means. Therefore, much attention is given o the theory of liquid spreading during squeezing of disks, using different boundary condi-tions for the rigid surface.[53]

The absence of friction (slip) on a solid wall (a case opposite to flow when fluid adheres to the solid wall), the equation for calculation of force takes a form:[54]

$$F(t) = \frac{3\pi\eta R_0^2}{h(t)}\frac{dh}{dt}$$

[5.4.7]

In many testing machines, the method of measurements is realized under the regime of $V_0 = dh/dt = \text{const.}$ Then integration of equation (5.4.7) gives:

$$F(t) = \frac{3\pi\eta R_0^2 V_0}{h(t)} \qquad\qquad [5.4.8]$$

Hence, the viscosity of liquid being investigated is determined from the measured dependence of F(t) and $h^{-1}(t)$.

The method of velocity measurement of the motion of the upper disk is the most critical element of the construction of squeezing flow plastometers, since its displacements are small. In the simplest case, this is achieved manually using an indicator of displacement of the dial-type and stopwatch. In more advanced instruments, displacement transducers of induction or capacitor type are used. For precision measurements, an interferometric method is used.[55] This method allows one to measure very low speeds, as low as 3×10^{-7} mm/s, although the usual operating range is higher by two orders of magnitude.

The critical element of the design of the squeezing flow plastometer is a strictly parallel installation of disks. Analysis[56] shows that parallel misalignment of even 1-3° can give significant errors during the processing of experimental data.

As a result of the simplicity of the device and comparatively low cost, squeezing flow plastometers found wide acceptance in the industry. Here, the *Williams plastometer* should be noted. It was proposed in 1924 and until now it is widely used in practice in different modifications. The use of this instrument is standardized for determining the conditional characteristic of viscometric properties of rubber compounds, called *plasticity*.[57]

The squeezing flow plastometers also received wide acceptance in control of the degree of scorching of rubber compounds, for which the *Goodrich plastometer* is used.

The variation of the squeezing flow plastometers are defometers – instruments in which the load, which causes deformation, is measured. By this method, the rigidity of natural rubbers and unvulcanized rubber compounds is evaluated.[58]

The diverse variants of plastometers of industrial purpose (first of all, for the rubber industry) are produced by a number of instrument-manufacturing companies in different countries.

5.4.3 METHOD OF TELESCOPIC SHEAR

The shear between parallel planes is shown in Fig. 5.4.3. Here, flow is accomplished between the coaxial cylinders, but in contrast to the rotary instruments, one of the cylinders does not revolve but moves along the axis. This flow is called *telescopic shear*, and instruments in which this flow is realized are called the *Pochettino viscometers*.[59]

If the clearance between coaxial cylinders is small in comparison with their radii, i.e., $[(R_o - R_i)/R_i] \ll 1$, then the curvature of the channel can be disregarded, and telescopic shear proves to be practically identical to shear in plastometers with parallel plates. But if clearance is not small, then the solution of telescopic flow is sufficiently simple only for Newtonian liquids. The rate of the lowering inner cylinder, V, which moves along axis under its own weight is expressed as follows:

$$V = \rho_1 g \frac{R_o^2 - R_i^2}{4\eta} + \frac{(\rho_s - \rho_1)gR_i^2}{2\eta} \ln\left(\frac{R_o}{R_i}\right) \tag{5.4.9}$$

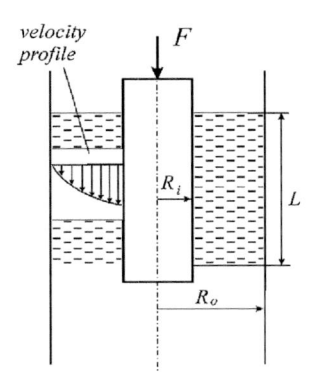

velocity profile

where ρ_1 and ρ_s, are, respectively, the density of the liquid being investigated and the material from which the inner cylinder is made, g is the acceleration due to gravity.

Hence, by measuring the speed of lowering of the inner cylinder, it is possible to find viscosity.

Usually, plastometers with telescopic shear are used for viscosity measurements at low deformation rates. This is due to a comparatively small force, created by the weight of the moving cylinder. At the same time, for viscosity measurements of high viscosity materials, such as polymer melts, rubber compounds, asphalts and bitumens (with viscosity in the range of 10^5-10^9 Pa*s), this force is insufficient for measurements within a convenient time range. Then, it is expedient to add weight to create an external force, F, as shown in Fig. 5.4.3.

Figure 5.4.3. Plastometer with telescoping flow.

This method was effectively used in laboratory installation, in which the external force was created by pressure of compressed gas.[60] With a radius of cylinders of the order of 25 mm and the clearance of 0.25 mm between them, a very rapid application of the force using aerodynamic drive created shear rates of the order of 10^5 s^{-1} for the duration of the experiment of about 0.01 s. The speed of movement of the cylinder was measured using an induction sensor, and the force, holding the external cylinder from movement, was measured with a very rigid piezoelectric transducer. The pressure of 1 MPa was reached, which made it possible to measure values of viscosity up to 150 Pa*s.

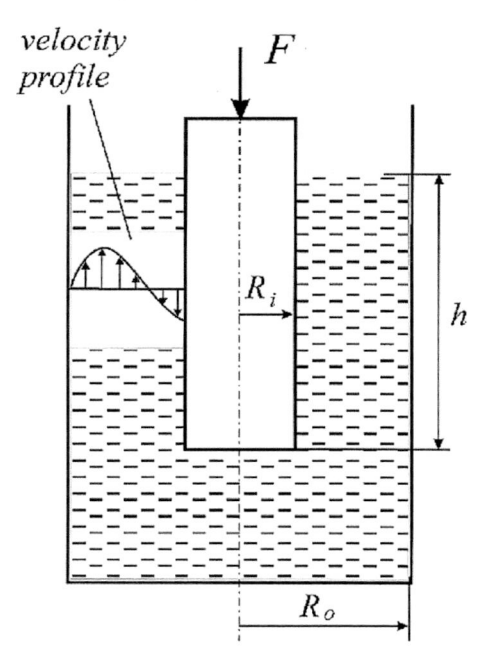

velocity profile

5.4.3.1 Telescopic shear penetrometer

A *Penetrometer* is a variation of plastometer utilizing a method of telescopic shear. Penetrometers are a separate group of instruments. The operating principle is based on the indentation (*penetration*) of a solid body by a tool called the indentor. The instrument, a schematic of which is shown in Fig. 5.4.4, occupies an intermediate position between plastometers and penetrometers. If the clearance between cylinders is small, then this measuring device is close to the classical plastometers. A difference in the diagram, shown in Fig. 5.4.4, from the basic principle of measurements during the telescopic flow,

Fig. 5.4.4. Telescoping shear in cylinder with closed bottom (Penetrometer).

lies in the fact that the outer cylinder is made with a closed bottom. Therefore, the telescopic shear between the cylinders is accomplished because the material under study is extruded between the cylinders from the bottom part of the instrument. Liquid resists a movement of the inner cylinder and the telescopic shear of liquid between cylinders occurs. Profiles of velocities of liquid flow between cylinders are different, as shown schematically in Figs. 5.4.3 and 5.4.4.

The basic kinematic feature of telescopic flow in the instrument, shown in Fig. 5.4.4, is that the height of the layer between cylinders changes with time. As a consequence, the rate of motion of the inner cylinder under the assigned force is variable.

The relationship between the force of insertion, F, and the rate of motion of the indentor (inner cylinder), $V = dy/dt$, is established by solving the hydrodynamic problem. Such a solution, known for Newtonian liquids, takes the form:

$$F = V \frac{2\pi y \eta \dfrac{R_o^2 + R_i^2}{R_o^2 - R_i^2}}{\dfrac{R_o^2 + R_i^2}{R_o^2 - R_i^2}\left[\ln\left(\dfrac{R_o}{R_i}\right) - 1\right]} = 2Ky\eta \qquad [5.4.10]$$

where K is the geometric (shape) factor, whose structure is evident from Eq. 5.4.10, and y is variable insertion depth.

Further, using equality $V = dy/dt$, it is possible to obtain the following dependence of insertion depth on time:

$$y(t) = \sqrt{\frac{Ft}{K\eta}} \qquad [5.4.11]$$

By plotting the dependence y versus $\sqrt{t}$, it is possible to find viscosity.

The possible method of viscosity measurement is to specify a certain assigned insertion depth, y_0, and measuring time, t_0, required to achieve this depth under assigned load, F. Then, it is pertinent that

$$\eta = kt_0 \qquad [5.4.12]$$

where coefficient $k = F/Ky_0^2$ is constant under selected (assigned) conditions of the experiment.

In some cases, the resistance, exerted by the bottom surface of the indentor during its penetration into a liquid, can be disregarded in comparison with resistance caused by telescopic flow. This happens, for example, when the indentor has the form of a needle with a sharp tip. With this form of the indentor, the shape factor is expressed by simpler means, namely

$$K = \frac{\pi}{\ln(R_o/R_i)} \qquad [5.4.13]$$

The above-written formulas cannot be used for non-Newtonian liquids. However, if the dependence of $\sigma(\dot{\gamma})$ is known, then the corresponding hydrodynamic problem for telescopic flow can be solved. In practice, the method of telescopic shear, especially in the version of combination with a penetrometer, is used only as a relative method of evaluating viscometric properties of the material.

Penetrometers have found widespread use in technological practice. They are produced by a number of companies in different modifications. An instrument of such type is known as the *Hoeppler consistometer* (Germany). In the USA, the penetrometers are produced by Gardner. The geometric shape of the indentor can be arbitrary. For determining the softness of natural rubber, plastic materials, and unvulcanized rubber compounds, the so-called *Humboldt penetrometer* is used. Tests on this instrument are done as follows. Under a load of 150 g, the needle of the diameter of 1 mm with a rounded apex is pressed into a sample of standard dimensions. Penetration continues for 10 min, after which the indentation depth is measured. The latter value serves as a measure of the rheological properties of the material.

In some technical applications, an indentor, prepared in the form of a cone, is used This is especially convenient for tests of high viscosity, filled, polymeric compositions. In this case, measurement of penetration velocity gives the characteristics of viscometric properties of the material, and the maximum indentation depth, reached during the application of a specific load, characterizes certain structural strengths of the material. In the simplest case, this strength, F_Y, is defined as

$$F_Y = \frac{F}{Kh^3} \qquad\qquad [5.4.14]$$

where the shape factor, K, is expressed as $K = \pi\cos^2\alpha\cot\alpha$ with α being half of the vertex angle of the cone.

The main methodological problem arises because of measurements of small deformations (displacements). This difficulty is especially significant when the application of large contact loads is inadmissible since measurements are carried out at low stresses and therefore small displacements. The general solution to such a problem is to utilize noncontact methods of measurements, for example, using optical methods (including interferometry), and also mechano-electronic lamp sensors.

5.5 METHOD OF FALLING SPHERE

5.5.1 PRINCIPLES

Experimental methods, considered in this section, are based on the measurement of resistance to motion of a solid body in liquid. In contrast to all methods discussed in previous sections, these methods are about external flow around solid bodies. The resistance to motion of a solid body in the liquid is determined by its viscosity. Two versions of a method can be envisioned: one is velocity measurement of motion under a given force and another is the force measurement under a given speed of motion. These two versions are completely analogous to methods used in the theory of capillary or rotational viscometry where preselection of one parameter and measurement of another parameter from a pair pressure/volumetric flow rate or torque/velocity of rotation is used.

A theory of the absolute method of viscosity measurements will be examined below. The theory is based on the simplest measuring scheme: a solid sphere is used as a moving body. The sphere falls along the axis of a cylindrical tube filled with the liquid being

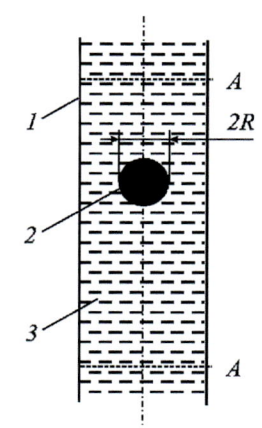

Figure 5.5.1. Scheme of measurements of viscosity by method of a falling sphere. 1 – tube; 2 – sphere; 3 – fluid; A – marks.

tested, and the velocity of the sphere is measured. This is shown in Fig. 5.5.1. It is assumed that the tube is vertical. The theory of motion of a solid sphere, which can be further refined, is valid as the first approximation under the following assumptions:

- the motion is steady, i.e., the rate of sphere descent is constant
- the inertia effects are considered negligible, i.e., motion occurs at low values of Reynolds number, Re (Re << 1)
- liquid is Newtonian
- the wall effect on a sphere motion is neglected, i.e., the sphere radius is much smaller than the tube radius
- the sphere moves strictly in a vertical direction.

During the sphere motion of a radius, R, made out of a material having a density, ρ_s, in a liquid having a density, ρ_l, under the action of the gravitational force, the driving force equals:

$$F = \frac{4}{3}\pi R^3 (\rho_s - \rho_l) g \qquad [5.5.1]$$

where g is the gravitational acceleration. The force resisting motion of a sphere in a viscous fluid is described by the *Stokes formula*:

$$F = 6\pi R U_\infty \eta \qquad [5.5.2]$$

where U_∞ is the rate of steady (indicated by subscript ∞) descent of sphere and η is the viscosity of the investigated liquid.

Under the assumption of the equality of the two above-written expressions (motion is assumed to be steady), the following formula for determining viscosity from the measured rate of sphere descent is obtained:

$$\eta = \frac{2}{9} \frac{(\rho_s - \rho_l) g R^2}{U_\infty} \qquad [5.5.3]$$

The measured viscosity is inversely proportional to the steady state-velocity of sphere descent and linearly proportional to the difference in densities of sphere and liquid. It is usually considered that this formula is applicable if Re < 0.1. The Reynolds number is calculated from

$$Re = \frac{2 R U_\infty \rho_l}{\eta} \qquad [5.5.4]$$

The shear stress on the surface of the moving sphere, σ_R, can be estimated according to the following equation:

$$\sigma_R = \frac{1}{3} R (\rho_s - \rho_l) g \qquad [5.5.5]$$

5.5.1.1 Corrections

5.5.1.1.1 Correction for inertia effect

If the condition Re < 0.1 is not fulfilled in this experiment, then it means that it is necessary to make the correction for the inertia effect using the following expression:[61]

$$F = 6\pi R U_\infty \eta \left[1 + \frac{3}{16} Re - \frac{19}{1280} Re^2 + ... \right] \qquad [5.5.6]$$

The expression in brackets is a correction for the inertia effect. Eq. 5.5.6 must be used instead of Eq. 5.5.2 to introduce correction. It is sufficient to include the second term in the correction factor. Taking into account this correction, an expression for calculating viscosity takes the form:

$$\eta = \frac{2}{9} \frac{(\rho_s - \rho_l) g R^2}{U_\infty} \left(1 - \frac{27 U_\infty^2}{16 g R} \frac{\rho_l}{\rho_s - \rho_l} \right) \qquad [5.5.7]$$

Correction for the inertia effect is negligible if inequality $U_\infty \ll \sqrt{gR}$ is fulfilled.

5.5.1.1.2 Correction for wall effect

In many practically important cases, the wall effect of a tube on the results of measurement cannot be disregarded, i.e., an inequality $R \ll R_0$ is not fulfilled. In this case, it is necessary to make the correction for the wall effect. If η is viscosity, found from the first approximation, i.e., according to Eq. 5.5.3, then the viscosity η_0, calculated by taking into account wall effect is determined as[62]

$$\eta_0 = \eta \left[1 - 2.104 \frac{R}{R_0} + 2.09 \left(\frac{R}{R_0} \right)^2 - 0.95 \left(\frac{R}{R_0} \right)^4 \right] \qquad [5.5.8]$$

The correction for the wall effect can be neglected only if $(R/R_0) < 0.001$, which is a sufficiently restrictive condition that is usually not satisfied. Therefore, an introduction of correction according to Eq. 5.5.8 is usually necessary.

5.5.1.1.3 Correction for non-Newtonian behavior of liquid

A method of viscosity measurement based on the velocity of sphere descent is intended for the analysis of Newtonian liquids. However, it is *a priori* unknown, if the liquid being investigated exhibits Newtonian or non-Newtonian properties. But if the experiment does show that apparent viscosity, calculated from the above-given formulas, is dependent on shear stress, then a basic method is to extrapolate the obtained dependence of apparent viscosity and the shear stress, η(σ), to a limit $\sigma \to 0$. For this purpose, it is assumed that the dependence η(σ) is described by known formulas. For example, it is possible to represent η(σ) by the following empirical dependence

$$\eta = \frac{\eta_0}{1 + K\sigma^2} \qquad [5.5.9]$$

where η_0 the initial Newtonian viscosity, K is the constant.

Then, after presenting experimental data in coordinates of η^{-1} vs. σ^2 and extrapolating them to $\sigma \to 0$, it is possible to find a value of η_0. In this case, a constant K is also determined. It can be related to the characteristic relaxation time of non-Newtonian liquid.

An analogous method is used if rheological properties of the liquid are described by a different equation than Eq. 5.5.9: a selection of the form of dependence of $\eta(\sigma)$ is determined by the selection of coordinates which permit linearization of experimental data in order to ensure their reliable extrapolation to $\sigma = 0$.

The use of the methodology of processing experimental data presented above makes it possible to obtain values of viscosity with an error not exceeding fractions of a percent.

Additionally, the increase in accuracy of measurements is achieved by the application of the following recommendations: velocity measurements should be conducted in the middle part of the tube (in order to eliminate inertia effect at the initial stage of sphere descent and the end effects at the bottom of a tube); averaging data, obtained in repeated measurements after turning a tube by 180° (in order to exclude a non-perpendicularity of tube axis); use of an automatic method of timing motion of a sphere between two markers (in order to exclude a subjective factor in measurements), etc. In the measurement of viscosity of Newtonian liquids, by varying material from which a sphere is made, or the sphere radius, it is possible to conduct measurements in the range of shear stresses from ~1 to 100 Pa and to measure viscosity in the range from $<10^{-3}$ to 10^2 Pa*s.

5.5.2 METHOD OF THE ROLLING SPHERE

A cylindrical tube can be used under a preselected angle, φ, in one of the variations of the method. The driving force, instead of Eq. 5.5.1, is written as follows:

$$F = \frac{4}{3}\pi R^3 (\rho_s - \rho_l)\cos\varphi \qquad [5.5.10]$$

The motion of the sphere occurs by rolling and sliding along a wall. The diameter of the tube is made slightly larger than the diameter of the sphere so that in reality flow of liquid occurs in a comparatively narrow clearance between the sphere and the tube wall. A solution of the flow of Newtonian liquid can be written as follows[63]

$$\eta = \frac{Cp}{U_\infty} \qquad [5.5.11]$$

where p is the pressure exerted by a sphere of weight, F, namely $p = F/\pi R^2$, U_∞ is the velocity of a steady motion of the sphere, C is the instrument constant. The theory gives the following formula for constant, C, expressed through the diameter of the sphere, D, and the clearance, δ, between the tube wall and the sphere:

$$C = 0.18\frac{(D/\delta - 1)^{2.5}}{1 + 0.956(D/\delta - 1)} \qquad [5.5.12]$$

Use of Eqs. 5.5.11 and 5.5.12 allows one to consider instruments with the rolling sphere as absolute instruments. However, usually, instruments of such type are used for comparative measurements of viscosity. In this case, Eq. 5.5.11 is written in a somewhat different form, namely

$$\eta = \frac{C_0(\rho_s - \rho_l)g\cos\varphi}{U_\infty} \qquad [5.5.13]$$

with the instrument constant, C_0, found during calibration of the instrument for each sphere, available in a complete set.

For evaluating shear rates and shear stresses, at which viscosity measurements in instruments with the rolling sphere are made, it is possible to use the following formulas:

$$\dot{\gamma} = \frac{2(U_\infty/\delta)}{(1 - \delta/D)^2} \qquad\qquad [5.5.14]$$

and

$$\sigma = 0.2(1 - \delta/D)^{1/2} \qquad\qquad [5.5.15]$$

Instruments of the type under consideration are rarely used for measurements of the viscosity of non-Newtonian liquids, although the theory of such measurements is known.[64]

5.5.3 VISCOMETERS WITH THE FALLING SPHERE

Production of instruments with a falling sphere is simple. It is sufficient to have a glass tube, stopwatch, and a sphere of a known diameter.

Commercial viscometers preferably do not use a free fall, but a rolling sphere. This helps in avoiding eccentricity during the free fall and gives reproducible results. Such instruments operate as viscometers for measuring relative viscosity, and their calibration is carried out on standard liquids.

Viscosity measurements by the falling sphere method are standardized in a number of countries (e.g., ISO/DIN 12058, DIN 53015, etc.).

The *Hoeppler viscometer* (Germany) is one of the commercial instruments. The instrument cylinder is made from heat-resistant glass. The cylinder has a diameter of 16 mm, a length of 200 mm. The cylinder is installed at an angle of 10° to vertical. The cylinder is immersed in a liquid thermostat, maintaining temperature in the range from -35 to 150°C. The experiment consists of the measurement of time for the sphere to move between two markers, located on the cylinder in its middle part. The viscometer is supplied with a collection of spheres, made of materials of different densities, e.g., steel, glass, tungsten. Viscosity can be measured from a few hundredths to 200 Pa*s.

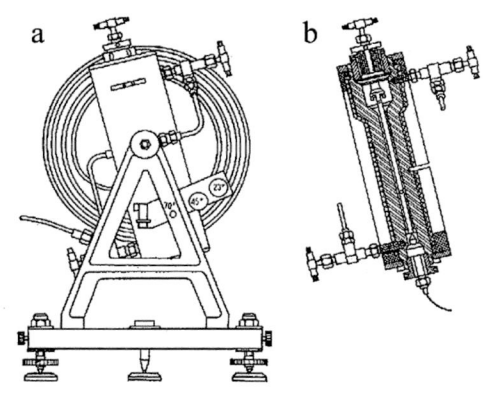

Many companies produce these instruments. *Physica* produces a portable viscometer, *Anton Paar AMVn*. Special features of this instrument are a very small volume of the test sample (to 150 ml), a possibility of changing the inclination angle of the cylinder from 15 to 80°, a possibility of conducting measurements in the

Figure 5.5.2. Rusk viscometer with a rolling ball. a – general view; b – tube placed in a high pressure chamber. (Rusk Instruments, USA).

range of temperatures from 10 to 100°C, automation of measurements and processing of the obtained results.

An interesting modification of instruments with a falling sphere is the *Rusk viscometer* manufactured by *Rusk Instruments*, USA (see Fig. 5.5.2). Measurements of viscosities can be carried out under high pressures. In a working cylinder of this instrument, a pressure of up to 84 MPa is created. The instrument can work at temperatures up to 350°C. The construction of the instrument makes it possible to vary the angle of inclination of the cylinder. Measurements of ball velocity are accomplished with an electrical detector, which makes it possible to automate the process of measurement.

Laboratory instruments, analogous in designs to the Rusk viscometer, are known in which pressure and temperature can, respectively, reach 500 MPa and 400°C.[65] A working cylinder in this instrument is prepared from a quartz tube with a diameter of 7.125 mm. The tube can be installed at different inclination angles. The velocity measurement is accomplished by a non-contact method using capacitance pickups, which are the Nichrome rings, attached to the tube. For obtaining clear signals, the liquid being investigated must have a specific resistance not lower than 10^6 Ohm*cm.

5.5.4 VISCOMETERS WITH FALLING CYLINDER

In the laminar motion of any body in a viscous, Newtonian liquid, a force resisting motion, F, is proportional to the viscosity of the medium. Therefore, by measuring F, it is possible to determine viscosity. In viscometers in which this principle is used, sometimes instead of a falling sphere, a body of the cylindrical form is used.

During the motion of a cylinder having a radius of R_i along its axis being coaxial with the cylindrical tube having a radius of R_o, a flow occurs in the space between two cylindrical surfaces. The velocity profile is described by a known formula, easily obtained from the equilibrium condition of a fluid element by taking into account corresponding boundary conditions. This formula takes the form:

$$V(r) = V_0 \frac{\ln(r/R_o)}{\ln(R_o/R_i)} \qquad [5.5.16]$$

where $V(r)$ is the velocity, which depends on the radial coordinate, r, V_0 is the velocity of motion of the inner cylinder.

Hence, the following expression for a force, f, resisting motion per unit length of the cylinder is obtained

$$f = 2\pi R_i \eta \frac{dV}{dr} = \frac{2\pi \eta V_0}{\ln(R_o/R_i)} \qquad [5.5.17]$$

and the shear rate (derivative dV/dr) is calculated on a surface of the inner (moving) cylinder.

Eq. 5.5.17 makes it possible to calculate a component of resistance to the motion of the cylinder, that acts on its lateral surface. A complete expression for the velocity of a body descending in a liquid under its own weight, V_0, with a difference in the material densities $(\rho_s - \rho_l)$ of the solid body and liquid, takes the form[66]

$$V_0 = \eta g(\rho_s - \rho_l)\frac{R_i^2}{2}\left(\ln k^{-1} + \frac{k^2 - 1}{k^2 + 1}\right)\chi \qquad\qquad [5.5.18]$$

where $k = R_i/R_o$, and χ the correction factor due to edge effects, the values of which can theoretically be computed.

Checking this formula for Newtonian liquids with viscosity from 0.08 to 5 Pa*s shows that viscometers under consideration permit one to determine viscosity with the error of up to 0.5% if the Reynolds number does not exceed 25.

In spite of the existence of theoretical background for viscometers with falling cylinders, they are usually used for relative measurements of viscosity, by calculating viscosity from a measured rate of fall of a cylinder, V_0, as

$$\eta = \frac{B V_0}{\rho_s - \rho_l} \qquad\qquad [5.5.19]$$

where B is the instrument constant, determined during its calibration using a standard liquid of known viscosity.

Viscometers with a falling cylinder are very simple to construct, but their use in practice is limited since they do not possess essential advantages of viscometers with the falling sphere. These instruments are used for measuring the viscosity of foaming liquids under conditions of high hydrostatic pressure.[67] For this application, the instrument has an enclosed high-pressure chamber with noncontact measurement (by induction sensor) of rod velocity (rod is rigidly connected to the falling cylinder). The cylinder, lifted to a given height falls under its own weight. Measurements of viscosity at pressures up to 15 MPa are possible.

5.6 EXTENSION

5.6.1 GENERAL CONSIDERATIONS

Important and unique methods of determining viscometric properties of liquids are measurements that are carried out by uniaxial extension. Mechanical properties of solid materials and rubbers under stresses are measured by uniaxial extension. A basic methodological difficulty exists for liquid measurements. The integrity of the stream has to be maintained in order to make successful measurements. This is difficult or even impossible with low viscosity liquids, such as water. But for high-viscosity fluid (e.g., polymer melts) integrity of a measured sample can be maintained for the duration of the measurement. The method is mostly used for polymeric materials.

General theoretical considerations, related to the extensional flow, were examined in Section 3.7. Total deformation has to be separated into irreversible and elastic components. For Newtonian liquids, there is a correlation between viscosity measured during shear flow and longitudinal viscosity (the *Trouton law*). The same is correct for a linear viscoelastic body, since viscoelastic characteristics, measured in shear flow, with factor 3 are equal to the analogous characteristics, measured by elongational flow (see Eq. 3.7.2). Results of testing elongational flow in the nonlinear region give new independent information about rheological properties of the material, which cannot be obtained from shear

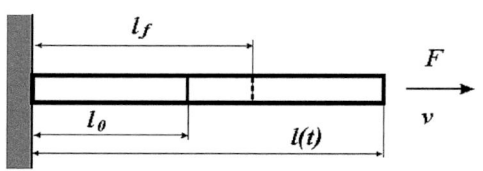

Figure 5.6.1. Typical diagram illustrating uniaxial extension.

flow measurements. For these reasons, the linear extension is an important independent type of experiment. The information has theoretical value in rheological studies and it is important for many practical applications, for example, in the analysis of processes of fiber spinning, film drawing, etc.

Uniformity of deformation is of fundamental importance in uniaxial extension experiments. During extension experiments with viscous fluids, the sample retains its integrity because viscosity is very high ($\sim 10^9$ Pa*s). But the sample decreases its surface of cross-section, which eventually leads to its rupture. With the extension of polymeric viscoelastic liquids, uniformity of sample shape along the length results from material elasticity, which is a stabilizing factor.

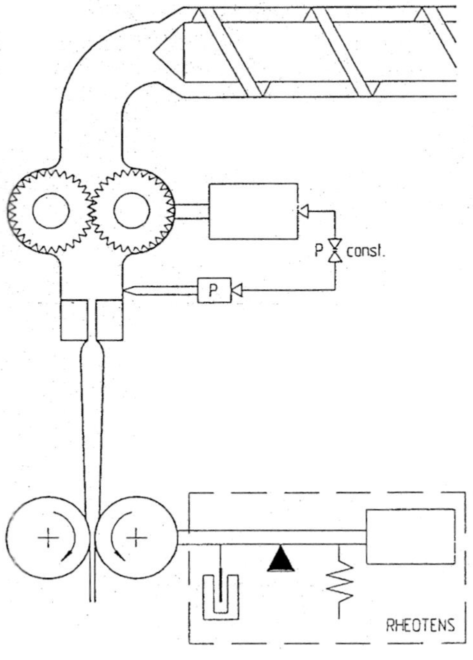

Let us examine a diagram of a sample, that, at one end is fixed in a clamp, and the other end is extended at a certain speed, v, by force, F, (Fig. 5.6.1). The initial length of the sample equals l_0. The length of the sample after extension for time, t, has length, l. Then, the external force is removed, and the sample restores its length to l_f. The difference between l_f and l_0 is a strain experienced during viscous flow. Measuring this strain in time, t, the strain rate is obtained. By measuring the reversible component of strain, i.e., the difference $(l − l_f)$, the elasticity of material under study can be determined.[68]

5.6.2 EXPERIMENTAL METHODS

5.6.2.1 The simplest measuring schemes

In a simple experiment, a sample is attached to a fixed clamp and a load is attached to another end of the sample. Measuring the rate of extension gives strain rate and known load gives stress. Using these data, apparent viscosity can be calculated as a measure of sample resistance to elongation, although, in a strict sense, calculated apparent viscosity is not

Figure 5.6.2. Instrument Rheotens for qualitative evaluation of extensibility and strength during elongation of polymer melts. [Adapted, by permission, from M. F. Wagner, B. Collington, J. Verbeke, *Rheol. Acta*, **35**, 117 (1996)].

true viscosity but a relative characteristic of rheological properties of the material in uniaxial extension. This characteristic is useful for the technological evaluation of materials.

The molecular weight of ultrahigh molecular weight polyethylene was estimated by uniaxial extension.[69] Stress at the extension of up to 600% was measured with different samples at 150°C for 10 min. The change of molecular weight from 2×10^5 to 10^7 (about 2 decimal orders) increases stress by 4 decimal orders.[70]

Rheotens[71] (Fig. 5.6.2) is a commercial instrument used for measurements of viscosity in extension. This instrument permits the determination of two basic parameters of polymer: apparent longitudinal viscosity and the maximum extension to rupture. Melt is supplied by an extruder into the head. Then, the extrudate is pulled from the head by rotating rollers. The speed of roller rotation can be varied. Pressure, P, at the entrance of the head and the force with which rollers pull the extrudate are measured (a block diagram of the latter is shown in the dash rectangle). By changing the feed rate of melt and roller rotation rate, the deformation rate can be varied. A maximum permissible speed of drawing characterizes the melt strength properties during the extension.

5.6.2.2 Tension in a controlled regime

Quantitative measurements of elongational viscosity in uniaxial extension consist of controlled extension through the imposition of constant strain and measurement of stress, or imposition of constant stress and measurement of strain rate. In essence, the methodology here is the same as in any other rheological experiment, in which one parameter (kinematic or dynamic) is maintained constant and the other is measured.

The end of the extrudate in Fig. 5.6.1 is moved with a velocity of $v(t)$, which is varied with time in such a manner that strain rate (or similarly a gradient of the longitudinal velocity) remains constant. At the initial length of the sample, l_0, the initial strain rate equals v_0/l_0, where v_0 is the initial extensional velocity.

As extension progresses, the length of the sample increases according to the following expression

$$l(t) = l_0 + \int_0^t v(t)dt \tag{5.6.1}$$

In order for the strain rate to remain constant and equal to the initial value of $\dot{\varepsilon}$, it is necessary that the rate of extension increases according to the equation

$$v(t) = \dot{\varepsilon}_0 l_0 e^{\dot{\varepsilon}_0 t} \tag{5.6.2}$$

If stress is constant, then the initial tensile force, F_0, must decrease proportionally to elongation, since elongation is accompanied by a decrease of the cross-section of the sample. Thus, if the sample was stretched λ times, then the tensile force must decrease to the value of F_0/λ.

Automatic control of force or rate of extension must be used to conduct extension under controlled conditions. Early attempts used various mechanical devices, but now only electronic control systems are used. Measurement of longitudinal viscosity is shown in Fig. 5.6.3. The extension is accomplished using clamps with the speed of their rotation (also the strain rate) regulated by a drive system. The force is measured by a sensor connected to an elastic element.[72] In order to avoid sample bending, the sample is placed on the surface of the liquid. This liquid bath also serves as a thermostat.

The imposed strain rate is controlled by the selection of rotation frequency of driving rollers. If it is necessary to conduct tests under constant stress, then the force transducer

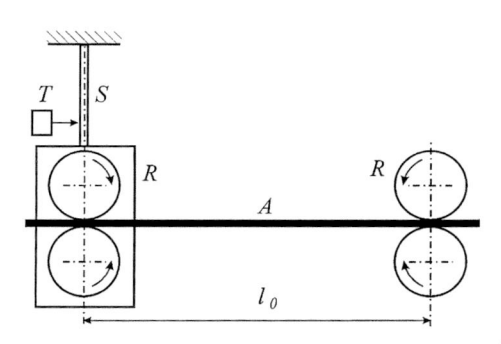

Figure 5.6.3. Principal scheme of testing of sample in uniaxial elongation under controlled conditions. A – sample; R – two pairs of drawing rollers; S – elastic element; T – force transducer.

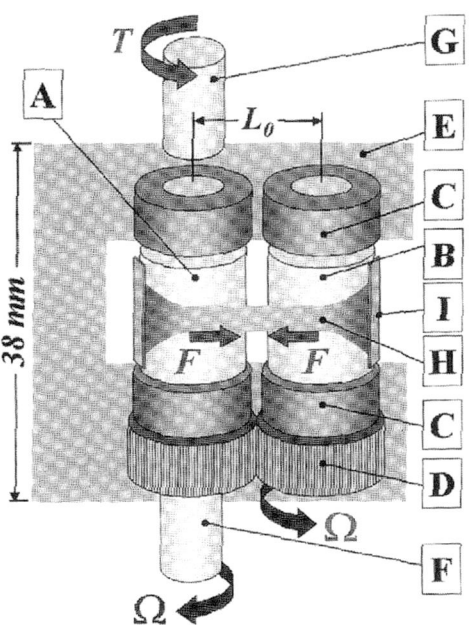

Figure 5.6.4. Scheme of the SER method of measuring extensional properties of a film. A and B – rotating shafts, C – cups, D - coupling, E - screen, F – drive, G – torque transducer, H – sample under study.

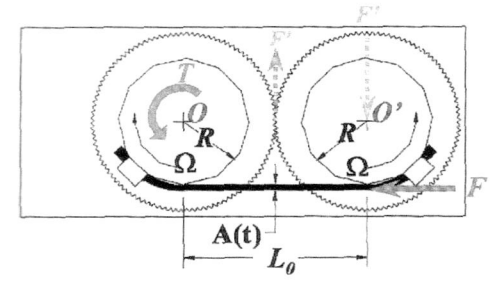

Figure 5.6.5. Section of a device showing the main parameter under measuring – time evolution of the width of a film A(t).

controls speed. Dimensions of the instrument permit measurements of extension up to 50 times the initial length of the sample.

These instruments can measure strain rates from $\sim 10^{-3}$ to 10 s^{-1}. Detailed analysis of errors confirmed the reliability of results and possible limitations of the method.[73]

The latest version of the extensional rheometer intended for high tensile stresses can create tensile stresses of up to 0.3 MPa for a time shorter than 0.02 s. The total deformation ratio can be as high as 100 (it corresponds to Hencky deformation of approximately 4.6). These parameters are achieved using a pneumatic loading system.[74]

An interesting and rather simple method for extensional rheometry of films is known as the Sentmanat Extensional Rheometer (SER).[75] The concept of the method is shown in Fig. 5.6.4 and the horizontal section of a sample is drawn in Fig. 5.6.5.

A film H (it can be rubbery-like or liquid film) is stretched due to the rotation of shafts A and B rotating with a given rate, Ω. The values under measuring are torque and thus stretching force F and the width of a film changing in time A(t) and thus the deformation of a sample. These values allowed us to calculate the rheological characteristic of a sample in extension. The Hencky strain rate applied to the sample is calculated as

$$\dot{\varepsilon}_H = \frac{2\Omega R}{L_0}$$

Then the transient elongational viscosity is calculated as

$$\eta_E^+(t) = \frac{F(t)}{\dot{\varepsilon}_H A(t)}$$

5.6.2.3 Tubeless siphon instruments

In a uniaxial extension of viscoelastic liquids a so-called *siphon effect*, described in Section 3.7.4 and shown in Fig. 3.7.10, is observed.

A complex velocity profile, arising during flow, does not permit us to consider this as an absolute method. However, direct measurements demonstrated that at a certain length of jet, the velocity gradient remains constant.[76] The latter gives the basis for quantitative calculations of elongational viscosity according to experimental data obtained in the regime of siphon flow. The strain rate is controlled by the variation of the speed of winding of the liquid jet. The force of extension is created either by connecting the free end of the capillary to a vacuum (as in Fig. 3.7.10) or by measuring the force acting on the winding roller (as in Fig. 3.7.10).

In known instruments of such type, the frequency of roller rotation is in the range from 0.1 to 170 s^{-1}, which gives velocity gradients of up to 200 s^{-1}. Such high rates of elongational deformation cannot be created in instruments with controlled regimes of extension. The attractiveness of this method is also related to the fact that with siphon flow it is possible to investigate comparatively low viscosity (but elastic), moderately dilute, polymer solutions that cannot be studied in instruments with controlled regimes of extension. At the same time, this method and instruments based on this principle should be considered as relative, since during the siphon flow it is difficult to separate viscous flow from the elastic deformation.

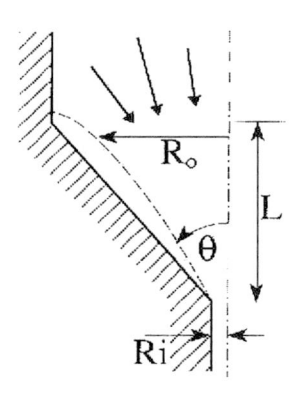

Figure 5.6.6. Flow in converging channel – longitudinal velocity gradient.

5.6.2.4 Flow in convergent channels

During an abrupt change in the cross-section of a channel, the longitudinal acceleration of flow is developed at the entrance to a small channel. It was proposed to use this special feature of flow for the evaluation of longitudinal viscosity.[77] The diagram of flow in a converging channel is shown in Fig. 5.6.6. It is evident that upon transfer from a tube having a radius, R_o, to a capillary having a radius, R_i, the cone-shaped flow is formed with an angle, θ, and length, L. In this zone, shear and elongational flows are superimposed.

In order to estimate the elongational viscosity, a pressure difference, Δp, at the entrance into a capillary should be measured. This difference is responsible for the elongational flow and determination of the normal stress. The rate of stretching deformation is found from rates of flow in the tube, V_o, and in a capillary, V_i, as

$$\dot{\varepsilon} = \frac{V_i - V_o}{L} = \frac{Q}{\pi L}\left(\frac{1}{R_i^2} - \frac{1}{R_o^2}\right) \qquad [5.6.3]$$

Hence, it is possible to find a relative value of the elongational viscosity, η_E.

However, a problem that remains undetermined during this evaluation is the contribution of shear strains. For stricter calculations of the velocity field and, correspondingly, the elongational viscosity, the approach of a lubricating film layer is used. According to this model, it is assumed that shear occurs in a very thin (lubricating) layer, and in the material bulk uniaxial extension occurs.[78] The problem with this concept is that correct theoretical calculation is possible only if the model of rheological behavior of investigated material is *a priori* known, which is not always possible.

5.6.2.5 High strain rate methods

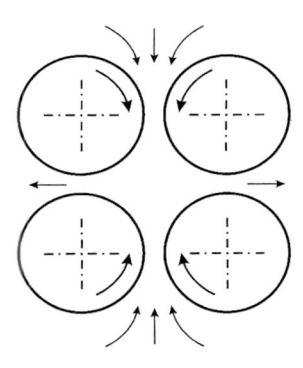

Figure 5.6.7. Four-rollers method for creation of longitudinal flow.

Several methods of studying elongational flow were proposed for the analysis of elongational deformations of dilute solutions, in the limit for studying the extension of individual macromolecules.[79] A diagram of the four-rollers method is shown in Fig. 5.6.7. A liquid is taken into the interroller space and then it is drawn out of it. There is a certain region where flow occurs in a pure shear mode that is equivalent to two-directional (planar) elongation (see Section 1.2.4.2). Using optical methods (birefringence) it is possible to estimate the gradient of longitudinal velocity and stress, and thus the elongational viscosity of polymeric solutions at deformation rates of the order of hundreds s^{-1}. This method is very informative for evaluating some physical properties of macromolecules (for example, see Fig. 3.7.11 and discussion in Section 3.7.5). This method also allows one to observe instabilities due to the elasticity of polymeric strands in dilute solutions.

Another optical method of investigation of the longitudinal flow of dilute polymer solutions includes the flow of liquid in two coaxial capillaries moving towards each other. Here, in a certain flow region, the uniaxial longitudinal flow is realized. Deformation rates can exceed 10^3 s^{-1}. Both methods can be used for studies in the field of polymer physics, in particular, to observe the coil-to-stretch transition of flexible macromolecules.

5.6.2.6 Capillary breakup elongational rheometry

The interest in extension properties of low viscosity substances (polymer solutions, emulsions) leads to the creation of a new type of instrument.[80] The method consists of an extension of a small liquid droplet placed between parallel plates. This liquid droplet has a cylindrical shape. Then, the plates are rapidly moved apart from each other. This type of deformation can be treated as a step strain. The evolution of the liquid bridge between plates is treated in the terms of viscous and viscoelastic properties of a matter. The final stage is the break-up of the sample. This method can be applied to liquids with viscosities down to several tens mPa*s.

The scheme of this method is presented in Figure 5.6.8 (first proposed in[81] and later widely used in many studies, including theoretical basing of the method, e.g.).[82]

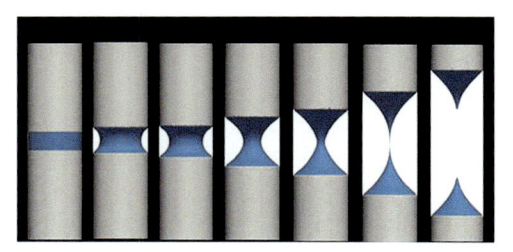

Figure 5.6.8. A scheme of the method for measuring deformability of a liquid droplet in extension.

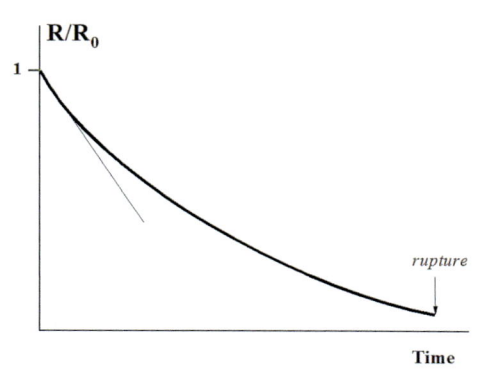

Figure 5.6.10. A typical time dependence of the sample radius at the most contracted position (at the middle of a sample).

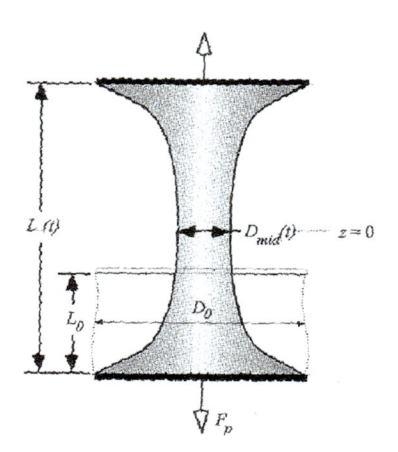

Figure 5.6.9. A sample with measured values.

A model of a sample under stretching is shown in Fig. 5.6.9. Measuring the profile of a sample surface allows for calculating the relaxation times, while the $L(t)$ function is determined by the driving device.

A typical view of the profile characterized by the changes in the radius at the most contracted position is shown in Fig. 5.6.10.

The complete analysis of the dynamic of stretching a viscoelastic filament was given elsewhere.[83] As the first approximation, the time dependence of the radius is described by the exponential function

$$R = R_0 e^{-t/3\theta}$$

This approach allows us to find the characteristic relaxation time, θ, of a visco-elastic liquid.

The important advantage of the described method is a possibility of its application to rather low viscosity liquids including colloidal systems and dilute polymer solutions. This opens definite promising aspects in characterizing liquids used in oil recovery industry and in other non-traditional areas. For example, this method can be used for estimation fiber spinning ability defined as $\ln (L_{max}/L_0)$, where L_0 is the initial length of a filament and L_{max} is the length at the point of rupture.[84]

5.6.3 BIAXIAL EXTENSION

Material testing using the biaxial extension (at equal or different strain rates in two mutually perpendicular axes), just as in a uniaxial extension, gives independent information about rheological properties of materials, and therefore, it is of interest. Biaxial extensional flow occurs in several polymer processing operations such as film blowing, blow

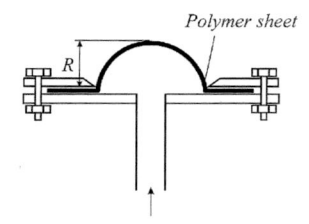

Polymer sheet

Figure 5.6.11. Schematic of bubble inflation rheometer. [Adapted, by permission, from C. D. Denson, R. J. Gallo, *Polym. Eng. Sci.*, **11**, 174 (1971)].

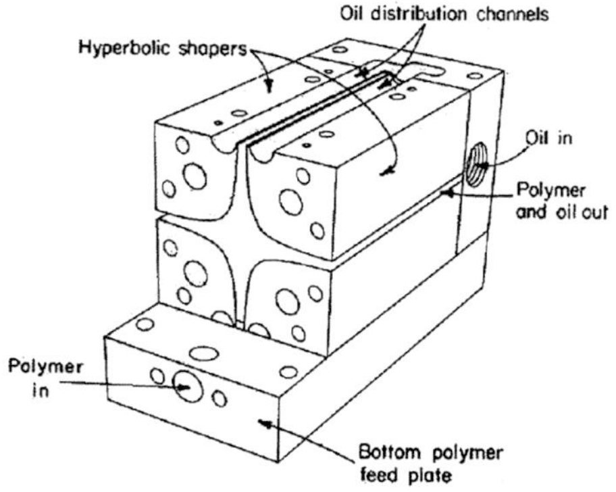

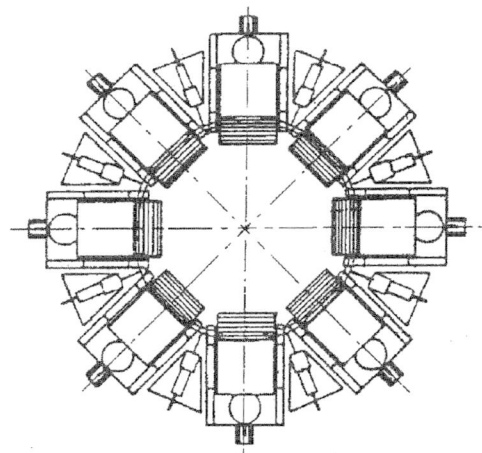

Figure 5.6.12. Schematic representation of die for creating lubricated planar stagnation flow of a polymer melt. [Adapted, by permission, from H. H. Winter, C. W. Macosko, K. E. Bennet, *Rheol. Acta*, **18**, 323 (1979)].

molding, and vacuum forming. Therefore, the rheological behavior of polymers subjected to biaxial extension is of prime importance in understanding and improving such processes. Significant research in biaxial rheology has occurred, primarily during the last two decades.

A number of methods are used to obtain a biaxial extension of polymer melts including sheet inflation,[85] axisymmetric and planar stagnation,[86] lubricated squeezing flow,[87] and sheet elongation.[88]

Sheet or bubble inflation involves a circular sample that is clamped around its perimeter (Fig. 5.6.11). Inert gas or silicone oil under pressure is introduced to one side of the sheet, causing inflation of the bubble. An equal biaxial elongation occurs at the top of the bubble. The strain rate is calculated by measuring the deformation of grid marks made on the sample. The stresses in the inflated bubble are related to the internal pressure by controlling which one can impose a constant stress or constant elongation rate.

An axisymmetric and planar stagnation flow involves impinging two melt streams through lubricated hyperbolic-

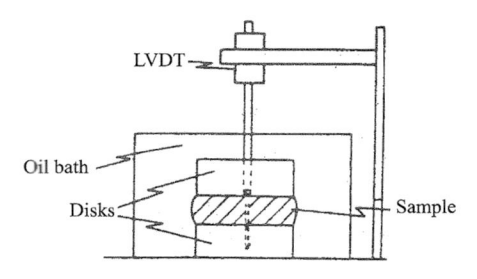

Figure 5.6.13. Schematic of lubricated squeezing flow instrument. [Adapted, by permission, from S. Chatraei, C. W. Macosko, H. H. Winter, *J. Rheol.*, **25**, 433 (1981)].

Figure 5.6.14. Instrument with rotary clamps for measurement of biaxial viscosity. [Adapted, by permission, from J. Meissner, S. E. Stevenson, A. Demarmels, P. Portmann, *J. Non-Newt. Fluid Mech.*, **11**, 221 (1979)].

shaped walls (Fig. 5.6.12). Constant biaxial elongation rates are obtained with stresses measured by the birefringence method or by measuring the pressure required to drive the polymer melt through the diverging and converging flow by means of a pressure transducer.

A lubricated squeezing flow is achieved by placing a polymer melt disk between two solid disks with their surfaces lubricated by a low viscosity fluid (Fig. 5.6.13). Squeezing is imposed by moving the top disk towards the lower disk at a controlled speed. Constant elongation rates can be achieved. The normal stress difference is calculated by measuring the normal force during squeezing flow and therefore biaxial viscosity can be measured.

A sheet elongation is achieved by stretching a sheet of polymer along its periphery using specially designed rotary clamps (Fig. 5.6.14). The rotary clamps are able to pull the sample at a controlled rate in two directions. The pulling force is measured on clamps. Depending on the arrangement of clamps, equal biaxial stretching or planar elongational flow can be achieved.

5.7 MEASUREMENT OF VISCOELASTIC PROPERTIES BY DYNAMIC (OSCILLATION) METHODS

5.7.1 PRINCIPLES OF MEASUREMENT – HOMOGENEOUS DEFORMATION

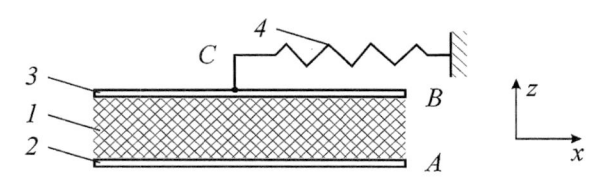

Figure 5.7.1. Principal scheme of measurements in forced oscillating mode. 1 – sample; 2 – driving plate A; 3 – plate B connected with a measuring device; 4 – spring of a measuring device with rigidity Z.

Let a uniform isotropic sample be placed between two parallel plates A and B (Fig. 5.7.1). The gap between the plates is small in comparison with the size of plates. The edge effects are assumed to be negligible.

Plate A is subjected to forced oscillations in accordance with a harmonic law:[89]

$$x_A(t) = x_{0A} e^{i\omega t} \qquad\qquad [5.7.1]$$

where x_A is a displacement of plate A depending on time t, x_{0A} is the amplitude, and ω is the frequency of oscillation.

It is assumed that slip on the boundaries of the plate is absent. Stresses appearing as a result of the movement of plate A are transferred *via* sample to plate B. The latter is joined to a stationary fixed frame of the measuring device through an elastic element – a spring of rigidity, Z.[90] The displacement of the upper plate is a measure of stresses in the sample which characterize the rheological properties of the material. It is assumed that a layer of material is thin enough to suppose that stresses inside the sample are uniform and the mass of the sample is negligible. The last suppositions will be formulated quantitatively below in Section 5.7.2.

If properties of material do not change in time, the movement of plate B will occur by a harmonic law with complex amplitude $x_B^* = x_{0B} e^{i\alpha}$ and with the same frequency, ω, as the movement of plate A:

$$x_B(t) = x_B^* e^{i\omega t} = x_{0B} e^{i\alpha} e^{i\omega t} = x_{0B} e^{i(\omega t + \alpha)} \qquad [5.7.2]$$

In this expression, x_{0B}, is the amplitude of oscillations of plate B and α is the *phase angle* – phase difference between displacements of both plates.

For material with arbitrary properties, the relationship between shear stresses, σ, and deformations, γ, can be written as

$$\sigma = G^* \gamma = G^* \left(\frac{dx}{dz}\right)$$

where G^* is the complex dynamic modulus depending on frequency.

The main experimental goal is to calculate $G^* = G' + iG''$ based on the pre-set value of x_{0A}, and measured values of x_{0B} and α, and known sizes of the measuring device at different frequencies.

In the ideal scheme of deformation considered here, the equation of motion of the point C in Fig. 5.7.1 is written as

$$m\frac{d^2 x_B}{dt^2} + Zx_B + \frac{S}{h}G^*(x_B - x_A) = 0 \qquad [5.7.3]$$

The value $(x_B - x_A)/h$ is deformation dx/dz, S is the surface area of plates, h is the gap between plates, the ratio $K = S/h$ is the form-factor, and m is the mass of plate B. This equation is solved by substitution of expressions for x_A (Eq. 5.7.1) and x_B (Eq. 5.7.2). The final result obtained after rearrangements is

$$G^* = \frac{(Z - m\omega^2)x_{0B}e^{i\alpha}}{K(x_{0A} - x_{0B}e^{i\alpha})} \qquad [5.7.4]$$

By separating real and imaginary parts of the expression for G^*, one obtains the following formulas for the components of the dynamic modulus:

$$G' = \frac{(Z - m\omega^2)(p\cos\alpha - 1)}{K[(p - \cos\alpha)^2 + \sin^2\alpha]} \qquad [5.7.5]$$

$$G'' = \frac{(Z - m\omega^2)p\sin\alpha}{K[(p - \cos\alpha)^2 + \sin^2\alpha]} \qquad [5.7.6]$$

and

$$\tan\delta = \frac{G''}{G'} = \frac{p\sin\alpha}{p\cos\alpha - 1} \qquad [5.7.7]$$

The factor p is the amplitude ratio: $p = x_{0A}/x_{0B}$.

Eqs. 5.7.5 and 5.7.6 are the solutions to the formulated problem. A particular case of this solution is important when $Z \gg m\omega^2$. The last inequality means that the inertia of the moving parts of plate B is negligible in comparison with the force of the deforming spring. The high rigidity of the spring also provides its small displacement. So it can be assumed that in this case p >> 1. This leads to the most simplified expressions for the components of dynamic modulus:

$$G' = \frac{Z}{Kp}\cos\alpha \qquad\qquad\qquad\qquad\qquad [5.7.8]$$

$$G'' = \frac{Z}{Kp}\sin\alpha \qquad\qquad\qquad\qquad\qquad [5.7.9]$$

The phase angle, α, equals loss angle, δ. However, this is not so in a general case, as seen from Eq. 5.7.7.

Eqs. 5.7.5 and 5.7.6 are valid in the whole frequency range except for the resonance frequency of the measuring device, ω_0, which is calculated as

$$\omega_0 = \sqrt{Z/m} \qquad\qquad\qquad\qquad\qquad [5.7.10]$$

The results of measurements in the vicinity of ω_0 are unreliable because even a slight error in measuring phase angle and amplitude ratio leads to a large error in calculating components of dynamic modulus.[91]

In the application of the principal scheme of the complex modulus calculation, a problem arises as to how to correctly measure small phase angles. One possible solution to the problem is to use controlled amplification of signals of a measuring device.[92] This method permits the measurement of phase angles as low as $1*10^{-4}$ rad. The other approach is based on the *correlation method* – removing noise of signals by measuring the angle in a number of cycles of oscillations. This improves reliability in calculating components of dynamic modulus.[93] In both cases, this problem is solved by the application of known electronic means and using computer techniques.

5.7.2 INHOMOGENEOUS DEFORMATIONS

An ideal scheme of measurements, as discussed in Section 5.7.1, did not take into account possible inhomogeneity in the sample deformation. However, a general case must include consideration of stress distribution inside a sample if a gap size between plates in Fig. 5.7.1 is not very small. Besides, it is necessary to give a strict definition of the limitation of the gap size in order to consider it to be small. The solution is based on the analysis of equilibrium of an infinitesimally thin layer of material of thickness, dz, which is parallel to the plates. The equilibrium equation for this layer is

$$\rho\frac{d^2x}{dt^2} - G*\frac{\partial^2 x}{\partial z^2} = 0 \qquad\qquad\qquad\qquad\qquad [5.7.11]$$

where ρ is the density of material under investigation.

The displacement x depends on time as well as on the distance from plate A designated by coordinate z (for plate A, z = 0). The general solution of Eq. 5.7.11 is

$$x(z, t) = x_0^*(z)e^{i\omega t} = x_0 e^{i\alpha}e^{i\omega t} \qquad\qquad\qquad\qquad\qquad [5.7.12]$$

where x_0^* is the complex amplitude of oscillations in a layer located at position z and α is the phase angle in this layer (depending on z).

Substituting Eq. 5.7.12 into Eq. 5.7.11 gives the linear second-order differential equation for x_0^*

$$G^* \frac{d^2 x_0^*}{dz^2} + \rho \omega^2 x_0^* = 0 \qquad [5.7.13]$$

The integration constants are found from boundary conditions:

$$x_0^* = x_{0A} \text{ at } z = 0$$

and the equilibrium conditions for forces acting at plate B:

$$\frac{\partial x_0^*}{\partial z} = \frac{x_B^*(m\omega^2 - Z)}{SG^*} \text{ at } z = h.$$

The solution of Eq. 5.7.13 with these boundary conditions gives the function $x(z, t)$ and its particular case at $z = h$: $x(h, t) \equiv x_B^*(t)$. Then, the ratio x_{0A}/x_B^* is found which is expressed in the following manner

$$\frac{x_{0A}}{x_B^*} = \cosh(kh) - \frac{m\omega^2 - Z}{SG^*}\cosh(kh) \qquad [5.7.14]$$

where $k = \omega\sqrt{-\rho/G^*}$

Eq. 5.7.14 is the complete solution of the problem under discussion because it is an equation for G^* including all necessary parameters – x_{0A} and x_B^*, the latter being determined if amplitude, x_{0B}, and phase angle, α, have been measured.

However, this equation is not convenient for practical applications because the unknown value G^* enters not only the coefficient of the second term at the right-hand side of Eq. 5.7.14 but also constant k. Therefore, approximations based on this equation are ordinarily used. For this purpose, the functions entering this equation are presented as the Taylor series, and the higher-order terms of G^*, other than $(G^*)^{-1}$, are omitted. Then, the *linear approximation* leads to

$$G^*\left(1 - \frac{p}{e^{i\alpha}}\right) = \frac{m\omega^2 - Z}{K} + \frac{\rho\omega^2 h^2}{2} \qquad [5.7.15]$$

A possibility to use Eq. 5.7.15 instead of the exact solution Eq. 5.7.14 is determined by the condition $(kh)^2 << 5$ or $G^* >> 0.2\rho\omega^2 h^2$, i.e., the gap h must not be too large.

The practical application of Eq. 5.7.15 is based on separating it into real and imaginary parts.

Eq. 5.7.15 can be written in a form equivalent to Eq. 5.7.4, if one writes the right-hand side of Eq. 5.7.15 as

$$\frac{m\omega^2 - Z}{K} + \frac{\rho\omega^2 h^2}{2} = \frac{m_0\omega^2 - Z}{K} \qquad [5.7.16]$$

where $m_0 = m[1 + (m_s/2m)]$ is called the *reduced mass* and the value $m_s = \rho Sh$ is the mass of the deforming sample. The value $m_s/2$ is called the *coupled mass*. Using m_0 instead of m in Eq. 5.7.4 accounts for the inertia of the sample.

The physical meaning of approximation, Eq. 5.7.15, and neglecting the stress distribution inside a sample, can provide a clear understanding of the meaning of the product kh. If the displacement of plate B is very small such that $x_{0B} \ll x_{0A}$, then it is possible to show that the initial amplitude of deformation, x_{0A}, damps by e times at a distance δ from plate A determined by equality $(k\delta) = 1$.[94] The value δ in the theory of oscillations is called attenuation. Damping is negligible (or deformations in the sample can be considered as uniform) if the following strong inequality is valid: $h \ll \delta$. This inequality shows that the term "small gap" means such a gap thickness along which damping oscillation inside the sample is negligible.

5.7.3 TORSION OSCILLATIONS

This type of deformation mode is realized for rigid materials if the sample is prepared in the form of a long rod that can maintain its shape. Also, this deformation mode is realized in all rotational devices, which were described in Section 5.3 for measuring rheological properties of fluid materials, if one uses oscillations instead of rotation. In this case, the sample fills a gap between coaxial cylinders, cone and plate, or two parallel disks in the rheometer. In all these cases, the angle of torsion is assumed to be small enough to exclude axial deformations.

An equation describing torsional oscillations of the cylindrical sample caused by twisting of one of its ends is written as

$$G^* \frac{d^2 \theta_0^*}{dt} + \rho \omega^2 \theta_0^* = 0 \qquad\qquad [5.7.17]$$

where $\theta_0^*(z) = \theta e^{i\alpha}$ is a complex amplitude of twisting. Both amplitude and phase angle are varied along the vertical axis z.

This equation is identical to Eq. 5.7.13 with evident changes of linear sizes for circular ones. Then, all details of the solution and discussion of results of calculations are the same, if one changes mass, m, to the moment of inertia, I, rigidity, Z, to the twisting rigidity of a torsion bar and the form-factor to $K = \pi R^4/2H$, where R is radius and H is the cylinder height.

If the experiment is carried out in a cone and plate device, then again nothing changes, though the form-factor becomes $K = 2\pi R^3/3\varphi$, where R is the radius of the cone and φ is the angle between the cone and the plate.

In measuring rheological properties of viscoelastic liquid-like materials, the deformation of a hollow cylindrical sample is often studied. Such a sample is formed by a material filling gap of height, H, between two coaxial cylinders. One of the cylinders (let it be the outer cylinder with radius R_o) is oscillating with a frequency, ω, and an angular amplitude, θ_{0A}. The angular displacement of the other cylinder, θ_B, (with radius R_i) is measured. If the gap between cylinders is small, the curvature of the sample can be neglected and this is similar to shear deformation shown in Fig. 5.7.1. This situation takes place when the strong inequality $(R_o - R_i) \ll R_i$ is satisfied. In this case, all discussion is the same as in Section 5.7.1 and the form-factor is expressed as

$$K = \frac{2\pi R_i^2 H}{R_o - R_1} \cong \frac{2\pi R_o^2 H}{R_o - R_i}$$

The arbitrary size of the gap between coaxial cylinders can be an interesting case. The theory of this experimental method in its principal features is the same as described in Section 5.7.2, though an equilibrium equation, written in cylindrical coordinates, looks somewhat different than Eq. 5.7.13:

$$\frac{d^2\theta_0^*}{dr^2} + \frac{3}{r}\frac{d\theta_0^*}{dr} + \frac{\rho\omega^2\theta_0^*}{G^*} = 0 \qquad [5.7.18]$$

where θ_0^* includes amplitude and a phase angle, both depending on the current radius, r.

The complete solution and analysis of this equation are possible[95] but practical interest is limited by the linear approximation, which gives:

$$\frac{\theta_{0A}}{\theta_B^*} = 1 + \frac{1}{G^*} \qquad [5.7.19]$$

Then, by separating complex numbers into real and imaginary parts, the following expressions for components of dynamic modulus are given by:

$$G' = \frac{L(p\cos\alpha - 1)}{(p - \cos\alpha)^2 + \sin^2\alpha} \qquad [5.7.20]$$

$$G'' = \frac{Lp\sin\alpha}{(p - \cos\alpha)^2 + \sin^2\alpha} \qquad [5.7.21]$$

The analogy between these equations and Eqs. 5.7.5 and 5.7.6 is quite evident if one takes into account that $p = \theta_{0A}/\theta_{0B}$, and coefficient L is expressed as

$$L = \left[\frac{I\omega^2 - Z}{4\pi H} + \frac{\rho\omega^2(R_o^2 - R_i^2)R_i^2}{8}\right]\frac{R_o^2 - R_i^2}{R_o^2 R_i^2} \qquad [5.7.22]$$

This coefficient is equivalent to the factor $(Z - m\omega^2)/K$ that enters Eqs. 5.7.5 and 5.7.6 with an appropriate choice of form-factor. Besides, the coefficient L includes the inertia of moving elements coupled with a torsion element and inertia of the moving material. Therefore, similar to Eq. 5.7.16 reduced moment of inertia must include the coupled inertial term.

An expression for loss tangent, derived from Eqs. 5.7.20 and 5.7.21, is equivalent to Eq. 5.7.7. It is essential that both equations for tanδ do not include any geometrical factors. Also, tanδ is calculated by measuring p and phase angle only. It is useful for measurements in some applied problems where the main interest is in finding mechanical losses, but not for dynamic modulus.

5.7.4 MEASURING THE IMPEDANCE OF A SYSTEM

One of the versions of oscillating measurements is fixing plate A and applying force F(t) to plate B varying as

$$F = F_0 e^{i\omega t} \qquad [5.7.23]$$

where F_0 is the force amplitude and ω is frequency.

Displacement of plate B, as well as its velocity, v, follows harmonic law. The characteristic of such a system is its *mechanical impedance*, Y_m^*, determined as

$$Y_m^* = \frac{F}{v} = R_m + iX_m \qquad [5.7.24]$$

and consisting of real (active), R_m, and imaginary (reactive), X_m, components.

In the scheme shown in Fig. 5.7.1

$$X_m = \frac{G''}{K\omega} \qquad [5.7.25]$$

and

$$R_m = -\frac{G'}{K\omega} - \frac{Z}{\omega} + m\omega \qquad [5.7.26]$$

Now it is reasonable not to measure force and velocity separately but to measure their ratio. It permits the calculation of components of dynamic modulus from the mechanical impedance. Some experimental devices directly measure mechanical impedance and thus this is a simple way to calculate G' and G".

This experimental scheme can also be realized in torsion deformation if one fixes one boundary surface and varies torque by a harmonic law. One of the possible versions of this method of measurement of viscoelastic properties is by subjecting a tube-like sample placed between two coaxial cylinders to an axial displacement. The velocity of oscillations is given by the equation:

$$v = v_0 e^{i\omega t} \qquad [5.7.27$$

Then, by measuring mechanical impedance it is possible to find components of dynamic modulus. The theory for the arbitrary shape of the measuring device is not difficult, but the final equations are rather bulky. However, in the linear approximation (which is applicable in the majority of real experimental schemes), the result of the calculation is very simple as indicated by the following formula:

$$Y_m^* = \frac{F_0^*}{v_0} = \frac{F_0 e^{i\alpha}}{v_0} = \frac{KG^*}{\omega} \qquad [5.7.28]$$

where the form-factor K is

$$K = \frac{2\pi H(R_o^2 + R_i^2)}{(R_o^2 + R_i^2)\ln\dfrac{R_o}{R_i} - (R_o^2 - R_i^2)} \qquad [5.7.29]$$

Then, very simple equations for G' and G" are derived:

$$G' = \frac{F_0 \omega}{K v_0} \cos \alpha \qquad [5.7.30]$$

and

$$G'' = \frac{F_0 \omega}{K v_0} \sin \alpha \qquad [5.7.31]$$

which are analogous in their structure to Eqs. 5.7.8 and 5.7.9, respectively.

5.7.5 RESONANCE OSCILLATIONS

There is a very special case of oscillation when the amplitude of deformation is at maximum. These are called *resonance oscillations*.

The theory of resonance oscillations is based on the analysis of the movement of the upper plate B in Fig. 5.7.1, which is loaded by the oscillating force, as described by Eq. 5.7.23. The equilibrium equation, in this case, is written as

$$m\frac{d^2 x}{dt^2} + Zx + KG^* x = f_0 e^{i\omega t} \qquad [5.7.32]$$

where f_0 is the amplitude of force, all other notations are similar to previously used.

The solution of this equation for components of dynamic modulus is

$$G' = \frac{1}{K}\left[\frac{f_0}{x_{0B}}\cos\alpha - (Z - m\omega^2)\right] \qquad [5.7.33]$$

$$G'' = \frac{1}{K}\left(\frac{f_0}{x_{0B}}\sin\alpha\right) \qquad [5.7.34]$$

Then, excluding the phase angle, it is possible to obtain the following expression for the amplitude of oscillations of plate B:

$$x_{0B} = \frac{f_0}{\sqrt{(KG' + Z - m\omega^2)^2 + (KG'')^2}} \qquad [5.7.35]$$

Resonance corresponds to the maximum of x_{0B} as a function of ω. However, it is impossible to obtain the exact solution for x_{0B}^{max} from Eq. 5.7.35 because G' and G" are unknown functions of ω. The analysis becomes easier if rigid viscoelastic materials with low losses are considered, i.e., if $G'' \ll G'$. In this case, the minimum value of the dominator (and consequently, maximum of x_{0B}) is reached at the *resonance frequency*, ω_0, which is calculated as

$$\omega_0 = \sqrt{\frac{KG' + Z}{m}} \qquad [5.7.36]$$

Then, the components of dynamic modulus are easily found at this frequency as

$$G' = \frac{m}{K}\left(\omega_0^2 - \frac{Z}{m}\right) \qquad [5.7.37]$$

and

$$G'' = \frac{F_0}{Kx_{0B}^{max}} \qquad [5.7.38]$$

The experimental procedure consists of varying frequency and measuring amplitude of oscillation until a maximum x_{0B}^{max} is reached. Additional information can be obtained by measuring the width of a resonance curve. This width is the difference of frequencies $\Delta\omega = \omega_1 - \omega_2$ at half of the height of the resonance value, x_{0B}^{max}. In the case of low losses and sharp resonance, this value characterizes the loss modulus, which is calculated as

$$G'' = \frac{\Delta\omega G'}{\sqrt{3}\,\omega_0}\frac{1}{1 - Z/(m\omega^2)} \qquad [5.7.39]$$

The resonance method is applicable for measuring G' and G" at a single resonance frequency for low-loss materials. In fact, the resonance frequency can be changed (though not within a wide range) by varying the front-factor, K, and mass, m. A possible version of this method consists of measuring modulus at overtones of the resonance frequency.

5.7.6 DAMPING (FREE) OSCILLATIONS

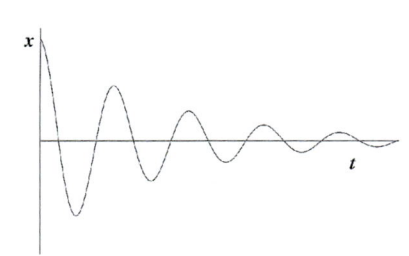

Figure 5.7.2. Damping oscillations.

The oscillations of plate B in Fig. 5.7.1 are supported by the applied force. If the force is imposed on a plate to initially shift its position from an equilibrium state, the plate will approach this state either monotonously or in the mode of *damping oscillations* (Fig. 5.7.2).

The equation of motion of point C in Fig. 5.7.1 is the same as in all previous cases with the difference in the boundary conditions used for the solution. The equilibrium equation can be written as

$$m\frac{d^2x}{dt^2} + Zx + K\left(G'x + \eta'\frac{dx}{dt}\right) = 0 \qquad [5.7.40]$$

where the term in the parenthesis reflects both components (elastic, G', and damping, η') of the reaction of liquid.

The solution of Eq. 5.7.40 has the form

$$x(t) = x_{0B}e^{(i\omega - \alpha)t} \qquad [5.7.41]$$

or

$$x(t) = x_{0B}e^{-\alpha t}\cos\omega t \qquad [5.7.41a]$$

where ω is the frequency of oscillations and α is a damping factor. The substitution of Eq. 5.7.41 into Eq. 5.7.40 and dividing the resulting expression into real and imaginary parts gives

$$G' = \frac{m}{K}\left[(\omega^2 + \alpha^2) - \frac{Z}{m}\right] \qquad [5.7.42]$$

$$G'' = \frac{2m\omega\alpha}{K} \qquad [5.7.43]$$

and

$$\tan\delta = \frac{2m\omega\alpha}{m(\omega^2 + \alpha^2) - Z} \qquad [5.7.44]$$

The important characteristic of a measuring system is its *natural frequency* $\omega_0 = \sqrt{Z/m}$, which is determined in an experiment without a sample. Using this value, the final expressions for the components of dynamic modulus are written as

$$G' = \frac{m\omega^2}{K}\left[1 + \left(\frac{\alpha}{\omega}\right)^2 - \left(\frac{\omega_0}{\omega}\right)^2\right] \qquad [5.7.45]$$

$$G'' = \frac{2m\omega^2}{K}\frac{\alpha}{\omega} \qquad [5.7.46]$$

and

$$\tan\delta = \frac{2(\alpha/\omega)}{(\alpha/\omega)^2 + [1 - (\omega_0/\omega)^2]} \qquad [5.7.47]$$

These equations allow one to find G' and G'' by determining the parameter of a measuring device, ω_0, and two experimental values ω and α.

The frequency of oscillations is easily observed and found from the experiment. The value of a damping factor, α, is related to the so-called *logarithmic decrement of damping*, Δ which is determined as

$$\Delta = \ln\frac{X_{n-1}}{X_n} = \alpha(t_n - t_{n-1}) \qquad [5.7.48]$$

where X_i are the maximum values of displacement (of the function x(t)) reached at the moments t_i. The difference $(t_n - t_{n-1})$ is the period of oscillations and equals $2\pi/\omega$.

In the simplest case (which is valid in many real experimental situations), $\alpha/\omega \ll 1$ and $Z/m\omega^2 \ll 1$. Then, the following equations for G', G'' and tanδ can be used

$$G' = \frac{m\omega^2}{K}\left(1 + \frac{\Delta^2}{4\pi^2}\right) = \frac{m}{K}(\omega^2 + \alpha^2) \qquad [5.7.49]$$

$$G'' = \frac{m\omega^2}{K}\frac{\Delta}{\pi}$$ [5.7.50]

$$\tan\delta = \frac{\Delta}{\pi}$$ [5.7.51]

Parameters of material determined by the damping oscillation method are measured at a single frequency. This frequency can be measured or roughly estimated from Eq. 5.7.49 as $\omega \approx \sqrt{KG'/m}$ assuming that $\alpha \ll \omega$. However, the frequency can be varied (though not within a wide range) by changing the parameters of the measuring system – form-factor, K, and moving mass, m.

The basic theory of damping oscillations does not take into account sample inertia. The situation here is quite the same as in the above-discussed cases. The equations derived in this *section* are valid, if $\omega \ll (1/h)\sqrt{G^*/\rho}$, where ρ is the density of material under study.

Damping oscillations are often realized using the torsion mode of deformations. If the sample is cylindrical, Eqs (5.7.45) – (5.7.47) as well as Eqs (5.7.49) – (5.7.51) are valid with a change of mass, m, for a moment of inertia, I, and form-factor i. $K = 4\pi R_o^2 R_i^2 H/(R_o^2 - R_i^2)$.

Samples of other geometrical forms can also be used in the damping oscillation experiment utilizing torsion. Formulas for calculating G' and G'' are much more complicated due to the contribution of out-of-plane bending and twisting.[96]

5.7.7 WAVE PROPAGATION

In these methods,[97] the propagation of waves is directly observed for samples of a large thickness such that several wave lengths are present within the material. If the damping characteristics of materials are not high, the wavelength and attenuation can be measured. These methods are different from the previously discussed methods. In the former methods, the effect of a wave is detected on the surface of instruments with sample thickness being much smaller than the wavelength. Typically, shear and longitudinal waves are used for measurements of the viscoelastic properties of materials.

5.7.7.1 Shear waves

In this case, a plate is placed in the material. The plate is subjected to oscillations in its own plane along the x axis, as indicated in Fig. 5.7.3. The wave propagates in the direction of the z-axis. The amplitude and frequency of oscillations are x_o and ω, respectively. At some distances z and z+dz from the plate, the displacements in the material are x and x+dx, respectively. Then, the shear strain and shear stress at the position z in the material is, respectively:

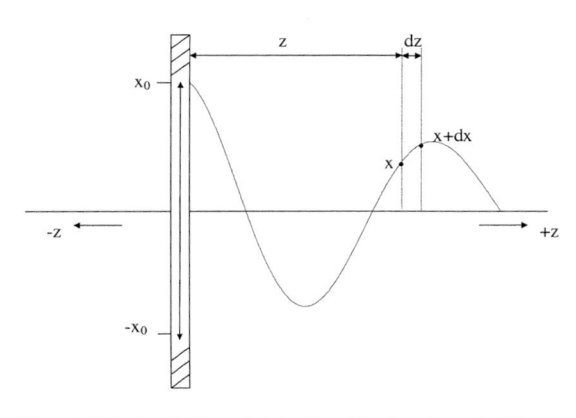

Figure 5.7.3. Oscillation of plate along its plane in material.

$$\gamma(t, z) = \frac{\partial x(t, z)}{\partial z}, \ d\sigma(t, z) = \frac{\partial \sigma(t;z)}{\partial z} dz \qquad [5.7.52]$$

Equation of motion is

$$\rho \frac{\partial^2 x}{\partial t^2} = \frac{\partial \sigma}{\partial z} \qquad [5.7.53]$$

with the shear stress, σ, being

$$\sigma = G'\gamma + \frac{G'' d\gamma}{\omega \, dt} \qquad [5.7.54]$$

Substitution of Eq. 5.7.54 in Eq. 5.7.53 leads to

$$\rho \frac{\partial^2 x}{\partial t^2} = G' \frac{\partial^2 x}{\partial z^2} + \frac{G''}{\omega} \frac{\partial^3 x}{\partial t \partial z^2} \qquad [5.7.55]$$

Since the solution of Eq. 5.7.55 is proportional to $e^{i\omega t}$, one obtains

$$\rho \omega^2 x = G' \frac{\partial^2 x}{\partial z^2} + iG'' \frac{\partial^2 x}{\partial z^2} = G^* \frac{\partial^2 x}{\partial z^2} \ \text{or} \ \frac{\partial^2 x}{\partial z^2} = \Gamma^2 x \qquad [5.7.56]$$

with $\Gamma = -(\rho \omega^2)/G^*$. The solution of Eq. 5.7.56 is

$$x(z, t) = (k_1 e^{\Gamma z} + k_2 e^{-\Gamma z}) e^{i\omega t} \qquad [5.7.57]$$

with k_1 and k_2 being integration constants that are determined based on the geometry of the experiment. In a semi-infinite medium, $k_1 = 0$ and Eq. 5.7.57 becomes

$$x(z, t) = x_0 e^{-\Gamma z} e^{i\omega t} \qquad [5.7.58]$$

It is convenient to replace Γ in Eq. 5.7.58 by

$$\Gamma = \frac{1}{z_0} + i2\pi\lambda \qquad [5.7.59]$$

Then, Eq. 5.7.58 becomes

$$x(z, t) = x_0 e^{i(\omega t - 2\pi z/\lambda) - \frac{z}{z_0}} \qquad [5.7.60]$$

where λ is the wavelength.

It is seen that at $z = z_0$, the amplitude of wave in material decreases by a factor e, and the value of $\alpha = 1/z_0$ is attenuation (Fig. 5.7.4). Then, the dynamic moduli are as follows:

$$G' = \frac{\rho \omega^2 (4\pi^2/\lambda^2 - \alpha^2)}{(4\pi^2/\lambda^2 + \alpha^2)^2} \qquad [5.7.61]$$

$$G'' = \frac{\rho\omega^2 4\pi\alpha/\lambda}{(4\pi^2/\lambda^2 + \alpha^2)^2} \qquad [5.7.62]$$

$$\tan\delta = \frac{G''}{G'} = \frac{4\pi\alpha\lambda}{4\pi^2 - \alpha^2\lambda^2} \qquad [5.7.63]$$

It is pertinent from Eqs. 5.7.61, 5.7.62, and 5.7.63 that, by measuring wavelength and attenuation, the dynamic properties of the material can be measured. However, if attenuation is small the wave propagates over a long distance, and reflection from walls may cause the measured damping to be magnified. At the other extreme, if attenuation is large, the shear wave decays over short distances and this causes difficulty in the measurement of wavelength. Thus, an upper limit is typically $\lambda\alpha = 3$. This technique can be used in the range of frequencies from 4 to 5000 Hz. At high frequencies from 3 kHz to 3 GHz, the reflection of propagating waves in a quartz crystal against the interface between the quartz and a thin film of liquid is used.

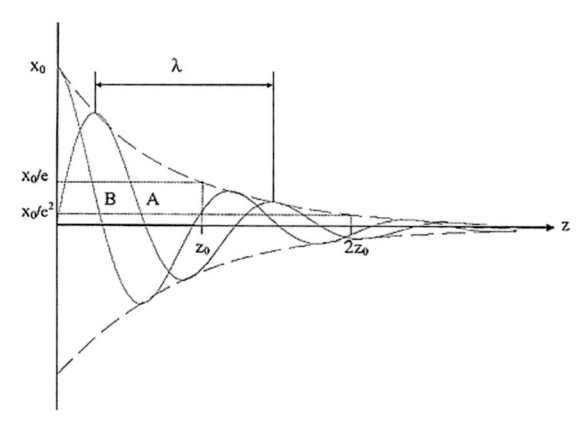

Figure 5.7.4. Propagation of a shear wave of wavelength, λ, and attenuation, $1/z_0$, in the direction z: A at time t $=2\pi n/\omega$ and B at t$=(2n\pi + \pi/2)/\omega$.

5.7.7.2 Longitudinal waves

In this case, the oscillation of the plate is in the z-direction. The material is subjected to oscillatory extension and compression. The dynamic longitudinal moduli, E' and E" are measured according to the following equations:

$$E' = B' + \frac{4}{3}G' \qquad [5.7.64]$$

$$E'' = B'' + \frac{4}{3}G'' \qquad [5.7.65]$$

$$E^* = B^* + \frac{4}{3}G^* \qquad [5.7.66]$$

where B' and B" are the components of the complex dynamic bulk modulus B*.

5.7.8 VIBRATION VISCOMETRY

Viscosity (more exactly – Newtonian viscosity) of inelastic liquids is frequently measured by various versions of the oscillation methods. They are based on a very well documented theory of oscillations and the simplicity of apparatus is their realization. The main field of application of *vibration viscometry* is relatively low viscous fluids.

The basic theory of viscosity measurements by oscillation methods is a particular case of a more general theory of measuring viscoelastic properties as discussed above in this chapter. However, the general theory can be substantially simplified if it is *a priori* known that $G' = 0$. Then, viscosity equals G''/ω, and, therefore, it is necessary to measure losses only.

The basic equations of equilibrium for viscous inelastic liquid, according to the scheme shown in Fig. 5.7.1, is

$$m\frac{d^2x}{dt^2} + Zx + \frac{S}{h}\eta\frac{d(x - x_0)}{dt} = 0 \qquad [5.7.67]$$

instead of Eq. 5.7.3, where the term $(1/h)[d(x - x_0)/dt]$ is deformation rate. The solution of this equation gives two equivalent expressions for viscosity

$$\eta = \frac{Z - m\omega^2}{K\omega}\frac{1}{\sqrt{p^2 - 1}} \qquad [5.7.68]$$

$$\eta = \frac{Z - m\omega^2}{K\omega}\cot\alpha \qquad [5.7.69]$$

where the notations are the same as before.

For the limiting case $Z \gg m\omega^2$ and $p \gg 1$, Eq. 5.7.68 leads to

$$\eta = \frac{Z}{K\omega p} \qquad [5.7.70]$$

which is equivalent to Eq. 5.7.9 because for inelastic liquid $\alpha = \pi/2$ and $\sin\alpha = 1$.

If oscillations of the plate take place in a vessel of arbitrary size, the equilibrium equation is

$$m_0\frac{d^2x}{dt^2} + Zx + k_u\frac{dx}{dt} = f_0 e^{i\omega t} \qquad [5.7.71]$$

where m_0 is a reduced mass, as in Eq. 5.7.16, and the coefficient $k_u = S\sqrt{\eta\rho\omega/2}$.

The solution of this equation is well known. It is an equation of harmonic oscillations with amplitude, x_{0A}, and phase angle, α, which are expressed as

$$x_{0A} = \frac{f_0}{\sqrt{(m_0\omega^2 - Z)^2 + k_u^2}} \qquad [5.7.72]$$

$$\tan\delta = \frac{\omega k_u}{m_0\omega^2 - Z} \qquad [5.7.73]$$

Any of these equations allows calculation of viscosity from measured values of x_{0A} or $\tan\delta$ because viscosity enters expression for k_u. However, these equations are not convenient in real practice. Therefore, the main interest is in the limiting cases. If a plate is vibrating in a large volume (such that $h \gg \delta$ or $h \gg 3.2\sqrt{\eta/\rho\omega}$; the meaning of the parameter, δ, was explained in Section 5.7.2, then Eqs. 5.7.72 and 5.7.73 have the following form:

$$\frac{A}{f_0} = \left\{ \left[\left(m + S\sqrt{\frac{\rho\eta}{2\omega}} \right)\omega^2 - Z \right]^2 + \frac{S^2\rho\eta\omega}{2} \right\}^{-1/2} \qquad [5.7.74]$$

$$\tan\alpha = \frac{\omega S\sqrt{\frac{\rho\eta\omega}{2}}}{\left(m + S\sqrt{\frac{\rho\eta}{2\omega}} \right)\omega^2 - Z} \qquad [5.7.75]$$

For further analysis of these equations it is convenient to use the dimensionless variables:

dimensionless frequency: $\lambda = \omega/\omega_0$, where $\omega_0 = \sqrt{Z/m}$ is the natural frequency of a measuring device
dimensionless amplitude: $a = AZ/f_0$
dimensionless viscosity: $\gamma = (S/m)\sqrt{\rho\eta/2\omega_0}$
After some rearrangements Eqs. 5.7.74 and 5.7.75 become:

$$a = \left[\left(\frac{1 + \gamma\sqrt{\lambda}}{\lambda^2} \right)^2 + \frac{\gamma^2}{\lambda^3} \right]^{-1/2} \qquad [5.7.76]$$

and

$$\tan\alpha = \frac{\gamma}{\gamma - \frac{\lambda^2 - 1}{\sqrt{\lambda}}} \qquad [5.7.77]$$

The dependencies of a and α on the dimensionless parameters γ and λ are called the *amplitude* and *phase characteristics* of a vibrating system. These dependencies, built in accordance with Eqs. 5.7.76 and 5.6.77, are presented in Fig. 5.7.5.

Some important particular cases are worthy special discussion.

Measurement of resonance amplitude. If viscosity is low ($\gamma \ll 1$), then the resonance in Fig. 5.7.5 is reached at $\lambda = 1$. Viscosity is directly related to the resonance amplitude, x_0^{max}, as

$$\rho\eta = k(x_0^{max})^{-2} \qquad [5.7.78]$$

where coefficient k can be calculated or found by calibrating a measuring system.

If viscosity is not low, the maximum is not sharp. Then, it is necessary to use the complete equation instead of Eq. 5.7.78.

Measuring amplitude at an arbitrary frequency. Viscosity can be calculated from the amplitude characteristics of a system at an arbitrary measured amplitude, though it is not convenient. Moreover, if the viscosity is not low this method becomes unreliable because, at large values of λ, an amplitude equals ~1 for all values of γ.

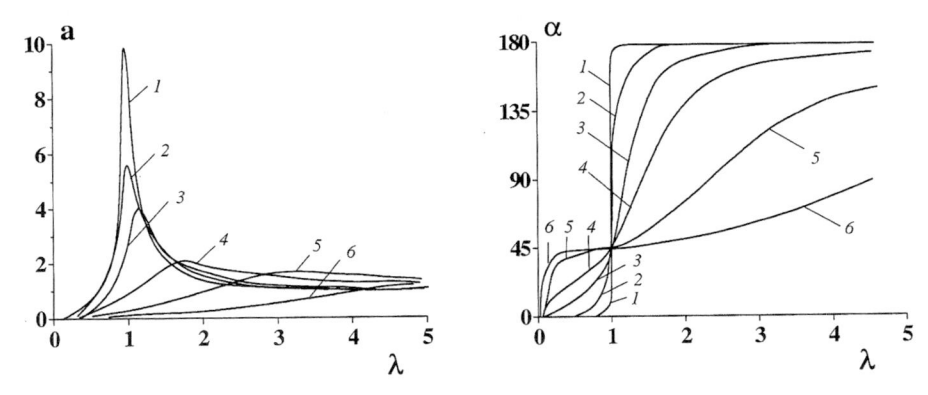

Figure 5.7.5. Amplitude a(λ,γ) (left) and phase $\alpha(\lambda,\gamma)$ (right) characteristics of oscillations of flat plate inside viscous liquid (in dimensionless variables). Values of γ are: 0.01 (curves 1); 0.1 (curves 2); 0.3 (curves 3); 1 (curves 4); 5 (curves 5); 10 (curves 6).

Measuring amplitude at a defined frequency. There exists such a value of frequency at which the relationship between a and γ is rather simple. Indeed, let $\alpha = \pi/2$. According to Eq. 5.7.77, $\gamma = (\lambda^2 - 1)/(\sqrt{\lambda})$ and a $= \lambda^{3/2}/\gamma$. Then

$$\gamma = [a^{1/3}(a-1)]^{-3/4} \qquad [5.7.79]$$

For low viscosity materials, this condition is close to the resonance ($\lambda \approx 1$), but in a general case, this method does not give reliable results because even a slight variation of frequency near the resonance results in large changes in viscosity.

Measuring frequency at a chosen phase angle. Viscosity is calculated from Eq. 5.7.77. If $\alpha = \pi/2$, then $\gamma = (\lambda^2 - 1)/\sqrt{\lambda}$ and

$$\rho\eta \propto \frac{\lambda^2 - 1}{\lambda} \qquad [5.7.80]$$

The dependence of viscosity on λ is strong, especially in the range of high values of viscosity, where condition $\alpha = \pi/2$ corresponds to high values of dimensionless frequency.

Measuring resonance frequency. This method is invalid for low γ because resonance is achieved at $\lambda = 1$ for liquid of any viscosity. However, this method can be used for high viscosity liquids, though it is not sensitive as the resonance maximum is not distinctly expressed.

Measuring phase angle. As seen from Fig. 5.7.5, this method can be used for high values of γ. A particular case of this method is measuring the phase angle at the resonance frequency. The condition $(\partial x_0/\partial\lambda) = 0$ leads to relationship between α and γ. Numerical analysis shows that noticeable changes in α take place in the range of γ between 0.1 and 10. It corresponds to the phase angle changes by about 20°. Beyond this range of γ this method is not applicable, due to the very small variation of α.

5.7.8.1 Torsion oscillations

Similar to the above-discussed cases of measuring dynamic modulus, the transition from plane shear to torsion does not change basic equations, except for the use of circular displacements instead of linear ones.

For example, if the liquid is placed between two coaxial cylinders and one of them is under torsion oscillations, viscosity can be calculated from one of the following equations:

$$(K_1\omega^2 - ZK)\frac{1}{\eta} = \omega p \sin\alpha \qquad [5.7.81]$$

$$(K_2\omega^2 - ZK)\frac{1}{\eta^2} = p \cos\alpha \qquad [5.7.82]$$

where p is the ratio of the amplitude of oscillations of inner and outer cylinders, Z is the rigidity of measuring device, and K_1, K_2, and K are form-factors. Their values can be calculated without any problems, though the final formulas are rather bulky. They can also be measured by calibrating a measuring device. By measuring either $\omega \sin\alpha$ or $(p\cos\alpha - 1)$ as a function of ω^2, viscosity can be found through an experimental procedure according to Eq. 5.7.81 or Eq. 5.7.82.

5.7.8.2 Oscillation of a disk in liquid

A plane disk of radius, R, is making torsional oscillation around its axis according to harmonic function $\theta(t) = \theta_0\cos\omega t$, where θ_0 is an angular amplitude and ω is its frequency. If a disk is moving in a sufficiently large vessel, an equilibrium equation is written as

$$T = k_\omega\Omega + I_s\frac{d\Omega}{dt} \qquad [5.7.83]$$

where M is torque and Ω is angular displacement. The coefficients in this formula are expressed as

coefficient of resistance: $\qquad k_\omega = \pi R^4\sqrt{\rho\eta\omega/2}$

jointed (coupled) moment of inertia: $\qquad I_s = \pi R^4\sqrt{\rho\eta/2\omega}$

If the liquid is placed on both sides of a disk, a multiplier 2 must be introduced in both coefficients, k_ω and I_s.

These equations permit us to find viscosity by measuring parameters of the disk movement. If the gap between an oscillating disk and stationary walls is small, the expressions for the coefficient should be modified but the scheme of calculations remains intact.

5.7.8.3 Oscillations of the sphere

In torsion oscillations of a sphere around its diameter, torque appears, due to the resistance to movement of the sphere. The equilibrium equation is the same as Eq. 5.7.83, though the expressions for the coefficients are different:

$$k_\omega = \frac{8\pi R^3(3 + 8a + 6a^2 + 2a^3)}{3(1 + 2a + 2a^2)}\eta \qquad [5.7.84]$$

$$I_s = \frac{16\pi R^3 a^2(1 + a^2)}{3\Omega(1 + 2a + 2a^2)}\eta \qquad [5.7.85]$$

where a = R/δ, R is the sphere radius and Ω is the angular velocity. The physical meaning of δ was explained in Section 5.7.2.

Of special interest is a particular limiting case when a >> 1. In this case, Eqs. 5.7.84 and 5.7.85 become:

$$k_\omega = 8.38R^4 \sqrt{\frac{\rho\eta\omega}{2}} \qquad [5.7.86]$$

$$I_s = 8.38R^4 \sqrt{\frac{\rho\eta}{2\omega}} \qquad [5.7.87]$$

An analogy between these expressions and the corresponding equations obtained for an oscillating plate is obvious with evident changes of coefficients.

A special case is the oscillation of a sphere filled with liquid. Liquid impedes the oscillation of the sphere and it can be used as a method of viscosity measurement. This method was proposed as a very sensitive, absolute method of measurement. In this case, many methodical contrivances were used.[98]

5.7.8.4 Damping oscillations

Damping oscillations in a liquid depend on its viscosity. Therefore, by measuring damping oscillations it is possible to find viscosity. The basic equation of equilibrium is Eq. 5.7.40 with G' = 0 and η' = η. Then viscosity can be found from Eq. 5.7.46 as

$$\eta = \frac{G''}{\omega} = \frac{2m\alpha}{K} \qquad [5.7.88]$$

This equation is valid for any mode of damping oscillations (including torsional oscillations) by appropriate choice of form-factor, K. The corresponding values of K for different geometrical shapes of the sample were discussed above (in this chapter in connection with rotational instruments of different types). All these expressions can be used in Eq. 5.7.88. However, an exact solution for damping oscillations of a sample of arbitrary shape in a vessel of arbitrary size may be complicated. However, there are no principal difficulties in obtaining such a solution. Nevertheless in real practice, it is preferable to take into account the limitations of applicability of Eq. 5.7.88 or to calibrate a measuring system using standard liquids.

5.7.9 MEASURING VISCOELASTIC PROPERTIES IN NON-SYMMETRICAL FLOWS

Viscoelastic properties of the material are measured during unsteady motion with deformation changing in time. A different approach was developed based on changes of deformation in space when small harmonic perturbations of velocity are superimposed on a steady flow.[99]

Let us consider circular flow with radial and axial velocity components equal to zero. Small angular periodic perturbations are superimposed on the main circular flow so that the velocity field is described as

$$u_r = u(r, z)e^{i\varphi} \qquad [5.7.89a]$$

$$u_{\varphi} = \omega r + v(r, z)e^{i\varphi} \qquad\qquad [5.7.89b]$$

$$u_z = w(r, z)e^{i\varphi} \qquad\qquad [5.7.89c]$$

and the values u, v, and w are small in comparison with ωr.

In this formulation, the equation of motion is exactly the same as for viscous liquid if Newtonian viscosity is replaced by a complex viscosity, η^*. Then, the problem consists of solving dynamic equations for a defined geometry of flow and calculating η^* for different experimental schemes.

Parallel disks with shifted axes. The small radial shift of axes leads to periodic changes of velocity. As a result, radial forces F_x and F_y appear. Analysis of possible approximation based on the exact solution of the problem gives the following (linear approximation) formulas for components of dynamic modulus

$$G' = KF_y/a \qquad\qquad [5.7.90]$$

$$G'' = KF_x/a \qquad\qquad [5.7.91]$$

where the form-factor K is

$$K = \frac{h}{2\pi R^2} \qquad\qquad [5.7.92]$$

R is the radius of disks, h is the distance between disks, and a is a radial shift of axes.

Parallel cylinders with shifted axes. Theoretical analysis of deformation gives (in linear approximation) the same equations as Eqs. 5.7.90 and 5.7.91 for components of dynamic modulus, where the form-factor K is

$$K = \frac{\ln(R_o/R_i) - \dfrac{R_o^2 - R_i^2}{R_o^2 + R_i^2}}{4\pi H} \qquad\qquad [5.7.93]$$

R_o and R_i are the radii of outer and inner cylinders, respectively, H is the height of a liquid layer in a gap between cylinders and a is the radial shift of cylinder axes.

Rotation between surfaces with a small angle between them. There are three types of the simplest geometries of such kind:
- coaxial disks with inclined surfaces
- cylinders with inclined axes
- cones with inclined axes.

In all these cases, the angle between the inclined axes, ε, is small.

Equations for calculating the components of dynamic modulus are the same in all cases. They are:

$$G' = KF_y/\varepsilon \qquad\qquad [5.7.94]$$

$$G'' = KF_x/\varepsilon \qquad\qquad [5.7.95]$$

and the form-factor depends on the geometry of deformation.

Rotation of spherical surfaces. There are two principal schemes of rotating spheres with small perturbations of the velocity field:
- flow between two spheres with slightly shifted centers
- flow between two spheres with the common center but slightly inclined axes of rotation.

For the first case, Eqs. 5.7.90 and 5.7.91 are valid, while in the second case equations analogous to Eqs. 5.7.94 and 5.7.95 can be used. The front factors are different in the cases under discussion.[100]

5.7.10 ABOUT EXPERIMENTAL TECHNIQUES

Many hundreds of experimental devices have been constructed for measuring the viscoelastic properties of different materials. In particular, many designs of vibration viscometry were described during the last century. It is impossible here to give even a short survey of all units. Therefore, only the main features of the design of experimental devices are presented below.

5.7.10.1 Rotational instruments

Essentially, these are rotational devices (rheogoniometers, rheometers, elastoviscometers, and so on) as described in Section 5.3. Instruments should be capable of oscillating measurement, in addition to the steady rotation. These types of devices are instruments with variable mechanical drives. In modern versions of such instruments, the frequency can be changed in the range of more than 7 decimal orders.[101] The amplitude of deformations can also be changed in a wide range, providing a possibility to follow the non-linear effects of the dependence of viscoelastic properties on deformation and to find the boundaries of linearity in the mechanical behavior of the material.

Rotational instruments produced by some companies (*Rheometrics, TA Instruments, mass*) are also equipped with techniques for measuring viscoelastic properties in non-symmetric flows, commonly using a scheme with shifted axes.[102] *Contraves* has made an instrument of this type that is called a *Balance Rheometer*, in which flow occurs between two spheres with inclined axes of rotation.

5.7.10.2 Devices with electromagnetic excitation

This type of experimental technique initially used a method of impedance measurements.[103] Instruments of this type were proposed primarily to study rigid materials.

Another approach is based on separate measurements of forces and deformations, and measurement of properties of various fluid materials, including dilute solutions with modulus varying in the range from 0.1 to 10^4 Pa and viscosity varying from 0.2 to 10^3 Pa*s.[104]

A promising development along this line of constructing instruments for measuring viscoelastic properties was the application of a multi-frequency resonator placed in liquid media.[105] This instrument makes measurements at ten different frequencies in the range from 10^2 to 8.3×10^8 Hz. The optical system of measurement was used to measure amplitudes as small as 10^{-3} degree. This instrument can measure modulus in the range from 0.1 to 10^3 Pa for liquids with the viscosity not less than 0.2 Pa*s.

Electromagnetic excitation is also convenient in instruments for torsion deformation, where different modes of deformation are possible – forced oscillations, damping oscillations, creep, elastic recoil, and steady flow.[106] An example of instruments of this type is

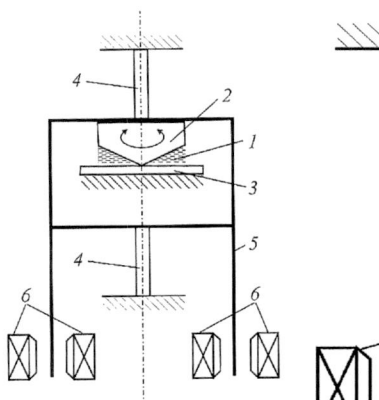

Figure 5.7.6. Rotational instrument with electromagnetic excitation. 1 – sample; 2 – cone surface; 3 – plate; 4 – two torsion rods; 5 – frame-rotor; 6 – force coils of electrodynamic drive.

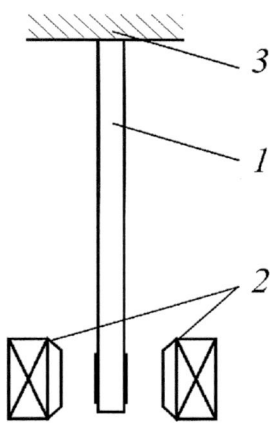

Figure 5.7.7. Vibrating reed method of measurement. 1 – sample; 2 – electro-magnets; 3 – sample holder.

shown in Fig. 5.7.6.[107] A sample is placed between a cone and a plate or between two disks. A moving part is fixed by torsions, which provide reliable alignment of the working unit. Frame is made of aluminum. It works as the rotor of an electrodynamic drive. By varying the electrical parameters, one can vary the regime of deformations.

Electromagnetic excitation is also widely used in the *vibrating-reed method* (shown in Fig. 5.7.7). Excitation of oscillation (periodic or damping) is created by means of a charge placed on the sample surface and an electromagnetic drive. The frequency may vary in a wide range, practically from several to hundreds Hz. Deformations are followed and registered by an optical method (not shown here). This version of the method is applicable to materials with modulus from 10^5 to 10^1 Pa. The devices of this type can be combined with very different measuring techniques. These instruments are also very conve-

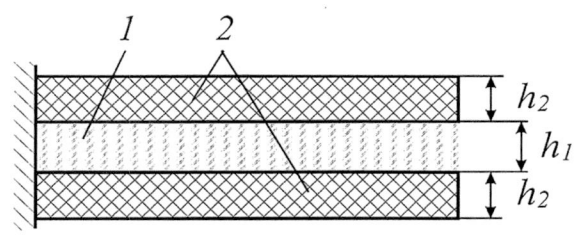

Figure 5.7.8. A sandwich-type sample: layers (2) of a soft material are placed on a hard support (1).

nient for the resonance method of measurement. These devices are simple and they are used either as home-made versions or produced by specialized companies.

This method is very convenient in different applications. In particular, the vibrating reed method was used at very low temperatures down to 4K and allowed measurement of loss tangent as low as 10^{-3}.[108]

Free bending oscillation (as in Fig. 5.7.7) can be applied to many materials. However, when testing mild materials an experimental scheme must be modified. An example of modification of a sample is shown in Fig. 5.7.8. Here, two layers (with the thickness h_2) made of soft material (it might be rubber, gel, or a viscoelastic solution) are placed on a surface of rigid support (with thickness, h_1). The whole sandwich construction oscillates together. Measurement of properties of support and sandwich sample separately permits calculation of rheological characteristics of a soft material. The reverse scheme – a soft material between two rigid layers – is also possible.

A sandwich construction can be used in a different system. For example, it was proposed to prepare samples as hollow cylinders and other cylindrical body, made of different materials. They were placed inside a hole.[109]

5.7.10.3 Torsion pendulums

Torsion pendulums are the most popular and widely used instruments for centuries to solve different fundamental physical problems.[110] In the practice of measuring viscoelastic properties, torsion pendulums were introduced in the 1950s,[111] though earlier they were widely used in viscometry. Now devices of this type are frequently used as home-made instruments as well as being produced by several companies.

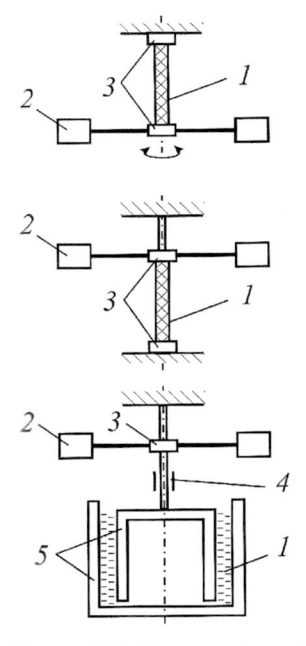

Figure 5.7.9. Three versions of torsion pendulum: direct (upper), reverse (middle), and for liquid samples (bottom). 1 – sample; 2 – inertial masses; 3 – holder; 4 – gas bearing; 5 – measuring cell of a cylinder-cylinder type.

Principal schemes of torsion pendulums are shown in Fig. 5.7.9 in three different versions: upper scheme – direct pendulum, central one – reverse pendulum, and bottom scheme – pendulum for liquid samples. There are a lot of different designs of torsion pendulums differing in methods of deformation measurement, mode of load application, construction of a holder, material of a torsion bar, design of thermo- and/or cryo-chamber, the geometry of measuring cell, and so on. However, all reproduce the main design feature of the instrument shown in Fig. 5.7.9. Depending on the design, special torsion pendulums can work at frequencies between 0.01 and 80 Hz, at temperatures from liquid nitrogen up to >1500K. They can measure modulus in the range from 10^2 up to 10^{12} Pa and tanδ from 10^{-4} (or even lower values, though the latter requires very accurate measurements) up to about 3.[112] Different optical or photo-electronic systems of measuring deformation are used.

Torsion pendulums can be used for the monitoring process of oligomer curing. This is realized by using a torsion element made as a braid (prepared from quartz, glass, silk, carbon fibers, and so on) impregnated with liquid. Then, the evolution of viscoelastic properties of such a complex sample is measured, which gives important technological information about relative changes in the sample.[113] The advantage of this method consists of the possibility to use a very small quantity of the sample, as low as 0.1 g, and to study unstable materials.

5.8 PHYSICAL METHODS

5.8.1 RHEO-OPTICAL METHODS

5.8.1.1 Basic remarks

The term the "*rheo-optics*" contains two words: rheo (flow) and optics. The rheo-optics is a method of study of deformations and stresses during the flow of transparent polymeric systems by means of optical techniques that measure the difference of the *refractive indi-*

ces, n_{ij}. The refractive index is determined by the polarizability of atomic groups and bonds in molecules.

When a parallel beam of light is incident to the surface separating two transparent media, part of the light is reflected back into the medium and part is transmitted into the second medium. These are the *reflected* and *transmitted rays*. The direction of the transmitted ray does not coincide with that of the incident ray and the transmitted ray is said to be *refracted*. The angles that the incident, reflected, and refracted rays make with a normal to surface at the point of incidence are known as the angles of incidence, reflection, and refraction and they are denoted as i, R, and r, respectively. Reflection and refraction in isotropic media obey *Snell's law* such that i = R and sini/sinr = n_{12}. If light enters a medium from a vacuum, the above ratio is called the absolute refractive index of the medium. If c is the velocity of light in a vacuum, v_1 and v_2 are velocities of light in media, the absolute refractive indices are $n_1 = c/v_1$, $n_2 = c/v_2$, and $n_{12} = v_1/v_2 = n_2/n_1$. The frequency f of waves, $f = v/\lambda$, is unchanged when light travels through various media. Therefore, the wavelength, λ, changes. If λ_1 and λ_2 denote the wavelengths in the two media, then

$$n_1 = \lambda_0/\lambda_1, \; n_2 = \lambda_0/\lambda_2, \; n_{12} = \lambda_1/\lambda_2 = v_1/v_2 \qquad [5.8.1]$$

Polymer molecules are typically anisotropic. However, if molecules are in a coiled state and randomly oriented in space, they form optically isotropic materials. If molecules are oriented under deformation of the material, the material becomes anisotropic, causing a phenomenon of double refraction (*birefringence*) due to dependence of the refractive index on the direction. The degree of anisotropy of the refractive index is characterized by symmetrical tensor of the refractive index, n_{ij}:

$$n_{ij} = \begin{bmatrix} n_{11} & n_{12} & n_{13} \\ n_{12} & n_{22} & n_{23} \\ n_{13} & n_{23} & n_{33} \end{bmatrix} \qquad [5.8.2]$$

This tensor is analogous to the stress or strain tensor. The components of the tensor of refractive indices follow the same rules as the components of other tensors during the transformation from one coordinate system to another. In particular, one can find three mutually perpendicular directions in which the diagonal components of tensor, n_{ij}, attain the maximum values of n_I, n_{II}, and n_{III}. These are the principal values of tensor n_{ij}.

Optical quantities related to this tensor can be determined by using a polarization optical technique. Through a transparent model (plate, channel, etc.), a transmitted beam of polarized light and reflected light give an *interference picture*, which characterizes the stress-strain state of the model.

Brewster[114] was the first to discover this phenomenon and suggested its use to study the stress state in glasses. Later it was proven that for a wide class of planar elastic problems, the distribution of stresses is independent of elastic constants of materials. Thus, the stress distribution can be determined using transparent models with elastic constant different from the material of objects. This fact opened an avenue for use of polarization-optical techniques to study stresses and deformations in solids. This area of study is called *photo-*

elasticity. Good examples of the use of photoelastic techniques are investigations of the stress state in models of bridges and dams made from glasses.

By means of interference, differences in the principal stresses and their directions, but not the stress components, can be determined. In some cases, this information is sufficient to solve practical problems, but usually one needs to determine separate values of principal stresses. For these purposes, methods of separation of stresses are used.[115]

At the beginning of the 20th century, work was initiated to study the relationship between the refractive index tensor and stresses during plastic deformation of transparent models. In particular, a method of coating received wide acceptance. According to this method, the photoelastic transparent coating covers a model. Coating deforms together with the surface and using the reflection technique, stresses can be determined on the studied object. This is a fertile area of research related to solid materials.

However, the main interest here is the application of optical techniques to study the flow of polymeric melts and solutions. The major boost of research in the application of these techniques to flowing polymeric melts and solutions was received in the middle of the 1950s. This was due to the theoretical work done by Lodge,[116] and the experimental work carried out by Philippoff,[117] Janeschitz-Kriegl,[118] and their co-workers.

At the present time, methods of the rheo-optics are used in various areas, such as studies of stresses and strains in solids, residual stresses, and structure in solids, stresses in fluids, rubbers, products made by polymer processing, etc.

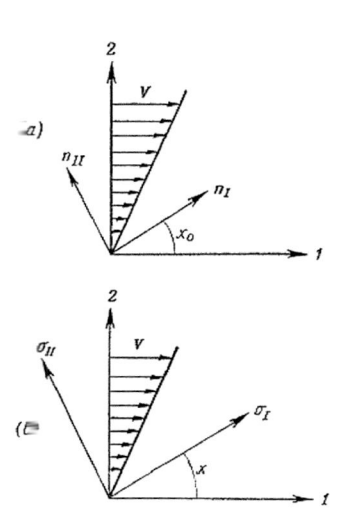

Figure 5.8.1. Coordinate system with axes of stresses and refractive indices in simple shear flow.

The most extensive information concerning the measurement of birefringence of polymeric fluids is available for simple shear, as shown in Fig. 5.8.1. Here 1 designates the direction of shear, 2 is the direction of the velocity gradient. Hence, shear occurs in plane 1-2. The third is known as a neutral direction, which is normal to the plane of drawing and corresponds to the direction of light propagation. For simple shear the principal directions of a birefringence tensor are defined by vectors n_I and n_{II}. They characterize the orientation of the birefringence ellipsoid in space.

Experimental determination of this orientation is carried out as follows. Use is made of a system of crossed polarizing devices – the planes of polarization of light in a polarizer and analyzer are arranged at an angle of 90° (*cross Nicols*); in this case, light does not reach an observer situated after the analyzer. The object under study is placed between the analyzer and the polarizer. The plane-polarized light of wavelength, λ, passing through a birefringent material of thickness, d, is split into two beams, which propagate along with the principal directions of the refractive index ellipsoid. Having passed through the analyzer, these two beams interfere with a phase shift of $\delta = 2\pi d(\Delta n/\lambda)$. Here $\Delta n = n_I - n_{II}$. It is this quantity that is a measure of birefringence.

Upon synchronous rotation of the crossed polarizer and analyzer, the light will be annihilated. The angle of complete light extinction (the *extinction angle*) is taken to be the smaller of the two angles between the plane of polarization and the direction of shear. Then, by definition, the angle, χ, is smaller than or equal to 45°. For low-molecular-mass liquids, the angle χ is 45° within ordinary values of shear stresses and shear rates. For polymer melts and solutions the angle, λ, equals 45° only at very low shear stresses. As the stress increases, the refractive index ellipsoid rotates relative to the neutral axis and the extinction angle diminishes. In the limiting case, the refractive index ellipsoid is found to be oriented in the direction of axis 1 and the angle equals zero.

5.8.1.2 Stress – optical rules for polymer melts

The method of quantitative measurement of stresses in polymeric fluids is based on a linear stress-optical relationship, which is sometimes called the *stress-optical law*. According to this rule, birefringence is linearly proportional to the difference of principal stresses and the directions of optical axes, χ_{op}, and mechanical stress axes, χ_{mech}, coincide such that

$$\Delta n = n_i - n_j = C\Delta\sigma = C(\sigma_i - \sigma_j), \ \chi_{op} = \chi_{mech} \quad\quad [5.8.3]$$

This rule holds true for a large group of industrially important polymers, such as polyethylene, polystyrene, and many other thermoplastic melts and solutions, uncured and randomly crosslinked (cured) elastomers. This rule is obeyed by polymeric systems consisting of linear flexible-chain molecules. The stress-optical rule means that the principal axes of stress and refractive index ellipsoids coincide. This relationship is expressed through a proportionality factor, C, called the stress-optical coefficient, which is a fundamental characteristic of a polymer:

$$C = \frac{2\pi}{45kT}\frac{(n^2 + 2)^2(\alpha_1 - \alpha_2)}{n} \quad\quad [5.8.4]$$

where n is the mean value of the refractive index, which is determined from the relation $3n = n_I + n_{II} + n_{III}$; α_1 and α_2 are the longitudinal and transverse polarizabilities of the Kuhn random link of a chain molecule, k is Boltzmann constant, T is the absolute temperature. The above expression for C and the stress-optical rule follow from the classical theory of an ideal network of rubber elasticity.[119] There are many examples[120] where the condition of C = const has been observed. It is measured in *Brewsters* (1 Br = 10^{-12} Pa^{-1}.

The use of the stress-optical rule for estimation of stresses in polymeric fluids is based on Lodge's idea of the existence of a fluctuating entanglement network, which manifests itself like a network formed by covalent bonds in crosslinked elastomers. Then the macromolecular chains between the entanglement points undergo numerous conformational transformations during the time required for small displacements of centers of gravity of macromolecules. The most important criterion of applicability of the stress-optical rule to uncured polymers must be the existence of a random entanglement network.

A remarkable feature of the stress-optical rule is that, over a wide range of stresses, the quantity C does not depend on deformation and rate of strain. At the same time, it is well-known that in simple shear, the shear rate depends strongly on shear stress. The constancy of C under simple shear for high-molecular-mass polymeric liquids may be explained by proceeding from the earlier discussed concept that the nonlinear relation between shear rate and shear stress for polydisperse polymers is due to the successive transition of the highest molecular mass fractions to the rubbery state (Section 3.3.5). This

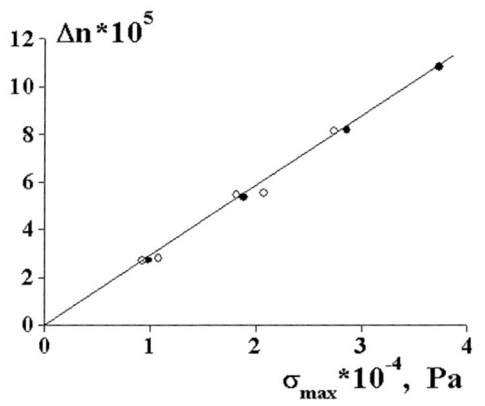

Figure 5.8.2. Birefringence vs. maximal tangential stress, σ_{max}, in a slit die during polyisobutylene flow. Stress-optical coefficient, C = 1414 Br. [Adapted, with permission, from A.I. Isayev and R.K. Upadhyay, *J. Non-Newt. Fluid Mech.*, **19**, 135 (1985)].

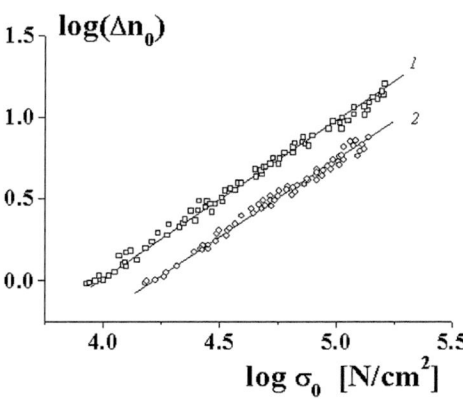

Figure 5.8.3. Dependence of birefringence amplitude, Δn_0, on shear stress amplitude, σ_0, for various polybutadienes (1) and polyisoprenes (2) at different frequencies. Lines are drawn according to the stress-optical coefficient of the respective polymers. [Adapted, with permission, from G.V. Vinogradov, A.I. Isayev, D.A. Mustafaev and Y.Y. Podolsky, *J. Appl. Polym. Sci.*, **22**, 665 (1978)].

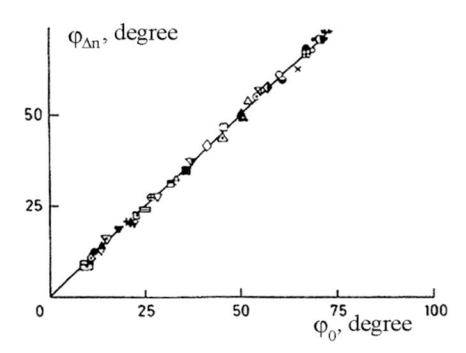

Figure 5.8.4. Phase angle by which oscillatory birefringence leads the oscillatory strain against phase angle. Line is drawn according the equality of these phase angles. [Adapted, with permission, from G.V. Vinogradov, A.I. Isayev, D.A. Mustafaev and Y.Y. Podolsky, *J. Appl. Polym. Sci.*, **22**, 665 (1978)].

reduces hydrodynamic losses, thereby changing the viscosity of polymers, but the transition of polymer from fluid to rubbery state is not accompanied by a change in the quantity C.

As an example, Fig. 5.8.2 shows the dependence of birefringence, Δn, on maximum tangential stress, σ_{max}, which equals half of the difference of the principal stresses during the flow of polyisobutylene melt in a slit die.[121] The linear relationship between birefringence and stress is fulfilled in a wide range of stresses indicating that C is constant. It should be noted that the linear stress-optical rule is also satisfied for the oscillatory flow of polymeric melts if the frequency of oscillations and temperature imposed are in fluid and rubbery states of polymer. These are shown in Figs. 5.8.3 and 5.8.4 for series of polybutadienes and polyisoprenes of various molecular weights.[120] In particular, Fig. 5.8.3 indicates that the birefringence amplitude as a function of the stress amplitude is linear. The lines drawn in this figure are based on the stress-optical constant of polymers obtained in steady shear flow. Fig. 5.8.4 compares phase shifts between oscillatory birefringence and strain, $\varphi_{\Delta n}$, and between oscillatory stress and strain, φ_0. It follows from this figure that $\varphi_{\Delta n} = \varphi_0$, i.e., phases of birefringence and stress with respect to

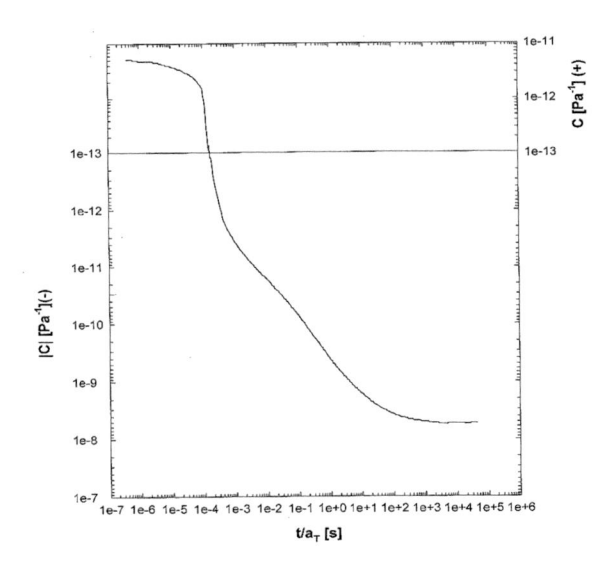

Figure 5.8.5. Master curve of stress-optical coefficient for polycarbonate at reference temperature of 147.5°C indicating the dependence of the coefficient on the time and temperature in the transition region from rubbery to glassy state. [Adapted, with permission, from G.D. Shyu, A.I. Isayev and C.T. Li, *J. Polym. Sci., Phys. Ed.*, **39**, 2252 (2001)].

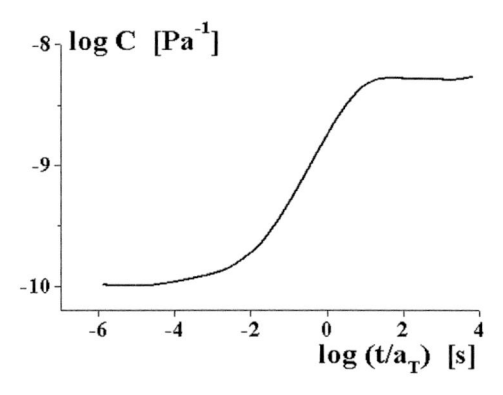

Figure 5.8.6. Master curve of the stress-optical coefficient for polystyrene at the reference temperature of 97°C indicating the dependence of the coefficient on the time and temperature in the transition region from rubber to glassy state. [Adapted, with permission, from G.D. Shyu, A.I. Isayev and C.T. Li, *J. Polym. Sci., Phys. Ed.*, **39**, 2252 (2001)].

the assigned deformation coincide, and hence they coincide with respect to one another as well.[122]

In the transition region from a rubbery to a glassy state, the stress-optical rule is no longer valid.[123,124] In this region, the stress-optical coefficient becomes a function of time and temperature, as shown in Fig. 5.8.5 for polycarbonate in the form of a master curve.[123] With increasing and decreasing time and temperature the stress-optical coefficient approaches respective constant values corresponding to their values in the fluid and rubbery states and the glassy state. This non-constancy of the stress-optical coefficient between the glassy and rubbery states is generally observed for many polymers. It should be noted that for polycarbonates, as seen from Fig. 5.8.5, the stress-optical coefficient exhibits its positive value at any time and temperature. However, such behavior is not generally true for all polymers. In particular, for polystyrene, the stress-optical coefficient is positive in the glassy state and negative in the rubbery and fluid states, as shown in Fig. 5.8.6.[123] Such a behavior of the stress-optical behavior of polystyrene was explained as follows. The birefringence of polystyrene is mainly determined by the orientation of the phenyl groups. Upon deformation in the glassy state, the motion of polymer chains is restricted. The phenyl groups are tilted toward the stretching direction, resulting in a positive birefringence. However in the rubbery and fluid state, polymer chains are able to move freely and tend to align along the stretching direction to some extent. Thus, the phenyl groups lie preferentially perpendicular to the stretching direction, leading to a negative birefringence. Therefore, the positive value of C comes primarily from the tilting of side phenyl groups, and the negative value of C results from the chain segmental orientation. The stress-optical coefficient of polymers is also affected by the strain level, especially when a polymer is in a glassy state. As shown for polycarbonates its stress-optical coefficient is a decreasing function of the strain.[123] This time, temperature and strain dependency of optical behavior is a general feature of various polymers.

In addition, the stress-optical rule is not applicable to filled polymers even in cases when they exhibit high optical transparency and the light depolarization is insignificant. This also applies to some copolymers with a heterogeneous microstructure.

Turning to the quantitative aspect of the stress-optical rule, it is useful to refer once more to Fig. 5.8.1 which, apart from the geometrical characteristic of relationships between simple shear and birefringence, shows a graph that gives an idea of a system of acting forces. Assuming a linear relationship between birefringence and stress tensors, one obtains the following system of equations:

$$\Delta n \sin 2\chi = 2n_{12} = 2C\sigma_{12}$$

$$\Delta n \cos 2\chi = n_{11} - n_{22} = C(\sigma_{11} - \sigma_{22})$$

$$n_{22} - n_{33} = C(\sigma_{22} - \sigma_{33}) \qquad [5.8.5]$$

$$n_{11} - n_{33} = C(\sigma_{11} - \sigma_{33})$$

$$\cot 2\chi = \frac{n_{11} - n_{22}}{2n_{12}} = \frac{\sigma_{11} - \sigma_{22}}{2\sigma_{12}} = \cot 2\chi_m$$

where χ is the extinction angle measured by the optical technique and defining the direction of the principal refractive indices, χ_m is the mechanical angle measured by mechanical means, and defining the direction of the principal stresses.

The determination of the stress optical coefficient, C, and tests of its constancy are usually carried out by using the relationship

$$\Delta n = \frac{2C\sigma_{12}}{\sin 2\chi} \qquad [5.8.6]$$

or

$$C = \Delta n \sin 2\chi / 2\sigma_{12} \qquad [5.8.6a]$$

Though measurements of stresses by the optical method are not direct, they are of great interest for the following reasons. The method makes it possible to conduct measurements in a polymer stream by a non-invasive method, without flow perturbations that could be caused by any measuring mechanical devices. The optical method is of special importance for the estimation of the first and the second normal stress differences. Such measurements can be made in rigid measuring devices. In the measurement of these values by means of mechanical devices, considerable difficulties arise, due to the deformation of various force transducers.

Let us consider briefly the latter method of measurement. Fig. 5.8.7 shows the scheme of measurements of the normal stress differences in a slit instrument. Measure-

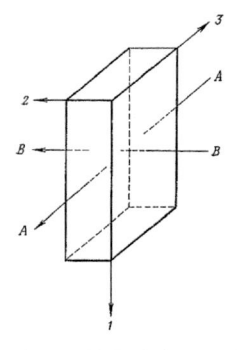

Figure 5.8.7. Scheme of birefringence measurement in slit die.

ments of the quantities $n_{11} - n_{22}$ and $n_{11} - n_{33}$ in the AA and BB directions, respectively, can be carried out. The second normal stress difference, N_2, can be found (see Section 3.4.2), which is proportional to $(n_{11} - n_{22}) - (n_{11} - n_{33})$. According to the results of these measurements, $N_2 < 0$ and is about 10% of N_1. This is in good agreement with data obtained by direct mechanical measurements of N_1 and N_2.[125]

The rheo-optical method can also be used for measuring quasi-equilibrium compliance, J_s^0, phase angle, δ, and recoverable deformation, γ, corresponding to the steady-state flow. This is based on the following equation:

$$\frac{\sigma_{11} - \sigma_{22}}{2\sigma_{12}} = J_s^0 \sigma_{12} = \cot 2\chi_m = \cot 2\chi = \cot \delta = \gamma$$

However, determination of recoverable deformation based on polarization-optical measurements can yield satisfactory results only at relatively low values of σ_{12}.

5.8.1.3 Stress-optical rule for polymer solutions

The quantity C remains constant over the entire range of polymer concentrations, from infinitely dilute solutions up to molten polymers.[126] At a low content of polymer in solution the properties of solvent begin to play an important role. Then, estimation of the observed optical properties must be carried out by taking into account the contribution of stresses on optical anisotropy of solvent compared with optical anisotropy of solution. As an example, it may be pointed out that, whereas the stress optical coefficient of polyisobutylene has a value close to 1,500-1,600 Br, solvents, such as cetane and methylnaphthalene, have C values equal to 1,100 and 1,900 Br, respectively. Therefore, depending on the ratio of optical properties of polymer and solvent, the stress optical coefficient may vary with the change of concentration of polymer solution.

The general method of determining normal stresses in solutions using optical measurements consists of separating the total observed birefringence into contributions of polymer and solvent. They are added up as vectors (in the case of biaxial stress) since the deformation of solvent is accompanied by the appearance of only shear stresses and that of the polymer by the development of both shear and normal stresses. The difference of the refractive indices of a system, Δn, is expressed in terms of the difference between the refractive indices of solvent, Δn_0, and polymer, Δn_1, using vector equality of $\Delta n_1 = \Delta n - \Delta n_0$. This is shown in Fig. 5.8.8a, which also gives an observed angle, χ, and angle, χ_1, associated with stresses that arise because the polymer is in solution. The angle χ due to deformation of solvent equals 45° because no normal stress arises during its flow. The diagram of stresses developing in solution is shown in Fig. 5.8.8b, where σ_{12} is the total shear stress acting in the system and σ_{12}^0 is the component of the total stress, which is due to the flow of pure solvent. The difference, $\Delta\sigma_{12} = \sigma_{12} - \sigma_{12}^0$, is the contribution introduced by the presence of polymer in solution to the shear stress (and, hence, to the viscosity of the system). The vector BC corresponds to the normal stress difference $\sigma_{11} - \sigma_{22}$ and the vector OC is the difference between the principal stresses acting in solution. It is important to

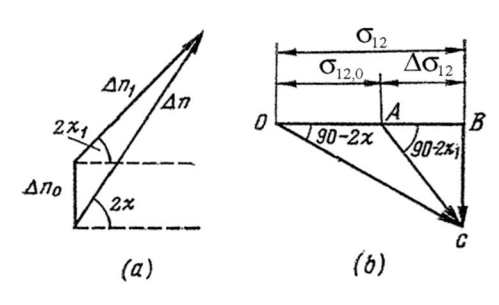

Figure 5.8.8. Vector diagram of the plane stress state for a polymer solution: the difference of refractive indices (a) and stresses (b) for solution and polymer.

note that the angle χ between the direction of shear and the direction of principal stresses in solution is not equal to χ_1. Of special interest in the case of dilute solutions is the question of the relation between the angle χ_1 and the stress-optical properties of the polymer. Experimental investigation of this problem has shown that the extinction angle of polymer melt $\chi_m = \chi_1$, but $\chi_m \neq \chi$. The stress-optical rule for the polymer melt is written in the following form:

$$\Delta n_1 = \frac{2C\Delta\sigma_{12}}{\sin 2\chi_1} \qquad [5.8.7]$$

where C is the stress-optical coefficient of the polymer melt, which is independent of the concentration of the solution and solvent type.

The obvious difference between Eqs. 5.8.7 and 5.8.6 is that, instead of the total difference between the refractive indices, Δn, measured experimentally, in Eq. 5.8.7 use is made of the quantity Δn_1, and the total shear stress, σ_{12}, is replaced by the quantity, $\Delta\sigma_{12}$, which refers to shear stresses that arise additionally because of the polymer presence in the system.

Calculations using the above relationships require knowledge of two parameters of solvent: viscosity, η_s, and stress-optical coefficient, C_0. Then

$$\sigma_{12}^o = \dot\gamma\eta_s, \; \Delta n_0 = 2C_0\sigma_{12}^o = 2C_0\dot\gamma\eta_s$$

The quantities Δn and χ are determined experimentally and the method of determining Δn_1 and χ_1 using vector triangle is pertinent from Fig. 5.8.8a. The difference of normal stresses arising during the shear flow of solution is then calculated from the equation:

$$N_1 = 2\Delta\sigma_{12}\cot 2\chi_1 \qquad [5.8.8]$$

In concentrated solutions, solvent makes a negligibly small contribution to stress and birefringence, and therefore $\chi_m = \chi = \chi_1$, and the decisive role is played by the stress-optical properties of the polymer.

If the stress-optical coefficient of polymer is unknown, then, according to Eq. 5.8.7, it can be found from the results of measurements carried out in dilute solutions.

5.3.1.4 Viscometers for optical observations

The study of structural transformations in liquid medium under the action of deformation represents a special field of research. The methodological basis of such studies is a combination of viscometers and optical measurements. Depending on the purpose, different methods are used.

Extensive research is devoted to the measurement of birefringence in the flow of dilute polymer solutions based on the so-called *dynamo-optic Maxwell effect*. Birefringence appearing in a shear field results from structural transformations of macromolecules, caused by deformation. During polymer solution flow, the Maxwell effect depends on geometrical, mechanical, and optical properties of dissolved macromolecules, i.e., on their structure. Therefore measurements of birefringence in the flow is an effective method of structural studies of macromolecules.[127]

Instruments, in which birefringence during flow is measured, are called *dynamo-optimeters* or *rheo-optical instruments*. The range of instruments of such type is distinguished by two basic elements of construction, namely viscometric and optical parts. The typical example, which illustrates the experimental measuring scheme is shown in Fig. 5.8.9. A rotational viscometer with a bell type intermediate revolving cylinder is used. The shear of liquid being investigated is accomplished in narrow clearances between the revolving cylinder and two fixed cylinders. For the optical measurements during flow in the outer clearance in the upper base of rotor perforation in the form of a system of annuluses is arranged along the circumference. In the lower and upper bases, observation windows, S, are made.

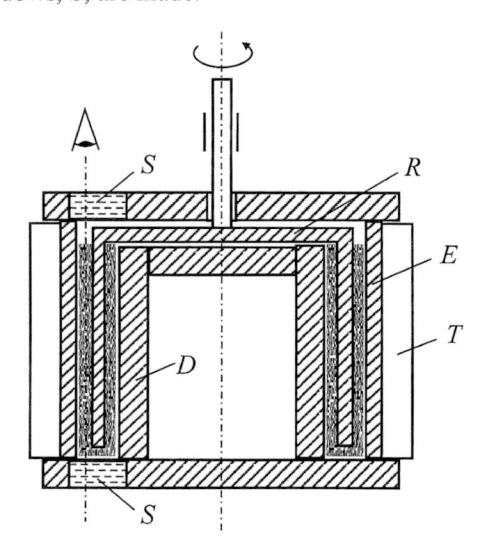

Figure 5.8.9. Bell-type viscometer as a rheo-optical instrument. R – rotating cylinder (rotor); D and E – fixed inner and outer cylinders; S – windows for visualization through the outer clearance. [Adapted, with permission, from E.B. Frisman, V.N. Tsvetkov, *Zh. Teor. Exp. Fiz.* (J. Theor. Experim. Phys. – in Russian), **23**, 690 (1952)].

The viscometric part of the rheo-optical apparatus does not differ from usual viscometers. The specific attribute of such instruments is optical measurement. The corresponding diagrams are described in the specialized monograph.[128] Essentially, these diagrams give the possibility of measuring two basic parameters – birefringence and orientation (extinction) angle during the flow of polymer melts and solutions.

Instruments in which the viscometric and optical observations are combined are also used for the study of shear-induced phase transition in polymers (see Section 3.5.3). For the realization of such studies, a rotational viscometer with coaxial cylinders is applied with a transparent outer cylinder.

The onset of phase transition is detected by the classical method of cloud point.[129] A monochromatic light beam passes tangentially through a transparent outer cylinder, then through a flowing solution, and the exiting beam is detected by a photomultiplier (Fig. 5.8.10).[130] In another version of this method, the light beam is sent along a radius of cylinders, then it is reflected from the polished surface of the inner cylinder, and, after passing twice through the solution being investigated, it is captured by a photomultiplier. The appearance of intensive light scattering (cloudiness) depends on shear rate.

A combination of viscometric and optical schemes of measurements is especially effective for observation of transient regimes of deformation when structural transforma-

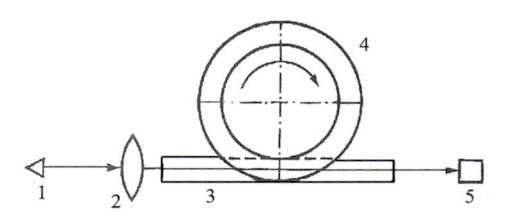

Figure 5.8.10. Diagram of measurement of cloud point of polymer solution flowing between coaxial cylinders of rotational viscometer. 1 – monochromatic light source; 2 – lens; 3 – optical tube; 4 – working cell; 5 – photocell. [Adapted, with permission, from A.Ya. Malkin, S.G. Kulichikhin, G.K. Shambilova, *Vysokomol. Soedin.* (Polymers – in Russian), **33B**, 228 (1991)].

tions due to deformation occur in multi-component systems.[131]

Commercial instruments supplied with optical devices, manufactured by specialized companies, can be used. Rheometer *Rheometrics RMS 800* (TA Instruments, USA) equipped with the optical system *Rheo-optical Analyzer* is equipped with a He-Ne laser. The working cell is equipped with cone and plate or parallel disks. The light beam is oriented in a direction parallel to the velocity gradient. The observed picture of light scattering at small angles gives a very good image of structural processes taking place during the deformation of a mixture of incompatible polymers.

5.8.1.5 Polarization methods for measuring stresses

Stress-optical measurements are the most effective and interesting in the study of polymer flow through channels and dies of different types, including capillaries. Experimental instruments are used for this purpose. As an example of the use of the polarization-optical method, let us consider measurement of normal stresses arising when polymer flows from a reservoir into a rectangular die.[132]

Fig. 5.8.11 shows the dependence of the first difference of normal stresses (extensional stress), N_1, on the dimensionless length, Z/H, i.e., the distance along the flow axis referred to the die width. Each curve corresponds to a constant value of shear stress on the die wall in the region of a fully developed velocity profile. The positive values of Z/H refer to the pre-entrance region, and their negative values refer directly to the die. The value of Z/H = 0 corresponds to the die edge. Inspection of Fig. 5.8.11 shows that in the pre-entrance region of die, the extensional stresses increase, reaching a maximum. The position of the maximum of extensional stresses is located over the die entrance at a distance of (0.2–0.3)H from the die edge. Then relaxation begins and terminates inside the die.

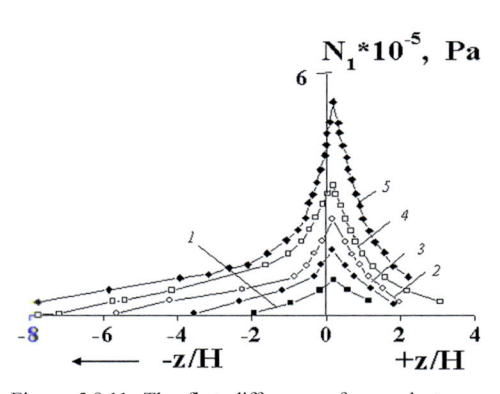

Figure 5.8.11. The first difference of normal stresses along flow axis at the entrance into a slit die for broad MMD polybutadiene (molecular mass 2.4 x 10⁵; 25°C). The arrow indicates the direction of flow. The curves correspond to shear stresses at wall in the zone of developed flow: 0.051 (1); 0.098 (2); 0.141 (3); 0.21 (4); 0.25 MPa (5). [Adapted, with permission, from V.I. Brizitsky, G.V. Vinogradov, A.I. Isayev and Y.Y. Podolsky, *J. Appl. Polym. Sci.*, **21**, 751 (1977)].

The rheo-optical method can also be used to measure stresses under conditions of sinusoidal oscillatory shear.[133] It has been shown that the normal stresses measured by the dynamic method consist of two components: the constant steady-state term, $N_{1,c}$, and the oscillating term of doubled frequency, $N_{1,2\omega}$, as compared with the frequency of specified shear stresses. The

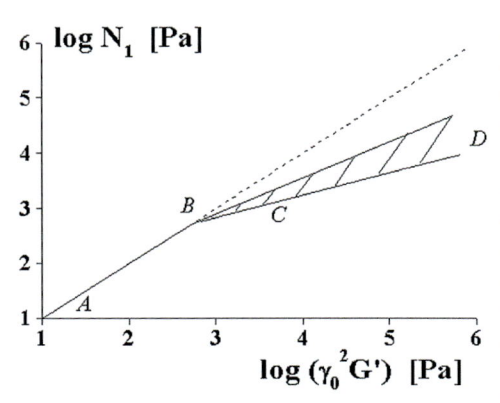

Figure 5.8.12. Master curve showing dependence of the steady state component of the first difference of normal stresses on the amplitude of strain in oscillation regime of deformations. The portion of the curve ABC refers to polystyrene solutions according to data by H. Endo and M. Nagasawa, *J. Polym. Sci.*, *A-2*, **8**, 371 (1970); the band CD refers to a series of polybutadienes and polyisoprenes. [Adapted, with permission, from G.V. Vinogradov, A.I. Isayev, D.A. Mustafayev and Y.Y. Podolsky, *J. Appl. Polym. Sci.*, **22**, 665 (1978)].

constant component reflects the long-time end of the relaxation spectrum since the stress has no time to relax during each deformation cycle.

The quantity $N_{1,c}$ in the linear deformation domain equals $\gamma_0^2 G'$, where γ_0 is the amplitude of deformation and G' is the storage modulus. Fig. 5.8.12 shows this relation for various polymeric materials. At low oscillatory amplitude, $N_{1,c}$ is proportional to $\gamma_0^2 G'$. At higher amplitudes, there is a deviation from proportionality between these two quantities, but a master curve can be constructed.

5.8.1.6 Visualization of polymer flow in dies

In Section 3.6.3, the critical deformation regimes for flexible-chain linear polymers and the associated spurt of their stream (loss of fluidity) in dies was discussed. It is very important to have experimental information explaining what happens to polymers during the spurt and under the above-critical regimes of occurrence of the elastic instability. The rheo-optical technique is useful for visualization of polymer stream. It is also particularly useful in the investigation of polymer flow through channels of complicated geometrical forms.[134]

What can be achieved by the method of visualization of polymer flow can be illustrated by the data presented in Fig. 5.8.13. The streamlines were recorded by observing a movement of 10-20 μm glass beads in the polymer. Flow at low shear rates is associated with the regular nature of streamlines in a die and at its entrance.

What can be achieved by observation of the flow of polymer under circularly polarized light was demonstrated in Figs 3.6.12 and 3.6.13. (see discussion of results in Section 3.6.3).

5.8.2. VELOCIMETRY

One evident consequence of non-Newtonian flow

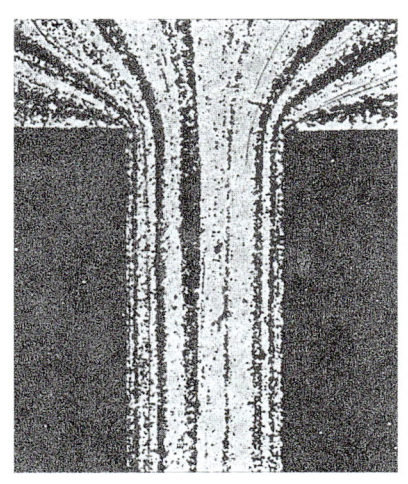

Figure 5.8.13. Flow lines of polybutadiene at shear stresses of 6.31×10^4 Pa (at a wall of rectangular channel). [Adapted, with permission, from G.V. Vinogradov, *Rheol. Acta*, **12**, 357 (1973)].

properties of liquid is a non-parabolic velocity profile in round channels. This statement can also be formulated in an inverse form: if a velocity profile along the channel radius is non-parabolic, then the liquid is non-Newtonian. Measurement of velocity near a channel wall in order to confirm the concept of polymer adhesion to a wall or to observe slip along the wall surface is a separate problem. Both effects require measurements of velocity profile in order to carry out calculations of rheological properties of the liquid.

This approach (called *velocimetry*) is successfully realized by utilizing several physical phenomena. The most popular phenomena used as a base for velocimetry are:
- the Doppler effect[135]
- Nuclear magnetic resonance (NMR) imaging.[136]

The *Doppler effect* is a change of frequency of vibration as a function of the velocity of relative movement of a source of vibrations and an observer. Measuring a shift in frequency, a motionless observer can measure the velocity of movement.[137]

NMR[138] flow imaging (used in different versions) is the most interesting method because it allows the experimenter to obtain velocity values at many points across the radius of a channel approaching the wall of the channel (about 0.2 mm). This is especially interesting for measuring near-wall effects. NMR technique of velocimetry has many applications in medical studies as well as in investigations of the flow of thixotropic suspensions.[139]

Other versions of optical methods for measuring velocity include creating an interference pattern by means of a laser system and measuring the intensity of reflected light that varies as small particles cross the interference pattern.[140]

Velocimetry can be useful to obtain data on velocity evolution in transient flow. The velocity profile in a steady laminar flow reflects shear rate distribution. The latter can be calculated if the velocity profile is measured. This is a typical inverse problem (analogous to some other mentioned in this book), which is solved by a standard procedure, e.g., by the Tikhonov regularization method.[141]

5.8.3 VISCOMETERS-CALORIMETERS

Measurement of thermal effects, which accompany viscous flow, can be of definite interest for studying some physical phenomena, for example, detection of phase transitions or other structural transformations caused by deformation. These measurements are carried out using a combination of viscometers with a calorimetric device, i.e., strictly viscometric measurements are supplemented by measurement of heat fluxes.

As mentioned above (see Section 3.2.5), thermal effects always accompany viscous flow in view of energy dissipation of external forces. Therefore, it is important to separate structural phenomena from a trivial energy dissipation of forces of viscous friction. In practice, this is achieved by a combination of a rotational viscometer with a calorimeter into which the working cell of viscometers is placed.[142]

Calorimetric measurements are accomplished by the installation of heat flux sensors on fixed and rotating cylinders. Calibration of the instrument is accomplished based on heat emission measured for standard Newtonian liquid. The intensity of heat emissions is proportional to the square of shearing rate: $W = k\dot{\gamma}^2$. Thus, the instrument constant, k, is determined. Then, heat emissions during the shear flow of non-Newtonian liquids are measured. The component of heat flux due to structural transformation is evaluated based on the deviation of observed dependence $W(\dot{\gamma})$ from quadratic dependence indicated by viscous flow.

REFERENCES

1   Flow of viscous liquid is always accompanied by heat output because work must be done and dissipated in creating flow, therefore, flow cannot be isothermal in principle. However, this effect cannot be weak and the basic theory of capillary viscometry neglects this influence. Non-isothermal effects can be essential in flow of high viscosity liquids, and the analysis of this problem is a separate task in the theory of capillary viscometry.

2   Hagen (1839) and independently Poiseuille (1846) have experimentally shown that volumetric flow rate is proportional to R^4. Eq. 5.2.10 was later independently obtained by Weidman (1856) and Hagenbach (1860).

3   A.I. Isayev, K.D. Vachagin, A.N. Naberezhnov, *J. Eng. Phys.*, **27**, 998 (1974).

4   Basic hypothesis for deriving this equation is an assumption about the absence of slip effect. This idea was suggested by K. Weissenberg. The equation was obtained by B. Rabinovitsch, *Z. Phys. Chem.*, **A145**, 1 (1929).

5   C.L.A. Berli, J.A. Deiber, *Rheol. Acta*, **40**, 272 (2001).

6   R.K. Upadhyay, A.I. Isayev, and S.F. Shen, *Rheol. Acta*, **20**, 443 (1981).

7   SigmaPlot Software, CambridgeSoft, Inc.

8   E.B. Bagley, *J. Appl. Phys.*, **28**, 624 (1957); *Trans. Soc. Rheol.*, **2**, 263 (1961).

9   A. I. Isayev, B. Chung, *Polym. Eng. Sci.*, **25**, 264 (1985).

10   F. P. La Mantia, A. Valenza, D. Acierno, *Rheol. Acta*, **22**, 299 (1983).

11   A.Ya. Malkin, V.G. Kulichikhin, I.V. Gumennyi, *Phys. Fluids*, **33**, 013105 (2021).

12   A.Ya. Malkin, S.O. Ilyin, G.B. Vasilyev, M.P. Arinina, V.G. Kulichikhin, *J. Rheol.*, **58**, 433 (2014).

13   This formula was obtained by A.M. Stolin (see Section 16.4) in: A. Malkin. A. Askadsky, A. Chalykh, V. Kovriga, **Experimental Methods of Polymer Physics**, *Mir Publishers - Prentice-Hall*, New York, 1983.

14   M. Mooney, *J. Rheol.*, **2**, 210 (1931). The problem of a surface slip in studies of rheological properties of raw elastomers (the most important case of wall slip) is discussed in: M. Mooney, **The Rheology of Raw Elastomers**, in Rheology, ed. F.R. Eirich, vol. 2, p. 181, *Acad. Press*, 1958.

15   C.P. Thomas, *Soc. Petrol. Eng. J.*, **16**, 130 (1976).

16   This instrument is called Saybolt Universal. It is standardized in the USA for measurement of the relative viscosity of oils.

17   See, for example, T. Tovrog, A. A. Krawetz, *Rev. Sci. Instr.*, **36**, 1261 (1965).

18   W.S. Park, W.W. Graessley, *J. Polym. Sci., Polym. Phys. Ed.*, **15**, 71 and 85 (1977).

19   This instrument was originally developed for measurements of the viscosity of oil products. Procedures of measurements are standardized according to ASTM D 1092.

20   An excellent review on the history of viscometers with coaxial cylinders and description of first viscometers of this type is given in: J.M. Piau, M. Bremond, L.M. Couette, M. Piau, *Rheol. Acta*, **33**, 357 (1994).

21   Obtained by Margules in 1892.

22   This formula was obtained in I. M. Krieger, M. E. Woods, *J. Appl. Phys.*, **37**, 4703 (1966).

23   Problem of determining the resistance of a thin disk to rotation in viscous infinite medium was first suggested and approximately solved by Th. Karman in 1921.

24   J.V. Kelkar, R.A. Mashelkar, J. Ulbrecht, *J. Appl. Polym. Sci.*, **17**, 3069 (1973).

25   A.I. Leonov, A.Ya. Malkin, G.V. Vinogradov, *Kolloid Zh* (Colloid. J. – in Russian), **26**, 335 (1964).

26   A.Ya. Malkin, A.M. Stolin, in **Rheology**, Eds: G. Astarita, G. Marrucci, L. Nicolais, *Plenum Press*, NY, Lnd, 1980, VIII Intern. Congress Rheol., Naples (Italy), v. 2, p. 223.

27   It should be noted that the phenomenon of super-fluidity (i.e., absence of viscosity) of quantum fluid was discovered by P.L. Kapitsa (Nobel Prize 1978) also with aid of an instrument analogous to a rotational viscometer – rotational element suspended on an elastic thin wire.

28   This instrument was manufactured with a number of modifications; the last modification was R19.

29   This company specializes in manufacturing testing machines for various materials that permit using experience and enormous capability of developed measuring techniques for the creation of high precision rheogoniometer.

30   ARES is an abbreviation for Advanced Rheometrics Expansion System.

31   RDA and RFS are abbreviations for Rheometrics Dynamic Analyzer and Rheometrics Fluid Spectrometer.

32   SR is an abbreviation for Stress Rheometer.

33   According to the ASTM, ISO 2555 and ISO 2319.

34   E.W. Merril et al., *J. Polym. Sci.*, **1A**, 1201 (1963); R.S. Porter, R.E. Klaver, J.F. Johnson, *Rev. Sci. Instr.*, **36**, 1846 (1965).

35   B.H. Zimm, D.M. Crothers, *Proc. Natl. Acad. Sci. USA*, **48**, 905 (1962).

36   F.S. Geata, *Rev. Sci. Instr.*, **37**, 844 (1966).
37   B. Lee, M.T. Shaw, *J. Rheol.*, **45**, 641 (2001).
38   H. Orihara, N. Miwa, M. Doi, *J. Rheol.*, **45**, 773 (2001).
39   Measuring scheme, used in these instruments, was proposed by M. Mooney in 1934.
40   Procedure of viscosity measurements using the Mooney viscometer appears in the following standards:
     USA ASTM D 1646 (in standard ASTM D 927 procedure for sample preparation is given),
     British BS 1673, German DIN 53523, DIN 53524, Russian 10722 and also in recommendations ISO P289.
41   This was done in earlier publications devoted to the Weissenberg effect. See e.g., K. Weissenberg, Proc.
     1st Intern. Rheol. Congress, Scheveningen, p. 1, p.29 (1948); F.H. Garner and A.H. Nissan, *Nature*, **158**,
     534 (1946); R.S. Rivlin and H.W. Greensmith, *Nature*, **168**, 664 (1951).
42   The detailed analysis of the reliability of force measuring cell in such devices is given in: J.M. Niemiec,
     J.-J. Pesce, G.M. McKenna, S. Skocypic and R.F. Garritano, *Trans. Soc. Rheol.*, **40**, 323 (1996).
43   S.-G. Baek, J.J. Magda, *J. Rheol.*, **47**, 1249 (2003).
44   N. Adams and A.S. Lodge, *Phil. Trans. Roy. Soc. London*, **A256**, 149 (1964).
45   S.F. Mall-Greissle, W. Gleissle, G.H. McKinley, and H. Buggish, *Rheol. Acta*, **41**, 61 (2002).
46   See a detailed analysis of this problem, though looking special but important for the application of methods
     under discussion in: M. Keentok and S.-C. Xue, *Rheol. Acta*, **38**, 121 (1999). The theoretical analysis is
     based on certain rheological models. However, it is interesting to mention that a possibility of the edge
     fracture (as a limit of the experimental ability of technique) is directly related to the value of N_2.
47   J.M.P Papenthuijzen and M. van den Temple, *Rheol. Acta*, **6**, 311 (1958); H. Markovitz, *J. Polymer Sci.*,
     3B, 3 (1965). See also: T. Kotaka, M. Kurata and M. Tamura, *J. Appl. Phys.*, **30**, 1705 (1959).
48   R.S. Rivlin, *J. Rat. Math. Anal.*, **5**, 179 (1956); J.W. Hayes and R.I. Tanner, Proc. 4th Intern. Congress on
     Rheology, v. 3, Interscience, New York - London - Sidney, p. 389 (1956).
49   The complete theoretical background of different methods of normal stress measurements based on
     capillary flow is discussed in: K. Walters, **Rheometry**, *Chapman and Hall*, Chapter 5, 1975.
50   R. Tanner and A.C. Pipkin, *Trans. Soc. Rheol.*, **13**, 471 (1969).
51   This was done in J.M. Broadbent, A. Kaye, A.S. Lodge and D.G. Vale, *Nature*, **271**, 35 (1968).
52   Compare, for example, publications: J.W. Hayes and R.I. Tanner, Proc. 4th Intern. Congress on Rheology,
     v. 3, Interscience, New York - London - Sidney, p. 389 (1956); and M.V. Keentok, A.G. Georgescu,
     A.A. Sherwood, R. I. Tanner, *J. Non-Newton., Fluid Mech.*, **6**, 303 (1980).
3   The main problem here is the formulation of the law of wall friction.
4   C.H. Chatraei, C.W. Macosko, H.H. Winter, *J. Rheol.*, **25**, 433 (1981); P.R. Soskey, H.H. Winter, *J. Rheol.*,
     29, 495 (1985); A. I. Isayev, A. D. Azari, *Rubber Chem. Technol.*, **59**, 868 (1986).
5   S. Kh. Peschanskaya, G. S. Pugachev, P. P. Yakushev, *Mekh. Polymerov* (Polymer Mech. – in Russian), # 2,
     357 (1977).
5   B. Hoffner, O. H. Camanella, M. Peleg, *Rheol. Acta*, **40**, 289 (2001).
57   See USA standard ASTM D 926.
5   See, for example, German standard DIN 53514.
59   A. Pochettino, *Nouvo Cimento*, **8**, 77 (1914). This is instrument after modification continued to be used for
     measuring rheological properties of highly viscous substances, e.g., A.W. Myers, J.A. Faucher, *Trans. Soc.
     Rheol.*, **12**, 183 (1968).
60   D.G. Williams, C.L. Garey, G.A. Hemstock, *Trans. Soc. Rheol.*, **6**, 93 (1962).
6   Formula obtained by S. Goldstein.
62   Theory of calculating correction for the wall effect was developed by H. Faxen and R. Ladenburg
     (see papers by R. Ladenburg, *Ann. D Phys.*, Paris, **22**, 287 (1907); **23**, 447 (1907).
63   Original solution of this problem was obtained by G. Duffing.
64   J. Sestak, F. Ambros, *Rheol. Acta*, **12**, 70 (1973).
65   D.E. Harrison, R.B. Croser, *Rev. Sci. Instr.*, **36**, 1840 (1965).
66   M.C.S. Chen, G.W. Swift, *Amer. Inst. Chem. Eng. J.*, **18**, 146 (1972).
67   F. Ramsteiner, *Rheol. Acta*, **15**, 427 (1976).
68   A method of measurement of the elongational viscosity based on the separation of the total strain into
     components was developed by G.V. Vinogradov, B.V. Raduskevitch, V.D, Fikhman, *J. Polym. Sci.,
     A-2*, **8**, 1 (1970); G.V. Vinogradov, A.Ya. Malkin, V.V. Volosevitch, *J. Polym. Sci., Polym. Phys. Ed.*,
     13, 1721 (1975).
69   J. Berzen, H. W. Bimkraut, G. Braun, *Brit. Polym. J.*, **10**, Dec., 281 (1978).
70   This method is essentially analogous to that described in standards: ASTM 1430 62T (USA) and
     DIN 53493 (Germany).
71   This instrument was developed by J. Meissner, *Rheol. Acta*, **10**, 230 (1971); later theory of this instrument
     and its capability were discussed in many papers, in particular, M. Laun, H. Schuch, *Rheol. Acta*, **33**, 119
     (1989); M. H. Wagner, B. Colognon, J. Verbecke, *Rheol. Acta*, **35**, 117 (1996).

72   Such scheme of measurements is realized, for example, in papers by J. Meissner, *Rheol. Acta*, **8**, 78 (1969); *Rheol. Acta*, **10**, 230 (1971); *Trans. Soc. Rheol.*, **16**, 505 (1972).

73   T. Schweizer, *Rheol. Acta*, **39**, 428 (2000).

74   M. Stadlbauer, H. Janeschitz-Kriegl, M. Lipp, G. Eder, R. Forstner, *J. Rheol.*, **48**, 611 (2004).

75   M.L. Sentmanat, *Rheol. Acta*, **43**, 657 (2004); M.L. Sentmanat, B. N. Wang, C.H. McKinley, *J. Rheol.*, **49**, 585 (2005).

76   W. C. McSporran, *J. Non-Newton. Fluid Mech.*, **8**, 119 (1981).

77   This method of evaluation of the longitudinal viscosity was first proposed by F. N. Cogswell, *Polym. Eng. Sci.*, **12**, 64 (1972). Later, a theory of the method was developed in: D.M. Binding, *J. Non-Newt. Fluid Mech.*, **27**, 173 (1988) and **41**, 27 (1991); D. M. Binding, K. Walters, *J. Non-Newt. Fluid Mech.*, **30**, 233 (1988) and also A. D. Gotsis, A. Odriozola, *Rheol. Acta*, **37**, 430 (1998).

78   M.E. Mackey, G. Astarita, *J. Non-Newt. Fluid Mech.*, **70**, 219 (1997).

79   The four-roll method was first suggested by A. Keller and described in series of papers, for example, F.C. Frank, A. Keller, M. R. Mackley, *Polymer*, **12**, 467 (1971); M.R. Mackley, *Phyl. Royal Soc., London*, **278**, 29(1975); D.P. Pope, A. Keller, *Colloid Polym. Sci.*, **255**, 633 (1975); M.J. Miles, A. Keller, *Polymer*, **21**, 1298 (1980); A. Keller, A. Odell, *Colloid Polym. Sci.*, **263**, 181 (1985). G.G. Fuller, L.G. Leal, *Rheol. Acta*, **19**, 580 (1980). The flow from opposite capillaries method was proposed in: M.R. Mackley, A. Keller, *Phil. Trans. Royal Soc (Lnd)*. **278**, 29 (1975).

80   M. Steller, G. Brenn, *J. Rheol.*, **44**, 595 (2000); A.V. Bazilevskii, V.M. Entov, A.N. Rozhkov, *Polymer Sci., A*, **43**, 716 (2001); M.S.N. Olivera, R. Yeh, G.N. McKinley, *J. Non-Newtonian Mech.*, **137**, 137 (2006); S. Kheirandish, I. Guybaidullin, W. Wohlleben, N. Willenbacher, *Rheol Acta*, **47**, 999 (2008); S. Kheirandish, I. Guybaidullin, N. Willenbacher, *Rheol Acta*, **48**, 397 (2009); K.Niedzwiedz, H.Buggisch, N. Willenbacher, *Rheol Acta*, **49**, 1103 (2010); O. Arnolds, H. Buggisch, D. Sachsenheimer, N. Willenbacher, *Rheol Acta*, **49**, 1207 (2010).

81   A.V. Bazilevsky, V.M. Entov, A.N. Rozhkov A.L. Yarin, in: Oliver DR, Ed.. Proc. 3rd Eur. Rheol. Conf. and golden jubilee meeting British Soc. Rheology. Amsterdam: Elsevier Ltd.; 1990. p. 41-3.

82   J.E. Matta, R.P. Tytus, *J. Non-Newton. Fluid Mech.*, **35**, 215 (1990); Tirtaatmadja, T. Sridhar. *J. Rheol.*, **37**, 1081 (1993); P. Szabo, *Rheol. Acta*, **36**, 277 (1997); S.L. Anna, Ch. Rogers, G.H. McKinley, *J. Non-Newton. Fluid Mech.*, **87**, 307 (1999); J.P. Rothstein, G.H. McKinley, *J. Rheol.*, **46**, 1419 (2002).

83   V.M. Entov, E. J. Hinch, *J. Non-Newton. Fluid Mech.*, **72**, 31 (1997).

84   D.F. James, N. Yogachandran, *Rheol. Acta*, **46**, 161 (2006).

85   C.D. Denson, R.J. Gallo, *Polym. Eng. Sci.*, **11**, 174 (1971); D.D. Joye, G.W. Poehlein, C.D. Denson, *Trans. Soc. Rheol.*, **16**, 421 (1972); **17**, 287 (1973); C.D. Denson, D.C. Hylton, *Polym. Eng. Sci.*, **20**, 535 (1980).

86   H.H. Winter, C.W. Macosko, K.E. Bennet, *Rheol. Acta*, **18**, 323 (1979); J.A. van Aken, H. Janeschitz-Kriegl, *Rheol. Acta*, **19**, 744 (1980); **20**, 419 (1981).

87   S. Chatraei, C.W. Macosko, H.H. Winter, *J. Rheol.*, **25**, 433 (1981); A.I. Isayev, A.D. Azari, *Rubber Chem. Technol.*, **59**, 868 (1986).

88   J. Meissner, S.E. Stevenson, A. Demarmels, P. Portmann, *J. Non-Newt. Fluid Mach.*, **11**, 221 (1979); J. Meissner, *Chem. Eng. Commun.*, **33**, 159 (1985); *Ann. Rev. Fluid Mech.*, **17**, 45 (1985).

89   In calculations of periodic processes, it is convenient to use complex variables. The same results are obtained if one uses real parts of complex numbers.

90   The rigidity is a displacement of spring related to force causing this displacement.

91   The detailed analysis of this problem is given in: T.E.R. Jones, K. Walters, *Rheol. Acta*, **10**, 365 (1971).

92   N.W. Tschoegl, *Rheol. Acta*, **13**, 897 (1974).

93   J.D. Watson, *Rheol. Acta*, **8**, 201 (1969).

94   Do not confuse this value with loss angle also designated as δ.

95   The detailed analysis of this equation in application to torsion oscillations was made by H. Markovitz, *J. Appl. Phys.*, **23**, 1070 (1952). See also: K. Oka, The principles of rheometry, in **Rheology. Theory and Applications**, 3, Ch. 2, p. 18-82, Ed. F.R. Eirich, *Acad. Press*, N.Y. London, 1960.

96   General theory of torsion damping oscillations for a sample of the arbitrary cross-section was developed in R.D. Glauz, *J. Polymer Sci., A-2*, **8**, 329 (1970). This publication also contains the results of calculations for a rectangular sample, which is widely used, especially in testing of rigid materials.

97   These techniques were extensively discussed in books: J. D. Ferry, **Viscoelastic Properties of Polymers**, 3rd edition, *Wiley*, New York, 1980; R.W. Whorlow, **Rheological Techniques**, *Ellis Harwood*, Chichester, 1980; see also K. ten Nijenhuis, in **Rheology**, vol.1: Principles, Eds. G. Astarita, G. Marrucci and L. Nicolais, *Plenum*, New York, 1980, pp. 263-282.

98   H.S. White, E.A. Kearsley, *J. Nat. Bur. Stand.*, **75A**, 541 (1971).

99   This idea was proposed and developed in a series of publications: T.N.G. Abbott, C.W. Bowen, K. Walters, *J. Phys. D: Appl. Phys.*, **4**, 190 (1971); T.N.G. Abbott, K. Walters, *J. Fluid Mech.*, **40**, 205 (1970); T.N.G. Abbott, K. Walters, *J. Fluid Mech.*, **43**, 257 (1970).

100 The analytical expressions for the front factors for the cases discussed in this section were obtained in the original publications and those who are especially interested in this experimental technique are referred to literature.

101 For a more detailed description of the instruments of this type see Section 5.3.

102 The first realization of this approach was described in B. Maxwell, R.P. Chartoff, *Trans. Soc. Rheol.*, **9**, 41 (1965).

103 This method was first proposed by J.D. Ferry and E. Fitzgerald (1952). See the classical monograph: J. Ferry, **Viscoelastic Properties of Polymers**, *Wiley*, New York, 3rd ed. 1980. Later the same basic principle of constructing experimental devices was used by many authors.

104 This approach was proposed by M. Birnboim (1961). The highly developed version of this experimental technique is described in: D.J. Massa, J.L. Schrag, *J. Polymer Sci., A-2,* **10**, 71 (1972).

105 J.L. Schrag, R.M. Johnson, *Rev. Sci. Instr.*, **42**, 224 (1971).

106 Instruments of this type were first proposed in D.J. Plazek, *J. Polymer Sci., A-2*, **6**, 621 (1968) and later were widely used by many authors.

107 G.C. Berry, C.-P. Wong, *J. Polymer Sci., Polymer Phys. Ed.*, **13**, 1761 (1975).

108 R. Bichdahl, R.J. Morgan, R.I. Nielsen, *Rev. Sci. Instr.*, **41**, 1342 (1970).

109 L. Szilágyi, G. Locati, *Rheol. Acta*, **9**, 535 (1970).

110 For example, torsion pendulum was used by C.A. Coulomb (1784) who measured the attractive force between magnet charges and by H. Cavendish (1797) for measuring the Newton gravitational constant (and made it with very high accuracy even according to the present standards).

111 It was first made by L.E. Nielsen (1951) and K. Schmieder and K. Wolf (1952).

112 Such very low losses are important in studying metals, ceramics, and other special engineering materials. See, e.g., A.E. Schwaneke, R.W. Nash, *Rev. Sci. Instr.*, **40**, 1450 (1969).

113 This method is known as "torsion braid analysis" (TBA). It was first proposed in A.F. Lewis, J.K. Gillham, *J. Appl. Polym. Sci.*, **7**, 685 (1963) and widely used by J.K. Gillham and coauthors.

114 D. Brewster, *Phil. Trans. Roy. Soc.*, **106**, 156 (1816).

115 M.M. Frocht, **Photoelasticity**, *Wiley*, New York, 1948.

116 A. Lodge; **Elastic Liquids**, *Academic Press*, New York, 1964.

117 J.G. Brodnyan, J.G.F.H. Gaskins, and W. Philippoff, *Trans. Soc. Rheol.*, **1**, 95 (1957); W. Philippoff, *ibid.*, **4**, 159 (1960); **5**, 149 and 163 (1961); F.D. Dexler, J.C. Miller, and W. Philippoff, *Trans. Soc. Rheol.*, **5**, 193 (1961); W. Philippoff and S.J. Gill, ibid., **7**, 33 (1963); W. Philippoff, *ibid.*, v. 7, 45 (1963); W. Philippoff, and R. A. Stratton, *ibid.*, **10**, 467 (1966); W. Philippoff, *J. Appl. Phys.*, **32**, 984 (1956); W. Philippoff, Proc. Fifth Intern. Congress on Rheology, 4, 3 (1968).

118 H. Janeschitz-Kriegl, **Polymer Melt Rheology and Flow Birefringence**, *Springer-Verlag*, Berlin, 1983.

119 L.Treloar, **The Physics of Rubbery Elasticity**, 3rd Edition, *Clarendon Press*, Oxford, 1975.

120 J. Wales, **The Application of Flow Birefringence to Rheological Studies of Polymer Melts**, Delft University Press, 1976.

121 A.I. Isayev and R.K. Upadhyay, *J. Non-Newt. Fluid Mech.*, **19**, 135 (1985).

122 G.V. Vinogradov, A.I. Isayev, D.A. Mustafaev and Y.Y. Podolsky, *J. Appl. Polym. Sci.*, **22**, 665 (1978).

123 G.D. Shyu, A.I. Isayev and C.T. Li, *J. Polym. Sci., Phys. Ed.*, **39**, 2252 (2001).

124 T.H. Lin and A.I. Isayev, *Rheol. Acta*, **47**, 977 (2008).

125 K. Walters, **Rheometry**, *Chapman and Hall*, London, 1975 (Table 4.2).

126 Tsvetkov, V. N., in: **Newer Methods of Polymer Characterization**, Ed. B. Ke, Chapter 14, *Interscience*, New York, 1965.

127 G.C. Fuller, **Optical Rheometry of Complex Fluids**, *Oxford University Press*, New York, 1995. See also Ref. 113.

128 See, for example, H. Janeschitz-Kriegl, **Polymer Melt Rheology and Flow Birefringence**, *Springer Verlag*, Berlin, 1983.

129 Method was first proposed by V. F. Alekseev (Russia) in 1877.

130 A.Ya. Malkin, S.G. Kulichikhin, G.K. Shambilova, *Polym. Sci. USSR*, **32A**, 228 (1991); A.Ya. Malkin, S.G. Kulichikhin, *Polym. Sci. USSR*, **38B**, 362 (1996).

131 H. Yang, H. Zhang, P. Moldenaers, J. Mevis, *Polymer*, **39**, 5731 (1998); M. Minale, P. Moldenaers, J. Mevis, *J. Rheol.*, **43**, 815 (1999); P. Van Puavelde, H. Yang, J. Mevis, P. Moldenaers, *J. Rheol.*, **44**, 1401 (2000).

132 V.I. Brizitsky, G.V. Vinogradov, A.I. Isayev and Yu.Ya. Podolsky, *J. Appl. Polymer Sci.*, **21**, 751 (1977); A.I. Isayev and R.K. Upadhyay, *J. Non Newt. Fluid Mech.,* **19**, 135 (1985).

133 G.V. Vinogradov, A.I. Isayev, D.A. Mustafayev and Yu.Ya. Podolsky, *J. Appl. Polym. Sci.*, **22**, 665 (1978).
134 G.V. Vinogradov, *Rheol. Acta*, **12**, 357 (1973).
135 This effect was theoretically predicted and described by Austrian physicist Ch. Doppler (1842).
136 P.T. Callagham, Y. Xia, *J. Magn. Reson.*, **91**, 326 (1991); S.J. Gibbs, D. Xing, T.A. Carpenter, L.D. Hall, S. Ablett, I.D. Evans, W. Frith, D.E. Haycock, *J. Rheol.*, **38**, 1757 (1994).
137 Some interesting examples of the application of this method for measuring velocity profiles and wall slip velocity can be found in N.C. Sharpley, R.A. Brown, R.C. Armstrong, *J. Rheol.*, **48**, 255 (2004); L. Robert, Y. Demay, B. Vergnes, *Rheol. Acta,* **43**, 89 (2004).
138 NMR is one of the methods of radio-spectroscopy. Its principle is based on a selective interaction of the magnetic component of electromagnetic field with the system of nuclear magnetic moments of the substance. It is observed when high frequency sn electromagnetic field is orthogonally imposed on a constant magnetic field. The effect was first observed by American physicist I.I. Rabi in 1937 (Nobel Prize 1944).
139 J.S. Raynaud, P. Moucheront, J.C. Bodez, F. Bertrand, J.P. Guilbaud, P. Coussot, *J. Rheol.*, **46**, 709 (2002).
140 B. Ahmed, M.R. Mackley, *J Non-Newton. Fluid Mech.*, **56**, 127 (1995).
141 Y.L. Yeong, J.W. Taylor, *J. Rheol.*, **46**, 351 (2002).
142 V. F. Polyakov, A.Ya. Malkin, B.A. Arutyunov, V.A. Platonov, N.V. Vasil'eva, V.G. Kulichikhin., *Polym. Sci. USSR*, **18A**, 2134 (1976).

QUESTIONS FOR CHAPTER 5

QUESTION 5-1

Melt flow index, MFI, was measured using a capillary with the following dimensions: diameter $d = 2.16$ mm, length $L = 8.0$ mm. The diameter of the barrel was $D = 9.5$ mm. Density of melt was $\rho = 0.8$ g/cm^3. Let the weight of a load be $G = 2.16$ kg. Estimate the shear stress attained in this experiment and calculate the apparent viscosity if a measured value of MFI was 2 g/10min.

Additional question

Give a general formula for calculation of apparent viscosity for some arbitrary weight of a load, G, and melt flow index, MFI.

QUESTION 5-2

How do you calculate the shear rate at a wall for liquid with viscous properties described by a power-law type equation?

QUESTION 5-3

Experimental study of the tube flow of suspension gave the following results:

for tube I: $D_1 = 2$ cm, $L_1 = 20$ cm, output $G_1 = 42$ g/min under pressure $P = 4$ bar;

for tube II: $D_2 = 4$ cm, $L_1 = 40$ cm, output $G_2 = 294$ g/min (pressure $P = 4$ bar).

Density of suspension was $\rho = 1.4$ g/cm^3.

Explain the results and estimate the rheological parameters of the material.

QUESTION 5-4

Calculate the velocity profile in the flow of Newtonian liquid through an annulus produced by two coaxial cylinders of length, L. Flow is induced by a pressure gradient, ΔP, at the ends of the annulus. The radii of inner and outer cylinders are R_i and R_o, respectively.

Additional question

How does one obtain the form-factor for this type of annular flow?

QUESTION 5-5

For a rotational viscometer of a coaxial cylinder type, what should be the diameter of an inner cylinder if the diameter of an outer cylinder is 40 mm and the acceptable inhomogeneity of the stress field is 5%?

QUESTION 5-6

In Section 5.7.3, a portable viscometer was described that measures viscosity *via* time of the turn of a light cylinder from the initial position by a constant angle, the initial deformation is set by twisting of a torsion spring. What is the relationship between measured time and viscosity?

QUESTION 5-7

In the principal scheme of measuring viscoelastic properties of the material, a spring is used (Fig. 5.7.1) that is not ideal elastic but viscoelastic (has some losses in deformation). How do you calculate the viscoelastic properties of the material under investigation?

QUESTION 5-8

In Section 5.7.2, the condition of uniform deformation of the sample in oscillation was formulated as $h \ll \delta$. Analyze this condition for inelastic viscous liquid.

QUESTION 5-9
Prove that Eq. 5.7.39 is valid for materials exhibiting low losses.

QUESTION 5-10
The value of maximum displacement X_{0B} appears in a solution of Eq. 5.7.41 of an equilibrium Eq. 5.7.40, but this value does not appear in Eq. 5.7.44 and other equations for G' and G". Explain why?

QUESTION 5-11
Prove that Eqs. 5.7.5 and 5.7.6 give Eqs. 5.7.53 and 5.7.54 for inelastic liquid.

QUESTION 5-12
In Section 5.7.6 deformations of the sample were treated as damping oscillations. Is it a unique case of damping deformations? Explain the answer.

Additional question

In which case does the deformation of inelastic liquid become aperiodic but not oscillating?

QUESTION 5-13
Eq. 5.3.49 is valid as an approximation only. What should be the exact solution?

Answers can be found in a special section entitled Solutions.

APPLICATIONS OF RHEOLOGY

6.1 INTRODUCTION

Rheological measurements provide us with the *properties* of materials. Fundamental rheological theories show how to treat the results of measurements of these properties and how to use them in solving applied problems. This is similar to measurements of any other properties of materials. How to apply knowledge of rheological properties to solve real technological and applied problems depends mainly on our understanding of the physics of these problems. Rheological characteristics of materials determine their behavior in many applications and they are important, together with various other material properties.

With this background, it is possible to point out several main directions of application of rheology in practice:

- Rheology is a ***physical method*** of characterization of the *structure* of matter. Rheology gives unambiguous, physically meaningful, quantitative parameters of materials. These parameters can be correlated with the structure of matter, either chemical (molecular structure of a compound, length and architecture of a molecule, and so on) or physical (physical intermolecular interactions, phase state, size and distribution of components in multi-component systems, and so on) structure. Rheological parameters *correlate* with the structure of the material and can be used for structural characterization.
- Results of rheological characterization of various similar materials give a basis for ***comparison*** of these materials. Rheology does not answer questions as to whether materials under test are "good" or "bad". The answer depends on expert estimation and previous experiences in the application of similar materials. The latter allows us to establish what parameters characterize the "ideal" material, with which other materials of a similar type can be compared. This line of application results in a great number of *standards* and *standardized test methods* developed for the main types of commercial materials. Thus, rheology proposes methods for quality control of materials.
- Description (and ***modeling***) of the dynamic behavior of different materials, including their flow in technological equipment. The description is based on solving *field equations* and rheological (material) properties enter these equations as coefficients and/or functions. Predicted dynamic behavior strongly depends on the rheological properties of matter.
- Special ***rheological effects***, i.e., phenomena, which do not exist in Newtonian liquids, but can be used for some practical application. Such rheological phenom-

ena as, e.g., deformation-induced transition from liquid to rubbery state, the elasticity of flowing liquids might be of theoretical and engineering interest.

The most important and impressive examples of such effects are:

- The Toms effect (suppressing turbulence at high Reynolds numbers)
- The Weissenberg effect (rubbery elasticity and appearance of normal stresses in shear flow)
- expandability of liquid streams due to the superposition of elasticity and fluidity of the material
- superposition of liquid- and solid-like properties providing stability of shape at rest and fluidity at intensive loading
- memory of pre-history of loading and deformation
- thinning or thickening effects as a result of flow accompanied with deformation-induced structure rearrangements
- solidification (liquid-to-solid transition) in intensive deformations, which may lead to stabilization of a liquid stream in various applications
- electro- (or magneto-) rheological effects (influence of electric or magnetic fields on rheological properties of liquid).

All these effects can be used (and are used) in engineering practice. Their limitless applications depend on the inventiveness of engineers.

6.2 RHEOLOGICAL PROPERTIES OF REAL MATERIALS AND THEIR CHARACTERIZATION

Rheology as an independent branch of natural sciences has come into existence as a method of characterizing deformation properties of real materials, which are far from idealized models of Newtonian liquid and Hookean solid.

Properties of numerous real materials are so diverse that it is impossible to invent a single or even a few different models. However, it is reasonable to try to classify numerous real materials into several principal groups or classes, depending on their nature or the similarity of their rheological behavior. Such classification cannot be absolute, because the same material can be treated as belonging to different groups, depending on the approach to classification. For example, paint with a polymeric binder can be regarded as a polymeric material, or a colloid system, or dispersion. Blood is a special liquid, though it can be treated as a substance that belongs to a much wider class of dispersions. The same is true for numerous food pastes, creams, pharmaceuticals, and so on.

Below, the main peculiarities of rheological properties of different groups of materials will be discussed with reference to the above-mentioned directions of application of the results of rheological studies.

6.2.1 POLYMER MATERIALS

This is a large group of different materials including polymer melts, solutions, filled materials, and multi-component blends.[1] The main difficulty in the application of rheological methods for characterizing these materials is encountered because practically none of them is an individual material. In the best case, a real polymer consists of fractions of macromolecules of different lengths of the same chemical nature, i.e., real polymers are always *polydisperse*. This is the reason why possible correlations between rheological

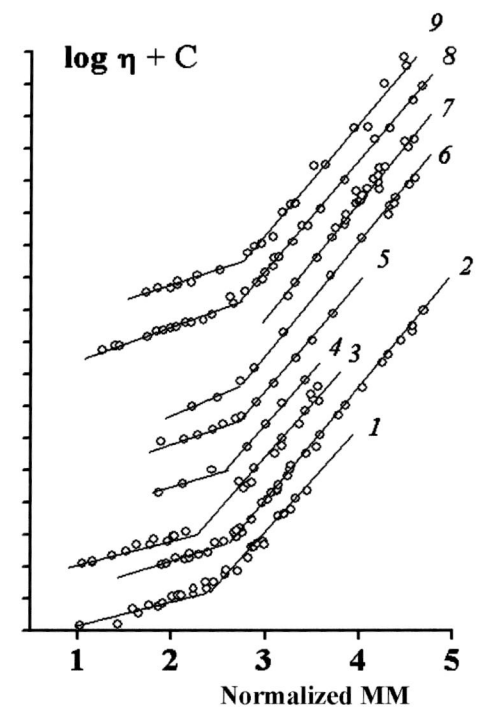

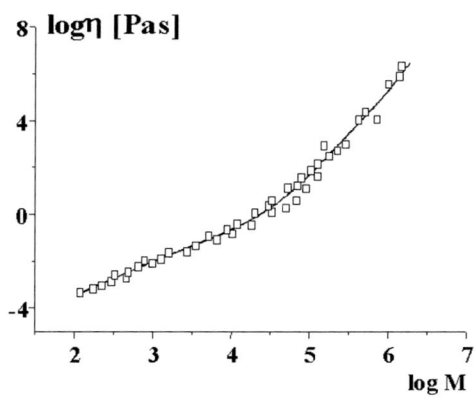

Figure 6.2.2. Molecular mass dependence of viscosity for polydimethylsiloxane. Collection of experimental data of T. Kataoka and S. Ueda. [Adapted, by permission, from *J. Polym. Sci., Polym. Lett.*, **4**, 317 (1966).]

Figure 6.2.1. Relationship of viscosity and length of molecular chain (normalized MM) for: 1 – polystyrene; 2 – polyvinylacetate; 3 – polyethyleneglycol; 4 – polymethylmethacrylate; 5 – poly(tetra-methyl p-silphenil siloxane); 6 – polybutadiene; 7 – polyethylene; 8 – polyisobutylene; 9 – polydimethylsiloxane. Viscosity values are shifted along the Y-axes by the arbitrary values. One division on the Y-axis corresponds to one logarithmic unit. [Adapted, with permission, from G.C. Berry, T.G. Fox, *Adv. Polymer Sci.*, **5**, 261 (1968)].

properties and molecular parameters are based on averaged characteristics of molecular mass distribution, MMD.

The most important correlations based on vast experimental data are the dependence of viscosity on average molecular mass, M, and between flow properties (non-Newtonian behavior) and MMD. This was discussed in detail in section 3.3.4 because analysis involves different aspects of rheology fundamentals. However, experimental results on which this analysis is based, including Fig. 3.3.2 and Eqs. 3.3.14, 3.3.15, 3.3.18, can be equally treated as a part of this section.

As was established for many polymers, the initial Newtonian viscosity, η_0, is very sensitive to the average molecular mass (see section 3.3.4). One illustration of this fact was shown in Fig. 3.3.2. The other illustration is the molecular mass dependence of initial Newtonian viscosity of different polymer melts (Fig. 6.2.1). The transition from a low-molecular-mass branch of the $\eta_0(\overline{M}_w)$ dependencies to a high-molecular-mass part of the curves is pertinent. The most important experimental fact here is the universality of a slope of a high-molecular-mass part of dependencies, which is close to 3.4-3.5.

Another impressive illustration of the relationship between viscosity and molecular mass is shown in Fig. 6.2.2 for polydimethylsiloxane covering 11 (!) decimal orders of viscosity changes. The slope of the curve changes more smoothly in this relationship.

It is pertinent from these examples that the viscosity of polymers depends on their molecular mass, but the exact form of this dependence may differ for a specific polymer.

This is important in practical applications because the "universal" value of the slope is only a rough approximation.

As was shown in section 3.4, the dependence of the coefficient of normal stresses, Ψ_0, on molecular mass is even more pronounced than the $\eta_0(M)$ dependence (see Eq. 3.4.7).

The dependence of intrinsic viscosity $[\eta]$ on molecular mass is usually expressed by a power-type equation (see Eq. 3.3.18). However, the exponent in the $[\eta](M)$ dependence is much smaller than the exponents in $\eta_0(M)$ and $\Psi_0(M)$ dependencies, i.e., the order of sensitivity of rheological parameters to molecular mass is:

$$\{\Psi_0\} > \{\eta_0\} > \{|\eta|\}$$

In practical technological applications, the direct determination of molecular mass is based primarily on measurements of intrinsic viscosity. The reasons for using the least sensitive quantity (in comparison with viscosity and normal stress variations) is partly because of tradition but mainly because of simpler experimental technique and the possibility of measurements of small size samples.

Rheological measurements can be used for the determination of the average molecular mass of the polymer. As discussed in section 3.3.5, it is possible to determine molecular mass distribution based on non-Newtonian flow curve measurements. However, this relates to the practical reliability of the method, which is always in doubt, because even strong variations of MMD are only slightly reflected by the shape of a flow curve. The ratio of viscosity values measured at two different shear stresses (i.e., measuring two points on a flow curve) can be used as a qualitative measure of the width of MMD and this method can be useful in comparing materials synthesized at analogous technological processes.

MMD of polymer is more effectively reflected in the elasticity of melt. A typical example was shown in Fig. 3.4.3. The addition of even small amount of high-molecular-mass fraction immediately causes a rapid increase of compliance (decrease of the rubbery modulus). No complete theory relating the elasticity of melt to MMD exists, though the correlation between these two properties does exist and can be used in practice for estimation of MMD. This correlation is expressed in the form of some empirical relationships relating steady-state compliance, J_s^0, to ratios of average molecular mass, including higher values of MMD. The following structure of such relationships is the most popular and can be used in practical applications:[2]

$$J_s^0 = K\left(\frac{\overline{M}_w}{\overline{M}_z}\right)^\beta \qquad\qquad [6.2.1]$$

However, it is more likely that the elasticity of polymer solutions and melts is determined by higher values of MMD, though it is difficult to separate the role of different MMDs. For a monodisperse sample ($\overline{M}_n = \overline{M}_w = \overline{M}_z = \overline{M}_{z+1}$) compliance does not depend on molecular mass. This is experimentally well-known.

According to these or similar empirical relationships, and based on measurements of melt elasticity, qualitative estimation of width of polymer MMD is possible. Using the relationship between molecular mass parameters and viscosity or elasticity, it is possible to regulate MMD to obtain the required properties of the material.

Rheological methods are also sensitive to the structure of the polymer chain. However, only qualitative estimations are possible. Also, a choice of experimental method depends on experience and its successful use by the experimenter. The central point here is a choice of parameter or dependence which is the most sensitive to structure details of interest.

Comparison of linear and branched polymers in uniaxial extension (see section 3.5.3 and Fig. 3.7.8) is an important example of the application of rheological methods for characterizing a structure of polymer chains. This is a qualitative comparison only. It is uncertain what quantitative measure of branching of polymer chains can be used because it depends on polydispersity, molecular mass, distribution of branches, their length, and their position along a chain.

The activation energy of viscous flow (or the temperature coefficient of viscosity) is a rheological parameter that is sensitive to the presence of side branching. There is almost a two-fold difference between activation energy for high density (linear) polyethylene and conventional low-density (branched) polyethylene. Intermediate cases were also observed.[3] Analogous effect was also described for linear and branched polystyrenes.[4]

The effect of long-chain branching can be characterized by following concentration dependence of viscosity of the polymer melt with variable solvent concentrations.[5] This dependence is different for linear and branched polyethylenes. The example demonstrates that the correlation between results of rheological measurements and structure parameters can be unusual and unexpected, but useful in practical applications if known.

Rheological properties are also sensitive to peculiarities of chain structure, such as the composition of copolymers, side group content, and so on. The results of rheological measurements can be used for polymer structure characterization. The kinetics of polyvinylacetate conversion to polyvinylalcohol is one example. It is discussed in section 6.3. However, this area of rheology application is still more a potential possibility than a real technological method.

Polymeric materials are processed with a variety of other materials. Use of filler is commonplace (see example in section 3.2, Fig. 3.2.4). The polymer melt is a continuous phase in this case, but polymeric substances may also be a disperse phase – latex is an example. Polymer dispersions are mainly stabilized by colloid size. The viscosity of such systems can be estimated as for any other dispersion (see section 3.3.4). The viscosity of dispersions depends not only on the total concentration of a dispersed (polymer) phase but on particle size distribution. This is illustrated in Fig. 6.2.3, where results of viscosity measurements of

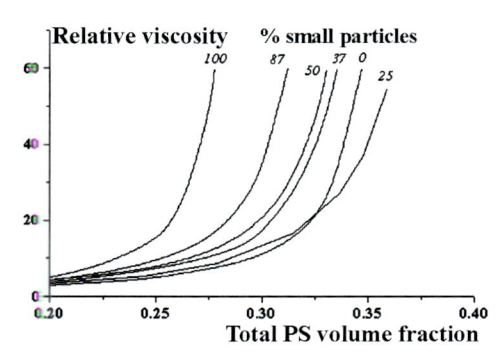

Figure 6.2.3. Particle size distribution effect on viscosity of polystyrene, PS, latex: relative zero shear viscosity as a function of the total PS volume fraction for a mixture of two monodisperse particles. Salt in water concentration [KCl] = 0.1 mM. Original experimental points are omitted. [Adapted, with permission, from F.M. Horn, W. Richtering, *J. Rheol.*, **44**, 1279 (2000)].

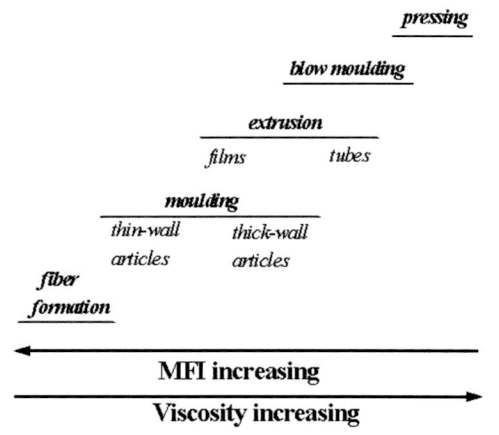

pressing

blow moulding

extrusion

films   *tubes*

moulding

thin-wall   *thick-wall*
articles   *articles*

fiber
formation

MFI increasing

Viscosity increasing

Figure 6.2.4. Primary correlation between viscosity (expressed by MFI) of a polymer melt and recommended processing method.

latex-containing mixtures having fractions of two sizes with the same total concentration are presented. Control of the particle size distribution of a dispersed phase is a sensitive method of changing the rheological properties of the material.

The above-discussed measurements are treated in terms of absolute parameters of rheological properties, such as viscosity and compliance. There is a lot of standardized rheological methods, which do no give *absolute* values of molecular parameters of polymer but some *technologica characteristics* related to molecular mass and (in some cases) concentration of poly mer in solution. All these rheological methods give valuable information concerning the technological properties and quality of polymeric material used in engineering practice.

Melt flow index, MFI, (see section 5.2.6) is a characteristic of viscosity directly related to the molecular mass of the polymer. However, MFI by itself is not suitable for determining any molecular parameter; it is only useful for a technical description of polymers produced by industry, and many polymers are classified according to their MFI. This rheological parameter determines the recommended method of polymer processing, and *vice versa*, selection of the suitable polymer grade for a particular processing method is based on a value of MFI. This is illustrated in Fig. 6.2.4.

A large number of different qualitative methods based on rheological measurements were proposed, standardized, and used in the technological practice of characterizing polymer melts and solutions.

A list of the most popular methods includes:

Melt Flow Index (MFI) – according to ISO 292; ASTM D 1238; BS 2782 (the amount of extruded melt in ten minutes under specified conditions, such as temperature, load, and size of the capillary)
Rossi-Peaks Test – according to ASTM D 569 (a thermoplastic material flows a specified distance under a specified pressure when tested in a standard apparatus)
Davenport Extrusion Rheometer (for shear rates up to 10^4 s^{-1})
Ubbelohde Viscometer (for dilute polymer solutions)
Brabender Plastograph (for materials with changing properties)
Castor-Severs Viscometer (for plastisols and organosols at high shear rates)
Brookfield Viscometer (the same but for low shear rates, also for other polymer solutions and gelling materials)
Cup Flow Method (for thermosetting polymeric materials)
Mooney Shearing Disk Viscometer – according to ISO 617 (a multi-purpose device for testing rubbery materials)
Humboldt Penetrometer
Hoeppler Consistometer
Meissner Drawability Tester
Williams Plastometer
Goodrich Plastometer
Defoe Plastometer
Convey Extrusion Machine

Agfa Vulcameter
Ring-and-Ball Method (for synthetic rubbers of Novolac type)
Hot Plate Method (for polyamide and fluoropolymers)
Rebound Method (ASTM D 1054; DIN 53512) (estimating elasticity)
Ford Cup Method (ASTM D 1200) (for paints and lacquers)
Stormer Viscometer (ASTM D 562) (for polymer solutions).

It is important to stress two terms always present in the description of any method – *specified conditions* and *standard apparatus*. Only because of these limitations is any such method useful for testing and qualifying polymeric materials.

Application of results of the rheological analysis of polymeric materials is based on the following concepts:

- polymers can be characterized by definite molecular values – molecular mass, molecular mass distribution, and branching. There are different empirical or model correlations between these values and rheological properties; therefore, when measuring rheological properties it is possible to make a prediction of molecular parameters of a polymer chain
- there is a large number of standardized methods which permit the characterization of technological properties of polymers and defining grades of industrial products; therefore, rheological methods are widely used for technological and processing control.

6.2.2 MINERAL OILS AND OIL-BASED PRODUCTS

Mineral oils are natural materials obtained from various sources. Oil is a mixture of liquid hydrocarbons and other chemical compounds. These are waxes, paraffins, sulfur derivatives, organic aromatic compounds, solid components, etc. The composition of crude oil differs depending on origin, see for example in Ref. 6. Consequently, the rheological properties of crude oils are different ranging from viscous liquids to viscoplastic materials with yield stress. Rheological properties of many oils are temperature dependent because the temperature of processing and transportation overlaps with a temperature of crystallization of some components, mainly waxes, and paraffins. Temperatures of crystallization of these components are in the range from 40 to 80°C. It means that under processing conditions (usually below 40°C) crude oils are either multiphase or unstable systems.

Measurement of rheological properties of crude oils has practical importance for their transport properties and pipeline and pump station design. The composition of crude oils of different origins dictates the necessity of measuring rheological properties and methods of their quantitative description for any oil well.

Investigation of oils demonstrates that diverse rheological behaviors can be observed. Viscosity characteristics (flow curve) are the most frequently applied rheological properties because oil transportation through pipelines is its main engineering problem. Crude oil is an unstable material and its rheological properties and transport characteristics also depend on material history, which determines the state of crystallizable components. Fig. 6.2.5 shows results of viscosity measurements for a model sample (containing 25% wax) cooled at different rates. At a starting temperature (48°C), the material is a homogeneous liquid whereas during cooling gelation of wax components takes place, and viscosity increases. The gelation temperature (viscosity attains some limiting high value) depends on the rate of cooling. This also depends on wax content.

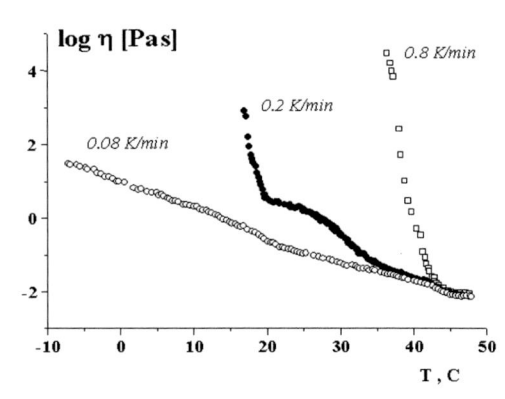

Figure 6.2.5. Influence of cooling rate (given on curves) on viscosity of a crude oil with 25% wax. Measurements were carried out at constant shear stress 7.4 Pa. [Adapted, with permission, from P. Sigh, H.S. Fogler, N. Nagarajan, *J. Rheol.*, **43**, 1437 (1999)].

Measurement of gelation temperature helps to estimate the wax content but also data similar to that presented in Fig. 6.2.5 can be used for the design of a pipeline for crude oil transportation.

Similar to many other structured systems, the thixotropic effect is often observed for waxy crude oils. Viscous properties of crude oil can be described by the Casson model (see Eq. 3.3.8), but upward and downward flow curves are different, and, consequently, preliminary deformation results in a decreased yield stress.[7]

No general equations or recommendations for the construction of a rheological model can be given for crude oils because of their variety. However, it is possible to obtain the parameters required for engineering calculations.

Oil is a raw material used in the production of numerous commercial products. This contributes to a large number of standardized methods used to control their quality. These methods usually measure parameters of material performance, which cannot be recalculated into absolute characteristics of rheological behavior.

A list of main standard methods is presented below.

> Seconds Saybolt Universal SSU
> Kinematic Viscosity, centistokes
> Kinematic Viscosity, ft/sec.
> Saybolt Furol, SSF
> Redwood 1 Standard
> Redwood 2 Admiralty
> Engler
> Barbey
> Parlin Cup
> Ford Cup
> Mac Michael
> Zahn Cup
> Demmier Cup
> Stormer Load
> Pratt and Lambert.

Light engine oils are usually treated as Newtonian liquids, though some viscoelastic effects can be observed at high rates of deformation. Therefore, they are characterized primarily by their viscous properties at low and high temperatures.[8] According to Standard SAE[9] J300, engine oils are marked as xWy, where the first index x is a low-temperature viscosity parameter (which is measured at −18°C). The lower the index, the lower the low-temperature limit of the practical application of oil.[10] The last index y is kinematic viscosity (in mm²/s) measured at 100°C. The letter W means that oil can be used in the wintertime.

The real performance of lubricating mineral oils is more complex, especially in low-temperature applications. Wax and paraffin crystallization in crude oils is also important for lubricating materials. The viscosity of mineral oils used as a base for lubricants depends on the rate of cooling and crystallization of wax.[11] This results in the appearance of yield stress at low temperatures. Certainly, this effect is very crucial for engine start-up in winter-time, and rheology is the method of choice for estimation lubricant quality.

Various *greases* are also oil-based products. The principal advantage of greases is the absence of fluidity in a stationary state, i.e., at low shear stress. From a rheological point of view, it means that these materials have a yield point and they are most likely obtained by the addition of solid components to oil. These additives form an inner structure characterized by a certain strength (and that is "yield stress"). Yield stress causes grease not to flow out of a bearing. Greases are viscoplastic materials and yielding is their important designed function.

The rheological behavior of viscoplastic materials was discussed in section 3.2.2. The main peculiarities of their flow properties are shown in Fig. 3.5.5 and discussed in reference to this figure. The low shear stress domain is not very important because viscosity in this part of a flow curve, η_0, is so high that behavior resembles a solid. Yield stress and viscosity, η_∞, in the high shear rate range are important characteristics of greases. The difference between η_0 and η_∞ is substantial (Fig. 3.5.5 contains only a schematic diagram and it is not meant to give quantitative information on a scale of difference between η_0 and η_∞ values). Low values of η_∞ provide low friction in bearing. The level of yielding is important for starting an engine. The structure of greases is thixotropic and its strength increases with time. This is especially important for winter applications. The engineering properties of greases are mainly determined by their rheological characteristics.

Bitumens (also known as *asphalts*) consist of another group of oil-based products. They are viscoplastic materials of high viscosity. Frequently, their elasticity (or viscoelasticity) is also important in their applications. Bitumens (natural or synthetic) are resin-like mixtures of higher hydrocarbons and their derivatives. Bitumens or asphalts, which are liquid binders, can be mixed with mineral fillers, such as sand, gravel, and other modifying components to form paving or roofing compounds and other construction products. They are composite materials with dispersed particles.

The behavior of these products in real applications depends on numerous factors. For example, in pavement applications, the amount and type of aggregate, and temperature properties of bitumen (especially low-temperature properties), play a dominant role in temperate climatic conditions. The geographic location of pavement also plays an essential role as well as many other factors. Rheological properties of compositions used in road construction are frequently modified to address local requirements. One example of such modification is shown in Fig. 6.2.6.

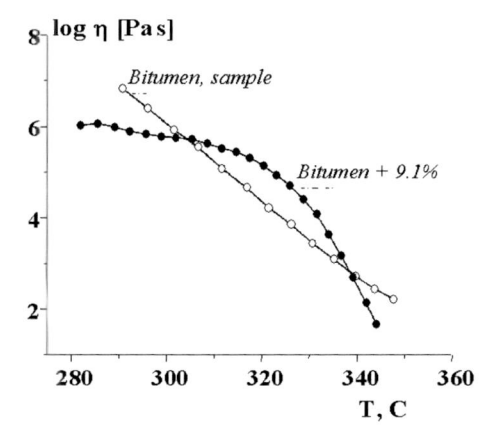

Figure 6.2.6. Example illustrating bitumen modification by addition of polymer (SBS – block copolymer of butadiene and styrene). [Adapted, with permission, from F. Martinez-Boza, P. Partal, F.J. Navarro, C. Galegos, *Rheol. Acta*, **40**, 135 (2001)].

Here, the effect of the addition of a block copolymer changes the relationship between viscosity and temperature. It is not immediately evident which one of these two materials is "better" or what else should be done in order "to improve" material. The answer depends on the requirements of the application. However, rheology gives a chance to compare materials of different compositions and suggests avenues to required modification of properties.

Viscous properties of asphalts are important for technological processes. Maintenance of characteristics of pavements or other compounded products depends on their viscoelastic properties and mechanical strength; the latter is most likely related to the viscoelasticity of material.

In real practice, measurements of viscoelastic properties can be carried out in a limited experimental frequency (or time) window, whereas such materials as asphalts are used in a wide temperature range and therefore relaxation times are changing by many decimal orders. As discussed in section 2.7, estimation of relaxation characteristics of the material in a wide time range can be solved by applying the time-temperature superposition principle. It was proven that this method may also be applied to asphalts[12] to estimate their relaxation properties and to calculate their strength and lifetime at different stresses.

In many other cases, it is possible to estimate the quality of asphalt based on the correlation between their structure and rheological properties, on one hand, and rheological properties and performance characteristics in product applications, on the other hand. Rheological measurements are also applied as standard methods in quality control and the results of these measurements are the necessary parameters for material characterization.

The following are the applications of rheological methods in the oil industry:
- measurement of rheological properties of oils and oil-based products to develop reliable engineering criteria for pipeline design
- estimation of temperature boundaries for transportation of crude oils and applicability of oil-based composite materials
- classification and grading of petroleum products for different applications
- development of new oil-based products on the basis of the correlation between applied properties of these products and their rheological characteristics.

6.2.3 FOOD PRODUCTS

The dough was one of the first objects of rheological studies, which demonstrated that dough is a viscoplastic material.[13] Dough testing in bread production is probably the most successful application of modern, sophisticated, rheological studies in control of product development and intermediate quality testing.[14]

Numerous food products were examined by rheological methods. The main problem in testing food products is their inhomogeneity. It is relatively easy to test such products as mayonnaise or cream but very difficult to experiment with a variety of products that have variable composition. Specific to food industry, rheological methods are widely used for product quality testing.

A list of some standard methods of testing food products is given below.[15]

Continuous Puree Consistometer
Denture Tenderometer
MIT Denture Tenderometer
General Foods Texturometer
Brabender Farinograph

Mixograph
Extensiograph
Alveograph
Chopin Alveograph
Cone Penetrometer
Bloom Gelometer
F.I.R.A. Gel Tester
Bostwick Consistometer
FMS and Adams Consistometer
Werner-Bratzler Shear
Zhan Viscometer
Kramer Shear Cell
Brookfield Disks and T-bars
Stephens Texture Analyzer
Simple Compression

Difficulties in applications of standard rheological experimental methods for food products lead to attempts to invent new approaches for characterizing these materials. Squeezing flow between two parallel plates was considered a useful method for measuring rheological properties of "semi-liquid" food products, such as tomato paste, low-fat mayonnaise, and mustard.[16]

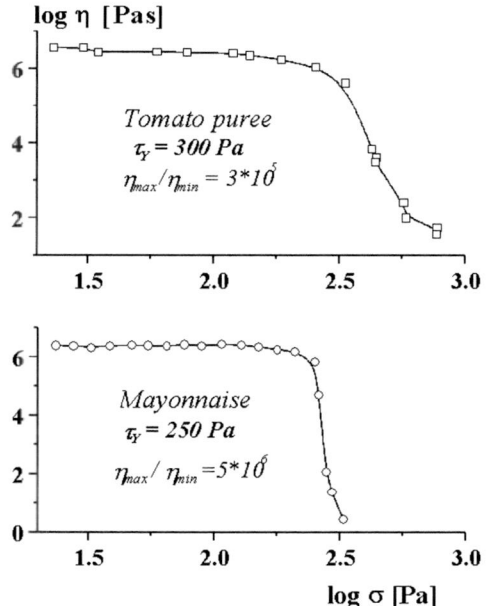

Analysis of rheological properties of numerous food products showed that they are mainly characterized using the following features:

- non-Newtonian flow properties
- yield stress
- thixotropy

while elasticity and viscoelastic properties, though they exist, are not very important.

It is also worth adding that the region of non-linear viscoelastic behavior of these products is reached at low deformation. This is explained by the weakness of the physical structure existing in these materials, which is easily broken by small stresses. It was found experimentally that shear stress must be as low as 0.2 Pa (for *yogurt*) in order to carry measurements in a linear viscoelastic range.[17]

A typical example of the viscous properties of some liquid-like food products is presented in Fig. 6.2.7 for *tomato puree* and *mayonnaise*.[18] The following characteristic features of the rheological behavior of these products (as well as some others) are seen in this figure: the rapid change from upper to lower branches of a flow curve is large but flow at low shear stresses occurs and flow at maximum viscosity in this stress range is of practical importance.

Figure 6.2.7. Flow curves of food products. Original experimental data obtained by C. Gallegos. [Adapted, with permission, from H.A. Barnes, *J. Non-Newt. Fluid Mech.*, **81**, 133 (1999)].

The measured level of yield stress is several hundred Pa, which is typical for paste-like food products. Yielding does not happen rapidly but viscosity is constant for some shear stress range. This type of rheological behavior was observed for many food materials including some unusual products, such as "*black cumin*".[19]

Properties of *chocolate* mass[20] are characterized by its rheological parameters. The recommended method[21] of measuring rheological properties of chocolate mass is based on the Casson equation and viscous properties are measured in the shear rate range from 5 to 60 s^{-1}. Typical values of rheological parameters (entering the Casson equation) are:

for milk chocolate: σ_Y = 0-20 Pa, η_p = 0.5-2.5 Pa*s

for chocolate bar grades: σ_Y = 10-200 Pa, η_p = 1-20 Pa*s.

The full range of the above-mentioned rheological effects was observed in testing *peanut butter*.[22] This is a suspension with micron-size particles. The yield stress in the range of 24-370 Pa (depending on composition) was measured. Its value correlated with stabilizing of a suspension structure. Strong non-linear effects in periodic oscillations were evident even at a very low-stress level. Time-dependent effects and non-Newtonian flow behavior were also observed in product testing.

Yield stress determines the quality and/or composition of the product. Sour cream is a good example. The yield stress depends on the amount of fat and it can be used as a quantitative measure of sour cream quality under standardized test conditions.

Qualitative description of steady-state (excluding time-dependent behavior) properties is based on the Hershel-Bulkley equation or similar ones (see Eq. 3.3.9). If the clearly pronounced yield stress is absent, it is possible to describe the flow properties of food products by other equations. For example, flow curves measured for aqueous dispersions of *spray-dried egg yolk* were successfully fitted into the Carreau-type equation by adding a limiting value of the minimum Newtonian viscosity at high shear rates (see Eq. 3.3.3).[23]

Many food products consist of polymeric substances. General approaches developed for polymer materials are also applied to food, as was demonstrated, for example, for *soy flour*.[24]

6.2.4 COSMETICS AND PHARMACEUTICALS

There is a great variety of cosmetic and pharmaceutical materials such as body lotions, face creams, toothpaste, liquid soaps, and so on. The difference between "creams" and "pastes" is not essential, and from a rheological point of view, it is more quantitative than qualitative. Many materials are emulsions, i.e., dispersions of liquid droplets in another continuous liquid phase. There is frequently some amount of solid component dispersed in a continuous liquid phase.

Two main classes of emulsions are distinguished: oil droplets dispersed in a water phase (*O/W emulsion*) or a water phase dispersed in a continuous oil phase (*W/O emulsion*). Emulsions with more complex structures also exist. Disperse phase is stabilized with an emulsifier covering the surface of droplets.

Pharmaceutical emulsions, like many other emulsions, are liquids and can flow at any stress. Disperse phase creates some "structure" appearing due to inter-particle interactions of a different nature. This structure contributes to a stress domain where viscosity drops rapidly. This stress region resembles yielding. At higher stresses, these systems are typical non-Newtonian liquids.

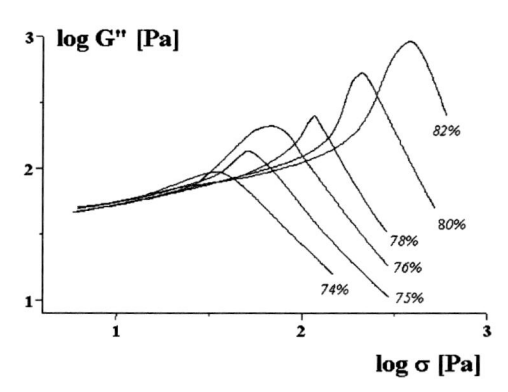

Figure 6.2.8. Loss modulus as a function of stress amplitude in testing highly concentrated W/O cosmetic emulsion. Concentrations of a disperse phase are shown at the curves. Curves are drawn instead of points in original publication. [Adapted, with permission, from A. Ponton, P. Clément, J.R. Grossiords, *J. Rheol.*, **45**, 521 (2001)].

Two main rheological characteristics are most important in pharmaceuticals:

- non-Newtonian flow properties
- yield stress.

The selection of a target value for these parameters is determined by human experience and goals of the application. Many cosmetics are viscoplastic materials (*"pastes"*) and yield stress is important because the material is expected to remain on the skin for a long time. The level of the yield stress depends on the application. The yield stress of body lotion is very low, about several Pa. The optimal level of σ_Y for body lotion is about 10 Pa, and for soft creams, σ_Y may reach 100 Pa. Thixotropic behavior is also important in some applications, and thixotropic effects are caused by an inner structure of some pharmaceuticals (e.g., see Fig. 3.5.7). This is a weak structure and it can be measured at low stresses.

Elasticity is not essential for the application of these materials, although viscoelastic properties and frequency dependence of elastic modulus are measured. These systems are strongly non-linear because inter-particle forces are weak and elastic in nature. Therefore, even at small deformation, amplitude the loss modulus becomes strain-dependent.

"Super-concentrated" cosmetic materials of the W/O type with a concentration of disperse phase exceeding 74 vol.% form an unusual group of emulsions.[25] The dispersed droplets are not spheres but polyhedrons. Such emulsions are viscoelastic liquids. The limit of linearity corresponds to very low stresses (see Fig. 6.2.8). Dependencies of this kind are typical for cosmetic emulsions[26] and the maximum of the loss modulus is treated as yield stress. Its value is low but the existence of yield stress in pharmaceuticals in many cases is crucially important for the application,[27] because creams should remain on a body surface without flowing.

Nature of components, the concentration of a disperse phase, and intimate details of the structure of an emulsion are the most important influences of rheological properties of pharmaceutical products and cosmetics. Measured rheological parameters by themselves do not express quality and application characteristics of the material. They only become valuable when compared with organoleptic features of materials, which cannot be expressed by the objectivity of measurement but must be compared with consumer acceptance of a product. Rheology helps to quantify these observations.

6.2.5 BIOLOGICAL FLUIDS

The central concept in the measurement of rheological properties of biological fluids is to establish the norm for a healthy individual and to compare the results of measurements for any person with such a norm. But norms may differ for various individuals in a broad range, and measures of this kind should be used with extreme caution in these applications.[28]

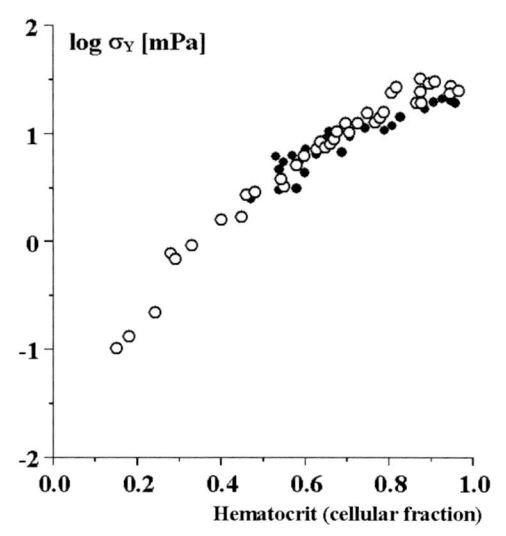

Figure 6.2.9. Yield stress as a function of the content of hematocrit (content of red blood cells). Open labels – original data; black labels – literature data. [Adapted, with permission, from C. Picart, J.-M. Piau, H. Gallard, P. Carpentier, *J. Rheol.*, **42**, 1 (1998)].

Blood is undoubtedly the most important biological fluid and its rheology is interesting from both theoretical and applied points of view. Other biological fluids such as synovial fluid and sweat are also interesting subjects for rheological analysis.

"*Blood is a juice of very special kind*"[28] – this statement is true in many aspects. By its structure, blood is a multi-component system (suspension) consisting of elastically deformable particles (mainly red blood cells) dispersed in a Newtonian liquid (plasma). Blood viscosity (at different velocities of flow), rate of coagulation (accompanied by the evolution of rheological properties), and influence of medicinal preparations are directly related to the health state of an individual. The main problem in the characterization of properties of blood is the absence of a "norm". Measurements are useful to compare properties of blood of the same person to study their evolution but are difficult to generalize and compare with the health status of an individual.

Another problem in the estimation of blood properties is its instability, caused by temperature and chemical reactions in the presence of air oxygen. These reasons require special measures to assure that the rheological characteristics are free of artifacts. It is doubtful whether blood shows real yield stress. If it has, it may indicate that blood was exposed for too long to low shear stress and it is objectionable from a biological point of view. Sometimes blood is treated as a "percolating physical gel" to determine conventional "yield stress" as its characteristic property.[29] As discussed in section 3.3.3, the determination of yield stress depends on the method of approximation. If a method is selected for the whole range of samples, their properties can be compared unambiguously. This is shown in Fig. 6.2.9, where stress at shear rate equal 10^{-3} s^{-1} was "taken as a realistic approximation of the yield stress of blood". Experiments were carried out in a Couette-type viscometer and deformations of red cells due to wall effects were absent. It is not a real, physically meaningful yield stress but a convenient measure of the rheological properties of blood. A good correlation between "yield stress" and the content of red cells was observed. The fitting curve for all points is well described by the cubic dependence:

$$\sigma_Y = A\varphi^3 \qquad\qquad\qquad [6.2.2]$$

where φ is the content of red blood cells and A is an empirical constant.

The yield stress data appear to be different for three donor blood samples, though the variability of data was not very large.[29] The authors of the publication emphasize that the results are strongly dependent on the details of the experiment. In particular, the measurements made with smooth or roughened capillary surface lead to very different conclusions. Variation of the method used for estimating yield stress can give quite different numerical values of this parameter. In the development of rheological methods for studies of blood properties, especially in clinical practice, it is very important to standardize conditions of measurement and use the same procedure for treating experimental data.

Application of rheological analysis was found useful for other biological materials, such as bones, tissues, arteries, etc.

6.2.6 CONCENTRATED SUSPENSIONS

These materials include a variety of compositions. Concentrated dispersions in low viscosity liquids, such as suspensions of minerals, sand, or stones in water, drilling compositions, slurries, slush, mud, coal suspensions, coating colors, paints, and so on, are interesting objects of rheological investigations. The following rheological properties of concentrated suspensions are typical:
- non-Newtonian flow including possible thickening at high concentrations and high shear rates
- yielding; the yield stress is a strong function of concentration
- critical concentration threshold corresponding to structure formation; at concentrations above this limit such effects as jamming, wall slip, and so on, take place
- thixotropy.

The attention here is drawn to the "structure" formed by solid particles.

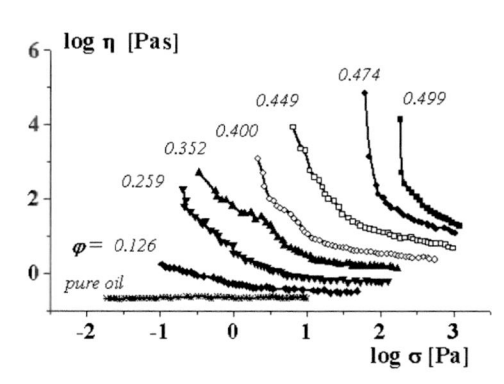

A typical example demonstrating the evolution of viscous properties in the transition from dilute to concentrated suspensions is shown in Fig. 6.2.10. The addition of solid particles leads to the appearance and gradual broadening of a non-Newtonian region in a flow curve, which eventually results in rapid yielding. Experimental data were obtained for numerous suspensions, though the boundary of concentration domains, corresponding to flow-to-yielding transition, depends on the nature of dispersed particles and their surface treatment. Contrary to the data in Fig. 3.2.3, the low shear stress domain, corresponding to very high constant values of viscosity,

Figure 6.2.10. Effect of concentration on flow curves of PMMA spherical particles (diameter 5 μm) in silicon oil. [Adapted, with permission, from L. Heymann, S. Peukert, N. Aksel, *Rheol. Acta*, **46**, 307 (2002)].

was not reached in Fig. 6.2.10, even though high shear rate constant values of viscosity are present in Fig. 6.2.10. But it is likely that such a domain exists at low shear stresses for all suspensions, regardless of concentrations.

Viscoelasticity of material is generally not as important because the structure is formed by hard solid particles. The structure can be modified by the viscoelastic matrix. Inter-particle interactions have a negligible influence on the viscoelastic properties of the material. Solid filler suppresses the elasticity of the matrix.

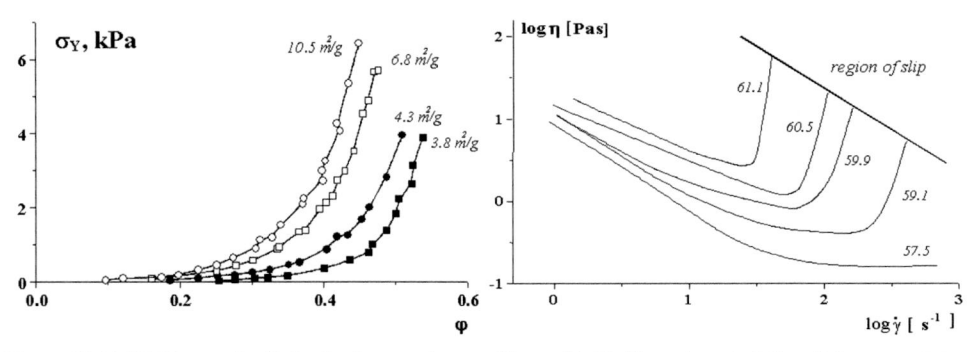

Figure 6.2.11. Yield stress in Al₂O₃ slurries as a function of volume fraction. Curves correspond to different sizes of particles presented by their surface area (shown on curves). [Adapted, with permission, from Z. Zhou, M.J. Solomon, P.J. Scales, D.V. Boger, *J. Rheol.*, **43**, 651 (1999)].

Figure 6.2.12. Flow characteristics of slurry consisting of 1 μm particles of silica in water at pH = 8.1 with 0.10 M added NaCl. Volume concentration (in%) of the solid phase is shown on curves. Points obtained in the original publication are omitted. [Adapted, with permission, from G.V. Franks, Z. Zhou, N.J. Duin, D.V. Boger, *J. Rheol.*, **44**, 759 (2000)].

The domination of a particular rheological effect depends on numerous factors, primarily on solid-phase concentration and inter-particle interactions which can be modified by different methods.

Yield stress, σ_Y, caused by a structure formed by solid components, does not depend on the viscosity of the matrix. This is shown for model materials in Fig. 3.2.5 and it applies to the majority of concentrated suspensions. At the same time, a matrix can influence the inter-particle interaction determining the strength of a solid-like structure formed by filler. The values of σ_Y depend on solid-phase concentration and the nature of solid particles, and their surface treatment.

The yield stress in concentrated suspensions in a low viscosity matrix depends on the concentration of solid particles. Fig. 3.2.4 and Fig. 6.2.11 demonstrate it very clearly. The role of particle sizes (surface area) is also evident from this experimental data. The concentration and particle size dependencies of yield stress, σ_Y, for experimental data of this figure can be represented by the generalized fitting function:

$$\sigma_Y d_s^2 = k\varphi^n \qquad\qquad [6.2.3]$$

where d_s is the average diameter of particles, k and n are fitting parameters. It was found that for the concentration range $\varphi < 0.42$, the exponent n = 4.2 (at higher concentration n rapidly increases). The analogous dependence of yield stress on concentration was observed for numerous suspensions, for example, for clay dispersions.[30] In fact, the exponential dependence $\sigma_Y(\varphi)$ discussed in section 3.3.3 may be approximated by two power-law functions with changing power-law index values.

It is reasonable to think that there is some threshold concentration in suspensions corresponding to the formation of solid-like structures. The exact value of concentration corresponding to this threshold depends on the composition of the system.

The formation of such a solid-like structure leads to the rapid increase of yield stress and to changes in the rheological behavior of the material. A typical example is shown in Fig. 6.2.12.[31] First of all, a very strong influence of concentration in a narrow range of its

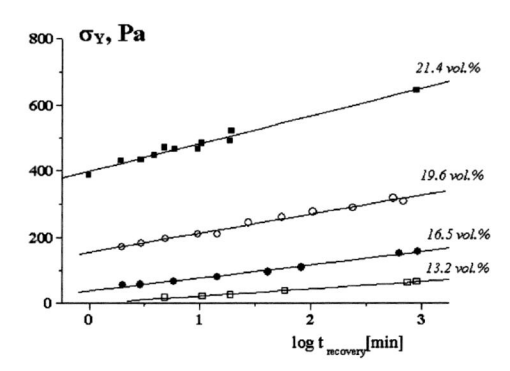

Figure 6.2.13. Thixotropy of concentrated suspensions: recovery of yield stress as function of rest time for Na-montmorillonite-based suspension. Concentration is shown on graph. [Adapted, with permission, from R.G. de Kretser, D. V. Boger, *Rheol. Acta*, **40**, 582 (2001)].

change is pertinent. A thickening effect is observed at a concentration above some threshold. The most popular interpretation of this effect is related to the formation of clusters.[32] This effect of volumetric dilatancy is similar to attempts to force dry sand through a tube at high applied force. Shear thickening can be attributed to increased inter-particle interaction.

Shear thickening at high shear rates is typical of many widely-used technological materials, for example, surfactant-stabilized concentrated kaolin-in-water suspensions.[33]

From an applied point of view, the existence of a critical concentration at which strong shear thickening and transition from flow to slip becomes dominant, preventing steady transportation of a suspension, is the most important finding.

Structure formation is a reversible process[34] and from a rheological point of view, it leads to the thixotropy of material. The existence of thixotropy and, as its manifestation, viscosity bifurcation, was demonstrated for such popular concentrated suspensions as bentonite-in-water suspensions.[35] Sometimes, thixotropy (structure formation and destruction) can be observed as an unusual effect of periodic oscillation of apparent viscosity.[36]

Thixotropy is mainly understood as reversible changes of viscosity (due to deformation and rest). In section 3.5.1 it was shown that the strength of the material is also time-dependent. For concentrated suspensions, it results in the time dependence of yield stress for materials that were initially destroyed by deformation. This effect is illustrated in Fig. 6.2.13.

Thixotropic effects are typical of concentrated suspensions of different types, though not widely investigated. Their quantitative description is important for some technological processes, in particular for mixing and transportation.

As was mentioned in Section 3.3.4.3, the concentration dependence of rheological properties of concentrated emulsions is similar (in many aspects) to the behavior of suspensions. For example, Fig. 3.2.9 discussed in Section 3.3.4.3 clearly demonstrates that the shape of flow curves changes in a rather narrow concentration range and this change is similar to the evolution of viscous properties of concentrated suspensions (see, for example, Fig. 6.2.10). At high concentrations (Fig. 3.2.9), flow curves can be treated as those of solid-like substance with very high viscosity at low shear stresses exhibiting the yield stress.

6.2.7 ELECTRO- AND MAGNETO-RHEOLOGICAL MATERIALS

Electro- and magneto-rheological materials (ER and MR materials, respectively) are a special type of concentrated dispersions of solid particles in a viscous medium.[37] Their solid phase can be formed from inorganic materials. But, it can also be formed from hydrated starch particles[37] or any other particles, which have induced dipole or dipole8-

dipole interaction creating a continuous structure. These suspensions consist of micron-level particles dispersed in a continuous phase. An inter-particle interaction in ER and MR liquids depends on the concentration of particles as well as on permittivity (for ER) and permeability (for MR) materials.

A structure formed after applying an electrical or magnetic field has some strength due to the interaction of these dispersed particles. Therefore, such a composition placed in an electrical or magnetic field loses fluidity at stresses lower than the strength of the structure and becomes a viscoplastic medium with yield stress corresponding to the strength of particle structure. Application of field results in drastic changes of rheological properties.

This effect can be used in numerous practical applications, such as the construction of switching devices, electrical valves, breaks, clutch dampers, and so on.

The rheological behavior of electro-sensitive suspensions can be described by equation used for viscoplastic medium, for example, the Bingham type, but with coefficients depending on the strength, E, of an electrical field:[39]

$$\sigma(\dot{\gamma}, E) = \sigma_Y(E) + \eta\dot{\gamma} \qquad [6.2.4]$$

Rheological properties of ER and MR materials can be changed from a purely viscous liquid to a viscoplastic medium. According to Eq. 6.2.4, the following characteristics are important for ER liquids: viscosity, η, in the absence of a field, and the dependence of the yield stress on E. In magneto-rheological materials, the yield stress depends on magnetic flux density, B, instead of E.

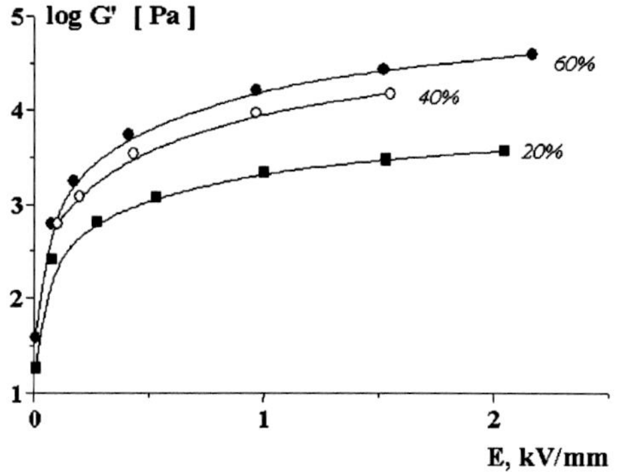

For practical applications, the characteristic time of switching, t_{sw}, i.e., time of formation and disappearance of solid-like structure, is an important factor. For systems of practical interest, this time must be in the range of 1-10 ms.

Structure formation on switching on an electrical field leads to yield stress and/or an increase in apparent viscosity. Material subjected to stresses below σ_Y is solid-like and this transition is of practical interest for applied science.

The transition to solid-like behavior can be demonstrated by measuring the elastic modulus under a superimposed electrical field. A typical example is shown in Fig. 6.2.14 for diatomite suspension in transformer oil. An increase of elastic modulus attains several decimal orders of magnitude.

Figure 6.2.14. Electro-rheological effect: influence of electrical field on storage modulus of diatomite suspensions at frequency of 0.16 Hz, amplitude of shear deformation of 0.027. Concentrations are shown at the curves. [Adapted, with permission, from E.V. Korobko, V.E. Dreval, Z.P. Shulman, V.G. Kulichikhin, *Rheol. Acta*, **33**, 117 (1994)].

The peculiarities of the rheological behavior of electro-rheological liquids depend on the concentration of the solid phase, though the strength of the structure (i.e., yield stress) is also sensitive to particle size distribution.[40] Analogous effect of particle size distribution on rheological properties is a characteristic feature of various emulsions and suspensions.

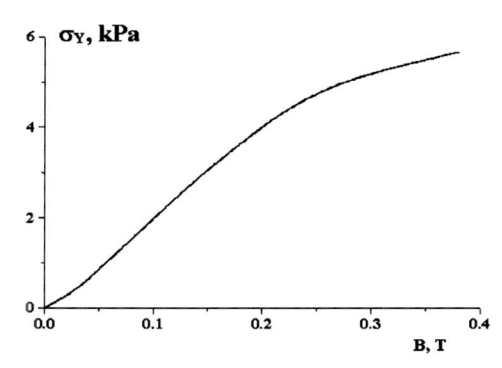

Figure 6.2.15. Magneto-rheological effect: dependence of yield stress on applied magnetic flux density for carbonyl iron suspension in different media. A line is an averaged curve drawn for the same particles dispersed in 7 different media. [Adapted, with permission, from P.J. Rankin, A.T. Horvath, D.J. Klingenberg, *Rheol. Acta*, **38**, 471 (1999)].

An example illustrating the magneto-rheological effect is shown in Fig. 6.2.15, where the dependence of yield stress on magnetic flux density, B, is shown. In an original study,[41] this dependence was measured for 10% suspension of carbonyl iron in silicone oil and in 6 different greases. Silicone oil was a Newtonian liquid and greases were viscoplastic materials with σ_Y values in the range of 1-37 Pa. The dependencies $\sigma_Y(B)$ for all these systems are within a narrow range, and the line in Fig. 6.2.15 is an averaged curve. The magnetic field produces a structure with strength several orders of magnitude higher than initial yield stress and in practice, this strength does not depend on the viscosity of the matrix.

Properties of ER materials can be characterized by a dimensionless group (known as the *Mason number*, Mn) equal to a ratio of viscous to polarization forces.[42] Polarization forces are proportional to the square of electrical field strength and electrical parameters of the system. However, experimentally observed $\sigma_Y(B)$ dependence is usually described by a power-law equation with an exponent close to 1.5.[43]

The transition from viscous to viscoplastic behavior takes place at some critical value of the Mason number, which depends on the concentration of the interacting particles. The linear dependence of σ_Y on the concentration of filler is a good approximation of real performance of some ER materials,[44] though more complicated concentration dependencies of σ_Y can be observed, as was shown in section 6.2.6.

The electro-rheological effect can also play a role in mixtures of two immiscible liquids, e.g., polymer blend, including a liquid crystal polymer component.[45] However, due to the high viscosity of the polymer system, transient switch-time in such systems increases up to several seconds, and viscosity jump is not sufficiently rapid as compared with suspensions of solid particles.

6.2.8 CONCLUDING REMARKS

The following is the summary of the concepts discussed so far in this chapter.

Rheological measurements provide a set of parameters characterizing properties of the material under test. It is supposed that these parameters are related to the molecular structure and composition of industrial materials. Establishing a correlation between rheological parameters and structure factors is the central research problem. The results of rheological measurements by themselves are not a direct indication of the quality and application properties of a product. The latter must be based on practical experience and expert evaluation.

First, there is a triangle: *"rheological properties"* – *"molecular structure and composition content"* – *"quality and application"*. Any side of this triangle is the subject of studies and only the complex results of these studies form a basis for the application of rheological methods in material control and characterization.

Rheology operates with different measures of resistance to deformation, primarily determined by viscosity and modulus, though these characteristics are being understood in a more complex sense than when they were initially introduced in basic theories. There is a difficult and in many cases impossible to solve the task of carrying out rigorous measurements based on a modern complicated continuum or molecular theories. Such measurements can be a goal of advanced academic investigations, which are not directly ready for technological applications. As an alternative to this rigorous approach, a great number of conventional industrial methods of material characterization, related to rheology, have been proposed and are used in the practice of testing various real materials.

Second, the principal direction of application of rheology in material characterization is the *standardization* of proposed experimental methods, including the design of the apparatus used and the testing procedure. The benefit of standardization depends on the correctness of the method selection for a specific field of application and its sensitivity to factors that are assumed by experts as determining product quality.

The choice of the most sensitive rheological parameter depends on the goals of testing. The traditional subject of rheology is the flow of materials. Then, the dominant share of experimental methods is related to viscosity measured under different conditions. Many real (either technological or biological) materials are viscoplastic, and their rheological properties in the low shear stress range can be useful for the assessment of their quality. In other cases, time-dependent behavior, primarily the thixotropic properties of the material, can play an essential role in the application.

Third, it is not enough to limit oneself to apparent viscosity as an applied measure of rheological properties of real materials but it is necessary to deal with a wider choice of material characteristics. Among them, the yield stress is very important for many materials and its reliable measurements are promising for practical applications.

6.3 RHEOKINETICS (CHEMORHEOLOGY) AND RHEOKINETIC LIQUIDS

6.3.1 FORMULATION OF THE PROBLEM

Rheological methods are used in practice to control chemical reactions in oligomeric and polymeric systems, physical transformations, and quality of products at different stages of the process. This branch of rheology is called *rheokinetics*, or sometimes the terms *chemorheology* and *kinetorheology* are also used.[46]

Its importance for polymeric materials (though not only these) is explained by the high sensitivity of rheological properties, primarily viscosity and elastic modulus, to the length of the macromolecular chain, its structure, and/or concentration of polymer in solution. This is contrary to reactions of low-molecular compounds, where rheological properties of reactive media are only slightly affected during the course of the reaction.

In rheokinetic measurements two different cases can be distinguished:
- linear polymerization, where the increase of a linear chain is a dominating process
- curing, where chemical reactions lead to the formation of a three-dimensional network.

In the first case, viscosity increases gradually (if the reaction takes place in a liquid phase). In the second case, when continuous network formation begins, the viscosity becomes unlimitedly high and material loses its ability to flow. The third case of interest for rheology is related to the chemical (or physical) transformation of polymeric materials without changes in chain length or its branching.

6.3.2. LINEAR POLYMERIZATION

The rheokinetics of linear polymerization depends on the process chemistry and/or the chain growth mechanism. Three main reactions can be realized and modeled, such as ionic polymerization, radical polymerization, and polycondensation.

According to the simplest model of *ionic polymerization*, the growth of chains happens on the active centers; both molecular mass, MM, of a newly-formed polymer and its concentration change.

The model of *free-radical polymerization* is based on the supposition that the MM of the forming polymer is constant and polymerization proceeds by an increase of its content in a reactive medium.

Polycondensation is a process of polymer formation by reaction of growing chain ends; concentration of polymer in the reactive medium is constant but its molecular weight increases.

Then, viscosity dependence on the concentration of polymer in a reactive medium (solution) and its MM is supposed to obey the standard rules discussed in section 3.3. Based on these simplest arguments, rheokinetic depends on viscosity, η, on the degree of conversion, β, for the three model schemes of polymer formation in the solution that can be obtained as follows:

for ionic polymerization:

$$\eta = K_1 \beta^{a+b} \tag{6.3.1}$$

for free-radical polymerization:

$$\eta = K_2 \beta^b \tag{6.3.2}$$

for polycondensation:

$$\eta = K_3 (1 - \beta)^{-1} \tag{6.3.3}$$

where K_1, K_2, and K_3 are constants, and the exponents a and b are the same as in the dependencies of viscosity on MM. As the first approximation, it can be assumed that for high MM compounds $a \approx 3.5$ and for concentrated solutions $b \approx 5$.

The types of $\eta(\beta)$ dependencies are different for various chemical processes. In addition, it is necessary to determine the kinetics of polymerization, i.e., the equation describing the time dependence of the degree of conversion, $\beta(t)$. Such equations are known for polymerization processes of different types.[47] Then, it is possible to calculate the time dependence of viscosity for polymerization processes of any type.

Fig. 6.3.1 is an illustration of the time dependencies of viscosity and the degree of conversion measured simultaneously.[48] Strong influence of temperature on rheokinetics of polymerization is pertinent. In fact, temperature influences rheokinetics in two ways.

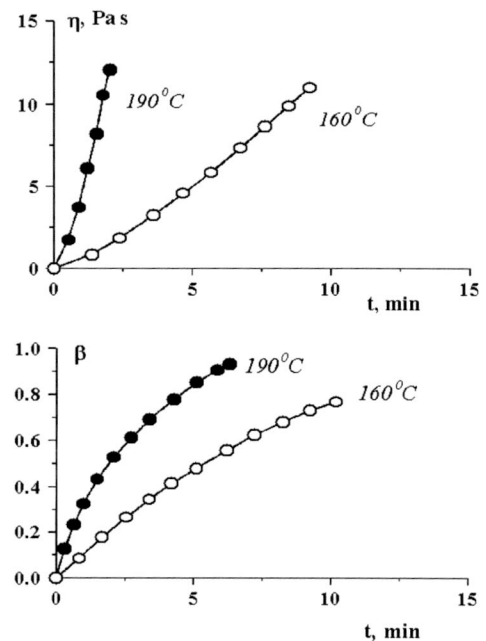

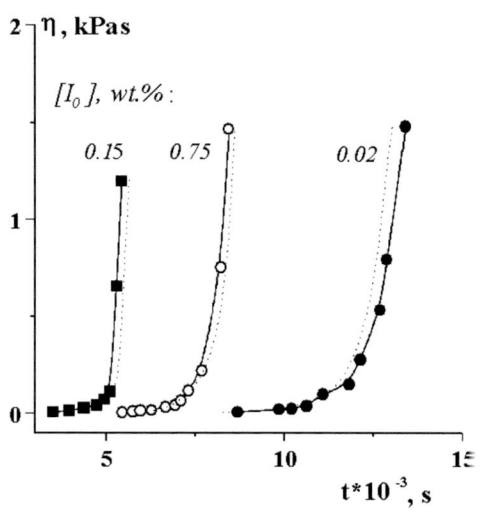

Figure 6.3.2. Comparison of experimental data (solid lines) and calculated viscosity profiles (dotted lines) in polymerization of methylmethacrylate at different content of initiator (its concentrations, $[I_0]$, are shown on curves). [Adapted, with permission, from A.Ya. Malkin, S.G. Kulichikhin, D.N. Emel'yanov, I.E. Smetanina, N.V. Ryabokon, *Polymer*, **28**, 778 (1984)].

Figure 6.3.1. Rheokinetics of ionic polymerization of dodecalactam. Evolution of the degree of conversion, β, and viscosity, η, at two temperatures. [Adapted, with permission, from A.Ya. Malkin, S.G. Kulichikhin, S.L. Ivanova, M.A. Korchaghina, *Vysokomol. Soedin.* (Polymers – in Russian), **22A**, 165 (1980)].

First, viscosity depends on temperature. Second, the kinetics of the chemical reaction is highly sensitive to temperature. Direct estimations show that the second factor always dominates.

Calculation of time dependence of viscosity based on pure chemical and kinetic arguments is illustrated in Fig. 6.3.2 for free radical polymerization of methylmethacrylate. Dotted (calculated) lines were found based on laboratory studies of the kinetics of polymerization and from independent data on viscous properties of polymethylmethacrylate solutions. It is seen that the model of rheokinetic calculations gives realistic results.

Numerous experimental data confirm that the simplest rheokinetic schemes of linear polymerization work satisfactorily when and if chemical processes take place in a homogeneous phase. However, in real technological practice, newly formed polymers can be insoluble (in contrast to their monomers) in a reactive medium, and, at a certain degree of conversion, phase transformation occurs. Rheokinetic data clearly demonstrate this effect, as shown in Fig. 6.3.3. The polymerization proceeds in a mixed solvent and the "quality" of solvent is improved for a newly formed polymer by increasing the cyclohexane fraction in relation to toluene (movement from the right curve to the left curve). Polymerization in "good" solvent proceeds in a homogeneous system and the smooth rheokinetic curve is observed (right curve in this figure). If the concentration of the "bad" solvent is higher, the phase separation occurs at earlier stages of the process.

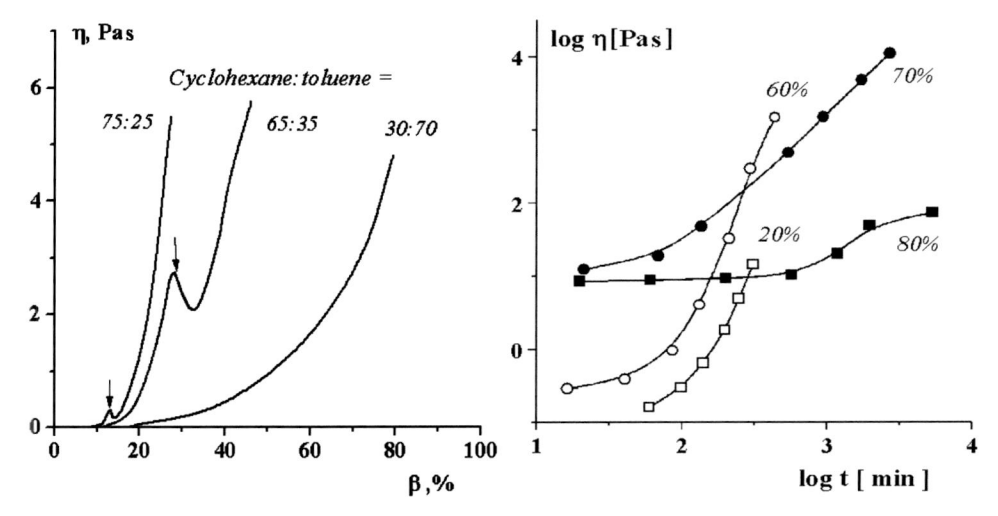

Figure 6.3.3. Polymerization of 50% solution of methylmethacrylate in a mixed solvent (ratios of solvents are shown on curves) – phase separation in the process of polymerization. Arrows show phase transitions. [Adapted, with permission, from A.Ya. Malkin, S.G. Kulichikhin, D.N. Emel'yanov, I.E. Smetanina, N.V. Ryabokon, *Polymer*, **25**, 778 (1984)].

Figure 6.3.4. Evolution of viscosity in the process of synthesis of polysulfone from reactive monomers. T=150°C. Concentrations of reactants are shown on curves. [Adapted, with permission, from A.Kh. Bulai, V.N. Klyuchnikov, Ya. G. Urman, I.Ya. Slonim, K.M. Bolotina, V.A. Kozhina, M.M. Gol'der, S.G. Kulichikhin, V.P. Begishev, A.Ya. Malkin, *Polymer*, **28**, 1047 (1987)].

The effects of phase separation in polymerization processes, regardless of cause, are always manifested by a maximum on the rheokinetic curve.

Rheokinetic measurements are, by necessity, accompanied by the shearing reactive medium. Then, it is natural to ask whether deformation influences the kinetics of the chemical process. Experiments have proven that the answer is "no" if the reaction takes place in a homogeneous system. However, the answer is the opposite if the system is heterogeneous. In this case, shearing is equivalent to mixing and averaging the content and properties throughout a reactive vessel, which influences the rate of reaction. However, the role of shearing can be even more impressive, if and when deformations shift the conditions of phase separation. This phenomenon has already been mentioned in section 3.5.3 and Figs. 3.5.14 and 3.5.15 are very clear demonstrations of the shear-induced rheokinetic effects.

Rheokinetic measurements also demonstrate the transition to the diffusion-controlled stage of reaction in linear polymerization, even if this transition is not accompanied by phase transition. This is shown in Fig. 6.3.4 for polycondensation synthesis of polysulfone. According to the general principles of chemical kinetics, the increase in the concentration of reactants in solution causes acceleration of reaction. This concept is valid and corresponds to the transition from 20 to 60% concentration. However, further increase of reactant concentration leads to a reverse effect – suppression of polymerization, which is clearly seen from the rheokinetic data. This effect is definitely related to the diffusion-controlled limitation of a chemical reaction. Rheological measurements are useful instruments in demonstrating this phenomenon.

As a summary of the present section, the following applications of results of the rheokinetic analysis of linear polymerization can be mentioned:

- calculation of viscosity change during the technological process with application to design of equipment (power of mixer motors, the output of pumps, and so on); the evolution of rheological properties of the reactive medium also influences the flow lines and efficiency of mixing[49]
- monitoring and control of the technological process with feedback to regulating parameters
- study of mechanisms of polymerization and estimation of their kinetic factors; this is achieved by comparison of theoretical predictions coming from different models with experimental data; a particular case of this line of investigation is based on using the Toms effect as the rheokinetic method.[50]

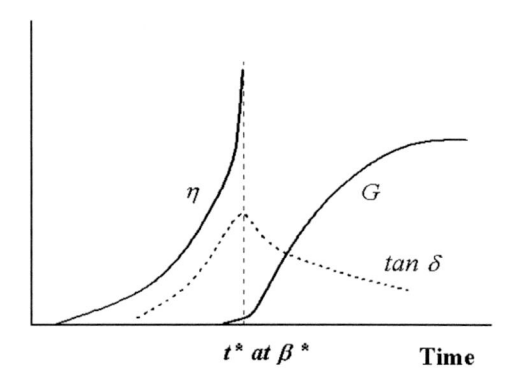

Figure 6.3.5. Evolution of main rheological parameters during oligomer curing.

6.3.3 OLIGOMER CURING

The evolution of two main parameters − viscosity and modulus of elasticity − in the process of oligomer curing is schematically shown in Fig. 6.3.5. Both factors are important in technological applications. The following general characteristics of the rheokinetic analysis of the oligomer curing process, based on experimental data obtained for numerous oligomeric systems, can be formulated:

- viscosity grows unlimitedly when approaching a "critical" point (of time or degree of conversion); this point is called a *gel-point*, t*; gelation occurs at a certain degree of transformation, β*
- noticeable values of the elastic modulus appear before the gel-point, though the main part of its growth takes place after the gel-point; however no special points can be marked on the G'(t) dependence, and the value of modulus at β* depends on the mechanism of three-dimensional network formation (more exactly, on the functionality of reactants)
- modulus reaches its constant limiting value at the end of the curing process
- loss tangent passes through a maximum at the gel-point, so it is a point of relaxation transition.

Let us discuss the behavior of different rheological characteristics of curing in more detail.

6.3.3.1 Viscosity change and a gel-point

The rate of viscosity growth is very important for technological applications at the stage of article formation. Viscosity should not be too high during wetting of reinforcing filler or forming a part's shape. However, an increase in viscosity is expected to be rapid after the

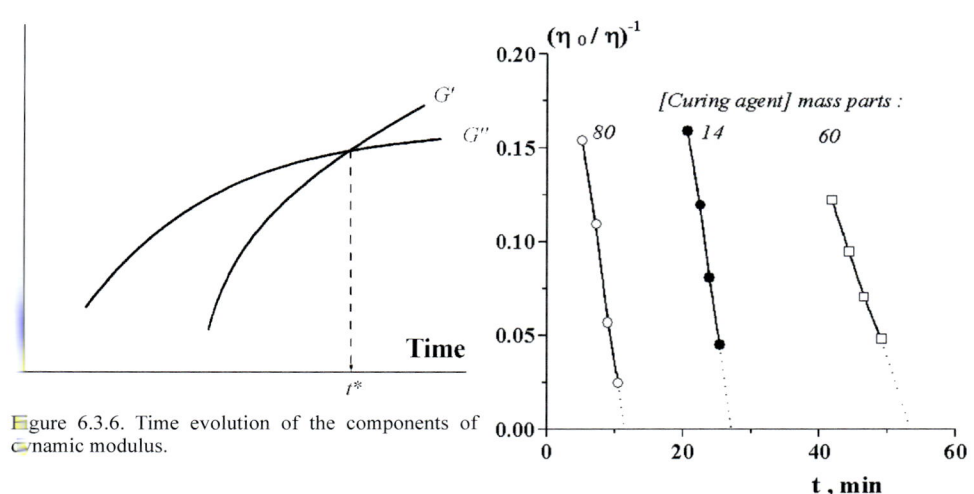

Figure 6.3.6. Time evolution of the components of dynamic modulus.

Figure 6.3.7. Finding a gel-time by extrapolation in coordinates $\eta^{-1} - t$ for an epoxy resin with different content of curing agent (shown on curves). [Adapted, with permission, from A.Ya. Malkin, S.G. Kulichikhin, M.L. Kerber, I.Yu. Gorbunova, E.A.Murashova, *Polym. Eng. Sci.*, **37**, 1322 (1997)].

completion of these technological operations in order to shorten the overall process. It is necessary to know the technological time, i.e., the time required to reach gel point when the oligomer being cured is still in a liquid form.

Two matters are important for technology – estimation of the gel-time, t* and formulation of the kinetic equation for viscosity evolution in time.

The important point in the discussion of materials with increasing (in time) viscosity is the transition from fluid to a solid-like state. This point is conditionally treated as the gel-time, t*. Different methods for finding t* exist. In many publications, the gel-point is determined as the time when G'(t) and G''(t) dependencies crossover[51] as shown in Fig. 6.3.6.

The sense of this definition is rather evident: in the fluid state (at t < t*) G'' > G' and losses dominate over elasticity. At t > t* the situation is reversed: G'' < G' and this is treated as the solid-like state.

However, material can remain in the fluid state even if G'' < G'. So, the gel-point determined by this method does not obligatory corresponds to the gelation of a material.

A technologically convenient definition of the gel point determines the gel-time and the time necessary for reaching some level of viscosity, e.g., 100 Pa*s. Formally, a material remains fluid, but not gel, below this time. This high viscosity limits the possibility of further processing material and thus can be used as the threshold of the transition to a solid-like state.

Meanwhile, gelation, by definition, is the transition to the real solid state when the fluid flow is impossible. Then the gel-time is a limiting time at which $\eta \to \infty$. It is usually found by the extrapolation procedure. One viscometric method of its determination is shown in Fig. 6.3.7. Constructing dependence of reciprocal viscosity (or ratio η_0/η, as in Fig. 6.3.7, where η_0 is the viscosity of the initial uncured system) versus time and extrapolating this dependence to $\eta^{-1} = 0$, it is easy to find t*. This method is satisfactory in many cases.

There are two limiting cases:

- macro-curing in which a three-dimensional network is formed throughout the whole volume of the sample
- micro-gelation in which curing proceeds in separate particles dispersed in the solvent.

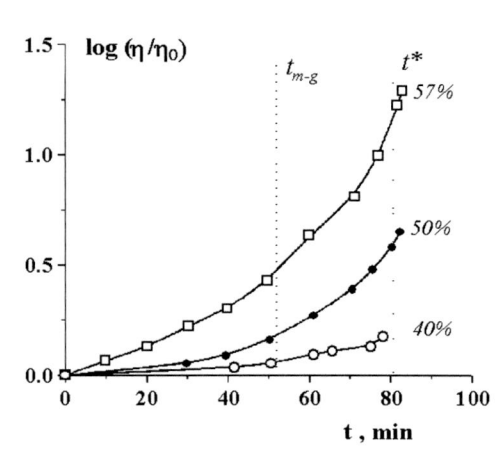

Figure 6.3.8. Viscosity evolution on curing melamine-formaldehyde resin in water solutions. Concentrations are shown at the curves. T = 80°C. Dotted line t_{m-g} – time of micro-segregation as observed by an optical method. Dashed line t* – gelation of system as a whole. [Adapted, with permission, from A.Ya. Malkin, S. G. Kulichikhin, *Adv. Polymer Sci.*, **101**, 217 (1991)].

Fig. 6.3.8 shows how micro-gelation (marked by a dotted line t_{m-g}) takes place before the loss of fluidity of material (marked by a dashed line t*) occurs. Micro-gelation, or micro-phase segregation, is observed by an optical method as the intensive increase of optical density of material because of the formation of insoluble cured micro-particles of colloidal size. This is the limiting case of heterogeneous curing. In fact, curing of real oligomeric products s more or less heterogeneous, and statistically homogeneous curing is a limiting ideal case only.

A quantitative description of viscosity growth, which is necessary for practical calculations in modeling technological processes, can be based on various concepts.

The very rapid increase in viscosity suggests the use of exponential formulas of various types. A number of $\eta(t)$ relationships can be described by the simplest exponential equation:

$$\eta = \eta_0 e^{t/t_c} \qquad\qquad [6.3.4]$$

where η_0 is the initial value of viscosity (at t=0), and t_c is some characteristic time constant (not to be confused with t*).

This equation does not fully fit experimental data, because it does not predict unlimited growth of viscosity on the approach of the gel-point. However, it describes a wide range of viscosity values. Generally, it is assumed that this equation can be applied up to viscosity $\sim 10^3$ Pa*s. This limit corresponds to a high viscosity level that is close to the limit of the fluidity of most materials. For these reasons, it gives an acceptable technological estimation of conditions of gelation.

A rigorous approach to $\eta(t)$ dependence is based on a scaling concept.[52] The following equation is expected to be valid only close to a gel-point:

$$\eta = \eta_0(1 - t/t^*)^{-s} \qquad\qquad [6.3.5]$$

where s is the "scaling factor". It equals 0.7+/-0.07.

Numerous experimental data show that an equation of an analogous type can describe $\eta(t)$ dependence even in a wider time domain, but the value of exponent s is not equal to the theoretical value but may change in a wide range, depending on the chemical composition of the curing system, temperature, and so on. Therefore, s is not a constant for various real systems, which means that the scaling concept is not valid in many practically important situations. This experimental fact can also be treated as indirect proof of heterogeneity of curing real oligomeric materials.

Another empirical equation for $\eta(t)$ dependence can be proposed:

$$\eta = \eta_0 \exp\left(1 - \frac{t}{t*}\right)^{-a} \qquad [6.3.6]$$

This equation contains a limited number of fitting parameters and correctly predicts important features of $\eta(t)$ dependence: existence of gel-point (approach to $\eta \to \infty$) and rapid growth of viscosity during curing.

Finally, it is worth mentioning that the rheokinetics of chemical processes of curing does not always coincide with results of a kinetic study made by chemical and/or other physical methods. Different methods emphasize different transformations and that is why even the form of kinetic equations obtained can be very different, depending on the method applied.[53]

6.3.3.2 Curing at high shear rates

Shearing may influence the rate of chemical reaction in several ways. *First*, shearing is a mixing. In this sense, shearing promotes contacts of reactive groups and *accelerates* a reaction. *Second*, shearing can be a kinetic factor by itself (see section 3.5 and Fig. 3.5.16). The kinetic effect of shearing can be observed in multi-phase systems only, but no direct evidence of this effect in curing processes is known. *Third*, deformation at high shear rates leads to intensive *heat dissipation* and thus to non-isothermal effects. An increase in temperature accelerates reaction and thus shortens oligomer processing.

Let the induction period, when oligomer can be treated as a low-viscosity liquid, be t_0 at low shear rates (in isothermal conditions). At high shear rates, due to heat dissipation, the induction period becomes shorter and equals t_n. Dimensionless value $\hat{T} = t_n/t_0 < 1$. Model calculations show[54] that $\hat{T}$ is a unique function of the dimensionless shear rate, Γ, which is expressed as

$$\Gamma = \frac{E\sigma_0\dot{\gamma}t_0}{c\rho RT_0^2} \qquad [6.3.7]$$

where E is the activation energy of viscous flow, σ_0 is the shear stress, at which the isothermal induction period t_0 is measured, $\dot{\gamma}$ is the shear rate at which the induction period of curing is measured, c is the heat capacity, ρ is density, R is the universal gas constant, and T_0 is the initial temperature at which t_0 is measured.

The final equation expressing this dependence is

$$\hat{T} = \frac{1}{\Gamma}\ln(1+\Gamma) \qquad [6.3.8]$$

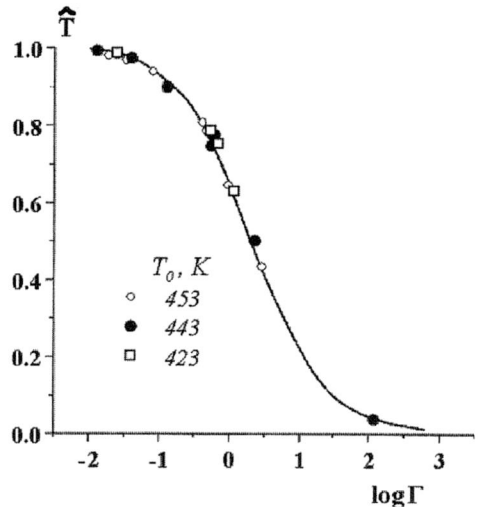

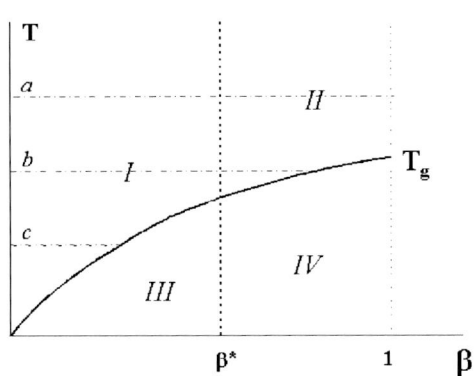

Figure 6.3.10. Dependence of the glass transition temperature on the degree of conversion.

Figure 6.3.9. Curing at high shear rates – dependence of the dimensionless induction period on dimensionless shear rate. Experimental data are presented for a silicon-organic oligomer at different initial temperatures, T_0. Solid line is calculated in accordance with Eq. 6.3.8. [Adapted, with permission, from A.Ya. Malkin, G.I. Shuvalova, *Vysokomol. Soedin.* (Polymers – in Russian), **27B**, 865 (1985)].

Experimental data obtained for various oligomeric systems, having a different chemical mechanism of curing, and at different initial temperatures, T_0, confirm that this equation completely describes experimental data (one example is shown in Fig. 6.3.9). The effect of shortening the induction period can be as high as a hundred times.

Comparison of experimental and theoretical results shows that a non-isothermal effect is responsible for shortening the induction period at high deformation rates and, therefore, deformation rate plays a minor (if any) role in kinetics. However, it would be premature to make this conclusion about a direct kinetic role of shearing if curing leads to phase segregation, as in the case of melamine-formaldehyde oligomers (Fig. 6.3.8).

In technological practice, high shear rates are common. That is why the results of measurement of the induction period performed at low shear rates should be used cautiously for estimating of lifetimes of oligomeric compositions used in real technology.

6.3.3.3 Curing after gel-point

Monitoring the process of curing after gel-point is possible on the basis of elastic modulus measurements.

The general understanding of the curing process is based on two basic concepts: formation of the three-dimensional network of chemical bonds at a certain degree of transformation (conversion), β^*, and possible transition to glassy state at curing temperature. This transition occurs at some degree of conversion because network density must be sufficiently high to restrict molecular mobility (at a given temperature). The transition to a glassy state occurs at isothermal conditions (in contrast to a trivial understanding of the glass transition as a phenomenon taking place on cooling).

The sequence of physical events which can happen on curing is presented in Fig. 6.3.10. The solid line is a dependence of the glass transition temperature, T_g, on the degree of transformation, β. The vertical dotted line, β^*, corresponds to the gel-transition. Four

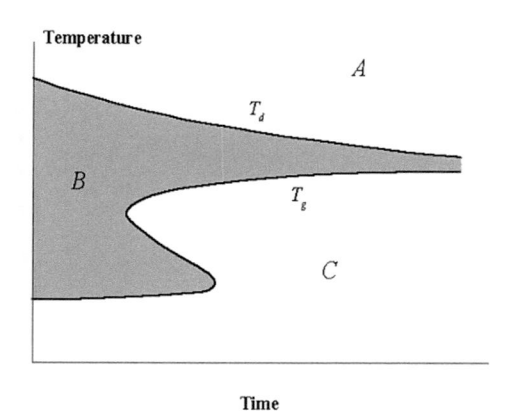

Figure 6.3.11. T-T-T diagram of curing. T_d – line of thermal degradation; T_g – line of the glass transition. A – domain of thermal degradation; B (shadowed zone) – domain of curing; C – domain of solid state. [Adapted, by permission, from J.K. Gillham, *Polym. Eng. Sci.*, **19**, 676 (1979)].

domains of different states of material are recognized and designated by Roman numbers: in domain I, the material is liquid (it can flow); in domain II, the material is an elastomer (cured rubber); in domain III, the material is vitrified liquid; and in domain IV, the material becomes a glassy polymer (a polymeric glass).

This scheme permits us to propose the primary analysis of the process of isothermal curing at different temperatures. Three temperatures, designated a, b, and c in Fig. 6.3.9 mark different stages and behaviors as discussed below.

If curing proceeds at a temperature corresponding to line a, initially the liquid oligomeric system changes to the rubbery state after crossing line β*. The process of curing continues to the end – to the complete consumption of reactive groups. Then the final value of β=1. Moving along line a, one maximum corresponding to passing through line β* can be observed on the tanδ-vs.-time dependence.

If curing takes place at a temperature corresponding to line b, a liquid system also passes to the rubbery state at β = β* and the process continues in a rubbery state to line T_g corresponding to the glass transition temperature. After this point, the chemical reactions stop because of freezing molecular motions, though in this case, when β < 1 two maxima on the tanδ-vs.-time dependence (corresponding to transition through β* and glass transition temperature) are observed.

If curing proceeds at a temperature corresponding to line c, glass transition takes place before β*, i.e., when the material is still in a liquid state. Glass transition also manifests itself as tanδ-vs.-time dependence passes through a maximum.

The kinetic factor is not reflected in the scheme in Fig. 6.3.10 at all, in spite of the fact that it is one of the determining factors in the rheokinetic studies, as becomes evident from the so-called T-T-T (Transformation-Time-Temperature) diagram in Fig. 6.3.11.[55] Curing is impossible above the line T_d, because of intensive thermal degradation at high temperatures. Curing is also impossible below the line T_g because chemical reactions are frozen at low temperatures. There is a domain between the lines T_d and T_g (shadowed zone in Fig. 6.3.11) in which curing is realized. The S-shape form of curve T_g is caused by the competition between the rate of curing and the distance of processing temperature from the glass transition temperature.

Diagrams such as those presented in Fig. 6.3.11 can be constructed for different curing systems, and they are the basis for the selection of technological parameters of curing.

Rheokinetic curves for the G'(t) dependence are different for various ranges of the T-T-T diagram. If curing takes place along the line a, as in Fig. 6.3.9, the final equivalent state of the material is reached at various temperatures, and this state is characterized by

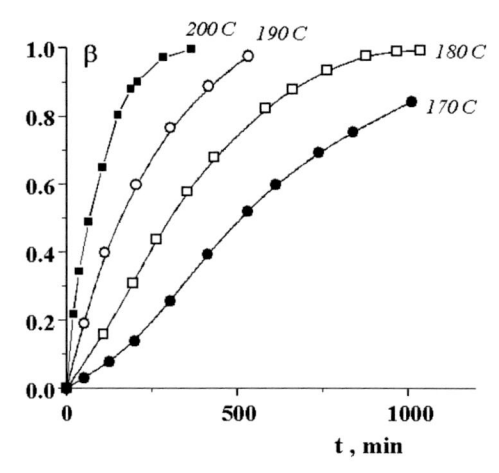

Figure 6.3.12. Kinetics of curing of an epoxy silicone organic oligomer at different temperatures. [Adapted, by permission, from S.G. Kulichikhin, P.A. Astakhov, Yu.P. Chernov,V.A. Kozhina, L.I. Golubenkoba, A.Ya. Malkin, *Vysokomol. Soedin.* (Polymers – in Russian), **28A**, 2115 (1986)].

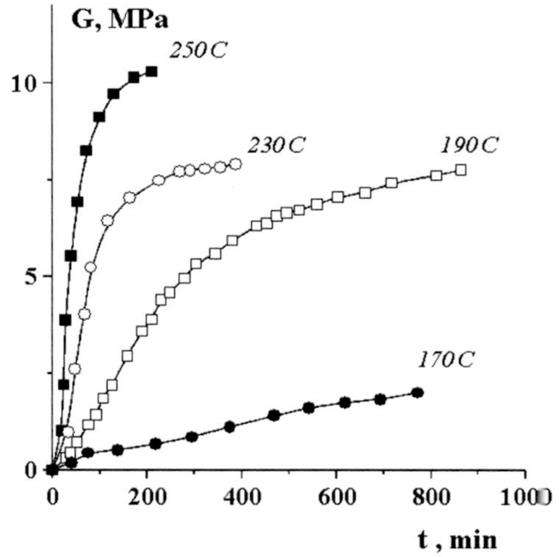

Figure 6.3.13. Kinetics of curing of phenylmethylsiloxane oligomer at different temperatures. [Adapted, by permission, from S.G. Kulichikhin, G.I. Shuvalova, V.A. Kozhina, Yu.P. Chernov, A.Ya. Malkin. *Vysokomol. Soedin.* (in Russian), **28**, 497 (1986)].

the limiting (or equilibrium) value of modulus, G'_∞, which in practice does not depend on temperature or frequency. Then, it is reasonable to determine the degree of transformation, β, in the following way:

$$\beta(t) = \frac{G'(t) - G'_0}{G'_\infty - G'_0} \qquad\qquad [6.3.9]$$

where $G'(t)$ are current values of elastic modulus and G'_0 is its initial value. In fact, in all cases $G'_0 \ll G'_\infty$ and therefore the following equation is valid:

$$\beta(t) = \frac{G'(t)}{G'_\infty} \qquad\qquad [6.3.10]$$

An example illustrating experimental results of curing of oligomer from the liquid through the rubbery state is shown in Fig. 6.3.12. The curve for the lowest temperature also reaches the limiting value $\beta = 1$, though this part of the curve is not shown in this figure because it corresponds to time as long as 10^4 min.

Incomplete curing is shown in Fig. 6.3.13. In this case, the limiting values of elastic modulus are different at various temperatures because they relate to different degrees of transformations. Eq. 6.3.10 can be used in this case too if G'_∞ is treated as the limiting value of modulus at the highest temperature, which is supposed to be above the glass transition line. The form of all curves is analogous to those shown in Fig. 6.3.13, but the ordinate axis are limited to 1. Practical application of results of curing analysis and comparison of different materials and conditions of curing are based on fitting equations for the rheokinetic curves. A general equation for kinetics, obtained by treatment of the results of calorimetric studies of curing, can be written in the following form:[56]

$$\dot{\beta} = (k_1 + k_2 \beta^m)(1 - \beta)^n \qquad [6.3.11]$$

or

$$\dot{\beta} = k_0 (1 - \beta)^n (1 + C\beta^m) \qquad [6.3.12]$$

where k_0, k_1, k_2, m, n, and C are empirical parameters.

The particular cases of these equations are of interest. If $C = 0$, it means that the kinetics of curing is described by the standard n^{th}-order equation (for example, $n = 1$ corresponds to the first-order kinetics, $n = 2$ to the second-order kinetics, and so on). If $C > 0$, it shows that the kinetic equation reflects the effect of self-acceleration.

The same equations were applied to the kinetics of curing, and many constants are necessary for fitting experimental data. For complete curing curves, as in Fig. 6.3.12, the kinetic equation of self-acceleration type can be written as:

$$\dot{\beta} = k_0 (1 - \beta)^n (1 + C\beta) \qquad [6.3.13]$$

and practically all experimental data can be described using $n = 1$ or $n = 2$. In this case, the constant k_0 is the initial rate of the curing process, and C reflects the effect of acceleration, regardless of the chemical mechanism of this phenomenon.

Incomplete curing, as in Fig. 6.3.13, requires the introduction of a special term into the kinetic equation which reflects this effect. A convenient rheokinetic equation for incomplete curing can be written as

$$\dot{\beta} = k_0 (1 - \beta)^n (1 + C\beta)(1 - \zeta\beta) \qquad [6.3.14]$$

where $n = 1$ or 2. The new factor ζ reflects the effect of limited curing. At $t \to \infty$, the limiting degree of curing $\beta_{lim} = \zeta^{-1}$ is reached.

Though the standard frequency of 1 Hz is mainly used for the rheokinetic monitoring of curing reactions by measuring G' as a function of time, the measurements can be carried out at different frequencies as well. It is difficult to compare results obtained at different frequencies because the sample is changing during measurements. This methodological problem can be solved by the mechanical Fourier transform spectroscopy (MFTS) method when several frequencies are superimposed and the output signal is analyzed by using the Fourier series.[57] This shows that measuring at different frequencies gives non-identical results. The evolution of different relaxation modes during oligomer curing may also be examined.

Rheokinetic studies of curing processes (in different versions of instruments) are widely applied in the technology of oligomers for the following purposes:
- selection of the optimal technological regimes (time-temperature evolution) for compositions of practical interest
- estimation and control of product quality
- qualitative comparison of different materials
- solving boundary problems in modeling different technological processes.

6.3.4 INTERMOLECULAR TRANSFORMATIONS

6.3.4.1 Polymeric reaction

Polymer-polymer transformations (chemical reactions occurring in the side groups of a macromolecular chain) are a natural part of chemical technology. Transformations of such kind are easily detected with rheokinetic methods because changes in chain structure lead to changes in its rigidity (flexibility) and, thus, of rheological properties of the material. Rheological measurements permit the detection of phase transitions caused by chemical transformations. Shearing influences the phase state of the system and thus the kinetics of transformations.

Both situations – polymer-polymer transformations in a homogeneous state and influence of shear stress on this transformation due to transition to the heterogeneous state – are presented in Fig. 6.3.14, showing results of rheokinetic studies of polyvinylacetate conversion to polyvinylalcohol induced by sulfuric acid. The descending (linear) branch of dependence corresponds to a homophase reaction and viscosity decrease accompanies polymer-polymer reactions. Superposition of shearing leads to phase transition and the reaction reaches a heterogeneous domain.

Rheokinetic study is a convenient and sensitive method of monitoring chemical reactions of polymer chains.[58]

6.3.4.2 Physical transformations

Changes in the material state can be caused by slow physical processes ("*aging*"), such as crystallization and the formation of intermolecular physical bonds. Any of these processes are reflected in changes in the rheological properties of the material. Rheological measurements are useful instruments for monitoring the state of material and estimating its technological quality. There are numerous materials for which these pro-

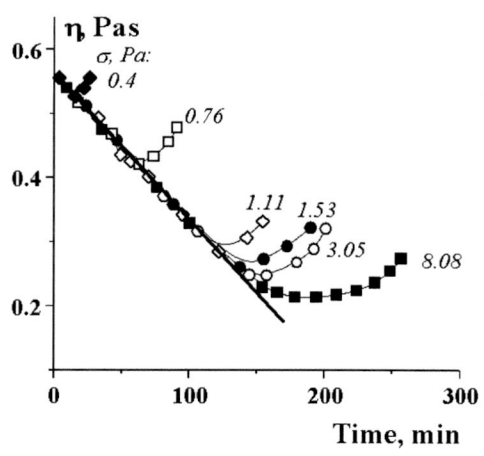

Figure 6.3.14. Viscosity evolution in conversion of polyvinylacetate to polyvinylalcohol in homogeneous (decreasing thick solid line) and heterogeneous (increasing parts of curves) domains. The influence of shear stress on transition to the heterophase system is shown. The reaction takes place in 20% solution in ethanol at 20°C. [Adapted, with permission, from A.Ya. Malkin, S.G. Kulichikhin, V.A. Kozhina, Z.D. Abenova, *Vysokomol. Soedin.* (Polymers – in Russian), **28B**, 408 (1986)].

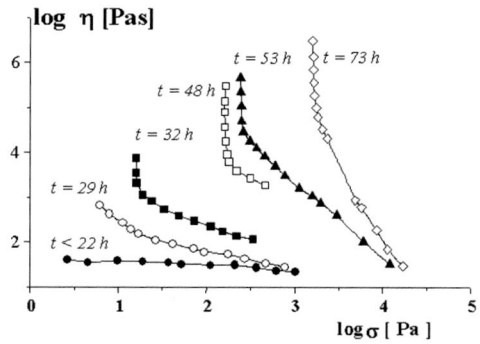

Figure 6.3.15. Solution-to-gel transition: evolution of rheological properties as a result of gelation process of a 12% solution of polysulfone in dimethylacetamide with 6% of water at T = 22°C. [Adapted, with permission, from A.Ya. Malkin, L.P. Braverman, E.P. Plotnikova, V.G. Kulichikhin, *Vysokomol. Soedin.* (Polymers – in Russian), **18A**, 2596 (1976)].

cesses are of practical importance. Fig. 6.3.15 shows a characteristic example. Aging causes a transition from fluid, through viscoplastic media (with a clearly expressed yield point), to a rubbery gel-like material.

6.4 SOLUTION OF DYNAMIC PROBLEMS

6.4.1 GENERAL FORMULATION

The previous parts of the book were devoted to the description of material properties "at a point", i.e., establishing relationships between local stresses and deformation rates related to a point. The central problem in measuring these properties (Chapter 5) was the transition from measured macro-values (forces, torques, velocities) to the relationships between local tensors related to dynamic and kinematic values. For solving any applied problems of movement of a rheologically-complex medium, the inverse problem must be solved – this is a transition from rheological properties measured in the laboratory to the prediction of material behavior in a real technological practice.[59]

The general formulation of this problem is: rheological properties of the material are known (measured) – how then to find the relationship between forces and velocities for an arbitrary geometry of deformation (flow)? The answer to this question is represented by the scheme in Fig. 6.4.1.

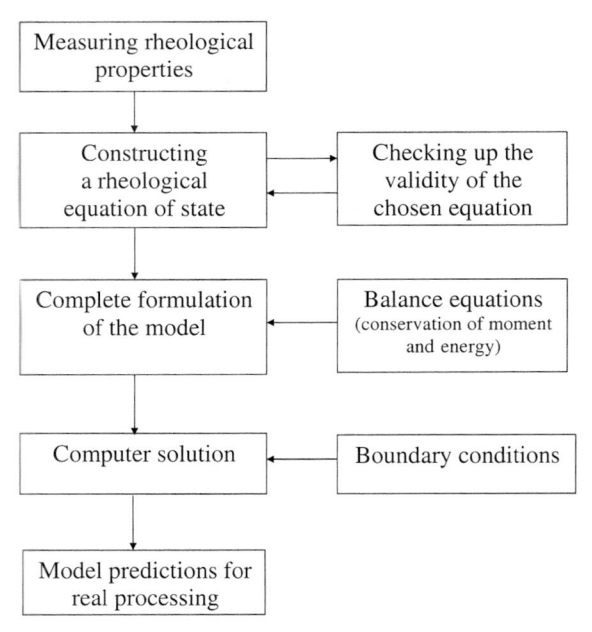

Figure 6.4.1. General scheme illustrating the method of solving applied dynamic problems.

The first line in this scheme is evident: it is relationships between stresses and deformations (and deformation rates) measured in different simple schemes of flow. Methods of measurements were discussed in Chapter 5 and numerous results of such measurements were presented in different parts of the book, mainly as dependencies of the shear stress on shear rates, as well as the elongational stress on uniaxial deformation rate.

The left side of the second line in Fig. 6.4.1 is the fitting of experimental data by a suitable empirical or theoretically-based equation. Experimental data are usually obtained in one-dimensional flow conditions in order to simplify their treatment. Then it is necessary to choose a method of their generalization for three-dimensional, 3D, deformations because the flow in a real processing environment takes place in 3D space. This general equation is called a *rheological equation of state* or *constitutive equation* (or a *rheological model*) and initial experimental data enter this equation as a particular case.

Therefore, a crucial step is listed on the right side of the second line. This is the checking of a chosen rheological equation of state for conditions quite different than the initial experimental scheme used for measuring the stress vs. deformation rate relationship.

For example, initial data are shear stress vs. shear rate. These data are fitted by one of the equations discussed in section 3.3. Can one use this equation for the prediction of material behavior in the uniaxial extension? To answer this question it is necessary to write the fitting equation in 3D form using ideas of continuum mechanics in order to receive an invariant presentation of experimental data (as discussed in Chapter 1). The principal point is that the methods of such presentations are ambiguous and it is possible to do it in a different manner. This leads to different rheological equations of state for the same initial experimental data. Consequently, the predictions of different rheological equations of state might appear different for a new dynamic or kinematic situation. The right side of the second line shows that it is necessary to confirm that at least in some new experimental situations a generalized rheological equation of state gives reliable predictions of material behavior.

There are several general rules (or principles) that must be fulfilled in the formulation of a rheological equation of state:[60]

- *Principle of coordinate invariance* requires that formulation of the rheological equation of state does not depend on the choice of a coordinate system while inertial systems are used. This requirement is realized by the formulation of any relationship between stress and deformation tensors *via* their invariants.
- *Principle of determinism* means that the stress state of material points can depend on the prehistory of its deformation but does not "feel" or forecast what will happen in the future. In fact, this principle is cast into the theory of viscoelasticity where the previous history of material is expressed *via* integrals with fading memory.
- *Principle of local action* supposes that only the closest neighboring points can influence the stress state at some chosen points. This principle denies the possibility of a long-distance action, though in some cases long-distance forces can exist but they are not of mechanical origin.
- *Principle of material objectivity* expresses the idea that behavior of material must be written in such a way that it would be independent of any motion of an observer, and in particular, of movement and rotation of a rigid body as a whole.[61] Based on this principle it is necessary to use some complicated laws to coordinate transformation and objective tensors of deformations such as the Rivlin-Ericksen and the White-Metzner tensors mentioned in section 1.3.1.

The solution of boundary problems, which is the final goal of modeling any technological process, begins with the complete formulation of a system of equations describing this process. The central step here is a complete formulation of the model (left side of the third line in Fig. 6.4.1). This is based on combining a rheological equation of state with balance (or conservation) equations (the right side of the third line in Fig. 6.4.1). The latter are the general laws of nature. Three of them are of interest to the problem under discussion.

Conservation of momentum

These are equations formulating the conditions of equilibrium of all forces acting at a point. They were written in section 1.1.6 as a system of Eqs. 1.1.18 and in a simplified form as Eqs. 1.1.19. They can be written in different coordinate systems, for example, in cylindrical coordinates (see Eqs. 1.1.20). The equations for conservation of momentum are written in components of the stress tensor, which are 6 unknown variable values.

Conservation of mass

This equation expresses the concept of constancy of mass. If we consider the flux of mass entering an elementary unit volume (i.e., at a point) and leaving the same volume and assuming that media is incompressible, then the mass of material inside this volume must be constant. This concept leads to the following balance equation:

$$\frac{\partial v_1}{\partial x_1} + \frac{\partial v_2}{\partial x_2} + \frac{\partial v_3}{\partial x_3} = 0 \qquad [6.4.1]$$

where the notation of velocity components, v_i, and axes, x_i, is related to the Cartesian coordinate system.

Conservation of energy

This is also a fundamental law of nature expressing the following concept: changes of energy inside some volume of space can happen, due to heat flux into this volume and heat dissipation inside this volume as a result of viscous flow (which is dissipative by its nature). In some cases, it is necessary to take into account the heat of phase transitions or reactions taking place in this volume. Sources of energy other than heat flux are usually not taken into consideration in formulating dynamic problems. Heat flux is possible due to the existence of temperature gradients. That is why the equation of conservation of energy is formulated for temperature, T.

Balance of heat fluxes in relation to a unit space volume leads to the following equation written in the Cartesian coordinate system:

$$\frac{\partial T}{\partial t} + v_1\frac{\partial T}{\partial x_1} + v_2\frac{\partial T}{\partial x_2} + v_3\frac{\partial T}{\partial x_3} = a\left(\frac{\partial^2 T}{\partial x_1} + \frac{\partial^2 T}{\partial x_2} + \frac{\partial^2 T}{\partial x_3}\right) + \frac{W}{\rho c_p} \qquad [6.4.2]$$

where v_i are the components of the velocity vector, a is the thermal diffusivity, ρ is density and c_p is the thermal capacity. The coefficients a, c_p, and ρ, as well as viscosity, are (potentially temperature-dependent) properties of matter.

The value, W, entering this equation is an intensity of energy dissipation. This value is the work required to sustain flow. For a viscous fluid, it is expressed as the product of stresses and deformation rates:

$$W = \sum_{i,j} \sigma_{ij} D_{ij} \qquad [6.4.3]$$

where D_{ij} are components of the deformation rate tensor and σ_{ij} are components of the stress tensor. It is evident that W is an invariant value (in respect to the choice of a coordinate system).

Based on the above concepts, it is possible to formulate a flow model for material under discussion. Two main possibilities exist: to consider the flow of time-independent or time-dependent media. Then, it is possible to move to the fourth line of the scheme in Fig. 6.4.1.

The solution of a system of equations describing the movement of a medium with known rheological properties is possible if boundary conditions are formulated. In fact, just these conditions specify peculiarities of flow in any arbitrary geometry, either through channels of a certain form or around solid bodies. Equations describing the flow of rheologically complex liquids are very complex and only the simplest cases can be treated as analytical functions. At present, there are no principal difficulties in solving any correctly formulated boundary problem using modern computer techniques and developed computational methods. That is why the fourth line in the scheme of Fig. 6.4.1 supposes that only technical difficulties can appear in modeling any real technological (processing) situation.

The solution of any concrete boundary problem gives quantitative model predictions of material behavior in a real processing operation. And this is the bottom line in Fig. 6.4.1, which is the final goal of the application of rheological methods for engineering design and operation.

It means that starting from rheological experiments it is possible to forecast all technological parameters of a process, such as dependence of output on force, stress and temperature fields, and so on, up to discussing (based on the obtained numerical solution) the influence of variation of initial technological factors (composition, initial temperature, etc.) and boundaries of stability of the real technological process.

As shown in Fig. 6.4.1, the boundary conditions are the necessary component for solving dynamic problems.

In most cases, it is assumed that a liquid moving in space and contacting solid body boundary sticks to this body and has the same velocity as the boundary, ether this boundary is stationary or moves. When the boundaries of the space where a liquid moves do not move, the velocity at the surface equals zero. This supposition is called the hypothesis of stick and in many cases this assumption is correct. However, this is not always true since the slip along a solid surface is quite possible.[62] In this regard, it is reasonable to distinguish two possibilities. First, this is a real slip with the detachment of a liquid from a wall. This phenomenon can be a consequence of the liquid-to-solid transition due to high deformation rate when characteristic time of the process exceeds relaxation times (high Deborah Numbers). This is typical for polymer melts and leads to spurt (Fig. 3.6.11). The wall slip can occur for visco-plastic media at low speeds when the shear stress is lower than the yield stress and a material behaves in a solid-like mode. The velocity of slip is not known beforehand and depends on the stress at the wall.

The second case is a quasi-slip that happens in the flow of multi-component materials. This phenomenon is due to phase separation when a low viscosity component segregates from a homogeneous liquid (e.g. in a concentrated solution) and forms a low-viscous layer at a wall. Then a jump in the velocity takes place and it looks like a wall slip. The same effect is observed in the flow of concentrated suspensions.

Measuring the rheological properties of different materials is a basis for engineering rheology, which is used for designing technological equipment and predicting processing characteristics of various real materials.

Engineering practice encounters two main types of flow: flow through tubes under applied pressure ("*pumping*") and flow caused by the movement of a solid boundary surface ("*drag*" flow). The latter case can be caused by flow produced by rotation of the screw, calibrating rolls, brush spreading paint, rotor inside a mixer, and so on.

6.4.2 FLOW THROUGH TUBES

This is an important engineering problem, which models many real situations in the transportation of rheologically complex liquids. The prime goal of calculations consists of establishing a relationship between the pressure gradient in a tube (or pipe) and output. This scheme is a good approximation of some other more complex situations encountered in processing equipment.

The simplest case is the flow through a tube of a circular cross-section.

The output, Q, vs. pressure, p, dependence is calculated based on the analytical presentation of a flow curve.

Some of the most important cases are considered below.

In the simplest case of flow of Newtonian liquid with viscosity, η, the Q(p) dependence is expressed by Poiseuille's well-known law:

$$Q = \frac{\pi R^4 p}{8 \eta L} \tag{6.4.4}$$

where R and L are the capillary radius and length, respectively.

The Q vs. p dependence for a power-law type non-Newtonian liquid (see Eq. 3.3.4)[63] is expressed as

$$Q = \frac{\pi K R^{m+3}}{m+3} \left(\frac{p}{2L}\right)^m \tag{6.4.5}$$

This equation can be rewritten as

$$Q = K_f p^m \tag{6.4.5a}$$

The last equation shows that for power-law liquid, Q is proportional to p^m, where m is the same exponent as in rheological law and K_f is a form- or geometrical-factor. Then it is possible to think that for channels with an arbitrary cross-section this relationship is also correct with its own value of a geometrical (front) factor depending on the form of a cross-section of the channel.

The Q vs. p dependence in the flow of a viscoplastic Bingham-type medium (see Eq. 3.3.9) through a cylindrical channel is expressed by the Buckingham-Reiner equation (see solution of Problem 3-9):

$$Q = \frac{\pi R^4 p}{8 \eta_p L} \left[1 - \frac{8 L \sigma_Y}{3 p R} + \frac{1}{3} \left(\frac{2 L \sigma_Y}{p R}\right)^4 \right] \tag{6.4.6}$$

where the rheological constants η_p and σ_Y are the same as in Eq. 3.3.7.

This equation can also be presented in the following equivalent form:

$$Q = \frac{\pi R^3 \sigma_R}{4 \eta_p} \left[1 - \frac{4}{3}\left(\frac{\sigma_Y}{\sigma_R}\right) + \frac{1}{3}\left(\frac{\sigma_Y}{\sigma_R}\right)^4 \right] \tag{6.4.6a}$$

Transport characteristics for the practically important case of flow of the Hershel-Bulkley viscoplastic medium are written as:

$$Q = \frac{n\pi R^3}{K^{1/n}\sigma_R^3}(\sigma_R - \sigma_Y)^{(1+n)/n}\left[\frac{(\sigma_R - \sigma_Y)^2}{1+3n} + \frac{2\sigma_Y(\sigma_R - \sigma_Y)}{1+2n} + \frac{\sigma_Y^2}{1+n}\right] \qquad [6.4.7]$$

where σ_R is the shear stress at the wall which is expressed in the usual manner as $\sigma_R = pR/2L$ and σ_Y, K and n are rheological constants, as in Eq. 3.3.9.

The above-written relationships are useful engineering equations for designing transport pipe systems, as well as channels in technological (processing) equipment.

However, there are the following limitations in applying these equations:
- they are written for steady flow in long tubes; therefore they are not correct for designing dies (for example, for channels with length-to-diameter ratio of the order of 1), especially in the flow of viscoelastic liquids; in the latter case the dominant role in resistance belongs to an elastic response
- time effects of thixotropic or rheokinetic nature might be important in the flow of some materials; these effects are not taken into consideration in the formulation of the above-written equations.

These situations must be treated using more complicated equations constructed for special time-dependent materials.

The solution of dynamic problems of flow through a channel of an arbitrary cross-section is achieved by introducing the form-factor characterizing the geometrical form of the cross-section. For Newtonian liquid, the basic linear relationship is given by Eq. 5.2.13 which is written as

$$Q = \frac{K}{\eta}p \qquad [6.4.8]$$

Analytical method of finding constant K consists of a rigorous solution of the dynamic problem of flow through a channel with arbitrary cross-section. The channel can be rectangular, elliptic, or have any other cross-section. For channels with complicated geometry, the solution is found by numerical methods.

Examples – form-factors for different cross-sections
Some important examples of the following geometries will be discussed:
- Flow through an elliptic channel with semi-axis a and b
- Flow through a flat channel. In this case, a channel is formed by two parallel flat plates; the gap, 2h between them is much smaller than their width $b \gg h$
- Flow through a circular channel formed by two co-axial cylinders along their axis; the radius of the outer cylinder is R_0 and of the inner cylinder is R_i.
- Flow through a channel with a cross-section having a form of an equilateral triangle with the length of a side a.

In all mentioned cases, the values of the form-factor can be calculated analytically. These values are listed below.

Cross-section	Form-factor, K
Round	$K = \dfrac{\pi R^4}{8L}$
Elliptic	$K = \dfrac{\pi}{4L}\dfrac{a^3 b^3}{a^2 + b^2}$

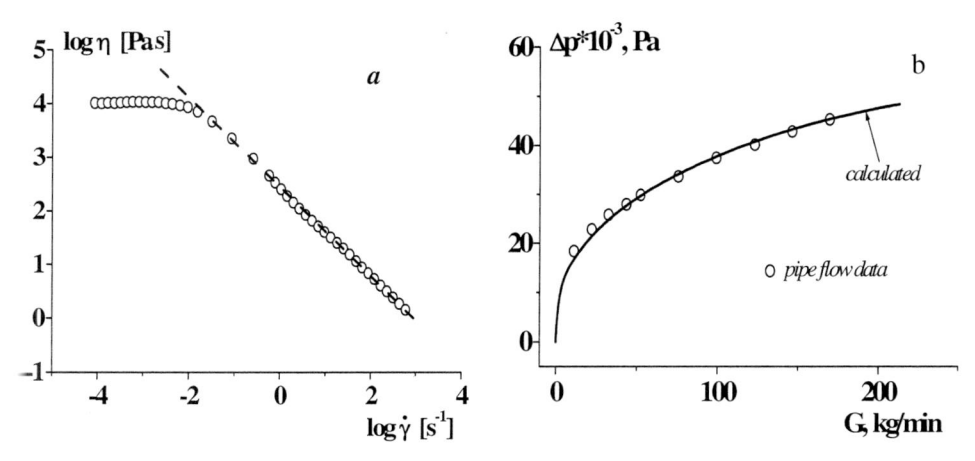

Figure 6.4.2. Flow curve of emulsion (a) and transport characteristic of a tube (b) calculated using power-law approximation of the flow curve. Average diameter of dispersed particles is 8.2 μm. 30°C. [Adapted, with permission, from A.Ya. Malkin, I. Masalova, D. Pavlovski P. Slatter, *Appl. Rheol.*, (2004)].

$$\text{Flat} \qquad K = \frac{2bh^3}{3L}$$

$$\text{Annular} \qquad K = \frac{\pi}{8L}\left[(R_o^4 - R_i^4) - \frac{(R_o^2 - R_i^2)^2}{\ln R_o / R_i}\right]$$

$$\text{Equilateral triangle} \qquad K = \frac{\sqrt{3}}{32\pi}\frac{a^4}{L}$$

Modern computer technique permits calculation of form-factor for channels with any arbitrary cross-section. An experimental approach for finding the form-factor is based on measuring Q–P pair of values for liquid of known viscosity. Then the form-factor is found using Eq. 6.4.8.

Tube flow of non-Newtonian liquid is more complex because there is no unique analytical solution for Q(p) dependence analogous to linear dependence for Newtonian liquid and analytical equations for this dependence cannot be presented using a single geometrical factor. However, there is a general solution for calculating the tube characteristic of any non-Newtonian liquid, for which a flow curve has been measured. This is known as the *Rabinowitch-Weissenberg equation* (see Chapter 5, Eq. 5.2.15). This equation can be written as

$$Q = \frac{\pi R^3}{\sigma_R^3}\int_0^{\sigma_R}\sigma^2 f(\sigma)d\sigma \qquad\qquad [6.4.9]$$

where $\sigma_R = pR/2L$ is the shear stress at a wall, and it is also normalized pressure, and $f(\sigma)$ is the flow curve, i.e., the dependence of shear rate and shear stress measured in a laboratory test. If $f(\sigma)$ has been measured, Q(p) dependence can be easily calculated from Eq. 6.4.9.

Function $f(\sigma)$ can be expressed by several appropriate equations. The choice of fitting method to describe function $f(\sigma)$ is an important problem in engineering rheology. An experimenter usually tries to fit experimental points as accurately as possible, sometimes using multi-constant equations for this purpose. However, it is not necessary in any case. As an example, Fig. 6.4.2a presents experimental points obtained for super-concentrated water-in-oil emulsion.[64] These points can be successfully fitted using the Cross-type equation:

$$\eta = \frac{\eta_0}{1 + (\lambda \dot{\gamma})^n}$$ [6.4.10]

which is a simplified form of the Cross equation, Eq. 3.3.1.

The flow curve in Fig. 6.4.2a comprises the clearly expressed domain of Newtonian flow at low shear stresses and therefore equations with yield stress are not appropriate for this flow curve.

Eq. 6.4.10 can be used for predicting Q-vs.-p dependence for tube flow and undoubtedly this approach gives good results. The high-shear-rate part of the flow curve can be described by a power-law equation, Eq. 3.3.4, as well as the Hershel-Bulkley equation Eq. 3.3.9. The straight line in Fig. 6.4.2a gives a power-law approximation. This type of fitting does not express properly a low stress domain. Fig. 6.4.2b presents Q-vs.-p predictions based on power-law approximation calculated from Eq. 6.4.5. Points in this figure are experimental data. Power-law type approximation gives accurate predictions (maximum error does not exceed 2.3%). Similar results were obtained when the Hershel-Bulkley approximation was used.

This result, as well as some other results obtained for different materials,[65] is explained by the fact that flow rates in real industrial transportation processes are high and the high-shear-rate domain of flow curves determines the total output. It is true in numerous cases because industrial engineers are interested in increasing transportation rates. However, many cases can also be pointed out in which behavior of liquid in low-shear-rate domain controls the process, e.g., deformation of greases in bearings. In such cases, freedom in the choice of fitting equation becomes invalid.

The choice of analytical approximation of laboratory-measured rheological properties should be made based on goals of applications, though in many cases there is no need for "exact' fitting of points obtained in the rheological experiment.

In discussion of engineering problems of tube transportation, it is necessary to take into account the following limitations of the above written equations:

* these equations are correct for steady flow through a long channel
* instabilities of various types may appear at high enough flow rates
* normal stresses in shear flow lead to circular fluxes in cross-sections of non-round channels. These fluxes do not give a large input into total energy consumption but can influence mixing processes and quality of final products.

6.4.3 FLOW IN TECHNOLOGICAL EQUIPMENT

These are primarily *drag flows*, i.e., flows caused by the movement of solid boundary surfaces in technological equipment.

6.4.3.1 Pumping screw

The complete theory of pumping screw machines (*extruders*) is not a subject of this book.[66] Here, only basic concepts related to the rheology of pumping materials are discussed.

The general engineering problem in designing a pumping screw extruder consists of establishing the relationship between the speed of screw rotation, power consumption, and output for equipment of known size, temperature regime of processing, and measured properties of the material. Rheological properties of the material are essential and necessary parts of analysis.

A pumping screw is often used in different technological processes. In particular, various extrusion machines are widely used in polymer technology and transportation devices.

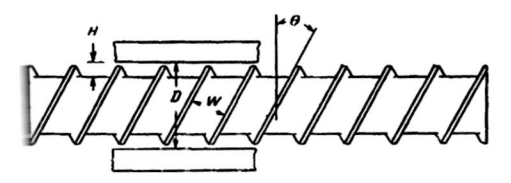

Figure 6.4.3. Screw. Geometrical parameters: D – diameter of barrel; W – distance between the neighboring flights; H – depth of channel; θ - helix angle (flight inclination).

A schematic diagram of a screw is shown in Fig. 6.4.3. Flow takes place in a narrow space between the body of the screw and a cylindrical barrel. Let us imagine that this channel is not cylindrical but transformed into a flat rectangular channel with a width, W, (equal to the distance between neighboring flights of the screw) and height, H, (distance between a body of screw and a barrel). This operation is correct if it is possible to neglect the curvature of the channel, which is acceptable if H << D. The most important point is the source of liquid flow in a channel. This is the relative movement of the upper side ("roof") of the channel, assuming that the channel is stationary. The velocity of movement has two components – along the channel V_z and in transverse direction due to inclination of the flight with an angle, θ. The velocity component, V_z, provides the output while the transverse component leads to the circulation of liquid inside the channel and it is responsible for the mixing effect.

The dynamics of flow in the channel of a screw pump are described by balance equations and a rheological equation of the state of liquid. Pumping screw extruders are usually used for plastic processing, and in this case, the theory[67] is based on a power-type viscosity law.

A more complete theory of flow in a pumping screw extruder includes the following important additional details.

Flow in a channel creates a longitudinal pressure gradient; so the pressure gradient exists between neighboring sides of flight. Therefore, a back (leakage) flow in a gap between a flight and a barrel emerges and the negative input of this back flow into the total output can be essential. The gap is narrow and non-Newtonian effects in flow through these gaps are pronounced.

Typically, flowing liquid is highly viscous, therefore it is necessary to take into account a dissipative effect. The temperature of the barrel is different than that of liquid. It means that it is necessary to consider the process as non-isothermal with a thermal exchange between walls of the channel and the flowing liquid.

The complete theory of processes in a *plasticating extruder* must also consider that material enters the channel in a solid form and the movement of solid particles and their melting are inherent components of the technological process.

Pumping of plastic melts proceeds, not into empty space but into a die, which forms the shape of the article (profiles of various types – films, tubes, sheets, and so on). The die plays a role of hydrodynamic resistance; therefore, the pressure at the end of the screw exits, and under this pressure the back-flow occurs. As a result, a real

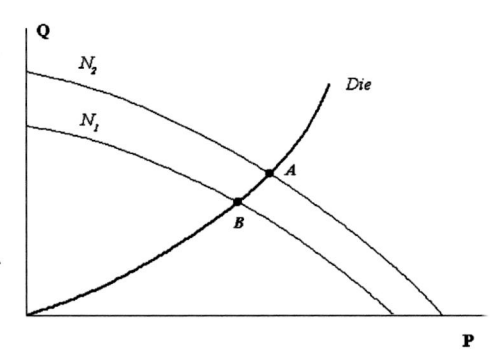

Figure 6.4.4. Performance characteristics of an extruder with a die.

flow consists of a superposition of two fluxes – drag flow produced by the screw and backpressure flow.

This is the most important feature of pumping extruders. Due to the superposition of two fluxes the total output, Q, can be (at the first approximation) written as

$$Q = AN - Bp^m \qquad [6.4.11]$$

where A and B are constants, N is the rotational speed of the screw and m is an exponent in a power-law equation describing rheological properties of liquid.

The real performance characteristics of an extruder in pumping polymer melts reflect two dependencies: performance characteristics of the screw as presented, for example, by Eq. 6.4.11, and resistance of die, as expressed, for example, by Eq. 6.4.5a. Combination of these dependencies results in the real performance of an extruder (*operating point*). This concept is illustrated in Fig. 6.4.4 in which two curves, N_1 and N_2, express Eq. 6.4.11 for two speeds of rotation ($N_1 < N_2$), and the curve "*Die*" expresses Eq. 6.4.5a. Points desig-nated as A and B are two pairs of real values of output and pressure (operating points) at two different speeds of screw rotation.

Curves are obtained from measuring rheological (and other physical) characteristics of the material. The solution of the problem, including all factors listed above and a full set of balance equations, is very complicated, though it can be realized using modern com-putation techniques. In fact, there is available standard software which permits calculation of necessary parameters of an extruder based on its characteristics as well as rheological and other physical properties of the material used in the technological process.

Screw extruders are also used as primary transportation machines, especially for concentrated suspensions (clay suspensions, mud, and so on). For this field of application – in contrast to polymer technology – a rheological equation of state must include yield stress. A power-law type equation is not suitable. However, in this case, non-isothermal effects and die pressure might be immaterial. But a complete theory of screw pumping is expected to take into consideration the effect of jamming at high concentrations and high rates of rotation, as was described in section 6.2.6.

Screw extruders can be used as mixing devices. In this case, the main factor is the transverse flow in channel cross-sections and the relative movement of layers of different materials. Rheological properties of both components play a dominating role in the mixing process and in formulation of process theory.

The extrusion process includes an important final stage: melt leaves a calibrating die and appears in free space. The crucial point here is a change of shape of an extruded profile due to elastic forces stored during flow inside calibrating channels of the die. This is a *swelling effect*, which was discussed in section 3.4.4. It is rather simple to measure die swell after flow through a channel with a round cross-section. In real technological practice, the extruded profiles are very often asymmetrical. Even if the profile is axisymmetric but not round, for example, in producing square cross-section profiles, there is no guarantee that the profile of the item is the same (square) as the cross-section of the die. The final profile is distorted in comparison with the profile of the die. In particular, it is very difficult to maintain sharp corners of the profile.

Calculation of profile evolution after leaving a forming (calibrating) die is a purely rheological problem.

There are two main mechanisms of die swell under isothermal conditions:
- velocity profile rearrangement observed for any liquid; this is a kinematic effect and its value is close to 1.12-1.13 for axisymmetric flow
- elastic unconstrained recovery; its value can be large and this effect is directly related to stored elastic energy in the flow of viscoelastic materials, as was discussed in Chapter 4 for rubbery solids; this mechanism dominates polymer solutions and melts.

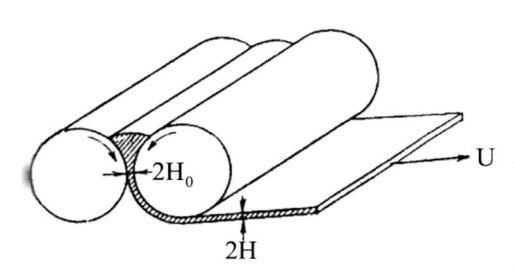

Figure 6.4.5. Calendering.

The majority of theoretical models of flow in extruder channels are based on non-linear flow curves only and do not take into account viscoelastic effects. It is incorrect in calculations related to the die swell because viscoelasticity and rubber elasticity are the main causes of this effect. Therefore, the rheological model becomes very complicated, the determination of its constants requires a more advanced experimental approach, and practical calculations can be carried out only by computational techniques.[68] The progress in this field is limited and it is restricted to model calculations of simple geometrical forms.

6.4.3.2 Calendering and related processes

A schematic diagram of calendering is shown in Fig. 6.4.5. Drag is achieved by rotation of two rollers, frequently with different speeds. Rollers pull material into a gap. Material deforms and flows. The main technological result is a decrease in thickness of a sheet, though some other effects also take place, especially if the speed of rollers rotation is different:
- intensive shearing in a narrow gap is accompanied by a significant heat release; rollers can also be heated up to a higher temperature to increase the temperature of the material (*heating* device)

- calendered material is composed of many components; intensive shearing in a narrow gap provides good *mixing* of components; flow in an entrance zone is two-directional, which improves mixing
- stresses in the calendering process can be so high that they may cause rupture of molecular chains (*mechano-chemical effect*); roller mills are used to regulate the rheological properties of the material.

The central goal in the analysis of calendering is to search for relationships between the speed of roller rotation, output, power of drive that rotates rollers, and forces acting in the transverse direction and causing separation of rollers.

Theoretical analysis of the process is based on analysis of conservation (balance) equations and rheological properties of the material.[69] The analysis of process involves dimensional arguments that are valid for any rheological model. The characteristic longitudinal length, L, (along the x-axis in Fig. 6.4.6) is much higher than the characteristic length h in the y-direction, i.e., h/L << 1. In this approximation, the momentum balance equation can be written as

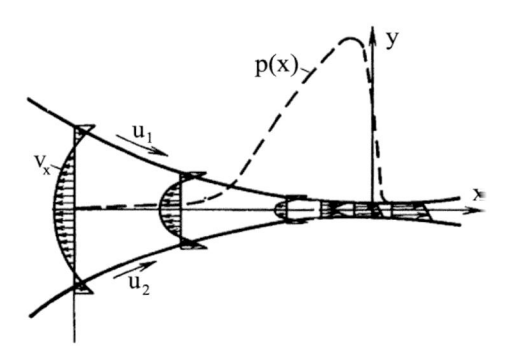

Figure 6.4.6. Velocity and pressure distribution in calendering. U_1 and U_2 – circumferential velocities of roller ; V_x – velocity in the x-direction, p – pressure.

$$-\frac{\partial p}{\partial x} + \frac{\partial \sigma_{xy}}{\partial y} = 0 \qquad [6.4.12]$$

Pressure p is assumed to be independent of y and it is a function of x, only: $p = p(x)$.

As a result, the following equation for stress distribution is valid for flow between rollers in the calendering process

$$\sigma_{xy} = y\frac{dp}{dx} + C(x) \qquad [6.4.13]$$

where C(x) is some function of x determined from boundary conditions (on the roller surfaces).

The condition of mass conservation is also valid for any medium and it is formalized by equation

$$\int_{y_2}^{y_1} u(x)dy = Q = const \qquad [6.4.14]$$

where y_1 and y_2 are coordinates corresponding to the roller surfaces, and u is velocity in the x-direction.

Rheology becomes of interest when Eq. 6.4.13 is used because the rheological equation of state relates stresses and velocities. In the simplest case of Newtonian liquid with viscosity, η, the following relationship for velocity distribution is obtained:

$$u(y) = U_2 + (U_2 - U_1)\frac{y}{h} - \frac{1}{2\eta}\frac{dp}{dx}(h-y)y \qquad [6.4.15]$$

where h is the distance between rollers (depending on x-coordinate).

Pressure distribution, p(x), is found from the condition of constant output, which is calculated from Eq. 6.4.14 and is written as

$$Q = \frac{U_1 + U_2}{2}h - \frac{h^3}{12\eta}\frac{dp}{dx} \qquad [6.4.16]$$

The force pushing rollers apart is calculated as the integral of pressure along the whole surface of rollers. The driving torque for rotating rollers is an integral of shear stresses acting on their surfaces.

A qualitative example illustrating the evolution of velocity distribution and pressure along the path of material in a gap between rollers is shown in Fig. 6.4.6. It shows characteristic features of the process:
- forward and backward flux in any cross-section
- rapid development of pressure along the material path with its maximum at some cross-section.

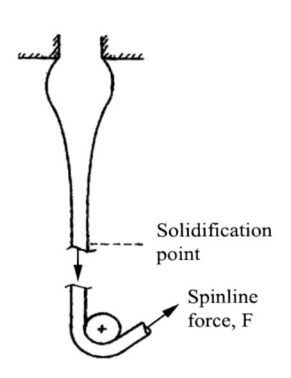

Solidification point

Spinline force, F

Figure 6.4.7. Schematic representation of the fiber spinning.

Extension of the calendering theory to flow of non-Newtonian liquids is made according to the same scheme as for Newtonian liquid, i.e., based on a balance equation with the introduction of appropriate rheological law. In this case, calculations become more complicated and require the application of numerical methods. The elasticity of rolled material is not involved in the discussion, though this rheological property plays an important role. Also mixing in calendering does not have definitive solution.

6.4.3.3 Extension-based technologies

There is a group of technological processes in polymer processing based on the application of stretching. The process creates the orientation of macromolecules, and as a result, increases the strength of the material.

Fiber spinning is a typical example. It has the following peculiarities:
- deformation (flow and elastic strains) is primarily extensional and shear stresses are neglected
- fiber is strongly inhomogeneous along its length due to the die swell at an outlet from the die (spinneret) and due to decrease of a cross-section of fiber caused by stretching
- flow is non-isothermal due to cooling of fiber until complete solidification (crystallization) of polymer; temperature is inhomogeneous along the radius of fiber due to slow heat exchange through the polymer
- tendency to increase the speed of spinning leads, in some cases, to various surface instabilities.

All these peculiarities make theoretical analysis of the fiber spinning process difficult and not directly related to a simple model discussed in section 3.7.[70]

A schematic diagram of fiber spinning is shown in Fig. 6.4.7.

Some basic qualitative relationships related to fiber spinning are as follows. The stretching force is constant along the length of the fiber and depends on normal stress, σ_E according to the equation

$$F = \pi \sigma_E R^2 = const \qquad [6.4.17]$$

Volume output, Q, is also constant along the length:

$$Q = \pi R^2 V = const \qquad [6.4.18]$$

where V is the longitudinal velocity that is dependent on the radius, R, changing along the length of stretching.

If the material being stretched is Newtonian liquid and the Trouton law is valid, then the gradient of elongational velocity, dV/dx, is expressed as

$$\frac{dV}{dx} = \frac{FV}{3\eta Q} \qquad [6.4.19]$$

At the starting point, at $x = 0$, $V = V_0$, and then the velocity profile along the fiber is expressed as

$$V = V_0 \exp\left(\frac{xF}{3\eta Q}\right) \qquad [6.4.20]$$

The process of stretching is characterized by the value of *draw ratio*, DR, which equals V/V_0, causing drawing along the length, L. Then, one can write

$$\frac{V}{V_0} = \exp\left(\frac{x}{L}\ln DR\right) \text{ and } \frac{R}{R_0} = \exp\left(-\frac{x}{2L}\ln DR\right) \qquad [6.4.21]$$

The above relationships provide some parameters determining the performance characteristics of the fiber spinning process. The Trouton law approximation is not good for real technological processes of fiber spinning. It is not reasonable to use any other flow curve equations, as is sometimes done. Rheological properties are modeled by the calculation of drawing force as a function of the kinematics of spinning. The complete theory must take into account the transient character of deformations. It is necessary to use a rheological model for the viscoelastic medium.

High rates of deformation in extension and the importance of orientation effects (a a technological goal of spinning) suggest that deformations in spinning processes are elastic (rubbery). It means that the rheological behavior of the material in spinning corresponds to zone III in Fig. 3.3.7, and it is preferable to search for a rheological equation of state from equations discussed in Chapter 4. The time factor can be excluded from the rheological equation of state and material can be treated as rubbery but not as a viscoelastic medium.

The rheological approach has to be combined with the kinetics of crystallization that proceeds under non-isothermal conditions. The crystalline phase influences constants of the rheological equation of state and the spinning process continues up to the line of crystallization (solidification) shown in Fig. 6.4.7. After that, the material solidifies and its deformation is negligible in comparison with stretching along with the distance from a die outlet to the solidification point.

The above physical description of the spinning process seems reasonable, though hitherto a model based on these rheological and kinetics arguments has not been developed.

In discussing the rheology of the fiber spinning process it is also necessary to remember possible instability effects in high-speed stretching (see section 3.5.7).

There are some other technological processes that are based primarily on the extension of polymer materials in a rubbery state; that is:

- formation of blown films
- vacuum and/or pressure thermoforming of plates, trays, cups, and analogous parts from sheets
- blowing of PET bottles from preform above the glass transition temperature of the polymer.

The technology of these processes is based on the same physical phenomena as fiber spinning, i.e., rheology of extension and kinetics of crystallization (or solidification for glassy materials).[71] However, in contrast to fiber spinning, bi-axial extension takes place. One invariant of rheological equations of state of rubbery material (as discussed in Chapter 4) supplemented with the kinetics of phase transition can be a base for an engineering calculation model for these technological processes.

6.4.3.4 Molding technologies

Molding processes are the major periodic manufacturing operations in the polymer industry. An enormous amount of plastic, rubber, and thermoset parts, ranging from automobile bumpers to personal computers, refrigerator housings to bottles and tires, are produced by molding processes. The automotive, appliance, computer, beverage, tire, and other industries are associated with molding.

Rheology plays a very important role in molding processes. In particular, the shear rate and temperature dependence of viscosity determine the resistance to the flow of polymer melts in dies and molds. As the flow rate or output of the process increases, higher shear rates and accordingly higher shear stresses are developed and higher forces and pressures are required to shape polymer products. The sensitivity of viscosity to variations in shear rate, characterized by the shear thinning effect, determines pressure increase during the mold filling process. This means that polymer melts exhibiting a lower power-law index and, therefore, more shear thinning, and show less increase in pressure with an increase of flow rate during processing. Therefore, they will have lower energy consumption. On the other hand, polymer melts exhibiting higher temperature sensitivity of viscosity, i.e., the higher activation energy of viscous flow, would cause a faster increase in pressures and forces required to carry out molding processes, which take place under fast cooling rates. Viscosity, and its shear rate and temperature dependence, determine whether a mold is fully filled or not.

Viscoelasticity of polymer melts has a significant influence on polymer molding and the performance of shaped products, especially with respect to a level of frozen-in molecular orientation introduced in them during cavity filling and packing stages and subsequent relaxation processes occurring upon the cooling stage of the molding cycle. In particular, polymer melts exhibiting a higher relaxation time would lead to a higher level of a frozen-in molecular orientation and residual flow stresses in final products. This strongly affects the performance characteristics of final products. Due to these effects, product properties become highly anisotropic. Depending on the type of products made, the molecular orientation effect may be beneficial or detrimental for product performance. For example, in the case of molding optical products, a higher level of molecular orientation introduces higher anisotropy of refractive index, leading to the deterioration of their optical quality. In particular, this effect of the frozen-in orientation is well known in the case of injection molding of optical products, such as compact disks, CD, and DVD substrates, and various lenses. A high level of frozen-in birefringence or retardation causes distortion of laser light propagation through CD and DVD media, leading to poor quality of reproduction of the sound of music and optical pictures. Therefore, very stringent specifications are established concerning the level of residual optical retardation in these molded products. It should also be noted that this optical retardation is strongly affected by process parameters during molding, as well as relaxation characteristics, such as the relaxation times and their distribution, and optical constants of polymer melts, such as the stress-optical and strain-optical coefficients.[72]

Plastics molding is an industry of enormous volume and scope. There are many variations of molding technology.[73] These processes include compression, injection, injection-compression, co-injection, transfer, resin transfer, blow, rotational molding, and thermoforming. Here we briefly describe some of these processes with the aim of indicating rheological relevance to the calculation of flow kinematics and dynamics.

6.4.3.5 Compression molding

It is one of the oldest techniques of manufacturing rubber, thermoset and plastic products. Compression molding dates back to the origin of the rubber industry. For many years, this has been a standard technique for molding, but recently it has been replaced to some extent by injection molding. By comparison, injection molding offers advantages in material handling and ease of automation. However, compression molding retains a distinct advantage when processing fiber-reinforced polymers. Moderate flow during compression molding helps to avoid high stresses and strains; therefore, reinforcing fibers are not damaged by flow during mold filling. Thus, a high concentration of reinforcing fibers and long fibers can be incorporated into composite materials.

Compression molding involves pressing (squeezing) of deformable material charge between two halves of a heated mold to fill and cure material in the mold, and subsequent part removal (Fig. 6.4.8). In manufacturing thermoset products, the transformation of flowable material into a solid product under elevated mold temperature takes place. Compression molding temperatures range from 140 to 200°C. Mold pressures can vary from 20 to 700 bars and curing times can vary from 1 min for thin parts to over 1 hour for very thick rubber parts. Recently, the development of thermoplastic matrix composites, to produce strong, lightweight structures, has increased interest in compression molding. In thermoplastic matrix composite molding, temperatures as high as 350°C are utilized.

Compression molding is carried out using compression molding presses. Two types of presses are used — down-stroking and up-stroking. Molds usually operate using a clamping ram or cylinder with clamping capacities ranging from a few tons to several

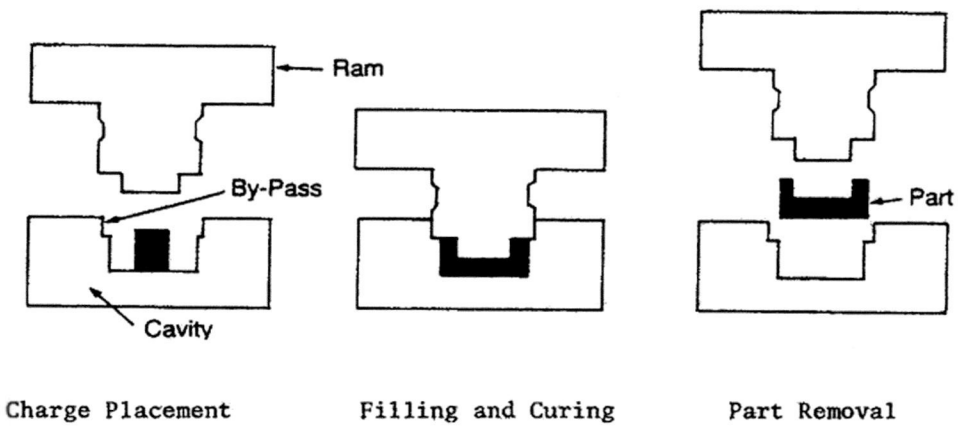

Charge Placement   **Filling and Curing**   **Part Removal**

Figure 6.4.8. Schematic representation of the compression molding process.

thousand tons. In addition to the clamping capacity, two other characteristics of a press are the daylight characterizing maximum platen separation, associated with stroke, and the platen size ranging from a few centimeters to several meters. The temperature of platens is controlled by built-in heating or cooling elements or by separate heaters.

There are five stages of the compression molding process:
- material preparation
- prefill heating
- mold filling
- in-mold curing
- part removal.

Material preparation includes compounding resin with fillers, fibers, and other ingredients or impregnating a reinforcing cloth or fibers with resin. This stage controls the rheology of material and the bonding between fibers and resin. The prefill heating stage is carried out to speed up the molding process. This stage can occur outside or inside the mold before the mold is closed and flow begins. Mold filling starts with material flow and ends when the mold is full. The effect of flow is critical for the quality and performance of the molded product. It controls the orientation of fibers, which has a direct effect on the mechanical properties of the part. In processes involving lamination of the long fiber-reinforced composites, there is little flow since the initial charges almost completely conform to the mold. In the case of a thermoset matrix, some curing may occur during the mold-filling stage. The in-mold-curing stage follows mold filling. In this stage, the part is cured in the mold while the final stage of cure may be completed during post-cure heating after part removal. In-mold curing converts the polymer from a liquid to a solid having rigidity sufficient for removal from the mold. Part removal and cool-down are the final stages. This stage plays an important role in the warpage of the part and residual stress development, which arise due to difference in thermal expansion in different portions of the part. Temperature distribution and rate of cooling affect these residual stresses.

Fig. 6.4.9 shows a typical curve of variation of plunger force required for mold closing as a function of time at a constant closing rate during molding of polymers not con-

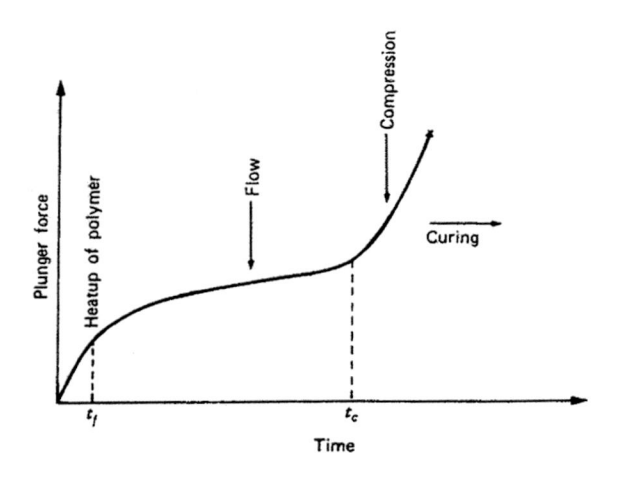

Figure 6.4.9. Schematic representation of the plunger force during compression molding at a constant mold closing speed.

taining fibers. In the first region, at time $t < t_f$, corresponding to softening of the material, the force increases rapidly as the preform is squeezed and heated. At t_f, the polymer in the molten state is forced to flow into the cavity and fill it. The filling is completed at t_c, corresponding to the initiation of curing. At this stage, compression of polymer melt occurs to compensate for the volume contraction due to curing.

To illustrate how the flow process can be described during compression molding, an idealized case of squeezing flow of a strip or disk blank is considered below. A *preform* in the form of a strip or disk is placed between parallel plates or disks.

Let us assume that in the case of a strip, fluid is confined between two sides in the width direction and the upper plate is moving toward the lower plate at a constant velocity $\dot{h} = dh/dt$, such that fluid is squeezed and forced to flow to fill the mold in the length direction. The fluid is Newtonian and the flow is laminar. The fluid adheres to the surface of the plates. The force required to fill the strip mold is

$$F = \frac{8\eta l^3 B \dot{h}}{h^3} \qquad [6.4.22]$$

where η is fluid viscosity, l is the filling length, B is the width of mold, and h is the current separation of plates. This equation can be used to determine force as a function of the closing velocity. If the squeezing process takes place under the force, F, being constant, then Eq. 6.4.22 is converted to a nonlinear ordinary differential equation for an unknown function of $h(t)$.

Let us assume that fluid in the form of a disk is confined between two parallel disks. The upper disk is moving under a constant velocity of $\dot{h} = dh/dt$ such that fluid is squeezed to fill the disk mold. The force required to fill the disk mold is

$$F = \frac{3\pi\eta R^4 \dot{h}}{8h^3} \qquad [6.4.23]$$

where R is a radius of the parallel disks.

This equation is known in the literature as the *Stefan equation*, indicating how much force is required to fill a disk cavity. If the process of filling a disk mold takes place under a constant force, F, then Eq. 6.4.23 is transferred to a nonlinear differential equation for an unknown function of $h(t)$.

E.4.3.6 Injection molding

It is one of the most widely employed molding processes. Injection molding is used for the processing of thermoplastics, elastomers, thermosets, ceramics, and metals to articles of various complexities. The advantages of injection molding are a high production rate, large volume manufacturing with little or no finishing operations, minimum scrap, and good dimensional tolerances.

Injection molding of thermoplastics includes automatic feeding of pellets into a hopper, melting, melt plasticizing, and feeding melt into an injection barrel at a temperature above the glass transition temperature, T_g, for amorphous polymers or melting point, T_m, for semi-crystalline polymers. The melt is then injected through a delivery system consisting of nozzle, sprue, runner system, and gate or gates into a mold having a temperature below T_g or T_m. The melt solidifies in the mold. Then, the mold is opened and the molded product is ejected.

Injection molding of elastomers includes automatic feeding of a preheated or plasticated rubber stock into an injection barrel at a temperature below the vulcanization temperature. Then, rubber is injected through a delivery system into a mold. The mold temperature is kept high enough to initiate vulcanization and subsequently vulcanize rubber inside the mold. After the rubber has been vulcanized, the mold is opened and the molded part is ejected.

Injection molding of thermosets and reactive fluids, which are able to form infusible crosslinked structures by irreversible chemical reactions, is also carried out using a hot mold. Reaction injection molding is characterized by in-mold polymerization from monomeric or oligomeric liquid components by a fast polymerization reaction. Thermosets are solid or highly viscous materials at ambient temperature. They are frequently highly filled.

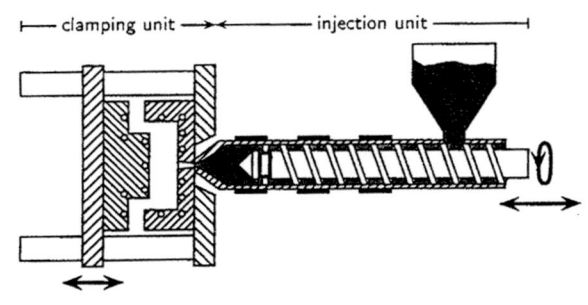

Figure 6.4.10. Schematic representation of the injection molding machine.

An injection molding machine consists of a clamping unit containing mold and an injection unit for feeding, melting, and metering thermoplastic material (Fig. 6.4.10). The most widely-used injection units utilize rotating screws to plasticize material. Rotation of the screw causes the plasticized material to accumulate in front of the screw, which is pushed back. The material is injected by the forward motion of the screw acting as a plunger, which pushes melt into the mold. The mold serves two functions: it imparts shape to the melt and cools the injection-molded part. The mold consists of cavities and cores and a base in which they are located (Fig. 6.4.11). The mold contains one or more cavities with stationary and moving mold halves. In many cases, molds may have multiple cavities. The latter is dictated by process economics.

The connection between runner and cavity is called a gate. In mold-making, the gate design is important. The size and the location of the gate are critical. The gate should allow the melt to fill the cavity and deliver additional melt to prevent shrinkage caused by

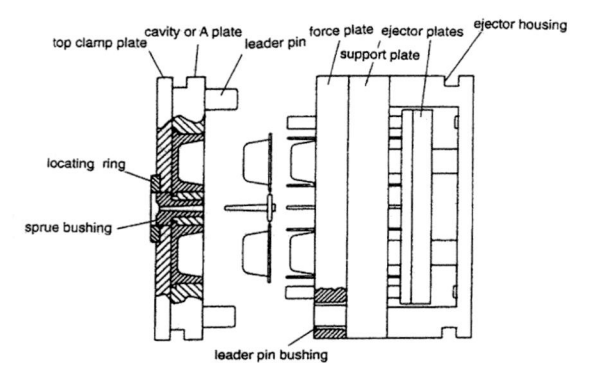

Figure 6.4.11. Schematic representation of a cold-runner, two-plate injection mold.

cooling. The material in the gate should freeze at an appropriate time during the molding cycle. Premature freezing will cause an undesirable phenomenon called underpacking, leading to excessive shrinkage and sink marks. The mold also requires cooling and/or heating system and venting to remove air during the cavity filling and rapid and uniform cooling. Venting is usually achieved by arranging small gaps in the parting line, which permit air to escape quickly. In some cases, forced removal of air is carried out by using vacuum venting. Mold cooling or heating is achieved by placing a number of channels in both halves of the mold through which cooling or heating liquid flows to remove heat from the melt or to add heat to the melt. Mold heating is also done by placing electric cartridge heaters in mold halves.

The injection molding cycle can be divided into three stages. These include cavity filling, packing (holding), and cooling. The three stages of the molding cycle can be easily seen from Fig. 6.4.12, indicating schematically the pressure variation with time. In the filling stage, the pressure rises as the melt propagates into the cavity. This stage is followed by the packing stage where a rapid increase (typically within 0.1 s) of pressure to its maximum is observed. Then, the cooling stage takes place at which pressure slowly decays.

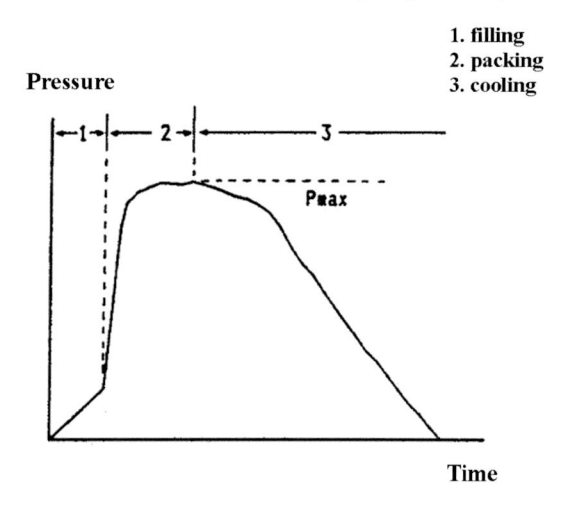

Figure 6.4.12. Schematic representation of the pressure-time curve during filing, packing and cooling stages of injection molding.

Molding variables such as injection speed, melt and mold temperatures, packing or holding pressure, and length of the packing stage have a strong influence on pressure development and properties of moldings. Frozen-in molecular orientation, residual stresses, polymer degradation, shrinkage, warpage, and weld line strength are influenced by process variables. In injection molding of semi-crystalline polymers, the molding variables strongly affect crystallinity and microstructure development in moldings, which influence their performance characteristics.

A simple isothermal analysis of the mold-filling process in cavities of simple geometries such as tubular and strip mold cavity is outlined below.

Consider a straight tubular mold of radius, R, and length, L. Melt is the Newtonian fluid that is injected at constant pressure or constant rate into the mold. The melt progresses along the mold until it reaches its end. We need to calculate the melt front position and the instantaneous flow rate when constant pressure is imposed at the mold entrance or pressure as a function of time when a constant velocity is imposed during the molding process. The fluid is incompressible and isothermal under a fully-developed flow. If constant pressure is imposed at the entrance of the tubular cavity, the penetration depth is

$$z(t) = \frac{R\sqrt{p_0 t}}{2\sqrt{\eta}}$$
[6.4.24]

where t is the filling time. It is seen that the penetration depth z(t) is proportional to the radius R.

When a constant velocity is imposed during the cavity filling process, the pressure required to fill mold is

$$p(t) = \frac{8\eta Q^2}{\pi^2 R^6} t$$
[6.4.25]

The pressure required to fill a tubular mold cavity is reciprocally proportional to its radius to the power of six.

For filling a strip cavity of thickness, h, under a constant injection pressure, p_0, the penetration depth is

$$z(t) = \frac{h}{2}\sqrt{\frac{p_0 t}{3\eta}}$$
[6.4.26]

The penetration depth z(t) is linearly proportional to the value of h.

For filling a strip cavity at a constant flow rate, the pressure variation with the filling time is

$$p(t) = \frac{12\eta Q^2}{B^2 h^4} t$$
[6.4.27]

The pressure varies linearly with the filling time and strongly depends on the width, B, and the cavity thickness, h, being reciprocally proportional to the width to the power of two and to the thickness to the power of four.

The process of calculation of filling patterns occurring in real molding processes is more complicated than simple geometries and flow of Newtonian fluid under isothermal conditions discussed here. Molding processes are more complicated because of the two- and three-dimensional flow of non-Newtonian viscoelastic fluid under non-isothermal conditions with solidification, crystallization, crosslinking occurring during processes. Various software is proposed for the calculation of flow in molding processes based on rheological properties of materials and designs of machines and molds.

The movement of material during mold filling can be so intensive that high deformations result in liquid-to-solid-like transition (see section 3.6.3). This problem is treated
sometimes from the point of view of the stability of the stream. The theoretical analysis
includes constitutive equations.[74]

6.4.3.7 Injection-compression molding

The injection-compression molding technique has been developed to utilize the advantages of both molding techniques. This technique utilizes the conventional injection molding machine and a compression attachment. At first, the polymer melt is injected in order
to partially fill the mold, which is partially open. Then, the compression stage is introduced which leads to the final closing of the mold by squeezing the flow of the melt. This
compression stage is introduced to replace the packing stage of conventional injection
molding. Since the pressure developed during the compression stage is significantly lower
than that in the packing stage of conventional injection molding, injection-compression
molding introduces lower residual stresses, lower molecular orientation and birefringence,
less and more even shrinkage, and better dimensional tolerances. At the same time, this
process maintains high output, good process control, and automation inherent to conventional injection molding. The process is especially useful for molding thin parts that
require high quality and accuracy. However, the process requires careful timing of injection clamp position and force. Injection-compression molding is presently employed in
making optical disks (DVD and CD) where requirements for dimensional tolerances and
optical retardation are very stringent. In the production of the optical disks, this process is
called *coining*. In comparison with injection molding, there are few little experimental and
theoretical studies in the literature on injection-compression molding.

REFERENCES

1   The number of publications devoted to the investigation of rheological properties of polymeric substances
     may be as high as several thousand. Therefore, publications discussed in this chapter include, in most
     instances, only the most recent data. There are several monographs reviewing the current status of this
     field, and for those who are especially interested in polymer rheology, these monographs may be of
     interest. The results of many early fundamental investigations are discussed in the book: G.V. Vinogradov.
     A.Ya. Malkin, **Rheology of Polymers**, *Springer*, Berlin, 1980. The most recent comprehensive reviews of
     this subject are included in book by C.W. Macosko, **Rheology: Principles. Measurements, and
     Applications**, *VCH*, NY, 1994.
2   H.J.M.A. Mieras, C.F.N. van Rijn, *Nature*, **218**, 865 (1968); N.J. Mills, *Nature*, **219**, 1249 (1968).
3   This effect is well documented for metallocene-catalyzed polyethylene, see e.g., E.E. Bin Vadud,
     D.G. Baird, *J. Rheol.*, **44**, 1151 (2000).
4   D. Ferri, P. Lomellini, *J. Rheol.*, **43**, 1355 (1999).
5   B.J. Crossby, M. Mangnus, W. De Groot, R. Daniels, T.C.B. McLeish, *J. Rheol.*, **46**, 401 (2002).
6   A. Ya. Malkin, G. S, Simon, S.O. Ilyin, M.P. Arinina, V.G Kulichikhin, J. Sjoblom, Some Compositional
     Viscosity Correlations for Crude Oils from Russia and Norway, *Energy and Fuels*, 2016,
     DOI: 10.1021/acs.energyfuels.6b02084
7   E.A. Kisanov, S.V. Remozov, *Rheol. Acta*, **38**, 172 (1999). E.A. Kisanov, S.V. Remozov, *Rheol. Acta*, **38**,
     172 (1999).
8   Quality of industrial engine oils is characterized by different methods, but not only their viscosity.
     However, the latter is the single most important rheological parameter used in engineering applications.
9   SAE – Society of Automotive Engineers (USA).
10   For example, if x = 25, oil can be used at temperatures higher than 0°C. If x = 5, it means that oil can be
     used at temperatures higher than -30°C.
11   R.M. Webber, *J. Rheol.*, **43**, 911 (1999).
12   This approach was studied in many works. Earlier publications along this line are R. Jongepier,
     B. Kuilman, *Rheol. Acta*, **9**, 460 (1970); A.I. Isayev, V.A. Zolotarev, G.V. Vinogradov, *Rheol. Acta*, **14**, 135

(1975). One of the latest is: L.-I. Palade, P. Attené, S. Camaro, *Rheol. Acta*, **39**, 180 (2000). The results in the last publication show that the time-temperature superposition principle has only qualitative meaning for asphalt and asphalt-based mastics.

13   M.P. Wolarowitch, K.I. Samarina, *Kolloid Z.*, **70**, 280 (1935).

14   N.M. Edwards, B. Peressini, J.E. Dexter, S.J. Mulvaney, *Rheol. Acta*, **40**, 142 (2001) and M. Keentok, M.P. Newberry, P. Gras, F. Bekes, R.I. Tanner, *Rheol. Acta*, **41**, 173 (2002).

15   This list of standard methods was referred by H.A. Barnes (3rd National Meeting of Sociedade Portuguesa de Reologia, 2001).

16   M.G. Corradini, V. Stern, T. Suwonsichon, M. Peleg, *Rheol. Acta*, **39**, 452 (2000).

17   D. Gabriele, B. De Cindio, P. D'Antona. *Rheol. Acta*, **40**, 120 (2001).

18   C. Bower, C. Gallegos, W.R. Mackley, J.M. Madiedo, *Rheol. Acta*, **38**, 145 (1999). This paper contains typical results of rheological characterization of oil-in-water emulsions in the linear and nonlinear viscoelastic regions. Many food products, including mayonnaise, are emulsions of this type.

19   B. Abu-Jdayil, *Rheol. Acta*, **41**, 441 (2002).

20   Chocolate mass is a suspension of various components, such as sugar, milk powder, cocoa, in a liquid phase (cocoa butter).

21   According to a working regulation of Office International du Cacao, du Chocolate et de Confiserie (1973).

22   G.P. Citerne, P.J. Carreau, M. Moan, *Rheol. Acta*, **40**, 86 (2001).

23   J. Muños, N.E. Hudson, G. Vélez, M. Del C. Alfaro, J. Ferguson, *Rheol. Acta*, **40**, 162 (2001).

24   M.E. Yildiz, J.L. Kokini, *J. Rheol.*, **45**, 903 (2001).

25   A. Ponton, O. Clement, J. L.Grossiord, *J. Rheol.*, **45**, 521 (2001).

26   N. Jager-Lazer, J.F. Tranchant, V. Alard, C. Vu, P.C. Tchoreloff, J.L. Grossiord, *Rheol. Acta*, **37**, 129 (1998).

27   Another types of "super-concentrated" emulsions are so-called "emulsion explosives"; see: H.A. Bamfield, J. Cooper, in **Encyclopedia of Emulsion Technology**, v. 7. *Marcel Dekker*, New York, 1985; some data concerning rheological properties of such unusual emulsions are described in: A.Ya. Malkin, I. Masalova, P. Slatter, K. Wilson, *J. Non-Newton. Fluid Mech.*, **112**, 101 (2003); A.Ya. Malkin, I. Masalova, P. Slatter, K. Wilson, *Rheol. Acta*, **43**, 584 (2004). See illustrations of rheological properties of such emulsions in section 3.5 (Figs. 3.5.12 and 3.5.13).

28   O.K. Baskurt, H.J. Meiselman, Blood rheology and hemodynamics, *Seminars in Thrombosis and Haemostatsis*, **29**, 5, 435-50 (2003); "Blut ist ganz bezonderer Saft" – Mephistopheles in Goethe's Faust.

29   C. Picart, J.-M. Piau. H. Galliard, P. Carpenter, *J. Rheol.*, **42**, 1 (1998).

30   Ch. Ancey, H. Jorrot, *J. Rheol.*, **45**, 297 (2001).

31   Experimental results of such kind can be found in numerous publications. This figure is taken from: G.V. Franks, Z. Zhou, N.J. Duin, D.V. Boger, *J. Rheol.*, **44**, 759 (2000). Analogous data were obtained in: J.-D. Lee, J.-H. So, S.-M. Yang, *J. Rheol.*, **43**, 1117 (1999).

32   M.K. Chow, C.F. Zukoski, *J. Rheol.*, **39**, 15 and 33 (1995).

33   V.T. O'Brient, M.E. Mackley, *J. Rheol.*, **46**, 557 (2002).

34   B.J. Maranzano, N.J. Wagner, *J. Rheol.*, **45**, 1205 (2001).

35   P. Coussot, Q.D. Nguyen, H.T. Huynh, D. Bonn, *J. Rheol.*, **46**, 573 (2002).

36   O.Seidal, F. Bagusat, H.-J.Mogel, *Rheol. Acta*, **38**, 305 (1999). This effect was described for 30% kaolin suspension and explained as a result of the agglomeration of kaolin particles.

37   The first publication devoted to electro-rheological liquids was by W.M. Winslow, *J. Appl. Phys.*, **20**, 1137 (1949) and to magneto-rheological effect was by J. Rabinow, *AIEE Trans.*, **67**, 1308 (1948).

38   S.L. Vieira, L.B.P. Neto, A.C.F. Aruda, *J. Rheol.*, **44**, 1139 (2000).

39   Y. Chen, A.F. Sprecher, *J. Appl. Phys.*, **70**, 6796 (1991).

40   H. See, A.Kawai, F. Ikazaki, *Rheol. Acta*, **41**, 55 (2002).

41   P.J. Rankin, A.T. Horvath, D.J. Klingenberg, *Rheol. Acta,* **38**, 471 (1999).

42   L. Marshal, C.F. Zukoski, IV, J.W. Goodwin, *J. Chem. Soc., Faraday Soc. Trans.*, **85**, 2785 (1989).

43   B.D. Chin, J.H. Park, M.H. Kwon, O.O. Park, *Rheol. Acta*, **40**, 211 (2001).

44   H.P. Gavin, *J. Non-Newton. Fluid Mech.*, **71**, 165 (1997).

45   A. Inoue, S. Maniwa, *J. Appl. Polymer Sci.*, **55**, 113 (1995); H. Orihava, A.Taki, M. Doi, *J. Rheol.*, **45**, 1479 (2001).

46   For comprehensive review of results of rheokinetic investigations, see monograph: A.Ya. Malkin, S.G. Kulichikhin, **Rheokinetics: Rheological Transformations in Synthesis and Reactions of Oligomers and Polymers**, *Hüthig & Wepf*, Germany, 1996.

47   See any standard textbook on polymer chemistry, for example, G. Odian, **Principles of Polymerization**, *Wiley*, New York, 1981.

48   These experimental data are related to anionic polymerization of ω-dodecalactame resulting in the synthesis of polyamide-12.

49   Y. Ide, J.L. White, *J. Appl. Polymer Sci.*, **18**, 2997 (1974).
50   See Ref. 94 in Chapter 3.
51   H.H. Winter, F. Chambron, *J. Rheol.*, **30**, 367 (1986); F. Chambron, H.H. Winter, *J. Rheol.*, **31**, 683 (1987);
     H.H. Winter, *Polym. Engng Sci.*, **27**, 1698 (1987); H.H. Winter, Gel point in **Encyclopedia of polymer
     science and technology** (1989).
52   M. Adam, M. Delsanti, D. Durand, G. Hild, J.P. Much, *Pure Appl. Chem.*, **53**, 1489 (1981); D. Stauffer,
     A. Coniglio, M. Adam, *Adv. Polymer Sci.*, **44**, 103 (1982).
53   A.Ya, Malkin, I.Yu. Gorbunova, M.L. Kerber, *Polym. Eng. Sci.*, **45**, 95 (2005).
54   A.Ya. Malkin, *Plastmasssy* (in Russian), No 4, 47 (1982); A.Ya. Malkin, V.P. Begishev, *Polym. Process
     Eng.*, **1**, 83 (1983); A.Ya. Malkin, G.I. Shuvalova, *Vysokomol. Soedin.*, (in Russian), **27**, 865 (1985).
55   T-T-T diagrams as a method of analysis of rheokinetic data in curing were proposed and developed in:
     P.G. Babayevsky, J.K. Gillham, *J. Appl. Polym. Sci.*, **17**, 2067 (1973); J.K. Gillham, *Polym. Eng. Sci.*, **19**,
     676 (1979); 26th Intern. Congress Macromol., Strasburg, Abstracts, 2, 1292 (1981) and *Polym. Mater. Sci.
     Eng.*, ACS Div. Polymer Materials, Spring Meeting, **54**, 4 and 8 (1986).
56   M.R. Kamal, S. Sourour, *Polym. Eng. Sci.*, **13**, 59 (1973); M.R. Kamal, M.R. Ryan, *Polym. Eng. Sci.*, **20**,
     859 (1980); J.S. Deng and A.I. Isayev, *Rubber Chem. Technol.*, **64**, 296 (1991).
57   This method was first proposed in A.Ya. Malkin, V.P. Begishev, V.A. Mansurov, *Vysokomol. Soedin.* (in
     Russian), **26**, 869 (1984) and then independently discussed in: S.N. Ganenvala, C.A. Rotz, *Polym. Eng.
     Sci.*, **27**, 165 (1987); E.K. Holby, S.K. Venkataraman, H.H. Winter, *J. Non-Newton. Fluid Mech.*, **27**, 17
     (1988); M. Wilhelm, D. Maring, H.-F. Spiess, *Rheol. Acta*, **37**, 399 (1998); M. Wilheim, P. Reinheimer,
     M. Ortseifer, T. Neidhofer, H.-W. Speiss, *Rheol. Acta*, **39**, 241 (2000). See: A.Ya. Malkin, *Rheol. Acta*, **43**,
     1 (2004).
58   See also the second publication in Ref. 96 in Chapter 3, where polymer transformations in very dilute
     solutions were proposed for study by a rheokinetic method based on the Toms effect.
59   Many important problems related to this line of application of rheology are discussed in details in the book
     R.I. Tanner, **Engineering Rheology**, 2nd Ed., *Oxford University Press*, New York, 2000.
60   These principles were first formulated in an important and fundamental paper of J.G. Oldroyd (1921-1982)
     J.G. Oldroyd, *Proc. Royal Soc. London*, **A200**, 523 (1950). Later the general principles of constructing
     rheological equations of state were discussed in review: C. Truesdell, W. Noll, The non-linear field theories
     of mechanics, in **Handbuch der Physik**, III/3, E. Fluegge (Ed.), *Springer*, Berlin, 1965, based on earlier
     publications of the authors.
61   This principle was first advanced by S.K. Zaremba, *Bull. Acad. Sci. Cracovie*, **85**, 380 (1903).
62   A.Ya. Malkin, V.A. Patlazhan, *Adv. Colloid Interface Sci.*, **257**, 42 (2018).
63   This equation was written as $\sigma = k\overset{\cdot}{\gamma}^{n}$. The slightly modified but equivalent form of this equation is more
     convenient $\dot{\gamma} = K\sigma^{m}$; evidently m = n^{-1}. This form of the power-law equation is used.
64   Detailed study of rheological properties of this emulsion can be found in: A.Ya. Malkin, I. Masalova,
     P. Slatter, K. Wilson, *Rheol. Acta*, **43**, (2004).
65   A.Ya. Malkin, I. Masalova, D. Pavlovski, P. Slatter, *Appl. Rheol.*, **14**, 89 (2004).
66   This theory is discussed in detail in many textbooks. See for example Z. Tadmor and C. Gogos, **Principles
     of Polymer Processing**, *Wiley*, New York, 1979.
67   The first simplest dynamic theory of pumping extruders was proposed in: J.F. Carley, R.S. Mallouk,
     J.M. McKelvey, *Ind. Eng. Chem.*, **45**, 974 (1953).
68   This problem is discussed in details in the book: R.R. Huilgol, N. Phan-Thien, **Fluid Mechanics of
     Viscoelasticity**, *Elsevier*, Amsterdam, 1997, as well as in numerous scientific publications.
69   Theory of plastic calendering is principally close to the theory of metal rolling based on a model of plastic
     media. The simplest theory of rolling of Newtonian liquids was proposed in R.F. Gaskell, *J. Appl. Mech.*,
     17, 334 (1950) and later was improved in many publications related to flow of more realistic models of
     non-Newtonian liquids.
70   The basic systematic analysis of complex processes taking place in fiber spinning was first summarized in
     the book: A. Ziabicki, **Fundamentals of Fiber Formation**, *Wiley*, New York, 1976.
71   A model of the film blowing process was proposed by I.A. Muslet and M.R. Kamal. *J. Rheol.*, **48**, 525
     (2004). This computer simulation clearly demonstrates a necessity to incorporate a complete description of
     rheological properties as well as physical processes (crystallization) for creating the realistic model of the
     technological process.
72   A.I. Isayev, C.A. Hieber, *Rheol. Acta*, **19**, 168 (1980); A.I. Isayev, *Polym. Eng. Sci.*, **23**, 271 (1983);
     A.I. Isayev, D.L. Crouthamel, *Polym. Plastics Technol. Eng.*, **22**, 177 (1984); G.D. Shyu, A.I. Isayev and
     C. T. Li, *J. Polym. Sci., Phys. Ed.*, **41**, 1850 (2003); G.D. Shyu, A.I. Isayev and H.S. Lee, *Korea-Australia
     Rheol. J*, **15**, 159 (2003).
73   A.I. Isayev, in **Handbook of Industrial Automation**, Ed. R.L. Shell and E.L. Hall, eds., *Marcel Dekker*,
     New York, 2000, Chapter 6.8; More extensive description of these technologies are given in various

references provided in Introduction.
74   A.C.B. Bogaerds, M.A. Hulsen, G.W.M. Peters, F.P.T Baaijens, *J. Rheol.*, **48**, 765 (2004).

QUESTIONS FOR CHAPTER 6

QUESTION 6–1

Compare sensitivity of different rheological methods to variation of molecular mass.

QUESTION 6–2

Explain why the MFI of polymer used for film extrusion must be higher than in tube extrusion (see Fig. 6.2.3).

QUESTION 6–3

Is it possible to vary the elasticity of melt without changing the average molecular mass of polymer?

QUESTION 6–4

Explain the advantage of synthetic lubricants in comparison with mineral oil-based lubricants.

QUESTION 6–5

Derive Eq. 6.4.5. As an intermediate result, obtain an equation for a radial velocity distribution.

QUESTION 6–6

How does the velocity profile during flow in a tube change during the transfer from Newtonian to a power-law type liquid? Make a comparison by analyzing the ratio of maximum to the average velocity.

QUESTION 6–7

Prove Eq. 6.4.6 for the Bingham liquid. Explain the necessary conditions required for the movement of the Bingham viscoplastic media through a tube.

QUESTION 6–8

Eq. 6.4.5a is convenient for solving engineering problems of the flow of non-Newtonian liquids through channels with non-circular cross-sections. Can a similar equation be formulated for viscoplastic liquids? Explain your answer.

QUESTION 6–9

Why does elasticity appear before gelation during polymer curing, i.e., the formation of a three-dimensional chemical network?

QUESTION 6–10

Describe what happens if an end of the pumping screw of an extruder is blocked? What pressure will be developed?

Answers can be found in a special section entitled Solutions.

NOTATION

A	intensity of dissipation (in a flow)
A	area of an ellipse in a dynamic experiment
A	constant in some equations
$A_{(n)}$	the Rivlin-Erickson tensor of the n-th order
a	thermal diffusivity
a	dimensionless amplitude in the theory of vibration viscometry
a	arbitrary coefficient, constant in different equations
a, b	semi-axes of an ellipse
a	length of a side of a triangle
a_T	temperature shift factor
a_1, a_2, a_3	components of the acceleration vector
a_{ij}	components of the gradient of velocity tensor (i = 1, 2, 3 and j = 1, 2, 3)
B	bulk modulus of elasticity (compressibility)
B^*	complex bulk modulus of elasticity
B	magnetic flux density
$B_{(n)}$	the White-Metzner tensor of the n-th order
B	width (of a slit channel)
B	constant in some equations
b	baric coefficient of viscosity
b	power factor in a molecular model
b	arbitrary coefficient, constant in different equations
C	electrical capacitance (in a model representation)
C	step in a discrete relaxation spectrum
C	(with different indexes) coefficients in different equations
C_{ij}	the Cauchy-Green tensor (i = 1, 2, 3 and j = 1, 2, 3)
C_{ij}^{-1}	the Finger tensor (i = 1, 2, 3 and j = 1, 2, 3)
$C_{I, inv}$	the first invariant of the Cauchy-Green tensor
$C_{I, inv}^{-1}$	the first invariant of the Finger tensor
C_k^{∇}	the Jaumann tensor derivative
c	concentration (in volume units)
c^*	critical concentration (in liquid-crystal solutions)
c, c_p	heat (thermal) capacity
c	arbitrary coefficient
D	diameter (of a tube, capillary)
D_{ij}	components of the deformation rate tensor (i = 1, 2, 3 and j = 1, 2, 3)
D_2	the second invariant of the deformation rate tensor

DR	draw ratio
d_{ij}	components of a small deformation tensor ($i = 1, 2, 3$ and $j = 1, 2, 3$)
d_1, d_2, d_3	principal values of a small deformation tensor
$d_{ij}^{(dev)}$	deviatoric part of the d_{ij} tensor
d_s	surface area based average diameter
E	functional of errors (in fitting experimental data)
E	number of entanglements per chain
E_1, E_2, E_3	invariants of a tensor of large deformations
E	measure of large non-linear deformations
E	Young's (elastic) modulus
E_0	instantaneous modulus
E^*	complex dynamic modulus (in extension)
E'	real part of complex modulus (storage modulus)
E''	imaginary part of complex modulus (loss modulus)
E_∞	equilibrium modulus
E_N^0	plateau modulus
E_{ijmn}	components of elastic modulus for an anisotropic material
E_a	activation energy
E	strength of an electrical field
F	kernel function in different models of mixing
F	force
F_x, F_y	components of forces acting along coordinate axes
F^*	limiting (critical) value of force
F_0	constant force; initial force
F_Y	strength of the material
F_{sp}	force acting on a spring in model representation
F_{pist}	force acting on a piston in model representation
F_1, F_2, F_3	components of force vector
F_n	normal force
F_σ	tangential force
f	coefficient of friction
f	frequency of oscillation (in Hz)
f_0	local coefficient of friction
f_E	engineering stress
f_M	neo-Hookean engineering stress
$f(x)$	arbitrary function of argument x
G	shear modulus
G	constants in the theories of large deformations (with different indices)
G	weight output
G_e	elastic (rubbery) modulus
G_i, G_n	partial shear modulus in a discrete relaxation spectrum
G_N^0	plateau shear modulus
G_r	relaxation modulus
G_∞	equilibrium shear modulus
G_∞	the final value of shear modulus in curing

G_0	instantaneous shear modulus
$G(\theta)$	relaxation time spectrum (measured in shear)
G^*	complex dynamic modulus (in shear)
G'	real component of complex shear modulus (storage shear modulus)
G''	imaginary component of complex shear modulus (loss shear modulus)
G'_{exp}, G''_{exp}	experimental values of G' and G'', respectively
G'_{cal}, G''_{cal}	calculated values of G' and G'', respectively
g	gravitational constant
g_i^σ	the measured value of a function in a regularization method
g_{ij}	components of tensor of relative displacements ($i = 1, 2, 3$ and $j = 1, 2, 3$)
H	distance; height; sagging
H	heat of transition
h	distance, gap, height
$h(\ln\theta)$	(with different indices) logarithmic relaxation spectrum
I	moment of inertia
[I]	concentration of an initiator
I_s	coupled moment of inertia
I_1, I_2, I_3	invariants of the stress tensor
I_E	the sum of the non-linear measures of deformations, E
I_{opt}	intensity of light transmission
J	compliance; viscoelastic compliance
J_c	creep compliance
J_0	instantaneous compliance
J_e	equilibrium shear compliance
J_s	steady state compliance
J_s^0	initial (linear) value of compliance
J^*	complex compliance (in shear)
J'	real part of complex compliance
J''	imaginary part of complex compliance
$J(\lambda)$	retardation time spectrum
J_i	partial compliance
J	electrical current
K	empirical parameters (with different indices)
K	kernel in an integral equation used in a regularization method
K	shape factor, geometrical factor, or form factor
K_{cr}	the Von Mises criterion of plasticity
K_H	the Huggins constant
K_M	the Martin constant
K_K	the Kraemer constant
k	kinetic rate constant (with different indices)
k	the Boltzmann constant
k	temperature coefficient of viscosity
k	coefficient in different equations (with different indices)
k	parameter in the theory of oscillations
$k(\ln\lambda)$	logarithmic retardation spectrum

k	ratio of radii
k_u	factor in the theory of vibration viscometry
k_ω	coefficient of resistance in vibrations
$L_{a,b}$	rate of formation of junctions in a network model
L	length
L	parameter in the theory of torsion oscillations
l	current length
l	coefficients in different equations (with different indices)
l_0	initial length
l_f	length of a sample after elongational flow
M or MM	molecular mass
$\overline{M}_n, \overline{M}_w, \overline{M}_z, \overline{M}_{z+1}$	number-, weight-, z- and (z+1)- molecular masses, respectively
$\overline{M}_{w,bl}$	the weight-averaged molecular mass of a blend
$\overline{M}_\eta$	viscometric-averaged molecular mass
M_0	molecular mass of a monomer unit
M_c	critical molecular mass (for entanglement formation)
M_e	molecular mass of chain segment between entanglements
M	number of elements in the model
MFI	melt flow index
m, m_e	reduced values of molecular mass
m_k	entrance correction factor
m	mass
m	memory function
m	arbitrary index
m	exponent in different empirical equations
m_0	reduced mass
m_s	doubled, coupled mass
N	number of elements in model or in a polymer chain
N	rotational speed of the screw
N_1	first normal stress function
N_1^+	first normal stress growth function
N_1^-	the first normal stress decay function
$N_{1,c}$	permanent component in the first normal stress function in oscillations
$N_{1,osc}$	amplitude of oscillations of the first normal stress function in periodic deformation
N_2	second normal stress function
N_c	intensity of nuclei formation
n_1, n_2, n_3	components of the normal to a surface
n	ordinary number in different sequences
n	exponent in different empirical equations
n	refraction index
n_∞	limiting value of refraction index
P, p	pressure at the entrance to a channel; current pressure
P_k	pressure drop responsible for kinetic energy losses; pressure drop responsible for end-correction

P_v	part of the pressure drop responsible for the resistance of a channel
p	amplitude ratio (in linear displacement or in torsion)
P_{max}	maximum pressure developed in a single-screw extruder
Q	volume output
Q	electrical charge
Q_0	initial charge of the capacitor in a model representation
Q_∞	equilibrium charge of the capacitor in a model representation
q	distributed load
q	parameter of a molecular model
R	universal gas constant
R	radius
R_i, R_o	inner and outer radii of cylinders, respectively
$\bar{R}$	average radius
R_0	initial radius of a sample in a squeezing plastometer
R	electrical resistance (in a model representation)
R_m	real (active) part of mechanical impedance
r	(current) radius; radial coordinate
r_0	initial radius
r_*	radius of a plasticity zone in the flow of a viscoplastic liquid
S	surface area in different equations: rectangular in dynamic measurements, under a relaxation curve and in other cases
S_R	swell ratio
s	distance between two points
s	scaling factor
T	torque
T^*	limiting value of torque
T	intensity of shear stresses
T	absolute temperature (in K)
T	period of oscillation
$\hat{T}$	dimensionless induction period (in curing)
T_{cef}	characteristic time of deformation
T_{ent}	life-time of entanglements
T_0	reference temperature
T_g	glass transition temperature
T_m	phase transition temperature, melting point
T_m^*	equilibrium phase transition temperature
t, t'	current and past time, respectively
t^*	critical time (in different applications)
t^*	gel-time
t_{m-g}	time of micro-gelation
t_{inh}	inherent time scale of material
t_{obs}	characteristic time of observation
t_{sw}	characteristic time of switching
t_n	non-isothermal induction period (in curing)
U	voltage

U_0	initial voltage
U_0	velocity of the movement of meniscus
U_∞	steady velocity (in different cases)
u	displacements of elements of a molecular model (with different indices)
u	current velocity
u_1, u_2, u_3	components of the displacement vector
u_0	speed of a jet flight
V	average deviation in a regularization method
V, V_0	velocity; initial velocity
V_s	slip velocity
V_∞	velocity of a steady movement
V_{max}	maximum velocity in the velocity distributions
$\overline{V}, V_{av}$	average velocity
V, V_0	volume; initial volume
V^*	volume after deformation
V_{ell}	volume of an ellipsoid (after deformation)
V_{sph}	volume of a sphere
v_1, v_2, v_3	components of the velocity vector
W	work; work per cycle in a dynamic loading; stored elastic energy; the intensity of energy dissipation
W_{sh}	energy responsible for shape changes
W	(with different indices) elastic potential (in non-linear models and theories of elasticity)
W	distance between the neighboring flights of screw
w_{ij}	components of the vorticity tensor ($i = 1, 2, 3$ and $j = 1, 2, 3$)
w	weight fraction in mixture
X^*	complex amplitude of displacement
X	displacement in a model representation
X_m	imaginary (reactive) part of mechanical impedance
X_0	initial displacement in a model representation
X_0	initial value of an arbitrary parameter X
X_{sp}	displacement of a spring in a model representation
X_{pist}	displacement of a piston in a model representation
X	arbitrary parameter
X_∞	the equilibrium value of an arbitrary parameter X
X_1, X_2, X_3	components of the body force
x_1, x_2, x_3	the Cartesian coordinate axes
x	variable in various equations
x_{0A}, x_{0B}	amplitudes of displacement of plates A and B, respectively
$x_{0, B}^{max}$	maximum (resonance) amplitude of displacement of plate B
Y	arbitrary variable
Y^*	complex mechanical impedance
y	coordinate axis
y	variable in some calculations
y_0	insertion depth in a viscometry

Z	rigidity of spring
z	coordinate axis
z_0	distance equal to reciprocal of attenuation

Greek letters

α	arbitrary angle; the angle in a cone-plate viscometer
α	(with different indexes) arbitrary factors in different equations
α	degree of crystallinity
α	coefficient of thermal expansion
α	phase angle
β	angle in a cone-plate viscometer device
β	angle of contact formed by meniscus
β	constant (coefficient) in different equations; scaling factor
β	degree of conversion
β^*	degree of conversion at the gel-point
Γ	parameter in the theory of wave propagation
Γ	dimensionless shear rate
γ	shear deformation
γ	dimensionless viscosity in the theory of vibration viscometry
γ_r	recoil shear strain or elastic (recoverable) deformation
γ_r^*	the critical value of elastic deformation (for the onset of instability in the flow)
γ_f	irreversible shear deformation (in the flow)
γ_m	shear deformation corresponding to maximum stress
$\dot{\gamma}$	deformation rate (in a simple shear)
$\dot{\gamma}_R$	deformation rate at the channel wall (in a simple shear)
$\dot{\gamma}_0$	constant shear rate; initial shear rate; average shear rate
$\dot{\gamma}_0^*$	apparent average shear rate (in the flow with slip)
$\dot{\gamma}_N$	quasi-Newtonian shear rate
$\dot{\gamma}^H$	the shear rate at the slit wall
$\dot{\gamma}_0^H$	the average shear rate in a slit
$\dot{\gamma}_m$	maximum shear rate (in viscometer)
$\dot{\gamma}_s$	critical shear rate (at the spurt point)
Δ	change (of something)
Δ	measure of the rate of damping in oscillations (logarithmic decrement of damping)
Δ	clearance between coaxial cylinders
δ	loss angle (in periodic oscillations)
δ	wall thickness of cylinder
δ	angle between plate and conical surface in viscometer
δ	clearance
δ	attenuation
δ_{ij}	the Kroneker Delta (unit tensor)
ε	ratio of stresses in a rotational viscometer
ε	relative change of distance (strain in tensile extension)
ε	deformation (of any type)

ε_0	initial deformation; constant deformation (in relaxation); amplitude of deformation (in periodic oscillation)
$\varepsilon_{r,0}$	characteristic recoverable deformation – empirical parameter (in extension)
$\varepsilon_{M,\,sp}; \varepsilon_K; \varepsilon_{M,\,pist}$	deformation of model elements in a model representation
ε_V	relative change of volume (volume deformation)
ε_{ij}	components of a tensor of large deformations (i = 1, 2, 3 and j = 1, 2, 3)
ε'_{ij}	deviatoric components of the deformation tensor
$\varepsilon_1, \varepsilon_2, \varepsilon_3$	principal values of the ij-tensor
ε^*	engineering measure of deformation
ε^H	the Hencky measure of deformation
ε_{inf}	residual deformation (in creep)
ε_{r_*}	tensile recoil (recoverable deformations stored in extension)
$\dot{\varepsilon}_r$	critical elastic deformation in extension (at the moment of rupture)
$\dot{\varepsilon}$	deformation rate (in extension)
ζ	constant in a kinetic equation
η	viscosity; apparent (non-Newtonian) viscosity
η_0	initial (Newtonian) viscosity; viscosity of a solution
η^0	viscosity at zero pressure
η_∞	upper (high-shear rate) Newtonian viscosity
η_s	solvent viscosity
η_M, η_K	viscosities of elements in a model representation
η^*	complex dynamic viscosity
η'	real part of complex dynamic viscosity
η''	imaginary part of complex dynamic viscosity
$\eta_{\parallel}$	viscosity of an anisotropic liquid measured in direction parallel to shear
$\eta_{\perp}$	viscosity of an anisotropic liquid measured in direction orthogonal to shear
η_B	biaxial stress growth coefficient
$[\eta]$	intrinsic viscosity
η_p	constant "plastic" viscosity
$\underline{\eta}_\infty$	limiting viscosity at high shear rates
$\underline{\eta}$	dimensionless viscosity
$\overline{\eta}$	average viscosity (in viscometer)
θ	angular coordinate
θ	angle in different expressions
θ_{ij}	components of the tensor of rotation (turn) (i = 1, 2, 3 and j = 1, 2, 3)
θ	characteristic time of process (with different indices); relaxation time in discrete spectrum
θ_c, θ_d	constants characteristic times in a tube model
$\theta_{max}, \theta_{min}$	maximum and minimum relaxation times in a spectrum
θ_K	the Kohlrausch relaxation time
θ_0	complex amplitude of twisting
θ_0	amplitude of twisting
θ_B	complex amplitude of twisting for body B
θ_{0A}	angular amplitude of twisting of body A
κ	coefficient of the thermal conductivity

λ	retardation time
λ	retardation time in a discrete spectrum (with different indices)
$\lambda_{max}, \lambda_{min}$	maximum and minimum values of retardation time in a discrete spectrum
λ	wavelength
λ	dimensionless frequency in the theory of vibration viscometry
λ	extension ratio
λ	heat transfer coefficient
λ	regularization parameter
$\lambda_1, \lambda_2, \lambda_3$	extension ratios along the principal axes
λ	coefficient of friction (of hydrodynamic resistance)
λ_R	coefficient of friction (of hydrodynamic resistance) expressed via radius
μ	Poisson's ratio (coefficient)
ν	empirical parameter
ξ	function characterizing non-linear effects
ρ	density
ρ_0	density at the reference temperature T_0
ρ_s	density of a solid body moving in liquid
ρ_1	density of liquid
Σ	sum
σ	various stresses; shear stress
σ	surface tension
σ_0	initial stress; constant stress (in creep); amplitude of stress (in periodic oscillation)
σ_A	shear stress at the capillary wall
$\sigma_{E,0}$	characteristic stress (empirical parameter) in extension
σ_E	normal (tensile) stress
σ_Y	yield stress
σ_R	shear stress at the tube wall of capillary; maximum shear stress on surface
σ_{ij}	components of stress tensor in the Cartesian coordinates (i = 1, 2, 3 and j = 1, 2, 3)
σ_{ij}	deviatoric components of the stress tensor
σ_{ij}	components of stress tensor in polar coordinates (i = θ, r, z and j = θ, r, z)
σ_{max}	maximum stress; limit of shear stress
$\sigma_1, \sigma_2, \sigma_3$	principal stresses
$\sigma_{1,max}, \sigma_{2,max}, \sigma_{3,max}$	maximum shear stresses (invariants, expressed in shear stresses)
σ_i, σ_o	stresses at the inner and outer surfaces of coaxial cylinders, respectively
$\bar{\sigma}, \sigma_{av}$	average shear stress
σ_C	stress at a surface in a cone-plate viscometer
σ_E	circumferential stress
σ_z	longitudinal stress
σ_{res}	residual stress (in relaxation)
$\sigma_{E, ext}, \sigma_{E, compr}$	normal stresses in extension and compression, respectively
σ^*	characteristic shear stress
σ^*, σ_E^*	stress limits of elasticity (critical stress for transition to plastic deformation in solids)

σ_H	shear rate at the slit wall
σ_s	critical stress (spurt stress)
σ^+	shear growth stress function
σ^-	shear stress decay function
σ_E^+	tensile growth stress function
σ_E^*	critical stress (at the moment of rupture)
φ	angle in the theory of break of solids
φ	angle of inclination in some instruments
φ	relaxation function
φ	concentration (in volume parts)
φ^*	critical concentration
χ	correction factor
Ψ_1	the first normal stress coefficient
Ψ_2	the second normal stress coefficient
$\Psi_{1,0}$	initial ("linear") value of the first normal stress coefficient
Ψ_0	permanent component of the first normal stress coefficient in oscillation
Ψ'	real part of the complex first normal stress coefficient in oscillation
Ψ''	imaginary part of the complex first normal stress coefficient in oscillation
ψ	creep function
Ω	angle of twisting
Ω	rotational speed
Ω_i, Ω_o	rotational speed of inner and outer cylinders, respectively
ω	angular velocity; frequency
ω_0	own (resonance) frequency
$\omega_{max}, \omega_{min}$	upper and lower boundaries of frequency window in the measurement of dynamic modulus
De	the Deborah number
Mn	the Mason number
Re	the Reynolds number
Re_R	the Reynolds number expressed via the radius
Re_c	the critical Reynolds number in rotational flows
We	the Weber number
Wi	the Weissenberg number

Vectors

A	acceleration vector
F	force
r	radius-vector
n	normal to the surface
u	displacement vector
v	velocity vector
X	body force

SOLUTIONS

CHAPTER 1

PROBLEM 1-1

What is the equilibrium state of a liquid and a solid in the absence of stresses?

Answer

There is a single equilibrium state for solid, determined by the absence of stresses. For liquid, any state in the absence of flow is equilibrium. The equilibrium state of liquid exists for any shape of the specimen.

PROBLEM 1-2

What are the possible limits of Poisson's ratio, μ? Can its value exceed 0.5? Can it be negative?

Answer

Poisson's ratio cannot exceed 0.5, because if it was so, hydrostatic pressure would lead to an increase in volume (see Eq. 1.2.18). This is physically impossible. Negative values of μ do not contradict fundamental laws. The value $\mu < 0$ means that in uniaxial extension the lateral size of a specimen increases. For real materials, this case is possibly realized for foams.

PROBLEM 1-3

What are the pressure and the shear stresses in the stress state created by the following normal stresses: $\sigma_{11} = 0$; $\sigma_{22} = -\sigma_0$ and $\sigma_{33} = 0$? What are shear stresses in this case?

Answer

Pressure

$$p = -\frac{1}{3}\sum_{i=1}^{3}\sigma_{ij}$$

(or the spherical part of the stress tensor) equals zero. Shear stresses, σ, acting on the plane inclined by an angle α are calculated as:

$$\sigma = \frac{1}{2}(\sigma_{11} - \sigma_{22})\sin 2\alpha = \sigma_0 \sin 2\alpha$$

Comment

This type of stress field leads to deformations known as "pure shear".

PROBLEM 1-4

Calculate stresses acting in a thread being suspended by its end and stretched by its own weight.

Answer

The normal (extensional) stress appears due to the gravitational force. The maximum stress, σ_{max}, acts on the cross-section of suspension. This stress equals to $\sigma_{max} = \rho g L$ (where ρ is the density, g is the gravitational acceleration constant, and L is the length of the specimen). So, σ_{max} increases as the sample becomes longer and there is a length of the thread at which σ_{max} exceeds the strength of a sample, σ^*.

Comment

This limiting length, corresponding to σ^*, can be treated as a measure of the strength of a material. Such a measure is used in engineering practice for characterizing the strength of fibers which is expressed as "*breaking length*".

PROBLEM 1-5

Analyze the situation where a horizontal long flexible engineering element (fiber, bar, etc.) is loaded along its length by a distributed force, q (i.e., force, normal to the bar, per the unit of a length).

Answer

Let the length of a bar be L and the sagging height be y. A reasonable assumption to make is that $y \ll L$. The balance equation (the sum of torques around the point of suspension is zero) is:

$$Fy - qx\frac{x}{2} = 0$$

where F is the normal force stretching a bar and x is the distance from a suspension point. Then the sagging is expressed as $y = qx^2/2F$ and the maximal sagging, H, is $H = qL^2/2F$.

Normal stress, σ_E, in a bar is $\sigma_E = F/S$, where S is the cross-section of the bar. Then the final relationship between stress in a bar and its sagging, H, is: $\sigma_E = qSL^2/2H$. The last equation shows that the decrease in H can be reached by an increase in stress, and H cannot be too small because stress is limited by the strength of the material.

PROBLEM 1-6

In section 1.3.1 the difference between the gradient of velocity and the rate of deformation is explained. What is the situation with these values for a uniaxial extension?

Answer

The uniaxial extension is a special and unique case in which the rate of deformation coincides with the gradient of velocity.

PROBLEM 1-7

Calculate the stresses in a hemispherical cup loaded by its own weight. Such a case is met in many engineering designs, for example, in a spherical roof covering a large area of a stadium or a warehouse.

Answer

The current radius of a hemisphere is expressed as $r = R\sin\alpha$, where r is its radius and the angle α is taken from a horizontal plane. Let the density of material be ρ and its (uniform) thickness be δ. It is evident that $\delta \ll R$ (such objects are called "*membranes*"). Then the analysis of the balance of acting forces gives the following resulting equation for the normal stress acting along the surface of a spherical membrane:

$$\sigma_E = \rho g R \frac{\sqrt{1-(r/R)^2}-1}{(r/R)^2}$$

where g is the gravitational acceleration.

The stresses on the supporting ring (at $r = R$) must balance the total weight of a hemisphere. Therefore, the maximum value of the principal stress, σ_{max}, is $\sigma_{max} = -\rho g R$.

PROBLEM 1-8

Let a liquid be placed between two coaxial cylinders with radii R_o (outer) and R_i (inner). The gap between cylinders $\Delta = R_o - R_i$ is small in comparison with the cylinder radii. Let the outer cylinder rotate with an angular velocity, Ω. Then the assembly of both cylinders begins to rotate with the same angular velocity, ω. What are the shear rates and gradients of velocity in these two cases?

Answer

The rate of shear in both cases is the same and equals $\Omega R_o/\Delta$. According to Eq.1.3.5, gradient of velocity in the first case is the same as the rate of shear and in the second case and it equals $\omega + \Omega R_o/\Delta$). The first member in this sum reflects quasi-solid rotation, and the second term is the shear rate.

PROBLEM 1-9

A cylindrical thread of length l_0 is fixed at one end and stretched at the other end. What must be the time dependence of velocity, $v(t)$, of stretching sufficient to maintain a constant deformation rate, $\dot{\varepsilon}_0 = const$?

Answer

Deformation rate is dl/ldt (but not $dl/l_0 dt$!). Then $\dot{\varepsilon}_0 = dl/ldt = const$ if $l = l_0 e^{\dot{\varepsilon}_0 t}$ and

$$v = \frac{dl}{dt} = l_0 \dot{\varepsilon}_0 e^{\dot{\varepsilon}_0 t}$$

i.e., velocity must increase exponentially.

PROBLEM 1-10

Advance your arguments proving a possibility of neglecting shear stresses in a thin-wall cylinder as in Example in section 1.1.4.

Answer

This can be proven by analysis of the second line in Eq. 1.1.2. The second and the third items equal zero because a cylinder is symmetrical along coordinates θ and z. There are no changes in stresses along with these coordinates. Then, the equation can be approximately written as:

$$\frac{\Delta\sigma_{\theta r}}{\Delta r} + \frac{2\sigma_{\theta r}}{r} = 0$$

The value Δr is small (because it is the width of a wall) and therefore the left side can be small (and neglected) only if $\Delta\sigma_{\theta r}$ and $\sigma_{\theta r}$ are also small. Shear stresses on a surface are absent (equal to zero). They cannot strongly vary across the wall and therefore they are close to zero and can be neglected.

CHAPTER 2

PROBLEM 2-1

For Maxwellian liquid with a relaxation time θ, what is the residual stress (in comparison with the initial stress σ_0), if the process of stress relaxation continues for a duration of time $t = 2\theta$?

Answer

For Maxwellian liquid $\sigma(t) = \sigma_0 \exp(-t/\theta)$. Therefore, for $t = 2\theta$ the ratio $\sigma/\sigma_0 = e^{-2} \approx 0.135$ i.e., the residual stress equals $\approx 13.5\%$ of the initial stress value.

PROBLEM 2-2

For a solid material with rheological properties described by the Kelvin-Voigt model, with a retardation time λ, what is the time necessary to reach 95% of its equilibrium (limiting) value?

Answer

The creep process is described by the equation $\varepsilon = \varepsilon_\infty(1 - e^{-t/\lambda})$, where ε_∞ is the limiting deformation. Then the value $\varepsilon/\varepsilon_\infty = 0.95$ is reached at $t \approx 3\lambda$.

PROBLEM 2-3

Viscoelastic properties of the liquid are described by two relaxation modes: the first with modulus G_1 and a relaxation time $\theta_1 = 1$ s and the second with modulus G_2 and a relaxation time $\theta_2 = 100$ s. Describe the evolution of stress in time. How do relaxation curves look if a linear time scale and a logarithmic stress scale are used?

Answer

Stress diminishes approximately according to an exponential curve from the initial value of $G_1 + G_2$ and after $t \approx 3$ s it will reach the plateau having a value of G_2. Then at $t \approx 10$ s a new relaxation process becomes visible, and stress practically diminishes according to an exponential curve but with different curvature.

In log(stress) versus time coordinates the relaxation process is presented by two linear branches with different slopes corresponding to two values of relaxation times with a short transient zone.

Additional question

What is the value of equilibrium stress in this case?

Answer

Equilibrium stresses are absent because the material under discussion is liquid.

PROBLEM 2-4

Explain why the value θ_K, entering the Kohlrausch function, Eq. 2.1.6 is not a relaxation time. How do you find relaxation times for this relaxation function?

Answer

Relaxation time, by definition, is a value entering an exponential relaxation function. So, θ_K is a relaxation time if an exponent n in the Kohlrausch function equals 1. In other cases, this function reflects the existence of a continuous relaxation spectrum, $E(\theta)$, found by means of Eq. 2.2.8. The equation for calculating $G(\theta)$ is:

$$e^{-(t/\theta)^{-n}} = \frac{1}{\varphi(0)}\int_0^\infty G(\theta)e^{-(t/\theta)}d\theta$$

where $\varphi(0)$ is the instantaneous modulus (see Eq. 2.2.4). The function $G(\theta)$, i.e., a relaxation spectrum, can be found from the last equation and written in an analytical form, using standard mathematical transformations. However, this solution is rather cumbersome and will not be cited here.

PROBLEM 2-5

Analyze the evolution of deformations in the following loading history: stress σ_0 was applied at the time $t = 0$; then additional stress σ_1 was added at the time t_1 and finally at the time $t*$ both stresses were taken away. The material is a linear viscoelastic solid.

Answer

The evolution of deformations (creep and recovery) is described by means of the Boltzmann-Volterra superposition principle.

In the time range $0 - t_0$ deformations follow function of time: $\gamma = \sigma_0\psi(t)$.

Then in the time range $t_0 - t*$, $\gamma = \sigma_0\psi(t) + \sigma_1\psi(t - t_1)$; and after unloading at $t = t*$ deformations diminish and the complete deformation is calculated as $\gamma = \sigma_0\psi(t) + \sigma_1\psi(t - t_1) - [\sigma_0 + \sigma_1](t - t*)$.

Additional question

What will be the final deformation at $t \to \infty$?

Answer

At large time values $t \gg t_1$ and $t \gg t*$. Therefore both differences, $\sigma_0\psi(t) - \sigma_0\psi(t - t*)$ and $\sigma_1\psi(t - t_1) - \sigma_1\psi(t - t*)$ are approaching zero, no residual deformation is expected, and the elastic recovery will be complete.

PROBLEM 2-6

What is the shape of the frequency dependencies of the components of dynamic modulus for a Maxwellian liquid?

Answer

Formal mathematical analysis of the functions $G'(\omega)$ and $G''(\omega)$ shows that the real part of dynamic modulus, G', increases from zero to G along with the growth of frequency and the loss modulus, G'', equals zero at very small and very high frequencies and passes through the maximum at $(\omega\theta) = 1$.

Comment

This result demonstrates that at a high-frequency limit, Maxwellian liquid behaves in a solid-like manner. In the low-frequency limits, such material is a typical liquid, and transition takes place at $\omega = \theta^{-1}$.

PROBLEM 2-7

An experimental relaxation curve was approximated with the sum of three exponential functions with the following parameters:

$G_1 = 2*10^3$ Pa, $\theta_1 = 100$ s; $G_2 = 10^4$ Pa, $\theta_2 = 20$ s; $G_3 = 10^5$ Pa, $\theta_3 = 6$ s.

What is the viscosity of this liquid?

Answer

According to Eq. 2.5.12, viscosity is the sum of relaxation modes calculated (for a discrete spectrum) as: $\eta = \sum_n G_n\theta_n$. Thus $\eta = 1$ MPa s

PROBLEM 2-8

Eq. 2.3.11 and its solution show that the Burgers model describes the behavior of a material with two relaxation times. The same behavior is represented by two parallel Maxwell elements with their relaxation times $\theta_1 = \eta_1/G_1$ and $\theta_2 = \eta_2/G_2$ where η_1 and G_1 are the viscosity and elastic modulus of the first and η_2 and G_2 of the second Maxwell elements joined in parallel. Calculate the values of the constants of the Burgers model expressed via constants of the two Maxwell elements.

Answer

First of all, it is necessary to prove that both models indeed describe the same type of rheological behavior. To do this, the relationship between stresses and deformations for the model consisting of two parallel Maxwell elements ought to be formulated.

This is achieved by the following procedure: the rheological behavior of each parallel element is given by the equations: $\dot{\varepsilon} = (\dot{\sigma}_1/G_1) + (\sigma_1/\eta_1)$ and $\dot{\varepsilon} = (\dot{\sigma}_2/G_2) + (\sigma_2/\eta_2)$ where indices 1 and 2 show the values related to both branches of a model. Then bearing in mind that $\sigma = \sigma_1 + \sigma_2$ and excluding both components σ_1 and σ_2, we come to the final rheological equation of state which is equivalent to Eq. 2.3.11.

A comparison of the coefficients of rheological equations of both models (the Burgers and the two-branch Maxwell) leads to the following relationships among the parameters of the models:

$$G_M = G_1 + G_2 \qquad \eta_M = \eta_1 + \eta_2 \qquad G_K = \frac{G_1 G_2 (\eta_1 + \eta_2)^2 (G_1 + G_2)}{(\eta_1 G_2 - \eta_2 G_1)^2}$$

$$\eta_K = \frac{\eta_1 \eta_2 (G_1 + G_2)^2 (\eta_1 + \eta_2)}{(\eta_1 G_2 - \eta_2 G_1)^2}$$

Comment

From this solution, it is seen that the presence of *two relaxation times* leads to the appearance of a *single retardation time*, and its value is

$$\lambda_K = \frac{\eta_K}{G_K} = \frac{\eta_1 \eta_2}{\eta_1 + \eta_2} \frac{G_1 + G_2}{G_1 G_2}$$

PROBLEM 2-9

Is it possible to measure dynamic modulus using non-harmonic periodic oscillations? How this is done?

Answer

It is possible to do it by decomposition of non-harmonic input and output waves by Fourier series expansions. Then, by comparison of the harmonics of the same frequencies, it is possible to calculate components of the dynamic modulus in an ordinary manner for different frequencies.

PROBLEM 2-10

In measuring a relaxation curve, it is assumed that the initial deformation is set instantaneously. In fact, it is impossible, and a transient period always exists. Estimate the role of this period for a single-relaxation mode ("Maxwellian") liquid.

Answer

Stress relaxation after an instantaneous set of deformation is expressed as $\sigma_{ins}(t) = \sigma_0 e^{-t/\theta} = G\gamma_0 e^{-t/\theta}$, where σ_0 is the initial stress, γ_0 is the constant deformation and G is elastic modulus.

If deformation is developed with a constant deformation rate until deformation γ_0 is reached, the evolution of stresses before relaxation can be written as (see Eq. 2.4.13)

$$\sigma_{trans}(t) = G\dot\gamma\theta(1 - e^{-t/\theta}) = \dot\gamma\eta(1 - e^{-t/\theta})$$

At $t_0 = \gamma_0/\dot\gamma$ deformation ceases and relaxation begins. Real stress evolution, $\sigma_{real}(t)$ in a relaxation range which can be described based on the Boltzmann-Volterra superposition principle, and the time dependence of stress is given by:

$$\sigma_{real} = G\dot\gamma t\left(\frac{\theta}{t}\right)(1 - e^{-t_0/\theta})e^{-(t - t_0)/\theta}$$

The difference between σ_{real} and σ_{inst} is evident. Estimations depend on the ratio of constants. In the range of measurements, $t \gg t_0$ and for a quick setting of necessary deformation γ_0, $\sigma_{real} \to \sigma_{inst}$. The correctness of this approximation is determined by the condition $t_0 \ll \theta$.

PROBLEM 2-11

Application of the theory of large deformations to a linear viscoelastic body leads to the following equation for the time evolution of the first normal stress difference, $N_1^+(t)$, at a constant shear rate, $\dot\gamma_0 = $ const:

$$N_1^+(t) = 2\dot\gamma_0^2\int_0^t t'\varphi(t')dt' \quad \text{(Can you prove this equation?)}$$

Calculate the function $N_1^+(t)$ for an arbitrary relaxation spectrum, $G(\theta)$.

Answer

Substitute a relaxation spectrum instead of a relaxation function $\varphi(t)$ and change the order of integration. This leads to the following formula:

$$N_1^+(t) = 2\dot\gamma^2\int_0^\infty \theta^2 G(\theta)\left[1 - e^{-t/\theta}\left(1 + \frac{t}{\theta}\right)\right]d\theta$$

The limiting value of $N_1(t)$ for steady flow is:

$$N_1(t \to \infty) = 2\dot\gamma_0^2\int_0^\infty t'\varphi(t')dt' = 2\dot\gamma_0^2\int_0^\infty \theta^2 G(\theta)d\theta$$

(compare with Eq. 2.5.14)

Additional question 1

Find the $N_1^-(t)$ dependence for stress relaxation after a sudden cessation of steady flow. Compare the rates of relaxation of shear and normal stresses.

Answer 1

Relaxation of normal stresses is described by

$$N_1^-(t) = 2\dot{\gamma}_0^2 \int_0^\infty \theta^2 G(\theta)e^{-t/\theta}\left(1 + \frac{t}{\theta}\right)d\theta$$

Relaxation of shear stresses after sudden cessation of steady flow is described by the formula (see Eq. 2.4.16):

$$\sigma(t) = \dot{\gamma}_0 \int_0^\infty \theta G(\theta)e^{-t/\theta}d\theta$$

The comparison of both functions related to their steady-state values: $\sigma(t)/\sigma_s$ and $N_1^-(t)/N_{1,s}$ (where the values σ_s and $N_{1,s}$ are shear and normal stress values at steady-state flow before the start of relaxation) depends on the comparison of the functions $e^{-t/\theta}$ and $e^{-t/\theta}[1 + (t/\theta)]$. It is evident that the second expression is always larger than the first one because $[1 + (t/\theta)] > 1$ (avoid the trivial case of $t = 0$). Therefore, the growth and relaxation of normal stresses always proceed slower than shear stresses.

Additional question 2

For a single-mode viscoelastic liquid with relaxation time, θ, calculate the relative residual shear and normal stresses after relaxation continuing for time 4θ.

Answer 2

Residual shear stress is found as $e^{-4} \approx 0.0185$ (1.85%), while relative residual normal stresses equal $5e^{-4} \approx 0.0927$ (9.27%), i.e., residual normal stresses are five times larger than shear stresses.

PROBLEM 2-12

Explain the procedure of transition from a discrete to a continuous relaxation spectrum (from Eq. 2.6.6 to Eqs. 2.6.7 and 2.2.8).

Answer

According to a spring-and-dashpot model, a discrete relaxation spectrum is written as a sum of delta-functions:

$$F(\theta) = C\sum_p \delta(\theta - \theta_p)$$

where C is a constant and θ_p are relaxation times. The distribution of relaxation times is expressed as: $\theta = \theta_{max}/p^2$. The transition from a discrete to a continuous spectrum is based on the change of a discrete argument p with a continuous one. According to the last relationship $p = (\theta_{max}/\theta_p)^{1/2}$ and for a continuous spectrum: $dp = k\theta^{-3/2}d\theta$, where k is a constant. The function $F(\theta) = k\theta^{-3/2}$ is a continuous relaxation spectrum.

Comment

If a logarithmic time-scale is used, the argument would be $\ln\theta$, then a logarithmic relaxation spectrum, $h(\ln\theta)$ is $h = \theta F(\theta) = k\theta^{-1/2}$.

PROBLEM 2-13

Let a small solid dead-weight of mass m be attached to a rod at its end and the rod fixed at the other end. Some initial displacement from the equilibrium position of the weight (deforming the rod) was created by an applied longitudinal force, and then the force was ceased.

Analyze the movement of the weight after the force is ceased. Is it possible to find the components of the dynamic modulus of a rod material following the movement of a weight?

Comment

A rod can be of different lengths and cross-sections. Not specifying the sizes and the geometrical form of a rod, the latter is characterized by the value of a "form-factor" k.

Answer

The movement of a dead-weight (approximated as a "point") is described by Newton's second law, which can be written as

$$m\frac{d^2x}{dt^2} + kE^*x = 0$$

where x is the displacement of the weight. The value E^* is the complex modulus. Using E^* instead of standard Young's modulus, it is assumed that the material of a rod is viscoelastic.

As the first approximation let E^* be the elastic modulus E, i.e., the material of a rod is ideally elastic. In this case, the solution of the dynamic equation is well known. It is a periodic harmonic oscillation with frequency $\omega = (kE/m)^{1/2}$. Measuring the frequency of oscillations, ω, it is easy to calculate modulus.

If the material is viscoelastic, it is supposed that oscillation will be damping and the solution of the equation for x is searched as $x = x_0 e^{i\omega t}e^{-\Delta t}$, where x_0 is the amplitude of oscillations, the first exponential factor represents periodic oscillations, and the second factor represents the effect of damping, which is characterized quantitatively by the value Δ. The substitution of this function into the dynamic equation and representing E^* as $E^* = E' + iE''$ leads to two equations (separate for real and imaginary components):

$$m(\Delta^2 - \omega^2) + kE' = 0 \text{ and } -2m\Delta\omega + kE'' = 0$$

These equations give the final expression for calculating the components of dynamic modulus:

$$E' = \frac{m}{k}(\omega^2 - \Delta^2) \text{ and } E'' = \frac{2m}{k}\omega\Delta$$

Measuring the frequency of oscillations and the rate of damping, it is possible to calculate E' and E" (see section 5.7 for a more detailed discussion).

Comment

Here, the geometry of movement is immaterial. It can be linear movement or rotation. In the latter case, mass in this equation is replaced by a moment of inertia and x is the angular twist.

CHAPTER 3

PROBLEM 3-1

Can viscosity be negative? Explain the answer.

Answer

Viscosity cannot be negative. If it were so, it would mean that the application of force in one direction leads to movement in the opposite direction. Also, heat would transform into mechanical energy and this contradicts the second law of thermodynamics.

PROBLEM 3-2

In measuring the viscous properties of the polymer solution, it appeared that the experimental data in the experimental range of shear rates can be fitted with the power-law equation (Eq. 3.3.4). Analyze the possibility of extrapolating this equation to the range of very high shear rates.

Answer

The power-law equation predicts that apparent viscosity at very high shear rates decreases to zero. It is physically impossible. The viscosity of a solution cannot be less than the viscosity of a solvent. Extrapolation of experimental data to the range of high shear rates is incorrect.

Additional question

Which kind of rheological behavior at high shear rates is expected in this case?

Answer

The following possibilities exist:
- apparent viscosity will reach its lowest (Newtonian) limiting value, slightly exceeding the viscosity value of a solvent
- instability of flow will appear due to inertial turbulence or the transition of a viscoelastic solution into the forced rubbery-like state; in both cases viscosity estimations becomes incorrect.

PROBLEM 3-3

What is the difference in stress relaxation of viscous liquids and viscoplastic materials?

Answer

Relaxation of stresses in liquids continues up to zero, whereas relaxation in viscoplastic materials stops at the level of the yield stress because at lower stresses viscoplastic materials behave in a solid-like manner.

PROBLEM 3-4

Can we expect that the values of the yield stress, σ_Y, found by treating a set of experimental data by means of Eqs. 3.3.7 to 3.3.9, are the same?

Answer

The general answer is "no", because any of these equations gives nothing more than the best fit of experimental data, and the constants of different empirical equations are not required to be the same. In this sense, the values of σ_Y found by fitting experimental data by any empirical approximation are not "true" yield stress but fitting constants only (compare with the example in section 3.3.3).

PROBLEM 3-5

Calculate shear stresses in the flow of liquid through a straight tube if the flow is created by the pressure gradient $\Delta p/L$ (L is the length of a tube).

Answer

Let us consider a force equilibrium of a cylindrical liquid element of the current radius, r. The pressure gradient is balanced by shear stresses acting on the surface of this element.

The balance is formulated as $\pi r^2 dp = 2\pi r \sigma dl$ or, changing dp/dl for $\Delta p/L$, the following final formula can be written: $\sigma(r) = \Delta p r/2L$.

Maximum shear stress acts at the wall of the tube and equals: $\sigma_R = \Delta p R/2L$. The following equation for the radial distribution of shear stresses is valid: $\sigma(r) = \sigma_R(r/R)$. This distribution is linear.

Additional question

Are the results valid for Newtonian liquid only?

Answer

No, this is a general result, because in proving the linear radial distribution of shear stresses no assumption about Newtonian flow was used. So, the result is valid for any liquid with properties described by an arbitrary rheological equation.

PROBLEM 3-6

Calculate the radial distribution of shear rates and flow velocity of Newtonian liquid (having viscosity η) through a straight tube with radius R.

Answer

The answer is based on the stress distribution obtained in Problem 3-5. For Newtonian liquid, shear rate is proportional to the shear stress. So, according to Newton's law, it is possible to write:

$$\dot{\gamma}(r) = \frac{1}{\eta}\sigma(r) = \frac{\sigma_R}{\eta}\left(\frac{r}{R}\right)$$

Maximum shear rate acts at the wall, where it equals $\dot{\gamma}_R = \sigma_R/\eta$. Evidently, the last formula is nothing more than Newton's law, written for the points at a tube wall.

Velocity distribution is found by integration of the $\dot{\gamma}(r)$ function:

$$u(r) = \int_r^0 \dot{\gamma}dr = \frac{\sigma_R}{\eta}\int_r^0 rdr = \frac{P}{4L\eta}(r^2 - C)$$

where C is the constant of integration, found from the following boundary condition: $u = 0$ at the wall (at $r = R$). The latter is the hypothesis of sticking (absence of slip at a solid boundary). Then, after formal rearrangements, the following equation for u(r) is obtained:

$$u(r) = \frac{PR^2}{4L\eta}\left[1 - \left(\frac{r}{R}\right)^2\right]$$

Additional question 1

Calculate the volume output, Q, for the flow of Newtonian liquid.

Answer 1

Volume output is found as:

$$Q = \int_0^R 2\pi r u(r)dr = \frac{\pi PR^4}{8L\eta}$$

The last expression is known as the *Poiseuille equation*.

Additional question 2
Express maximum shear rate, $\dot{\gamma}_R$, via volume output.

Answer 2
This value can be written as

$$\dot{\gamma}_R = \frac{\sigma_R}{\eta} = \frac{\Delta p R}{2L}$$

Then, the direct comparison of this formula with the equation for Q gives:
$\dot{\gamma}_R = (4Q)/(\pi R^3)$.

Additional question 3
Is the last expression valid for a liquid with arbitrary rheological properties?

Answer 3
No, for a Newtonian liquid only, because this equation was obtained based on Newton's law relating shear rate and shear stress.

PROBLEM 3-7
Calculate the velocity profile in the flow of a power-law type liquid through a straight tube with a round cross-section. The radius of a tube is R.

Answer
In the solution of Problem 3-5, it was proven that the radial distribution of shear stress is expressed as $\sigma(r)/\sigma_R = r/R$, where r is the current radius and σ_R is the shear stress at the tube wall. The flow curve of a power-law type liquid is written as $\sigma = k\dot{\gamma}^n$ (see Eq. 3.3.4), or $\dot{\gamma} = \sigma^{1/n}/k$. Then, the radial distribution of shear rate is presented as:

$$\dot{\gamma} = \frac{1}{k}\sigma_R^{1/n}\left(\frac{r}{R}\right)^{1/n}$$

The last step in the solution is the integration of this equation along the radius because $\dot{\gamma} = du/dr$. Therefore

$$u(r) = \frac{1}{k}\sigma_R^{1/n}\frac{1}{R^{1/n}}\int_R^0 r^{1/n}dr$$

with the evident boundary condition: u = 0 at r = R (the concept of liquid sticking to the wall or absence of slip at the wall). After simple rearrangements the following final equation is obtained:

$$u(r) = \frac{1}{k}\sigma_R^{1/n}\frac{n}{1+n}R\left[1-\left(\frac{r}{R}\right)^{\frac{1+n}{n}}\right]$$

This equation shows that for a power-law type liquid with an arbitrary value of the exponent n (not equal to 1), the velocity profile appears non-parabolic, unlike for Newtonian liquid.

If to introduce the expression $\sigma_R = PR/2L$ (see Problem 3-5), the dependence of velocity profile on pressure gradient is obtained.

Comment

The u is a nonlinear function of P (if $n \neq 1$), unlike for Newtonian liquid.

Additional question

Calculate the volume output, Q, as a function of P for a power-law type liquid.

Answer

Volume output is calculated from

$$Q = \int_{0}^{R} 2\pi r u(r) dr$$

and the final equation is obtained after substituting the formula for σ_R by P.

PROBLEM 3-8

An experimenter obtained two pairs of data: at $\dot{\gamma}_1 = 1*10^{-3}$ s^{-1} $\sigma_1 = 100$ Pa and at $\dot{\gamma}_2 = 1*10^{-2}$ s^{-1} $\sigma_2 = 600$ Pa.

Assuming that the flow curve is described by a power-law type equation, find the constants of this equation for a liquid under study.

Answer

The constants are found as the solution of the system of two equations:

$$\begin{cases} 100 = k(1*10^{-3})^{n} \\ 600 = k(1*10^{-2})^{n} \end{cases}$$

By dividing left- and right-hand-side terms, the following equation for n is obtained: $\log 600 - \log 100 = n$. Then, $n \approx 0.778$. Then k is found from any of the equations: $k \approx 2.16*10^{4}$ $Pa*s^{0.778}$.

Additional question

How does one find the constants of the power-law type equation if an experimenter obtained three or four pairs of experimental points?

Answer

This can be done by the least-mean-square-root procedure. In order to linearize the problem, it is preferable to present experimental points using the log-log coordinates.

PROBLEM 3-9

Analyze flow of a viscoplastic ("Bingham-type") liquid through a straight tube of radius, R. Find radial stress and velocity distributions and calculate volume output as a function of the pressure gradient.

Answer

Rheological properties of the liquid are described by Eq. 3.3.7, which includes two characteristic rheological constants: yield stress σ_Y and plastic viscosity η_p.

Stress distribution does not depend on the type of rheological properties of liquid and is linear (see Problem 3-5). However, flow is possible at $\sigma > \sigma_Y$. This stress is reached at the radius $r_Y = (2L)/(\Delta p \sigma_Y)$. In the central zone of the tube, at $r < r_Y$, there is no flow and the central core is moving as a solid, while the layer near the wall is flowing in a shear-like mode. Direct calculations based on Eq. 3.3.7 give the equation for a velocity profile:

$$u(r) = \frac{(R^2 - r^2)P}{4L\eta_p} - \frac{\sigma_Y}{\eta_p}(R - r)$$

which is valid at $r > r_Y$. At $r < r_Y$ velocity is constant and equals to $u(r_Y)$.

Volume output is found by integration of the function $u(r)$ along the radius:

$$Q = \int_0^R 2\pi r u(r) dr$$

Direct calculations lead to the following final result:

$$Q = \frac{\pi R^4 P}{8\eta_p L}\left[1 - \frac{4}{3}\left(\frac{2L\sigma_Y}{RP}\right) + \frac{1}{3}\left(\frac{2L\sigma_Y}{RP}\right)^4\right]$$

This solution is known as the *Buckingham-Reiner equation*.

At $\sigma_Y = 0$, the expressions for $u(r)$ and $Q(P)$ transform to the equations obtained for a Newtonian liquid.

PROBLEM 3-10

A ball with a radius R is falling in a Newtonian liquid having viscosity η. After some transient period, the velocity of the ball movement becomes constant. Find the velocity of steady movement, U_∞.

Answer

A ball falls under the action of gravitational force and this force is balanced by viscous resistance. The balance equation is written as

$$\frac{4}{3}\pi R^3(\rho_s - \rho_l)g = 6\pi R U_\infty \eta$$

where ρ_s is the density of ball, ρ_l is the density of the liquid, g is the gravitational acceleration at free falling. The right-hand side of this equality is the Stokes equation taken from a textbook on fluid dynamics. The velocity of steady movement is calculated as

$$U_\infty = \frac{2R^2(\rho_s - \rho_l)g}{9\eta}$$

Comment

The last equation can be used for the determination of viscosity in an apparatus where velocity U_∞ is measured (see section 5.5 for more details). In this case, viscosity is calculated as

$$\eta = \frac{2R^2(\rho_s - \rho_l)g}{9U_\infty}$$

PROBLEM 3-11

An experimenter measured the viscous properties of the material at different shear rates and obtained a flow curve. What can he say concerning the viscous properties of his material in the uniaxial extension? Explain the answer.

Answer

Generally speaking, nothing. Only if the material under investigation is Newtonian liquid is it possible to say that uniaxial elongation viscosity obeys the Trouton law (Eq. 3.1.6). In all other cases, shearing experiments give no information concerning the behavior of a material in extension. The extension is principally a different type of deformation than shearing. The connection between the two cases is provided by a general (three-dimensional) equation of state. However, shear experiments do not give sufficient ground for the construction of such an equation; experiments in extension mode need to be performed. Then, the generalization of different types of experiments will give shear properties, on one hand, and extensional properties, on the other hand.

However, if a general rheological model is known beforehand (from *a priori* model arguments or previous experimental data), then shear experiments, possibly with experimental data of another type, will give the necessary values of the model parameters. The latter can be used for predicting the rheological behavior of the material in deformation modes of any type.

PROBLEM 3-12

Prove the validity of Eq. 3.1.7 – the dependence between normal stress and deformation rate for a Newtonian liquid in two-dimensional (biaxial) extension.

Answer

This problem is a two-dimensional analogue of the Trouton viscosity. It is necessary to compare deviatoric stress and deformation rate tensors for the deformation mode under discussion.

The deviatoric the stress tensor for a two-dimensional extension is calculated as

$$\sigma = \begin{bmatrix} \sigma_E & 0 & 0 \\ 0 & \sigma_E & 0 \\ 0 & 0 & 0 \end{bmatrix} = -p\delta_{ij} + \frac{\sigma_E}{3}\begin{bmatrix} 1 & 0 & 0 \\ 0 & 1 & 0 \\ 0 & 0 & -2 \end{bmatrix}$$

where $p = (2\sigma_E)/3$ (compare with Eq. 1.1.17). The deformation rate tensor is deviatoric and it is written as

$$\dot{d} = \dot{\varepsilon}\begin{bmatrix} 1 & 0 & 0 \\ 0 & 1 & 0 \\ 0 & 0 & -2 \end{bmatrix}$$

Then, based on the generalized Newton law and comparing the deviatoric stress tensor and the deformation rate tensor, it is seen that $\sigma_E = 6\eta\dot{\varepsilon}$.

The biaxial elongational viscosity η_B equals 6η (Eq. 3.1.7).

PROBLEM 3-13

Normal stresses in shear appear as a second-order effect. However, at high shear rates, they exceed shear stresses. Estimate the condition when it becomes possible.

Answer

The following relationship (based on Eq. 3.4.4) can be written as a possible approximation:

$$N_1 = \left(\frac{\Psi_0}{\eta_0^2}\right)\sigma^2 = \theta\left(\frac{\sigma}{\eta_0}\right)\sigma = (\theta\dot{\gamma})\sigma$$

where the ratio (Ψ_0/η_0) can be treated as a characteristic relaxation time and $\dot{\gamma} = \sigma/\eta_.$. The product $(\theta\dot{\gamma})$ is the dimensionless Weissenberg number, Wi. So, $(N_1/\sigma) > 1$ at such high shear rates where Wi > 1.

PROBLEM 3-14

Can normal stresses appear in the shear flow of suspension of solid particles? Explain the answer.

Answer

If solid particles are anisotropic, some equilibrium distribution of orientation exists in a steady-state. Shear flow destroys this distribution and three-dimensional forces intending to restore the equilibrium state appear. This is the reason for the appearance of normal stresses in simple shear.

Additional question

Estimate the characteristic time ("relaxation time"), θ, of this process.

Answer

The driving force of this process is the Brownian motion of particles. Then it is possible to construct the following time-dimensional criterion of this process:

$$\theta = (kT/D^2)^{-1}$$

where k is the Boltzmann constant, T is the absolute temperature, and D is the characteristic dimension of particles.

PROBLEM 3-15

An experiment was carried out in shear at the constant shear rate, $\dot{\gamma}$ = const, and the curve similar to shown in Fig. 3.5.1 or Fig. 3.5.2 was obtained. Can the ratio $\sigma(t)/\dot{\gamma}$ be treated as the evolution of viscosity of liquid? Explain the answer.

Answer

No, because viscosity is the ratio of stress to the *irreversible part* of the deformation rate. In the experimental data under discussion, transient deformation is a combination of elastic (reversible) and plastic (irreversible) components. In order to find apparent viscosity, it is necessary to separate the deformation rate into parts and viscosity can be calculated using the irreversible part of deformation rate only after this procedure is done. Without doing so, false conclusions might be reached that at $t \to 0$ (where $\sigma \to 0$) viscosity is close to zero and then increases along the $\sigma(t)$ curve.

Comment

The same arguments are true for normal stresses applied in extension: only the irreversible (plastic) part of the total deformation rate can be used for calculating elongational viscosity, as discussed in section 3.7.

PROBLEM 3-16

A liquid layer is intensively sheared at shear rate $\dot{\gamma} = 1*10^2$ s^{-1}. A liquid is Newtonian and its viscosity $\eta = 500$ Pa*s. Shearing continued for 10 s. Temperature dependence of viscosity is neglected; density is assumed to be 1 g/cm^3 and heat capacity is 0.5 J/(g*K).

What temperature rise is expected?

Answer

Shearing continued for a short time. It can be assumed that flow proceeds under adiabatic conditions and the heat removal is negligible. It means that the whole heat dissipation leads to an increase in temperature. The work of deformation, W, is calculated as $W = \sigma\dot{\gamma}t = \eta\dot{\gamma}^2t = 50$ J/cm^2 and it leads to the temperature rise equal to 100K.

Additional question

If shearing proceeds for a longer time, what physical phenomena must be taken into consideration and what final thermal effect of shearing can be expected?

Answer

The following physical phenomena become important in prolonged shearing accompanied by intensive heat production:

- decrease of viscosity due to temperature growth and suppression of intensive heat production
- increase of heat removal due to heat exchange with surroundings due to increase of temperature gradient and therefore decrease of temperature of a flowing liquid.

Two possibilities can be realized;

- thermal stabilization at some temperature level at which heat production equals heat removal
- uncontrolled increase in temperature with unpredictable consequences (the so-called "*thermal explosion*").

PROBLEM 3-17

Analyze the Mooney equation (3.3.27) for the concentration dependence of viscosity for limiting case and, in particular, calculate the intrinsic viscosity of dilute suspensions.

Answer

The Mooney equation is written as

$$\eta = \eta_s\exp\left[\frac{2.5\varphi}{1 - (\varphi/\varphi^*)}\right]$$

Then, at very low concentration ($\varphi \to 0$) the factor (φ/φ^*) $\ll$ 1 and the exponent member can be decomposed into the power series. Conserving the first member of the series and neglecting the higher members, the Einstein equation is obtained: $\eta = \eta_s(1 + 2.5\varphi)$. Then, according to the definition of the intrinsic viscosity:

$$[\eta] = \lim\left(\frac{\eta - \eta_s}{\eta_s\varphi}\right) = 2.5$$

i.e., the intrinsic viscosity of a suspension of solid spheres always equals 2.5.

In the high concentration range, at $\varphi \to \varphi^*$ viscosity is increasing unlimitedly. It means that φ^* is the limit of filling, and flow at higher degrees of filling is not possible.

PROBLEM 3-18

Newtonian viscosity of a polymer sample with molecular mass $M_1 = 3*10^5$ is $\eta_1 = 5*10^5$ Pa*s. There is also another polymer of the same chemical structure with molecular mass $M_2 = 4*10^4$. How can one decrease the viscosity of a polymer by 10 times?

Answer

It can be done by blending two fractions of a polymer sample. Shares of different molecular mass components are determined based on the following arguments. Let viscosity of blend be (as expected) $\eta_{bl} = 5*10^4$ Pa*s. Then, (see Eq. 3.3.15):

$$\frac{\eta_{bl}}{\eta_1} = \left(\frac{\overline{M}_{w,bl}}{M_1}\right)^{3.4} \quad \text{or} \quad \overline{M}_{w,bl} = M_1 10^{-1/3.4}$$

Then $\overline{M}_{w,bl} \approx 1.5*10^5$. Such material can be prepared by adding a low-molecular-mass fraction, w_2, to the high-molecular-mass component calculated from the following relationship: $\overline{M}_{w,bl} = M_1(1-w_2) + M_2 w_2$ or $1.5*10^5 = 3*10^5(1-w_2) + 4*10^4 w_2$. The last equality gives $w_2 \approx 0.57$, i.e., mixing of both shares in approximately equal fractions decreases viscosity by 10 times.

PROBLEM 3-19

Experiments show that an electrical charge appears on the surface of a polymer stream leaving a capillary in an unstable or spurt regime. Explain the origin of the charge.

Answer

The appearance of an electrical charge can be treated as the consequence of a slip of material along the wall in unsteady or spurt movement. In this case an electrical charge appears as a result of surface friction of dielectric material (polymer) at the wall.

CHAPTER 4

PROBLEM 4-1

Values of Young's modulus, E, and the bulk modulus of compressibility, B, are measured. Find shear modulus for a Hookean solid.

Answer

Beginning with Eq. 4.2.1, after rearrangements, the following equation is obtained:

$$G = \frac{3BE}{9B - E}$$

Additional question

Show that for an incompressible material the last equation transfers to the relationship between extensional and shear modulus known for rubber-like materials.

Answer

The limit of the right-hand part of the last equation at B >> E equals E/3. So, G = E/3.

PROBLEM 4-2

A bar is placed between two rigid walls. Its temperature is 20°C. Then the bar is heated to 200°C. What are the stresses that appear in the bar?

Answer

During heating, material extends but the walls prevent its extension, leading to the creation of compressive stresses, σ. Let the linear coefficient of thermal expansion be α, the material behaves as Hookean body with Young's modulus, E. Then stresses are calculated as

$$\sigma_E = -\frac{\Delta l}{l_0}E = -\alpha \Delta T E$$

where ΔT is the temperature increase.

Additional question

Using standard values of parameters for steel, estimate the level of stresses. Ordinary values of the parameters of material are: $\alpha = 1.2*10^{-5}$ K^{-1}, $E = 2.1*10^5$ MPa. It was assumed that $\Delta T = 180$K. Then direct calculation gives $\sigma \approx -450$ MPa.

Comment

It is worth mentioning that this value is close to the strength limit of the material, which is about 500-600 MPa. Thermal stresses can lead to rupture of material even with moderate temperature increase.

PROBLEM 4-3

Analyze the stress field in torsion of a cylindrical shaft caused by torque, T. This occurs in transmitting torque in a gearbox.

Answer

Let the radius of a shaft be R. Then the moment of forces balance equation is formulated as

$$\int_0^R 2\pi r^2 \sigma(r)dr = T$$

where r is the current radius and σ is the radius dependent shear stress.

The equation for shear deformations is written as

$$\sigma = rG\frac{d\varphi}{dx}$$

where G is the shear modulus, and the angle φ is the twist angle depending on the position along the shaft, x, but not on radius.

Shear stress is a linear function of radius and its maximum value, σ_R, is reached on the shaft surface. The combination of both equations gives:

$$T = 2\pi G\frac{d\varphi}{dx}\int_0^R r^3 dr = \frac{\pi R^4}{2}G\frac{d\varphi}{dx}$$

After excluding the value $d\varphi/dx$ from both equations, one comes to the expression for the radial distribution of shear stresses:

$$\sigma = \frac{2T}{\pi R^3}\frac{r}{R}$$

The maximum shear stress is:

$$\sigma_R = \frac{2T}{\pi R^3} \text{ and } \sigma(r) = \sigma_R \frac{r}{R}$$

PROBLEM 4-4

Compare the stress state in torsion of a solid cylindrical bar of radius R_o and a tube with the same outer radius and the inner radius equal R_i. What is the increase of the maximum shear stress produced by a decrease in cross-section of bar caused by changing the solid cross-section to a tube?

Answer

The radial stress distribution in a bar under torsion was calculated in Problem 4-3. It is:

$$\sigma(r) = \frac{2Tr}{\pi R^4}. \text{ For a solid bar, torque is } T = \frac{1}{2}\pi R_0^3 \sigma_{max}$$

where σ_{max} is a shear stress at the outer surface of the bar. For tube:

$$T = \frac{1}{2}\pi R_0^3 (1 - \beta^4)\sigma_{max, tube}$$

where $\beta = R_0/R_i$. At the same torque

$$\sigma_{max, tube} = \sigma_{max} \frac{1}{1 - \beta^4}$$

It means that at $R_i = R_o/2$, the increase of the maximum shear stress is only 2.5%.

Comment

These calculations demonstrate that it is reasonable to save materials in construction elements working in torsion by using tubes instead of solid bars.

PROBLEM 4-5

Calculate the principal stresses and maximum shear stress, if torque, T, and the stretching force, F, act simultaneously on the shaft of radius R.

Answer

Maximum shear stress was calculated in Problem 4-3 and it equals to

$$\sigma = \frac{2T}{\pi R^3}$$

Normal stresses at all sections are the same and equal $\sigma_E = F/\pi R^2$. Based on the equations derived in Chapter 1 for a plane stress state, it is possible to find the following expressions for two principal (normal) stresses:

$$\sigma_{1,2} = \frac{1}{2}[\sigma_E \pm \sqrt{\sigma_E^2 + 4\sigma_R^2}] = \frac{F}{2\pi R^2}\left[1 \pm \sqrt{1 + \left(\frac{4T}{RF}\right)^2}\right]$$

Maximum shear stress, σ_{max}, is:

$$\sigma_{max} = \frac{\sigma_1 - \sigma_2}{2} = \frac{1}{2}\sqrt{\sigma_E^2 + 4\sigma^2} = \frac{F}{2\pi R^2}\sqrt{1 + \left(\frac{4T}{RF}\right)^2}$$

Additional question

Are these results valid for shafts made out of rubber?

Answer

No, because the expression for shear stress adopted from the solution of Problem 4-3 is based on Hooke's law, which is not valid for rubbers.

PROBLEM 4-6

A shaft is twisted with a torque T, as in Problem 4-3. However, the torque is high enough to produce stresses exceeding the yield stress, σ_Y, of material. Describe the stress situation along the radius of a shaft.

Answer

Yielding conditions are achieved in the outer layers of a shaft where stresses are maximum. The central part of a shaft works as an elastic body. Stress distribution in an elastic part of the material is linear. The yield stress is reached at some radius r_Y. It means that at r $> r_Y$ stress is constant and equals σ_Y.

The balance equation is:

$$T = \int_0^{r_Y} 2\pi r^2 \sigma(r)\,dr + \sigma_Y\int_{r_Y}^R 2\pi r^2\,dr$$

The first integral term gives the moment of forces in the inner (elastic) part of a shaft, at $r < r_Y$, where $\sigma(r) = \sigma_Y(r/r_Y)$. The second integral term presents the moment of forces in the outer (plastic) zone, at $r > r_Y$, and R is the radius of the shaft.

After the integration the following solution of the equation for r_Y is found:

$$r_Y = \left(4R^3 - \frac{6T}{\pi\sigma_Y}\right)^{1/3}$$

If $r_Y = R$, i.e., the yielding condition is reached at the outer radius of shaft, T = $\pi R^3 \sigma_Y/2$ (compare with a solution for σ_R in Problem 4-3).

Additional question

What will be the deformations after unloading the shaft?

Answer

The central (elastic) part of the shaft, at $r \le r_Y$, will go back to its initial state, while the outer part of the shaft (at $r > r_Y$) will store plastic deformation.

PROBLEM 4-7

Prove that at small deformation, Hooke's law is the limit of the rubber elasticity equation.

Answer

The rubber elasticity equation is written as:

$$\sigma_E = G\left(\lambda^2 - \frac{1}{\lambda}\right)$$

The elongation ratio, λ, is

$$\lambda = \frac{1}{l_0} = \frac{l_0 + \Delta}{l_0}$$

where l_0 is the initial length of the specimen, l is the current length, and Δ is the displace-ment of the specimen length. In the limit of small deformations $\Delta \ll l_0$.

After substitution and neglecting the terms of the order lower than Δ^3, it is possible to make the following rearrangements:

$$\sigma_E = G\left[\left(\frac{l_0 + \Delta}{l_0}\right)^2 - \frac{l_0}{l_0 + \Delta}\right] = G\frac{l_0^3 + 3l_0^2\Delta + 3l_0\Delta^2 + \Delta^3 + l_0^3}{l(1 + \Delta)} \approx G\frac{3l_0\Delta}{l_0^2} = \frac{3G\Delta}{l_0} = 3G\varepsilon$$

The last expression is Hooke's law. Young's modulus equals 3G, where G is the rubber elasticity modulus.

PROBLEM 4-8

A rubber-like strip is stretched by the applied force F = 0.2N. The area of the cross-section of the strip is S = 1 mm^2. The elastic modulus, E, was measured at small deformations and it equals $3*10^5$ Pa. What is the elongation of strip? What would be the estimated elonga-tion if one would use Hooke's law for calculations?

Answer

The initial normal stress $\sigma = F/S = 2*10^5$ Pa. The estimated elongation is $\varepsilon_H = \sigma_E/E = 0.67$ or 67%. However, for the calculation of elongation of a rubber-like material, it is neces-sary to use Eq. 4.4.17. Then, the solution of this equation gives the answer: $\lambda = 2.2$, or $\varepsilon_{r\,b-ber} = 120\%$. The difference between ε_H and ε_{rubber} is evident.

Additional question 1

Why does the coefficient 1/3 appear in this equation?

Answer 1

This is because the equation of rubber elasticity includes shear modulus but not Young's modulus and they are related by coefficient 3 for an incompressible material.

Additional question 2

Why was Eq. 4.4.17 used for calculations but not Eq. 4.4.16?

Answer 2

The stress was known for the initial cross-section and it is not real stress increasing along with elongation of the strip.

PROBLEM 4-9

According to Hooke's law, the use of compression instead of extension leads to the sym-metrical change of normal stresses. Is it the same for a rubbery material with rheological properties characterized by Eq.4.4.17?

Answer

No. The stress-deformation curve for a rubbery material is not symmetrical with respect to the zero point. For example, in extension at $\lambda = 2$, engineering stress $\sigma_{E,ext}$ equals

$$G\left(2 - \frac{1}{2^2}\right) = 1.75G$$

and in compression at $\lambda = -2$

$$\sigma_{E,\,compr} = G\left(-2 - \frac{1}{2^2}\right) = -2.25G, \text{ i.e., } \sigma_{E,\,ext} \neq \left|\sigma_{E,\,compr}\right|$$

Additional question

Can the last result be treated as proof of the anisotropy of material, i.e., the existence of different values of elastic modulus in extension and compression, as is known for some other engineering materials, for example, concrete?

Answer

No. The rheological properties of the material are characterized by the same constant, G, in extension, as well as in compression. However, asymmetry of the stress-deformation curve is the consequence of the special rheological equation describing the behavior of rubbery materials.

PROBLEM 4-10

How are time effects taken into account in the formulation of the constitutive equation for large deformations, e.g., Eqs. 4.4.7 or 4.4.20?

Answer

Time effects are not taken into account in these equations at all. These relationships describe equilibrium states of material under stress, assuming that the reaction of material on stress application is instantaneous, quite as is assumed in Hooke's law.

PROBLEM 4-11

A cylindrical rod of radius R was studied in uniaxial extension. It was found that it can work below the critical force F*. Then, this rod was used as a shaft working at torsion deformation mode. What is the limiting value of the torque, T*, that can be applied to the shaft?

Answer

The basic relationship between limiting normal stress in extension, σ_E^*, and shear stress in torsion σ^* is established by the criteria of strength. The general form of this criterion is $\sigma^* = k\sigma_E^*$, where the constant k depends on the choice of the most reliable theory of strength (see Eqs. 4.5.1 and 4.5.7). Assuming the constant k = 0.5, and using the expressions for $\sigma_E = F/\pi R^2$ in extension and $\sigma_{max} = 2T/\pi R^3$ in torsion, it is easy to obtain the following relationship between F* and T*: T* = F*R/4.

CHAPTER 5

PROBLEM 5-1

Melt flow index, MFI, was measured using a capillary with the following dimensions: diameter d = 2.16 mm, length L = 8.0 mm. The diameter of the barrel was D = 9.5 mm. Density of melt was $\rho = 0.8$ g/cm³. Let the weight of a load be G = 2.16 kg. Estimate the shear stress attained in this experiment and calculate the apparent viscosity if a measured value of MFI was 2 g/10min.

Answer

Shear stress is estimated from Eq. 5.2.26. End correction is unknown and we assume the reasonable value $m_K = 2$ (the Couette correction plus elastic component). It gives shear stress of $\sigma_R = 1.57 \times 10^4$ Pa.

Newtonian shear rate is calculated as $\dot{\gamma}_N = 4Q/\pi R^3$. For MFI = 2 g/10 min, Q=4.16x10^{-3} cm^3/s, and Newtonian shear rate is calculated as $\dot{\gamma}_N = 4Q/\pi R^3 = 4.58$ s^{-1}. Then apparent viscosity (Newtonian liquid) is $\eta = \sigma_R/\dot{\gamma}_N \approx 3.4\times10^3$ Pa*s.

Comment

There is only one experimental point. Therefore, it is impossible to introduce a correction for non-Newtonian flow and the value of η must be treated as a rough estimation.

Additional question

Give a general formula for calculation of apparent viscosity for some arbitrary weight of a load, G, and melt flow index, MFI.

Answer

Calculating σ_R and $\dot{\gamma}_N$ as written above for given geometrical parameters of instrument, one obtains $\sigma_R = 727.5G$ and $\dot{\gamma}_N = 1.834$MFI/ρ. It gives $\eta_a \approx 400G\rho/$MFI, where G is expressed in Newtons, MFI in g/10 min, and ρ in g/cm^3.

PROBLEM 5-2

How does one calculate the shear rate at a wall for liquid with viscous properties described by a power-law type equation?

Answer

For a liquid of a power-law type, a relationship between shear stress and shear rate is written as in Eq. 3.3.4. Then dlogσ/dlog$\dot{\gamma}_N = n^{-1}$ and according to Eq. 5.2. 8 $\dot{\gamma}_R = [\dot{\gamma}_0(3n + 1)]/n$. The application of this equation simplifies practical calculations

Comment

For Newtonian liquid n = 1 and $\dot{\gamma}_R = 4\dot{\gamma}_0 = \dot{\gamma}_N$

PROBLEM 5-3

Experimental study of the tube flow of suspension gave the following results:

for tube I: $D_1 = 2$ cm, $L_1 = 20$ cm, output $G_1 = 42$ g/min under pressure P = 4 bar;
for tube II: $D_2 = 4$ cm, $L_1 = 40$ cm, output $G_2 = 294$ g/min (pressure P = 4 bar).
Density of suspension was $\rho = 1.4$ g/cm^3.

Explain the results and estimate the rheological parameters of the material.

Answer

Shear stress, σ_R, in both cases is the same and equals 10^4 Pa.

Apparent shear rates are: for tube I: $\dot{\gamma}_1 = 0.637$ s^{-1}; for tube II: $\dot{\gamma}_2 = 0.557$ s^{-1}. The same stress produces different apparent shear rates depending on the tube diameter. It can be explained by wall slip. Let the wall slip velocity be V_s. Then the equation for flow rate is as follows:

$$\frac{32G}{\pi D^3 \rho} = \frac{8V_s}{D} + \frac{\sigma}{\eta}$$

This equation includes two unknown constants: V_s and η. Solving an equation for these two parameters using two experimental points, the following values are obtained: $V_s = 0.04$ cm/s and $\eta = 2.24$x10^4 Pa*s.

PROBLEM 5-4

Calculate the velocity profile in the flow of Newtonian liquid through an annulus produced by two coaxial cylinders of length, L. Flow is induced by a pressure difference, P, at the ends of the annulus. The radii of inner and outer cylinders are R_i and R_o, respectively.

Answer

The balance equation for a cylindrical elementary volume is written as

$$\frac{1}{r}\frac{d}{dr}\left(r\frac{du}{dr}\right) = -\frac{P}{\eta L}$$

The integral of this equation is

$$u(r) = -\frac{P}{4\eta L}r^2 + A\ln r + B$$

where constants A and B are determined by means of boundary conditions. These conditions are: $u = 0$ at $r = R_i$ and at $r = R_o$. The substitution gives the final formula for $u(r)$

$$u(r) = \frac{P}{4\eta L}\left[(R_o^2 - r^2) + \frac{R_o^2 - R_i^2}{\ln R_o/R_i}\ln\frac{r}{R_o}\right]$$

Additional question

How does one obtain the form-factor for this type of annular flow?

Answer

The form-factor can be obtained by calculating the integral

$$\int_{R_i}^{R_o} 2\pi r u(r)dr$$

where the function $u(r)$ is taken from the previous equation. Direct integration leads to the expression for the form-factor included in the table of section 5.2.1.

PROBLEM 5-5

For a rotational viscometer of a coaxial cylinders type, what should be the diameter of an inner cylinder if the diameter of an outer cylinder is 40 mm and the acceptable inhomogeneity of the stress field is 5%?

Answer

According to Eqs. 5.7.37 and 5.7.38 this degree of homogeneity of the stress field corresponds to the ratio of diameters above 0.975. Then the diameter of the inner cylinder must be no less than 39 mm.

PROBLEM 5-6

In section 5.7.3, a portable viscometer was described that measures viscosity *via* time of the turn of a light cylinder from the initial position by a constant angle, the initial deformation is set by twisting of a torsion spring. What is the relationship between measured time and viscosity?

Answer

Since the turning cylinder is light, its moment of inertia is negligible. Torque appears as a result of viscous resistance. It is expressed by Eq. 5.3.16. Then an equilibrium equation is:

$$K\eta \frac{d\varphi}{dt} + Z\varphi = 0$$

where the form factor is (according to Eq. 5.3.16) $K = 4\pi HR^2$, Z is rigidity of a torsion, η is viscosity, and φ is the angle of rotation. Initial condition is: at $t = 0$, $\varphi = \varphi_0$. The solution of this equation is: $\varphi = \varphi_0 e^{-t/\upsilon}$, where $\theta = K\eta/Z$. If t^* is the time interval from the beginning of the experiment (at a certain angle φ^*), then the following relationship is valid:

$$\ln\left(1 - \frac{\varphi_0 - \varphi^*}{\varphi_0}\right) = -\frac{t^*}{\theta}$$

The left-hand side of this relationship is constant (due to the peculiarity of the measuring scheme) and therefore $\eta = kt^*$, where k is constant.

PROBLEM 5-7

In the principal scheme of measuring viscoelastic properties of material spring is used (Fig. 5.7.1) that is not ideal elastic but viscoelastic (has some losses in deformation). How do you calculate the viscoelastic properties of the material under investigation?

Answer

Viscoelasticity of spring means that value Z in Eq. 5.7.4 must be replaced by complex rigidity $Z^* = Z' + iZ''$. Then Eq. 5.7.4 must be written as

$$G^* = \frac{(Z^* - m\omega^2)Be^{i\alpha}}{K(A - Be^{i\alpha})}$$

The same formal rearrangements as were done for Eq. 5.7.4, i.e., separating it into real and imaginary parts gives the following expressions for the components of dynamic modulus:

$$G' = \frac{(Z'' - m\omega^2)(p\cos\alpha - 1) - Z''p\sin\alpha}{K[(p - \cos\alpha)^2 + \sin^2\alpha]}$$

$$G'' = \frac{(Z' - m\omega^2)p\sin\alpha + Z'(p\cos\alpha - 1)}{K[(p - \cos\alpha)^2 + \sin^2\alpha]}$$

The values Z' and Z'' are found by calibrating a measuring device.

PROBLEM 5-8

In section 5.7.2, the condition of uniform deformation of the sample in oscillation was formulated as $h \ll \delta$. Analyze this condition for inelastic viscous liquid.

Answer

According to definitions given in section 5.7.2, this condition is the same as $(kh) \ll 1$ because $k = \delta^{-1}$. The value of k is defined as $\omega\sqrt{-\rho/G^*}$. $G^* = G' + iG''$, and for inelastic liquid $G' = 0$, $G'' = \eta\omega$, k is expressed as $k = \sqrt{i\rho\omega/\eta}$. Then the real part of k equals

$\sqrt{\rho\omega/2\eta}$. Then finally, the condition of the uniformity of deformation in a gap is $h \ll \sqrt{2\eta/\rho\omega}$, which is the depth of propagation.

PROBLEM 5-9

Prove that Eq. 5.7.39 is valid for materials exhibiting low losses.

Answer

Eq. 5.7.38 gives the following expression for X_{0B}^{max} : $X_{0B}^{max} = f_0/KG''$. Then Eq. 5.7.35 leads to the following equation for the half-height of a resonance peak, $X_{0B}^{max}/2$:

$$\frac{f_0}{2K\eta'\omega} = \frac{f_0}{\sqrt{(KG' + Z - m\omega^2)^2 + (KG'')^2}}$$

where $G'' = \eta'\omega$.

The last relationship is rearranged into a fourth-order equation for frequencies, ω, corresponding to the half-height of the resonance curve:

$$\omega^4 - (2\omega_0^2 - a)\omega^2 + (\omega_0^4 + 4a^2\omega_0^2) = 0$$

where constant a is given by:

$$a = \frac{K\eta'}{m} = \omega_0\frac{G''}{G'}\left(1 - \frac{z}{m\omega_0^2}\right)$$

If a low loss material is considered ($G''/G' \ll 1$), it means that $a/\omega_0 \ll 1$. The solution of this equation for ω and calculation of the difference $\omega_1^2 - \omega_2^2$ (where ω_1 and ω_2 are frequencies at the half-height of a resonance curve) gives

$$\omega_1^2 - \omega_2^2 = \sqrt{12a^2\omega_0^2 + a^4}$$

The difference $\omega_1^2 - \omega_2^2$ is presented as $\omega_1^2 - \omega_2^2 = 2(\Delta\omega)\omega_0$ and this presentation is approximately correct for resonance because in this case, it is possible to change ω_1 and ω_2 for ω_0 and

$$\Delta\omega = \frac{1}{2\omega_0}\sqrt{12a^2\omega_0^2 + a^4}$$

Then the expression for $\Delta\omega$ is

$$\Delta\omega = \sqrt{3a^2(1 + a/12\omega_0^2)}$$

Finally, because $a/\omega_0 \ll 1$, the resulting equation is $\Delta\omega = \sqrt{3}a$, which is equivalent to Eq. 5.7.39.

PROBLEM 5-10

The value of maximum displacement X_{0B} appears in a solution of Eq. 5.7.41 of an equilibrium Eq. 5.7.40, but this value does not appear in Eq. 5.7.44 and other equations for G' and G''. Explain why?

Answer

The values of G' and G" are the inherent characteristics of the material, which (for the linear viscoelastic domain) must not depend on the amplitude of deformation, i.e., on X_{03}. The basic theory of damped oscillations is developed for the linear region of viscoelastic materials.

PROBLEM 5-11

Prove that Eqs. 5.7.5 and 5.7.6 give Eqs. 5.7.53 and 5.7.54 for inelastic liquid.

Answer

For inelastic liquid G' = 0. It means that in Eq. 5.7.5 $p\cos\alpha = 1$. Then, excluding either α or p in Eq. 5.7.6, one obtains Eq. 5.7.53 or Eq. 5.7.54, respectively.

PROBLEM 5-12

In section 5.7.6 deformations of the sample were treated as damping oscillations. Is it a unique case of damping deformations? Explain the answer.

Answer

Not necessary. Eq. 5.7.40 is the second-order differential equation with constant coefficients. It can be re-written as

$$\frac{d^2x}{dt^2} + \frac{K\eta'}{m}\frac{dx}{dt} + \frac{Z+KG'}{m}x = 0$$

From the theory of equations of such type, it is known that the solution is a sum of two exponents (but not damping oscillating function), if

$$\left(\frac{K\eta'}{2m}\right)^2 < \frac{Z+KG'}{m}$$

Additional question

In which case does the deformation of inelastic liquid become aperiodic but not oscillating?

Answer

For inelastic liquid, G'=0 and $\eta' = \eta$. Then the last inequality gives:

$$Z > \frac{K^2\eta^2}{4m} \quad \text{or} \quad \eta < \frac{1}{K}\sqrt{4mZ}$$

PROBLEM 5-13

Eq. 5.3.49 is valid as an approximation only. What should be the exact solution?

Answer

The expression in the square brackets in Eq. 5.3.48 is rearranged in the following manner:

$$(\sigma_{11} - \sigma_{33}) + (\sigma_{22} - \sigma_{33}) = (\sigma_{11} - \sigma_{22}) + 2(\sigma_{22} - \sigma_{33}) = N_1 + 2N_2$$

Therefore the exact solution of Eq. 5.3.48 should be

$$\sigma_{22}(r) = (N_1 + 2N_2)\ln\frac{r}{R}$$

It means that an approximate solution, Eq. 5.3.49 is valid only if $N_1 \gg N_2$.

CHAPTER 6

PROBLEM 6–1

Compare sensitivity of different rheological methods to the estimation of molecular mass.

Answer

Let two samples of the polymer have a two-fold difference in molecular mass, i.e., $M_1/M_2 = 2$. The ordinary values of exponents in the power-type dependencies of intrinsic viscosity, initial Newtonian viscosity, and coefficient of normal stresses are 0.7, 3.5, and 7.0, respectively. Then, the ratio of their intrinsic viscosities is $[\eta]_1/[\eta]_2 = 2^{0.7} \approx 1.75$, the ratio of initial viscosities is $\eta_{0.1}/\eta_{0.2} = 2^{2.35} \approx 11$ and the ratio of coefficients of normal stresses is $\Psi_{0,1}/\Psi_{0,2} = 2^7 = 128$.

Indices 1 and 2 in the above equations relate to samples with molecular masses M_1 and M_2.

PROBLEM 6–2

Explain why the MFI of polymer used for film extrusion must be higher than in tube extrusion (see Fig. 6.2.3).

Answer

Calibrating channels in a film-forming die of an extruder have much a smaller cross-section than in tube extrusion. Then in order to maintain the necessary output at the same pressure created by a screw, viscosity must be lower (i.e., MFI must be higher) for a smaller cross-section corresponding to a thin-film in comparison with a tube.

PROBLEM 6–3

Is it possible to vary the elasticity of melt without changing the average molecular mass of polymer?

Answer

Yes. It can be done by varying molecular-weight distribution while maintaining the average molecular mass of the polymer.

PROBLEM 6–4

Explain the advantage of synthetic lubricants in comparison with mineral oil-based lubricants.

Answer

Natural (mineral) oils contain crystallizable components which form a solid phase upon cooling. Therefore, in a low-temperature range, a solid-like structure appears in lubricant and its strength depends on the rate of cooling and the storage time. It makes starting characteristics of an engine worse. A synthetic lubricant oil of the equivalent viscosity does not contain such components due to its permanent and controlled composition.

PROBLEM 6–5

Derive Eq. 6.4.5. As an intermediate result, obtain an equation for a radial velocity distribution.

Answer

Eq. 6.4.5 can be obtained beginning with a basic definition of power-law type liquid: $\dot\gamma = K\sigma^m$ (see section 3.3; pay attention that m = 1/n). In section 5.2, it was shown (Eq. 5.2.18) that shear rate at a wall is expressed as:

$$\dot\gamma_R = \frac{Q}{\pi R^3}\left[3 + \frac{d\log(Q/\pi R^3)}{d\log\sigma_R}\right]$$

For power-law liquid $\dot\gamma_R$ can be written as:

$$\dot\gamma_R = \frac{Q}{\pi R^3}(3 + m)$$

Shear stress $\sigma_R = PR/2L$. The substitution of these expressions for $\dot\gamma$ and σ into a definition of power-law type liquid results, after some evident rearrangements, in Eq. 6.4.5.

The velocity profile is also obtained from a basic definition of power-law type liquid having a linear distribution of stresses along a radius regardless of liquid type:

$$u(r) = \int_0^r \dot\gamma dr = \int_0^r K\left(\sigma_R\frac{r}{R}\right)^m dr = \frac{KR^{m+1}}{m+1}\left(\frac{P}{2L}\right)^m\left[1 - \left(\frac{r}{R}\right)^{m+1}\right]$$

Eq. 6.4.5 is easily obtained from the last equation by calculation of the integral:

$$Q = \int_0^R 2\pi r u(r)dr$$

It directly leads to Eq. 6.4.5.

PROBLEM 6–6

How does the velocity profile during flow in a tube change during the transfer from a Newtonian to a power-law type liquid? Make a comparison by analyzing the ratio of maximum to the average velocity.

Answer

Maximum velocity, V_{max}, is calculated from the velocity profile, as found in Problem 6–5 at r = 0. It equals:

$$u(0) = \frac{KR^{m+1}}{m+1}\left(\frac{P}{2L}\right)^m = \frac{KR}{m+1}\sigma_R^m$$

Average velocity, $\overline{V}$, is defined as $\overline{V} = Q/\pi R^2$. According to Eq. 6.4.5

$$Q = \frac{\pi KR^3}{m+3}\left(\frac{pR}{2L}\right)^m \quad \text{and} \quad \overline{V} = \frac{KR}{m+3}\sigma_R^m$$

Finally,

$$\frac{V_{max}}{\overline{V}} = \frac{m+3}{m+1}; \ \overline{V} = \frac{Q}{\pi R^3}$$

For Newtonian liquid m = 1 and $(V_{max}/\overline{V}) = 2$. With increasing m, the velocity profile becomes more and more flat, approaching plug-like flow.

PROBLEM 6–7

Prove Eq. 6.4.6 for a Bingham liquid. Explain the necessary conditions required for the movement of Bingham viscoplastic media through a tube.

Answer

Bingham-type viscoplastic media can move through a channel only if shear stress at a wall exceeds the yield stress: $\sigma_R > \sigma_Y$, where as usual $\sigma_R = pR/2L$.

Shear flow is realized in the stress range $\sigma > \sigma_Y$. It corresponds to the near-wall ring zone at $r > r_Y = 2L\sigma_Y/p$. At $r < r_Y$, i.e., around the central axis of the channel, $\dot{\gamma} = 0$ and consequently u = const, i.e., velocity profile in the central part of the flux (around the channel axis) is flat and material in this zone moves in a solid-like manner as a plug.

Calculations are based on the definition of Bingham media (Eq. 3.3.9) written in the following form:

$$\dot{\gamma} = \frac{\sigma - \sigma_Y}{\eta_p}$$

Then, the velocity profile is found by its integration:

$$u(r) = \int_0^r \dot{\gamma}(r)dr$$

substituting the expression for $\dot{\gamma}$ from a rheological definition of Bingham media and assuring that the radial distribution of shear stresses is always linear. Finally, it leads to the following velocity distribution along a radius:

$$u(r) = \frac{PR^2}{4L\eta_p}\left[1 - \left(\frac{r}{R}\right)^2\right] - \frac{\sigma_Y R}{\eta_p}\left(1 - \frac{r}{R}\right)$$

This equation is valid in the range $r_Y < r < R$.

As was mentioned above, the velocity at $r < r_Y$ is constant, u_Y. It can be found from the last distribution by substituting $r = r_Y = 2L\sigma_Y/P$. The final expression for u_Y is

$$u_Y = \frac{PR^2}{4L\eta_p}\left[1 - \left(\frac{\sigma_Y}{\sigma_R}\right)^2\right] - \frac{\sigma_Y R}{\eta_p}\left(1 - \frac{\sigma_Y}{\sigma_R}\right)$$

Additional question

Prove Eq. 6.4.6.

Answer

It can be done by calculating the sum:

$$Q = \int_{r_Y}^R 2\pi r u(r)dr + \pi r_Y^2 u_Y$$

where the expressions for u(r) and u_Y are in the above equations.

PROBLEM 6–8

Eq. 6.4.5a is convenient for solving engineering problems of the flow of non-Newtonian liquids through channels with non-circular cross-sections. Can a similar equation be formulated for viscoplastic liquids? Explain your answer.

Answer

No. Eqs 6.4.6 and 6.4.7 show that Q-vs.-p dependence in the flow of viscoplastic media is more complicated and pressure enters various terms of these equations. Therefore, it is not possible to introduce a single geometrical factor for non-circular cross-sections.

Additional question

How do you calculate Q(p) dependence for viscoplastic media moving through tubes of non-circular cross-section?

Answer

As a general rule, an analytical solution is not possible, and it appears necessary to apply computational methods for engineering calculations.

PROBLEM 6–9

Why does elasticity appear before gelation during polymer curing, i.e., the formation of a three-dimensional chemical network?

Answer

In the process of curing:

- long flexible macromolecules are produced which, first, have their own inherent relaxation properties, and, second, can form temporary entanglements with viscoelastic interaction
- inhomogeneity of curing results in the formation of micro-gel particles before gelation of the system as a whole; these particles also contribute to the viscoelasticity of the medium.

PROBLEM 6–10

Describe what happens if the end of a pumping screw of an extruder is blocked? What pressure will be developed?

Answer

An extruder will work "for itself" providing the circulation of liquid inside a channel. Rotation of the screw will create a pressure at its end which is calculated from Eq. 6.4.8 at $Q = 0$. In this limiting case, maximum pressure equals $p_{max} = (AN/B)^{1/m}$.

INDEX

Printed in the United States
by Baker & Taylor Publisher Services